ISBN 978-0-484-60217-4
PIBN 10859704

THE

MECHANICS' MAGAZINE,

MUSEUM,

Register, Journal,

AND

GAZETTE,

JULY 5TH — DECEMBER 27TH, 1851.

EDITED BY J. C. ROBERTSON.

VOL. LV.

"The triumph of the industrial arts will advance the cause of civilization more rapidly than its warmest advocates could have hoped, and contribute to the permanent strength and prosperity of the country far more than the most splendid victories of successful war."

BABBAGE.

London:
ROBERTSON AND CO.,
MECHANICS' MAGAZINE OFFICE,
(No. 166, FLEET-STREET.)
AGENTS:—EDINBURGH, J. SUTHERLAND;
GLASGOW, W. R. M'PHUN AND DAVID ROBERTSON;
DUBLIN, MACHIN AND CO., 8, D'OLIER STREET;
PARIS, A. & W. GALIGNANI, RUE VIVIENNE;
HAMBURGH, W. CAMPBELL.

1851.

ALPHABETICAL LIST OF NEW PATENTS GRANTED FOR ENGLAND, SCOTLAND, AND IRELAND.

Name.	Subject.	England.	Scotland.	Ireland.	Page.
Adcock	Pipes, chimney pots, &c.	23 Oct.	....	....	340
Albright	Phosphorus	17 July	....	....	60
Alexander	Preparing cheese	8 Dec.	....	....	480
Allan	Electric telegraphs	....	2 July	23 July	100,180
Alliott	Cleaning, dyeing, and drying-machines	....	31 July	....	179
Applegarth	Printing machinery	24 Dec.	....	....	520
Armand	Distilling organic substances	10 Dec.	....	....	480
Aston	Buttons, &c.	3 July	....	....	20
Baildon	Writing, printing, or marking	7 July	....	....	40
Bailey and Bailey	Preparing, combing, and spinning wool	20 Nov.	....	....	420
Bale	Ornamenting buildings, &c	17 July	20 Oct	....	60,400
Barclay	Fatty and oily matters	....	....	11 Aug.	180
Barker	Chipping and shaving dye wood	....	6 Oct.	....	360
Barlow	Rotary engines	3 July	....	....	20
Barlow	Saws	31 July	....	....	99
Beadon	Roofing houses and structures	....	....	2 Aug.	180
Beattie	Railways, locomotives, &c.	22 Oct.	....	....	340
Bernard	Leather	13 Nov.	....	....	399
Bessemer	Producing ornamental surfaces	19 Nov.	....	....	420
Bewick	Bricks and tiles	6 Nov.	....	....	380
Betjemann	Bedsteads	....	16 Oct.	....	400
Biddell & Green	Moulding and casting	29 Oct.	....	....	359
Bissell	Sustaining carriages	5 Aug.	....	....	119
Blair	Beds and couches	11 Sept.	....	....	220
Blundell	Sweeping roads	14 Aug.	....	....	140
Boggett and Palmer	Heat and light	22 Oct.	14 Nov.	....	340,439
Booth	Generating and applying heat	....	15 Oct.	....	360
Borden	Treating animal and vegetable substances	11 Sept.	....	....	220
Bousfield	Manure	19 Dec.	....	....	520
Bower	Preparing flax, line and grasses	....	20 Aug.	8 Sept.	180,280
Bramwell	Steam-engine valves and paddle wheels	20 Nov.	12 Nov.	....	420,439
Brazil	Dyeing and preparing dye woods	....	21 July	....	100
Briggs	Oil lamps and lubricators	9 Oct.	....	....	300
Brooman	*Presses and pressing*	25 Sept.	....	....	259

...me.	Subject.	England.	Scotland.	Ireland.	Page.
....	Horticultural construction	15 Nov.	....	....	419
......	Agricultural implements, steam engines and boilers	1 Dec.	....	....	460
er.......	Motive power.........	21 Aug.	....	....	160
tt	Carpets	....	21 Oct.	....	400
......	Candles, mortars and night lights.........	22 July	....	....	80
....	Oxalate of potass........	....	25 Aug.	....	280
r	Motive power	....	29 Aug	....	280
nemoreau.	Gas lighting...........	28 Aug.	....	....	179
nemoreau	Preserving animal substances	4 Sept.	....	....	200
nemoreau.	Kneading and baking bread	8 Dec.	....	....	479
n	Casting, shaping and pressing metals	8 Dec.	....	....	480
ngs......	Forge hammers	16 Oct.	....	....	320
ay......	Steam engines..........	....	30 Sept.	....	360
er.......	Metal pipes, shafts, &c...	8 Dec.	....	....	480
th.......	Locomotive engines......	25 Sept.	....	....	260
....	Manures	16 Dec.	....	....	500
ier.....	Castors and legs of furniture	22 Nov.	....	....	440
....	Pulverising land..	16 Oct.	7 Nov.	....	320,439
......	Fatty and oleaginous matters	22 Nov.	....	....	440
and el	Treating paper, to prevent the taking of impressions............	14 Aug.	....	....	140
n	Odometers ,............	1 Dec.	....	....	460
....	Brass tubes	25 Sept.	....	....	259
rood & ourton .	Drawing and combing wool.............	3 Nov.	....	....	380
ad Bow-	Weavers' healds or harness...............	7 Aug.	4 Nov.	....	120,439
s	Steam engines and propelling	....	21 Oct.	....	400
od....	Communicating motion; and rudders........	14 Aug.	....	....	140
ell	Drying malt............	4 Sept.	5 Nov.	....	200,439
....	Preparing and spinning ..	22 Oct.	....	....	340
....	Looms	3 July	....	....	20
....	Weaving textile fabrics ..	15 Nov.	....	....	420
en....	Sawing, boring and shaping wood	....	....	22 May	20
....	Scythes................	....	....	6 Oct.	439
....	Barometers,............	....	....	10 June	20
n	Steam engines	8 Dec.	....	....	479
s.......	Brushes	....	16 July	....	100
ner ..	Obtaining instantaneous light..	19 Dec.	....	....	520
rson.....	Generating gas.........	23 Oct.	22 Oct.	....	340,400
....	Steam generators........	17 July	....	....	60
....	Woollen cloth	19 Dec.	....	....	519

...me.	Subject.	England.	Scotland.	Ireland.	Page.
...a.......	Motive power	....	....	9 Aug.	180
...s	Aërated liquids	11 Dec.	....	....	480
...s and ...lowsky	Washing, steaming and drying fabrics	....	....	5 Sept.	280
...e.......	Carpets	....	....	10 July	80
...n	Steam boilers	19 Dec.	....	....	520
....	Steam boilers and propelling machinery ..	22 Nov.	....	....	440
...es	Sugar	7 July	8 Aug.	....	40,180
...y	Crushing sugar canes....	5 Aug.	....	....	119
...n	Gutta percha and caoutchouc	14 Aug.	....	....	140
...ray......	Weaving machinery	....	21 Oct.	....	400
....	Melting furnaces........	....	2 July	....	100
...l........	Locks	....	24 Sept.	....	360
...n	Railways..............	....	4 Nov.	....	439
...n	Dyeing textile fabrics....	10 Dec	....	....	480
...n	Separating substances....	24 Dec,	....	....	520
..., Ball ...Bagley.	Knitted, looped and elastic fabrics	24 Dec.	....	....	520
...ndy ...Fell ...	Obtaining fresh water and concentrating sulphuric acid	7 Aug.	13 Aug.	....	120,180
....	Plain and figured fabrics..	16 Dec.	....	....	500
....	*Bricks*, tiles, quarries, etc.	9 Oct.	....	....	300
....	Spinning machinery	....	11 Aug.	....	180
......	Nuts, bolts, steps, bearings, &c.	16 Oct.	20 Oct.	....	320,400
....	Steel.	....	15 Oct.	....	360
...l and ...nd....	Sugar	....	6 Aug.	3 Nov.	180,439
....	Delineating objects......	23 Aug.	....	....	179
....	Ploughs	23 Oct.	....	....	340
....	Copper and alloys	18 Sept.	....	....	240
....	Locks	6 Nov.	....	....	380
...s	Cranes and railways	....	21 July	18 Nov.	100,439
....	Drying, heating & cooling	3 July	....	....	20
....	Capstans	31 July	....	....	100
....	Looped fabrics	16 Oct.	....	....	320
......	Preventing injurious effects of smoking	18 Sept.	....	....	240
....	Building materials	8 Dec.	....	....	480
....	Regulating engines	2 Oct.	24 Sept.	....	279,360
....	Textile fabrics..........	14 Aug.	....	....	140
...d Burch.	Looms..............	3 July	21 July	....	20,100
...d Schiele	Preparing fibrous materials	22 Oct.	....	....	340
....	Railway-axle boxes......	23 Oct.	....	....	340
......	Ventilating railway carriages	19 Dec.	....	....	519
......	Leather, and boots and shoes	....	17 Nov.	....	439
...ttom ..	Measuring flow of fluids..	22 Oct.	....	....	340
....	Grinding wheat	19 Dec.	....	....	520
....	Locks or fastenings	8 Dec.	....	....	480

...se.	Subject.	England.	Scotland.	Ireland.	Page.
......	Stopping railway carriages	22 July	....	....	80
......	Buffers, grease boxes, &c.	4 Nov.	....	....	380
......	Nickel and cobalt	4 Nov.	....	....	380
......	Knives and forks	21 Aug.	....	....	160
......	Railways and carriages ..	2 Oct.	....	....	279
....	Decomposing saline substances	25 Sept.	29 Sept.	....	259,360
......	Carriage and other springs	....	....	3 Nov.	439
......	Dyeing gloves	10 Dec.	....	....	480
....	Surgical instruments, scissors, &c.	22 Nov.	....	....	440
edt ..	Preserving substances....	4 Sept.	....	....	200
r......	Steam engines	31 July	1 Aug...	22 g.	99,180 280
:......	Printing and weaving	27 Nov.	....	....	440
......	Kettles, saucepans, &c....	19 Dec.	....	....	520
......	Rails for railways	13 Nov.	19 Nov.	....	400,439
Wil- Childs ckson	Presses, matting, candles, &c..........	3 Nov.	22 Oct.	....	380,400
'st ..	Steam engines and propelling	....	....	30 Sept.	439
......	Carpets, rugs, &c.	4 Dec.	....	....	460
a	Bricks and tiles	31 July	....	....	99
......	Spinning and doubling ..	18 Sept.	29 Sept.	....	240,360

Mechanics' Magazine,

MUSEUM, REGISTER, JOURNAL, AND GAZETTE.

No. 1456.] SATURDAY, JULY 5, 1851. [Price 3*d*., Stamped, 4*d*.

Edited by J. C. Robertson, 166, Fleet-street.

MR. WILLIAMS'S PATENT STEAM-BOILER FURNACES.

Fig. 1.

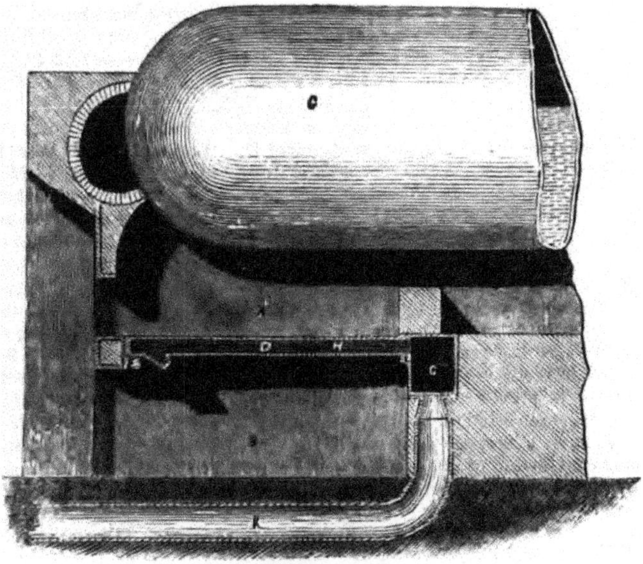

Fig. 2.

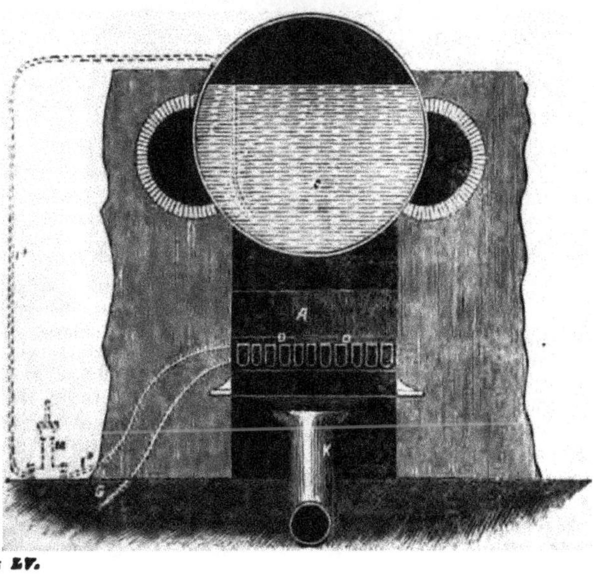

MR. WILLIAMS'S PATENT STEAM-BOILER FURNACES.

(Mr. D. L. Williams, of Thornhill, Llandilo, Carmarthen, Patentee. Specification Enrolled June 7, 1850.)

Specification.

My improvements in furnaces have for their object the construction of furnaces, particularly those employed for generating steam, in such manner that the furnace bars shall be always kept in a comparatively cool state, and that the air (employed to support combustion) or water (employed to feed the boiler) as the case may be, shall be heated previous to their introduction into, or subjection to, the direct action of the furnace. Fig. 1 is a longitudinal section of a steam-boiler furnace thus constructed, fig. 2 a cross-section of the same, and fig. 3 a plan. A is the fire-place; B the ash-pit; C a portion of the steam boiler; DD the furnace bars, which rest at front upon the cross-bearer E, and at the back or further end of the furnace upon a bar F, which forms one side of a hollow chamber G. The fire-bars D D are hollow, each having a channel H passing through it from end to end; at the back of the furnace, these different channels open into the chamber G (as represented in the sections of the bars in figs. 1 and 2), while in front they terminate in openings I I formed in the lower side of the bars. K is a pipe through which a constant supply of atmospheric air is kept flowing into the chamber G, and thence into the channels inside of the furnace bars, whence the air, in a heated state, issues through the openings I I, passes directly through, between the furnace bars, into the fire, and tends to support the combustion of the fuel. By the arrangements just described, the cold air passing through the bars, keeps them from becoming too much heated, and they therefore remain much longer in a good working condition, while the air supplied to the furnace is previous to its entry amongst the fuel raised to a temperature exceedingly favourable to combustion, and a considerable saving of fuel is thereby effected. In some cases the circulation of the air, which is thus employed for keeping the bars cool instead of being kept up by rarefaction alone, may be produced or assisted by means of a fan or other mechanical contrivance.

Fig. 3.

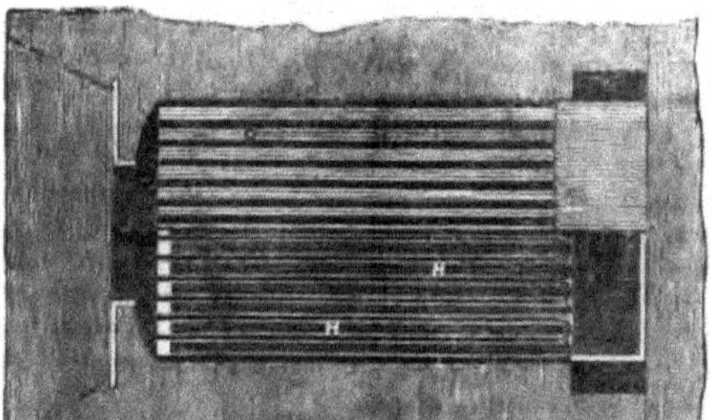

The arrangements, which are shown in the engravings, for heating air to supply a steam furnace may, with slight modifications, be applied to heating the water intended to feed boilers, or for other purposes. In that case the openings I I in the bars at the front of the furnace are connected to the pipe I^2 (indicated by dotted lines) and a constant supply of cold water is made to communicate with the chamber G by means of a pipe G^2, smaller than that for the air. The fire bars are thus kept constantly filled with water, which, as it gets heated, is drawn off by means of the feed-pump M (connected to the pipe I^2), and is forced by it into the boiler through the pipe I^3. By the force pump being thus interposed between the boiler

furnace bars, the pressure exerted by the steam upon the surface of the
the boiler is prevented from in any way being exerted upon that contained
urnace bars, so as to cause any disrupture to take place, either in them or in
ts. Instead of having the whole set of bars applied to either of the pur-
ove described, part of them may be employed for heating the air, and the
ler for heating the feed water; in which case the chamber G must be par-
off, and the connections of the different pipes disposed so as to suit such an
ment, or a continuous stream of water may be allowed to flow through the
keep them from becoming overheated, the heated water being permitted to
y instead of being forced into the boiler.

THE ROTATION OF THE EARTH.

been said that, in political mat-
lie, uncontradicted for twenty-
ars, is as good as a truth. How
alse theory in physics requires to
his prescription has not been
ut thinking it high time the pre-
pinion on the pendulum experi-
hould be further inquired into, I
u with much diffidence a few
. The English mathematicians
to have derived their solution of
lem from the French. It does
ear that any men of note here
blished original or complete in-
ions themselves. It is to be
d that men so able to form ori-
eories themselves as Professors
and Powell, should have been
d with merely illustrating those
s. If it should appear, on their
investigations, that the theory is
l, they will participate in this
Meantime, without any preten-
be able to form another complete
I am anxious to state the views I
theirs, which, if they have no
modifying it, may make some
h to a more popular appreciation
henomena.

e outset I wish to guard myself
being supposed in the least de-
call in question the reality of the
ents. I think they most assur-
n the basis of a demonstration
otation of the earth of a highly
ng character, and peculiarly fitted
the popular mind. The doubts
d upon the subject from various
, are obviously the result of most
: arrangements for performing
riments. No such doubts have
om the experiments performed
lace under the direction of your
l correspondent, Mr. Uriah
one of the apparent anomalies
having arisen.* In fact, the experiment is a new one, requiring great care to adjust the parts; but when the necessary care is taken, the result is uniformly decisive of the effect being caused by the rotation of the earth.

The theory we wish to examine is thus propounded:—

Mr. Silvester, in his letter to the *Times*, April 25th, inserted in your No. 1447, says—"M. Binet's investigation leaves nothing to be desired in point of vigour of demonstration; the same result has been obtained, after a more compendious method, by two English analysts, the correctness of which I can attest. For the sake of those thinkers who form an intermediate class between those who are incapable of any proof except what appeals directly to the senses, and the elevated few who can comprehend the form of an analytical investigation, I offer a brief recapitulation of the argument only hinted at in my former letter.

"At the pole it is obvious the plane of vibration remains fixed in space, and therefore appears to revolve at the rate of 15° per hour. At the equator the 'law of sufficient reason' shows there can be no apparent rotation either way. ...... As regards places intermediate

* A paper explanatory of the experiments referred to, by our esteemed friend Mr. Clarke, appeared in the *Leicester Mercury* about six weeks ago. We meant to have republished it in our pages, but it has been excluded by a pressure of other matters till its appearance now would be out of season. According to the theoretical rule for ascertaining the apparent deflection—namely, 15° per hour at the pole, and for other places as the sine of the latitude—the deflection should be somewhat less than 12° for the latitude of Leicester; but Mr. Clarke found, "from actual experiments many times repeated," that it is more than 12°. This discrepancy is supposed by Mr. Clarke to arise from the theoretical calculations having been made on the supposition of the earth being a spherical instead of a spheroidal body.

between the pole and the equator, a rough but substantially correct view of the phenomenon may be reasoned out thus:—By a process known to geometers, the rotation of the earth about its pole may be supposed to be replaced by two *simultaneous* rotations about two ideal poles, one running straight up under the place of observation, the other passing through the earth's centre at right angles to the former. This latter pole will produce no effect. The *principal, and I may say total* observed effect, will be due solely to the rotation about the ideal pole, which is on a line with the point of suspension, and the rate of its vibration is the entire and true rate diminished as the sine of the latitude to unity. The motion is the result of a rational and mathematical investigation in which two things have to be considered—the motion of the earth and the motion of the plane itself, if the mode of explanation be attempted to be kept up; and the difference of these two motions will become apparent to observation."

Now as at the pole the apparent motion is due to the invariability of the pendulum's plane, coupled with the earth's motion, at every other place the same effect would be produced in a complete revolution in twenty-four hours if the plane remained invariable, and therefore any variation in the time of a revolution could only be effected by an actual motion of the plane,—a mode of explanation, however, which Mr. Silvester seems to deprecate.

To say nothing about the easy way in which Mr. Silvester disposes of one of the two ideal rotations into which, for solving the problem, he supposes the rotation of the earth to be resolved, nothing can be clearer than that it is impossible to neglect either, unless it has its whole effect expended, in verily and indeed causing the plane of vibration to move in the same direction. Whether any such effect is due to the rejected rotation, and if so, in what manner it is accomplished, all the writers upon this subject are entirely silent; and yet they all follow M. Binet in making this artifice of geometry first used by Euler the foundation of their solution. It is of course difficult for a person of small *attainments in calculations* to say the *process is inapplicable to the question altogether, but it is not too* much to ask those better informed to point out some reasons why they suppose it to be applicable. The very loose terms in which Mr. Silvester states his proposition are such as not to create any great confidence in it; and your very intelligent correspondent, "S. Y.," who has on former occasions wielded no weak weapons against the undue pretensions of the mere mathematicians, is so struck by the insufficiency of the ground for first resolving the real rotation into two ideal ones, thus treating both as real, and afterwards arbitrarily rejecting one of them as having no effect, that he seems to doubt the reality of the phenomenon altogether. However, in the imperfect report of Professor Powell's lecture, we are told that the Astronomer Royal agrees with him in opinion that the exact determination of the direction of the plane cannot be made on *any general considerations*, but must be the result of detailed mathematical investigation.

Let us therefore return to the inquiry. At the pole the plane of vibration continues invariable—it maintains its first position, 1st, in direction; 2nd, in its relation to the plane of rotation; 3rd, in the horizontal direction of the line joining the centres of the ball at the two extreme distances. At the equator the plane of vibration is equally invariable in *direction*; for if the pendulum be set vibrating north and south, although the plane will revolve round with the earth, the direction of the plane north and south will be unchangeable. In like manner, if set vibrating east and west, the plane will remain unchanged, though the direction of the line joining the centres of the ball at the two extremes will vary through the whole circle. At any intermediate latitude, the plane will vary in inclination from the perpendicular on the plane of rotation to an angle equal to the latitude of the plane; and the line joining the centres of the ball at the two extremes will vary from a parallel with the plane of rotation to an inclination equal to the same angle; but it is not equally obvious, though it may be true, that the *direction* of the plane varies at all. Now this is an important consideration; for from the fact of the table under the pendulum maintaining always the same position with respect to the axis of the pendulum, and consequently with relation to the line where the plane pendulum cuts it, this *direction* of ane is the only quantity affecting blem at all; and if that is inva- the table under it will make a te revolution in twenty-four hours y part of the earth except at the ; however much the other condi- ay vary. Before proceeding to gate in what way this direction of ne may be liable to vary, let us w this view of the matter affects ory under examination. We have concluded that if Euler's process leable to this question, it will be of the rotations having its whole xpended in making the plane of on really move in the same direc- the earth. We can now substi- *direction* of the plane instead of ne itself.

as in travelling from the pole to ator, this theory assumes that the y of apparent motion of the pen- as caused by the effective rota- constantly decreasing, the effect other rotation is continually in- g till, in approaching the equator, it has become indefinitively near mplete revolution. How comes , when it arrives at the equator it we have just seen, have no effect et,—the direction of the plane remaining absolutely unchange-

great respect for Professor Young, der the explanation given by him true nor consistent with the of M. Binet. The professor not nsiders the plane of vibration at udes invariable, but he even con- he line joining the centres of the the two extremes as absolutely ble in the direction of its inclina- ong any one assumed meridian. pears a strange misconception— whole theory is built upon it. s this line, during the motion of th from one meridian to another, ake an angle, not of the distance meridians, but of their inclinations apex of a cone formed by a pro- on of the tangents at the latitude ation. Now as a mathematical ion from the immediate premises, unquestionable; but these pre- re not the conditions of the pro- When the horizontal line, as he , has arrived at the second posi- d assumed a different direction, it must still be horizontal; and as the horizontal plane has changed the direc- tion of its inclination, the line has no longer the inclination of the tangent to the meridian it had at first, and an addi- tion will have to be made to the angle of its motion of a quantity due to the change of its inclination. In its motion to each succeeding meridian, the angle to be added will be an increasing quantity; so that, on his own supposition thus cor- rected, when the earth has revolved 90°, the motion will have accomplished 90° also.

There is a mode of viewing Professor Young's theory which would render it intelligible, but which shows its utter inapplicability to solve the problem. By supposing the line to retain the first in- clination, and not to follow the horizontal plane, it would truly have the motion he ascribes to it, and would form a double cone during one revolution of the earth, each exactly similar to the cone formed by the prolonged tangents to the meri- dians; and in this manner might be con- sidered as making a complete revolution without going through 360°; but that the professor could conceive this as sup- porting Mr. Silvester's views, which he considered abundantly satisfactory, I am at a loss to conceive.

Let us now, however, endeavour to investigate, in an ordinary manner suited to common minds, by what means the direction of the plane at any latitude could be made to vary, and in the outset let us divest ourselves of the prevalent, though vague, notion that some of the earth's motions could of themselves effect this object. No truth in physics is better established than that any num- ber of motions impressed on a body are perfectly independent of each other, it is the very principle of the second law of motion which it is too late in the pro- gress of science to expect to see re- pealed. The only way in which one motion of a body can be instrumental in causing any alteration of another motion, is by bringing it within the action of a force, or by altering its relation to a force to which it is incident. In this instance, the motion of the earth can bring the pendulum within the sphere of no new force. Can it, therefore, alter its relation to the force of gravity to which it is already incident? Let us see. In the oscillations of a pendulum the ball is alternately on each side of the axis, in an exactly symmetrical manner; any effect, therefore, that can be produced on the ball in its passage from the axis to the extreme distance and back again, on one side, would be exactly balanced by a corresponding effect on the other side, and, as far as the present question is concerned, the whole effect on each side might be considered as concentrated in that on the ball when at each extreme. It will not, perhaps, therefore, involve any error if we suppose the force of gravity on the pendulum represented by a line from the centre of the earth to the ball of the pendulum at each of the two extremes. And if we consider the plane of the pendulum, and its axis extended to the centre of the earth within these two lines, we shall be able easily to see what effect any alteration of its position may be likely to cause. In the first place, this extended plane will at all times be in the section of a great circle of the sphere, that at all times *when at rest* gravity can have no effect whatever in causing any deflection of the plane, its action on the plane being in the *direction* of it, and symmetrical on each side its axis. Moreover, it might to some minds be sufficient to say that as the rotation of the earth in altering the position of this plane acts on a single point, that of suspension of the pendulum, or the extremity of the axis prolonged from the centre, the force of gravity from the centre can have no effect whatever in causing the plane to deflect or twist in any direction. But a little detailed consideration may make it clearer; first, the motion of the point of suspension is in a circle on the outside of a sphere, which motion may obviously be resolved into two motions,—one in a great circle in the direction of the plane itself, and the other in a great circle at right angles to the former. And this would in fact be true in its motion from one point to another, independent of the two points lying in a circle. Now, the motion of the plane in its own direction obviously never alters its relation to the central force, and its motion at right angles to it, as it is in a great circle, never alters its position as regards the centre, that remaining perfectly symmetrical to *it every moment.* Therefore the united *action, simultaneously,* in these two directions cannot have any effect. Consequently, if the assumptions I have started with are allowable, the direction of the plane of vibration at any latitude is absolute, invariable. If the assumptions are not allowable to the extent I have taken them, the simplicity of the process, perhaps, may make the corrections that would be necessary easy to be understood, and I may be informed by others more competent to such inquiries, what the amount due to such corrections is.

Some general considerations seem to strengthen these conclusions.— The second law of motion says, that a body in motion is acted on by a force, in *amount* and *direction* the same as a body at rest. Now, in actuating the motion of a pendulum, it cannot be said that any motion is imparted at all, but only a direction given to a motion derived from gravity; now, if direction only is the thing imparted, the force for the motion existing before, it is this direction only that is independent and likely to remain unchangeable. Besides, I understand the nicest experiments have been made long ago, proving that a pendulum vibrates in exactly the same time in whatever azimuth it oscillates; and as the earth's rotation, if it could have any effect in deflecting the plane of a pendulum hanging free, would have an equivalent effect in retarding the motion of one which, for such experiments, must be constructed to move in a prescribed direction, we may presume no such effect belongs to the rotation. Still, on this last point, it must be conceded, that it is perhaps not known whether the times of a pendulum hanging quite free would be the same as one constrained to vibrate in one direction, as all those attached to common clocks are.

Another point may be just mentioned; it seems probable that any deflection of the plane produced by gravity during the motion of the point of suspension in a circle, would not be all one way—that what might be produced in one direction through the first quadrant would be reversed in the second—what might be produced in the third would be reversed in the fourth, or *vice versâ.* If these views raise the presumption that the direction of the plane may be invariable at all latitudes, or that at any rate the compendious method of the two cannot be applied, it becomes next to inquire in what man- apparent motion of the plane ifest itself on the graduated any latitude. This inquiry, necessary, has not yet been nto by any of the writers on it,—their mode of investigation r its object to show the exact f the plane on the horizontal f. In an inquiry into the effect otion of an invariable plane on ontal table, it is necessary to the calculation the inclination rizontal plane to the plane of and the direction of its inclina- is the *projection* of the appa- ible motion of the plane of on the equatorial plane upon ontal plane that is required. have no occasion to calculate, ready provided to our hands in m of the shadow of an hori- n-dial: the number of degrees ites for every hour of the dial distances for every correspond- of the apparent motion of the i's plane. It will not be neces- illustrate this further, as the consideration will be sufficient the analogy of any invariable ersecting the plane of rotation shadow of the stile of a sun- ch is always parallel with the e earth. For the latitude of the distances will be as fol-

from the North both ways		11° 51′
....	....	12° 28′
	....	13° 44′
....	....	15° 33′
....	....	17° 30′
East and West	......	18° 54′

parent motion of the plane of , if invariable, on an horizontal each direction, as stated above, ue quantity derived from the notion of the earth of 15° per ertainly as 15° itself would be ity at the poles.

now remains to inquire how riments sanction either view. r case are they decisive. Gene- r are represented as giving an motion rather more than the he latitude require; and if the nts have been mostly confined ction nearly north and south, s reason to think probable, it would be slightly in favour of these views. Then the experiments at Dublin showed that the rate of motion varied, being least between the north and east and greatest between north and west, but the amount of difference is not stated; this also in some slight degree supports the views here stated. It must, however, be clearly admitted that no experiments have been reported that sanction in any degree the rate of motion in an east and west direction which these views require. Perhaps more accurate measurements of the angular motion in all directions may open up some further views, either more clearly to confirm the received theory or to point out more distinctly where it fails. If any experiments could show decisively the effect said to be produced by the little instrument invented by Professor Wheatstone, or even if full reliance could be placed upon the reported performance of that instrument itself, in all positions, the question would be at an end. In matters of this sort any theory not supported in all points by experiments applied successively to meet the difficulties of the question must be given up. No deduction of the reason from other facts and laws can stand comparison with the induction of a sufficient number of the special facts of the case. My wish being only to elicit truth, I have abstained from any harsh expressions or anything like a sneer at what I may consider unseemly airs assumed by some writers on this subject.

B. ROSSELL.

Leicester, June 24, 1851.

THE PATENT-LAW AMENDMENT BILL—"NO. III."

On Tuesday last Earl Granville laid before the House of Lords, from the Select Committee on the Law of Patents, a new bill, dubbed "No. 3." Our readers will recollect that there were already two bills in the field; namely, "No. 1," proposed by Lord Brougham, and "No. 2," brought forward under Government auspices. In the course of the discussion which took place on the present occasion, Lord Brougham observed, that "letters had reached him by scores from persons complaining that he had abandoned *his* bill about patents. Now his noble friend (Lord Granville) knew that he had not abandoned his bill, any more than he

(Earl Granville) had abandoned his. Both bills were referred to the Committee, and the principal provisions of each — with some important additions suggested in the Committee — had been mixed up and amalgamated, as it were, in the measure before the House."

This we find, on a comparison of the three bills, to be about the real state of the case; and as we have already laid the principal features of bills "No. 1" and "No. 2" before our readers, we shall now confine ourselves to the "important additions" introduced by the Select Committee.

These are; first, the entire exemption of the colonies from the operation of patents; and second, a provision for the advertisement of every application for a patent, in order that all and sundry may have due incitement to oppose the grant of it. Lord Brougham characterized these as "improvements," and the Select Committee are stated to have approved of them "unanimously." We humbly presume to think that they are, on the contrary, prodigious blemishes, and such as, if persisted in, must prove fatal not only to the bill No. 3, but to the whole set of bills. We did not altogether like either No. 1 or No. 2, but as both provided for the consolidation into one, of the three patents for England, Scotland, and Ireland (now so absurdly separated), we were willing, for the sake of so great an improvement as that, to overlook all minor objections, and well content that either bill should pass. If, however, the consolidation of the three patents is to be accompanied with the exemption of the whole of our colonies and foreign possessions from the operation of patents altogether,—*if the rights of inventors are only to be protected at home, and piracy of inventions is to be legalized throughout the whole of the foreign possessions and dependencies of the British Crown*,—and if, moreover, a legal obstruction unheard of before is to be interposed against the grant of patents even for the mother country,—then we must say that the advantage of consolidation would be dearly earned at such a price.

The exemption of the colonies is one of the most whimsical and irrational legislative projects which ever came under our observation. *Lord Granville did not assign a single reason for it*, and we defy him to produce a good one.

The plan of *advertising for opposition to patents* is not much better. It is a plan conceived in the interest of the law-officers of the Crown (*or rather of certain new functionaries proposed to be appointed* by the *Tertium Quid* Bill), —but wholly unjustifiable on grounds either of public policy or of private justice.

Both clauses are manifestly dictated by a spirit altogether adverse to patents and to the rights of inventors. The exemption of the colonies and foreign possessions is but a first step towards the *utter extinction of patent rights throughout the whole of the British Empire.* Indeed, Earl Granville and other lords frankly confessed that they were opposed on principle, to the system of patents altogether, and that in trying to amend the existing system, they but yielded to a conviction that the general sense of the country was so far against them, and called for amendment only—not abolition.

Inventors and patentees will now see, therefore, what they have to hope for at the hands of Government. The Bill, "No. 3," has still to go through the Commons, and it must be therefore opposed most promptly and vigorously, if inventive genius is still to have a place in this country among the sources of individual advancement and national prosperity. We shall return to the subject in our next.

THE FIRE AT LONDON-BRIDGE.

A person of the name of Finlay was the first who discovered the recent disastrous conflagration at London-bridge, and the first to enter the building after the fire had broken out. According to the *Times* of the 26th instant, he in evidence said, "His opinion was, that had he had a few pails full of water at hand when he first entered, he could have extinguished the flames, or have kept them under till the arrival of the engines." Evidence to the same effect has been given on many previous occasions when the outbreak of fire has been inquired into, and the simple means by which a ready supply of water could be insured in the interior of large buildings has been for half a century exemplified in Portsmouth Dockyard. Disregard of the example thus affords here amongst many instances the immense length of time requisite duce any general adoption of useful ntions, and hence the need arises for eated calls of attention to them, as done to the fire-extinguishing in the same dockyard. It is not possible to say that property to the of millions of money would ere have been saved, had similar ones very generally adopted for the protection of buildings from the ravages g. of

his disregard of the Portsmouth ex-has not arisen from any want of ity of the means employed, since, no industrial establishment—Crystal Palace excepted—has ever resorted to by so great a multitude as the wood-mills in that Arsenal, and visitants must have had little obser-while failed to notice the water-full, the hose, the buckets, conveniently distributed in every ready for instant use. In the of the recent conflagration, had a single such water-pipe been pro-with appropriate hose and buckets, communicating, as at Portsmouth, a cistern of water on the roof, there is every reason to believe, have been enabled to save pro-value, it is said, of 200,000*l*. either has disregard of the Ports-fire-extinguishing works arisen want of public information re-them through the press. Though al communications be usually con-to office, yet Sir Samuel Bentham, the Portsmouth works origi-having given a short description in his official "Statement of 1813, printed and largely dis-copies of that communication and afterwards published it; descriptions of them, more or have been given in the *Mech.* Nos. 1268, 1355, 1869, 1439, and *Builder*.

of Sir Samuel's applications of facts to useful purposes was of the fusibility of a compounded the heat of boiling water. As *fusible* metal appears in No. of the *Mechanics' Magazine*; he as a safety-valve, in the water-distilling apparatus of 1827; and in the *Builder* it proposed for plugs to close water pipes, as an important part of arrangements for the preservation of deeds and other manuscripts in cases of fire. So in buildings, especially those containing stores liable to spontaneous inflammation, it might be desirable to form portions of interior water-pipes of fusible metal, which, without the intervention of man, would melt, and introduce a body of water immediately when the surrounding heat rose to 212° Fahr. The *Times*, some months ago, mentioned a fact which proves that such a provision for the immediate application of water is far from a visionary speculation. The paper related, that on the outbreak of fire in the basement storey of some house in the city, the leaden water-pipe in the back of the building was found to have melted by the surrounding heat, and that the water thus admitted from the house cistern had been sufficient to extinguish the flames.

Should fusible metal be introduced for the purposes indicated, experiment would be requisite to determine its power of resisting pressure, as on this the necessary thickness of water-pipes would depend. Indeed, so little use has hitherto been made of this composition, that its properties, other than that of fusibility, are little known: nor does it seem that mixtures in different proportions of the several metals composing it have been tried with a view to ascertain at what different low temperatures a compound metal might be made to melt.

It will be seen, in some of the numbers of the Magazine above referred to, that by slow degrees the fire-extinguishing works at Portsmouth are being imitated. They have lately been so, to a certain extent, at the British Museum,—Mr. Braidwood (it is understood) having visited the dockyard there to examine them: they are said to be provided for the security of the Crystal Palace, and are thus likely to attract attention; but their adoption would be most likely to become general were some private civil engineer to devote himself to the fire-extinguishing branch of the profession; it would lead as much to his own emolument and credit as it certainly would to public benefit. Private interest is far more efficacious for the introduction of improvements than any representation of general good ever has been, or seems likely ever to be.

There are various precautions in respect to the construction and application of buildings, which would, if observed, doubtless be productive of much security against conflagration; some have been already particularized in the above-quoted numbers of the Magazine. One measure which Sir Samuel was exceedingly desirous of introducing in naval arsenals was, that of storing all particularly inflammable matters in *under*-ground compartments, and, when near a river or the sea, below the level of the water, so that it might be let into and submerge any one compartment, without damage to goods in other stores. The store cellars of his contrivance, over the reservoir in Portsmouth Dockyard, were perfectly well ventilated and dry. He would probably have recommended the stowage of rags in such compartments, appropriating the upper floors of a warehouse to less inflammable substances. He had at different times long official controversy as to a more appropriate arrangement of stores in naval arsenals; in them it was merely official habit that had to be overcome; in the case of private storehouses their proprietors naturally look to the appropriation that will best pay, yet still a classification of stores, in respect to inflammability, would be very useful, having regard also to other sources of deterioration than fire. There would be the head of highly inflammable stores—stores not easily ignited—stores injured by moisture—those which bear dampness with impunity, and so forth. Merchants, it is true, possess general ideas as to the kind of storehouse that suits well their commodities respectively, yet such a classification would lead to beneficial results; it would probably be found that many articles now consigned to cellars would be better above ground;—that existing warehouses are not provided with a sufficiency of underground accommodation for inflammable stores;—that little or no regard is had to the due ventilation of cellars, still less to their being made dry—a condition much to the present purpose, since neither rags nor hops could be kept in a moist atmosphere, consequently must continue to be stored in upper floors, though it should lead to such destructive conflagrations as that which is now deplored.

M. S. B.

July 2, 1851.

BUTCHER'S REGISTERED SELF-ACTING CHIMNEY-GUARD.

(Mr. William Butcher, St. James's-place, Bermondsey, Proprietor.)

Fig. 1. Fig. 3. Fig. 2.

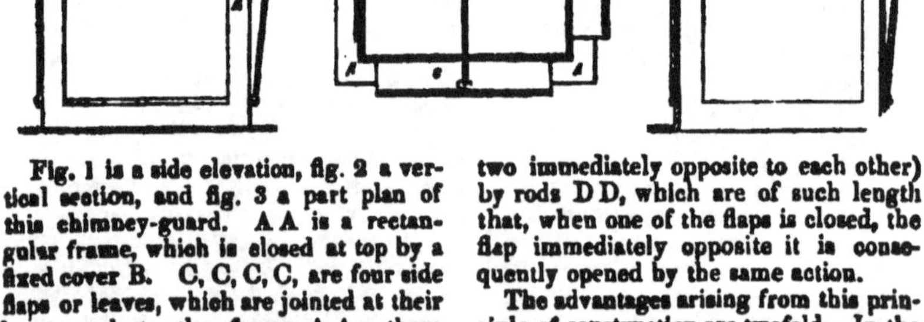

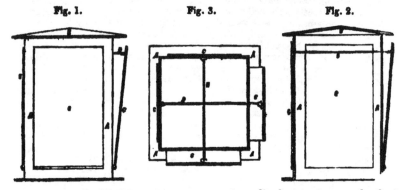

Fig. 1 is a side elevation, fig. 2 a vertical section, and fig. 3 a part plan of this chimney-guard. A A is a rectangular frame, which is closed at top by a fixed cover B. C, C, C, C, are four side flaps or leaves, which are jointed at their *lower ends to* the frame A A: these *flaps are connected in pairs* (that is, the two immediately opposite to each other) by rods D D, which are of such length that, when one of the flaps is closed, the flap immediately opposite it is consequently opened by the same action.

The advantages arising from this principle of construction are twofold. In the first place, when the wind blows against

n of the guard, it shuts the flaps are exposed to its action, and hose on the opposite side or sides escape of smoke for maintaining ght of the chimney; and in the place, the close cover at top at s prevents any down draught and g from that cause.

OSPECTS OF ELECTRO-MAGNETISM AS A MOTIVE POWER.

Letter of Professor Page to the *Scientific American*.)

ter, referring to my preference for ry form of the engine, says, I have back upon Davidson's and Avery's As to Davidson's engine, it was ted by myself on a large scale in Bos- 1837, and it was invented and tried imore by Dr. Edmondson, in 1834. *lliman's Journal*.] But the writer rehends the case. I have "fallen upon no one. The rotary form of l engine, as well as the reciprocating, most essentially from any engines fore tried. In my reciprocating , the *magnetic piston*, if I may so is impelled with nearly an equal hroughout the stroke, and this for *gth of stroke* desired. The rotary the perfection of the improvement, es not seem to involve the difficulties it in rotary steam-engines, for my require no *packing*. When the de- n of my engine is published, which ere long, I think the writer referred l others will appreciate its peculi- I have never claimed for electro- ic power that it is or would be supe- steam; that is, in every respect; nor ecessary that it should be, to answer rposes of my investigations. The the power has been with me a sub- e question, knowing full well that more important questions had to be first, before ever the cost could be ascertained. The abstract rule laid y M. Joule, Messrs. Hunt, Scoresby, d, and others, of the absolute duty ned by a given quantity of zinc, is nough as far as their experiments but is of little or no value in the al question of the availability of this To illustrate my meaning, take the t duty of coal in the best condensing s in the world; will any one pretend that there is no room for improve- even there? Why, in the Cornish s, within a few years, the expense of e-power has been reduced from 10d. per diem. But suppose it be admitted that the minimum cost has been attained, how many engines in the world can be worked as cheap as those engines? In reality, M. Joule's calculation makes the expense of magnetic power less than is steam power at the present day in some of our locomotive engines. The cost, therefore, I say, is not the practical question; and if the magnetic power will cost more than the dearest steam power, still, if we render it an *available* power in other respects, it must come into use for many and perhaps most purposes, by reason of its great advantages over steam in point of safety, simplicity of construction, readiness for operation, compactness of machinery, and, lastly, one very important condition, namely, there need be no consumption of material when power is not wanted for use.

* * * *

Magnetism, it is yet in its infancy, and steam is full grown. The proper appreciation of magnetic power is to be had by comparing it with steam in an equal stage of its development, when it will be seen that the magnetic power rather carries the palm. Steam power has not yet reached its climax, but it seems as if it were approaching its culmination, as its march seems to be comparatively slow; while magnetic power, evidently in its inception, is progressing rapidly. The first steam locomotive applied in England, in 1804, made, on a level plain, five miles an hour with about 15 tons; and ten years after, the celebrated Mr. Stephenson constructed a locomotive which was considered a great improvement, and carried eight carriages, about 30 tons, four miles an hour; and in 1829, after twenty-five years of experience (and all the while "invention was stimulated by necessity"), Mr. Stephenson produced his locomotive, the "Rocket," which made an average speed of fifteen miles an hour, with 17 tons, consuming about 1 lb. of coke per mile to a ton, as in the two trips of seventy miles 1,085 lbs. of coke were consumed. With my magnetic locomotive just as it is, I would willingly have entered the *list* with the "Rocket" in point of power, speed, and expense of working. I feel confident, however, that the magnetic locomotive is capable of carrying two loaded passenger-cars to Baltimore, at the rate of twenty miles an hour, as soon as some of the very great and obvious defects are remedied.

I had lately an opportunity of seeing how great was the friction of the machinery of the locomotive. They have at our station here, one of the largest and strongest horses I ever saw, and he is well trained to the work of pulling cars. In removing the magnetic car from its station, this horse was attached to it, but was found to be unable to pull it up the grade over which the car was propelled by magnetism six miles an hour. It required five men and this horse to get the car over this grade, and it was lighter by two tons than when driven by magnetic power; and moreover, when it ascended this grade at six miles an hour, the power of the battery was not fully up; and I have discovered a cause of great additional friction when the engine was in action, the remedy for which is obvious.

In regard to the doctrine of Liebig, that the zinc cannot give out more power than the coal required to smelt it, it is unfortunate, and though entertaining the highest respect for his reputation and ability, I must pronounce it a *practical absurdity*. It is reasonable to suppose that a given amount of zinc combining with oxygen, would not eliminate more heat than would be required to overcome this affinity, but we have no proof of any such relation of electricity to heat as to make the mechanical power of the one the measure of the mechanical power of the other. Whatever may be the connection and analogy between heat and electricity, we must consider them as distinct forces in their mechanical relations. In the combustion of coal we develope heat as the motive force, and no electricity; in the oxidation of zinc in the battery, we develope both heat and electricity, the latter only being the motive force. The *equivalents* of forces regarding affinities may be interesting as a matter of speculation, but, as furnishing a practical estimate for the amount of mechanical or available power, it cannot stand, and necessarily involves the unwarrantable assumption that the whole power or inherent force may be eliminated and rendered available in each case. But Liebig goes still further; he maintains that the heating power of the current is the equivalent of its mechanical power through electro-magnetism; or, in other words, that the heat developed by the passage of the current ought to raise steam enough to furnish a power equivalent to the electro-magnetic power of the same current; and from the fact that the mechanical force derived from steam raised by the heating power of the current is so small compared with that obtained by the combustion of coal, he arrives at the conclusion that electro-magnetic power "can never be useful." The speculation is thus pursued up to a point where facts are brought in to its support, and fortunately where facts enough can be adduced to subvert the whole doctrine. I will take but one, and one that can be easily admitted; or, rather, I will propound a question, "How many pairs of *plates would be required to operate through* their calorific or steam power the lever of the receiving magnet in Morse's Telegraph, say through a circuit of 40 miles?" I saw an experiment some years ago at the Capitol, when gunpowder was fired through this length of circuit, the powder being at the Capitol and the battery at Baltimore. Fifty pairs of Grove's battery, such as they used for the telegraph, would not ignite a platina wire one-thousandth of an inch in diameter. It finally required seventy-five pairs to fire the powder. Ten pairs of such plates will work the receiving magnet through this circuit vigorously. I leave it to mechanical minds here to form their own conclusions. The truth is, that the cost of electro-magnetic power, or any other power, is circumstantial, and the attempt to predicate the whole economy of magnetic power upon the cost of coal and cost of zinc, and the fact that coal is found native and zinc not, is, in effect, to make Nature's laws and operations amenable to market prices and other contingencies.

ON THE CONSTRUCTION OF STEAM BOILERS AND THE CAUSES OF THEIR EXPLOSIONS.

BY WM. FAIRBAIRN, ESQ., C.E., F.R.S.

(Continued from vol. iv., p. [illegible].)

In the flat ends of cylindrical boilers, and those of the marine construction, the same rule applies as regards construction; and the due proportions of the parts, as in those of the locomotive boilers, must be closely adhered to. Every description of boiler used in manufactories, and also those on board of steamers, should in my opinion be constructed to a bursting pressure of 400 to 500 lbs. on the square inch, and locomotive engine boilers, which are subjected to a much severer duty, to a bursting pressure of 600 to 700 lbs.

It now only remains for me to state that internal flues, such as contain the furnace in the interior of the boiler, should be kept as near as possible to the cylindrical form; and as wrought iron will yield to a force tending to crush it of about one-half of what would tear it asunder, the flue should in no case exceed one-half the diameter of the boiler, and with the same thickness of plates it may be considered equally safe to the other parts. In fact, the force of compression is so different to that of tension, that I should advise the diameter of the internal flues to be in the ratio of 1 to 2½ instead of 1 to 2 of the diameter of the boiler.

Various notions are entertained as to the causes of boiler explosions, and scientific men are not always agreed as to whether

ent from excessive pressure due to
mulation of heat, or to some other
uch as the explosion of hydrogen
erated by the decomposition of
ddenly thrown on heated plates, of
where an exceedingly indefinite spe-
That of the decomposition of
, I believe, a somewhat prevalent,
but I apprehend it cannot be the
le cause, inasmuch as in that case
assume the boiler to be nearly
of water, and the plates over the
red-hot.
not unreasonable to suppose that a
such sudden origin, and so imme-
destructive in its effects, should
the presence of an explosive mix-
I think it will be difficult, if not
to account for the accumulation
cient quantity of hydrogen without
ence of oxygen and other gases in
proportions to form an explosive
Now, as these equivalents are
enerated all at once by the simple
sition of water (admitting for the
that the water is decomposed) we
k for some other cause for the fatal
ructive accidents which of late years
so prevalent.
ating of this subject, I hope to
only what are the probable causes
sion, but what appears equally im-
what are not the causes. So many
(some of them exceedingly prob-
) have been brought forward on the
of disastrous explosions, that it
the utmost care and attention to
tances before they are generally ad-
To acquire satisfactory evidence as
precise condition of the boiler and
before an explosion is next to im-
as most frequently the parties in-
and from whose mismanagement and
we may in many cases date the ori-
the occurrence, are the first to
the victims of their own indiscre-
we can only judge from the havoc
tation that ensues as to the imme-
use of the event.
this it follows, that in many of the
no record, few, if any, of the
umstances of the case are made
and we are left to draw conclusions
appearances of the ruptured parts
terrific consequences which too fre-
follow as a result. This want of
n as to the precise condition of a
with all its valves and mountings,
g an explosion, is much to be
d, as it causes a degree of mystery
und the whole transaction; and the
d, sometimes inaccurate, testimony
ence, but too often baffles all

attempts at research, and creates additional cause of alarm to all those exposed to the occurrence of similar dangers.

In the discussion of this subject I shall however, endeavour to trace from a number of examples in which I have been personally engaged, and from others which have come to my knowledge,—the causes which have led to the disastrous effects.

In my attempts to ascertain facts by a course of reasoning which I shall have to follow in this investigation, I could wish it to be understood that it is not my intention to raise doubts and fears in the public mind, calculated to arrest the progress of commercial enterprise, or to cripple the energies of mechanical skill. On the contrary, I am most anxious to promote the advancement of the useful arts, to increase our confidence in the application of increased pressure, and to secure within moderate bounds the economical and useful employment of one of the most powerful agents ever known in the history of practical science. My object in this inquiry will, therefore, be to enlarge our sphere of action by a more comprehensive knowledge of the subject on which it treats; to induce greater caution along with improved construction; and to insure confidence in all those requirements essential to the public security.

For the attainment of these objects it will be necessary to divide the subject into the following heads:—

1st. Boiler explosions arising from accumulated internal pressure.

2nd. Explosions from deficiency of water.

3rd. Explosions produced from collapse.

4th. Explosions from defective construction.

5th. Explosions arising from mismanagement or ignorance; and

6th. The remedies applicable for the prevention of these accidents.

1st. *Boiler explosions arising from accumulated internal pressure.*—In nine cases out of ten a continuous increasing pressure of steam, without the means of escape is probably the immediate cause of explosion; in some instances it arises from deficiency of water, but accidents of this kind are comparatively few in number, as we often find in tracing the causes that they have their origin in undue pressure, emanating from progressive accumulation of steam of great force and intensity. Let us take an example—to some of which I am able to refer—and we shall find that a boiler under the influence of a furnace in active combustion (as the recipient of heat) will generate an immense quantity of steam, and, unless this is carried off by the safety valve or the usual channels when generated, the greatest danger may be apprehended by the continuous increase of pressure that is taking place within the boiler. Suppose that from some cause the steam thus accumulated does not escape with the same rapidity with which it is generated, that the safety valves are either inadequate to the full discharge of the surplus steam, or that they are entirely inoperative, which is sometimes the case, and we have at once the clue to the injurious consequences which, as a matter of fact, are sure to follow. The event maybe procrastinated,and repeated trials of the antagonist forces from within and the resistance of the plates from without may occur without any apparent danger, but these experiments often repeated will at length injure the resisting powers of the material, and the ultimatum will be the arrival of the fatal moment when the balance of the two forces is destroyed and explosion ensues. How very often do we find this to be the true cause of accidents arising from extreme internal pressure, and how very easily these accidents might be avoided by the attachment of proper safety valves to allow the steam to escape and relieve the boiler of those severe trials which ultimately lead to destruction. If a boiler, whose generative power be equal to 100, be worked at a pressure of 10 lbs. on the square inch, the area of the safety valve should also be equal to 100, in order to prevent a continuous increase of pressure; or in case of the adhesion of any of the valves, it is desirable that their areas should, collectively, be equal to 100. If two or more valves are used, 100 or 120 would then be the measure of outlet. Under these precautions, and a boiler so constructed, the risk of accident is greatly diminished; and provided one of the valves is kept in working order, beyond the reach of interference by the engineer, *or any other person*, we may venture to assume that the means of escape are at hand, irrespective of the temporary stoppage of the usual channels for carrying off the steam.

So many accidents have occurred from this cause—the defective state of the safety valves—that I must request attention whilst I enumerate a few of the most prominent cases that have come before me. In the year 1845 a tremendous explosion took place at a cotton mill in Bolton. The boilers, three in number, were situated under the mill, and from unequal capacity in the safety valves, and even those imperfect, as they were probably fast, a terrific explosion of the weakest boiler took place, which tore up the plates along the bottom, and the steam having no outlet at the top, not only burst out the end next the furnace, demolishing the building in that direction, but tearing up the top on the opposite side, the boiler was projected upwards in an oblique direction, carrying the floors, walls, and every other obstruction before it; ultimately it lodged itself across the railway at some distance from the building. Looking at the disastrous consequences of this accident, and the number of persons—from sixteen to eighteen—who lost their lives on the occasion, it became a subject of deep interest to the community that a close investigation should immediately be instituted, and a recommendation followed that every precaution should be used in the construction as well as the management of boilers.

The next fatal occurrence on record in this district was a boiler at Ashton-under-Lyne, which exploded under similar circumstances, namely, from excessive interior pressure, when four or five lives were lost—and again at Hyde, where a similar accident occurred from the same cause, which was afterwards traced to the insane act of the stoker or engineer, who prevented all means for the steam to escape by tying down the safety valve.

There was a boiler exploded at Malaga, in Spain, some years since, and my reason for noticing it in this place is to show that explosions may be apprehended from other causes than those enumerated in the divisions of this inquiry, and that is *incrustation*. Dr. Ritterbandt says—in a paper read before the Institution of Civil Engineers, by an eminent chemist, Mr. West—"That a sudden evolution of steam under circumstances of incrustation is no uncommon occurrence." In several instances I have known this to be the case, particularly in marine boilers, where the incrustation from salt water becomes a serious grievance, either as regards the duration of the boiler, or the economy of fuel.

If it were supposed, as Dr. Ritterbandt observes, that the boiler was incrusted to the extent of half an inch, it would at once be seen that nothing was more easy than to heat the boiler strongly, even to a red heat, without the immediate contact of water. Under these circumstances, the hardened deposits being firmly attached to the plates, and forming an imperfect conductor of heat, would greatly increase the temperature of the iron, and the great difference of temperature thus induced between the material—and the greater expansibility of the iron—would cause the incrustation to separate from the plates, and the water rushing in between them would generate a considerable charge of highly elastic steam, and thus endanger the security of the boiler.

These phenomena were singularly exemplified in the Malaga explosion, which is thus described by Mr. Hick:—"I have

...ined that a very thick incrustation of ...as found on the lower part of the ... immediately over the fire, and so far extended, the plates appear to have ...ed hot, thereby much weakened, and the explosion. The ordinary working ...re of the boiler is 130 lbs. per square ...and perhaps at the time of the explo... ...ry much above that pressure, as there ...nly one small safety-valve of 2¼ ins. ...ter. The boiler was only 2 feet 6 ins. ...ter, and 20 feet long.

...rustation, exclusive of being dange... ...is attended with great expense and ... to the boiler by its removal. In the ...f the transatlantic, oriental, or other ...sea-going vessels—even after the use of pumps, blowing out, &c.—a very large ...t of incrustation is formed, and con... ...ble sums of money are expended each ... to remove it.

...er explosions of a more recent date ...ose which occurred at Bradford and ...x. They are still fresh in the recol... ...of the public mind, and are so well ... as not to require notice in this ...

...nnot, however, leave this part of the ...t without reverting to an accident ...occurred on the Lancashire and York... Railway, which had its origin in the ...cause—excessive internal pressure. ...ccident is the more peculiar as it led ...ng mathematical disquisition as to the ... of the forces, which produced results ...e curious and interesting. The con... ...s which I arrived at, although *prac... right, were, however, considered by mathematically wrong*, as they were ... combatted by several eminent mathe... ...ians; and notwithstanding the number ...gebraic formulas, and the learned dis... ...ns of my friends on that occasion, I ...been unable to change the opinions I ...ormed to others more conclusive.

...e accident here alluded to, occurred to ...*rk* locomotive engine, which in Fe... ...y, 1845, blew up and killed the driver, ...r, and another person who was stand... ...ear the spot at the time. A great dif... ...e of opinion as to the cause of this ...nt was prevalent in the minds of those ...witnessed the explosion, some attribut... ...to a crack in the copper fire-box, and ...s to the weakness of the stays over the ... neither of these opinions were, how... ...correct, as it was afterwards demon... ...d that the material was not only entirely ...rom cracks and flaws, but the stays were ...d sufficient to resist a pressure of 150 ...0 lbs. on the square inch. The true ... was afterwards ascertained to arise ... *the fastening down of the safety valve ... engine (an active fire being* in operation under the boiler at the time), which was under the shed, with the steam up, ready to start with the early morning train.

The effect of this was the forcing down of the top of the copper fire-box upon the blazing embers of the furnace, which, acting upon the principle of the rocket, elevated the boiler and engine of 20 tons weight to a height of 30 feet, which, in its ascent, made a summerset in the air, passed through the roof of the shed, and ultimately landed at a distance of 60 yards from its original position. The question which excited most interest, was the absolute force required to fracture the fire-box, its peculiar properties, when once liberated, and the elastic or continuous powers in operation, which forced the engine from its place to an elevation of 30 feet from the position on which it stood. An elaborate mathematical discussion ensued relative to the nature of these forces, which ended in the opinion that a pressure sufficient to rupture the fire-box was, by its continuous action, sufficient to elevate the boiler and produce the results which followed. Another reason was assigned—namely, that an accumulated force of elastic vapour, at a high temperature, with no outlet through the valves, having suddenly burst upon the glowing embers of the furnace, would charge the products of combustion with their equivalents of oxygen, and hence explosion followed. Whether one or both of these two causes were in operation is probably difficult to determine; at all events, we have in many instances precisely the same results produced from similar causes, and unless greater precaution is used in the prevention of excessive pressure, we may naturally expect a repetition of the same fatal results.

(To be continued.)

SPECIFICATIONS OF ENGLISH PATENTS ENROLLED DURING THE WEEK ENDING JULY 3, 1851.

WILLIAM HODGSON GRATRIX, of Salford, engineer. *For certain improvements in the method of producing or manufacturing velvets or other piled fabrics.* Patent dated December 26, 1850.

These improvements are applicable only to those descriptions of velvet and other piled fabrics in which the pile is produced by the weft thread, and the cut is consequently made in the direction of the warp.

Claims.—1. A method of producing or manufacturing velvets or other piled fabrics by weaving the pile threads over a series of fine longitudinal knives with elongated points of wire—such knives being stationary, and having their cutting ends or extremities attached to a suitable holding frame (and being kept in a state of tension by weights attached to their points), so that simultaneously with the weaving, the portions of the weft intended to form the pile slides consecutively upon the points of the knives as the cloth is woven, until it arrives at the cutting portions of the knives, by which the weft thread is severed, the velvet or other fabric being then wound upon the cloth beam as usual.

2. A method of producing or manufacturing velvets and other piled fabrics, by weaving in the cloth, a series of fine wires, passing through the spaces between the pile threads and the cloth, which wires are subsequently separated or removed from the cloth by passing the fabric between two rollers or bowls (one composed of a yielding material, such as paper—the second of metal), acting by friction or pressure, so as to sever the pile threads, and leave the velvet or other fabric complete.

George Edward Dering, of Lockleys, Herts, Esq. *For improvements in the means of and apparatus for communicating intelligence by electricity.* Patent dated December 27, 1850.

The improvements claimed under this patent comprehend—

1. The use of elastic supports in place of axes of suspension for telegraphic arrangements. [This system is particularly applicable to the indicators and other similar moveable parts, whether such moving parts are permanent magnets, or composed of metal rendered temporarily magnetic.]

2. A method or methods of applying the force of gravity to restore magnetic needles and other similar telegraphic arrangements to their position of rest, after having been set in motion, by the passage of currents through coils placed in juxtaposition with them. [In one arrangement for this purpose, the centre of gravity of the needle is placed immediately below the centre of motion, by attaching to the needle (which is itself equally balanced on either side of the centre of motion), a button of metal of sufficient weight to bring the needle quickly to rest—the principle here adopted being that of the quick beat of a short pendulum as compared with that of a long one. Another method of effecting the same object is, to suspend the needle from its upper end, which is provided with a triangular aperture, through which is passed a hook of circular wire, which forms the suspender. According to a third method, the needles are suspended by magnetic attraction.]

3. A method of applying electro-magnetic coils to produce motion for telegraphic purposes. *[In this arrangement the needle is* placed or suspended in a vertical position, and instead of being influenced by deflection, is caused to vibrate by the direct attractive and repulsive action of coils, which are so placed that their axes are parallel to or in the plane of vibration of the needle. The same method of producing motion is equally applicable to needles moving on a central pivot or suspended by their upper extremity.]

4. An arrangement for telegraphic alarums, by which, when included in the same circuit with the signalling apparatus, they are prevented sounding during the ordinary working of the instruments, though ready to be acted on at any moment when required. (This is effected in double needle telegraphs by reserving from use in sending messages, one particular signal or motion of one or both needles, and employing that motion only when it is desired to sound the alarum.)

5. A method of transmitting secret intelligence to any one or more stations on a line at choice without the use of an extra wire. [Each station is provided with a metallic disc, capable of revolving by a step by-step motion worked at pleasure by a current in the ordinary manner, and having pieces of ivory or non-conducting material, let into the surface of its periphery which revolves in contact with a metallic spring.] The bearings of the disc and the spring are both in connection with the extremity of the coil which actuates the indicator motion, the coil being in the same circuit with the line wire. When the discs are caused to revolve, and the springs rest on the conducting surfaces, a short circuit is established which permits the passage of the electricity from end to end of the line wires without affecting the indicating instruments; but when the springs rest on the non-conducting materials, then the short circuit will be broken, the indicator brought into the line wire circuit, and the messages delivered as in the ordinary working of the telegraph. In order to prevent the working of the discs under ordinary circumstances, it will be requisite to adopt some such as the following system: that is, to employ the reverse current to produce the step-by-step motion of the discs when a current in one direction only is used for the conveyance of messages, or to arrange the step-by-step motion so that it shall remain unaffected, except when increased battery power is applied.

6. A method of counteracting the effects of currents of atmospheric electricity by introducing into the circuit an opposing current of equal force [produced by a galvanic battery or other suitable means].

7. The use of pairs of conducting surfaces, for the purpose of carrying off atmo-

electricity collected in suspended [These surfaces are two brass plates ed with a file, or with grooves cut them in opposite directions, and a linen cloth interposed. One plate ected with an extremity of the line d the second with the earth, or the re attached at each terminal of the es between them and the indicating ents. By these means the atmo- electricity contained in the sus- wires instead of passing through the r coil to the injury of the appara- nected therewith, is conducted at o the earth.]

self-acting arrangement for the same . [Two brass balls are suspended by om a brass bar which is in connec- h the line wires, in such a manner y rest lightly against each other, when a sufficient amount of electri- collected in the wires they may be , and coming in contact with two ates, be held in that position until ipe of the electricity through the nto the earth shall allow them to their original state of rest.]

e use for the same purpose of strips llic leaf introduced into the circuit ich under the influence of a strong of electricity (such as is produced h of lightning) in the wires may be d thus break the circuit, and permit pe of such electricity into the earth of passing through the coil into the r].

, method of improving the insula- telegraphic wires by applying an second bell of insulating material he outer one.

STE MENOTTI, of Rue de la Paix, *For certain chemical compositions dering cotton, linen, woollen, silk, er fabrics impervious to water, fixing colours in dyeing."* Patent ecember 27, 1850.

vords of the title contained between commas have been disclaimed. The adopted by the patentee in manu- g his improved waterproofing com- , which he calls "hydrofugine," is s:—1. In a vessel capable of hold- e gallons, he places 22 lbs. of sul- alumina, or of potash and alumina, umina and ammonia, or sulphate of opper, or chloride of tin, reduced to (with the exception of the sulphate na, which is cut in slices, and the of tin, which is employed in a crys- state.) 2. In a second similar ves- places 14 ozs. of oleic, stearic, or acids (the two last of which sub- re known in commerce as stearine), e soap or saponaceous matter. 3. He mixes the oleic acid, or dissolves the other substances, by the aid of heat, in two gallons of alcohol at 30° Cartier, and pours this compound on the salt employed; then submits the whole to a temperature of about 30° Reaumur, and obtains the hydrofugine in a dry, powdered, or moulded state. For waterproofing cotton and linen cloth, one part of the composition is dissolved in 100 parts of water, and the fabric, after having been soaked in the solution, is hung up to dry, and is then in a finished state. For silk and woollen, one part of the composi- tion is added to 200 parts of water. Paste- board, cords, and other materials, may be also treated with this composition, which possesses the advantage of permitting the passage of air, whilst it effectually excludes wet and moisture.

Claim.—The processes described for the production of the composition called by the inventor hydrofugine, and its application for rendering linen, cotton, silk, woollen, and other fabrics impervious to water, but per- vious to air.

JOHN MATHISON FRASER, of Mark-lane, merchant. *For improvements in the manu- facture of sugar.* (A communication.) Pa- tent dated December 27, 1850.

This invention consists in a method of treating the expressed cane or beetroot juice by adding quicklime, and subsequently a saturated aqueous solution of sulphurous acid, in certain specified proportions. The acid for this purpose may be procured by the direct combustion of sulphur in atmo- spheric air, or by heating metallic oxides (such as that of manganese) and sulphur, or by decomposing sulphuric acid by the aid of heat, and of certain metals (such as mercury or copper), or carbonaceous materials (such as wood charcoal.) The last is the process preferred by the patentee, and in carrying it into effect he employs an apparatus consist- ing of three closed vessels communicating with each other by pipes, the first to contain the wood charcoal operated on (which is coarsely powdered and imbued with eight times its volume of sulphuric acid;) the second, water to wash the acids evolved; and the third, the water to be saturated. The application of heat to the first vessel in the range causes the evolution of sulphurous and carbonic acid gases, which pass through the water in the second vessel into the third, the water in which is saturated with the sulphurous acid, whilst the carbonic acid gas is allowed to escape, together with the super- abundance of the sulphurous acid. This loss may, however, be prevented by employ- ing a larger number of receivers. The "sulphurous solution" thus obtained should be bottled off into carboys, and corked up for use. Its specific gravity should be about

1·05, and it should contain thirty volumes of gas to one of water.

The method in which quicklime and the solution are employed is as follows:—The expressed cane-juice is poured into an open vessel through a sieve containing 0·312 lbs. Netherlands, of quicklime, and a similar quantity of lime is mixed with four cans Netherlands (about seven-eighths of a gallon) of juice, to about the consistence of cream, and held ready for use. This quantity is sufficient for 1000 cans, or about 220 gallons English measure. When about 500 cans of juice have run into the vessel, the above-named mixture is added, together with three cans of solution, and the whole well incorporated by stirring. After this the remainder of the juice is run in, another can of solution added, and the mixture stirred and allowed to settle. The clear liquor is drawn off when subsidence of the suspended particles has taken place, and is boiled in an open pan, the scum and flakes which rise during the operation being carefully and completely removed. The liquor will at first have an olive-brown colour and a peculiar odour, and will throw up brown flakes; but as the boiling proceeds the smell lessens, and the colour gradually changes to a green, and finally assumes a rich golden hue, throwing up at this time yellow-coloured flakes. When the colour is quite clear the boiling is discontinued, and the liquor is then fit for evaporation and crystallization in the ordinary manner. The boiling may be conducted in the vacuum pan, care being taken to remove the scum in this as in the former operation. This may be conveniently done about the time when the density of the liquor is about 38° Beaumé, equal to a specific gravity of 1·3.

The patentee does not claim generally the employment of quicklime and sulphurous acid in the purification of saccharine juices, but he claims the treatment of such juices in the manner and for the purposes set forth.

Edward Dunn, of New York, master mariner. *For an improved engine for producing motive power by the dilatation or expansion of certain fluids or gases caused by the application of caloric.* Patent dated December 26, 1850.

The patentee describes two arrangements of machinery for obtaining motive power by the dilatation of air or fixed gases, the caloric absorbed by the air during the heating process being surrendered, and again employed to heat a fresh supply. By this means a constant circulation of caloric is maintained, and the furnace employed only to supply the amount of heat lost during its transference from one supply of air to another.

Claims.—1. A regenerator, whereby the caloric in the air or other circulating medium, as it passes from the cylinder, is transferred to a series of discs of wire net or minute mineral or metallic particles, and again delivered to the working medium, either at stated intervals or at each successive stroke of the piston.

2. The combination of an expansion heater with the working cylinder, by which the fall of temperature consequent on the expansion of the circulating medium during the upward stroke of the piston is restored, and the force of the piston augmented to a greater extent than if no such retransfer of heat took place.

3. A heat-intercepting vessel attached to the working piston, by which any injuriously high temperature is prevented from reaching the packing of the piston, and by which also the very desirable end is obtained of presenting always surfaces of uniform high temperature to the acting medium under the working piston.

4. Placing the working and supplying cylinders in an inverted position, and leaving their ends open.

5. The direct attachment of the working and supplying pistons, by which arrangement the acting and reacting forces are uniformly distributed, and the maximum working effect of the pistons obtained.

Thomas Symes Prideaux, of Southampton, gentleman. *For improvements in generating and condensing steam, and in fire-places and furnaces.* Patent dated December 28, 1850.

Claims.—1. Coating the inside of steam boiler furnaces, and any desired part of the flue with fire-bricks, or any stone or composition having analogous qualities, and also placing one or more rows of water tube, cased outside with fire-brick, or any equivalent substitute at or behind the flue bridge. Also the use of a closed ash-pit under pressure in combination with this form of furnace.

2. A method of constructing tubular boilers [with a single tier of tubes laid in an inclined direction, and perpendicular conical tubes rising from them; the intermediate spaces forming the flue.]

3. The application of a lining of loosely woven porous or fibrous material to the inside of a steam chamber of steam boilers. [The lining is kept in its place by a wire frame, and should extend downwards into the water of the boiler to such a depth as to be kept constantly saturated with moisture by capillary attraction.]

4. The addition to any apparatus for heating steam in its passage from the boiler to the cylinder, or for heating the cylinder, of an automatic contrivance for regulating the heat applied according to the temperature produced.

5. The combination of a fan-blower and steam jet for producing a draught for the furnaces of locomotives and other non-condensing engines.

6. A method of constructing condensers by employing corrugated metal arranged to produce parallel spaces [which are occupied alternately by the steam to be condensed and the water employed for that purpose, which is supplied by a centrifugal pump or other means of causing a rapid circulation].

7. The construction of fire-places with bars made to open [for the supply of fuel under the incandescent mass] and the application of reflecting surfaces to grates so arranged with respect to their angle of inclination as to produce a maximum of useful effect.

8. The application of furnaces of an air-supply valve to be closed by any suitable apparatus at a determinate period after the supply of fuel.

9. The combined use in the ash-pit of reverberatory furnaces and steam boiler furnaces of water and air under pressure in closed ash-pits.

ALFRED VINCENT NEWTON, of Chancery-lane, mechanical draughtsman. *For improvements in the construction of metal shutters.* (A communication.) Patent dated December 27, 1850.

These improvements consist in forming the connecting joints to the strips of metal of which metallic shutters are composed by curving or bending the meeting edges in opposite directions, and sliding one curve within the other, the hollow being filled by inserting a rod of wire which serves to prevent indentation of the joint, and when bent down at the ends, increases its security, and prevents the parts from being disconnected. In order to prevent the forcing open of these shutters when closed, the ends may be turned in at right angles, and an L-shaped groove be provided for them to slide up and down in.

Claim.—The construction of metal shutters by the employment of the connecting joint formed as above described.

WILLIAM HENRY JONES, M.A., of Queen's College, and Chorley, Sussex, clerk. *For improvements in apparatus to be used when burning candles.* Patent dated December 28, 1850.

This improved apparatus, which the patentee calls an "acolyte," consists of a cap composed partly of metal and partly of glass, porcelain, plaster of Paris, or other non-conductor of heat, which fits loosely on the top of the candle, and descends gradually as it burns away, thus preventing swaling or guttering, at the same time that it causes a regular supply of melted wax, &c., on every side of the wick.

Caps composed wholly of metal or glass have been in use before, but the distinguishing features of the "acolyte" are a cap of metal, or heat-conducting material, combined with a guide-ring of non-conducting material.

The acolyte may be constructed with a collar, for the purpose of receiving a glass or paper shade.

JOHN RANSOM ST. JOHN, of New York, engineer. *For improvements in the construction of compasses and apparatus for ascertaining and registering the velocity of ships or vessels through the water.* Patent dated December 27, 1850.

The improvements claimed under this patent comprehend—

1. The application to mariners' compasses of satellites, or auxiliary needles, for indicating the amount and direction of any variation from the true meridian caused by local attraction.

2. A method of and apparatus for ascertaining the velocity of a ship through the water, by the action of the water on the blades of a fan-wheel.

3. An arrangement of log-glass, lever, and auxilliary contrivances, whereby the motion given to the clockwork by the reel is communicated to an index-hand during a definite period of time, the different parts of the apparatus being so proportioned, and the dial so divided, that the index, moving while the sand is running in the glass, may show the rate of speed at which the vessel is moving during a given time.

4. The application to logs of a parachute, and cylindrical wedge for keeping it extended in the water, and so arranged as to admit of its being readily withdrawn when the log-ship is drawn on board the vessel.

JAMES SLATER and JOHN NUTTALL SLATER, of Dunscar, bleachers. *For certain improvements in machinery or apparatus for the purpose of stretching and opening textile or woven fabrics.* Patent dated December 28, 1850.

The apparatus here specified is applicable to the opening and stretching of fabrics which have been attached together in continuous lengths for convenience in the dyeing and bleaching operations. The fabric, in a twisted state, is first passed through a tube, to which motion is communicated in a direction opposite to that of the twist in the fabric, and is thence conducted between two vibrating bars, which effect the opening, to a taking-up roller.

Claims.—1. The stretching and opening of textile fabrics, by machinery or apparatus arranged as described.

2. The opening of textile fabrics by a vibrating motion.

WEEKLY LIST OF NEW ENGLISH PATENTS.

John Platt, of Oldham, Lancaster, engineer, and Richard Burch, of Heywood, Lancaster, manager, for certain improvements in looms for weaving. July 3; six months.

James Howard, of the Britannia Iron Works, Bedford, agricultural implement maker, for improvements in ploughs, and other implements or machines used in the cultivation of the soil. July 3: six months.

John Aston, of Birmingham, manufacturer, for improvements in buttons and ornaments for dress, and the machinery for making the same respectively. (A communication.) July 3; six months.

Charles Payne, of Wandsworth-road, Surrey, gent., for improvements in drying animal and vegetable substances, and in heating and cooling liquids. July 3; six months.

Robert Haynes Easum, of Commercial-road, Stepney, Middlesex, rope-maker, for improvements in the manufacture of rope. July 3; six months.

William Hamer, of Manchester, for certain improvements in looms for weaving. July 3; six months.

George Kemp, of Carnarvon, North Wales, doctor of medicine, for a new method of obtaining power by means of electro-magnetism. July 3; six months.

Richard Jex Crickmer and Frederick William Crickmer, of Page's-walk, Bermondsey, engineers and co-partners, for improvements in packing stuffing-boxes and pistons. July 3; six months.

Charles Cowper, of Southampton-buildings, Chancery-lane, Middlesex, patent agent, for improvements in the preparation of cotton for dyeing, and bleaching. July 3; six months.

Charles Barlow, esq., of Chancery-lane, London, for improvements in rotary engines. (A communication.) July 3; six months.

LIST OF IRISH PATENTS FROM 21ST OF MAY TO THE 19TH OF JUNE, 1851.

James Hamilton, of London, engineer, for improvements in machinery for sawing, boring, and shaping wood. May 22.

Adolphus Oliver Harris, of High Holborn, Middlesex, philosophical instrument-maker, for improvements in barometers. June 10.

William Becket Johnson, of Manchester, manager for Messrs. Ormerod and Son, engineers, for certain improvements in steam engines, and in apparatus for generating steam; such improvements in engines being wholly or in part applicable where other vapour and gases are used as the motive power. June 10.

WEEKLY LIST OF DESIGNS FOR ARTICLES OF UTILITY REGISTERED.

Date of Registration.	No. in the Register.	Proprietors' Names.	Addresses.	Subjects of Design.
June 26	2861	Henry Squire	Willenhall	Lock.
"	2862	D. Hulett	Holborn	Apparatus for working valves of dry gas-meters.
"	2863	W. Higginbottom	Manchester	Stand pipe and valve.
27	2864	John Hall and Son	Lombard-street	Powder canister.
"	2865	George Thomson	Stirling	Apparatus to keep a carriage body in equilibrium under different circumstances of use.
28	2866	Longin Gantart	Glasgow	Machine for tramping and squeezing yarn and cloth by the dyeing and bleaching process.
"	2867	A. Marion	Regent-street	Index or book marker.
30	2868	Will Bishop and Robert Cooke	Boston; Huntingdon	Elastic tightener for trousers.
"	2869	Edward M'Keon, Thomas MacAnaspie, N. D. Maillard	Dublin	Spinning-wheel.
"	2870	Timothy Lonasgan	Ludlow	Spring-lock staple.
"	2871	G. Spill	London and Bristol	Thorough metallic ventilated coat.

CONTENTS OF THIS NUMBER.

LONDON: Edited, Printed, and Published by Joseph Clinton Robertson, of No. 166, Fleet-street, *in the City of London—Sold* by A. and W. Galignani, Rue Vivienne, Paris; Machin and Co., *Dublin; W. C. Campbell and Co., Hamburg.*

Mechanics' Magazine,

MUSEUM, REGISTER, JOURNAL, AND GAZETTE.

No. 1457.] SATURDAY, JULY 12, 1851. [Price 3*d*., Stamped, 4*d*.

Edited by J. C. Robertson, 166, Fleet-street.

WALKER'S REGISTERED DOUBLE-ACTING SCREW PRESS.

Fig. 1.

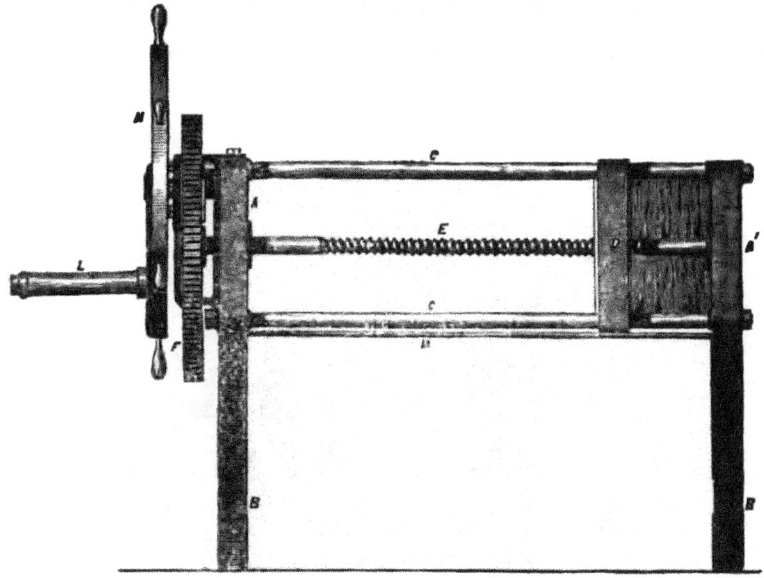

Fig. 3.

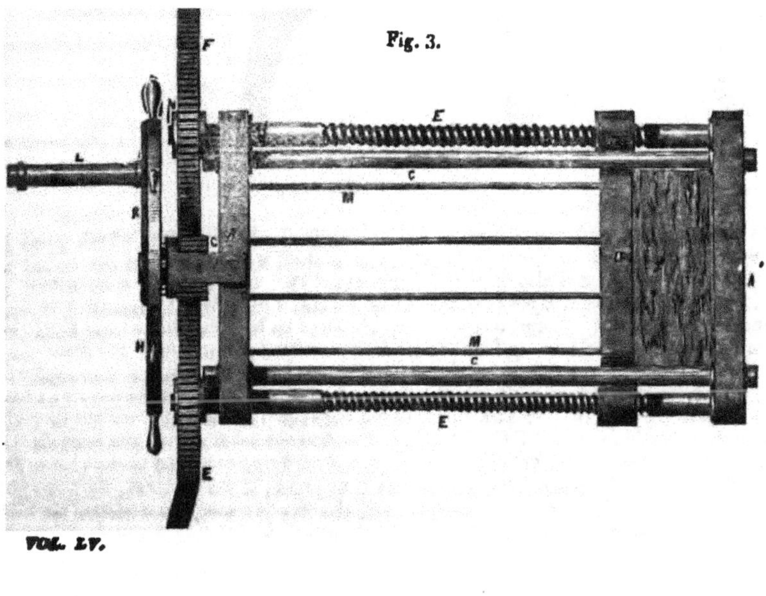

WALKER'S REGISTERED DOUBLE-ACTING SCREW PRESS.

(John Walker, of the City-road, Hydraulic Engineer, Proprietor.)

Description.

FIGURE 1 of the prefixed engraving is a side elevation, fig. 2 a front elevation, and fig. 3 a plan of this press. A A′ are the two sills or end blocks, which are cast in one piece with the frames or supports B B, and connected together by the four tie-rods and nuts C C. D is the ram, which is made to travel backwards and forwards upon the tie-rods or guides C C by means of two double-threaded screws E E, which are tapped through nuts placed in the two ends of the ram. One end of each of these screws is centred in the sill A′, and the other end passed through the sill A.

Fig. 2.

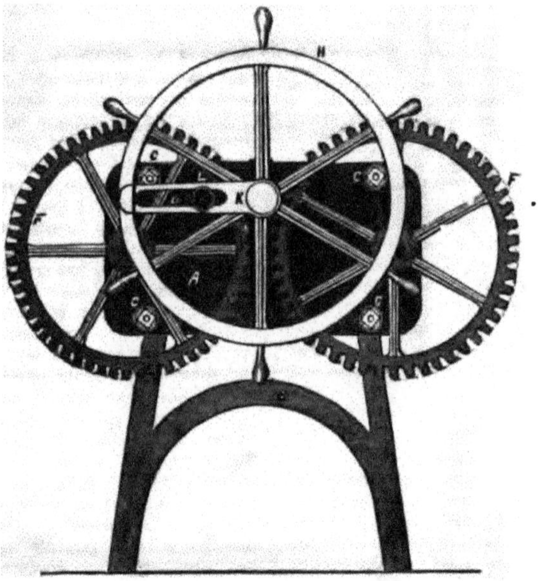

On the A end of each screw there is keyed a toothed wheel F, into which gears a pinion G, which pinion again is keyed upon a shaft which turns in the socket I, bolted to the sill A, and also carries the fly-wheel H. K is a slotted arm, which is securely attached to the fly-wheel H, and in the slot *a* of which the handle L is fastened by means of a nut and screw, which allows of its being shifted to and from the centre as required. M M are four rods, which pass from end to end of the press, and serve as a frame for the support of the materials to be pressed. When it is required to use this press, the ram D is drawn to either end of it, and the material—such as leather, cotton goods, hay, or straw—is placed between the ram and the sill farthest from it, and within the rods C C; the handle L is then shifted along the slot towards the centre of the wheel H, which allows of a quick motion being imparted to the pinion G, and through it to the wheels F F and screws E E, which, as they revolve, act upon the ram D, and gradually force it towards the sill, thereby pressing the materials between them. As the materials become compressed, and more power is required, the handle

L is shifted along the slot towards the periphery of the wheel H, which thereby gives a greater length of leverage, and consequently increases the power exerted upon the ram. Should this, however, not be sufficient, there are a series of handles placed around the circumference of the wheel H, by which the additional force of two, or even three men may be applied to it. When the ram has arrived at the end of its travel, as shown in figs. 1 and 3, and the material is sufficiently pressed, a fresh supply of goods is placed between the reverse side of the ram and the opposite sill, and the handle being shifted to the centre again, and turned in a contrary direction, the ram is caused to travel back, and the whole process of pressing is again repeated.

By the arrangements above described, a very great power is comprehended within a small compass, and the press is rendered double-acting, whereby a very considerable saving in time is effected.

WHITE'S PATENT HYDRO-CARBON GAS.—REPORT ON ITS MANUFACTURE, COMPOSITION, AND ILLUMINATING POWER. BY DR. FRANKLAND, F.C.S., PROFESSOR OF CHEMISTRY, OWEN'S COLLEGE, MANCHESTER.

(Having given insertion to a paper by Dr. Fyfe, the eminent chemical professor (vol. liii., p. 92), in which the reality of the advantages claimed for this gas was gravely impugned, we deem it but an act of justice also to lay before our readers the principal portions of a Report just made on the subject by Dr. Frankland, of Manchester, who has arrived experimentally at conclusions diametrically opposed to those of Dr. Fyfe.—Ed. M.M.)

The experiments detailed in the following pages were conducted at the gas-works attached to the mill of Messrs. Geo. Clarke & Co., Pollard-street, Ancoats, Manchester. These works consist of a bench containing two of Mr. White's resin-gas retorts and two of his water-gas retorts of the largest size, all of which had been in use for four months previously. The water retorts discharge themselves into the resin retorts, and these last work into an hydraulic main, from which the gas passes successively through a refrigerator and wet lime purifier to the gas-holder—a vessel of the ordinary construction, and capable of containing about 20,000 cubic feet. The volume of gas produced was measured by a meter placed between the last purifier and the holder. A copper for melting the resin, and an oil cistern for collecting the residual oil, condensed in the hydraulic main and refrigerator during the process, complete the apparatus, the whole of which was placed under my own entire control, and every facility given me for correctly ascertaining the weight and measure of the materials used, and the products of the operations.

Before commencing each day's experiments, the cubical contents of the holder were carefully determined, and a specimen of the gas contained in it withdrawn for analysis: the charcoal retorts were then filled, the resin melted *in the oil of a former working—about 7½ gallons being* used for each 112 lbs. of resin, and the water and oil tanks being first accurately gauged, the process of gas-making was commenced.

It was of importance to ascertain the temperature at which the gas passed through the meter, since, if not sufficiently cooled, the indications of this instrument might be far from correct. I found, however, that the gas streaming through the pipes previously to entering the meter had been so perfectly cooled in the refrigerator that its temperature never exceeded 60° Fahr., and was frequently much below this point, thus affording a sufficient guarantee for the correctness of the numbers read off. The illuminating power of the gases was taken by the shadow test, and is expressed in the quantity of gas per hour, which is equivalent to the light of one s. 6. composite candle. The specimens of gas for analysis were drawn from the holder on the following morning, in order to ensure a perfect mixture and a fair sample. I have made the analyses of these specimens with great care, and in all cases over mercury: the olefiant gas was determined by absorption with strongly fuming sulphuric acid—the only method by which an accurate estimation of this gas can be attained: the carbonic acid was determined by a bullet of caustic potash, and the rest of the gases by explosion with excess of oxygen.

The prices attached to the several articles employed in the manufacture, are those charged in Manchester, and the value assigned to the residual oil is derived from the price which the hydro-carbon gas patentees give for that article, usually exclusive of carriage.

[Here follow the details of five days' experiments.]

The foregoing analytical results furnish

us with a satisfactory explanation of the processes which go on both in the water and resin gas retorts. In the water retorts two distinct decompositions take place; viz., first, the decomposition of steam by charcoal with the production of equal volumes of hydrogen and carbonic oxide gases; and second, the decomposition of steam by charcoal with the formation of two volumes of hydrogen, and one volume of carbonic acid.

This mixture of hydrogen, carbonic oxide and carbonic acid along with a large excess of steam then passes into the resin retort, where, mixing with the decomposing resin vapour, it twice traverses the whole length of the red-hot vessel. There is no doubt that the greater portion of water gas is produced by the decomposition of this excess of steam in the resin retort, since the weight of charcoal required for the formation of the volume of water gas generated in each of the above experiments is more than twice as great as that which disappeared from the water retort. This circumstance elucidates the advantages arising from the passage of this gas mixed with steam through the resin retort, the fuliginous matter which would otherwise accumulate and block up this retort and its exit pipe, as is well known to be the case when resin alone is used, is converted into permanent combustible gas, which, although possessing no illuminating power, yields valuable service in a manner presently to be described.

It has been maintained that the hydrogen of the water gas enters into combination with the carbonaceous matters formed in the resin retort, and produces hydro-carbon gases, possessing highly illuminating properties; this opinion originated from the fact that if the mixture of the resin and water gases be not made in the resin retort, but on the contrary, the two gases be conducted separately into the holder, the resulting gas will be greatly inferior both in quantity and quality. The foregoing experiments, however, afford no foundation for this opinion, but, on the contrary, prove that no portion of the hydrogen of the water gas enters into any chemical combination whatever, for, as before mentioned, when steam acts upon charcoal at an elevated temperature, every cubic foot of carbonic oxide produced generates an equal volume of hydrogen, whilst every cubic foot of carbonic acid formed liberates two cubic feet of hydrogen; hence, if the volume of hydrogen contained in the hydro-carbon gas equals the volume of carbonic oxide plus twice the volume of carbonic acid, we have the strongest evidence that no hydrogen has *entered into combination*, and this is proved *to be the case by the results of* the first and second days' experiments; for the gas produced on the first day contained:

	Per cent.
Hydrogen	39·38
Carbonic Oxide	28·98
Carbonic Acid	6·31

Hence

Hydrogen : Carbonic Oxide + 2 Carbonic Acid = 39·38 : 41·60.

The Gas produced on the Second Day contained—

	Per cent.
Hydrogen	33·54
Carbonic Oxide	8·40
Carbonic Acid	10·78

Hence

Hydrogen : Carbonic Oxide + 2 Carbonic Acid = 33·54 : 29·96.

On the third day a large excess of hydrogen was produced, owing, no doubt, to the decomposition of light carburetted hydrogen by the much higher heat which was employed on that day.

But although it be thus proved that the water gas does not in any way enter into chemical union with the constituents of the resin gas, yet I do not conceive that its value in the process is diminished by this consideration. I have already shown its use in taking up and converting into permanent gas much of the fuliginous matter which would otherwise choke up the resin retorts and exit pipes; but even this is of very secondary importance compared with the service it performs in rapidly sweeping out of the red-hot retort the permanent illuminating gases produced by the decomposition of the resin, and in saturating itself with the various volatile hydro-carbons upon which so much of the illuminating power of all gas depends, and which would otherwise, to a great extent, be left behind with the tar and water in the condensers. It is well known how rapidly olefiant gas and all rich hydro-carbons are decomposed into charcoal and gases possessing little or no illuminating power when in contact with the walls of a red-hot retort, and therefore the value of the water gas in thus rapidly removing them from this destructive influence and retaining them in a permanently gaseous form can scarcely be overrated; indeed, this principle has not been entirely neglected by coal gas manufacturers, several companies having attached exhausters to their retorts, which, however, perform their work very imperfectly compared with the water gas.

The generation of water gas free from carbonic acid is a problem of great importance, and one which deserves the best attention of the hydro-carbon as manufacturer. The relative quantity of this gas produced varies so considerably (from 10·78 to 4·72 per cent.), owing no doubt to the degree of heat at which the decomposition takes place, and also probably to the rapidity with which the water is admitted into the retorts, that it is not impossible, by varying the conditions, to get rid of it altogether; its quantity appears to decrease as the temperature increases; but I have hitherto been unable entirely to prevent its formation: it is therefore requisite to have an efficient means for removing it from the gaseous mixture before it arrives at the holder, since this gas is not only entirely useless, being perfectly incombustible, but has a decidedly injurious influence on the combustion of the gas, by cooling the flame, and thus greatly diminishing the illuminating power. Lime, both in its wet and dry state, is quite inefficient for the removal of this carbonic acid, since the carbonate of lime first formed prevents further contact between the gas and the purifying agent. I therefore recommend caustic soda, produced by mixing lime with a solution of common soda, as a most efficient and inexpensive purifying agent when applied in the following manner:—Let 1 cwt. of soda be dissolved in not less than 120 gallons of water (and proportionately for smaller quantities); add to this 70 lbs. or 80 lbs. of quicklime; mix the whole well together, and transfer it to the purifier, where it should be occasionally well agitated: after about 8,000 cubic feet of gas have passed through, the mixture should be run off and allowed to settle in a suitable tank, from which the clear liquor floating above the sediment of carbonate of lime must be pumped up into the supply tank for the purifier, and being again mixed with the same quantity of lime, used as before. Thus little or no loss of soda occurs, this base being simply used as a carrier of the carbonic acid from the gas to the lime. The sediment of carbonate of lime may be thrown away between each operation. The cost of purification by this method would not exceed ½d. per 1,000 cubic feet.

The following experiment was made with the purifier charged in the manner described, except that only 75 lbs. of soda were employed:

	cwt.	qr.	lb.
Resin used..........	2	0	7
Coal	1	2	0
Charcoal............	0	0	10
Lime, &c	0	1	0
Water..............	0	2	6½
Residual oil produced..	8 75 gallons.		
Gas produced........	3090 cub. ft.		
„ per 112 lbs. resin	1520 „		

Cost of Production.

				s.	d.
Resin..........	2 cwt. 0 qr. 7 lbs. at	3s. 6d...........		7	2¼
Coal	1 „ 2 „ 0 „	6s. per ton.......		0	5½
Charcoal	0 „ 0 „ 10 „	....		0	2¼
Purification				0	2
				8	0¼
			s. d.		
Deduct 8·75 gallons oil at 7d.........			5 1¼		
„ Cask			0 5		
				5	6¼
				2	6

Hence cost of 1,000 cubic feet = 9¾d.

Per Centage Composition of Gas.

Olefiant gas	8·22
Light carburetted hydrogen	31·09
Hydrogen...............	42·06
Carbonic oxide..........	15·04
Carbonic acid	3·59
	100·00

It is therefore evident that, whilst the whole of the carbonic acid can be readily removed by this method, if the caustic soda be employed in sufficient quantity, and the gas brought in contact with a large surface of it, the quality of the gas is not in the least deteriorated in its passage through the liquid, as is proved by the increased per centage of olefiant gas in the above analysis.

The following Table shows the quantity and composition of the gas obtained in the first, second, third, and sixth days' experiments, when perfectly purified, as just described:

	1st Day.	2nd Day.	3rd Day.	6th Day.	Average.
Volume of Gas produced from 1 cwt. Resin......................	1300	1406	1841	1465	1503
PER CENTAGE COMPOSITION.					
Olefiant Gas......................	8·27	7·94	7·78	8·53	8·13
Light Carburetted Hydrogen	18·76	45·06	22·79	32·25	29·71
Hydrogen	42·03	37·59	50·27	43·62	43·38
Carbonic Oxide	30·93	9·41	19·16	15·60	18·78
	100·00	100·00	100·00	100·00	100·00

In order to establish a practical comparison between the partially purified hydrocarbon gas and the Manchester coal gas, as supplied to the town, the latter was conducted from a house adjoining Messrs. Clarke's mills into the dark room set apart for testing their illuminating powers. By a simple arrangement, either of the gases could be passed through a meter which indicated the consumption per hour in observations of one minute. Care was always taken to displace the whole of one gas from the meter before the illuminating power of the other was taken. The following results were obtained:

First Experiment.

Pressure of Gases, ½ inch.

Consumption per hour equivalent to the light of a S. 6 Composite Candle.

Manchester Coal Gas.

7½ tenths of 1 cubic foot.

Hydro-carbon Gas.

7½ tenths of 1 cubic foot.

Second Experiment.

Pressure of Gases, 1 inch.

Consumption per hour equivalent to the light of a S. 6 Composite Candle.

Manchester Coal Gas.

8 tenths of 1 cubic foot.

Hydro carbon Gas.

7¾ tenths of 1 cubic foot.

This method of determining the relative value of the two gases did not seem to me entirely free from objection, since some of the lighter hydro-carbons contained in the coal gas might possibly become condensed in its passage through the great length of cold piping intervening between the gas-works and the illuminating room; a circumstance which might thus give an undue advantage to the hydro-carbon gas which had to pass through a comparatively much shorter length of pipe. I therefore procured a specimen of the coal gas from the immediate neighbourhood of the works of St. George's-road station, and submitted it to analysis; it yielded me the following numbers, which I have placed side by side with the average composition of the hydro-carbon gas *before* and *after* purification:

	Manchester Coal Gas.	Hydro-carbon Gas before Purification.	Hydro-carbon Gas after Purification.
Olefiant Gas	5 50	7·41	8·13
Light Carburetted Hydrogen	40·12	26·50	29·71
Hydrogen	45·74	40·27	43·38
Carbonic Oxide................	8·23	18·55	18·78
Carbonic Acid	·41	7·27	....
Nitrogen	trace	trace	trace
	100·00	100·00	100·00

As the illuminating power of both coal and resin gases depends almost exclusively upon the quantity of their constituents condensible by fuming sulphuric acid, and which generally appear in all gas analyses under the somewhat inappropriate title of olefiant gas, the superiority of the hydro-carbon over the Manchester coal gas is sufficiently apparent from the above comparison, yet a question might arise relative to the quality of the substances composing the olefiant gas in each of the specimens, and an answer to this question was therefore imperative before a definite conclusion respecting the merits of the two gases could be arrived at.

The illuminating power of the hydro-carbons grouped together in the above analysis under the term "olefiant gas" depends directly upon the weight of carbon contained in a given volume, and hence by ascertaining the quantities of carbonic acid which this portion of the two gases yields on explosion with excess of oxygen, it is easy to calculate their relative value. My experiments have led to the following proportion:

Illuminating Power of Equal Volumes of the Olefiant Gases.

Manchester coal olefiant gas : hydro-carbon olefiant gas=3·62 : 2·8.

Hence the olefiant principles in the Manchester coal gas have a higher value than the same volume of the olefiant gas contained in the hydro-carbon gas, and this must therefore be allowed for in calculating the relative value of the two gases.

According to the above proportion the true value of the 5·5 per cent. of olefiant gas will be 7·11, and hence the illuminating power of the average hydro-carbon gas will be to the Manchester coal gas as follows:

I. Relative illuminating power of hydro-carbon gas *unpurified* and Manchester coal gas:

Hydro-carbon Gas.		Manchester Coal Gas.
7·41	.	7·11
or, 100	.	95·9

II. Relative illuminating power of hydro-carbon gas *purified* and Manchester coal gas:

Hydro-carbon Gas.		Manchester Coal Gas.
8	.	7·11
or, 100	.	88·9

numbers which exactly confirm the results obtained by the previous experiments on the illuminating powers of the two gases.

The above facts prove that 1,000 cubic feet of hydro-carbon gas *before purification*, are equal to 1,042 cubic feet of the Manchester coal gas, and that 1,000 cubic feet hydro-carbon gas *after* purification are equivalent to 1,125 cubic feet of the Manchester gas, and further, that at the present market price of the articles consumed and produced, 1,000 cubic feet of average hydro-carbon gas *before purification* can be produced, exclusive of rent, taxes, wages, wear and tear, at the cost of 9½d. to 1s. 1½d., according to the mode of working, whilst 1,000 cubic feet of the same gas purified will cost from 10½d. to 1s. 2½d.

A distinction must be made between unpurified coal gas and unpurified hydro-carbon gas: the former contains many deleterious ingredients, which entirely prevent its use; the latter does not contain any noxious principle, but simply has its illuminating power diminished by the presence of carbonic acid.

It is also evident that a moderate heat is better adapted for producing good gas economically than a stronger one, which, although it produces much more gas, yet does so at the expense of the oil, which becomes much diminished in quantity, and thus the cost of the gas is greatly increased, whilst its quality also appears slightly to suffer. This is seen in the produce of the first, second, and sixth days, when a more moderate heat was employed compared with that of the third, fourth, and fifth days, when the heat was much higher; and although the yield in gas was considerably less than on the latter days, yet its quality was somewhat better, and the yield in oil being much greater, its price was reduced in a proportionate degree.

It appears to me a mistake to suppose that the hydro-carbon gas requires a different form of burner from those used for coal gas, as I could find no difference in the powers of the gas when coal gas burners were substituted for those generally used with the hydro-carbon; possibly, however, a larger burner might be required if the gas contained much carbonic acid. A careful determination of the specific gravities of the hydro-carbon and Manchester coal gases, which I here append, shows that they do not materially differ in this respect, and thus confirms the opinion that no difference of burner is necessary if the gas be properly manufactured:

Specific Gravity.

Hydro-carbon gas before purification......	·65886.	Manchester coal gas ·52364.
Hydro-carbon gas after purification	·59133.	

In conclusion, its purity of composition and freedom from all substances which can, during combustion, produce compounds injurious to furniture, drapery goods, &c., gives the hydro-carbon gas great advantages *over coal gas, which always contains* more or less bisulphuret of carbon, a volatile substance that has hitherto defied all attempts to remove it or diminish its quantity by any process of purification, and which, during combustion, generates sulphurous acid, the compound to which all the mischief produced by coal gas is probably owing. The odour of the hydro-carbon gas while it is sufficiently strong to give warning of any escape, is far less nauseous than that of the coal gas, and might even by some persons be deemed pleasant, whilst the process of manufacture is so simple that any person of moderate intellect can at once conduct it.

E. Frankland,
Ph. D., F.C.S.

Owen's College, Manchester, June 23, 1851.

ON THE CONSTRUCTION OF STEAM BOILERS AND THE CAUSES OF THEIR EXPLOSIONS. BY WM. FAIRBAIRN, ESQ., C.E., F.R.S.

(Continued from page 15.)

The preventives against accidents of this kind are well-constructed boilers of the strongest form, and duly proportioned safety valves—one under the immediate control of the engineer, and the other, as a reserve, under the keeping of some competent authority.

2nd. *Explosions from deficiency of water.* This division of the subject requires the utmost care and attention, as the circumstance of boilers being short of water is no unusual occurrence. Imminent danger frequently arises from this cause, and it cannot be too forcibly impressed upon the minds of engineers, that there is no part of the apparatus which constitute the mountings of a a boiler which require greater attention—probably the safety valves not excepted—than that which supplies it with water. A well-constructed pump, and self-acting feeders—when boilers are worked at a low pressure—are indispensable, and where the latter cannot be applied, the glass tubular gauge steam and water-cocks must have more than ordinary attention.

In a properly constructed boiler, every part of the metal exposed to the direct action of the fire should be in immediate contact with the water, and when proper provision is made to maintain the water at a uniform height and depth above the plates, accidents can never occur from this cause.

Should the water, however, get low from defects in the pump, or any stoppage of the regulating feed valves, and the plates over the furnace become red hot, we then risk the bursting of the boiler, even at the ordinary working pressure. We have no occasion, under such circumstances, to search for another cause, from the fact that the material when raised to a red heat has lost about five-sixths of its strength, and a force of less than one-sixth will be found amply sufficient to bear down the plates direct upon the fire, or to burst the boiler.

When a boiler becomes short of water, the first, and perhaps the most natural action is to run to the feed valve, and pull it wide open. This certainly remedies the deficiency, but increases the danger, by suddenly pouring upon the incandescent plates a large body of water, which, coming in contact with a reservoir of intense heat, is calculated to produce highly elastic steam. This has been hitherto controverted by several eminent chemists and philosophers; but I make no doubt such is the case, unless the pressure has forced the plates into a concave shape, which for a time would retard the evaporization of the water when suddenly thrown upon them. Some curious experimental facts have been elicited on this subject, and those of M. Boutigny and Professor Bowman, of King's College, London, show that a small quantity of water projected upon a hot plate does not touch it; that it forms itself into a globule surrounded with a thin film, and rolls about upon the plate without the least appearance of evaporation. A repulsive action takes place, and these phenomena are explained upon the supposition that the spheroid has a perfectly reflecting surface, and consequently the heat of the incandescent plate is reflected back upon it. What is, however, the most extraordinary in these experiments is, the fact that the globule, whilst rolling upon a red hot plate, never exceeds a temperature of about 204° Fahr.; and in order to produce ebullition, it is necessary to cool the plate until the water begins to boil, when it is rapidly dissipated in steam.

The experiments by the committee of the Franklin Institute, on this subject, give some interesting and useful results. That committee found that the temperature of clear iron, at which it vaporized drops of water, was 334° Fahr. The development of a repulsive force which I have endeavoured to describe was, however, so rapid above that temperature, that drops which required but one second of time to disappear at the temperature of maximum vaporization, required 152 seconds when the metal was heated to 395° of Fahr. The committee goes on to state that—"One ounce of water introduced into an iron bowl three-sixteenths of an inch thick, and supplied with heat by an oil-bath, at the temperature of 546°, was vaporized in fifteen seconds, while at the initial temperature of 507°, that of the most rapid evaporization was thirteen seconds."

The cooling effect of the metal is here strikingly exemplified, by the increased rapidity of the evaporization, which, at a reduced temperature of 38°, is effected in thirteen instead of fifteen seconds.

This does not, however, hold good in every case, as an increased quantity of water, say from one-eighth of an ounce to two ounces, thrown upon heated plates, raised the temperature of its evaporization from 460° to 600° Fahr.: thus clearly showing that the time required for the generation of explosive steam under these circumstances is attended with danger, and it may be doubted whether the ordinary safety valves may not be wholly inadequate for its escape.

Numerous examples may be quoted to show that explosions from deficiency of water, although less frequent than those arising from undue pressure, are by no means uncommon—they are nevertheless comparatively fewer in number, and the preventatives are good pumps, self-acting feeders (when they can be applied), and all those conveniences, such as water-cocks, water-gauges, floats, alarms, and other indicators of the loss and reduction of water in the boiler.

3rd. Explosions produced from collapse.

Accidents from this cause can scarcely be called explosions, as they arise, not from internal force which bursts the boiler, but from the sudden action of a vacuum within it. In high-pressure boilers, from their superior strength and circular form, these accidents seldom occur, and the low pressure boiler is effectually guarded against it by a valve which opens inwards by the pressure of the atmosphere whenever a vacuum occurs. In some cases a collapse of the internal flues of boilers has been known to take place, from a partial vacuum within, which, united to the pressure of the steam, has forced down the top and sides of the flue, and with fatal effect discharged the contents of the boiler into the ash-pit, and destroyed and scalded everything before it. A circumstance of this kind occurred on the Thames on board the steamer *Victoria*, some years since, when a number of persons lost their lives, and serious injury was sustained in all parts of the vessel within its reach. This accident could not however be called an explosion, but a collapse of the internal flues, which were of large dimensions, and the consequent discharge of large quantities of steam and water into the space occupied by the engines.

One or two cases which bear more directly on this point are however on record, and one of them, which took place in the Mold mines, in Flintshire, was attended with explosion. The particulars, as given by Mr. John Taylor, will be found circumstantially recorded in the first volume of the *Philosophical Magazine*. This occurrence seems to prove that rarefication produced in the flues of a high pressure boiler may determine an explosion. The boiler which exploded belonged to a set of three feeding the same engine; the fuel used was bituminous coal. The furnace doors of all three of the boilers *had been opened, and the dampers of two* had been closed, when a gust of flame was seen to issue from the mouth of the furnace of these latter, and was immediately followed by an explosion. The interior flue of this boiler was flattened from the sides, the flue and shell of the boiler remaining in their places, and the safety-valve upon the latter not being injured.

Other similar cases of collapse might be stated, but as most of them have been attended by a defective supply of water in the boiler, the plates over the fire having become heated, they can scarcely be included in the category of this class of accidents, and more properly belong to those of which we have just treated—explosions from a deficiency of water in the boiler.

It is nevertheless necessary to observe that cases of collapse should be carefully guarded against, as the great source of danger is in the escape of hot water, which, with the steam generated by it, produces death in one of its worst and most painful forms.

The remedies for these accidents will be found in the vacuum valve, and careful construction in the form and strength of the flues.

4th. *Explosions from Defective Construction.*

This is, perhaps, one of the most important divisions that can possibly engage our attention, and on which it shall be my duty to enlarge. In a previous inquiry, I have already shown the nature of the strain and the ultimate resistance which the material used in the construction of boilers is able to bear. We have not, however, in all cases, shown the distribution and position in which that material should be placed in order to attain the maximum of strength, and afford to the public greater security in the resisting powers of vessels subject to so severe and sometimes a ruinous pressure. This is a subject of such importance that I shall be under the necessity of trespassing upon your time, in endeavouring to point out the advantages peculiar to form, and the use of a sound and perfect system of construction.

For a number of years the haycock, hemispherical, and wagon-shaped boilers were those generally in use, and it was not until high-pressure steam was first introduced into Cornwall, that the cylindrical form with hemispherical ends, and the furnace under the boiler, came into use; subsequently, this gave way to the introduction of a large internal flue extending the whole length of the boiler, and in this the furnace was placed. For many years this was the best and most economical boiler in Cornwall, and its introduction into this country has effected great improvements in the economy of fuel, as well as the strength of the boiler. Several attempts have been made to improve this boiler by cutting away one-half of the end, in order to admit a larger furnace. This was first done by the Butterley Company, and it has since gone by the name of the Butterley boiler. This construction has the same defects as the haycock or *hemispherical* and wagon-shaped boilers; it is weak over the fire-place, and cannot well be strengthened without injury to the other parts of the boiler, from the vast number of stays necessary to suspend the part which forms the canopy of the furnace. Of late years a much greater improvement has, however, been effected by the double flue and double furnace boiler, which is now in general use, and has nearly superseded all the other constructions. It consists of the cylindrical form, varying from five to seven feet in diameter, with two flues which extend the whole length of the boiler; they are perfectly cylindrical, and of sufficient magnitude to admit a furnace in each. The boiler is the simplest, and probably the most effective that has yet been constructed. It presents a large flue surface as the recipient of heat, and the double flues, when riveted to the flat ends, add greatly to the security and strength of those parts. It moreover admits of the new process of alternate firing, so highly conducive to perfect combustion, and the prevention of the nuisance of smoke.

5th. *Explosions arising from mismanagement or ignorance.*

To mismanagement, ignorance, and the misapplication of a few leading principles in connection with the use and application of steam, may be traced the great majority of accidents which from time to time occur. Many of these accidents, so fruitful of the destruction of property and human life, might be prevented, if we had well constructed vessels judiciously united to skill and competency in the management. To convey a few practical instructions to engineers, stokers, and engine-men, would be an undertaking of no great difficulty. A young man of ordinary capacity would learn all that is necessary in a few months; and if placed under competent instructors, he might be made acquainted with the properties of steam, its elastic force at different degrees of pressure, the advantages peculiar to sensitive and easy working safety valves, the necessity for cleanliness and keeping them in good working condition; the use of water gauges, fusion plugs, indicators, signals, &c., &c., connected with the supply and height of water in the boiler. The dangers to be apprehended from a scarcity of water, the danger of explosion when the engine is standing, or when the usual channels for relieving the boiler of its surplus steam are stopped,—all these are parts of elementary instruction which the stoker, as well as the engineer, should be acquainted with, and no proprietor of a mill, captain of a steam-ship, or superintendent of locomotive, should give employment to any person unless they can produce certificates of good behaviour, and a knowledge of the elementary principles of their profession.

If these precautions were adopted, greater care observed in a selection of men of skill and responsibility in the construction of boilers, and a more strict and rigid code of laws in the management, we might look forward with greater certainty to a considerable diminution, if not a prevention, of those calamitous events which so frequently plunge whole families into mourning by unexpected and instantaneous death.

As an individual I would cheerfully lend my best assistance to the development of a principle of instruction, calculated to relieve the country of the ignorance which pervades that part of the community on which the lives of so many depend. A resolution on the part of those who employ persons of this description, and whose interests are so much at stake, to take none whose knowledge and character does not come up to the requisite standard, and *pay for it*, would soon find from the economy of the management and the increased security of their property, a very important change in all the requirements of the economy, as well as the application of steam. How often do we find implements of danger, and vessels containing the elements of destruction, in the hands of the most ignorant and reckless practitioners, whose insensibility to danger, and total incompetency to judge of its presence, renders them above all others the most unfit to be employed. And why? because they are the very persons, from their defective knowledge, to increase the danger and aggravate the evils they were selected to prevent. It is not the first time that engineers, to secure (if I may use the expression) an insane pressure, have fastened the safety valves, and screwed down the steam valve, closing every outlet, without ever thinking of the fire that was blazing under the boiler.

(*To be continued.*)

PANNELL'S REGISTERED RETORT CALORIFERE, FOR CONSERVATORIES, GREENHOUSES, ETC.

(John Pannell, of 5c, Fetter-lane, London, Agricultural and Horticultural Engineer, Proprietor.)

The engraving represents a vertical section of this apparatus. A A is the boiler, which is built into the furnace like a retort, being completely surrounded by the fire, which is supplied with fuel through the door N. B is a partition, which divides the boiler into two compartments C^1 C^2, D^1 D^2 are the two cold water or inflow pipes, and E^1 E^2 the outflow or hot water pipes. The upper compartment C^1, with its pipes D^1 and E^1, are employed for the purpose of producing a surface heat, while the compartment C^2, with its pipes D^2 and E^2, are for heating the bed. When a surface heat is not required, the whole circulation is sent under the bed by shutting the taps F F, and opening G G. H H is an air-heating chamber; the cold

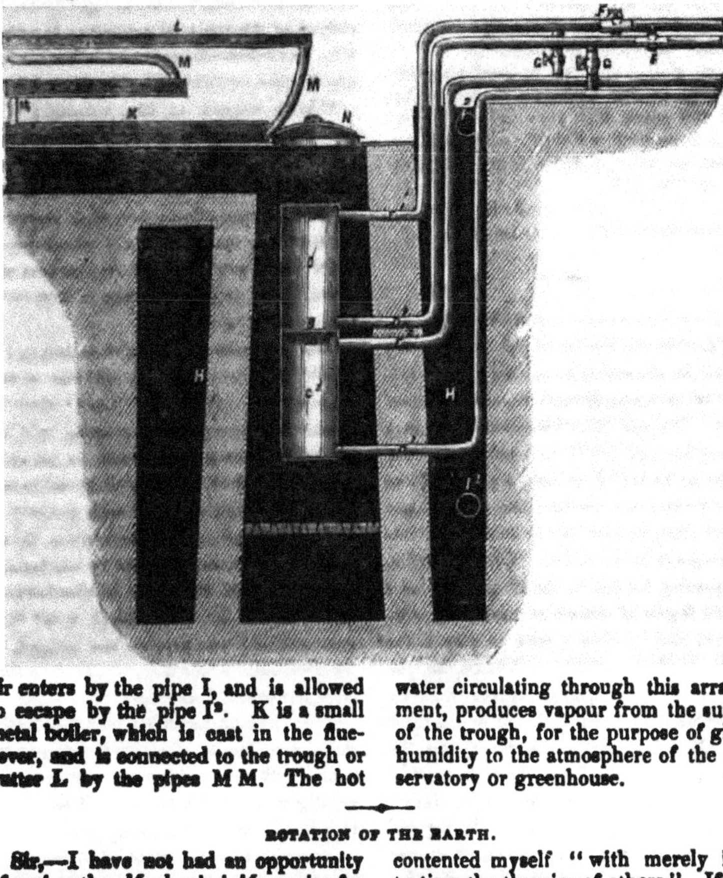

air enters by the pipe I, and is allowed to escape by the pipe I^2. K is a small metal boiler, which is cast in the flue-cover, and is connected to the trough or gutter L by the pipes M M. The hot water circulating through this arrangement, produces vapour from the surface of the trough, for the purpose of giving humidity to the atmosphere of the conservatory or greenhouse.

ROTATION OF THE EARTH.

Sir,—I have not had an opportunity of seeing the *Mechanics' Magazine* for several weeks, and consequently am not even yet in possession of the information forwarded to you by the Rev. Mr. Reynolds, and other correspondents on the rotation of the earth. I have, however, just seen the article on this subject by Mr. B. Rozzell, printed in the Magazine of last week, from which I regret to find that there is *one* reader of your journal to whom my discussion of that interesting topic is unintelligible. I am sorry that it is out of my power to make the matter plainer. Your correspondent expresses *his regret that I should have* contented myself "with merely illustrating the theories of others." If such be the fact, I shall be thankful for an opportunity of restoring to the rightful claimant whatever there may be, in the trifle referred to, which belongs to another. I can only say that, up to this moment, I have never seen a single line of the investigations of Binet and the other Parisian analysts: indeed, I have never seen any mathematical details on the subject at all. It would be quite futile for me to attempt anything like a reply to Mr. Rozzell: each of us, I see, is quite unintelligible to the other. Mr. R. reads in my paper what I never wrote, and he writes in his what I can't read: he says, I "consider the plane of vibration at all latitudes invariable!" Now, I don't think I ever mentioned the plane of vibration at all: simply because I had nothing to do with it: but sure I am that he will nowhere find, in my communication, the "strange misconception" on which he comments.

I am glad to see that so good a mathematician as Mr. Osborne Reynolds has read my paper with very different eyes, and I hope that my short additional explanations were in some degree satisfactory to "S. Y."

J. R. YOUNG.

London, July 8, 1851.

LORD GRANVILLE'S PATENT BILL.

Whatever the merits of this bill may be, it must be allowed to have been a great surprise on all whom it more immediately concerns. No one ever imagined that, in a Committee appointed to consider how the Patent Laws might be best *amended*, there was a project secretly hatching for the ABOLITION altogether of the system of protecting property in inventions. Lord Granville, in reporting the Bill to the House, affected a certain degree of shame at having so treacherous and ignoble a part to play in the affair. "He was afraid it might be thought he was taking a strange course when he supported the present Bill, and at the same time avowed himself of opinion "that the whole system of the Patent Laws was unjust to the public, disadvantageous to inventors, and wrong in principle." If the bill in its *ensemble* could be considered as one for the "*amendment* of the laws touching Letters Patent for Inventions," the course taken by his lordship would have been "strange" indeed; but being, as it really is, a bill calculated to do more harm than good to patentees and inventors,—a bill to obstruct as much as possible the protection of property in inventions,—there was nothing at all strange in the case. Lord Granville, on taking the bill under his charge, was but acting in perfect consistency with his (now) declared views on the subject; and all that we have to complain of is, that his lordship should pretend to be enacting the part of the inventor's friend and patron, while, in fact, aiming a deadly blow at his very existence.

Of the state of public opinion—that is, of opinion out of the House of Lords—on the subject of the right of inventors to protection, Lord Granville gave, we believe, a very exact representation when he observed:

"With respect to the necessity of a patent law, he believed it would have been easy for him, as Chairman of the Committee, to get *one hundred* sensible persons to give evidence to that effect; but with respect to the injurious tendency of the whole system, *there were probably not six persons who could be got to give evidence in support of that view*"!'!

And so, because opinion is as 100 to 6 in favour of protecting (by all fair means) property in inventions, Lord Granville thinks it only right to promote, with the whole influence of Government, a bill which has for its object, or at least its entire tendency, the abolition of all such property.

The witnesses against the system, though few in number, are paraded by his lordship as if he thought they were in themselves a host. Let us see. We detest at all times personalities; but here we are obliged, by the necessities of the case, to deviate a little from our established usage.

1. "The first (witness) was Mr. Cubitt, the civil engineer, whose evidence was entirely worthy of consideration, because, very greatly to his credit, he had raised himself by gradual steps from being a working or journeyman to his present high position of President of (the Institution of) Civil Engineers, and was held in great consideration throughout the country for his personal integrity and professional attainments. His opinion was *conclusive* against the whole system of patents."

Mr. Cubitt may be all that is here averred (we have no personal knowledge whatever on the subject); but how does that affect the value of his opinion on the point immediately in question? What has successful rising in the world—which may be by ways innumerable, from the selling of old clothes

ng old inventions—to do necess- experience or knowledge of the s of an inventor?—Or what "per- egrity?"—Or what even "profes- ttainments," if such professional nts embrace nothing in the way of ? If Mr. Cubitt had been known rld as an inventor or projector,—if imself ever invented, or devised, or anything new or original, worthy ng his name down to posterity,— leed, his opinion would have been mething; but as it is, it is worth than that of any other unit of the

The next was Mr. Brunel, of whom t necessary to say that his evidence reat weight and importance."

lence it wasn't? The rashest, the esumptuous, the most injudicious, entifically regarded) the least suc- f all the engineers of his time: the , too, of a name which was founded uttressed round by patents* beyond most men; and to which inherited owes fully as much as to his own he reputation which he enjoys.

The next witness was Mr. Ricardo, ho also gave valuable evidence on side of the question."

real value of this gentleman's tes- our readers may judge from a spe- it furnished by Lord Granville (we lised the whole by-and-bye.) Mr. hes, it seems, found out—by much research, no doubt—that two of test acquirements ever made by —writing and arithmetic—belong ed when there were no such things to encourage and reward inventive We wonder he did not go still back, and, parodying Southey, de-

When Adam delved and Eve span,
Where was then your *patent man?*"

The next witness was Colonel Reid, r of a work on the Law of Storms, and Chairman of the Executive Committee of the Exhibition. He also gave evidence to the same effect, *although his opinion was derived from reading, and was not founded on a practical knowledge of the question.*"

The discoverer of the law of storms, and therefore a good judge of the law of patents! Has no "practical knowledge of the question," and therefore to be regarded as one of the best of all possible authorities upon it!

5. "Another witness was Mr. Farrie, a sugar-refiner; and although he held the opinion that it was unjust to exempt the Colonies, if we were to have patent laws, he spoke most strongly of the injury which was done to this country by the existence of patent laws at all."

Mr. Sugar Refiner Farrie has had to pay, we suppose, for the use of some valuable patent invention relating to his particular business—for the piracy of it, perchance—and what is injury (*damnum sine injuria?*) to himself, he denounces as injury "*done to the country.*" The old story!

6. "The last witness was the Master of the Rolls, who, notwithstanding the experience he had had as one of the law-officers of the Crown in administering the Patent Laws, and although he took charge of the first bill which the Government proposed on the subject, was decidedly of opinion that Patent Laws were bad in principle, and were of no advantage either to the public or inventors."

Sir John Romilly is as estimable a gentleman as ever filled the offices of either Attorney-general or Master of the Rolls; but he filled the former post for only a short time, and was antecedently but little engaged as a counsel in patent cases. His "experience" of this particular branch of the law has therefore been but small; and he has no more knowledge of what patents have done for "the public and inventors" than any one else may obtain from reading or from hearsay.

Of such, then, is Lord Granville's half-dozen minority composed—such the new lights of the age whose authority is put for-

. I. Brunel took out in his time no less ian patents.

ward by way of apology for this insidious attempt to undermine, in order to ultimately overthrow, the established policy of the country for the last two centuries and more.

A minority, not including more than one name which we can, by the utmost stretch of courtesy, consider as of weight on this particular matter, and confessedly outnumbered in the proportion of more than 100 to 1, by the "sensible" portion of the community.

But there is Lord Granville's own opinion, in corroboration of the precious half-dozen. "He (Lord Granville) thought it impossible to hold any innate right of property in ideas, and that the only reasonable ground upon which the Patent Laws could be supported was, that they stimulated inventors, and encouraged them to discover their inventions."

So far we quite agree with his lordship; and are glad to see that he has so completely escaped being infected by the nonsense propagated on the subject of "the innate right," by the Society of Arts and others. But his lordship ought in all fairness to have added, that the Statute of Monopolies (21 James 1, c. 3), which may be called the Inventor's Charter, is expressly founded on that "only reasonable ground," and is upheld by "sensible" men solely because of the stimulus it affords to inventors. Lord Granville, however, "entirely disbelieved that in the present state of the world—even if it was different at the earlier stages of society—it was at all necessary to stimulate inventors...... He found that scientific men were in the habit of making known their discoveries with great alacrity without seeking any protection from patents."

If the case really were so, there would indeed be an end to all defence of the system of patents. But where is the proof that the occasion for the stimulus has vanished? It is not a case which can be taken for true on the mere affirmance of Lord Granville or of any one else; it must be established by abundant and incontestible evidences before it can be accepted as a groundwork for so serious a step as the abolition of the existing law. We know not what evidence on the subject has been adduced before the Select Committee (for that evidence has not yet been printed); but if the scope of that evidence is in favour of the positions laid down by his lordship—then sure we are that this can only have arisen from some system of inquiry being pursued before the Committee by which the truth has been either wholly excluded or abominably obscured. All who are intimate with the real history of the arts and sciences, and with the views and feelings of those who cultivate them, know well that the hope of individual gain continues as much as ever to be the prime mover in the march of improvement. Of all the more remarkable inventions of modern times there is scarce one which may not be traced distinctly to this source, and to the protection to individual enterprise which patents afford. Witness the electric telegraph, the railway system, lighting by gas, steam navigation—all of which had their origin in patents, and not one of which would have made the wonderful progress it has done but for the patent system. If there is a probability that but one of these great improvements would have been missed, or even delayed a single year from the want of such protection as the patent system affords, that is of itself reason enough for keeping up the system. We cannot afford to throw away a single chance of this sort, for the sake of all that your mere philosopher may do for us from love of science alone.—To talk, indeed, of the discoveries which scientific men are "*in the habit* of making known with alacrity without seeking any protection from patents" is mere dreamwork. We know of no such men, nor of any such free-gift revelations.

Lord Granville urged farther, that it was "quite clear that the tendency of the patent system was to raise the price of the commodity during the fourteen years while the patent existed, and it was often worth the while of a rich company to keep the sale of a patented article exclusively in their own hands by the exorbitant price which they

...enses, to prevent any other ...use of the patent during the ...

...we should be disposed to ...his lordship. We deny that ...the system is as he alleges. ...ase, the tendency of all im... ...the arts is to lower the cost ...and consequently the price ...; and there are many patents ...or their validity on no other ...e reduction in time, labour, ...e, which they accomplish. ...whether price was ever in a ...nced by the operation of a ...re entitled, at all events, to ...f the fact before we admit it. ...so most decidedly the authen... ...rich company" story, and ...iberty of setting it aside as a ...ion, till we are favoured with ...e company, and the particu... ...ent right which it is said to ...y abused. There are many ...his story cannot be true; ...we have not now time or ... Besides, if it were even the ...y patent makes the article ...bject of it dearer for the four... ...if it make it ever afterwards ...etter (one or other of which ...necessity always ensue), then ...ser, but very much the reverse. ...ville expressed fears that he ...t by the case of copyright of ...denied that the cases were ...

...his denial he supported by ...which are so silly that they ...h repetition—and not being ...ion are not worth quoting. ...thing clearer or more univer... ...edged than that property in ...roperty in inventions rest on ...same grounds;— both are ...ffspring of intellect, yet both ...statutary enactment alone, for ...he benefits they may confer, ...duals, but the public. As the ...his charter in the Statute of Monopolies, so have authors theirs in the Copyright Act of Queen Anne, with the supplementary extension Act of William the Fourth. You cannot impeach the one charter without endangering the other; nor (justly) abolish either without abolishing both. The two rights must, if equal justice is to be done, stand or fall together.

In the Proceedings before the House of Lords (1774) in the celebrated case of Donaldson against Becket and others, in which the question at issue was, whether authors had at common law a perpetual property in their productions? (for there were "innate right" Quixotes in those days as there are now), the perfect identity in principle between copyrights and patent rights was admitted on all hands.

Thus, Chief Baron Eyre (a much respected authority) "considered a book precisely on the same footing with any mechanical invention. In the case of mechanical inventions, ideas were in a manner embodied, so as to render them tangible and visible; now a book was no more than a transcript of ideas; and whether ideas were rendered cognizable to any of the senses by the means of this or that art, of this or that contrivance, was altogether immaterial.... *The clothing may be dissimilar, the essences clothed are identically the same.*"*

We have not space left to go into the grounds (briefly set forth in our last) upon which this Bill of Lord Granville's is, in truth, intended to pave the way for the overthrow of the Patent Law system, under the pretext of simply amending it; but cannot quit the subject for the present without earnestly recommending to inventors and patentees throughout the kingdom to bestir themselves instantly and zealously, to do what in them lies to defeat the measure.—Every mechanics' and literary institution ought to send up a petition against it; every constituency which includes any considerable number of the industrial classes should solicit its representatives to oppose it. Not a moment is to be lost. Already

* Parliamentary History, vol. xvii., p. 974.

the Bill has been read a first time in the House of Commons, and the legal advisers of the Crown do not conceal, that it is their intention to pass it through this Session with all the expedition which the forms of the House will allow.

All that is for the present asked is a fair hearing for the invention interest. That it *has confessedly not yet had*; the whole of the proceedings before the Select Committee of the Lords, having been contrived so as to bring out only the defects of the existing patent system, and no one (beyond the clique of which Lord Granville is the organ) being aware that any hostility was entertained against the general policy of the system, or suspecting that there was the least occasion for offering a word in its defence. Let but a fair opportunity be given to inventors and patentees to vindicate their title to a continuance of the protection which they have enjoyed for nearly two centuries and a half at the hands of the State, and we have no fears whatever for the result.

THE PROPOSED NEW PATENT BILL— "NO. III."

Sir,—I have just read in your last Number, the impending fate which awaits all the inventive skill that may be hereafter directed to improvements in the arts and manufactures in and for the British Colonies.

This monstrous scheme of disfranchising the Colonies, or those interested in their manufactures, is so flagrant an error, and so grossly unjust, as to demand the interference of all who are likely to be affected by this masterpiece of Legislative blundering.

I am one of those who entertained the hope (*now* a slender one), of seeing at least, some of the evils and errors so long complained of in the old state of things—swept away; such as excessive costs,—want of due security,—the dangerous liability of litigation through informal technicalities; separate patents for England, Ireland, Scotland and the Colonies, &c., &c. These defects, so universally admitted, are, however, in my opinion, better to stand as they do, in all their *primitive rottenness*, than that any *improvement in the present* law should be based upon the annihilation of patents for the Colonies.

Let us suppose such inventions or improvements in machinery as the following, to be required for the Colonies, and applicable for the Colonies only; viz., An improved sugar-cane mill, suitable for the West Indies;—a saw-frame for the better cutting of the immense forest timbers of Australia;—a furnace and machinery, suitable for reducing the rich steel-iron ore of New Zealand, so as to render England independent of Sweden and Russia for a similar article;—or a machine for effectually dressing the "Phormium tenax," the indigenous flax of New Zealand, remembering that "*one only*, of our four royal rope-yards paid to Russia eleven millions sterling, in fourteen years, for hemp!"

Now, in the absence of those advantages, which a cheap and efficient Patent Law would secure, where is the incentive to invention? Who will devote money and time in the experiments necessary to success, where there is no hope for an equitable return?

Being already a patentee, and intending to secure by patent some of the items hinted at above, I can speak *feelingly* as to the importance of retaining the Colonies on the list, as offering a remunerative premium to inventive talent; and in doing this I humbly hope to be aiding in some degree the present effort to *improve*, and not to annihilate, the law of patent-right.

I am, Sir, yours respectfully,

J. STENSON.

Northampton, July 7, 1851.

DR. RUTHERFORD'S TRACT ON EQUATIONS.

Sir,—Your correspondent, "R. S." (*Mech. Mag*, No. 1455, p. 502), gives me credit for more influence than I really possess. I have not the honour to be acquainted with Dr. Rutherford, either personally or otherwise, and am therefore not in a position to request the information your correspondent desires. Such information, however, as I happen to possess is entirely at the service of your readers, and as it appears to fix, within a trifle, the date of the publication of Dr. Rutherford's tract on equations, it will probably serve your correspondent's purpose in this respect.

he *Mathematician* for March,
. Rutherford inserts his "New
le Process for determining all
roots of a Cubic Equation;"
er is dated "February 2nd,
d contains a reference to the
quations "*recently published*."
the *Phil. Mag.* for "March,
Professor Young refers to Dr.
rd's tract; and as this paper is
elfast, Feb. 15th, 1849," the
must have been in possession
previously to this date.
m a private note to me dated
th, 1849," it appears that Pro-
avies had *then* read the pam-
ordered the work for myself,
Longman's, on "January 23rd,
he answer returned was "*not*
did I succeed in obtaining a
il "March 9th, 1849," as
rom a reference to my book-
ay-book for that period.
fessor Young claims for Mr.
e "essential principle" of Dr.
rd's method in the *Phil. Mag.*
, 1849, p. 282; and Mr. Davies
maintains his *priority* in the
s of the principle in a "*Note*
erical Transformation," in-
the *Phil. Mag.* for May, 1849.
er communication bears date
, March 8th, 1849," and the
Royal Military Academy,
h, April 6th, 1849."
ce appears that copies of Dr.
rd's tract were in the hands of
ticians *early in February*, 1849,
the publication, most probably,
e in "January, 1849," as the
ntimates. My copy contains
ress errors *apparently corrected*
uthor himself, and hence we
onably account for the difficulty
ing *early* copies.

I am, Sir, yours, &c.,

THOS. T. WILKINSON.

ncashire, July 9th, 1851.

PENDULUM EXPERIMENT.

As the pendulum experiment
ed considerable interest, and as
e your correspondent Mr. Ros-
fully understand how the plane
ion apparently rotates 15° per
e pole and remains stationary
uator, but cannot comprehend
uld vary in intermediate latitudes, I send you the following experiment, by which I think they will be able to practically demonstrate the fact:

Construct a parallel ruler of two pieces of fine steel wire, four or five inches long, and capable of opening out to two or three inches wide. Fix on a globe, at any latitude (say 30°) a circular piece of paper representing the table and rule, a line across the centre to represent the plane of vibration, say from north to south. Place one arm of the parallel ruler directly over and in a line with the pencil-mark, hold it firmly, and turn the globe through 15°; bring the other arm over the centre of the table, and rule a line. Now remove the ruler, and placing it over the last mark, turn the globe 15° more, and draw another line. Repeat this operation every 15° until the globe has revolved 90°, and you will have a paper with seven lines on it thus:

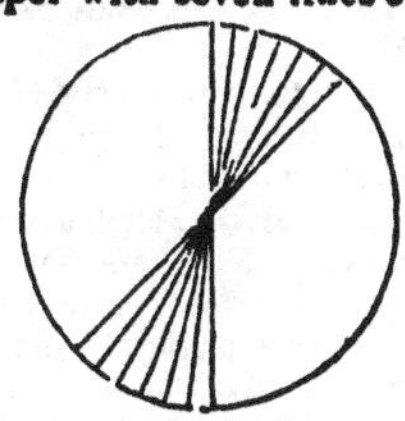

Now repeat this process with the table affixed to another latitude (say 70°) and you will have a paper with the seven lines thus:

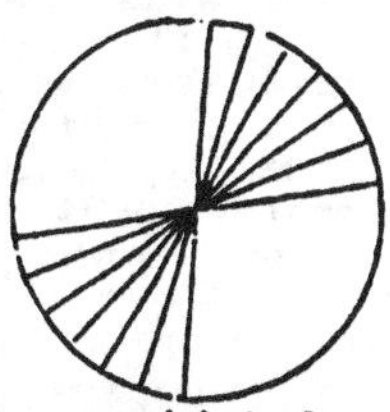

Try it in several latitudes, and it will be found that, whereas at the pole the lines marked on the table will correspond with the number of degrees through which the globe has been turned, so they will gradually describe a less and less angle as you approach the equator, where there will be no deviation whatever.

It is not necessary, though perhaps more convenient, to commence with the plane of vibration in a line with the meridian; and the process, if wished, may be continued during an entire revolution of the globe.

Although this experiment will prove that the period of rotation varies with the latitude, yet it is not strictly correct, because that which is here made to occur every 15°, actually takes place every vibration of the pendulum, but cannot be shown on so small a scale. The correct demonstration must be left for the higher powers of mathematical analysis, and has been proved to vary with the sine of the latitude.

I am, Sir, yours, &c.,
R. Webster, Jun.

74, Cornhill.

SPECIFICATIONS OF ENGLISH PATENTS ENROLLED DURING THE WEEK ENDING JULY 10, 1851.

William Robinson, of Holsham, York, machinist and agricultural implement maker. *For improved machinery for separating corn from straw.* Patent dated January 11, 1851.

This machine, which is intended to be used in conjunction with an ordinary thrashing machine, consists of a rectangular framework, across which extend a series of axes parallel to each other, some having two and others three radial flaps (thin strips or leaves of wood or other material), projecting from them, and of equal length with the axes. The axes, which are placed at such distance apart as to allow room for them to revolve without bringing the flaps in contact, have mounted on them pinions gearing into each other, and are caused to revolve at a rapid rate, and with the same linear velocity. By the shaking action thus produced the corn is effectually separated from the straw, and falls through the spaces between the axes, whilst at the same time the straw is gradually carried forward and delivered from the machine.

Claim.—A series of flaps or flies in a frame, arranged to receive the straw from the thrashing machine, and separate the corn therefrom whilst the straw is carried forward.

John Corry, of Belfast, damask manufacturer. *For improvements in machinery or apparatus for weaving figured fabrics; which machinery or apparatus is also applicable to other purposes for which Jacquard apparatus is or may be employed.* Patent dated January 2, 1851.

The improvements here claimed have relation to Jacquard apparatus, and comprehend:

1. The application as a substitute for the cards ordinarily employed in such machines, of a perforated metallic sheet, covered with paper, parchment, or other similar material, on which is painted, drawn, or traced, the pattern or design intended to be produced on the figured fabric, which pattern or design so painted, drawn, or traced, is afterwards punched out by means of suitable punches, and thereby holes or perforations produced which answer the same purpose as the holes punched in the ordinary cards.

2. A peculiar arrangement of Jacquard apparatus, in which the bent wires to which the harness is attached are arranged in one, or not more than two rows, instead of being in several rows, as at present; in which also the needles are combined with the bent wires in such a way as to act thereon by means of their heads or ends, and being kept in contact by the bent ends of the wires, which act as springs.

3. The exclusive use of needles which act on the bent wires by means of their heads or ends striking against them, the heads or ends being enlarged for the purpose, instead of being provided with an eye or loop through which to pass the needle, as in the present arrangement.

Joshua Horton, of Ætna Works, Smethwick, steam-engine boiler and gas-holder manufacturer. *For improvements in the construction of gas-holders.* Patent dated January 2, 1851.

The main object of the present improvements is to dispense with guide standards of the usual great height, and to employ instead standards of about half the height of the holder when afloat. The claims are—

1. The use of vertical travelling guide standards fixed to the lower or outer gasholder, to which standards friction rollers, wheels, or pulleys are attached, which work against guide-plates or bars, for the purpose of keeping the holder in a steady and upright position during its rising and falling, caused by the increase or diminution of the quantity of gas contained therein.

2. The application of guide-bars or plates fixed to and forming part of the sides of the gas-holder, instead of being loosely attached thereto, as in the ordinary construction of gas-holder.

3. A peculiar method of forming the hydraulic cups for the water joints of the different parts of the holder, by bending the metal of which the top and bottom are composed into a cup shape, so that the joint may be made complete, and attached to the sides of the holder without the necessity of employing angle iron and double sets of rivets for the purpose.

John Tatham and David Cheetham, of Rochdale, machine-makers. *For certain improvements in steam engines, in apparatus for generating and indicating the pres-*

and for filtering water to be lers: also improvements ap- am vessels or ships. Patent 2, 1851.

ments here claimed are—

ication to locomotive engines ial cylinder or cylinders, in m exerts its expansive force tuated the piston in the first the arrangements described, ylinder is placed either at the t, and has its piston attached d as the main cylinder, or it it, and the two pistons are a crosshead, so as in both simultaneously in the same lso, a method of transferring ttachment of the crank con- ith the driving cylinder to a he end of the cylinder from ton rod or rods issue. [In nt the patentees employ an ler, with a piston of corre- . The piston is connected by s to a crosshead, at the centre nother rod, which slides in ircular space at the centre of The connecting rod of the shaft is attached to the end ad rod which works in the these means the object above ttained, and the driving power it within a very short distance aft.]

lication to steam boilers of s to connect the flues, when constructed in the direction of the boiler.

ication to steam boilers of a th, and extending about half r the purpose of receiving mpurities. [The chamber is a cock for drawing off the quired intervals.]

loyment of a fan for causing of combustion, after passing iler flue, to be drawn through es, and facilitate the genera- by heating the feed-water for

ng of the fire-bars of furnaces transversal to the fire-door. this arrangement, a part of e replaced when worn by ex- fire, which cannot be done to n the bars are placed in the ion.]

loyment in apparatus for mea- icating the pressure of steam f mercury, which is caused to a undue pressure of steam, *ity opens the safety-valve.*

7. The application of a filter for purifying the water employed for injection into the condenser. [The patentees describe two arrangements of filter adapted for this purpose. The first of these consists of a casing, within which is a double cylinder of perforated zinc or metal mounted in horizontal bearings, and capable of revolving. The space between the perforated cylinders is filled with wool or fibrous materials, through which the water passes, and after being thus purified, it is drawn off by a pump and injected into the condenser. An air pipe is provided leading to the interior of the cylinder, in order that in certain cases the pressure of the atmosphere may be allowed to be exerted on the water contained therein, and thus the action of the exhausting pump be facilitated. According to the second arrangement, the cylindrical filter occupies a vertical position, and is made stationary instead of being moveable]

8. The employment of steam, conducted through suitable pipes, for extinguishing fire on board ships. [The arrangement of the pipes, which should in all cases be provided with suitable cocks, must be varied with circumstances, such as the build of the vessel and the situation in which they are required to be placed.]

THOMAS LAWES, of the City-road. *For improvements in generating and applying steam for certain purposes.* Patent dated July 4, 1851.

The improvements sought to be secured under this patent comprehend—

1. A peculiar construction of tubular boiler, in which the tubes are arranged in alternate horizontal and vertical lines, and provided with plugs to admit of their being cleared out when necessary.

2. An apparatus for drying feathers, hair, wool, &c., by the aid of steam pipes enclosed within a cylinder, to which a rotary motion is communicated when the materials have been placed therein.

3. The application to locomotive engines of an intermediate toothed driving wheel gearing into a rack or toothed rail placed between the ordinary rails, and provided with suitable means for lifting it out of gear.

Specifications Due, but not Enrolled.

BENJAMIN COOK, of Birmingham, manufacturer. *For a certain improvement or certain improvements in the manufacture of metallic tubes.* Patent dated January 3, 1851.

JOHN PERCY, of Birmingham, doctor of medicine, and HENRY WIGGIN, of the same place, manufacturer. *For a new metallic alloy, or new metallic alloys.* Patent dated January 3, 1851.

Cornish Engines.—The number of pumping engines reported this month is 27. They have consumed 2081 tons of coal, and lifted 19,000,000 tons of water 10 fathoms high. The average duty of the whole is, therefore, 52,000,000 lbs. lifted 1 foot high by the consumption of 94 lbs. of coal.—*Lean's Engine Reporter, Jan. 8*, 1851.

WEEKLY LIST OF NEW ENGLISH PATENTS.

Frederick Rosenborg, of the Albany, Middlesex, Esq., for improvements in the manufacture of casks, barrels, and other like articles, and the machinery employed therein. July 5; six months.

Henry Craven Baildon, of Edinburgh, chemist, for improvements in writing, printing, or marking letters, characters, or figures upon paper, parchment, or other materials properly prepared for that purpose. July 7; six months.

James Buchanan Mirlees, of Glasgow, Lanark, North Britain, engineer, for certain improvements in machinery, apparatus, or means for the manufacture or production of sugar. July 7; six months.

WEEKLY LIST OF DESIGNS FOR ARTICLES OF UTILITY REGISTERED.

Date of Registration.	No. in the Register.	Proprietors' Names.	Addresses.	Subjects of Design.
July 3	2872	George Orpwood	Bishopsgate-street	Register or book mark.
,,	2873	George Mallock	Carpenter-street, Berkeley-sq.	Suspending hook.
4	2874	James Kimberley	Birmingham	Stay and fastener for windows, doors, and shutters.
,,	2875	Nicholas Stead and Son,	Hulme, Manchester	Ventilating chimney-top.
,,	2876	Bathgate and Wilson	Canning Foundry, Liverpool	Metallic cask.
,,	2877	John Pannell	Fetter-lane	The retort calorifere for conservatories, green-houses, &c.
5	2878	Thomas Foxall Griffiths,	Birmingham	Portable cooking stove.
,,	2879	George Chambers and Co.	Priory-mills, Studley, and Gresham-street	Needle eye.
7	2880	Samuel Last	New Bond-street, and Oxford-street	Prend-tout, or railway portmanteau.
8	2881	Simcox and Pemberton,	Birmingham	Blind roller, and swing-glass axle.
9	2882	Robert Smith Bartlett	Redditch	Part of a watch key.

WEEKLY LIST OF PROVISIONAL REGISTRATIONS.

July 7	256	Harrild and Sons	Gt. Distaff-lane	Printer's mitring guard.
8	257	George Pigall	St. Martin's-court, Leicester-square	Watch-guard.
,,	258	Richard Timmins and Sons	Pershore-street, Birmingham	Loose heater or Italian-iron curling tongs.

CONTENTS OF THIS NUMBER.

LONDON: Edited, Printed, and Published by Joseph Clinton Robertson, of No. 166, Fleet-street, *in the City of London*—Sold by A. and W. Galignani, Rue Vivienne, Paris; Machin and Co., *Dublin; W. C. Campbell* and Co., Hamburg.

Mechanics' Magazine,

MUSEUM, REGISTER, JOURNAL, AND GAZETTE.

No. 1458.] SATURDAY, JULY 19, 1851. [Price 3*d*., Stamped, 4*d*.

Edited by J. C. Robertson, 166, Fleet-street.

DUNN'S PATENT CALORIC ENGINE.

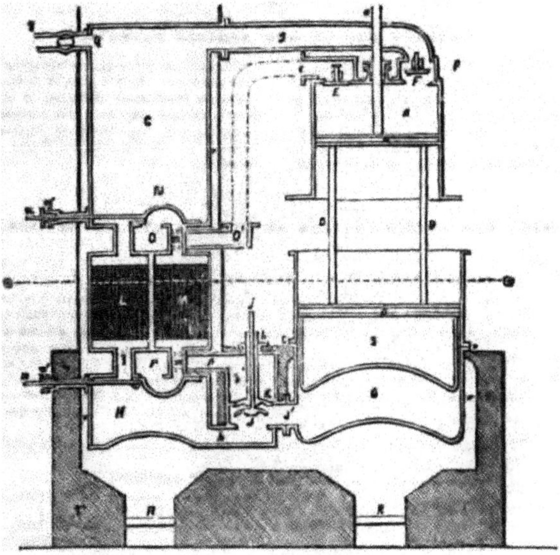

Fig. 1.

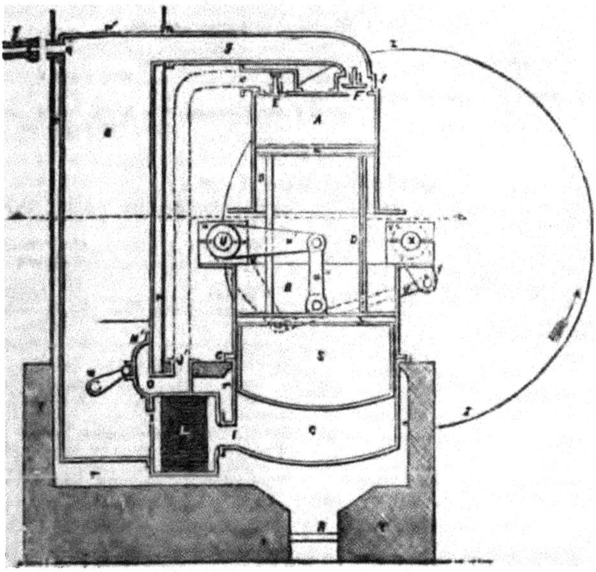

Fig. 2.

DUNN'S PATENT CALORIC ENGINE.

(Patent dated December 26, 1850. Specification enrolled June 26, 1851.)

Specification.

THIS invention consists in producing motive power by the application of caloric to atmospheric air, or other permanent gases or fluids susceptible of considerable expansion by the increase of temperature,—the mode of applying the caloric being such that, after having caused the expansion or dilatation which produces the motive power, the caloric is transferred to certain metallic substances, and again retransferred from these substances to the acting medium at certain intervals, or at each successive stroke of the motive engine—the principal supply of caloric being thereby rendered independent of combustion or consumption of fuel. Accordingly, whilst in the steam engine the caloric is constantly wasted by being passed into the condenser, or by being carried off into the atmosphere, in the improved engine, the caloric is employed over and over again, enabling me to dispense with the employment of combustibles, excepting for the purposes of restoring the heat lost by the expansion of the acting medium, and that lost by radiation also, and for the purpose of making good the small deficiency unavoidable in the transfer of the caloric.

Having thus stated the object and general character, I will now proceed to describe the structure of the improved engine for producing motive power, reference being had to the engraving. Figs. 1 and 2 represent longitudinal sections of the engine, both being alike in all essential points, differing only in part of the detail, as will be clearly seen by the following description. I will first describe fig. 1:—A and B are two cylinders of unequal diameter, accurately bored and provided with pistons *a* and *b*, the latter having air-tight metallic packing rings inserted at their circumferences. I call A the supply cylinder, and B the working cylinder; *a'* piston rod attached to the piston *a*, working through a stuffing box in the cover of the supply cylinder. C is a cylinder with a spherical bottom attached to the working cylinder at *cc:* I call this vessel the expansion heater; DD rods or braces connecting together the supply piston *a* and the working piston *b*. E is a self-acting valve opening inwards to the supply cylinder; F a similar valve opening outwards from said cylinder, and contained within the valve box *f*. G is a cylindrical vessel, which I call the receiver, connected to the valve box *f* by means of the pipe *g;* H a cylindrical vessel with an inverted spherical bottom: I call this vessel the heater. J a conical valve supported by the valve stem *j*, and working in the valve chamber J', which chamber also forms a communication between the expansion heater C and heater H by means of the passage *h*. K is another conical valve supported by the hollow valve stem *k*, and contained within the valve chamber *k'*. L and M two vessels of cubical form, filled to their utmost capacity, excepting small spaces at top and bottom, with discs of wire-net or straight wires closely packed, or with other small metallic substances, or mineral substances such as asbestos, so arranged as to have minute channels running up and down. I call these vessels L and M, with their contents, regenerators. *l l*, *m m*, pipes forming a direct communication between the receiver G and the heater H, through the regenerators. N N two ordinary slide valves arranged to form alternate communications between the pipes *l l* and *m m* and the exhaust chambers O and P, on the principle of the valves of ordinary high-pressure steam engines. *n n* valve stems working through stuffing boxes *n' n'*. *p* pipe communicating between the valve chamber *k* and exhaust chamber P. *o'* pipe leading from exhaust chamber O. Q pipe leading into the receiver G, provided with a stop-cock *q*. R R fire-places for heating the vessels H and C. *r*, *r*, *r*, *r*, flues leading from said fire-places, and terminating at *r'*. S a cylindrical vessel attached to the working piston *b*, having a spherical bottom corresponding to the expansion vessel C. This vessel S, which I call the heat-intercepting vessel, is to be filled with fire clay at the bottom, and ashes, charcoal, or other non-conducting substances towards the top, its object being to prevent any intense or injurious heat from reaching the working piston and cylinder. T T brickwork or other fireproof material surrounding the fire-places and heaters. I now proceed to describe fig. 2. All corresponding parts in this figure are marked by similar letters of reference as in fig. 1; it will, however, be well briefly to repeat the description of the same:—A supply cylinder; *a* supply piston; B working cylinder; *b* working piston; C expansion heater; C junction of working cylinder and expansion heater; D rods connecting the supply and working pistons; E inlet valve of supply cylinder; *e* valve chamber of the same; F outlet valve of supply cylinder, and *f* its chamber; G receiver, *g* pipe connecting the same to outlet valve chamber *f;* L regenerator, *l* passage between the same and receiver; *l'* passage between the regenerator and the expansion heater; N' slide valve, *n* stem or spindle for working the same; O exhaust chamber under the slide valve; O' outlet pipe; Q pipe leading into receiver, *g* stop-cock in the same; R fire-place; *r, r, r, r,* flues leading from said fire-place; *r'* exit of said flues; T brickwork surrounding the fire-place and flues; U rock shaft, supported at both ends by appropriate pillar-blocks; *u u'* crank lever or arm attached to the said rock shaft; *u''* link connecting said arm to the working piston *b*; V another crank lever or arm attached to the extreme end of the rock shaft; *x* crank shaft or axle, having a crank Y firmly attached; *v* connecting rod connecting the arm V to the crank pin *y* of crank Y; *y'* pillar blocks supporting the crank shaft *x*; Z Z represent the circumference of a fly-wheel, paddle-wheel, propeller, or other rotary instrument to be worked by the engine. Fig. 3 represents a sectional plan of the fig. 1, and fig. 4 a sectional plan of fig. 2. Before describing the operation of the improved engine, it will be proper to observe that the piston rod *a'* only receives and transmits the differential force of the piston *b*, viz.,

Fig. 3.

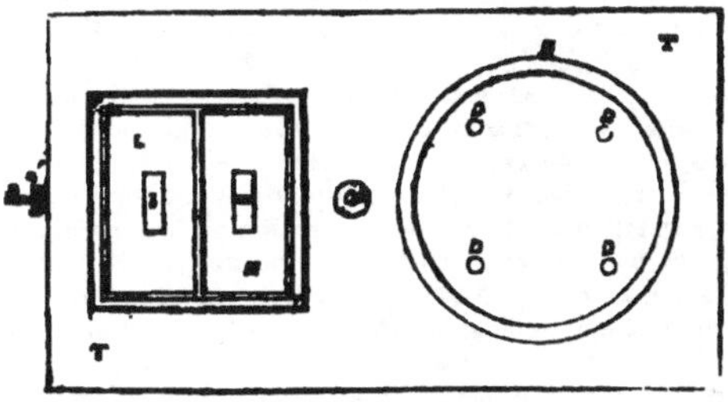

Fig. 4.

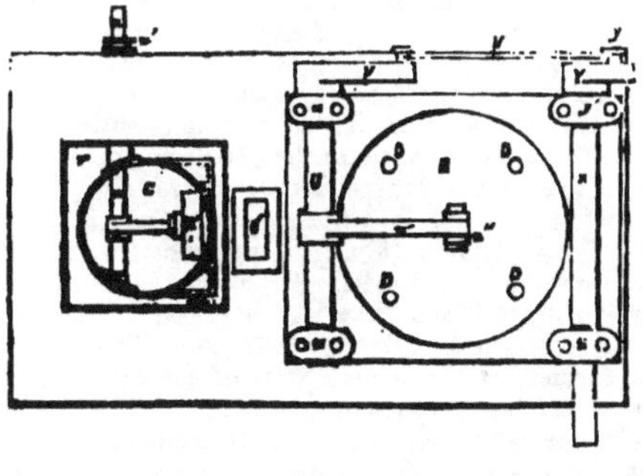

the excess of its acting force over the reacting force of piston *a'*: it will also be proper to observe that this differential force imparted to said piston rod may be communicated to machinery by any of the ordinary means, such as links, connecting rods, and cranks; or it may be transmitted directly, for such purposes as pumping or blowing. I have further to observe that the conical valves K and J may be worked by any of the ordinary means, such as eccentrics or cams, provided the means adopted be so arranged that the valve K will commence to open the instant that the piston *b* arrives at the full up-stroke, and be again closed the instant the piston arrives at full down-stroke, whilst the valve J is made to open at the same moment, and to close shortly before or at the termination of the up-stroke. In like manner the slide valve N' is to open and close as the piston *b* arrives respectively at its up and down stroke, similar to the slide valve of an ordinary high-pressure engine. It will be seen that the link *u''*, like the piston rod *a'*, only transmits the differential or useful force of the piston *b*.

Having thus described the construction of the engine, I will now proceed to describe the manner in which the same is to be put into operation, reference being first had to fig. 1. Before starting, fuel is put into the fireplaces RR, and ignited, a slow combustion being kept up until the heaters and lower parts of the regenerators shall have been brought to a temperature of about 500°. By means of a hand pump, or other similar means, atmospheric air is then to be forced into the receiver G, through the pipe Q, until there is an internal pressure of some eight or ten pounds to the square inch. The valve J is then to be opened, as shown in the engraving, the pressure entering under the piston *b* will cause the same to move upwards, and the air contained in A will be forced through the valve F into the receiver. The slide valves NN being by means of the two stems *nn*, previously so placed that the passages *l l* are open, the air from the receiver will pass through the wires in L into the heater H, and further into C, the temperature of the air augmenting, and its volume increasing as it passes through the heated wires and heaters. The smaller volume forced from A will in consequence thereof suffice to fill the larger space in C. Before the piston arrives at the top stroke, the valve J will be closed, and at the termination of the stroke the valve K will be opened; the pressure from below being thus removed, the piston will descend, and the heated air in C will pass through *k'*, *p*, P, and *m* into the regenerator M and in its passage through the numerous small spaces or cells formed between the wires part with the caloric, gradually falling in temperature until it passes off at O', nearly deprived of all its caloric. The commencement of the descent of the piston *a* will cause the valve F to close, and the valve E to open, by which a fresh charge of atmospheric air is taken into the cylinder A. At the termination of the fall-down stroke, the valve K is closed and the valve J again opened, and thus a continued reciprocating motion kept up. It will be evident that after a certain number of strokes the temperature of the wires or other matter contained in the regenerators will change,—that of M will become gradually increased, and that of L diminished. The position of the slide valves NN should therefore be reversed at the termination of every fifty strokes of the engine, more or less, which may be effected either by hand or by a suitable connection to the engine. The position being by either of these means accordingly reversed to that represented in the engraving, the heated air or other medium passing off from C, will now pass through the partially cooled wires in L, whilst the cold medium from the receiver will pass through the heated wires of M, and on entering H will have attained nearly the desired working temperature. In this manner the regenerators will alternately take up and give out caloric, whereby the circulating medium will principally become heated independently of any combustion after the engine shall have been once put in motion. Having thus with special reference to fig. 1, described the manner of putting the improved engine into operation, I have now to notice that the said engine, as represented in fig. 2, is operated precisely in the same manner, excepting that the regenerator is arranged in a single vessel, and that the metallic substances therein take up the caloric from the circulating medium that leaves the working cylinder or vessel C, and returns the same to the circulating medium that enters the working cylinder at each stroke of the engine, instead of transferring and retransferring the caloric at intervals, as shown in fig. 1. The manner in which the differential or useful upward force of the working piston *b* (fig. 2) in conjunction with its descending power, caused by gravity, are made to impart rotary movement to the crank-shaft *x* becomes self-evident on examining the disposition of the working gear of the engine, as shown in the engraving. It is particularly worthy of notice that the relative diameter of the supply and working cylinder will depend on the expansibility of the acting medium employed; thus in using atmospheric air or other permanent gases, the difference of the area of the pistons may be nearly as two to one, whilst in using fluids (such as oils, which dilate but slightly), the difference of area should not much exceed one-tenth. I have next to notice that in employing any other medium than atmospheric air, it becomes indispensable to connect the outlet pipe O', and the valve-box *e* of the outlet valve E, as indicated by dotted lines in both figures, these dotted lines representing the requisite connecting pipe. The escaping air or fluid at O' will, when such a connecting pipe has been applied, furnish the supply cylinder independently of other external communication, and the acting medium will perform a continuous circuit through the machine under this arrangement, the operation being in other respects as before described. It is evident that the several parts composing the improved engine may be arranged in various ways, and that the external form thereof may be greatly changed whilst its principle of operation remains substantially as I have ascertained and described. It is also evident that the working cylinder may be placed horizontally or otherwise, and that it may be made double acting, and that a heat-intercepting vessel may be applied at each end of the working piston, and also an expansion heater at each end of the working cylinder. I do not, therefore, confine myself to the exact form represented in the engraving, but I claim as the invention the substantial features of the devices I have described as new, and by which I secure great and beneficial results; and particularly I claim as the invention:

First. I claim the structure which I call the regenerator, by which I effect a transfer of the caloric contained in the air or other circulating medium that passes off from the working cylinder to a series of discs of wire net, or to other minute metallic or mineral sub-for the purpose of being again retransferred to the air or other circulating medium ers the working cylinder, whether said structure be arranged that the metallic or substances are made to take up and again return the caloric at each successive f the working piston, as in fig. 2; or whether it is so arranged that the transfer unsfer of the caloric takes place at intervals, as in fig. 1; or whether said structure ged in any other manner for the purpose of accomplishing substantially the purposes fore mentioned.

d. I claim the combination of the expansion heater with the working cylinder, by e fall of temperature consequent upon the expansion of the air or other circulating during the upward movement of the working piston becomes restored, and by so the force of said piston becomes augmented beyond what it would be if the acces- caloric effected by the expansion heater did not take place.

. I claim the heat intercepting vessel attached to the working piston, by which any ly high temperature is prevented from reaching the packing of said piston and by so the very desirable end is attained of presenting at all times surfaces of uniform perature to the acting medium under the working piston.

h. I claim the inverted position and open ends of the working and supply cylin- represented in the engraving.

I claim the direct attachments of the working and supply pistons by which not acting and re-acting forces may be uniformly distributed over the area of each out by which also the entire differential power of the working piston is rendered , less only the friction of the packings.

CONSTRUCTION OF STEAM BOILERS AND THE CAUSES OF THEIR EXPLOSIONS.

BY WM. FAIRBAIRN, ESQ., C.E., F.R.S.

(Concluded from page 30.)

r such circumstances what could be l but a blow up? A madman rush- a lighted match into a powder ma- ould not act with greater insanity. owever, has been the case, and that from want of thought, or, what is from the total absence of know- ich it was the duty of his employer s himself to have possessed.

e on former occasions stated that I an advocate for Legislative interfer- er in the construction or manage- boilers; but seeing the dangerous of these vessels when placed under rol of ignorance and incapacity, I orego many considerations to en- a more judicious and intelligent men than has hitherto been employed are and management of steam and m engine. The reforms necessary roduced may be done by the owners engines, steam boats, railways, and gaged in the use and application of ortant element. A desire to enforce dicious and stringent regulations, erate talent, and to employ only hose good conduct and superior ge entitle them to confidence, is the guarantee of public safety and the ty of the employer.

The remedies applicable for the on of accidents arising from explo-

g noticed in the foregoing remarks the *causes incident to boiler ex-* plosions, it now only remains to draw such inferences as will point out the circumstances which it is desirable to cultivate, and others which it is desirable to avoid. These circumstances I have endeavoured to class in such way as to bring the subject prominently forward, and to point out under each head; first, the causes which lead to accident; and secondly, the means necessary to be observed in avoiding it. In a general summary it may not be inexpedient briefly to recapitulate these statements, in order to impress more forcibly upon the mind of those concerned the necessity for care and consideration in the use of one of the most powerful agents ever placed at our disposal.

One of the most scientific nations of Europe places the greatest confidence, as a means of safety, on the use of a fusible metal plate over the furnace. These plates are alloys of tin and lead with a small portion of bismuth, in such proportions as will ensure fusion at a temperature something below that of molten lead. In France, the greatest importance is attached to these alloys, and in order to ensure certainty as to the definite proportions, the plates are prepared at the royal mint, where they may be purchased duly prepared for use. In this country these alloys are not generally in use; but in this respect I think we are wrong, as boiler explosions are not so frequent in France as in this country, and high-pressure steam, from its superior economy, is more extensively used in France than in England. In my own practice I invariably insert a lead rivet, 1 inch in diameter, immediately over the fire-place; and as lead melts at 640 degrees, I have invariably found these metallic plugs a great security in the event of a scarcity of water in the boiler. I am persuaded many dangerous explosions may be avoided by the use of this simple and effective precaution; and as pure lead melts at 600 degrees, we may infer from this circumstance that notice will be given and relief obtained before the internal pressure of the steam exceeds that of the resisting power of the heated plates. As this simple precaution is so easily accomplished, I would advise its general adoption. It can do no harm to the boiler, and may be the means of averting explosions and the destruction of many valuable lives.

The fusible metal plates as used in France are generally covered by a perforated metallic disc, which protects the alloy of which the plate is composed, and allows it to ooze through as soon as the steam has attained the temperature necessary to insure the fusion of the plate. The nature of the alloy is however somewhat curious, as the different equivalents have different degrees of fluidity, and the portion which is the first to melt is found out by the pressure of the steam causing the adhesion of the less fusible parts in a most imperfect state incapable of resistance to the internal force of the steam. The result of these compounds is, the fusion of one portion of the alloy and the fracture of the other, which is generally burst by pressure.

This latter description of fusible plates is different to the lead plug over the fire, as the one is fused at 600 degrees by the heat of the furnace, and the other, by the temperature of the steam, raised to the fusible point of the alloy, which varies from 280 to 350 degrees.

Another method is the bursting plate, fixed in a frame, and attached to some convenient part of the upper side of the boiler; this plate to be of such thickness and of such ductility as to cause rupture whenever the pressure exceeds that of the weight on the safety valve. There can be no doubt that such an apparatus, if made with a sufficiently large opening, would relieve the boiler: but the objection to this and several other devices is the frequent bursting of those plates, and the effect every change of pressure has upon the material in reducing its powers of resistance, and thus increasing uncertainty as to the amount of pressure in the boiler, as well as the constant renewal of the plates.

It has already been noticed that one of *the most important securities* against explosions is a duly proportioned boiler, well constructed, and to this must be added ample means for the escape of the steam on every occasion when the usual channels have been suddenly stopped. The only legitimate outlets under these circumstances appear to me to be the safety valves, which, connected with this inquiry, are indispensable to security. Every boiler should, therefore, have two safety valves, of sufficient capacity to carry off the quantity of steam generated by the boiler. One of these valves should be of the common construction, and the other beyond the reach of the engineer or any other person.

Whilst tracing the causes of explosions from a deficiency of water in the boiler, I have recommended as the usual precautions; good pumps, self-acting feeders, water cocks, glass gauges, float alarms, and other indicators which mark the changes and variations in the height of the water. To these may be added the steam whistle; but, above all, the constant inspection of a careful, sober, and judicious engineer. Above all other means, however ingeniously devised, this is the most essential to security, and on that official depends, not only the security of the property under his charge, but also the interests of his family, and the lives of all those within the immediate influence of his operations. One of the most important considerations in this and every other department of management is cleanliness and the careful attention of a good engineer.

Explosions produced from collapse have their origin in different causes to those arising from a deficiency of water, and the only remedy that can be applied is the vacuum valve and the cylindrical or spheroidal form of boiler.

Defective construction is unquestionably one of the greatest sources of the frightful accidents which we are so frequently called upon to witness. No man should be allowed unlimited exercise of judgment on a question of such vital importance as the construction of a boiler, unless duly qualified by matured experience in the theoretical and practical knowledge of form, strength of materials, and other requirements requisite to insure the maximum of sound construction. It appears to be equally important that we should have the same proofs and acknowledged system of operations in the construction of boilers, as we have in the strength and proportions of ordnance. In both cases we have to deal with a powerful and dangerous element, and I have yet to learn why the same security should not be given to the general public as we find so liberally extended in an important branch of the public service. In the Ordnance

: at Woolwich (with which I have or less connected for many years) care and precision is observed in cture of guns, and the proofs are made under the superintendence nt officers as to render every gun perfect safety to the extent of 200 rounds of shot.

nd artillery are equally exposed , and it appears to me of little hether the one is burst by the unpowder, or the other by the of steam.

nto consideration all the circum- nnected with the bursting of the bursting of guns, and looking re competition which exists, and be extended, in manufactories, fic, and steam navigation, where every day more desirable to coit by an extended use of steam higher pressure, it surely becomes m to secure the public safety by ction of some generally acknow- es of construction that will bear xperience and involve a maximum istance. The most elaborate dis- ave taken place, by the most dis- men of all ages since the inven- powder, to discover the strength guns of every description—surely equally if not more important, fice of human life appears to me r in the one case than the other: fore, a subject of paramount im- the public to know that the facts inquiry, and the knowledge of till, have combined to give *secu- rity as well as confidence*, that properly constructed and capable at least *three times* their working

question of explosions arising nagement and ignorance, we further to add; and it now only state that the subject of security explosions is of such importance for more able exponents than have endeavoured to trace the less lamentable occurrences, and h deductions therefrom as I trust ful in at least mitigating, if not ly averting, the danger.

he means of prevention and the necessary to be observed in the and management of boilers.

avoid explosions from internal ylindrical boilers of maximum rength must be used, including cessary appendages of safety

plosions arising from deficiency of water may be prevented by the fusible alloys bursting plates, good feed pumps, water gauge alarms, and other marks of indication; but, above all, the experienced eye and careful attention of the engineer is the greatest security.

3rd. Explosions from collapse are generally produced from imperfect construction, which can only be remedied by adopting the cylindrical form of boiler, and a valve to prevent the formation of a vacuum in the boiler.

4th. Explosion from defective construction admits of only one simple remedy, and that is, the adoption of those forms which embody the maximum powers of resistance to internal pressure, and such as we have already recommended for general use.

Lastly. Good and efficient management, a respectable and considerate engineer, and the introduction of such improvements, precautions, and securities as we have been enabled to recommend, will not only ensure confidence, but create a better system of management in all the requirements necessary to be observed in preventing steam boiler explosions.

DEEP SEA SOUNDINGS.

(From the "Scientific American.")

An Act of Congress authorises the vessels of the navy to co-operate with the scientific Lieutenant Maury, in procuring materials for his investigations into the phenomena of the "great deep." An order of the chief of the Bureau of Ordinance requires the commanders of our public cruisers to get a deep sea sounding whenever it is calm. Heretofore this had been a difficult object. The difficulty was in getting a line long enough, and in knowing when the plummet had reached the bottom

Recourse had been had by other navies to wire of great length and tenuity, and the greatest depth ever known to have been reached, before the subject was taken up here, was the sounding, by an officer of the English Navy, in 4,000 fathoms, which was by no means satisfactory. Lieutenant Walsh, in the United States schooner *Taney*, has reported a sounding without bottom, more than a mile deeper than this.

Instead of costly implements used for

sounding the depths of the ocean, our vessels are simply supplied with twine, to which they attach a weight, and when the weight ceases to sink they know it is on the bottom; and thus the depths of the ocean, in the deepest parts, may, without trouble or inconvenience, be ascertained in every calm of a few minutes' continuance.

With this simple contrivance the *Albany*, Captain Platt, has run a line of deep sea soundings across the Gulf of Mexico, from Tampico to the Straits of Florida.

The basin which holds the waters of this Gulf has thus been ascertained to be about a mile deep, and the Gulf stream in the Florida Pass about 3,000 feet deep.

Captain Barron, of the *John Adams*, has been sounding the Atlantic Basin, between the Capes of Virginia and the Island of Maderia, belonging to Portugal. He got bottom with a line of 5,500 fathoms, the deepest, and 1,040 fathoms, the shallowest.

Men of science will recognise in these results some of the most interesting and valuable physical discoveries of the day. They reflect the highest credit upon our navy and those who planned and set on foot these simple and beautiful arrangements, which have cleared away the difficulties with which all have found themselves beset who heretofore have undertaken to fathom the sea at great depths.

AMERICAN PROGRESS IN PRACTICAL ASTRONOMY.

At the recent Meeting at Ipswich of the British Association, a description was read of "An Apparatus for Making Astronomical Observations by means of Electro-Magnetism," by G. P. and R. F. Bond, of the Cambridge (U.S.) Observatory.—The apparatus exhibited has been in use at the Harvard Observatory, Cambridge (U.S.), and is the property of the United States Coast Survey. It consists of an electric break-circuit clock, a galvanic battery of a single Grove's cup, and the spring governor, by which uniform motion is given to the paper. Two wires pass from the clock, one direct to the battery, and the other, through the break-circuit key used by the observe through the recording magnet, back t battery. The length of the wire is of immaterial. When the battery is in nection, the circuit is broken by the leaving the tooth of the wheel, and stored at the instant of the beat of the which is, in fact, the sound produced b completion of the contact restoring th cuit,—the passage of the current through the pallet and the escap wheel alone. With the exception o connecting wires, and the insulation of parts, the clock is like those in comm for astronomical purposes. Several have been proposed by different pe for interrupting, mechanically, the ga circuit at intervals precisely equal. present instance the clock is of the proposed by Mr. Bond. Professor W stone, Professor Mitchell, Dr. Locke Saxton, and others, have contrived di modes of effecting this object:—the f several years since, but for a purpo tinct from the present. The cylinder a single rotation in a minute. The s marks, and the observations succeed other in a continuous spiral. When a is filled, and is taken from the cylinde second marks and observations appe parallel columns, as in a table of d entry, the minutes and seconds bein two arguments at the head and side sheet. The observer, with the break- key in his hand or at his side, at the i of the transit of a star over the wire telescope, touches the key with his f The record is made at the same insta the paper, which may be at any dis many hundred miles, if required, fro observer. It is a well-established fact not only may observations be increa number by this process, but that the of error of each individual result are narrowed. As far as comparisons hav been made, the *personal* equation bet different observers, if not entirely in ble, is at least confined to a few hundr of a second. It is through the faciliti means furnished by the Coast Survey partment of the United States, unde superintendence of Dr. A. D. Bache, individuals there have been enabled to to its present stage the application of tro-magnetism to the purposes of ge and of astronomy, it having been a expense of that department, and frequ by its officers, that nearly all the ex ments have been conducted.

Daguerreotypes of the Moon were sh taken by Messrs. Whipple and Jone

Boston, from the image formed in the focus of the great equatoreal of the Cambridge (U.S.) Observatory.

The Astronomer Royal (Professor Airy) said, that the principle of the method was entirely the discovery of the Americans, and Professor Bond had the merit of originating what he had no doubt would prove of the utmost importance in the practice of astronomy: for besides the distraction of the attention of the observer at present having to listen to and to count the beats of the clock, and having then to occupy many seconds in recording his observations when made,—he could not often repeat these observations at as short intervals as would be desirable. But by this method he might even repeat an observation within the compass of one second, if required. It was also believed that there was a more direct connection between the senses of sight and touch, the senses that he required the aid of in this mode of observing, than there was between the senses of hearing and seeing, the senses called into united operation in the present mode of observing; and if this were so, what was at present known to practical observers under the name of personal equation, would be got rid of, if not entirely, yet to a great degree. These and other considerations had made him determine to give this method of observing the most mature consideration and the fullest trial. He had a cylinder constructed of twenty inches length and one foot diameter, and of which a fair conception of the size might be formed when he stated that it would gauge to about a bushel. This cylinder he hoped to be able to cause to revolve with something of an approach to astronomical uniformity. For this purpose, it was his intention to dispense with the fly-wheel which regulated the motion in Mr. Bond's apparatus, and to depend on a large conical pendulum revolving in a circle, the diameter of which would be about equal to the arc of vibration of an ordinary seconds pendulum. This he intended should be a well-made mercurial compensating pendulum; and thus he hoped to be able to dispense with the clock used by Mr. Bond. The construction of the conical pendulum he intended to use was also peculiar. He intended to take advantage of the principle of the chronometric governor of the steam engine invented by a Prussian, and which the members of the section might see at work in Mr. Ransome's factory; but without such actual inspection, he feared he could not make himself understood in an attempt to explain this curious governor. Suffice it to say, that this governor was made to revolve by a bevel wheel, which engaged another bevel wheel attached to the governor, not directly, but through the intervention of a third, which worked upon a centre that was not entirely fixed. The moving of this intermediate wheel was made to work the valve which admitted or shut off steam, and thus equalize the motion of the machine as the resistance varied. In the apparatus he proposed to use, the resistance would occasionally vary from many causes, for instance, at the changing of the cylinder; and as this would affect the rate of the clock, if not provided against, he proposed to use the principle of the foregoing governor, by causing it to produce a varying by moving further out or nearer to the fulcrum of a steelyard a weight, which would thus increase or diminish, as was requisite, the friction caused by a point connected with the steelyard on a wheel kept revolving by the machine. In this way he hoped to be able to produce a motion which, under all changes to which the machine should be exposed, would remain uniform to the extreme accuracy required.

Mr. Bond exhibited daguerreotypes of the moon, taken with the 23-feet equatoreal of Cambridge (U. S.) Observatory. These daguerreotypes were very beautiful, and admitted of being very considerably magnified. But Mr. Bond stated that the motion of the equatoreal, although very steady, was yet not sufficiently so to admit of their being examined by very high magnifying powers. Sir David Brewster stated that, if these daguerreotype impressions were taken on transparent sheets of gelatine paper, and so placed before a telescope as to subtend accurately thirty minutes of a degree, they would assume all the appearance of the moon itself.

THE HAMMER SUPERSEDED IN BLOOMING IRON.

At the last Meeting of the Birmingham Institution of Mechanical Engineers, a paper was read "On a New Machine for Blooming Iron." A complete working model of this machine in brass was exhibited, and also a sectional model in wood, which very clearly elucidated its action. The working portion of the machine consists of three eccentric, cuspidated, semilunar-shaped cams, working simultaneously, and all kept rotating in one direction by wheels and pinions, firmly connected together in a strong frame, and set in motion by a steam engine. The convex sides of these semi-cylindrical cams are deeply grooved and serrated, and their peculiar form is such, that on dropping a bloom of iron into the concavity of the upper cam, as it presents itself, it is immediately drawn into the vortex, or centre of motion, of the three cams at the instant when that opening is the largest. As they rotate, the convexities, in consequence of the eccentricity of the centres, approach nearer and nearer—the ridges and rough surfaces squeezing, rolling, and kneading the iron in all directions, like squeezing a sponge in the hand. The cinders and impurities are thus ejected, and fall out beneath the machine; and the cams, in the latter part of their rotation, having closed the space between them to the smallest dimensions in the revolution, the bloom is elongated and ejected in the form of an iron cylinder. The paper stated that the machine was the invention of Mr. Jeremiah Brown, late of the Oak Farm Iron-works, and that its use was calculated to form a new era in the iron trade. For the production of superior iron, it had hitherto been considered that the hammer was indispensable; but for all purposes of efficiency, rapidity of action, and economy, this machine, it was assumed, would come into general use. From its strength and simplicity, it would not cost in repairs 20*l.* a year; while a hammer involved expenses of ten times that amount, and the cost of replacing a broken hammer was well known in the iron trade to be a serious *item.* It turned out a finished bloom, entirely free from cinder, in twelve seconds, the engine working moderately; while under the hammer it could not be completed under eighty seconds. Thus, by the machine, the cylindrical bloom, when ejected, was still at welding heat, and could be at once passed through the rolls, while from the hammer it had again to pass through the furnace.

In the discussion which followed the reading of the paper, Mr. Beazley, of Smethwick, the author of the paper, stated that, from some comparative experiments he had made, as to the strength of the same iron finished by hammer and by the machine, he considered the quality about equal; on different-sized bars, in some cases, they were a trifle in favour of the hammer, and in others of the machine; but he considered the economy highly important. In labour there was a saving of 1s. 3d. per ton; in tools of 1s. per ton; and the saving in time was equally worthy of consideration. That a more perfect ejection of the cinder was effected by the machine than by the hammer, was clear from the fact that the same quantity of iron weighed less after passing the former than from the operation of the latter; and Mr. Beazley said that he had taken two blooms direct from the machine successively, passed them together through the rolls; and the result was a perfectly welded joint.

Mr. Adams bore testimony to the efficiency of the machine; but he had seen a bloom passed through the rolls from it, and noticed that a considerable quantity of cinder still oozed from the ends. He thought, after leaving the machine, the iron might be subjected to a few blows of the hammer with advantage, and thus aim at the production of a highly superior article, rather than at saving 1s. a ton. Mr. Beazley thought the hammer would be superfluous, as the rolls effected what the machine had left undone. Mr. Cowper had often seen the machine in operation, and had not noticed the cinder in the iron at the rolls, as represented by Mr. Adams. Mr. Williams said, if iron was imperfectly puddled, the hammer would knock it to pieces and show the defect; but he feared the machine would roll the iron up, whether good or bad. From the rolling action, the cinder would be lapped up in the iron. He considered the cost of the machine and repairs would be an important consideration. Mr. Beazley assured Mr. Williams he was in error; it had been repeatedly proved that if the iron was imperfectly puddled, the machine instantly tore it in fragments; that, as to complexity, it was as simple as the ordinary rolls, and no more likely to get out of repair. It had worked four months with only one trifling accident, which arose from faulty construction at first. Mr. Siemens, Mr. Slate, and Professor Hodgkinson, also bore testimony to the efficiency of the machine.

THE GOVERNMENT PATENT-LAW AMENDMENT BILL.

We contemplated last week, following up our strictures on the general character of this Bill, by a detailed examination of its provisions; but we are spared, to a considerable extent, the trouble of doing so by the appearance of a paper headed "Reasons against the Bill, issued by the *Association of Patentees for the Protection and Registration of Patent Property.*" We give these "Reasons" at length.

Reasons.

I. For more than two centuries it has been the settled policy of this country to encourage the production of new inventions and improvements, by granting patents for the exclusive use of the same for a brief period (fourteen years) to the authors and their assigns or licensees.

II. The system has worked well; for to its stimulating influence may be distinctly traced the unequalled progress which Great Britain has made in arts and manufactures, and all the wealth and power of which these have been the acknowledged source. Not a single branch of our industry can be named that has not had its origin in, or been materially aided by, patents. The steam engine, the cotton-spinning machine, the stocking-frame, the power-loom, steam navigation, railways, gas lighting, the hot blast, the electric telegraph, may be cited as notable and familiar examples.

III. The maintenance of this system, so worthy of all respect for its antiquity, and for the important services which it has rendered to the country, is now (for the first time) hostilely assailed. The Bill (No 3), though professing only to amend the existing law, would in fact introduce so many new obstructions to the granting of patents, and so large a limitation of the privileges conferred by them, as practically to put an end to the encouragement and protection which they have hitherto afforded to men of inventive genius. For—

Firstly. It is proposed by the bill to grant no patent for an invention until a preliminary inquiry has been made by a Government examiner into the novelty of the invention, instead of leaving the novelty, as heretofore, open to subsequent impeachment. In all cases where the invention claimed is of no value,—which is probably in nine out of every ten,—any such inquiry would be wholly superfluous.

Secondly. No patent is to be granted until all the world has been invited, by advertisement, to come forward and oppose it; thus wantonly stirring up opposition where none might otherwise arise, tempting agents and servants to betray the secrets of their employers, and subjecting the honest and meritorious inventor to a system of *espionnage* at once offensive to his feelings, perilous to his interests, and wholly unnecessary for the public safety.

Thirdly. The proposed preliminary examination would be of the most loose and unsatisfactory description, it not being proposed to invest the examiners with power to take evidence on oath, or to enforce the attendance of witnesses or production of books and papers. There is to be a right of appeal from the examiners to the law-officers of the Crown, but neither are these officers to have any such judicial authority.

Fourthly. The patent, when granted, is to extend only to England, Scotland, and Ireland, to the exclusion of the whole of the Colonial and Foreign possessions of the *British Crown,—a distinction which* would have the double effect of injuriously limiting the field of encouragement to the home inventor, and of depriving the excepted portions of much of that assistance which they might reasonably hope to derive from the continued application of scientific knowledge and mechanical skill to the development of their industrial resources.

Fifthly. It is proposed by the Bill to increase in a majority of cases the cost, already enormous, of patents. The cost of a patent for England and the Colonies is at present 110*l.*; the bill will raise it, *without the Colonies being included*, to at least 175*l.* Patents for Scotland and Ireland now cost respectively 75*l.* and 135*l.*; but under this bill, neither could be obtained for less than 175*l.* The only case in which the cost would be reduced would be where an inventor desired to patent his invention for all the three kingdoms; then instead of 310*l.*, he would have 175*l.* only to pay. But the cases are numerous where an inventor requires protection from one or two only of the three kingdoms.

And *sixthly*. The bill proposes to make the use or publication of an invention in any part of the world previous to the grant of a patent for it, a ground for avoiding the patent; whereas, hitherto that penalty has attached only to previous use or publication in some part of the British dominions. A man may be reasonably presumed to be aware of what has been promulgated in his own country or its dependencies, but if he is to be held bound to know all that has previously been invented anywhere, so wide a door will be opened to pretexts for infringement, that it will be next to impossible to sustain any patent whatever; and a degree of uncertainty will hence result, which will be most disheartening to inventors, and be a great obstacle to the investment of capital in the promotion of patent inventions.

IV. The sweeping and hazardous changes thus proposed to be made in the existing system for the encouragement of inventions, are wholly uncalled for by any pressure of public opinion, or by any proof whatever of its injurious tendency.

Lord Granville, who was Chairman of the Select Committee of the House of Lords, from which the bill emanated, has frankly owned that public opinion is in favour of the system, in proportion of at least one hundred to six.

The evidence taken on the subject before the Lords' Select Committee, has not yet (12th July) been printed; but when it appears, it will be found, it is believed, wholly insufficient to justify the conclusions founded upon it. The witnesses whose testimony is most relied on, are understood to have been mostly persons who have had no practical knowledge or experience of the working of the existing system. Of the hundreds of living persons who have been patentees, and more or less benefactors to their country by their inventions, and who could speak best to the influence which the law, as it exists, exercises in stimulating the inventive faculties, *not one of any note has been examined.* Some pains would seem to have been taken to shut out all evidence that might tell in favour of the system, and a willing ear given to such persons only as could testify against it. The bill, according to all appearance, has not grown out of the inquiry, but is the result of a foregone conclusion, conceived in ignorance of facts, and based on some exceedingly false theory of the indifference of inventors to remuneration for their labours.

These "Reasons," abundant, just, and forcible as they are, do not form the whole of the case against the Bill. We subjoin some

Additional Reasons.

(A). The Bill proposes to appoint the Lord Chancellor, the Master of the Rolls, the Attorney-general, the Solicitor-general, the Lord Advocate and Solicitor-general for Scotland, the Attorney-general and Solicitor-general for Ireland for the time being respectively, "*together with* SUCH OTHER PERSON OR PERSONS AS MAY BE FROM TIME TO TIME *appointed by Her Majesty*," to be "*Commissioners for Patents.*"

The duties of these Commissioners are to be—1. To appoint an "examiner or examiners," "officers, clerks, and servants," to execute the powers of the Act. 2. To provide an "office." 3. To make rules for the regulation of the said office, and "all matters and things which may appear to them necessary and expedient for the purposes of this Act."

Now what is there in these duties which makes the addition to the eight law functionaries named of any "other person or persons" as Commissioners, necessary?—What is there which the Attorney and Solicitor-general for the time being cannot do of themselves, without help from any one, especially if it be provided, as proposed, that "all the powers vested in the Commissioners *may be exercised* by any three or more of them, the Lord Chancellor or Master of the Rolls being one?" We can see no occasion whatever for the appointment of any additional Commissioners, unless, indeed, it be such occasion as may arise from the powers given by clause 29 to the Lord Chancellor and Master of the Rolls to saddle the country with a salary to every such extra official functionary. Where there is a salary going, there must, of course, be somebody to receive it,—services required or to be rendered being equally out of the question.

We are not sure that there is any need for a Commission at all; but as long as it is composed of such officials only as those named, we shall apprehend no harm from it, and shall therefore not oppose it. *Let, however, but one paid and superfluous Commissioner be added, and the affair will be made a job of.*

(B.) The Bill proposes to perpetuate what is universally allowed to be one of the worst features of the existing system, the granting of separate patents for the three principal divisions of the United Kingdom —England, Scotland, and Ireland. Why should not one patent for the whole suffice? One application, one warrant, one grant? What sensible reason can be given against the consolidation? The Bill proposes (though, to be sure, in rather an indirect and obscure way) to do away with the old three sets of fees; but that only makes the keeping up of the three sets of forms the more absurd. As long as there is money to be extracted from the pockets of the lieges by a multiplicity of useless forms, that may be some reason—though not a very good one—for keeping them up; but take away the money apology, and reason there is none of any kind, good or bad, left for the abuse.

The history of this Bill, and of the private and sinister influences under which it has sprung so suddenly into existence and Government favour, has yet to be brought to light. Lord Granville, in introducing the Bill into the House of Lords, made this observation:

"Before concluding, he must be permitted to do an act of justice, by declaring that the Committee were much indebted to Mr. Thomas Webster, a barrister in great practice, *who attended their meetings constantly, and aided them by his valuable suggestions.*"—*Times Report.*

Concurring with his Lordship in opinion that it is but right that inventors know to whom they are indebted for the manifold *benefits* the Bill will confer upon them, if passed into a law, we beg to observe that the Mr. Webster who was in this constant attendance on the Lords' Committee, and aiding them with his "valuable suggestions," was himself the author of sundry proposals for the amendment of the law of patents, which were pretty extensively circulated just previous to the meeting of Parliament, and that in these proposals will be found traces of some of the worst provisions of Lord Granville's Bill. For example—

Mr. Webster advocated the appointment of Commissioners, after having, in common with all reflecting and considerate persons, deprecated anything of the kind.

Mr. Webster proposed the system of preliminary examination—and was (we believe) the first to do so.

Mr. Webster proposed the exception of the Colonial and Foreign possessions of the Crown from the operation of patents.

The similarity of feature is remarkable, and strongly illustrates the advantage of being, as the Scotch say, "at the lug of the law."

But there is much which your "constant" ear-wigger can do in the way of omission as of commission. We beg, therefore, to add a couple of questions, which Mr. Webster may answer or not, as he pleases, but which will, at any rate, be allowed to be of abundant significance.

First. Did Mr. Webster suggest to Lord Granville, as he might have done, that there were other barristers of as high standing and of as extensive practice as himself in patent cases, whose opinions it might be worth having?—Mr. Rotch, for instance, or Mr. Hindmarch? What reasonable excuse can be made for omitting both of those *gentlemen from the list of witnesses?* And, *Second.* Did Mr. Webster suggest to Lord Granville, as he ought to have done, that there were practical men—themselves inventors and patentees—such as your Heathcotes, Neilsons, Hancocks, Brockedons—the evidence of any one of whom as to the influence of the existing system of encouraging inventions would have been worth that of all the lawyers in the world?*

From the "Caledonian Mercury."

For the last twelve months the attention of the public has been anxiously drawn to the reform of the patent laws, the subject, independent of its intrinsic importance, having been in a manner forced into the arena in the progress of the Great Exhibition, with the anticipated results of which it is so intimately associated. The expectation was, that the Government would introduce a measure of reform, and there could be no difficulty as to the general scope of that measure; for there have not been wanting several and well-digested schemes, all tending to the same object — the diminution of the present extravagant costs, and of the cumbrous and inefficient forms which patentees have to encounter under the existing law. Economy—simplicity of preliminary procedure — an improved system of registration—so that the interests of discoverers or patentees, and those of the public, which two interests ought not to be at variance, might be defined with precision, and amply secured, were the grand *desiderata.* What, then, must be the surprise of the public after all this discussion, to find the Government introducing a bill at the fag-end of the session, which bill has been well designated "a masterpiece of legislative blundering." Instead of seeking to amend the Patent Laws, the Government, judging by the observations of Lord Granville, the sponsor of the bill, have made the discovery that the general principle of the Patent Laws is inimical to the interests of society, and hence that they ought to pave the way for the overthrow of the whole patent system! Accordingly, in the present bill, the improvement in the law at home is based upon the annihilation of patents for the Colonies.

A greater and more inexpedient, and, let us add, though in no vicious sense, more unprincipled delusion never entered into the

* *Erratum in last week's Article.*—P. 35, col. 1, l. 10 and 11 from the bottom, for "repetition," read "refutation."

mind of any statesman. We take our cue from the sentiments of Lord Granville in defending his measure before the House of Lords, and these we consider so short-sighted, puerile, and withal so complete a deviation from moral justice, that we do not wonder at the strictures in the *Mechanics' Magazine*, the able article from which we subjoin. His lordship plainly lays it down, that it is not at all necessary to stimulate inventors, because he found that scientific men were in the habit of making known their discoveries with great alacrity, without seeking any protection from patents; and as a meet corollary from this notable and very novel idea, his lordship farther urged, "it was quite clear that the tendency of the Patent Laws was to raise the price of the commodity during the fourteen years while the patent existed." Of course, the tendency is to give the inventor a fair remuneration for what he has given to the public, and which but for the labour of his brain the public would have been *without*. Nothing more. If the discovery is valuable, abridging labour, and diminishing cost, those in the particular trade pay that tribute, not for the benefit of the inventor, though that is an indirect effect, but for the primary end of enhancing their own profits. If, again, the discovery is not worth that cost, then it remains dormant, and the patentee loses his pains. In fact the reasoning —the very shallow reasoning—of Lord Granville comes to this, that the public have lost a great deal because the works of Sir Walter Scott, Dickens, and Macaulay, can only be issued by the respective publishers who have paid for the privilege, and thereby remunerated those great luminaries for labours to the fruits of which they are as well entitled as Mr. Cubitt (of whom significant mention is made in the annexed article) is for his labour in raising any edifice of less enduring materials. This principle, so consonant with every idea of moral justice, we would maintain against all the world. Lord Granville, anticipating an illustration so palpable, denies its analogy; but Chief-Justice Eyre—an authority at least equal to this renowned Whig Statesman of 1851, whose fame has in an unlucky hour come in contact with the patent laws,—long since, and in felicitous and memorable language, expressly laid down the identity of principle between the two descriptions of property. The *dictum* of Chief-Justice Eyre, which is sustained by the common sense of mankind, has not as yet been shaken by that of Lord Granville.

We contend that the present patent law, *defective as it is, and easily capable of remedy, has still even almost in its rottenness conferred inestimable benefits on mankind,* and which but for it might have slumbered in the womb of time. It supplies the great stimulus to inventive energy, the hope of distinction, and also of substantial reward; and in this practically moving world, those two aims will be found to move in great harmony, and ought never to be disunited. The sound philosophy and social wisdom are to blend the same. But Lord Granville, on this question at least, appears to disregard all these broad, liberal, and wise maxims which have their source deep in our moral constitution, and are sanctioned by the highest code which man ever received.

We sincerely trust that the present measure, resting altogether on a false principle, will be thrown out; but, at the same time, it is necessary that the public opinion, and especially of our municipal bodies and chambers of commerce, should be speedily conveyed to the Legislature; for, strange to say, the Government, as will be seen from the brief discussion in the House of Commons on Friday, 11th inst., have intimated their intention to carry this bill during the present session. The Ministry must be greatly pressed for matter to fill up the chronicle of their achievements, when they take such a crude measure under their wing. But we refer our readers to the following article:—(Here follows the article in the last Number of the *Mechanics' Magazine.*)

EFFECT OF THE EARTH'S ROTATION ON PROJECTILES.

It happens, generally, when two persons differ concerning the result of a mathematical investigation, that one, at least, of the two, has mistaken the meaning which the other attaches to the expressions he has used; and it will probably be admitted that this is the case with respect to the law of the deviation of military projectiles, in consequence of the diurnal rotation of the earth, which is given in a former Number of the *Mech. Mag.* (p, 474, col. 2) It is there stated by our correspondent, Mr. Reynolds, that the deviations vary *directly as the range and inversely as the speed of the projectile;* and we, in answering (p. 495) the question which that gentleman had proposed (p. 474, col. 2), after giving reference to a *Mémoire* on the subject by a most distinguished mathematician, made a passing remark that the law was not *exactly* that which Mr. R. had stated. In that remark we did not mean to imply that our correspondent was unacquainted with the true law; in fact, we considered that his expression, without attention to the mathematical signification of the terms used, may have been intended only as a popular approximation to the law, founded on the fact, which is undoubtedly true, that an increase of horizontal range gives rise to an increase of deviation, and that an increase of speed is attended by a decrease of deviation. Now the exact formula for the deviation, including the effects of the air's resistance, is

$$-\frac{2\, n \sin.\, \gamma}{a c^3}\left(\begin{matrix} e\ \omega \\ e - c\ \omega - 1, \end{matrix}\right)$$

in which a is the initial velocity and ω is the range; the other terms are constants. The expression is, of course, transcendental; but, if the exponential quantity be developed, the deviation might be expressed, approximately, by two terms, one of which depends on the range simply, and the other on *the square of the range*, while a in the denominator shows that the deviation is inversely proportional to the *initial* velocity, not the mean velocity (p. 506, *ante*) of the spot in its trajectory. The law of the deflection is therefore very different from that mentioned above (p. 474), and the discrepancy is a sufficient justification of our remark that the latter is not *exact*.

THE ROTATION OF THE EARTH.

Sir,—Allow me to tender my thanks to your correspondent, Mr. Webster, for the ingenious experiment he has devised, for elucidating the pendulum experiments. I think it calculated to assist some minds to comprehend the subject, in one point of view, with perfect clearness. I cannot, however, accept it at present, as the complete demonstration of the problem, which it is claimed to be. At the most, it can only meet some of the difficulties of the subject. Assuming that the line in which the plane of vibration cuts the table, *continues parallel to itself*, without which the experiment has no relevancy, but which requires proof, the line drawn after the globe has been moved 15°, is no doubt the correct position of it, at the end of the first hour; but as the second line has now the position the first had, at the commencement, and we are directed to move the ruler to the first position, and repeat the first *step, the third line will not show the distance travelled from the* second, but only the repetition of the first quantity,—that due to the motion in the first hour. Your correspondent evidently assumes that this is a matter of indifference. If it has attracted his attention at all, he must consider each succeeding step as of the same magnitude as the first. I submit that this stands in need of proof, and that the experiment proposed is quite capable of deciding it, and that it would decide it in the negative. Your correspondent says, "It is not necessary, though perhaps more convenient to commence with the plain of vibration in a line with the meridian." He has accordingly made the experiment, of which he has sent you diagrams, in this direction,—and in this direction they appear perfectly correct. Now, I beg to suggest to him, to try the experiment in two more ways; first, with the plane of vibration east and west; and secondly, in a direction intermediate between the two. If he tries these at the latitude of 30°, and supplies you with two corresponding diagrams to the one, at that latitude, in his last communication, your readers will be able fully to determine the conclusions to be drawn from the experiment. I should have been ready to have sent the diagrams myself, but think it better to obviate carping, which this subject unfortunately seems calculated to provoke, that your correspondent should complete the experiment himself. I trust to his candour, at the same time, to state if his own views should have then undergone any modification.

B. ROZZELL.

Leicester, July 15, 1851.

LORD ROSSE'S SPECULUMS OF SILVER.

It is well known that in reflexion by silver much less light is lost than by any of the other metals,—but, unfortunately, this metal is so soft, that great difficulties present themselves in giving it the requisite degree of high polish. Lord Rosse had tried by the electrotype process to procure a surface with a high polish by depositing silver on a surface of speculum metal, and treating it by the same process as that used by the distinguished officers engaged in the Trigonometrical Survey of Ireland. But, unfortunately, he soon found that, use what precautions he would, either there took place an adherence of the deposited silver to the surface on which it had been deposited, or the polish was rendered imperfect by the means resorted to to prevent this. He tried copper similarly, which did not adhere, but produced a high degree of polish,—but of course its colour and other properties rendered it inadmissible as a reflector. He then determined to endeavour to grind and polish a plane surface of silver, the softness of that metal having, however, heretofore caused the attempt to fail in the hands of the most experienced who had tried it. The processes of grinding and polishing are essentially different:—in grinding, the substance, whether emery or other powder, must run loose between the substance which is used to rub it against the other and that which is to be ground; and he soon found that he could not use emery or any other grinding powder for bringing a surface of silver to a correct form,—for, from the softness of the metal and the unequal hardness of its parts, the emery was found to confine its action to the softer parts, leaving the harder portions in elevated ridges and prominences, something in the way that the iron handle of a pump which has been long and much used may be observed to be worn away. Hard steel he found he could bring to a very true surface, and even impart to it a high degree of polish; but the quantity of light it was capable of reflecting was by no means sufficient, nor could he succeed in imparting to the surface of silver by compression with highly-polished steel surfaces the evenly and high-polished surface requisite for his purpose. At length, he found that he could by the use of good German hones grind surfaces of silver perfectly true, and he had now no doubt that he could with safety recommend for that purpose as the best material the blue variety of the German hone. The next point was, to polish the surface to a true optical plane reflecting surface. This was by no means so easy a task as may be supposed;—for although our eminent silversmiths do produce surfaces of silver of an extremely brilliant polish, as in the magnificent plateaux and other articles which they turn out, yet if any one will take the trouble to examine these surfaces, they will be found to be so irregular, though highly polished, as to be entirely unfit for producing correct images by reflection. And it is a singular fact, that although in the first part of the process of polishing chamois leather of the finest kind was used to rub the rouge on the silver surface, yet the finer finishing polish had always to be communicated by the human hand. Nor would the hand of every individual answer;—the manufacturer had to select those with the very softest and finest grain,—nor would the hand of perhaps one in every twenty of the persons employed answer for thus giving the final finish. But it was obvious that the irregular action of the human hand would by no means answer the end he had in view. Suffice it to say, that at length, after many fruitless trials, he had succeeded in producing a polishing surface which seemed fully to answer the purpose by exposing spirits of turpentine to the continued action of air, or by dissolving a proper quantity of resin in the spirits of turpentine, and by means of this varnish applying the rouge to the same description of polishing substance which he used in polishing the speculum metal, and which he had heretofore so frequently described. By the use of this polishing substance he had produced a plane surface of silver, which, as far as the photometric means he had within his reach would enable him to measure the light before and after reflexion, did not lose in that action seven parts out of the hundred, and which, tested in the manner which he usually adopted, defined admirably.

The Astronomer Royal begged to know how Lord Rosse secured the plane form of the surface in grinding and polishing?—The Earl of Rosse replied that, as to the mode of grinding, it was that commonly adopted for producing accurately flat surfaces. But the mode in which he tested it was peculiar. It was this:—A watch-dial was placed before a good telescope, and as soon as the eyepiece was accurately adjusted to the position of most distinct vision, the plane mirror was placed in front of it at an angle of 45°, and the watch-dial was moved round by a simple contrivance to such a position as that its image should very nearly occupy the place it had been just removed from. If now the adjustment of the telescope for distinct vision remained unchanged, the proper form had been attained: but, if by drawing out the eye-piece more distinct vision was obtained, it was concluded it had received a convex form,—if on pushing it farther in it gave the image more distinct, then it was concluded the mirror had received a concave form.—Prof. Chevalier wished to know whether Lord Rosse intended to form the great speculum similarly of silver.—The Earl of Rosse replied that he had at present no expectation of doing so. That it was a very different matter to grind and polish a speculum of a few square inches surface—which could be done by a small machine worked by hand, and from which to the eye-piece the light had to travel but about 3 feet—to executing the same operations over the surface of a speculum 6 feet in diameter, and from which the light after reflexion had to travel a distance of 53 feet. For Newton, with his usual sagacity, had long since shown that any error in the form of the object speculum of a reflector was a much more serious injury to the performance of the instrument than an equal error would be in the plane speculum,—and that for the identical reason he had just pointed out.—*Athenæum Report of the Ipswich Proceedings of the British Association.*

NEWTON'S LAW OF THE RESISTANCE TO THE MOTION OF BODIES THROUGH FLUIDS.*

Sir,—Although I never could perceive the "incomparable" value of Mr. Frost's "stame," I do think that he does not deserve all the scolding of "J. B. G." in your Number for 26th April last. He is there taken to task severely for saying that Sir Isaac Newton and himself agree in an opinion that, to produce double speed, double power only is required, and that he has proved this by an experiment which he relates. I beg to add my name as that of a humble, but earnest convert to the same opinion, and should be glad to see this most important question submitted to the test of intelligent experiment and dispassionate reasoning. Permit me, therefore, to lay before your readers my view of the matter, and in as simple and untechnical a way as I can.

In No. 1429, "A. B. C." asks for an explanation of the dictum of scientific men, that the velocity of a vessel's motion is in proportion to the cube root of the propelling power, while the resistance is in proportion to the square of the speed; or, as I understand him, he wants to know how it can be that, if when a vessel is at a speed of five miles per hour, it is made to increase its velocity to ten miles per hour, the resistance is quadrupled, and yet the power required to overcome that resistance is eight times that at first used? To this "J. B. G.," and also "S. G. S.," replied in No. 1432, endeavouring to maintain the said dictum, but by means of arguments which are to me by no means satisfactory. "J. B. G." adduces the experiment of a boat, drawn by a cord having a weight attached to it, and passed over a pulley; he says that *four times* the weight will be required to produce double speed; while Mr Frost, in trying the same very simple experiment, found that double the weight produced double the speed. I believe Mr. Frost to be correct in his report; for though I have not tried this identical experiment, from want of opportunity to do so, I have made some rather similar ones, and with coinciding results. It is to be observed, too, that Mr. Frost has actually performed the experiment "often," while "J. B. G." appears to rest on hypothesis; how far that is worthy of being opposed to facts may be perhaps illustrated by a statement made in a highly commended work on the steam engine, which was lately published,—to the effect that the atmospheric resistance to the advance of a train increases so greatly with a doubling of the speed, that to overcome it an engine is required that is capable of exerting eight times the power, "so far as regards the atmospheric resistance," "which constitutes the greatest impediment to motion at high speeds;" while Mr. Gooch's carefully-made dynamometer experiments had already shown that the power exerted by the engine at double speed was not far from "double." Let "J. B. G." actually try the experiment, and he will see at once whether his foundation is sound or not; and so will the opposite party. In both his letters "J. B. G." argues upon the point of double or quadruple, as the case may be, the weight descending twice as fast when the speed is doubled, and evidently considers the pith of the matter to lie in that. I entertain a different opinion; and will endeavour to prove it by means of an illustration which is, I think, the best that we can adopt in treating of this subject. Imagine that the boat, in his letter in No. 1432, is drawn forward by a cord attached to a piston rod, working in a cylinder fixed at the lockhead of a canal, and that it has been caused to move the distance (say 40 feet) of the length of the cylinder in four minutes, by a charge of 20 lbs. steam. Let a charge of 40 lbs. be in the next experiment used, and if it causes the boat to travel the same distance in two minutes, will it not follow that twice the quantity of power has caused it to travel the same distance twice as fast? "J. B. G." appears to me to mistake an effect for a cause, when he supposes that the speed with which the weight falls is a reason or cause for the more rapid motion of the boat; and the last lines of his first letter imply, that he thinks that eight times the power is exerted in moving the boat *the same distance* at twice the speed, which is in opposition to "S. G. S.;" both can't be

---

* After the frequent demonstrations of this law which have been given in our pages, it will not be supposed that this paper is published from any idea that further proof is required. We insert it simply as a clever exercise on the subject by one of a class worthy of all respect, and happily becoming every day more and more numerous—the class of self-improvers and independent thinkers,—persons who, with every inclination to put faith in the laws which science has developed for the guidance and direction of practice, are nevertheless eager to work out for themselves the sufficient *reasons* on which they are founded.—ED. M. M.

right, though it is possible both may be wrong.

If we admit, for the moment, that four times the pressure would be required to cause the boat to move twice as fast, what would follow then?—Not that the speed at which the piston travelled was any measure of either the power or the resistance. The speed of its travel is, I conceive, a measure of the speed with which the power overcomes the resistance,—but not of the quantities of either: I think that its speed is an incident (like the fall of the weight) depending upon how effectively the power acts upon the resistance, and not in any way causing that action to be slow or fast, little or great—in short, to be an occurrence taking place *after* the action of the weight upon the resistance: and if it is so, it cannot be a cause operating in that action; whereas, in both his letters, "J. B. G." builds upon its being a cause, *i. e.*, part of the reason why the boat's speed is increased.

I think that much of the confusion which I believe exists concerning this question, arises from the persons treating of it allowing the point of the amount of power which is at any one moment being exerted to be confused with the amount of power which is expended per hour. If the piston before instanced travelled twice as fast, at twice the pressure, four times the quantity of steam would be consumed per hour; but it would not follow from that that it was overcoming four times the resistance, as is explained by "S. G. S." It would be overcoming twice the resistance, twice as fast; but the power being exerted at any one moment would not be quadruple, although the power exerted per hour was; yet "J. B. G." would tell us that the power was to be reckoned by multiplying the pressure by the velocity, though how the velocity of the motion could increase or diminish the *power*, I cannot conceive: that it would increase the amount expended per hour, I can see; but he appears to have confused the two things, and to have supposed (in his second letter) that Mr. Frost was exerting quadruple power when twice the weight "descended twice as fast." Suppose it had been a piston instead, would the power have been quadrupled by its descending twice as fast as thrice the pressure? Decidedly the quantity expended per hour would, but not the power being exerted upon the boat at any one instant. And what we want to know is, if double pressure on the piston, or double weight, will produce double speed of the boat, can the resistance of the boat be four times what it was at single speed? "J. B. G.'s" principle would make out that the piston was exerting a power four times as great by only doubling the pressure; for he says "the power" is the weight (here represented by the pressure) multiplied by the velocity. Now the pressure being double, and the velocity being double, we have, according to him, used quadruple the power; but it would be easily seen that only twice as much steam, and therefore only twice the power, have been used in each movement of the length of the cylinder: therefore, has not "J. B. G.," in trying to calculate what the power was, multiplied it by something (the velocity) by which it ought not to be multiplied? unless your object was to see how much steam you used per hour, not to find what power you were exerting; for that you knew already when you knew the pressure. The power and the resistance must, I imagine, from the law of action and reaction, always be just equal; and yet "J. B. G." tells us that the weight represents the resistance, while he says the weight, multiplied by the velocity, is the power. It appears to me to be clear that, directly the power was greater than the resistance, an increase of speed would result sufficient to raise the resistance to an equality with the power. And is there not a contradiction of this great principle of mechanics contained in the statement in "J. B. G.'s" first letter, and apparently accepted by the other writers (except Mr. Frost), that the velocity is as the cube root of the power while the resistance is only as the square of the velocity? or as I understand it, and as the paragraph beginning with "whereas," in "J. B. G.'s" first letter lays it down, that to produce double speed, eight times the power must be exerted, while the resistance is only four times as great. I cannot imagine how the power required can ever be greater or less than the resistance; if the resistance being one is increased to four, and a power of eight is applied, what becomes of the second half of the power? it must have something to act against; yet "J. B. G." evidently does not reckon on any increase of speed; and if he did, how could his argument stand? Neither is he speaking of the quantity of power expended per hour, but of that employed in traversing a certain space at a certain speed.. In opening the question, "J. B. G." makes an assertion which is frequently made, but which is, I think, utterly incorrect; he says, "if we double the speed, double the quantity of water is displaced at double the velocity." He is here speaking of the resistance, and argues from the position which has been just laid down, that "thus we see the resistance is as the squares of the velocity." But to this

I reply, that I conceive that at double the speed a vessel is not displacing double the quantity of water (actually, rather less), but is simply displacing at any moment the same quantity, twice as fast. "J. B. G." seems to me to have here confused the quantity displaced per unit of time, with the amount of the resistance which there actually is at any one time. "S. G. S." has well explained how the power may be given out twice as fast without the resistance being increased, by instancing a horse towing on a canal. If it follows from "J. B. G.'s" assumed position, that the resistance is as the squares of the velocity, it will, if his assumption is correct, follow that the resistance is not as the square of the velocity. The doctrine of "J. B. G.," that in the example which he gives the power is eight times as great for the same space traversed at double speed, is contradicted by "S. G. S.;" who says, that it is only four times for "equal distances passed over." Such a contradiction in the pages of an established Magazine must tend to confuse the student. I have tried in vain to reconcile the statements of these gentlemen, and have long thought differently to them. If "J B. G." is right, sixteen times the power, or fuel, must be expended per hour when double speed was used; and even if "S. G. S." is right, it would follow, that if a steamer used a ton of coal per hour, when going two miles per hour, she would use eight tons per hour at four miles per hour, and sixty-four tons per hour at eight miles. Now, is this really the case? Yet I cannot see how "S. G. S." can escape from this conclusion. If in any calculation on this subject, we substitute the piston and cylinders for the string and weight, I apprehend we shall not thereby introduce any element of error into it, and by reckoning the quantity of steam consumed per mile shall be able to see exactly how much power per mile is exerted; and by calculating the number of strokes per hour, also obtain a knowledge of the power consumed per hour, without mixing up things, which (in my opinion) are not naturally connected; and I also think we shall see from experiments so made, that the velocity of the weight or piston is really no cause at all in the case, but simply an effect which will always be found to depend upon the proportion of the power to the resistance.

"J. B. G." says, that before Mr. Frost writes of this dictum as an "enormous error," he ought to study the elements of mechanics; I have done so, and beg to subjoin, the following extract from a work on the "Elements of Mechanics" which was published some years ago, and had an extensive sale:

"A vessel, if it moves twice as fast of course displaces twice as many particles in the same time, and requires to be moved by twice the force on that account; but it also displaces every particle with a double velocity, and requires another doubling of the power on this account. The power thus twice doubled becomes a power of four."

This clearly states, that *for the same time*, at double velocity, only four times the power is expended.

If the rule as laid down by "J. B. G." were correct, it would operate backwards as well as forwards; and if a vessel usually travelled at eight miles an hour, only an eighth (or sixteenth) of the fuel would be required to propel it four miles an hour, and only a 64th at a speed of two miles an hour,—and it would be indeed the cheapest plan to carry goods that went by steamers slowly.

If I do not much mistake, I have shown a flat contradiction between the doctrines of "J. B. G." and "S. G. S.," although I imagine they think they agree; but if I am in the wrong respecting this interesting point in mechanical science, I hope some more competent person than myself will clearly set me right.

I am, Sir, yours truly,

AN EARNEST INQUIRER.

May, 1851.

SPECIFICATIONS OF ENGLISH PATENTS ENROLLED DURING THE WEEK ENDING JULY 13, 1851.

JOHN HARCOURT BROWN, of Fir Cottage, Putney, Gent. *For certain improvements in the manufacture of wafers.* Patent dated January 7, 1851.

Mr. Brown's improved wafers are composed of very thin sheets of metal or metallic alloys of any description, and either plain or ornamented by gilding, varnishing, or lackering, coated with an adhesive compound. The compound which the patentee employs, consists of sixteen parts strong glue, four parts clean gum arabic, five parts treacle or other saccharine substance, three parts spirits of wine, one part camphor, one part virgin wax, and twelve parts distilled water. These ingredients are heated in a closed vessel by means of a sand bath, and maintained for about eight hours at a temperature of 210° Fahr.; then filtered and diluted to fluidity with water (in which alum has been dissolved, in the proportion of one part to every fifteen parts water), being kept during the time, just below the boiling point. The sheets of metal to be coated are passed between a pair of metal rollers, one having a plain polished, and the other a slightly-roughened surface, and after the composition has been applied to the roughened side of the sheets, and allowed to dry, they are cut or punched into wafers of any form or pattern.

Claims.—1. The composition for rendering metals or metallic alloys adhesive for the purpose of securing letters and envelopes, whereby they are rendered impervious to damp and unaffected by wet, which is not the case with sealing-wax, wafers, and adhesive compounds at present in use.

2. The employment of machinery for roughing the surface of the sheets of metal and any other compound, in addition to that above described, for rendering them adhesive.

WEEKLY LIST OF NEW ENGLISH PATENTS.

John Hick, of Bolton-le-Moors, Lancaster, engineer, for certain improvements in steam boilers or generators. July 17; six months.

William Dickinson, of Blackburn, Lancaster, machine-maker, and Robert Willan, of the same place, mechanic, for certain improvements in machinery or apparatus for manufacturing textile fabrics. July 17; six months.

Thomas Wilks Lord, of Leeds, York, flax and tow machine maker, and George Wilson, director of the flax-works of John Fergus, Esq., M.P., of Prenlaws, Fife, North Britain, for a machine to open and clean tow, and tow waste from flax and hemp and other similar fibrous substances, and an improved mode of piecing straps and belts for driving machinery, and a machine for effecting the same. (A communication.) July 17; six months.

John M'Nab, of Midtownfield, Renfrew, North Britain, for certain improvements in stretching and drying textile fabrics or materials, and in the machinery or apparatus employed therein. July 17; six months.

Arthur Albright, of Birmingham, manufacturing chemist, for improvements in the manufacture of phosphorus, and in the apparatus to be used therein. (Being a communication) July 17; six months.

Thomas Sanders Bale, of Cauldon-place, Stafford, china manufacturer, for certain improvements in the method of treating, ornamenting, and preserving buildings and edifices, which said improvements are also applicable to other similar purposes. July 17; six months.

WEEKLY LIST OF DESIGNS FOR ARTICLES OF UTILITY REGISTERED.

Date of Registration.	No. in the Register.	Proprietors' Names.	Addresses.	Subjects of Design.
July 10	2883	W. Pearson	Maryport	Cooking apparatus or caboose.
,,	2884	J. Walker	City-road	Double-acting screw press.
12	2885	P. Nicholas	Thann, France	Machine for engraving rollers.
15	2886	J. Hall and Son	Lombard-street	Safety stopper and measure for powder canisters and flasks.
16	2887	A. Saxon	Middleton, Lancaster	Throstle bobbin [246 provisional.]
,,	2888	E. Leach and Sons	Rochdale	Adjustable traverse slot.
,,	2889	W. Mabon	Manchester	Double angle iron for the cup and dip of gas-holders.
,,	2890	J. Strutt	Belper, Derbyshire	A cheese-turning machine.

CONTENTS OF THIS NUMBER.

LONDON: Edited, Printed, and Published by Joseph Clinton Robertson, of No. 166, Fleet-street, in the City of London—Sold by A. and W. Galignani, Rue Vivienne, Paris; Machin and Co., *Dublin; W. C. Campbell* and Co., Hamburg.

Mechanics' Magazine,

MUSEUM, REGISTER, JOURNAL, AND GAZETTE.

No. 1459.] SATURDAY, JULY 26, 1851. [Price 3*d*., Stamped, 4*d*.

Edited by J. C. Robertson, 166, Fleet-street.

ERICSSON'S RECIPROCATING FLUID METER.

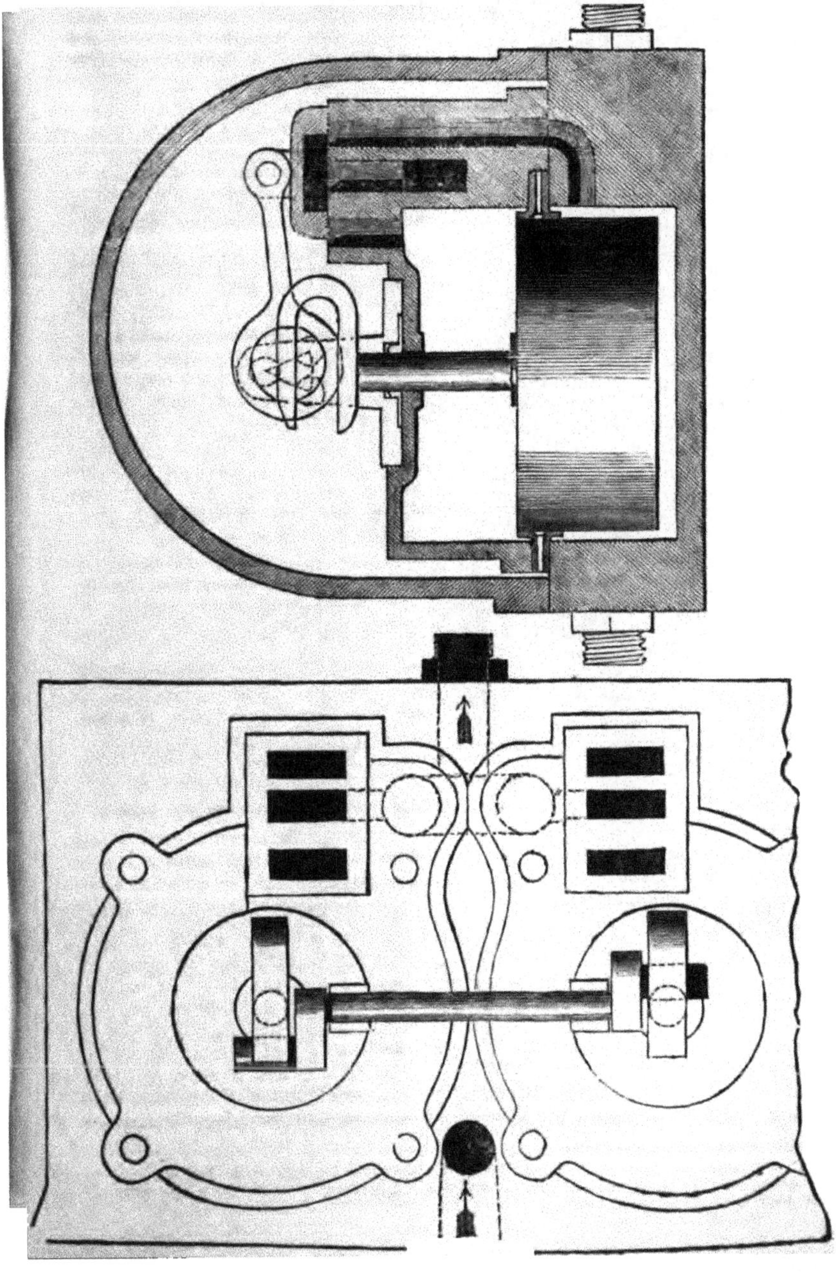

ERICSSON'S RECIPROCATING FLUID METER.*

The principal object of this meter is that of measuring the quantity of water which passes through pipes during definite periods. The accuracy of the instrument having been well established by actual trial, the directors of the celebrated Croton Aqueduct, New York, have recently adopted it for measuring the quantity of water consumed by the principal manufacturing establishments, the intention being to apply this meter wherever the Croton water is consumed, with a view of changing the present mode of levying water-rate, as well as preventing the existing extraordinary waste of water. By reference to the engravings (in vertical section and top view), it will be seen that this meter works in the medium to be measured. Numerous practical difficulties are avoided by this expedient; such as tight joints, lubrication, strain on the working parts and packings, &c. The admeasurement is effected by two double-acting plungers, connected to cranks placed at right angles. The most important feature of the movement consists in checking the motion of the plungers before the cranks reach full up-and-down stroke, which is effected by stops operating directly on the plungers.

The oval slots in the piston-rods, as will be seen by the engravings, are made so much wider than the diameter of the crank-pins, that these latter may move through an arc of 20 degrees whilst the pistons remain stationary. The advantage of reversing the position of the slide valves, whilst the pistons remain stationary, will be readily perceived by any one practically acquainted with the action of hydraulic engines. The advantage of working the pistons between definite stops is self-evident, the wear of crank-pins, journals, &c., having no effect whatever on the accuracy of admeasurement. It is particularly worthy of remark, that since the meter works within the medium to be measured—its operation being at the same time simply that of a scapement—an equilibrium of pressure (or nearly so) is established, which prevents all injurious friction and wear of the moving parts. The simplicity and strength of the entire mechanism will be best seen by inspecting the instrument exhibited. Three different sizes of meters are now being made for the Croton Works; viz., 5, 9, and 21-inch plungers, the quantity measured being respectively one-fifteenth, one-third, and 4 cubic feet for each turn of the crank—stroke about *one-fourth* of the diameter of plunger. The speed being about twelve turns per minute, each meter measures off the following qualities; viz., the 5-inch, 1,200; the 9-inch, 6,000; and the 21-inch, 70,000 cubic feet in twenty-four hours. An ordinary register is applied to these meters, locked up and set once a year, indicating at all times the quantity of water passed through. It may be asserted that no previous mode of measuring fluids in large quantities can compare in accuracy to this meter. The waste and inaccuracy attending the employment of the gallon measure is manifest; measuring by such means is but a rough approximation, when compared to the process here described, of determining the quantities by the spaces occupied by plungers of definite size working between definite stops; insuring, as it does, almost absolute precision.

INDUSTRIAL SCHOOLS.

Observers can hardly fail to have noticed the mischievous consequences that attend the leaving children of the industrious classes to themselves during the intervals between school-hours. It was a perception of such evils that led, in No. 1431, of the *Mech. Mag.*, to the suggestion of providing in industrial schools for the continuance of pupils in them during the whole of the time their parents might be employed in their respective callings. Further investigation has but strengthened the conviction that such a measure would at once greatly contribute to the comfort and convenience of the parent, and to the present well-being and permanent morality of the child. But it must be understood that continuance of a child at the school during the whole day should not be im-

* *An English patent for this invention has been taken out in the name of* Mr. Edward Dunn, on behalf of *Captain Ericsson, but it has not been yet specified.* A model of the apparatus is to be seen in the *American quarter of the Great Exhibition.*

ve; on the contrary, it should be ntirely to the option of the parent. examination into the actual state ngs, it would be found that fathers nilies of the industrial classes are, the exception of mealtimes, absent home during the whole day, and nother also, if employed elsewhere dustrial occupations — upholstery, aundry, charing. The child has in cases literally no home wherein to his lessons, to enjoy relaxation, or usefully employed during six of welve hours of the day; his sole rce is consequently the street; as-ion there with children of all de-ions—some good, more bad, leads oo frequently to evil courses. To the more favourable view—that the er is at home, keeps her child within, encourages him to learn his lessons, e has rarely opportunity to do so; infant's wailings disturb him—young child entices him to play; r other distractions occur, all equally e prejudice of profitable study. There frequent instances, too, where the er, rather than be interrupted in business of the hour, turns her chil-out of doors to play in the streets, marvels that they become unruly. e case of absence of parents, the children are often greatly to be l when not allowed to remain in d. The girls of a national school some of them been seen at dinner-sitting on the door-step, too cold to he provision in their little basket, rying from actual suffering,—others em crouching under a wall, drenched in to the skin, though partly shel-by trees. Happily it is not thus in chools; the mistress of an infant-l in the same parish receives the ones as early as their mothers go e day's work, and retains the chil-till the parents' return from labour. e infants are happy, orderly, healthy, rkably advanced in reading, writing, metic, and needlework; for they their mistress, and learn what is red of them for her sake. These girls fall back when taken into the nal school of the same place.*

The great objection likely to be made to retaining pupils the whole day in schools is, that of undue absorption of the master's time; indeed this would be an insurmountable obstacle were his attendance requisite for more than the usual hours of school. French schools (*colleges*, so called) afford examples of appropriate superintendence of the children when not in school. The masters (*professeurs*) attend only during school hours; at other times a different set of persons, *maitres d'études*, come on duty: they see that the pupils are undisturbed whilst studying their lessons, and that no quarreling or mischief takes place during the times of recreation. This is widely different from the practice in English boarding-schools; for in these the ushers are expected to do the duty of the French *maitres d'études;* but English ushers are also tutors during school-time, consequently are too much wearied and worn out to pay due attention to the pupils when out of school. *Maitres d'études* are usually young men looking forward to become professors in public, or masters in private schools: leisure for study is afforded to these young men during school-time. So in English schools, steady young men, of an age sufficiently advanced to insure respect, and intending to become school-masters, might be employed as superintendents of studies; or otherwise discreet, though unlearned persons, might be intrusted with this duty, and, for a very moderate salary, would suffice to keep order during the intervals of school, and take care that pupils should not be disturbed whilst learning their lessons.

Were children to be whole-day pupils, it might be advisable to introduce a few expedients for affording the petty advantages obtainable at home. One important point would be the inuring a child to cold, to wind, to rain,—to all of which he must be exposed in future life, if embracing any other than sedentary occupations. A walk to school would in this climate be but too much exposure to all of those vicissitudes, yet farther weather-hardening should be provided for. In an industrial school, carpentry, for ex-

* the infant-school in question there were t this day (15th July) 138 scholars, about 50 *dined in the school. For the manage-and instruction of the pupils in this numerously*-attended and successful establishment, there is a schoolmistress, one assistant, and some little monitors and monitresses of about five and six years old.

ample, might be carried on in the open air, or during rain, in a cool shed. Variation of the employment of a child during the day, alternately in a greater or lesser degree of heat, would well prepare him for exposure in after life to the vicissitudes of this variable climate.

Over and above such school-house arrangements, walking and running exercise—so congenial to childish temperament, so conducive to health—should be provided. Children enjoy country rambles; and, in the way of reward, a whole school might be frequently indulged in them under the care of an assistant master—say an out-of-school superintendent; not conveyed in carriages of any description, but walking expeditions. Whilst in streets, children should be required to proceed in a manner the least annoying to other passengers,—no trundling of hoops against a woman's dress, no throwing of stones or spattering of mud or water, no spreading across the whole footway; but once out of the course of general traffic, children might be allowed to run and jump, trundle hoops, fly kites, play at leap-frog, skip-rope, or what not, to their heart's content. Those who have witnessed the summer excursions which sometimes town children are afforded, can hardly fail to have noticed their orderly behaviour on such occasions, and that even the tiniest urchins seem to have acquired unwonted strength, keeping up with the elder children.

Though no head-master need as a duty accompany pupils in any such excursions, yet should he sometimes voluntarily join them, occasionally in a country expedition, where many opportunities would doubtless be availed of for exemplifying facts in natural history or rural economy that had been enunciated in school; and opportunity would be afforded him of witnessing the impression his lessons had made as to the out-of-door conduct of his pupils.

But though children should be inured to vicissitudes of climate, it seems important that they should be early habituated to the use of precautions diminishing danger from its variations; such, for example, as the changing wet shoes for dry ones before commencing a sedentary *employment. The individual* must be *poor indeed who could* not afford his *child a second pair of* shoes or slippers,
though they were but old ones;
blains and colds that so tormen
young arise most frequently from
down in wet shoes or damp clot
indeed, many of the most serio
nesses of after-life generally origin
the same cause, or from exposure
hot, to draughts of cold air.

An industrial school, where ch
might remain the whole day, cou
made a powerful aid in overcomi
great bane to health, to competen
domestic happiness—*strong drink*
is a most unfortunate circumstan
all schools it may be said, from th
the pauper to those of the wea
classes—to provide beer, so calle
the beverage of children at dinner
To say nothing of the deleterious
of such sour mixtures as are
served, the furnishing any drink
the name of beer gives a child the
that it is a necessary of life; it i
that in early years water comes
considered as prejudicial, where
fact, as an accompaniment to food i
least altogether harmless. It is thi
step that leads to inebriety in the
man—the poor small-beer is first ch
to that of better quality, then to p
strong ale, spirits. It might be a
regulation that water only shou
drank at dinner—no beer or spir
addition to water at any time; mill
water, tea, coffee, are now happily
the place of strong drinks in the i
trial classes, as they already have
in the superior ones; it is proof
that the afternoon meal formerly
tippling-time, is now denominate
time. Doubtless one of the less
morality in all schools is that of
nence from intoxicating drinks;
practical proof that they are ne
would go farther than could any
towards impressing this importan
on the juvenile mind.*

* It has has been often asserted that i
shortens life, but it is only of late that the
been verified from certain data. The *Athen*
the 28th ult. gives extracts from a paper lai
the Statistical Society on the 16th of June,
though not treating of mechanical matters,
from its importance to mechanists as well
others, to merit particular notice in the *Me*
Magazine. The paper in question was pr
by Mr. F. G. N. Nelson; it stated that
noticed only well-marked cases of intemp
that his observations were on 6,111-5 years
in which 357 deaths had taken place,
according to the general rate of mortalit

Travellers have rarely opportunity of becoming acquainted with the real habits of the people in foreign countries, hence it is little known that women in many parts of the south of France drink water only; yet these females of the industrial classes are robust, and competent to severer labour than the generality of English women could endure. French washerwomen, for example, will stand in a stream of water washing for twelve or fourteen hours of the day; on the Saturday walk six or eight miles, beside the donkey, carrying a load of clean linen, and will walk as many miles more on returning home with the soiled linen of the week. Peasant women labour in the field often with the same tools that are used by men in digging fields or vineyards, and in harvest-time females are frequently as laboriously employed with the sickle as the men, yet even when reaping, drink only *water*. Men drink wine, but are never seen intoxicated; great as is the temptation where good wine can be bought retail at from a penny to twopence a quart. But without going so far for examples, they may be found at home amongst *tee-totalers* and others. It is much to be regretted that instances in proof of this are not brought prominently to view, example being so far more convincing than argument.

To question the expediency of periodical school vacations is a bold adventure —yet in what rank of life is not holiday-time looked forward to with apprehension? In the upper-classes mothers and servants dread the unruliness of children coming home with the notion that all must be subservient to their amusement. A constant round of sight-seeing and pleasure parties, it is true, usually absorbs the greater portion of their time; yet part of it hangs heavy on them, unless parents have had the prudence to engage temporary tutors. In humble stations of life no such sight-seeing can be afforded, no such provision of masters at home during holidays; the children feel persuaded that during the vacation they, like the wealthy, have nothing to do but to amuse themselves, and indulge in every whim—debarred of expensive pleasures, little diversion to be found in the confined home, they seek it without door—thus, unfortunately, the idle and the vicious entice the good child to join in mischievous pranks, and too often in corrupt schemes. The evils resulting from indiscriminate association with children in the streets seems rarely to be adverted to by the many who interest themselves in juvenile education.

Holiday habits of idleness and disorder, even where greater mischiefs do not result, are really confined to periods of vacation. Listlessness is the least of the discouragements the master has to combat on the return of his pupils to school. On a late visit to a numerous one, the mistress observed to the stranger that the school would be seen to great disadvantage, for it was so soon after the holidays that the children had not yet recovered from the mischievous effects of the vacation. Indeed, it is not to be expected that parents of the industrial classes should have either ability or leisure to keep up the desirable habits of a good school. It would seem, in fact, that vacations, to the injury of pupils, have been invented for the benefit of masters solely.

This leads to a consideration of the quantum of the time a master should, or could, devote to school business. It is urged by some that even the usual six hours of school attendance is beyond the amount of mental application that can be expected from the generality of men, not to speak of the additional time which, where there are pupil teachers, a master must devote to their instruction. In this view the abolition of holidays would be an intolerable surcharge on his labours. Without referring to the much longer time many clerks daily sit at their desks, or to the long labours of literary and professional men, or to the incessant mental labour of the clergy in well-cared-for populous parishes, it may be affirmed that abundant relaxation might be pro-

land and Wales "the number of deaths would have been 110 only, or less than one third. At the term of life, twenty to thirty, the mortality was upwards of *five* times that of the general community; and in the succeeding years it was four times greater." The duration of life in beer-drinkers averages 21·7 years, spirit-drinkers 16·7 years, those who drink beer and spirits indiscriminately 16·1 years. There are of male drunkards in England and Wales 53,583, of females 11,223; total 64,806—that is of confirmed drunkards.

From the above, it appears how greatly mistaken is the notion that strong drinks do not shorten life till an advanced period of it, whereas, on the contrary, it is between the ages of twenty and thirty that they are most deleterious; so that five persons die at that age from indulgence in such beverages for every one person of those who abstain from them.

vided for the schoolmaster, though there were no vacations in his school. The Sabbath should to him be made a day of rest, instead of its being, as now in national schools, the most laborious of all the days of the week. By some such means] as those suggested in No. 1435 of the *Mechanics' Magazine*, on the seventh day he might be left at liberty both to worship his Creator, and to have besides some hours for needful relaxation of the mind. Liberation of the master from attendance in whole-day schools, excepting during school-hours, has been provided for above; by another arrangement occasional *holidays* might be allowed the *master* without much prejudice to the pupils, for during short intervals a school might be governed and taught by some assistant. In schools on private account, the master would of course nominate as his *locum tenens* that one whom he considered most competent to the duty; but where the establishment might be under other direction,—as a national school, for instance—the master should have liberty to fix upon that person as his substitute on whom he could most depend, even though it were but some well-advanced pupil teacher, since he (the master) should be made responsible for due conduct of the school during his short absence. It seems remarkable that provision is not generally made for the carrying on of school business in every school during temporary absences of the master; seeing that he, like all of the human race, is liable to disease. In small village schools there might not be funds sufficient for the payment of any assistant whatever; but even in such circumstances, the deficiency might be supplied by some hired parishioner—some one doubtless might be found willing to inform himself of the ordinary routine of business, and assemble the children as usual during the illness, or authorized short holiday of the master; thus keeping the pupils out of mischief at least, though he were absent.

(*To be continued.*)

THE GOVERNMENT PATENT LAW AMENDMENT BILL.

The Evidence.

The Evidence taken before the Select *Committee of the House of Lords* has now been printed. We expressed some doubts (*ante* p. 51), of the possibility of making out a case for the Government Bill (No. 3), by any evidence which it was in the power of its busy concoctors to adduce; and the big blue-book before us proves we were right. The Bill has followed upon the Evidence, but *not been founded upon it.* On the contrary, it is in its general tenor is as direct contradiction to the opinions of the majority of the witnesses as it is possible to imagine.

The existence of a foregone conclusion against the policy of the patent system in the mind of the Chairman of the Committee is made abundantly, and, indeed, painfully manifest throughout the whole of the inquiry. The questions put by his lordship to the witnesses are much too often, of that insidious and entrapping character known by the name of *leading* questions, and to which no honourable examiner, having truth only in view, ever has recourse. Take a few examples.*

Wm. Cubitt, Esq., examined.—Do you not think that the system has acted in an injurious way in stimulating persons to attempt to invent who had no peculiar inventive power?

Mr. John Fairie examined.—Is it not a hardship, if in the Colonies, to be obliged to pay for an expensive machine when they can get a similar one very cheap?

[*i. e.* "a similar" *and a pirated one.*]

In the case of mechanical improvements *is not it* very difficult to specify what is the invention contained in each of them? *is not it* generally the mere application, and mere dextrous manipulation of wheels to regulate, direct, and control forces *previously known?*

"It has been stated to the Committee that patents operate as a great encouragement to inventors and to invention; *is not* it your opinion that in the event of there being no patent to protect an inventor, he

* We assume here that Earl Granville was the questioner throughout, but possibly some of the questions may have been put by other peers. The "Minutes" furnish no information on this head, giving the questions only, without the names of the lords by whom they were put. We venture to ascribe the whole to Earl Granville, because at every sitting "Earl Granville (was) in the chair," and because the questions are all very much of a piece.

might still count, in most cases of useful inventions, upon remuneration for the communication of his improvement to the trade?"

It will be seen from these samples, that if the evidence was not all in favour of the *foregone conclusion*, it was no fault of Lord Granville's.

The noble lord in announcing his hostility to the patent system, confessed that he could muster but 6 witnesses against it (*i. e.* six to 100!) and it now appears from the printed Minutes that his lordship considerably exaggerated the force of the evidence of at least three of the six.

Mr. Cubitt, whose evidence was specially dwelt upon by Lord Granville as being "quite conclusive" against the system, qualifies it by an admission which takes away all value from it:—"It is a subject I NEVER WENT INTO, *or applied my mind to much*, having rather a dislike to the system, NOT BECAUSE IT MAY NOT BE GOOD IN PRINCIPLE, but because I never saw it act well."

The Master of the Rolls, who was represented as irreconcilably opposed to "the principle" of the system, is shown by the Evidence to have assigned a reason for his opposition, which *does not touch the principle at all*:—"In the greater number of cases I believe that the person who takes out a patent, and who makes the patent useful, is some one who finds out the little thing at the end, and which just makes it applicable and useful."

Sir John Romilly's objection, it will be seen, is not, that to encourage invention by rewards is pernicious (the affirmative of this being the principle involved), but that the rewards do not always fall to the lot of the most deserving (an accident, at best, which has nothing to do with the principle). However, we quite dispute the correctness of Sir John's representation of the usual course of things. We may safely challenge him to sustain his assertion by instances. All the most remarkable cases of patent success run quite the other way. Most assuredly it was not by "little things" that either Watt, or Arkwright, or Cartwright, or Brunel (the elder), or Howard, or Neilson, or Wheatstone, or Hancock, gained those great prizes in the invention lottery with which their names are identified. Nay, the instances of the "little thing" sort are so few that we cannot, for our own parts, call to mind at the moment a single notable case of the kind.

Mr. Ricardo's evidence exhibits simply a case of profound ignorance combined with its usual accompaniment, inordinate conceit. It appears to have been from this gentleman that Lord Granville derived his extraordinary delusion that scientific men would go on inventing just as actively, without as with the stimulus of a patent system. But testimony more in the teeth of known facts we never met with.

Mr. Ricardo loquitur.—"All the great inventions *now* are made without patents. I have alluded to the scientific discoveries which have been made without any stimulus of the kind (viz., arithmetic, writing, paper, glass, &c.!!!) I may say, again, that the great mechanical discoveries are made *without patents.* There is no patent for Stephenson's tube across the Menai Straits, but there is a patent for Nicholl's paletôt: *this last is the sort of thing which is encouraged.*"

Now the *real truth* is just the very opposite of all this. Of the great inventions "*now* made without patents," the number is exceedingly small, and Mr. Ricardo's instances to the contrary are mere fictions. *There is a patent for the tube*, though not in Stephenson's name, but in the name of Fairbairn, who was the real inventor, as is undeniably established by the fact that Stephenson was a party to the taking out of that patent *in Fairbairn's name*, and paid half the expense on condition of receiving half the profits—(see Appendix to Clark's History;)—and *there is not any patent* (though a registration) *for the "paletôt."*

Mr. Ricardo again.—"You stimulate the making of bedsteads and beer, and belts and bands, and blocks and bedding, and all that kind of thing, but you will seldom find a patent taken out for any wonderful and extraordinary public improvement; it is simply *these trivial things which patents are obtained for.*"

The answer again is—*not true*, and *impossible to be true*. For, of necessity, the same general system of protection which leads to the improvement of one class of inventions must tend, more or less, to the improvement of all others.

Mr. Sugar-Refiner Fairie we suspected (*ante*, p. 33) to be an opponent of the system on the same principle that a certain class of persons dislike laws against thievery, and the fact turns out to be even so. Mr. Fairie was the defendant in the well-known case of *Derosne v. Fairie*. He found Derosne's prepared charcoal well worth adopting in his manufacture, but did not like paying for it. He endeavoured to upset the patent, and had, no doubt, a heavy bill of costs to pay, besides being compelled at last to do justice to the meritorious inventor. *Hinc illæ lachrymæ.*

So much, then, for Lord Granville's phalanx of authorities. One has never gone much into the subject; a second objects to the "principle" for a reason which does not touch it at all; a third mistates almost every fact he brings forward; a fourth is a convicted infringer of patent rights; and the fifth and sixth are, as we pointed out in our last—one a philosophical colonel, who has confessedly had "no practical knowledge of the question," and the other a civil engineer, eminent only for differing generally from every one else, for the mere difference sake.

Is it fit that a system which has subsisted and flourished for nearly two centuries and a half, should be brought into jeopardy by such testimony as this?

Among the witnesses who spoke to the beneficial operation of the system—as they mostly all did with the exception of the precious six—we notice with pleasure Sir David Brewster, from whose evidence we cannot resist the temptation of quoting two or three striking passages:

"It appears to me that it is the duty of a wise Government to use every means to induce inventors to bring forward their ideas, especially in a country like this, where *so much depends upon the progress of the* useful arts, and at a time when foreigners are making such exertions—and often successful ones—to rival and to outstrip our manufacturers, both in the quality and cheapness of their productions.

"Every idea is of value, and every encouragement should be given to a man to come forward and take out a patent, for this reason:—When a man takes out a patent, he makes experiments; he enters into new researches; he very often makes in that way apparently a frivolous idea into a great and valuable one. I have never been able to see what evil could arise from the multiplication of patents. I have never heard any one state a reason which could not be answered why they should not be multiplied indefinitely."

Lord Granville.—"If you multiply patents very much, as soon as an idea is patented, does it not set other people to work, not to improve that idea in the best way, but to try and arrive at the same end by inferior means?"

"Never by inferior means, but by superior means, because, in order to overrule a patent, you must do something better."

Lord Granville.—"You do not think a great many workmen now are, by the hope of obtaining a patent, induced to sacrifice a great deal of their time to work privately at inventions which turn out to be of no use?"

"I am satisfied that no man of any ingenuity would work at all, unless he had some chance of obtaining reputation or wealth. The work of an ingenious workman never can be useless, even though it may be unprofitable."

* * "I cannot conceive how any person can be injured by there being a number of patents, and still less how any person can be so selfish as to complain of them, and so ignorant as not to see the national importance of encouraging the development of new ideas. If a patent appears to be frivolous, which I hold no patent can be (because a patentee makes new experiments in order to bring his invention into practical and beneficial use), I cannot see how ever a frivolous patent can affect injuriously the interests of any individual; such a patent

falls to the ground immediately. Hundreds of apparently useless patents now fall to the ground, because no person values them or desires to make use of them. But they contain ideas which suggest others more useful and practical; and what is a simple and amusing experiment in one age becomes a great invention in another."

Mixed with Sir David Brewster's very just views of the policy of giving State encouragement to inventions, there is, however, a very considerable fallacy with respect to the expense. He holds that, because the public must be gainers by all new ideas, however crude, the public ought, therefore, to be at all the cost. He would charge the patentee nothing whatever. Now, certainly, that would be going much farther than the premises warrant. To grant patents for nothing would be tantamount to granting a bounty on patents; and bounties have been long since proscribed as against all sound policy. The growth and culture of ideas ought no more to be exempted from the difficulties and risks, natural and common to all human enterprise, than the growth and culture of anything else. It is good that there should be such difficulties and risks in the path of industry; for the individual, in order that his mind may be quickened and invigorated by the struggle to overcome them, and for the public, in order that improvements may grow out of, rather than overshoot, the progress of society. It may be that there is no patent so bad but that it contains the germ of something good; but the same may be said of every book that ever was printed; and if you are to grant inventors patents at the expense of the State, for the sake of the good they contain or may produce, you ought, by a parity of reasoning, to print at the expense of the State whatever any one may choose to write.

The distance is vast, however, between granting patents for nothing, and charging so enormously for them as has hitherto been the practice in this country. Bounties are not more unwise on the one hand than disabling taxes and burdens on the other. None of the witnesses examined ventured to defend the costliness of the existing system;—all were for a large reduction, and some for reducing the expense much below what we should deem expedient or safe. The Bill, as we pointed out in our last, would in a majority of cases make the present enormous cost still more enormous—being in this, as in many other respects, directly at variance with the evidence on which it is pretended to be founded. The money, however, is to be paid by three instalments—the first of no great amount, and the two others of larger amount, but payable at the respective periods of three years and seven years after the date of the patent; and a patent may be thrown up after the first payment, so that the amount of that payment (about £20) would in future form the whole of the pecuniary difficulty in the way of an inventor desirous of a patent. No doubt this would, *in effect*, be much the same thing as reducing the cost of all patents to £20, since few probably would make the second and third payments, except out of profits realized, or pretty sure to be so. There is a convenience in this arrangement which makes it exceedingly tempting to inventors; and we fear much that it may be the means of carrying the Bill, in spite of its manifold defects. But accompanied, as it is, with so stringent and vexatious a system of preliminary inquisition as that proposed by the Bill, we abide by our opinion that, rather than take the two things together, it would be better far to leave matters as they are, and even bad as they are. In vain will the story of Esau have been written, if inventors are beguiled by so gross an artifice into the sacrifice of their birthright; (meaning by that, good reader, not the "innate right" of the Adelphi enthusiasts, but the statutory right of inventors, as established by the Act of James, and handed down unimpaired from father to son till the present time). They may think that Lord Granville and his brother lords are acting in a friendly spirit towards them in proposing this (in effect) cheapening of patents; but they will excuse us for reminding them that the gifts of avowed enemies are things to be aware of. Mr. Brunel, one of Lord Granville's select phalanx of anti-patent supporters, may be said, in homely phrase, to have *let the cat out of the bag* when he concluded his evidence in these terms:

"The result of your evidence is, that you are very decidedly of opinion that the *whole patent system should be abolished?*"

"Yes; I think it would be an immense benefit to the country, and a very great benefit to that unfortunate class of men whom we call inventors, who are at present ruined, and their families ruined, and who are, I believe, a great injury to society."

"And you think that those consequences—such as ruin to inventors and evils of that description—would subsist equally though the patent laws were made simple and more effective?"

"Yes; I think that they would be much increased; and if patents are continued, I hope *that the principle will be carried out thoroughly*, AND THEN IT WILL NOT STAND FOR TWO YEARS; I *think that* would put the principle to the test: if it is a right principle, it should be carried out fully.—MAKE PATENTS CHEAP, AND HAVE PLENTY OF THEM"!!!

Do, pray, inventors, mark the detestable glee with which this pet witness of Lord Granville's anticipates your utter ruin from the moneyboon which the Bill proposes to confer upon you. We do not think that the cheapening of patents to the extent proposed would have any such result (though certainly the tendency would be that way); but it should be enough that the Granville party entertain such an expectation to make you look well to any measure of pretended reform emanating from so hostile a quarter.

(*To be continued in our next.*)

ON AN IMPORTANT GEOMETRICAL THEOREM.—NO. II. BY T. T. WILKINSON, ESQ., F.R.A.S., BURNLEY.

In No. 1430 of the *Mech. Mag.*, pp. 12-13, may be seen a paper which attributes the *first* formal enunciation of the property, that "the three perpendiculars of any triangle intersect in the same *point*" to an Arabian commentator on the "*Lemmata Archimedis.*" This may be correct so far as the *formal enunciation* is concerned, but an inspection of "*Pappi Alexandrini Mathematicæ Collectiones. A Frederico Commandino. Venetiis.* 1589," has convinced me that the property was known to the ancient geometers, since it follows as an easy inference from "*Theorema* LVII., *Propositio* LX.," which may be thus literally translated:

"Let there be a triangle ABC, and through any point G draw AD, BE, CF, such that AD may be perpendicular to BC, and the points A, F, G, E in the circumference of the same circle: then the angles at F and E will be right angles." (*Pappi Math. Coll.*, folio 195, b; *Smellin's de Lectione Determinata*, translated by Lawson, p. 29; *Simsoni Opera*, p. 55, Prop. 40, *Lemma* 21.)

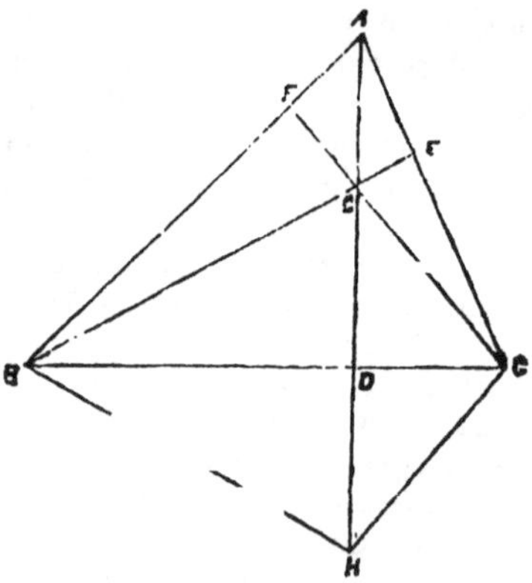

The demonstration given by Pappus is somewhat amended by Dr. Simson, and consists in producing GD to H making DH=DG. The points A, B, H, C are then shown to lie in the same circle, and hence that the <GCD=<GAF. Whence the <AFG=<GDC=a right angle; and by a similar process it follows that <AEG=a right angle.

Now since the lines AD, BE, CF are respectively perpendicular to BC, AC, AB; it follows conversely that they pass through the same point G, which, when formally enunciated, is the property of the triangle noticed at the commencement of this paper, and is besides so obvious an inference that it could not possibly have escaped the Greek geometers. The same property immediately follows from, and in fact is involved in, Pappus's demonstration of "*Theorema* LIX., *Prop.* LXII.," and is actually deduced from it by Commandine as a lemma in note "M, folio 198 b," where a demonstration may be seen much more complex than that given by the Arabian commentator on Archimedes. Dr. Simson demonstrates the property in p. 171 of his "*Opera Reliqua*" by a method very similar to that in the "*Lemmata*," and instances another investigation by Herigonius in his "*Cursus Mathematicus*," tome I., p. 318, Paris, 1644. This now well-known property of the triangle has therefore not escaped the attention of geometers, either in ancient or modern times, although it does not appear to have found its way into any *elementary* treatise previously to the publication of "Emerson's Geometry," in 1763. The preceding notices, however, will afford a satisfactory reason why the property had become so familiar as to be taken for granted, without proof, by English mathematicians in 1743.

Burnley, Lancashire,
July 14, 1851.

LEACH'S REGISTERED ADJUSTABLE TRAVERSE SLOT.

(Edmund Leach and Sons, Castle Works, Rochdale, Proprietors.)

Fig. 2. Fig. 3. Fig. 4.

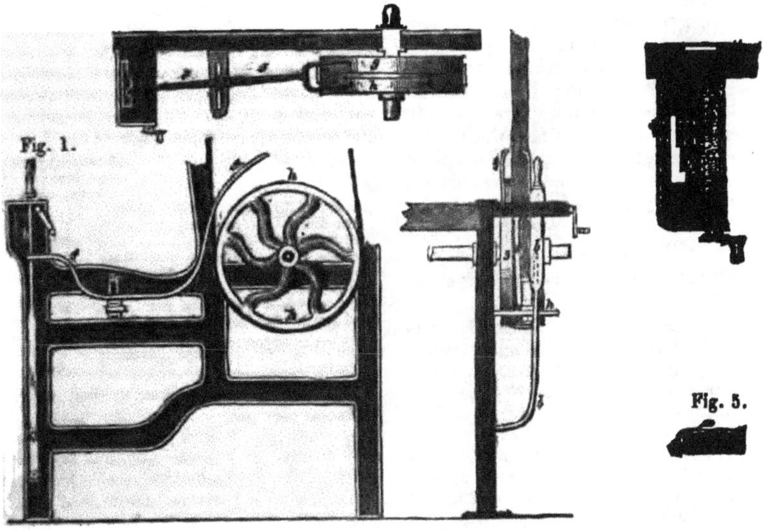

Fig. 1.

Fig. 5.

The object of this invention is the better adjustment of the length of the traverse slots of strap guides, so as to permit the strap being thrown more or less on to the fast pulley, at pleasure, whereby more or less friction will be produced without deranging the position of the strap when thrown on to the loose pulley.

Fig. 1, is an end elevation; fig. 2, a plan, and fig. 3, a *side elevation* of the adjustable traverse slot apparatu plied to a loom.

Figs. 4 and 5 are, an enlarged plan, and section of the box containing the traverse slot. *a* is the slot, in which the spring lever *b*, works. *c* and *d* are moveable pieces, by means of which the slot *a* is lengthened or shortened. *e*, an adjusting screw for setting the piece *c*; *f*, the strap guide; *g*, the fast pulley, and *h*, the loose pulley.

THE ROTATION OF THE EARTH.

Sir,—From Mr. Rozzell's remarks in your last Number, I fear that he has not followed the directions given in my former letter. If I understand him correctly he has ruled the third line parallel with the first, instead of parallel with the second line. He says, that "the line drawn after the globe has been moved 15° is no doubt the correct position at the end of the first hour," it must then naturally follow, that in drawing No. 3 line parallel to No. 2, No. 4 to No. 3, and so on, that each is only a repetition of the first process, and if the first be right all the others must be also.

He will find that if he begin with the plane east or west, or intermediately, that it will not make the slightest difference, and the diagrams already published will answer for the experiment however commenced, by merely turning them to agree with the original position. He must be careful to keep the ruler always in a horizontal position over the paper, and perhaps he will succeed better if he rule the lines every 7° 30′ instead of every 15°, and continue it for half a revolution of the globe, ruling each new line parallel to that last drawn.

I am, Sir, your obedient Servant,

74, Cornhill. R. Webster, Jun.

STIRLING'S PATENT IMPROVEMENTS IN IRON.

The Report of the Government Commissioners on the application of iron to railway purposes, and the extracts from it which have appeared in this and other journals have made workers in iron of all classes, familiar with the great improvements effected by Mr. Morries Stirling in the manufacture of iron, and led to their extensive adoption. These improvements it may be remembered relate to both *cast* and *wrought* iron. In the former case a mixture of cast and wrought iron in certain proportions, which has the effect of giving a fibrous nature to the cast metal, and thereby greatly increasing its strength and tenacity; for all kinds of beams, girders, and other castings, where strength is required, its use is found very advantageous and economical. Beams cast of the toughened iron may be made of less dimensions, and, consequently, of less weight than if cast of common metal, to support the same load; and they have the advantage of deflecting to a greater extent than common iron, and, therefore, are not so liable to sudden failure.

At page 101 of the Commissioners' Report an abstract is given of a series of trials, from which it will be seen that Mr. Stirling's iron is nearly fifty per cent. superior to sixteen other sorts of iron experimented upon. Various other experiments have been made by Mr. Owen for the Admiralty, and by Messrs. Rennie, and others, all with the same results, showing the great increase of strength obtained from the patent iron. The economy of its use is apparent; for instance, —common Scotch pig toughened can be had now for about 2*l*. 10*s*. per ton; and this is at least fifty per cent. stronger than the best Blaenavon, which costs three guineas per ton.

The improvements in the manufacture of *wrought* iron are; *first*, the admixture of a certain alloy in the puddling furnace, by which all malleable iron is rendered much more fibrous and stronger than common wrought iron, so much so that common or merchant bar becomes equal to best bar, thus saving one process to the manufacturer. Also very ordinary iron, which can scarcely be used at all, is made equal to the best. The following abstracts of experiments are given in the Report of the Commissioners appointed to inquire into the application of iron to railway purposes, page 417:

	Breaking Strain in Tons per square inch.
Average of Mr. Jesse Hartley's Experiments at Liverpool on many sorts of malleable iron	23·23
Average of S. C. Crown Iron from numerous trials at Woolwich Dockyard	24·47
Average of best Dundyvan Bar	24·33
Average of Mr. Stirling's best quality	27·81
Average of Mr. Stirling's, another quality	27·70

The cost of the process is only a few shillings per ton. When Mr. Stirling's toughened pig is used in the puddling furnace instead of common pig, and the alloy added, an iron is produced of a very superior quality, of a very fibrous nature, and much finer in the fibre than the iron mentioned above; this will be found very advantageous in the manufacture of thin plates and sheets.

Second. The admixture of a different alloy in the puddling furnace, whereby a quality of iron is produced quite opposite in its character to the last; instead of being fibrous, it becomes hard and crystalline—approaching to the nature of steel. The average stretch of common round bars 1 inch diameter, is about 3 inches per foot; whereas the average of Mr. Stirling's hardened iron is from one-eighth to three-eighths of an inch per foot. This shows the great stiffness obtained by his method. The crystalline nature of this description of iron causes it to resist compression, lamination, and abrasion. Thus, for the top portions of wrought-iron girders, it is precisely what is required to resist the compression force,—the fibrous iron being used for the bottom portion to resist the tension. For rails and tyres for wheels this sort of iron is peculiarly adapted; the top of the rails and the outside of the tyres, being made with it, will resist the wear and tear and lamination so universally complained of; and rails made of the patent iron are found to answer remarkably well. They have been used on the East Lancashire, Caledonian, Edinburgh and Glasgow, and other Railways with great success. The extra cost of rails made of this iron being only from 7s. 6d. to 10s. per ton.

SPECIFICATIONS OF ENGLISH PATENTS ENROLLED DURING THE WEEK ENDING JULY 22, 1851.

HENRY GRISSELL, of the Regent's-canal Iron Works, and THEOPHILUS REDWOOD, of Montague-street, professor of chemistry. *For improvements in coating metals with other metals.* Patent dated January 11, 1851.

1. For coating iron with zinc, the patentees melt the zinc in an open bath or vessel of suitable size, and place upon its surface a stratum of chloride of zinc (obtained by dissolving zinc in muriatic acid, and evaporating the product to dryness), or of a mixture of chloride of zinc and chloride of potassium in the proportion of eight parts of the former to ten of the latter, or of a mixture of equal parts of chloride of zinc with chloride of sodium, or anhydrous sulphate of zinc with chloride of potassium or sodium. When the salt is in a state of fusion the metal to be coated is placed in the bath, and allowed to remain there till a coating of sufficient thickness has been obtained; it is then withdrawn, and any parts of its surface imperfectly covered are sprinkled with sal-ammoniac, and the sheet of iron again immersed in the bath.

2. For coating zinc, or iron coated with zinc or other metal, with metallic alloys, the alloy employed is melted in an open vessel, and a stratum of chloride of zinc mixed with equal parts of sal-ammoniac deposited on its surface. The iron or other metal is then plunged into the molten bath, and allowed to remain there until a coating has been deposited in it.

One of the alloys which the patentees employ consists of 26 parts of tin, 10 parts of zinc, and 5 parts of lead, but they do not confine themselves to these proportions. They also use *the alloy known as "fusible metal,"* the proportions of which may be varied; provided always a melting point less than 400° Fahr. be ensured. A "fusible metal," suitable for the purpose is composed of 8 parts bismuth, 5 parts lead, and 3 parts tin.

3. To coat iron or other metals with tin or tin alloyed with lead, the tin or alloy is first melted as before directed, with a stratum of chloride of zinc and sal-ammoniac on its surface, and the iron or metal to be coated is immersed in the molten metal until sufficiently covered. The patentees here observe, that it will be found advantageous to remove the sheet of metal undergoing coating from the bath at intervals, and again immerse it in order that its surface may be brought well in contact with the fused salt. They also recommend that the sheet of metal should in the first instance be dipped into a hot solution of chloride of zinc acidulated slightly with muriatic acid.

4. For coating iron or other metals with silver, or any alloy of silver and copper, the metal to be coated must be first amalgamated with mercury. In the case of iron, the following is the process adopted:—12 parts of mercury, 1 of zinc, 2 of sulphate of iron, 2 of muriatic acid, and 12 of water, are mixed together and heated in an open vessel to about 200° Fahr., the iron is then immersed, and the mercury rubbed into its surfaces until amalgamation is effected. The silver or alloy is to be melted in a crucible, and the amalgamated iron placed therein, when a coating of silver or alloy will be deposited.

5. To coat iron with copper or brass, or with an alloy of copper, with zinc, lead, or tin, the copper or other coating is to be melted in a suitable vessel, and a stratum of borosilicate of lead (obtained by calcining together 24 parts boracic acid, 112 parts oxide of lead, and 16 parts silica) placed on its surface; the iron is then to be plunged into the molten metal, and retained there until a coating is deposited on it. Iron coated with tin or lead, or amalgamated as before described, may be also treated in a similar manner. Another method of coating iron with copper is, to place in a crucible a quantity of chloride of copper, upon which is laid the iron to be coated, and over that a quantity of charcoal. The crucible is then submitted to a red heat, and the chloride of copper fused, and a coating of copper deposited on the iron. Or, the vapour of chloride of copper may be employed for the same purpose. The coating of copper thus obtained may be converted to one of brass by exposing the sheet of metal to the vapour of zinc in a closed vessel.

Claims.—1. The use of chloride of zinc,

applied as described—in a previously fused state; also the use of chloride of zinc mixed with chloride of potassium or chloride of sodium, and of sulphate of zinc in mixture with chloride of potassium or chloride of sodium, in manner described.

2. The use of the alloys specified or referred to, and of the process described for coating metals with such alloys.

3. The use of a mixture of chloride of zinc and sal-ammoniac, which is kept in a state of fusion over the surface of the melted tin, or alloy of tin and lead, in effecting the deposition of these metals on the surface of other metals.

4. The process of coating iron or other metals with silver, or silver alloyed with copper, by amalgamating the surface of the metal to be coated, and then putting it into the melted silver or alloy.

5. The use of borosilicate of lead kept in a fluid state on the surface of the melted copper or brass, or alloy of copper with zinc, tin, or lead, used for coating iron by immersion.

6. The process of coating iron with copper by exposing the iron at a red heat to the action of fused chloride of copper or its vapour, and of subsequently converting the coating of copper to a coating of brass, by exposing the former to the action of vapours of zinc.

Charles Barlow, of Chancery-lane. *For improvements in propelling.* (Communicated by Abner Chapman, of Fairfax, Franklin, Vermont, U.S.) Patent dated January 11, 1851.

Claims.—1. A paddle wheel formed with winding curved paddles, or blades, set at an angle between the rims of the wheel.

2. A paddle-wheel constructed with double sets of winding-curved blades, set at opposite angles, in each of the sections of which the wheel is composed, and with a space left between the sections.

Samuel Hall, late of Basford, near Nottingham, civil engineer. *For improvements in the manufacture of starch and gums.* Patent dated January 11, 1851,

Claims.—1. The use of chlorine alone, or in combination with water, but not combined with lime, for bleaching starch and rendering it white. Also, the use of sulphurous acid in its fluid or gaseous state, for the same purpose.

2. The use of chlorine for bleaching gums. Also the use of sulphurous acid in any suitable state for acting on gums with the exception of "all the varieties of gum arabic, including that brought from Senegal."

3. The combining or commixing of starch and gums in any proportions that may be necessary for the purposes to which they are applied, but which proportions, owing to the various qualities of starch and gums, and the variety of uses for which they are required, experience alone can determine. Also the use of alkaline, saccharine or saponaceous matters, or lime, or hydrate of lime, or lime-water, previous to, or during the evaporation or desiccation of gum or of starch and gums conjointly. Also, the evaporation or desiccation of starch and gums (inclusive of gum arabic), or the mixtures of both, by means of a vacuum, or under a pressure less than that of the atmosphere. Also, the process of the inspissation of gums by the evaporation of the water in which they are dissolved, in vacuo, independent of or combined with the process of bleaching, by means of chlorine or sulphurous acid.

Thomas Allan, of Glasgow, iron-founder. *For certain improvements in paving or covering roads, streets, and other surfaces of a similar nature.* Patent dated January 11, 1851.

These improvements consist in the employment of paving constructed of iron plates and blocks, in lieu of that of wood or stone.

Claims.—1. The employment of plates of iron as a substitute for, and cast in imitation of ordinary stone paving.

2. The employment of separate hollow boxes or blocks in the formation of iron pavement.

3. A method of connecting the parts of which the paving is composed, by means of ribs or flanges.

4. A method of forming the side channels of iron paving.

John Alexander Archer, of the Broadway, Westminster. *For improvements in the manufacture of tobacco.* Patent dated January 11, 1851.

These improvements have relation to the cutting of tobacco, and consist in a combination of machinery for that purpose, in which the tobacco to be cut is fed in between two endless chains, by which it is carried forward and delivered through a mouthpiece to be acted on by revolving cutters. The tobacco in its passage between the chains is subjected to the action of free steam, and parts of the apparatus are heated by steam. The free action of the cutting blades is ensured, and clogging prevented by causing them to revolve in contact with a sponge kept constantly saturated with water.

Claim. The mode of combining machinery as herein described.

William Melville, of Roe Bank-works, Lochwinnock, calico printer. *For certain improvements in manufacturing and printing carpets and other fabrics.* Patent dated January 11, 1851.

Claims.—1. A peculiar method of throwing or actuating the needles used in raising the pile in weaving carpets and other fabrics.

2. A method of raising or forming the pile by actuating or throwing a needle from each side of the fabric.

3. A method of guiding the needles by an elastic guide or spring.

4. A method of working the needle action by treadles, or other actuating apparatus, working in connection or concert with the actuating apparatus of the shed or sheds of the fabric wherein one needle is inserted whilst the wool or yarn of the fabric is brought over or under the other.

5 & 6. The use of expanding or contracting needles for the purpose of facilitating the withdrawal of the needles from the pile.

7. The use of a double reed for working in connection with two shuttles.

8. The employment of a reed formed with a ledge or race on the dents or slips thereof for carrying the second shuttle.

9. A method of printing devices or patterns on yarns or woven fabrics by means of surface rollers revolving in or dipping into colour troughs or servers, or supplied by a feed roller revolving in such troughs.

10. A method of constructing printing rollers with the periphery formed or adjusted in sections for the purpose of facilitating the variation of the pattern.

11. The employment of wire cloth or other permeable fabric in printing yarns.

12. A method of holding up the yarn and supporting it, at a uniform tension by the employment of a separate fabric on one or both sides thereof.

13. A method of printing yarns by applying the colour through a permeable fabric.

14. A system of steaming printed yarns and woven fabrics directly as delivered from the printing machine, by passing the same through a steaming apparatus.

15. A method of printing woven fabrics without the use of an upper pressure roller.

16. A method of printing woven fabrics without the use of a blanket, by the employment of an upper printing surface pressure roller.

17. A method of forming or producing the pattern on the surface of the printing rollers. [A plain roller of the same size as the printing roller has its surface covered with gutta percha and flannel lined out in squares, on which the pattern is drawn, and these squares are then cut out and attached to the surface of the sections of the printing roller, their position thereon being determined by pattern lines drawn for the purpose.]

JOHN CLARKSON MILNS and SAMUEL PICKSTONE, of Radcliffe-bridge, manufacturers. *For certain improvements in machinery or apparatus used in spinning, doubling, and weaving cotton, flax, or other fibrous substances.* Patent dated January 11, 1851.

Claims.—1. The application of water to cotton and other fibrous materials as it passes through the spinning machine.

2. The application of a revolving surface or surfaces, for the purpose of conveying moisture to cotton and other material in its passage through the spinning or doubling machine, or for revolving in contact with moistened yarn.

3. The application of starch or other gummy or stiffening material to cotton or other substance, as it passes through the spinning or doubling machine by means of a roller or rollers immersed in such the starch or stiffening material, and revolving in contact with the cotton or other substance.

4. The application of artificial heat to cotton and other substance, for the purpose of drying such substance after having been moistened in the spinning or doubling machine.

5. The employment of cork for constituting the drag-washers in spinning machines.

6. The construction of the guides of the rising shuttle boxes of looms in such manner that each side thereof may be capable of independent adjustment.

GEORGE ANSTEY, of Brighton, gent. *For certain improvements in consuming smoke and in regulating the draught in chimneys.* Patent dated January 11, 1851.

The improvements here claimed comprehend,

1. A method of consuming the carbonaceous inflammable matter contained in smoke, by causing it to pass through a series of apertures, and thereby retaining it longer in contact with the flame and heat of the fireplace or furnace to which such apparatus may be attached.

2. A method of maintaining an equable temperature at the top of a chimney, by attaching thereto an apparatus for preventing the too sudden ingress of wind.

ALEXANDER SPEID LIVINGSTONE, of Swansea, engineer. *For improvements in the manufacture of fuel.* Patent dated January 11, 1851.

The patentee describes and claims,

1. An arrangement of apparatus for moulding and compressing fuel into blocks. [In this machine the moulds are formed in

an annular table, which revolves over a fixed bed plate, and the fuel mixed with bituminous material to effect its cohesion, when subjected to pressure, is delivered into the moulds from a shoot above the table, and is compressed by means of rollers having loosely hung projecting arms of a size equal to the apertures of the one above and the other below the table moulds, and revolving simultaneously with the table, the arms of the rollers taking into the moulds as they are brought round in succession by the revolution of the table.]

2. The employment of a long oven or tube heated to different degrees of temperature at different parts of its length, for drying the blocks of fuel, which are carried by tables running on rails, and caused to progress gradually through the oven.

3. the drying of the blocks of fuel by subjecting them to degrees of temperature gradually increasing to about 400° Fahrenheit.

CHARLES BARLOW, of Chancery-lane. *For improvements in machinery for the manufacture of railway chairs.* (A communication.) Patent dated January 14, 1851.

The machine claimed under this patent is intended to effect the manufacture of railway chairs from plates or bars of metal at a single operation. A piece of metal of sufficient size for the purpose is first sheared off from the plate and is pressed between two dies, one being stationary and the other attached to a vertical-sliding head, by which the edges are bevelled off, and the holes for the spikes punched out; the partially-formed chair then passes under the action of two other dies, attached by arms to a second sliding head, by which the lips are formed and bent up and the operation completed; each revolution of the mainshaft turning out a chair in a finished state.

JEAN MARIE TAURINES, of Paris, engineer. *For certain improvements in the machinery and apparatus for measuring and regulating the working of engines.* Patent dated January 16, 1851.

This invention relates to the construction of dynamometers, in which the flexure of a spring or springs, is made to indicate the amount of power exerted by a first mover to which the apparatus is attached, but owing to the unintelligible nature of the specification, which is apparently a literal translation by a Frenchman from the French original, we are altogether precluded from entering into details.

CHARLES WILLIAM LANCASTER, of New Bond-street, gun-maker. *For improvements in the manufacture of firearms and cannons, and of projectiles.* Patent dated January 16, 1851.

The improvements described under this patent are of such a nature as to require the aid of drawings to render a description of them intelligible.

CHARLES COWPER, of Southampton-buildings, Chancery-lane. *For improvements in the construction of apparatus for manufacturing, and apparatus for retaining and drawing off soda-water and other aërated liquors.* Patent dated January 16, 1851.

Claims.—1. The construction of apparatus for manufacturing aërated waters, by placing within the generator an acid vessel furnished with a cock or valve, and so arranged that the valve may be opened and closed from the exterior.

2. The general arrangement of the apparatus, in which the generator is surmounted by the receiver, and inclosing the acid vessel and washing vessel—both the generator and receiver being furnished with agitators on horizontal axes.

3. The construction of the valves of bottles for retaining and drawing off aërated liquors by means of a disc of vulcanized caoutchouc or other suitable elastic material, whose edges enter a recess in a cap, and which is pressed against a rounded or flat edge, as described. Also, the construction of such valves of a disc of vulcanized caoutchouc or other elastic material, contained in a cap so arranged as to confine the disc.

4. The application of a flexible metallic capsule for securing the valvular stoppers of bottles for containing aërated liquors, the junction being made good by cement, or by a ring or collar of vulcanized caoutchouc or other elastic material.

5. The construction of the valvular stoppers of bottles for containing aërated liquors in such manner that they may embrace the neck of the bottle, and be compressed closely around it, the junction being made good, as in the former case, by cement, or by a ring or collar of vulcanized caoutchouc, or other elastic material.

6. The application of a cap screw on the neck of the bottle for securing the valvular stopper of bottles for containing aërated liquors, the junction being made good by a ring of vulcanized caoutchouc, or other elastic material.

7. Constructing a stopper with a valve, and affixing the same to a bottle for containing aërated liquors by securing it into the neck of the bottle, the junction being made good by a ring of vulcanized caoutchouc, or other elastic material.

8. The application of a collar or tube of

vulcanized caoutchouc as a spring to close the valve, and serving at the same time as a packing to prevent the escape of liquid round the stem of the valvular stopper.

9. The construction of the valvular stoppers for bottles for retaining and drawing off aërated liquids with the stem of the valve made hollow, and with perforations therein for the purpose of admitting air.

10. The construction of valvular stoppers for retaining and drawing off aërated liquors with a tube which acts as the spout, and by pressing down which the valve is opened.

11. A method of securing the capsules and stoppers of bottles for retaining and drawing off aërated liquors.

Gustav Adolph Buckholz, of Agar-street, Strand, civil engineer. *For improvements in printing, and in the manufacture of printing apparatus, and also in folding and cutting apparatus.* Patent dated January 16, 1851.

Claims.—1. Certain machinery or apparatus for, and methods of manufacturing printing rollers or cylinders. [The patentee employs a hollow cylindrical matrix of gutta percha, and compresses thereinto gutta percha in a plastic state, in order to obtain a printing surface for the roller or cylinder.]

2. The construction of printing machinery in such manner that the middle roller (where three rollers only are employed), or all the rollers, with the exception of the first and last (where more than three rollers are employed), may produce impressions on two or more webs of paper or other material simultaneously, or nearly so, and so also that both sides of the paper or material may be printed during its passage through the machine.

3. A method of forming or producing the surfaces of inking or colour-serving rollers for printing paper or other material, by the employment of gutta percha softened by the action of nitric acid.

4. An arrangement of apparatus for folding continuous lengths of paper or other material.

Robert Cogan, Leicester-square, glass merchant. *For improvements in the application of plain or ornamental glass, alone or in combination with other suitable materials, to new and useful purposes of construction or manufacture.* Patent dated January 16, 1851.

The improvements claimed under this patent comprehend—

1. A method of constructing portable show-fronts for shops, verandahs, greenhouses, pavilions, tents, and other similar erections [*by the employment of* plates, panes, or panels of glass set in frames of wood or iron, which frames are constructed in detached parts, and may be so connected together as to form, when differently combined, the erections above mentioned.]

2. An improved glass ventilator, or any modification thereof. [A glass frame is attached inside of a window inclining inwards from the bottom, and having an open space at top; a narrow opening or slit is made in the top of the window, of the whole width thereof, to admit of the escape of heated and vitiated air from the room, and this is further assisted by taking out a pane from the bottom of the window, and allowing the external air to enter between the window and inclined frame.]

3. The construction of striking and propagating glasses, with a hollow conical space or inverted tube in the centre, for receiving and communicating heat to the earth contained in the pan.

4. The application of glass for making the barrels of churns in combination with dashers and beaters of wood or other non-corrodible material. [The patentee also specifies another application of glass to domestic purposes; viz., for making dish-covers, which possess the advantages over the wire cloth ones hitherto in use, of being easily cleaned, and not liable to corrosion or deterioration from dampness.]

5. The use of tubular glass candlesticks graduated on the outside, for the purpose of showing the hour [by burning a candle which is raised as consumed by a coiled spring].

6. The application to pens and pencils of a hollow moveable glass case, which may be made also to answer the purpose of a handle.

7. A method of constructing monuments "in commemoration of illustrious men and remarkable events," of blocks or slabs of glass, which may be cast or moulded to any design or part of a design.

Frederick Watson, of Moss-lane, Hulme, gentleman. *For improvements in sails, rigging, and ships' fittings, and machinery and apparatus employed therein.* Patent dated January 16, 1851.

These improvements consist in applying to the propulsion of ships or vessels the power derived from the action of the wind on sails arranged in a circular framing, like those of a windmill, and supported above the deck of the vessel,—the motion of the sails being communicated by suitable gearing to paddle-wheels or other propellers.

The claim is for the application to ships, boats, and vessels, of sails and rigging arranged on the above principle, for commu nicating motion to the hulls thereof by the agency of paddle-wheels, screws, or other propellers,—the sails being acted on by wind, instead of steam or other motive agent.

George Norman, of Shoreditch, cabinet-maker. *For an improved cooking and boiling apparatus.* Patent dated January 18, 1851.

These improvements in apparatus for cooking and boiling have for their object the application of heat derivable from steam, in combination with that radiated from or transmitted through the internal surface of the steam generator, to cooking purposes, whereby the apparatus is rendered more economic, convenient, and suitable for the preparation of various articles of food than any of the cooking apparatuses now in use. Fig. 1 is an external elevation, fig. 2 a longitudinal section, and fig. 3 a plan of this improved apparatus. A A and B B are two similar-shaped vessels, which are connected together near to their upper edges by a flange or strip of metal C C. The inner and smaller vessel is made of such size as to leave an intervening space D D of about three-quarters of an inch, more or less, between the sides and bottom of the two. When the article is in use, this intermediate space is partly filled with water, which is introduced through the plug E fitted into the bottom of the trough or channel formed by the upper edges of the vessels A B and the flange or strip of metal C by which they are connected together. F is a tray or pan, into which meat, fish, or poultry to be cooked is placed. G is a perforated tray, which is placed over the tray F, and is employed for holding potatoes, carrots, or other vegetables. F F are two pipes which, at their upper ends, are open to the interior of the space D D, between the vessels A and B; at their lower ends these pipes communicate with the interior of the inner vessel B, just immediately above the upper edge of the tray F, so that when the apparatus is set upon the fire, or heated by any other means to cause the water to boil, the steam generated within the chamber D is projected from the lower ends of the pipes I I directly upon the contents of the lower tray, by which they are speedily cooked, while the steam at the same time passing up through the upper tray amongst the vegetables placed upon it, prepares them for use. K is a gauge-tap for regulating the quantity of water put into the apparatus in the first instance, the chamber D being filled until it runs out at the tap. L is the cover. In the engravings the openings of the steam pipes I I into the interior of the vessel B

Fig. 1.

Fig. 2.

Fig. 3.

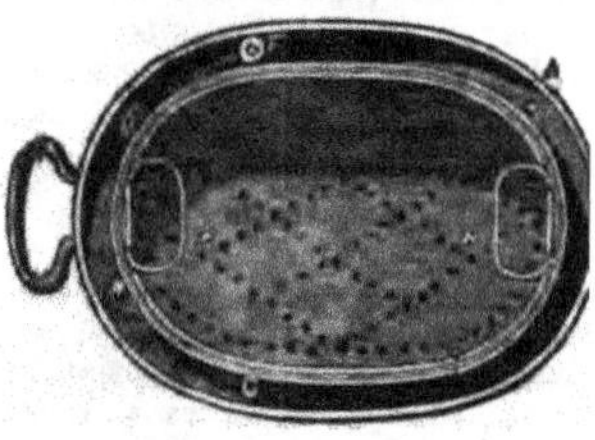

are shown as being placed immedi[ately at] the top of the lower tray, but it wi[ll be ob]vious that they may be placed furt[her] lower down at pleasure, accordi[ng to the] uses to which the apparatus is to b[e applied.] In some cases, such as the melting and other like substances, the ste[am pipes] may be carried into the interio[r of the] chamber B, so as to terminate am[ongst and] project the steam directly into [the sub]stances in that chamber; and the t[rays may] be altogether dispensed with. Th[e melted] substances would in such cases b[e drawn off] by a tap communicating with the i[nner ves]sel B. M M are caps, in which t[he lower] ends of the pipes I I are placed.

Claim.—The construction of cooking and boiling apparatus in such manner that the substances to be cooked or boiled are simultaneously subjected to the action of steam, and of radiated or of transmitted heat, as before described.

GEORGE FREDERICK MUNTZ, junior. *For improvements in furnaces applicable to the melting of metals for making brass, yellow metal, and other compound metals.* Patent dated January 18, 1851.

These improvements have for their object the prevention of the loss from volatilization which occurs when melting and mixing metals (especially when zinc is employed) for the manufacture of brass and other similar compound metals, and consists in the adaptation to the melting furnaces of two additional dampers, one in the bridge of the furnace, to shut off communication between the fire and the metal; and the second between the melting-pot and the chimney. There is also an additional flue (provided with a damper) between the fire and the chimney, for carrying off the smoke and products of combustion when the bridge damper is closed. The mixing operation will be thus performed in a close chamber, and the loss from volatilization much lessened, if not entirely prevented.

Claim. — The construction of furnaces for melting and mixing metals, for making brass and other compound metals, in which zinc forms a part, which will allow such metals when melted, and whilst being mixed, to be confined or nearly so from the air, by the furnace being converted into a close, or nearly close chamber; thereby preventing a great deal of the loss which occurred from volatilization in mixing such metals in the furnaces in use previous to the date of this invention.

WILLIAM REES, of Pembrey, Carmarthen, coal agent. *For certain improvements in the preparation of fuel.* Patent dated January 18, 1851.

The object of these improvements is to effect the consolidation of small coal, or of coal admixed with coke, into a compact mass without the aid of extraneous adhesive matter.

In carrying out his invention, the patentee prefers to use bituminous coal reduced to small pieces and dried by a gentle heat, which he places in moulds, and submits to a low continuous heat, of such a nature as to effect the partial fusion of the coal without volatilizing the gases contained therein. By these means blocks may be obtained of an even fracture, and the same specific gravity as the coal previous to the heating operation. When blocks of a very compact character are required, the moulds are submitted to and retained under compression during the heating process.

Claim.—The confining of coal or a mixture of coal and coke in moulds or vessels that are air-tight, or nearly so, and exposing the same to heat and pressure in manner described, thus retaining the gaseous constituents, and causing them to aid in the fusion and ultimate cohesion of the fuel into a compact mass.

JOHN LIENAU, junior, of Wharf-road, City-road, merchant. *For improvements in purifying or filtering oils or other liquids* Patent dated January 18, 1851.

The object of these improvements is to produce an oil of a superior quality for lighting and lubricating purposes, and this is effected by the addition to the oil under treatment (whether animal, mineral, or vegetable), of a sufficient quantity of creosote, or of the essential oils of cummin, aniseed, or rosemary, to produce fermentation, after which the oil is passed through a flannel filter (the action of an exhaust pump being employed to facilitate this part of the operation), and thereby obtained in a state of great purity.

RICHARD BYCROFT, of Paradise Wolsoken, gentleman. *For improvements in apparatus to be used by persons to secure warmth and dryness when travelling.* Patent dated January 18, 1851.

The apparatus here specified is specially adapted for the use of persons travelling in carriages, or otherwise in a sitting posture. It consists of a light metal framing forming the back and sides, to which is attached a covering folding over the front, and with a strap to attach it to the person using it. The wrapper is lined with some fabric having a long pile, or with sheepskin or fur, for the purpose of securing warmth when travelling in open carriages; and in this latter case, it is also provided with a hood for entirely enveloping the user.

Claim.—The construction of apparatus in the manner described, for the purpose of securing dryness and warmth when travelling.

EDMUND PACE, of the firm of Tylor and Pace, of Queen-street, iron bedstead-makers. *For certain improvements in bedsteads, couches, chairs, and other like articles of furniture.* Patent dated January 21, 1851.

Claims. — 1. The constructing of the frames of bedsteads, couches, and other like articles of furniture of sheet iron, drawn into angular or other suitable forms by passing the same through dies or rollers shaped to produce such forms.

2. A peculiar method of constructing the joints of beds, couches, chairs, and other like articles of furniture.

3. A method of forming the laths for beds, couches, chairs, and other seats, and also two several modifications of the same.

4. A peculiar method of constructing the rails of bedsteads and cots.

Specification Due, but not Enrolled.

WILLIAM ELLIOTT, of St. Helen's, Lancashire, chemist. *For improvements in the manufacture of alkali.* Patent dated January 21, 1851.

WEEKLY LIST OF NEW ENGLISH PATENTS.

Arthur Field, of Lambeth, gentleman, for improvements in the manufacture of candles, night-lights, and mortars. July 22; six months.

Samuel Varley, of Sheffield, engineer, for improvements in retarding and stopping railway carriages, and in making communications between the guards and engine-drivers on railways. July 22; six months.

Thomas, Earl of Dundonald, admiral in Her Majesty's Navy, of Chesterfield-street, Middlesex, for improvements in the construction and manufacture of sewers, drains, water-ways, pipes, reservoirs, and receptacles for liquids or solids, and for the making of columns, pillars, capitals, pedestals, vases, and other useful and ornamental objects from a substance never heretofore employed for such manufactures. July 22; six months.

LIST OF IRISH PATENTS FROM 21ST OF JUNE TO THE 19TH OF JULY, 1851.

Thomas Marsden, of Salford, for improvements in machinery for hackling and combing flax, and other fibrous materials. July 2.

William Melville, of Roebank-works, Lochwinnoch, Renfrew, North Britain, calico printer, for certain improvements in manufacturing and printing carpets and other fabrics. July 10.

WEEKLY LIST OF PROVISIONAL REGISTRATIONS.

Date of Registration.	No. in the Register.	Proprietors' Names.	Addresses.	Subjects of Design.
July 19	259	T. W. Stephens	Dublin	Wheat cleaning-machine.
,,	260	W. H. Dupré	Jersey	Albert roof-light and economical ventilator.
,,	261	R. Anderson	Westor, South Shields	Life boat.
,,	262	R. Powell	Great Pultney-street	Military coat or cloak.
22	263	J. Plimsoll	Sheffield	Moulding or surface file-holder.
,,	264	J. Plimsoll	Sheffield	Moulding or surface file-holder.

CONTENTS OF THIS NUMBER.

LONDON: Edited, Printed, and Published by Joseph Clinton Robertson, of No. 166, Fleet-street, in the City of London—Sold by A. and W. Galignani, Rue Vivienne, Paris; Machin and Co., Dublin; W. C. Campbell and Co., Hamburg.

Mechanics' Magazine,

MUSEUM, REGISTER, JOURNAL, AND GAZETTE.

No. 1460.] SATURDAY, AUGUST 2, 1851. [Price 3*d*., Stamped, 4*d*.

Edited by J. C. Robertson, 166, Fleet-street.

ALLAN'S IMPROVEMENTS IN ELECTRIC TELEGRAPHS.

Fig. 4.

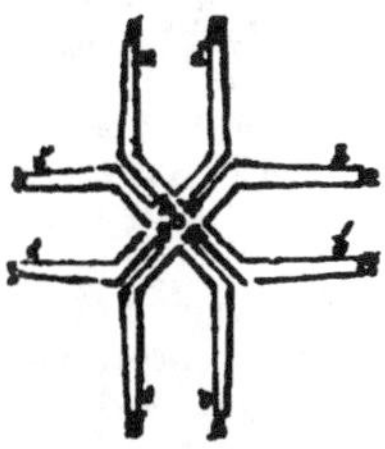

Fig. 5.

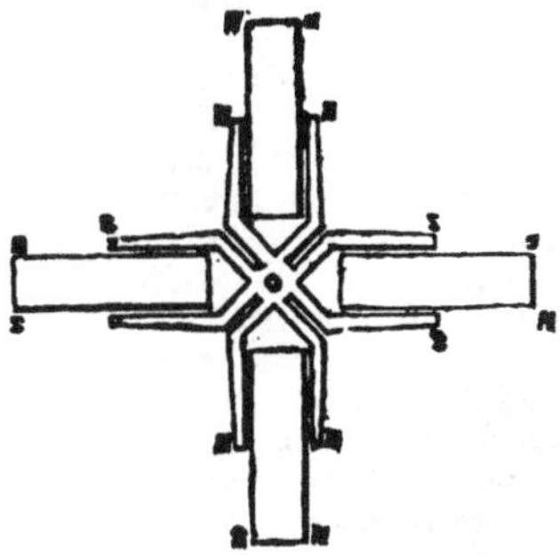

Fig. 6.

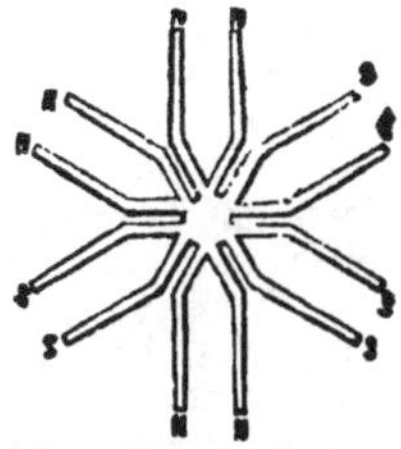

Fig. 9.

Fig. 10

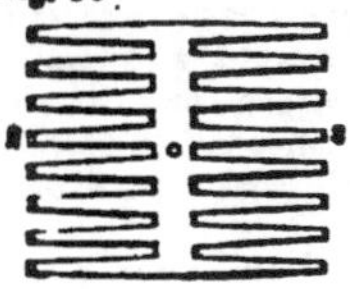

Fig. 12.

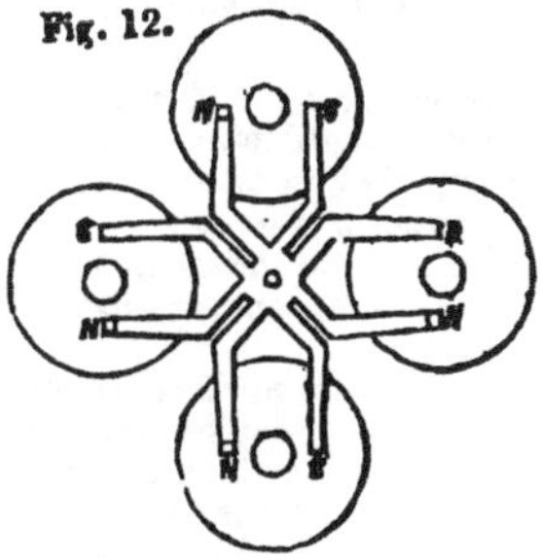

Fig. 11.

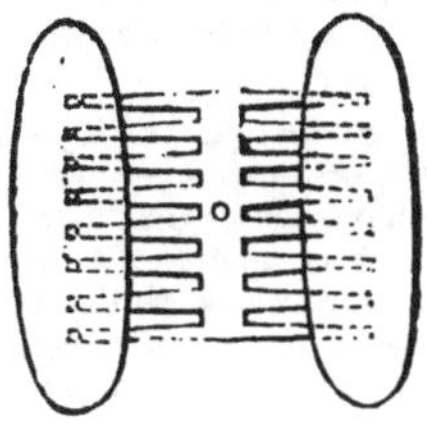

Fig. 17

Fig. 14

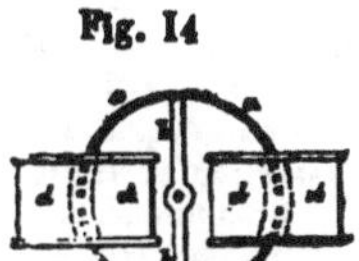

Fig. 15

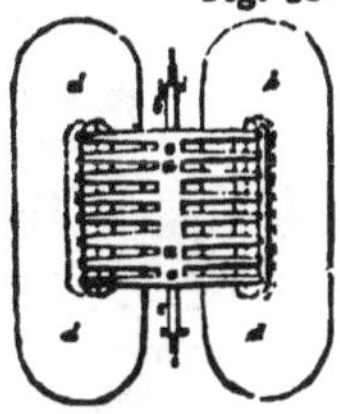

Fig. 18.

Fig. 19.

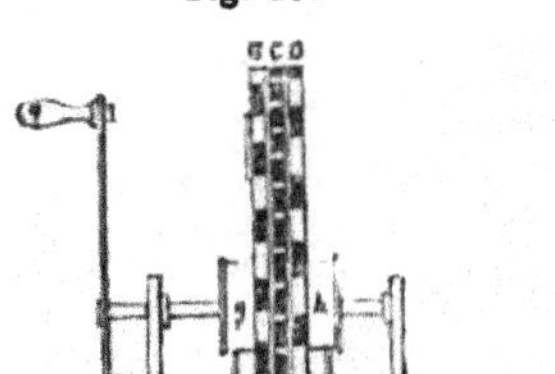

ALLAN'S IMPROVEMENTS IN ELECTRIC TELEGRAPHS.

Our readers may have noticed the progress through Parliament of a Bill for incorporating a new Company, to be called "The United Kingdom Electric Telegraph Company," for the purpose of purchasing and working certain Letters Patent granted to Mr. Thomas Allan, of Edinburgh (see vol. liv., p. 416), and the fruitless opposition which has been made to the Bill by certain rival parties. We are not surprised at the alarm evinced at the entrance of this new competitor into the field; for if we rightly comprehend the scope of Mr. Allan's improvements, they will, if fairly carried out in practice, *compel* a large reduction in the present high charges for wire postage (we give it this name for brevity's sake), and introduce besides a degree of precision and certainty which has not been realised by any of the plans of electric communication hitherto in use.

Mr. Allan prefers the original needle system of indication as, on the whole, the best yet devised, and it is to that exclusively his improvements have reference. These improvements he arranges in his specification under six distinct heads; of which the following are the leading features:

I. The first improvement described consists of a "mode of arranging the whole or a portion of the letters of the alphabet, or a series of words or sentences, into sections, so that each of the letters, words, and sentences so arranged may be telegraphed or signalled by *means of the number of the section to which it belongs, and the number or position of it in such section.*"

II. The second, which seems to us the most original and valuable improvement of the whole, we shall give in the inventor's own words (with but slight abridgment):

Another part of my said invention consists of a mode or modes of constructing what I call compound permanent magnets, or combining several simple magnets so as to form such compound magnets, which are intended to be used for electric telegraphs, and also of a mode or modes of deflecting or giving motion to such compound magnets, and thereby to deflect or give motion to needles, or other telegraphic apparatus connected therewith by means of the united action of such compound magnets and electric coils or cylinders of insulated wire or electro magnets.

I make my compound magnets for these purposes either in one and the same piece or in several pieces, and then combine or unite the pieces together, and I fix a compound magnet made in any such manner upon a suitable axle or spindle mounted in bearings, so that the axle or spindle will turn or move freely in its bearings to the extent to which it is intended to be capable of being deflected in either direction.

These compound magnets are made of various forms, and have their component parts arranged in various ways for the more conveniently adapting them to the purposes to which they are to be applied, and the manner in which they are to be used, and I construct every such magnet, and fix it upon an axle in such a way that the action of each of the arms or poles of the magnet (either by attracting or repelling) may be simultaneously exerted, and a combination of their several forms applied in deflecting, turning or moving the compound magnet upon its axle, and thus through the medium of the axle or spindle deflect or give motion to a needle or other description of electric telegraphic apparatus.

One description of my compound magnets I form in such manner that they may be used in conjunction with such electric coils or cylinders, as are or may be called electro-dynamic cylinders, and by varying the construction or arrangement of the arms of such magnets, or the polarities of their arms, so that they may be used in conjunction with electro-magnets, and others of my compound magnets are formed so as to act in conjunction with hollow electric coils or cylinders called galvanometer coils for the purpose of deflecting or giving motion to such compound magnets, and thereby to the needles or other descriptions of electric telegraph apparatus connected therewith.

My compound magnets which are intended to be used in conjunction with electro-dynamic cylinders are made in such a manner that each arm of every pair of arms, between which I place such a cylinder, shall have either a north or a south polarization, that is to say, so that the two arms at the sides or ends of every coil or cylinder shall both have the same polarity.

The cylinders which I use for the purposes of this part of my invention may either be made of round wire or of ribbon wire, and I make such cylinders consist of several yards *of wire,* insulated by gutta percha or any other convenient material, and wound in such a *manner as to form a cylinder,* very short in comparison to its diameter.

I place a cylinder of this description between each pair of arms of one of these compound magnets, each of the arms having (as before mentioned) the same polarity, and I fix each cylinder with a portion of its periphery towards the axle of the magnet, and with the plane of the end of the coil at right angles, or nearly so, with the plane of the side of the compound magnet.

Fig. 20.

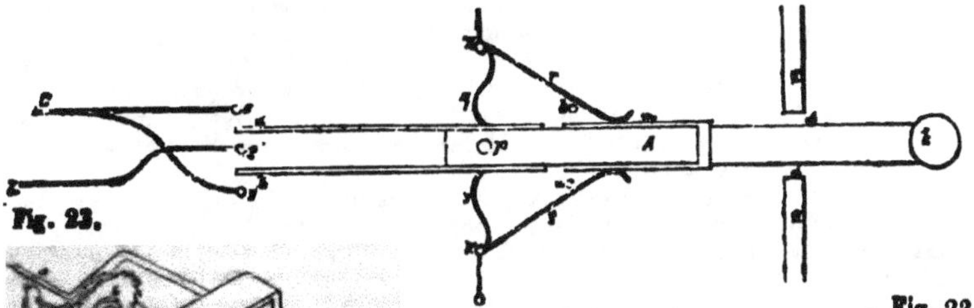

Fig. 23.

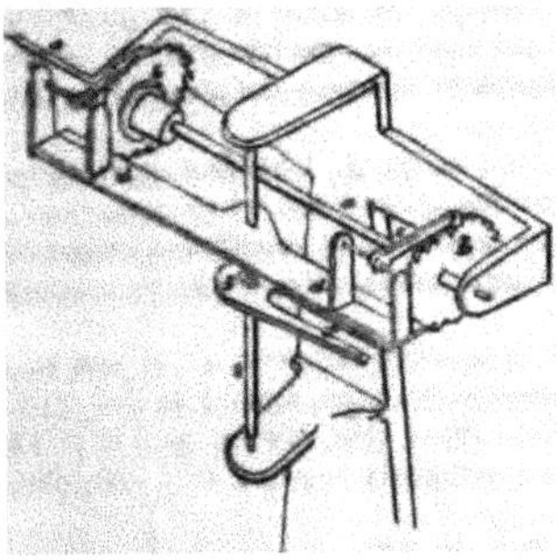

Fig. 22.

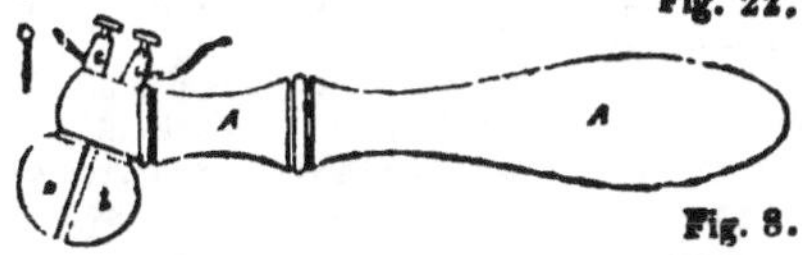

Fig. 13.

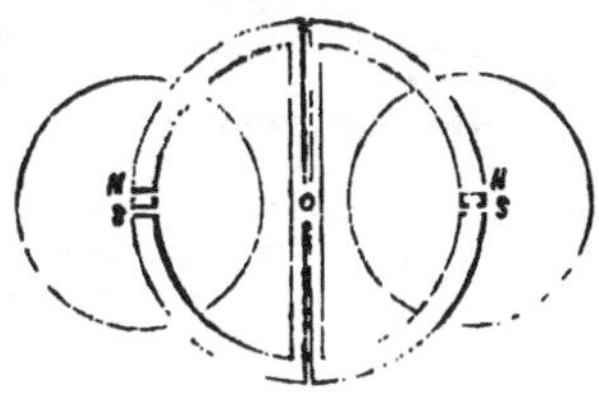

Fig. 16.

Fig. 8.

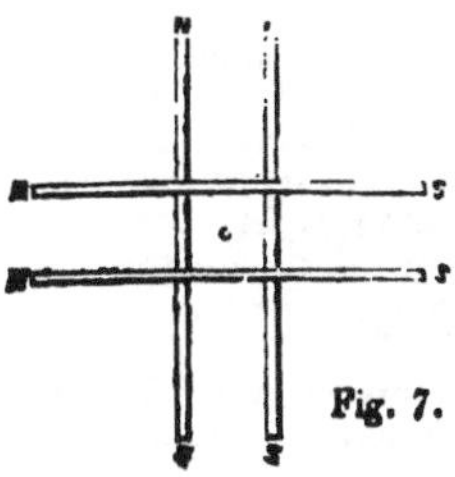

Fig. 7.

The sizes of the cylinders, the distances between each arm of every pair of arms, and the fixing of the cylinders, must be so regulated as to allow of a slight motion or turning of the magnet upon its axle: one half (in number) of the arms of the magnet resting against the ends of the cylinders when the magnet has been moved or turned in one direction, and the other half (in number) of the arms of the magnet resting against or coming in contact with the opposite ends of the cylinders when the magnet has been moved or turned in the other direction, so that the needle or other telegraphic apparatus connected with the axle or spindle of the magnet may be thus deflected, or have motion given to it.

In figs. *4 and 6 of the said engravings* I have shown two forms of this description of compound magnets, each of which may be either made of one piece of steel, or of several pieces, as may be preferred.

Fig. 4 is a side view of a compound magnet which may be used for deflecting the needle of an electric-needle telegraph mounted upon the end of the axle or spindle of the magnet, or for giving motion in a similar manner to any other description of electric telegraph apparatus.

This compound magnet is shown in the figure as having four pairs of poles projecting from the centre through which an orifice is made (as shown in the figure) to receive the axle or spindle upon which the magnet is to be mounted. Each arm of each of the four pairs of arms *aa'*, *bb'*, *cc'*, *d'd'*, is magnetised so as to have the same polarity, each of the arms *a*, *a'*, and also each of the arms *c*, *c'*, having a north polarity, and each of the arms *b*, *b'*, and also each of the arms *d*, *d'*, having a south polarity, and I magnetise the arms so that the two arms *a'*, *e*, and *b*, *e*, shall together form a horse shoe or bent magnet *a'*, *e*, *b*, and so that the other arms of the compound magnet shall form three other horse-shoe or bent magnets *b'*, *f*, *c*,—*c' g' d*, and *d' h' a*.

When I make this magnet all in one piece (which I prefer), I take a piece of steel plate of the thickness which I desire the magnet to possess, and cut it so that it shall have eight projecting arms, as shown in the figure, the object of cutting the arms of the particular angular forms shown in the figure is not only to admit as large an electric coil as practicable between each pair of arms of the same polarity, but also to make the magnet as light as conveniently may be.

When a piece of steel has been cut into the shape shown in fig. 4, I harden each of the arms in the usual manner, so as to enable them to retain the magnetism when magnetised, taking care not harden the central part of the magnet, so that what is sometimes called the neutral point, which is at or near to *e*, *f*, *g*, and *h* of these horse-shoe, or bent magnets, may be soft.

A piece of steel having been formed in this way, I magnetise the arms *a'*, *e*, and *b*, *e*, by drawing the ends of a pair of magnets, or of a horse-shoe magnet, from *e* to the points *a'* and *b*, first on one side of the arms and then on the other side, taking care that the points of the magnet or magnets used for magnetising are properly applied for producing the required polarities of the arms.

When these two arms have been magnetised, they will in effect form a horse-shoe or bent magnet; and in order to prevent any demagnetising of these arms by the magnetising of the other arms of the compound magnet, I place an armature or piece of iron across between the points *a'* and *b*, whilst I am magnetising the other arms. I then magnetise the other arms of the compound magnet in a similar manner, so that the arms *b' f* and *c f* may, in effect, form a horse-shoe, or bent magnet, the arms *c' g* and *d g*, another similar magnet, and the arms *g' h* and *a h*, another similar magnet, taking care to place an armature or piece of iron across each of the first three pairs of arms magnetised in manner and for the purposes I have already mentioned. The compound magnet thus completed, may now be mounted upon its intended axle or spindle, upon which it may be secured in any convenient manner, and it is then ready for acting in conjunction with cylinders, so that the magnet may be deflected or have an alternating or reciprocating motion given to it for the purpose of deflecting or giving motion to the needle or other apparatus fixed upon or connected with the axle or spindle.

If a compound magnet, such as I have just described, be made of several pieces of steel, I fashion the parts and solder them together so that they will, when thus united into one piece of the required shape, and when magnetized in the manner I have described, form a compound magnet such as shown in fig. 4.

In fig. 5, I have shown a compound magnet constructed in the manner I have described, with electric coils placed between each pair of arms of the same polarity, in the manner I have before mentioned.

The wires used in making these coils (whether round or flat) are coiled from their centres so as to make them solid, or nearly solid, cylinders, and get the length of wire used in forming a coil compressed into the smallest practicable space.

It is well known that when a cylinder of insulated wire such as I have just mentioned has a current of electricity sent through the wire in one direction, the cylinder acquires a north polarity at one end and a south polarity at the other, and that when the direction of the current is reversed, the polarities of the ends of the cylinders are also reversed.

When, therefore, a cylinder such as I have mentioned is placed between two arms of a compound magnet having the same polarities, and a current of electricity is sent through the wire so that the ends of the cylinders acquire polarities, the one end of the cylinder will attract one of the pair of arms, and the other end of the cylinder will repel the other of the pair, and the forces of the attraction and of the repulsion thus called into action will both be exerted in the same direction, or in such a way as to turn or move the compound magnet upon its own axis in one direction.

Thus if the two arms between which a coil is placed have both of them north polarities, the end of the cylinder which has acquired a south polarity will attract towards it the arm which is placed opposite that end of the cylinder, whilst the other end of the cylinder, which has acquired a north polarity, will repel the other arm of the pair. And thus both of the arms will—the one by attraction and the other by repulsion—be deflected or moved in the same direction.

By reversing the direction of the current of electricity sent through the wire composing the cylinder, the polarities of the ends of the cylinder will be reversed also; and the end of the cylinder which before attracted one of the arms, will now repel it, and the other end of the cylinder, which before repelled the other arm, will now attract it; and thus the arms will, by the reversed forces, be deflected or moved in a direction the reverse of that of the previous motion.

The wires forming the cylinders placed between the arms of this compound magnet are connected together in such a way that the same current of electricity passes through each of the cylinders; but the connection is made in such a manner that, when the current of electricity passing through a cylinder placed between two arms having north polarities, passes round the cylinders (to or from the centre of the compound magnet) in one direction, the current of electricity passing through a cylinder placed between two arms having south polarities, shall pass round the cylinder in an opposite direction, either to or from the centre of the compound magnet, so that the cylinders and the magnetic arms of the compound magnet may all concur in simultaneously exerting their attractive or repulsive forces in deflecting, turning, or moving the compound magnet upon its axle in the same direction.

The cylinders may be fixed in any convenient manner in proper positions between the arms between which they are to act, in the manner I have described. A compound magnet such as I have described being fixed upon an axle, and connected in any convenient manner, and cylinders being also fixed in proper positions between the arms of the magnet, the magnet may be deflected, turned, or moved at pleasure, in either direction, to the extent permitted by the distances between the cylinders and the arms, according to the direction in which the current of electricity shall be transmitted; and in this manner a needle at the end of the axle or spindle of the compound magnet may be deflected in either direction at pleasure.

In fig 6 I have shown a side view of a compound magnet similar to that shown in fig. 4, with the exception of the number of projecting arms, of which this magnet has six, instead of four pairs, the polarities of the arms being as shown in the figure. This magnet will require six electric coils in lieu of four; and they are to be fixed and used in like manner as I have described with reference to a magnet such as shown in fig. 4.

In fig. 8 I have shown one pair of straight simple magnets combined together (or made in one piece), so as to form a compound magnet capable of being mounted upon an axle. The north pole of each simple magnet composing this compound magnet is placed in the same direction, and, of course, each of the south poles is placed in the contrary direction, so that an electric coil may be placed between each pair of arms having the same polarity as shown in the figure.

In lieu of making a compound magnet of a pair of straight simple magnets, as shown in fig. 8, two compound magnets, such as shown in fig. 10, may be united together or fixed upon the same axle or spindle—the one at one side of the axle, and the other at the other side of the axle; for which purpose I connect them together by brass cross arms; one of which is seen by the top view of this compound magnet shown in fig. 9. In making electro-dynamic coils to be used in conjunction with this compound magnet, I coil the wire in such a manner that each end of a coil will be of a shape resembling a cylindroid; and I have shown a pair of those cylindroids in fig. 11, with one portion of the compound magnet (such as shown in fig. 10) placed behind them in the position it would occupy if the whole were mounted and in action.

Instead of mounting only one such compound magnet, as shown in either of the figs. 4, 6, or 7, upon an axle or spindle, several of such magnets may be so mounted, and placed as near to each other as conveniently may be, and so that the electro-dynamic coils may be placed between the arms of each compound magnet in like manner as if there were only one such compound magnet mounted upon the axles.

In order to make the electro-dynamic coils act more effectually in attracting and repelling the arms of all the magnets, I use coils made in the shape of cylindroids, as shown in fig. 11.

Compound magnets, such as I have described and shown in the figs. 4, 6, 7, and 8, may also be used in conjunction with electro-magnets, for the purpose of deflecting or giving motion to such magnets, and thereby to deflect or give motion to needles or other telegraphic apparatus, in manner aforesaid—the polarities of the arms of the magnet only being different when such a magnet is used in this way.

Thus in fig. 12 I have shown a compound magnet such as represented in fig. 4, placed so as to act in conjunction with two electro-magnets.

The mode of arranging the polarities of the several arms of the magnet, when thus used, are marked in fig. 12 (one of each pair of arms being north, and the other south), and the ends of each of the magnets are to protrude above the electric coils by which they are magnetised, so as the better to attract or repel the two arms of the magnet between which it is placed.

When the compound magnet used to act in conjunction with an electro-magnet is composed of a pair only of single magnets, the form instead of being, as seen in fig. 8, may be made as shown in fig. 13.

In figs. 14 and 15, I have shown a mode of constructing a compound magnet, which is intended to be used in conjunction with hollow coils or cylinders, called galvanometer coils.

Fig. 14 is an end view of this compound magnet, showing also a pair of coils in conjunction with which it is to be used; and fig. 15 is a side view of the same magnet and its coils, showing also the axle or spindle upon which the compound magnet is mounted; and in those figures similar parts are marked with similar letters of reference. One half of the compound magnet *a*, *a*, *a*, *a*, is made in one piece, and of thin steel plate, deeply serrated or cut in a similar manner to a comb, so as to form in effect, when magnetized, several magnets, all the north poles of which shall lie in one direction, and the south in the other direction. Each half of the compound magnet is bent into the shape of a semicircle, as shown in fig. 14; and by means of two arms *b b*, fig. 14, they are fixed upon an axle or spindle C C, fig. 15, so as to form, when thus secured, a hollow cylinder, but with small spaces between the ends of the magnets, as shown by dotted lines in fig. 14. The north poles of one half of the compound magnet are placed so that they shall be opposite to the north poles of the other half of the compound magnet, in order that all the north poles of he compound magnet may be placed and work within one of the coils, and all the south poles within the other coil.

Each of the coils *d d* is made of a shape nearly resembling a cylindroid, and with a hollow space in its centre of such a form that the compound magnet may be free to turn or move upon its axle or spindle to the intended extent in either direction. The direction of the current of electricity through the wires forming the coils, for the purpose of turning or moving this magnet and its axle or spindle, and so deflecting a needle, or giving motion to a ratchet-wheel or other telegraphic apparatus, must be regulated in manner before described.

In figs. 16 and 17, I have shown another mode of constructing a compound magnet, which is intended to be used in conjunction with either a single hollow coil or cylinder (as shown in fig. 17), or with two such coils or cylinders.

Fig. 16 represents this compound magnet, made of one piece of thin steel plate, cut out in the manner shown in the figure, and magnetized so that all the north poles shall be at one side of the circle, and all the south poles at the opposite side.

This compound magnet may be fixed upon an axle or spindle, and mounted upon bearings within the coil (if only one), in conjunction with which it is to act in the manner shown in fig. 17. This coil is made of a similar form to those shown in fig. 15, the hollow space in the centre of this coil being made sufficiently large to receive the magnet and its bearings. A magnet, thus mounted, may have a pointer attached to it, as shown by dotted lines in fig. 17, so that the pointer may, by the motions of the magnet, be deflected in the manner required for telegraphic purposes.

III. To impart greater rapidity of movement to the pointer, forms the subject of Mr. Allan's next improvement. He effects this by means of an exceedingly ingenious construction of insulated metallic discs. Fig. 18 is a front view of this arrangement, and fig. 19 a side or edge view.

A A are two standards, in which the axle of the compound disc has its bearings. B B, C C, and D D are three circular discs, which are held together by means of bolts and nuts, but with small spaces between them.

For the purpose of keeping the discs apart from each other, I interpose pieces of gutta *percha or other non-conducting* material between them, and I also interpose a piece of gutta percha, or of such other material as aforesaid, between the heads and nuts of the bolts and the discs, and also between the shanks of the bolts and the holes in the discs (made large for that purpose) through which they pass.

For the purpose of causing this apparatus, by means of one revolution of the compound disc, to produce as many motions of the pointer as there are letters of the alphabet, I make thirteen notches in the periphery of each of the two outer discs B B and D D, as shown in the disc B B, fig. 18; and the axle of the compound disc is in two pieces, one piece of it being attached to the centre of the disc B B, and the other to the centre of the disc D D; and those portions of the axle are mounted in the bearings as shown in fig. 19. The portion of axle attached to the disc B B is, however, made longer than the other, so that it may pass through a dial, and have mounted upon it a pointer *p*, furnished with a handle *q*, intended to be used in conjunction with a dial-plate, not shown in the figures.

I make the intermediate disc C C with twenty-six notches in its periphery, so that it may have exactly twice as many projections as either of the other discs.

When the discs are fixed together, these notches and projections are so placed that the centre of each projection upon the periphery of the disc B shall be opposite to the centre of a notch in the disc D, and the centre of each projection upon the periphery of the disc D shall be opposite to the centre of a notch in the disc B.

The notches or recesses in the peripheries of each of the discs B and D are made longer than the projections, so that there may be an intermediate space in the periphery of the compound disc between each projection on either of the discs B and D, and each of the two next projections on the other of those discs, and each of the projections upon the periphery of the disc C is of such a size, and so placed, that it will occupy rather more than one of the intermediate spaces in the compound disc first mentioned.

The two flat springs *a* and *b* (fig. 18) communicate, the one with the earth wire, and the other with the line wire of the telegraph—each of those springs forming part of the circuit. The nozzle or end of each of those springs is of the same width as the compound disc, or nearly so, in order that it may alternately come in contact with each of the projections of the discs B, C, and D, during the revolution of the compound disc. The nozzles or ends of those springs are made to pass gently against the periphery of the compound disc, at points upon the periphery at such a distance apart as shall be equal to the distance upon the same periphery between the centre of any projection upon either of the discs B and D, and the centre of the next succeeding projection upon the other of those discs.

The flat spring *e* communicates with a wire leading from the copper or positive pole of the galvanic battery, and is made to press gently upon the cogs *g* of the disc B. The flat spring *f* communicates with a wire leading to the zinc or negative pole of the battery, and is made to press gently upon the cogs *h* of the disc D D—those two springs thus forming part of the galvanic circuit.

When the compound disc is turned round, it will be seen that when the nozzle of the spring *a* presses against a projection of the disc B B, the nozzle of the spring *b* will press against a projection of the disc D D, and thus the galvanic current will be sent in one direction through the whole of the current. When the spring *a* presses against a projection of the disc D D, the nozzle of the spring *b* will press against a projection of the disc B B, and the galvanic current will then be sent in the reverse direction; and when the nozzle of each spring presses upon a projection of the disc C C, the galvanic current is broken.

By these means, when the compound disc is turned round, the galvanic current will be alternately sent in each direction throughout the current, and the galvanic current thus used for giving motion to the pointer of a step-by-step telegraph, as before mentioned.

IV. A new break or pole-charger is next described, by means of which the direction of the current of electricity may be changed with the greatest celerity as often as required.

A top view of this apparatus is shown in fig. 20 of the said engravings, in which AA is a wooden lever mounted upon a pivot *p*, within the case containing the apparatus, the lever projecting through an orifice *dd*, in the case EE, and being furnished with a knob *k*, so as to fit the projecting part for being used as a handle. The orifice *dd* is made just sufficiently large to allow the lever AA to be moved to the right or left hand to the extent which may be requisite for working the apparatus. To the end of the lever AA two straight flat brass plates *a* and *b* are attached, one being fixed upon one side of the lever, and the other upon the other side, and these two plates are made of such lengths, that their ends will come in contact with two of the three metallic pins *e*, *f*, and *g*, when the lever is moved either to the right or the left hand. The metallic pins *e*, *f*, and *g*, are fixed in such positions, that when the lever, and consequently the metallic plates *a* and *b*, are moved in one direction, the ends of the plates shall come in contact at the same moment with, and slightly press against two of the pins, the one plate pressing against the middle pin *f*, and the other plate against the pin *e* or *g*, according to the direction in which the plates shall have been moved. To the middle pin *f*, the end of a wire leading to one of the poles of the battery is to be secured, and to each of the other two pins *e* and *g* is to be secured the end of a wire leading to the other pole of the battery.

The wire secured to the pin *f* is represented in the figure as leading to the zinc or negative pole, and the wire secured to each of the pins *e* and *g* as leading to the copper or positive pole; but this arrangement may be reversed if thought fit.

It will be seen, therefore, that by moving the plates *a* and *b* in one direction, the plate *a* will be brought in contact with the pin *f*, connected with the zinc pole of the battery; the other plate being at the same time in contact with the pin connected with the other pole; and by moving the plates in the other direction, the plate *b* will be brought into contact with the pin *f* connected with the zinc pole, the other plate being at the same time in contact with the pin connected with the other pole, and thus the current of electricity transmitted along the two plates *a* and *b* may be varied at pleasure by moving the lever, and consequently the plate *a* and *b* in the requisite direction. Upon one side of the lever AA is also fixed a metallic plate *m* to receive the loose end of the spring *r*, and upon the opposite side of the lever is fixed another similar plate *n*, to receive the loose end of the spring *s*. These plates *m* and *n* are connected together by means of a cross-piece *i*, so as to enable a current of electricity to pass from either of the plates to the other. The spring *r* is fixed upon the pin T, so that it may press slightly upon the plate *m*, or upon the stop pin *t*, when the lever is moved to the left hand, and the spring *r* will therefore be capable of forming part of the electric circuit when it is in contact with the plate *m*.

The spring *s* is also fixed upon the pin X, so that it may press slightly upon the plate *n*, or upon the stop pin *u*, when the lever is moved to the right hand, and the spring *s* will therefore be capable of forming part of the electric circuit when it is in contact with the plate *n*. But when either of the springs *r* or *s* ceases to be in contact with its plate *m* or *n*, neither of them will form part of the electric circuit.

The wire *q* has one of its ends attached to the plate *a*, and the other to the pin T, so as to complete the circuit between the wire of the electric circuit connected with that pin and the plate *a*.

The wire *y* has one of its ends attached to the plate *b*, and the other to the pin X, so as to complete the circuit between the wire of the electric circuit connected with that pin and the plate *b*.

When this apparatus is in a state of rest (as shown in the figure) and the electric current is in action, the springs *r* and *s* will form part of the circuit through which the current of electricity, sent from a distant station, will pass in the one direction or the other, to or from the telegraph, which may happen to be used, and the battery wires connected with the pins *e*, *f*, and *g* will of course form no part of the circuit.

But when this apparatus is to be used for communicating signals to a distant station, the battery wire shown in the figure, and their battery, will be brought into action as the lever AA is moved to the right hand or to the left for the purpose of sending the electric current in the intended direction.

Thus, if the handle is moved to the right, the plate *b* will be brought into contact with the pin *g*, and the plate *a* into contact with the pin *f*, and then the current of electricity from the copper pole of the battery will pass along the plate *b* and the wires connected with it, returning by the wires connected with the plate *a*, pass along that plate to the pin *f*, and the battery wire connected with it.

By moving the handle of the lever to the left hand, the plate *a* will be brought into contact with the pin *e*, and the pin *b* into contact with the pin *f*, and thus the current of electricity transmitted will be reversed.

V. A description follows of an apparatus which Mr. Allan calls (not very happily) "a slotted frame apparatus," which he has invented "for the purpose of making, breaking, and reforming or reversing the electric current of a telegraph any number of times successively, and also, when desired, varying the continuance of the electric current each time the current is formed."

a, *a*, *a*, *a*, is a flat piece of dry wood, upon which several framed or slotted pieces of wood are fixed, there being as many of such slotted pieces as there are letters in the alphabet. Upon each side of each of those slots, and a little apart from each other, are placed square buttons or pieces of metal, let down into the wood so that the top of the wood and of the buttons may be level or flush with each other, and make the edges of the slots smooth for the instrument nextly described to pass along it. The under sides of some of those buttons are connected with one of the wires of the circuit, and the others with the other wire of the circuit; or otherwise, if thought better, some of the buttons may be connected with one of the wires and the battery, the others being connected with the other of the battery wires. The number of buttons placed along each side of a slot, and the spaces between the buttons, as well as the particular battery or line wire with which each button is made to communicate, must be determined or regulated according to the number of times and the manner in which the electric current is intended to be formed, broken, and reversed or repeated for the intended purpose.

Thus supposing the slot C to be intended for signalling the letter C, that may be done by two motions of the electric current in one direction, and one motion of it in the other direction; and for effecting this purpose, the first button on the other side of that slot (commencing from the left hand), marked with the letter *n*, may be made to communicate with one wire of the circuit or battery—the first button on the lower side of the slot marked with the letter *p* being made to communicate with the opposite circuit or battery wire. The second button on the upper side, also marked *n*, is to be made to communicate with the same wire as the first button marked *n*, and the second button on the lower side, also marked *p*, is to be made to communicate with the same wire as the first button marked *p*. The third button on the upper side, marked *p*, is to be made to communicate with the opposite wire to that with which the two first buttons, marked *n* and *n*, communicate, and the third button on the lower side, marked *n*, must also communicate with the opposite circuit or battery wire from that with which the third button on the upper side, marked *p*, communicates. The parts of this slot being thus arranged, it will be seen that when the instrument nextly described is used, and passed along the slot from left to right, the electric circuit will be formed and broken three times, the current being firstly sent twice in one direction, and then once in the opposite direction.

The buttons upon the upper side of the slot D, it will be seen, are all made to communicate with one of the battery or circuit wires, and the buttons on the lower side to communicate with the other of those wires, so that upon each of the three times when the electric circuit shall be formed the electric current will be transmitted in the same direction. It will be seen, also, that the space between the second and third of the upper as well as the lower buttons, there is a greater space than between the first and second; and this slot, arranged as shown, may be used for signalling the letter D, which is marked opposite to it.

In this way the direction in which the electric current is sent every time the circuit is formed, by means of buttons placed along such slots, may be varied at pleasure; and I have marked buttons on the upper and lower sides of the several slots in the figure in such manner as to show several variations which may be made by this apparatus in the forming, breaking, and reversing the electric current, which may be used either for telegraphing the letters of the alphabet, as marked, or for any other telegraphic purposes. To make the figure more clearly indicate the direction in which the electric current is intended to be sent each time when the circuit is formed, by means of the buttons along each of the slots, I have marked each button with a letter *p* or *n*, to show that those pieces are to be connected with one or other of the two wires of the circuit or battery, the one of which wires may, for the sake of distinction, be called the positive wire, and the other the negative wire.

In fig. 22, I have shown a side view of an instrument to be used in conjunction with the apparatus (fig. 21) before described, for making, breaking, and reforming or reversing currents for telegraphic purposes.

In this fig. (22) A A is a wooden handle, upon the end of which are fixed two semi-globular nozzles *b b*, which are secured to the handle by a screw passing down from the pins *c c*, through the head of the handle into those nozzles, each of the nozzles, with its screw and pin, being kept insulated from the other. One of the pins, *c c*, is connected (by means of a wire) with one of the battery or circuit wires, the other pin being connected (in a similar manner) with the other wire. If the buttons of the apparatus fig. 21 be connected with the circuit wires, the pins of the instrument fig. 22 must be connected with the battery wires, and *vice versâ*.

In working the apparatus fig. 21, by means of the instrument fig. 22, supposing the buttons of the apparatus to be connected with the line or current wires, and the nozzles of the instrument fig. 22 with the battery wires, and the battery in action, the operator is to take hold of the handle of the instrument, and to place its nozzles at the left hand end of one of the slots of the apparatus intended to be used, in such a position that the upper nozzle will touch the upper side or edge of the slot, and the lower nozzle will touch the lower side or edge of the slot, and then the nozzles are to be drawn along the whole of the slot, from left to right, taking care that the nozzles shall be kept in contact with the edges of the slots during the whole of the operation, so as to ensure the nozzles coming in contact with each of the buttons as they pass along. In this way it will be seen that whenever the nozzles come in contact with a pair of buttons, the one at the upper and the other at the lower side of the slot, the electric circuit will be completely formed, and the electric current sent in one direction or the other, according to the arrangement made for that purpose; and the moment the nozzles quit their contact with the buttons the circuit will be broken. The nozzles proceeding along the slot, and coming into contact with another pair of buttons, the electric circuit will again be made complete, and the current sent in the same or the reverse direction, according to the arrangement; and this operation of making, breaking, and reforming or reversing the circuit and current will be repeated as often as there are pairs of buttons in a slot, and according to the arrangements made for effecting those purposes.

VI. Mr. Allan's last improvement consists of an apparatus for giving motion to the pointer of a circular dial. A view of this is given in fig. 28.

A a is a portion of the spindle or axle of the pointer, upon which are fixed two ratchet wheels *b b*. Two palls or clicks *c c* are mounted upon the alternating frame *d d*, and take into the teeth of the ratchet wheels. The frame *d d* is fixed upon an upright spindle *e e*, which is mounted in suitable bearings, so as to allow the alternating frame to move freely in either direction. When one end of the frame *d d* is drawn back from the ratchet wheel opposite to it, the pall having hold of one of the teeth of the wheel will draw it partially round in one direction, the pall at the opposite end of the frame *d d* at the same time advancing so as to catch hold of another tooth of the wheel at that end which has been turned in the same manner as the other wheel; the motion of the frame being then reversed, the last-mentioned pall will, by means of the tooth which it has just caught hold of, draw the wheel at that end partially round in the same direction as before, the pall or click at the other end of the frame advancing towards the wheel at that end and catching hold of another tooth, and then the frame may be again moved in the same direction as before, and then by the continuous alternating motion of the frame and the two palls in the manner described the wheels *b b* and their spindle, together with the pointer mounted at the end of the spindle may be turned round step by step in the same direction. To the upright spindle *e e* and under the frame *d d* is fixed a slotted piece *f f*, and by means of a pin working in that slot the requisite alternating motion may be given to the frame, and this motion may be given by means of any description of electric telegraphic apparatus usually adopted for similar purposes.

Each motion of the frame *d d* will cause the wheels to be turned a portion of a revolution equal to one-half of one of the teeth, and to prevent the wheels turning further by any motion than is intended, the edge of the frame may be made to come in contact with each wheel alternately, and so prevent any overrunning.

To recapitulate.—The great commercial advantages which Mr. Allan's system offers are these:

First. A large accumulation of the magnetic power within a small compass.

Second. A great diminution of the frictional resistance to the electric current, and proportionate saving of battery power.

Third. An abbreviation of the signals necessary to be used, and great increase of celerity and speed in their transmission. And

Fourth. Simplicity in the mechanism, rendering it not only easy of comprehension in all its parts, but cheap to work and maintain.

HODGES' "PATENT SILENT HARPOON PROJECTOR."

Sir,—I am desirous of calling the attention of your nautical readers to the "Patent Silent Harpoon Projector," manufactured by me under the patent of Mr. Hodges, and placed for inspection in the Great Exhibition (Class viii., No. 38); but previous to entering into the details of it, I may be permtted to offer *a few explanatory remarks*. The oldest and still most frequent method of killing whales is that of throwing the harpoon by hand; this is attended with immi. nent risk, because the boat must be within a fathom or so of the fish, and it is very well known that they are often dashed to pieces or upset, the men being thrown out and sometimes killed. It was to obviate this danger that the harpoon powder-gun was invented. This kills at a great distance; but trusting too much to its long range, the fish is sometimes missed and lost. Although it places the men comparatively out of danger, it is objectionable on account of its report, as if the whale is not struck at the first shot, she takes alarm and immediately disappears. If, as is sometimes the case, there are more than one in sight, they are so startled by the explosion that they dive, and are not again seen. To overcome these difficulties, the harpoon air-gun was introduced as a third method. In theory, it was all that could be desired; it placed the men as much out of danger as the powder-gun does, and it possessed the further advantage of making no report. In practice it was found to become soon useless, because being composed of parts necessarily so very delicate, as all pneumatic apparatus is, it soon got out of order; in fact, it cannot be adapted to the rough usage it unavoidably meets with at sea. Being once deranged, it cannot be repaired,—for two reasons; first, from not having the requisite means at hand; and, second, from not having any one on board sufficiently acquainted with its principles and construction to undertake the task, which is indeed somewhat difficult under the most favourable circumstances. It is now very seldom, if ever, used. Both these guns require harpoons of a peculiar make, which cannot be used by hand—an obvious disadvantage.

Fig. 1.

Fig. 3.

Fig. 2.

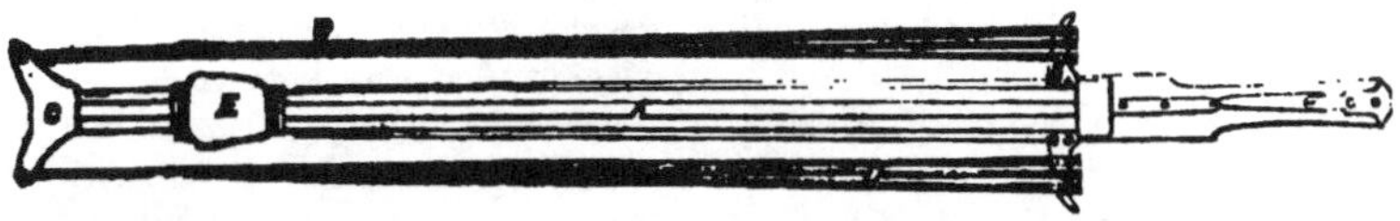

The "Patent Silent Harpoon Projector," meets every difficulty, inasmuch as it is effective at 10 fathoms, it makes no report, is not affected by wet or change of temperature, nor is it liable to be easily injured, and any part of it may very readily be repaired if it should be damaged. In describing the apparatus, it is important that it should first be clearly understood that neither powder nor air is the projecting agent but Hodges' Patent India-rubber purchases are employed for that purpose. Fig. 1 is a side view of the gun; and fig. 2, a plan. The projector, or gun, consists of a slotted barrel, A, in which a plunger, B, travels freely. On the muzzle of the gun is a crosshead C, to which are attached four of the patent purchases on each side (more may be added, if thought desirable). One of these purchases is shown in its unextended state in fig. 3. The lock, which is of the simplest construction, is provided with a catch which drops into a loop in the tail of the plunger, B, and holds it securely. The plunger being thus fixed, the purchases, DD, are elongated and looped one by one to each side of it, as shown in fig. 2. The harpoon is placed in the gun, the plunger is liberated by pulling the trigger, and the contractile power of the purchases causes them to fly up to the crosshead with great violence, carrying the plunger with them, and thus projecting the harpoon. E is a buffer provided to receive the shock of the plunger. Each purchase requires about the strength of one man to draw it down from the crosshead to the plunger; thus we have a very considerable accumulated power,—the original idea of which was suggested to the patentee by observing the blacks bend down young trees one by one, and secure them to larger baulks of timber in the woods; the power of several trees being sufficient to raise the block to enable them to carry it away. The hand harpoon of the ordinary make is used in this silent projector, thus requiring no extra stock of expensive instruments, and it is more effective at 10 fathoms projected by 8 purchases only, than when thrown by hand within 1 fathom. With regard to injury, the gun itself can receive no damage; the lock is strong and very simple, so that it can easily be repaired on board, in the event of its being out of order; and should the purchases be accidentally injured, they can be replaced by others at a nominal cost.

The "Silent Harpoon Projector" may be used on board yachts and vessels generally for defence, service, or amusement, in throwing lines, killing sharks, etc., etc.

I am, Sir, yours, &c.

WILL. MURRAY.

Sole Licensee.

21, University-street, Bedford-square,
July 25, 1851.

P.S. The "Projector" in the Exhibition is the first ever manufactured, and consequently a very rough and inferior production, being about a third larger and heavier than those now made; which may be seen at Messrs. Brown and Redpath's, Commercial-road, West India Docks.

W. M.

THE GOVERNMENT PATENT LAW AMENDMENT BILL.—"THE EVIDENCE."

(*Continued from p.* 70.)

We come now to the provision contained in the Bill for the previous examination into the novelty of inventions, before patents for them are granted. The weight of the evidence is decidedly against it; a majority of the witnesses, both in number and respectability, being of opinion, with the Protection Society, that it is "altogether unnecessary, of a nature ill calculated to elicit the truth, and likely to subject inventors to grievous risk and difficulty in the establishment of their claims." If a reader were, however, to be guided by the General Index to the evidence (prepared, of course, under the superintendence of the Chairman, Lord Granville, or his prompter, Mr. Webster), he would be led to believe that the testimony *was all the other way.* Under the head of "PATENTS—remedies proposed for, have this item:

"(5.) Previous examination as a upon repatenting of same subject;' opposite to that seven witnesses are red to as being of the opinion indi Now neither in this nor in any other the Index is reference made to the exi of any opposite testimony on the s Here we have the *suppressio veri* and *gestio falsi* most flagrantly combined dence ignored, because it does not s preconceived notions of the managers inquiry, and the false conclusion sugg that because no such counter testim referred to in the index, there was non adduced. Another striking specime of the partial and unfair spirit in whi whole of this inquiry has been cond Nor is this all; for of the seven wit referred to as being of opinion that vious examination" is necessary as "a upon the repatenting of some subject, of the most deserving of respect, Pro Woodcroft, did, in fact, give evidenc directly contrary description. We quote the Professor's own words:

Question.—"Do you adduce thi recent case of Steiner) for the purp illustrating the advantage that would from a commission consisting of sci and experienced persons?"

"*It is the reverse of this;* it is t *that those men who are supposed to b competent, are really quite incompet decide upon these points.*"

Question.—"You think that neithe yers nor scientific men could form a that would be *competent to decide up question whether a patent should be g or not?*"

"I AM MOST DECIDEDLY OF THA NION."

The reader will please to observe th the necessity for this previous examin depends almost entirely the appointm commissioners and examiners; for one were done away with, no decent p could be left for the other. Hen doubt, the pertinacity with which th as been persisted in, in the teeth nce, and, as we shall presently xisting experience on the sub·

ountries in which a preliminary (more than formal) takes place, and the United States. It was mportance, therefore, to ascer- how the system works in those Let us see, then, in what fashion ommittee addressed themselves ch of the inquiry. The only examined, as regards Prussia, 'addinge, who is "a member of n) Patent Commission," and f course, in upholding the use- e Commission of which he is a hy not have examined some less authority?—some of the many ssian engineers and manufac- ample, who are now in this coun- it of the Great Exhibition. Was her Prussians, *not of the Com-* ld have testified that the sys- led with universal dissatisfac- the inventive classes of their ? Yes,—that and no other was With respect, again, to the ' the system in the United Committee did not examine a of the States; but contented ith the evidence of one Paul ge (a Cornishman we believe), had practised as a patent agent ears in America (but of whom, ity, we never heard before), and system of Examiners to "work The Committee might have the evidence on the subject of n Consul in London, which been unimpeachable, and who ubt, have been able to supply copies of the latest official the Washington Patent-office; was the American Consul or the official evidence he could ed, sought after. Why? For loubtless, that any native Ame- have told them that the system ry examination is just as un-

popular among American as Prussian inventors, and because the Official Reports of the American Patent-office prove beyond all dispute (Paul Rapsey Hodge's evidence to the contrary, notwithstanding,) that the system does, in truth, work most abominably? We happen to have in our possession a piece of evidence on the American part of the question, which is of more recent date than even the latest published Official Report of the Washington Patent-office, and which being very much to the point, we shall here lay before our readers. It is the following statement furnished to the *Scientific American* by a late chief-clerk of the Patent-office, of the fees received number of applications made, and number of patents granted during each month for the past year, 1850:

	Cash Received.	Applications.	Patents.
January	8,777 47	239	60
February	7,239 26	176	60
March	8,119 43	196	38
April	6,683 72	177	48
May	7,589 43	196	60
June	8,847 88	191	44
July	6,188 23	161	31
August	6,287 93	174	49
September	6,984 00	181	34
October	6,095 57	166	61
November	6,392 81	165	52
December	7,721 32	199	65
	Dols. 86,927 05	2,193	602

Applications in 1848, 1,628; 1849 1,955; 1850, 2,193. Cases granted in 1848, 607; 1849, 595; 1850, 602. Patents issued in 1848, 660; 1849, 1,076; 1850, 995. Cash received in 1848, dols. 67,576 69; 1849, dols. 80,752 78; 1850, dols. 86,927 05. Cash expended in 1848, dols. 58,905 84; 1849, dols. 77,716 44; 1850, dols. 80,000 95.

The reader will see from this document that of 2,193 applications for Patents made under the previous examination system in the United States, during the past year, only 995 (less than one-half) were rejected. It is but fair to presume that, were the same system adopted in this country, the approved and rejected would be very much in the same proportion. It is further equally fair to presume, from the well-known self-sufficient temper of inventors, and from the want of authority which must necessarily attach to any body of Examiners however selected, that of the rejected applications, one-half at least (say 500) would be made the subject of appeal from the Examiners to the Law-officer of the Crown, (for which provision is made in the Bill).

Behold, then, Inventors! the sea of litigation and trouble on which you are about to be launched by the false friends who have taken your case under their treacherous care.

You have been dazzled into acquiescence in the Bill by the prospect it affords of cheap patents; that is, cheap in the first instance, but dear in the long run; but mark, that those only are to have them who can pass through the ordeal of an inquisitorial and incompetent Board of Examiners. One-half of your number at least will not have the patents they are reckoning upon *at any price*. And of a surety there will be not a few inventors of really good things, by which both the country and themselves would be largely benefited, who will be irremediably the victims of the ignorance and incapacity of the Examiners set over them.

(*To be continued.*)

Second Reading of the Bill. **July 30.**

The Attorney General, in moving the second reading of the Patent Laws Amendment Bill, reminded the House of the complaints made respecting the delay, uncertainty, and expense which attended the operation of the existing patent laws. Much inquiry had, in consequence, taken place on the subject; and, a committee having been appointed in the other House of Parliament, the result had been a bill, the fundamental principle of which was that patents of inventions should be issued to persons who should make discoveries for the benefit of society. Individuals whose views were entitled to considerable weight were of opinion that patents might be altogether dispensed with—an opinion in which he could not concur. The object of legislation ought to be to develope to the utmost possible extent the discovery of inventions calculated to meet the wants and promote the comforts and enjoyments of the community, and, though the desire of benefiting their species, and of obtaining reputation, might be strong incentives, no one acquainted with human nature would deny that these motives must be far less efficacious than that which arose from the expectation of individual reward. Long labour and much expense were to be incurred in making discoveries. Nay, more, to perfect and introduce them required further labour and expense. The bill proceeded on the principle, that when persons invented what was new and useful to society, they should have the reward which the patent laws gave. The most appropriate mode of giving that reward was by giving a monopoly for a limited period. If the notion lately started were

to be acted on, that there should be no in
adjudication on the question whether a man
have a patent or not, Westminster could
more effectually compensated for the de
business which was said to prevail. The
ought rather to be guarded against the ass
of rights in regard to inventions, where th
tions were neither new nor useful; and, i
cases where rights were conceded to in
these might be protected by a short and ch
cess. The next question was, by what syst
would determine the point whether a pers
entitled to patent rights. All were agreed t
present system was a bad one; for as matte
stood, every application for a patent mu
through no fewer than seven different office
It was proposed *in the bill that there should*
common patent for the three kingdoms, w
trouble and expense of application would
materially reduced. (Hear.) It was also p
to distribute the sum expended into three s
payments—one when the patent was first g
another at the end of three years, and the t
largest at the end of seven years. Th[illegible]
poor men an opportunity, wh[illegible] they [illegible]
present possess, of protecting inventi[illegible]
time as they might be able to deriv[illegible]
from them, and at a cost adap[illegible]
It was proposed then that the [illegible] pay[illegible]
be limited to 20*l*., with addition[illegible]
the end of three years the pate[illegible]
renewal for four years more [illegible] payment
with 10*l*. stamps. This gave h[illegible] a patent
seven years, and at the end of that period, w
had sufficiently tested his patent, and fo[illegible]
would benefit the public and hi[illegible]self, he
tend his right for seven other ye[illegible]rs by [illegible]
and a stamp duty of 20*l*. (Hear.) [illegible]
be a reduction in point of exp[illegible]nse from
about 170*l*. (Hear, hear.) As they were a
abolish five out of the seven stages through
applicants must now pass, there wou'd be c
sation given to certain parties, and [illegible] oblig
to take a large margin at starting; but he
that as they felt their way, and [illegible] the ne
expenses by actual experiment they would
in time to lower still more th[illegible] amount of t
of procuring a patent. (Hear, hear.) The
the mode of determining whe[illegible]er a p[illegible]rty w
tled to a patent or not, th[illegible] duty [illegible] h
devolved on the law-officers [illegible] the C[illegible]own.
there were two objections to [illegible] trib[illegible]nal,
thought them both well found[illegible] In th[illegible] first
the tribunal was a secret one though [illegible]
considered that this secrecy [illegible] not ar[illegible] fr
will or inclination of the law-[illegible]ficers, but f
necessity of the case; for it [illegible] made [illegible]
consequence of applicants be[illegible]g anxio[illegible]s to
from objectors and rivals wh[illegible] the nature o
inventions were till after th[illegible] patent was g
(Hear, hear.) To do away wit[illegible] this it was pr
to enable a man, at the same [illegible]me that he
for a patent, to lodge in the [illegible] of the c
sioners to be appointed unde[illegible] the bill a prov
specification—that was, to s[illegible] statemen
sponding with the deposit no[illegible] required by t
officers, of the precise nature [illegible] invention
protected. As soon as he dep[illegible] that pro
specification, they would g[illegible] the po
patent for six months, and h[illegible] therefor
no risk in using his inventio[illegible] making it
to the public. This protec[illegible] ould be f
and in certain cases nine mon[illegible] and it w
found a most valuable portion [illegible] the bill.
other hand, when a man obje[illegible] to a pat
must lodge the particulars of [illegible] objection
a way that the case could be f[illegible]
parties entitled to decide. [illegible]
objection to the present trib[illegible] that th
officers might be incompet[illegible] the
aright for want of proper scie[illegible] knowledg
therefore it was proposed to [illegible]nstitute a bo

omposed of persons of scientific know- ar, hear.) To them, in the first in- uestions about patents would be refer- any case of dispute, and the decision d not being satisfactory, it was pro- rry the matter before the law-officers n. In such cases the law-officers would report of the examiners, and the ulti- n would lie with them. (Hear, hear.) of application would not be increased nce of this arrangement. The law- perfectly willing to make a personal that respect, in order that the public a the full benefit of the proposed ar- (Hear, hear.) He omitted to state, at benefit of the provisional specifica- ch he had referred would be, that it e a man to bring his invention into the l to get the assistance of capitalists, slightest apprehension — a benefit r the present system, was wholly im- Hear, hear.) The next point provided ill was the proper classification of all filing and copying of all specifications, for the publication of a register of of inventors, all of which should be ection at any time. (Hear.) At pre- en happened that when a man had time, his patience, his money to an nd imagined that he was on the eve of fruits of his labours, he found that se had anticipated him, that a patent in force for the particular invention he and that therefore all his labour and it for naught. (Hear.) Now, in order s, two provisions were introduced into d it was also provided that when s had to a court of law, the court enter- question should be invested with an risdiction, so that a court once seized ter would be entitled to decide it on inciples, without putting the parties to of further litigation. It was also pro- if an invention had been practised in a itry that circumstance should be fatal As the law now stood, if a man intro- vention that was new in this country ecure a patent right for it. This was bsurd; but, on the other hand, he rong to say that because an invention scovered in some remote country the mind led him to the same dis- should have no claim to any reward at two extremes, he should say that a ought an invention into this country ire some right in it, but not a right t of the original inventor. He thought, middle course should be taken. Leav- use, however, to be decided upon in he would not now further discuss that question. He had now stated the re of the bill, and, though it might be as a great deal to be done hereafter in reforming the present system, yet he ouse would accept the measure as one ; direction, and a great improvement isting mode. (Hear, hear.) The hon- learned gentleman then moved the ng of the bill.

RICARDO complained that they were to consider the bill before they had had nine the evidence taken in committee bject. He agreed with the noble lord esident of the Board of Trade, that so fording facilities for securing patents, to proceed upon the ground that the iple of the system was wrong. (Hear, Attorney-General said the fundamental the bill was to give a stimulus to in- affording inventors a monopoly for a ber of years. Such a course might be inventors, but, at all events, the mono- poly itself thus granted was an evil to the public. He was supported by the greatest authorities in the view he had took of this question The whole principle of the patent system was bad, and ought to be abolished. (Hear, hear.) All the great inventions and discoveries by such men as Liebig, Brunel, and Stephenson were made without any encouragement from patents; and it was only small inventions in the making of sealing-wax, great-coats, or paletots, that seemed to require the protection of a patent. Such inventions as those of paper, of glass, of printing, were all made without the stimulus of patents (Hear, hear.) A great deal had been said in that House about the monopoly of landowners, but he thought they should carry out the principle a great deal further and apply it to the question now before the House. (Hear.)

Mr. M'GREGOR did not admit that the bill was an improvement on the law of patents. He denied that that law conferred any monopoly. He was far from wishing to give any great extension of the right, but still he maintained that a man was entitled to the fruits of his intellect as much as to any other property he might possess. He should support the bill.

Mr. LABOUCHERE observed, that it was quite unnecessary for him to occupy the time of the House after the very able and lucid statement which had been made by the Attorney-General. There were important reasons why the present system relating to patents should not continue any longer. In consequence of the Great Exhibition now taking place in this country, there were a number of inventions to be registered provisionally this year, and it would be monstrous to oblige the inventors to register their inventions in the expensive and complicated way which was required by the existing law, when it could be done in the simple and economical manner which was provided for by this bill. The present system was manifestly defective, and they had no alternative but to amend it, or sweep it away altogether. He would not express an opinion whether or not it was desirable to abolish the patent laws, but he believed the country was not prepared to pursue such a course; the only mode of proceeding, therefore, was to effect an improvement of the system.

Sir D. L. EVANS supported the bill.

Mr. ROUNDELL PALMER expressed his regret that the right honourable gentleman the President of the Board of Trade should not have declared a more decided opinion on the principle of the law. He (Mr. R. Palmer) distinctly dissented from the views of the honourable member for Stoke-upon-Trent (Mr. Ricardo), and most cordially concurred in those expressed by the Attorney-general. It appeared to him to be a fallacy to say that by the law of patents they were granting a monopoly for a certain number of years, to the prejudice of the public. What was it that was taken away from the public? It was that which the public would never have possessed but for the invention of the individual to whom the supposed monopoly was given. One of his own constituents, Sir W. Snow Harris, the inventor of conductors of lightning for ships, by means of which there had been a great saving of human life, had never derived any benefit from his invention, in consequence of the impediments occasioned by the law of patents. He hoped the merits of that gentleman would be considered, and that a more substantial remuneration would be made to him than that which he had hitherto received. He thought the present bill hardly went far enough in favour of the principle on which it was based. If the principle of the bill was in itself good, they might in his opinion go much further, and take the consequences of it. It was said that the present law did not act as a check to frivolous inventions; but the principle of the law was not to inquire whether the inventions were frivolous or not; the question was, whether they were new and useful, and such as the Attorney-general ought not to refuse a patent for, unless there were grounds to believe that they infringed upon some previously existing right.

Mr. Alderman SIDNEY remarked on the desire which had been expressed to regard this measure as one of economy, and observed that in the year 1847 there were 493 English patents granted, 168 Scotch, and 76 Irish, and that the fees paid to the Government and the official personages amounted to 87,640*l.* He (Alderman Sydney) presumed they would, under the improved state of the law, not have less than 742 patents registered next year; and, assuming that they were to be had at the reduced cost proposed by this measure, the amount would be, not 87,640*l.*, but 129,850*l.* (Hear.) He should consider it his duty to propose certain resolutions in committee with the view of modifying the fees. He did not think they had any right to tax an inventor 175*l.* for the completion of a patent. In many continental states an inventor would not have to pay more than from 2*l.* 10*s.* to 20*l.*, for which he would obtain real protection, while in this country if a patent was infringed the patentee was in many cases ruined in maintaining his right.

Mr. W. WILLIAMS wished to know from what source the commissioners who were to award compensation under this bill were to be paid? As to the necessity of the present measure, no doubt could be entertained about it. He understood there were not less than 700 persons waiting to take out patents under the cheap plan to be established by this bill.

Mr. MUNTZ had been for some twenty years engaged in litigation on the subject of inventions; he felt therefore competent to give an opinion upon the subject. He did not consider the patent law to have any connection whatever with a monopoly. It was a sort of recompense to a man for his ingenuity. No greater blow could be struck against the inventive genius of this country than by the abolition of the law of patents.

The SOLICITOR-GENERAL observed, that under the present measure the patentee, instead of paying 175*l.* for a patent for England, would obtain for 25*l.* a patent for the three kingdoms, the expense of which would previously have been about 300*l.* This patent, obtained for 25*l.*, would be in force for three years, and in that time the inventor would be able to ascertain whether the patent was worth keeping or not. The consequence would be, probably, that at least one-half of the patents would be relinquished at the end of the third year. The patentee might, if he chose, renew the patent for four years more, on the payment of 40*l.*, in addition to the stamp duty, which would be 10*l.* By the expiration of the first three years, however, the value of the invention would have been fully tested. It was well known that a great number of patents were useless for the whole period of 14 years, because in the course of time improvements were made and new patents were taken out for such improvements, which entirely superseded the old patents. The bill would enable an inventor to take out a patent which would be in force in the three kingdoms for a period of seven years, at a cost of 75*l.*, instead of 300*l.*, which would be the expense at present. The honourable member for Stafford (Mr. Alderman Sidney) had complained of the heavy fees paid to official personages, which he said amounted to 80,000*l.*, and he (the Solicitor-General) did not know whether the hon. gentleman meant the House to infer that the law-officers of the Crown received that amount. A very large proportion of that sum—from 45,000*l.* to 50,000*l.*—was expended upon stamps, or went to the consolidated fund; and the sum received by the law-officers upon each English patent, which cost 109*l.*, was between 9*l.* and 10*l.* When the cost of a patent was reduced to 25*l.*, he thought it would be found that there would be a very considerably *diminished payment to official personages.* He considered it most important that this bill should be passed as speedily as possible, for no less than 500 persons had taken out protections under the Designs Act passed in the present session for the protection of designs and inventions deposited in the Great Exhibition, and he believed that a large number of the persons so protected were waiting to apply for patents as soon as this measure became law.

Mr. WAKLEY thought they had arrived at a period of the session when they could scarcely legislate satisfactorily upon this very important subject. There was no doubt the present state of the law was extremely defective, but he was not quite satisfied that this measure was the best that could be proposed. He considered that the principle which should govern legislation with reference to patents and to copyright ought to be the same. The process with regard to copyright was as simple as possible, and he could not understand why the same simplicity should not be observed with respect to inventions. The machinery contemplated by this bill was by no means simple, and it seemed that even when a person had got his patent he would not have secured his right. The Commissioners to be appointed under the bill were to nominate examiners, who would carry on their inquiries in secret. That would, he thought, be a most unsatisfactory proceeding, and it would be infinitely better to refer examinations to one man, who should be compelled to give his decision and his reasons for it in public.

The bill was read a second time.

In the speech of the Attorney-general there is a statement which, as it seriously affects our own character for veracity, we are particularly called upon to notice. "It was *proposed*," he said, "*in the Bill, that there should be one common patent for the Three Kingdoms.*" Now the reader may remember that, among the "Additional Reasons" which we gave for concurring with the Protection Society in disapproving of the Bill, there was the following (see *ante*, p. 52):—"*The Bill proposes to perpetuate what is universally allowed to be one of the worst features of the existing system—the granting of separate patents for the three principal divisions of the United Kingdom, England, Scotland, and Ireland.*" Let us now see which of these contradictory statements is the true one—Mr. Attorney-General's or ours.

We have now before us a copy of the Bill as it came from the Lords to the Commons, and which copy came into our hands *direct from Lord Granville.* Clause XII. is in these words:—

"The Clerk of the Patents, so soon after the receipt of the said warrant as required by the applicant for the said Letters Patent, shall issue *three* transcripts of Letters Pa-be passed (1), under the Great Seal; Seal appointed to be used instead of eat Seal of Scotland; and (3), the eal of Ireland respectively: and the Patent (1) to be passed under the eat Seal shall relate and extend to d, the dominion of Wales, the town wick-upon-Tweed, the Channel Is-and Isle of Man; and the Letters (2) to be passed under the Seal ap-to be used instead of the Great Seal land shall relate and extend to Scot-and the Letters Patent (3) to be under the Great Seal of Ireland shall nd extend to Ireland."

Attorney-General's statement, there-thus proved to be the false one— s. The Bill as passed by the Lords, opose to perpetuate the old stupid of three separate patents. It may the Attorney-General intends to get ered in Committee, and to have the atents consolidated into one; but if would have been only fair to the to make them acquainted with the ate of the case. His description of l, of which he proposed the second , was not a true description of it; far, therefore, the second reading said to have been obtained under pretences. Besides, supposing such ment should be introduced in Com-, the Bill must go back to the Lords ir concurrence in it; and how can torney make sure that the Lords will ?

he Bill in Committee. July 31.

CHANCELLOR of the EXCHEQUER moved is bill be committed *pro forma*.

. THESIGER expressed his regret that a of such importance should be pressed at period of the session, when it was impos-e subject could receive the attention it d, and when many hon. gentlemen who doubtless have taken part in the discussion sent. He thought there were many serious ns to the details of the measure.

M'GREGOR said, that though this was far ing such a measure as he would like to see upon the subject, he believed it would effect aterial improvements in the existing law. esent state of the patent law was a disgrace country. In France there was scarcely any in obtaining a patent, and in Prussia no-hatever was paid by way of tax, provided invention were an original one. He trusted progress of the bill would not be stopped.

. L. EVANS approved of the bill. It had uch considered, not only by the Lords, but Board of Trade also, and it was too hard now ppose any objections. It was calculated to ameliorate the present patent system, s loudly called for. He begged to express his thanks to the two honourable and learned gentlemen, the Attorney and Solicitor-General, who had taken up the subject with great earnestness, and had pressed it forward, although it would probably affect their incomes to the extent of 1,000*l*. or 2,000*l*. a year.

The SOLICITOR-GENERAL deprecated discussion, which would longer delay the bill. There was at this time 509 persons who had taken out protection for articles in the Exhibition, and he believed that more than one half of those persons were waiting with great anxiety for the purpose of knowing whether they could take out their patents under this bill, and scarcely a day passed in which he did not receive applications from patent agents on the subject. (Hear, hear.)

Mr. CARDWELL hoped that the bill would be allowed to proceed at once, on the understanding that there should be a searching inquiry into the whole matter before a select committee next year.

Lord J. RUSSELL said that he should take into consideration the suggestion of the honourable member for Liverpool. In the mean time he asked the House to accept the bill as an improvement on the patent law. (Hear.)

SPECIFICATIONS OF ENGLISH PATENTS ENROLLED DURING THE WEEK ENDING JULY 24, 1851.

CHARLES ROPER MEAD, of Old Kent-road, mechanical engineer. *For improvements in apparatus for measuring gas, water, and other fluids.* Patent dated January 21, 1851.

The improvements claimed under this patent comprehend,

1. A method of supporting or counterbalancing the tumbling lever of wet gas-meters by a mechanical arrangement for bringing it into a vertical position, and whereby the resistance offered to the gas in working the meter is greatly reduced.

2. The application to meters of that description in which water or other fluid is employed as the measuring medium of two or more vibrating measuring vessels, communicating motion by one or more revolving shafts to two or more slide valves.

3. A method of actuating and arranging a series of drums or wheels, having figures on their circumference, and placed side by side in such manner that the quantity of gas, water, or other fluid passing through the meter or machine may be readily ascertained.

4. The use of a vibrating vessel having a bucket at each end for the purpose of tilting over the same, and a partition in the centre for changing the direction of the fluid from one side to the other of the measuring vessel, by which the use of slides for admitting and discharging the fluid is dispensed with.

WILLIAM BURGESS, of Newgate-street, gutta percha dealer. *For improvements in machinery for cutting turnips and other substances.* Patent dated January 21, 1851.

In the machine which forms the subject

of this patent, the knives are arranged horizontally in a cutter frame, to which reciprocating motion is communicated by a crank worked by hand or power. The bottom of the hopper is provided with several transverse blocks or stops, against which the turnips or other substances rest whilst being cut. When the turnips are required to be divided into small pieces vertical knives are attached to the front of the cutter-frame, which come into operation before the horizontal knives make their cut. In addition to the stop-pieces above mentioned, there are also plates, some of which are perforated to admit of the passage of stones or pebbles, and which prevent the turnips passing through the machine until they have been subjected to the action of the cutters.

Claim.—Combining mechanical parts into a machine for cutting turnips and other substances in manner described.

JOHN RANSOM ST. JOHN, of New York, engineer. *For improvements in the process of and apparatus for manufacturing soap.* Patent dated January 21, 1851.

In carrying into effect that part of his invention which relates to "the process of manufacturing soap," the patentee prepares a lye from soda-ash and lime, using 100 parts of the former to 40 parts of the latter, with sufficient water to bring the mixture to 10° proof. To every pound of this lye, in a hot state, he adds 6 lbs. of tallow or lard stearine, or its equivalent in palm oil, cocoa-nut oil, or lard, and boils the whole well, skimming off the upper portion from time to time, and thus obtaining "fullers' soap" for subsequent treatment. Or, instead of adopting the above process, the patentee employs the ordinary "fullers' soap" of commerce. In either case, he heats the "fullers' soap," and adds gradually to every 100 lbs. thereof about 50 lbs. of Paris white, or of common whiting, or silicate of magnesia, previously well mixed with 15 gallons of water. He then boils and stirs the mixture, and adds to it 12 lbs. of common resin, and continues the application of heat until the resin is dissolved, taking care not to raise the temperature to the boiling point lest bubbles should be formed, by which the soap would be deteriorated: or the resin may be used as the last ingredient in preparing the fullers' soap. He next mixes with the heated preparation 14 lbs. of sal-soda, and after stirring for a short time, 2 lbs. of biborate of soda, or borax, or its equivalent of tincal or other compound of borax, and finally a mixture of 3 lbs. of starch or boiled potatoes with 1 lb. of water, in the proportion of 20 lbs. of the mixture to every 100 lbs. of soap. In order to ascertain whether the boiling has been continued long enough, a trial should now be made of the soap by pouring some of it into a vessel—when, if it sets or hardens quickly, the whole of the contents of the soappan may be run off into shallow coolers, or into frames; in which latter case it may be found necessary to crotch or stir the soap during cooling, to prevent the separation of the ingredients. In order to facilitate the hardening, 1 lb. per cent. of rice flour may be added to the soap while in the pan; but when silicate of magnesia is employed in its preparation, the addition of rice flour is not necessary, as this material possesses in it the requisite hardening properties.

The "improved apparatus" consists of a pan or kettle, with a casing or jacket, between which and the bottom of the pan the steam employed for boiling circulates. A number of vertical pipes, closed at the top, and communicating at bottom with the steam space, rise into the kettle, and as the steam occupies the interior of these pipes, the boiling operation is facilitated, and the heat uniformly diffused through the contents of the kettle. A horizontal arm, mounted on a vertical shaft in the centre of the kettle, to which rotary motion is communicated by any suitable means, and provided with scrapers extending downwards to the bottom of the kettle, and so arranged as to pass between the vertical pipes, is employed to effect the stirring.

Claims.—1. The combination of materials above set forth, for the purpose of producing a soap which may be used with hard or soft, fresh or salt water; and particularly the use of biborate of soda, borax, tincal, or any substance containing these elements, in combination with the above specified or other materials.

2. The employment of the apparatus above described, when used for boiling soap.

AUGUSTE LORADOUX, of Bedford-street. *For certain improvements in machinery or apparatus for raising water and other fluids.* (Being a communication.) Patent dated January 21, 1851.

The improvements claimed under this patent have relation to pumps, and consist,

1. In certain means "to raise liquids without the help of a piston in producing a vacuum by means of the oscillating tube and the speed required in working."

2. In the application of an air-chamber to pumps constructed on this principle.

3. In a peculiar formation of barrel, in which a funnel-shaped piece is employed, as a connection between the upper and lower parts thereof.

ROBERT WILLIAM SIEVIER, of Upper Holloway, gentleman. *For improvements in weaving, and printing or staining textile or fabrics.* Patent dated January 51.

ims.—1. A method of forming the l surface of looped fabrics, such as ls carpets, terry velvets, and other r goods or fabrics from the weft s, by means of longitudinal wires or quivalent means, in combination with hreads, which, by being made to de- one after the other in rapid succession, l of simultaneously, as is usually the ith warp threads, thereby draw off he quill or bobbin in the shuttle a nt quantity of weft or shoot to form ps on such longitudinal wires.

A method of cutting the pile of looped . [The cutters employed are plain f steel with sharpened edges, which nted, at suitable distances apart, on ontal spindle extending in a direction rsal to the fabric, and revolve in s formed in the wires, being sup- in bearings formed in the free ends f. The cutters are worked by the of the loom, so that the loops of the are cut by them immediately after ormation, and the fabric then passes taking-up roller.]

A method of printing patterns on warp s or woven fabrics by the employment, it purpose, of a series of plates va- perforated, according to the number des of each colour which exist in the to be printed.

XANDER SAMUELSON, of Banbury, tural implement-maker. *For im- ents in apparatus for cutting turnips, , mangold wurzel, and other vege-* Patent dated January 23, 1851.

se improvements are based on the es known as Gardner's turnip cut- n which the knives are made of a form, and are arranged in one or two a circular barrel. They consist,

n the combination of wrought or cast rames and hoppers (both of which have been hitherto made of wood with knives and barrels of the de- n above mentioned.

n the application to the interior of lame boards" of a projecting piece, which the vegetable rests while in of being cut, and which ensures the of the last piece.

n a method of constructing and ap- the hoppers and hopper sides to such es with axes and hinges, for the pur- f allowing the hopper to be raised, s sides thereof opened, for cleaning or ng the barrel.

n arranging or adjusting the knives ting one of two widths of pieces, in such manner that no obstruction shall be offered to the free working of one set of knives, whilst the other set is not in action.

JOSEPH BUNNETT, of Deptford, engineer. *For certain improvements in public carriages for the conveyance of passengers.* Patent dated January 23, 1851.

Mr. Bunnett describes and claims—

1. A method of providing additional outside seats, easy of access, for omnibus passengers, without elevating them above the roof, by slightly recessing the sides and roof of the omnibus. Also, the application of guards, composed of a metal frame and panel of perforated zinc, to divide the seats of omnibuses into a given number of spaces, thus separating the passengers from each other, and preventing robbery.

2. The application of leather in its unprepared state, or prepared by immersion in melted tallow, and subsequent compression, for making the bearing parts of omnibus and other carriage axle-boxes.

3. A method of preventing lateral shaking of railway carriage axle-boxes by the introduction between the parts of which the axle-box is composed, of a wedge-shaped piece acted on by a screw or other mechanical contrivance.

Specification due, but not Enrolled.

SAMUEL CLIFT, of Bradford, near Manchester, manufacturing chemist. *For improvements in the manufacture of soda, potash, and glass.* Patent dated January 21, 1851.

WEEKLY LIST OF NEW ENGLISH PATENTS.

James Timmins Chance, of Birmingham, gentleman, for improvements in the manufacture of glass. (Being a communication.) July 28; six months.

Richard Lloyd, of Paris, in the republic of France, engineer, for improvements in steam engines and in treating steam. (Being a communication.) July 28; six months.

Peter Robert Drummond, of Perth, for improvements in churns. July 29; six months.

John Workman, of Stamford-hill, Middlesex, fumist and furnace-builder, for improvements in the manufacture of bricks, tiles, and other articles made of like materials. July 31; six months.

Charles Barlow, of Chancery-lane, London, for improvements in saws. (Being a communication.) July 31; six months.

Victor Lemoign, of Cette, Departement de l'Herault, France, for certain improvements in rotary and other engines. July 31; six months.

Charles Cowper, of 20, Southampton-buildings, Chancery-lane, Middlesex, for improvements in locomotive engines, and boilers, and carriages, part of which improvements are applicable to other similar purposes. (Being a communication.) July 31; six months.

James Whitelaw, of Johnstone, Renfrew, North Britain, engineer, for certain improvements in steam engines. July 31; six months.

Joseph Mansell, of Red Lion-square, Middlesex,

manufacturing fancy stationer, for improvements in ornamenting paper and other fabrics. July 31; six months.

Charles Perley, of the City of New York, United States of America, machinist, for certai useful improvements in the construction for nautical and general purposes. Jul months.

LIST OF SCOTCH PATENTS FROM 22ND OF JUNE TO THE 22ND OF JULY, 1

John Swindells, of Manchester, Lancaster, manufacturing chemist, for certain improvements in obtaining products from ores and other matters containing metals, and in the preparation and application of such products for the purposes of bleaching, printing, dyeing, and colour making. June 25; four months.

John Emmanuel Lightfoot, of Broad-oak, Accrington, Lancaster, calico printer, and James Higgin, of Cobourg-terrace, Stratford-road, in the same county, chemist, for improvements in treating and preparing certain colouring matters, to be used in dyeing and printing. June 26: six months.

Robert Hayes Easum, of Commercial-road, Stepney, Middlesex, rope maker, for improvements in the manufacture of rope. July 1; six months.

George Frederick Muntz, the younger, of Birmingham, Warwick, gentleman, for improvements in furnaces applicable to the melting of metals for making brass, yellow metal, and other metals. July 2; six months.

Thomas Allan, of Edinburgh, gentl certain improvements in electric telegrap apparatus connected therewith. July 2; si

Thomas Hawkins, of Inverness-terrace, road, Bayswater, Middlesex, oilman, for ments in brushes. July 16; six months.

John Brazil, of Manchester, Lancaste man, for certain improvements in dyeir the preparation of dye woods. July 21 · si:

John Platt, of Oldham, Lancaster, engi Richard Burch, of Heywood, in the sam manager, for certain improvements in weaving. July 21; six months.

Percival Moses Parsons, of Robert-str phi, Middlesex, civil engineer, for impr in cranes, capable of being used on railv in parts of railways. July 21; six montl

WEEKLY LIST OF DESIGNS OF ARTICLES OF UTILITY REGISTERED.

Date of Registration.	No. in the Register.	Proprietors' Names.	Addresses.	Subjects of Desi
July 23	2891	Benjamin Nickels, jun.	Albany-road, Camberwell	Draft or chess board
26	2892	John Gatliff	King's Arms-yard, Moorgate-street	Shawl and other pin
28	2893	George Holcroft	Manchester	Steam generator.
30	2894	John Bellerby	St. George's Saw-mills, York	Cart.

WEEKLY LIST OF PROVISIONAL REGISTRATIONS.

July 23	265	Gregory Kane	Dublin	Cabinet to contain furniture.
25	266	W. S. Adams	Haymarket	Sponging pan or bat
29	267	W. & J. Harcourt	Birmingham	Portable cylindrical case.

CONTENTS OF THIS NUMBER.

LONDON: Edited, Printed, and Published by Joseph Clinton Robertson, of No. 166, Flee in the City of London—Sold by A. and W. Galignani, Rue Vivienne, Paris; Machin Dublin; W. C. Campbell and Co., Hamburg.

Mechanics' Magazine,
MUSEUM, REGISTER, JOURNAL, AND GAZETTE.

No. 1461.] SATURDAY, AUGUST 9, 1851. [Price 3*d.*, Stamped, 4*d.*
Edited by J. C. Robertson, 166, Fleet-street.

BLACK'S PATENT FOLDING MACHINE.

Fig. 1.

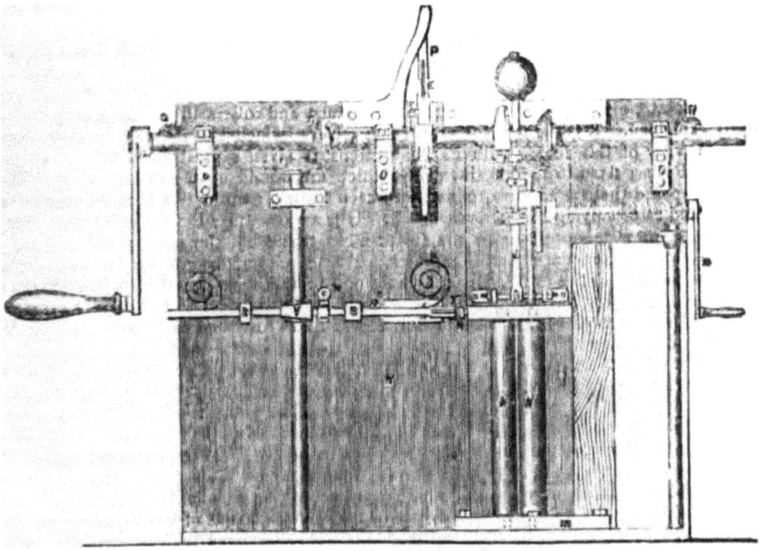

Fig. 2.

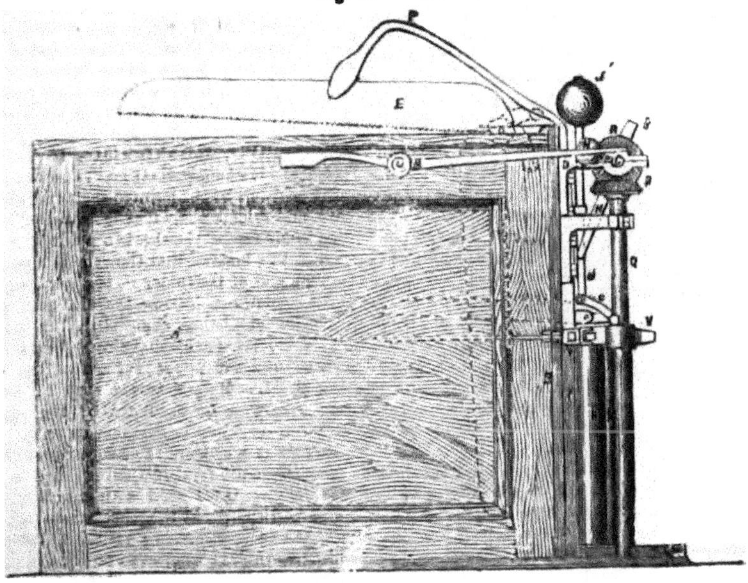

BLACK'S PATENT FOLDING MACHINE.

(Patent dated November 7, 1850. Specification enrolled May 7, 1851.)

A machine constructed according to this patent, for the purpose of book-work, has attracted a great deal of attention at the Exhibition, and been deservedly much admired for its simplicity and efficiency. Figs. 1, 2, and 3, of the accompanying engravings, are views of this machine; fig. 1 being a front elevation, fig. 2 an end elevation, and fig. 3 a plan. The following description of it we extract from Mr. Black's specification:

A A is a box or case, which forms the framework; B a metal plate, which forms one end of the box, and serves as a basement to which the moveable parts of the machine are attached. C is a main shaft, which has its bearings on the brackets D, D, D, and, when made to rotate, gives motion to certain folding blades and rollers in the manner to be presently explained. E is the first folder or folding blade, which has its axis in the brackets F F. A view of this folder, with its connections, is given separately in fig. 4. On the brackets F F are fixed two spiral rings G G, which are wound upon the axis of the folder, and so disposed that they have a tendency to keep the blade up in the position which it is represented as occupying in figs. 1, 2, and 3. H is an arm affixed to the main shaft, which is placed immediately opposite to the folder E, so that when the main shaft revolves, it comes against the short arm J of that folder, and causes it suddenly to assume the position indicated by dotted lines in fig. 2. The outer end of the short arm J is furnished with a friction roller, for the purpose of enabling the two arms H and J to slide freely over each other. The movement of the folding blade E, which has just been explained, produces the first fold or doubling of the sheet of paper, with the aid of the following supplementary arrangements. In the top of the box or case A, and immediately underneath the blade E, there is formed a long slot K K, which is continued down to the bottom of the box by means of partitions inside, so that there is formed thereby a chamber of the same length and depth as the interior of the case A, but of only about one-fourth of an inch in width. The paper to be folded is placed by hand upon the upper surface of the lid of the case and underneath the folding blade E, with the line in which the fold is to be made immediately below the blade, and consequently over the slot K K, in which position it is held from shifting during the interval of time intervening between the removal of the hands of the attendant and the coming down of the folding blade, by means of two sharp-pointed instruments L L, which project about one-sixteenth of an inch above the top surface of the case, and upon which the paper is slightly pressed by the finger of the person supplying the sheets to the machine. The pointed instruments L L are connected to two levers M M, which are centred upon axes affixed to the sides of the case; the outer or free ends of these levers are loaded with counterbalance weights N N, which serve to keep those ends of the levers down upon the main shaft upon which they rest, and also cause the points L L to maintain their position above the surface of the case A. Supposing a sheet of paper to be placed in the position directed, the main shaft being in motion, the instant the folder E comes down upon the paper so as just to take hold of it, then the two tappets O O, which are affixed to the main shaft, and immediately under the ends of the levers M M, come against and raise the outer ends of the levers, and thereby cause the projecting portions of the pointed instruments L L to descend below the surface of the top of the case, and release the paper from being longer held by them. The further descent of the folding blade forces the sheet in a doubled state into the thin or narrow chamber inside of the case. The instant that the arm H upon the main shaft passes beyond reach of the arm J upon the end of the folder E, the spiral springs G G cause the folder to fly up and resume its former position, leaving the sheet of paper inside of the narrow chamber: the first or single fold being completed, the tappets O O having also passed the ends of the levers M M, the counterbalance weights again bring the pointed instruments L L to protrude beyond the upper surface of the case, so as to be ready for the reception of another sheet. P is a spring stop or buffer, which serves to counteract the jar which would otherwise be produced by the sudden rising of the folder E by the action of the spiral springs mounted upon its axis. The folder ought to be made smooth on both sides, so that it may leave the folded sheet inside of the chamber very freely; but I have found that when the face or folding edge is serrated somewhat similar to a fine saw, that the folding is effected with much greater accuracy than if that edge is quite smooth. The teeth prevent the sheet from sliding, not only longitudinally upon the blade, but also in the cross direction; casualties, which if allowed to take place during the operation of folding to the extent of even one quarter of an *inch in either direction,* would destroy the usefulness of the machine as applied to "book-work," or the folding of printed papers of any description. Other methods may be adopted for the purpose of securing the proper "register" of the sheets than th e use of the pins LL; as, for instance, lines raised upon the surface of the case, on one side of and parallel with the folding blade. So far as the machine and its action have been described, one fold only has been accomplished, and the sheet is supposed to be now left in the nar-

Fig. 3.

Fig. 5.

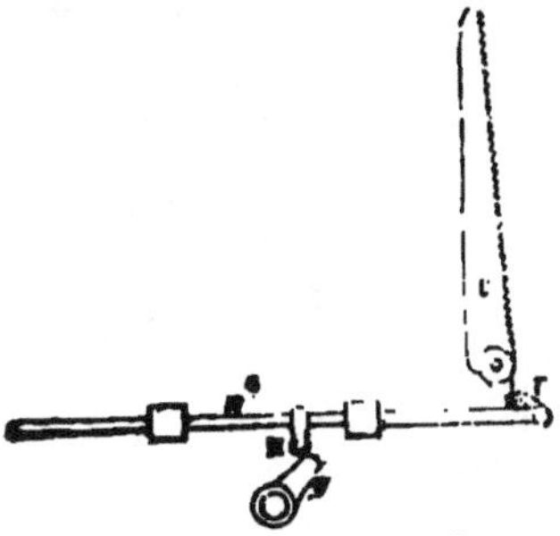

Fig. 6.

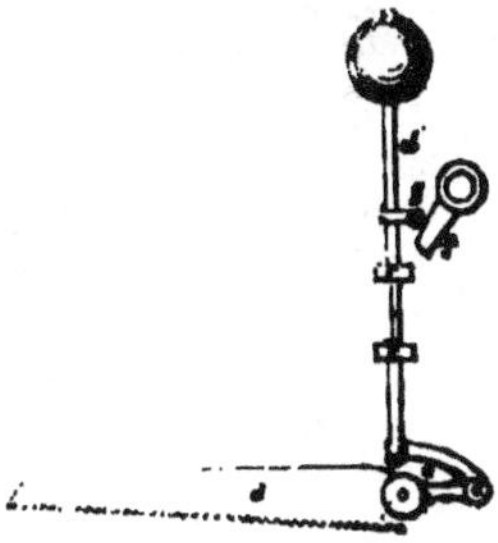

Fig. 4.

row chamber, into which it has been introduced through the slot KK. RR are a pair of bevel wheels of equal size, one of which is keyed on the main shaft, and the other upon an upright spindle Q, whereby the revolutions of the main shaft and the upright spindle are rendered synchronous. R^2 is a parallel sided bar which slides in bearings SS, and is connected by a link T to the second folder or folding blade U, as shown in fig. 3, and also in the separate view with the parts in immediate connection with it given in fig. 5. V is an arm keyed upon the upright spindle Q, which, as the spindle revolves, comes in contact with the arm W, affixed to the bar R^2, and causes the bar to be impelled forward so as to travel from right to left of the machine, which actuates the folder, and causes it to move through one quarter of a circle towards the right. This movement of the second folder takes the sheet from the narrow chamber, previously described, into a second and similar chamber formed in the side of the first narrow chamber; and the instant the arm V loses its hold of the arm W, the bar R^2 and the folder are again brought back to their former positions by the action of the spiral springs XX; the second fold (which forms the sheet into four leaves) is thus completed, the folded paper being left in the second narrow chamber, the sides of which are at right angles to the sides of the first narrow chamber, and parrallel with the upper surface of the case A.

Fig. 6 is a detached view of the third folder *a*, which has its axis or bearings at *bb*, and is connected by a link *c* to a parallel sided bar *d*, which slides up and down within the bearings *ee*. *f* is an arm fixed to the main shaft C, which as the shaft revolves, comes against the arm *g* affixed to the bar *d*, and raises that bar, which again through the intervention of the link *c*, causes the folder *a* to travel in a downward direction through a quarter of a circle, whereby the doubly folded sheet in the second narrow chamber is made to enter or pass through a third narrow chamber formed in the side of the second one, and the throw of the third folder *a*, brings the sheet folded three times (that is into eight leaves) within the grasp of the first of two pairs of rollers *h h'*. These rollers are kept in continuous motion by a pair of bevel wheels *i i*, which are of different sizes, so that the rollers may have their speed increased, so as to free the machine more readily from the papers passing through it. Motion is communicated from the first roller *h'*, upon the spindle of which the bevel-wheel is placed, to the other rollers, simply by the frictional contact of their surfaces. The two outer rollers are covered with cloth, and are made to bear against each other with a considerable degree of pressure by means of regulating screws *kk*, while the innermost pair of rollers are kept from touching each other. By this arrangement the inner pair of rollers grasp the folded paper without laying hold of the folder, and convey it to the outer pair of rollers, where the folded paper is more or less pressed according as the nature of the work may require. Again; the instant the arm *f* clears the arm *g*, the parallel bar *d* and the folder *a* are made by the action of the counter-balance weight *d'* to assume their original positions.

During the interval of time in which these various foldings have been effected, another sheet is applied to the machine by the attendant, and thus a continuous operation of folding is carried on;—one attendant alone being able to supply the machine at a rate varying from one to two thousand sheets per hour. The mainshaft of the machine may be caused to rotate by manual, or by any other power.

In the machine which has just been described, all the parts for effecting the movements of the first and second folders are fixed, and seldom or ever require adjustment after being once placed; the third folder, however, and the parts by which it is immediately put in motion, are connected to a moveable-plate *l*, which slides within two guides *m m*, and is capable of being acted upon by a screw and handle *n n*; so that the folder, the rollers, &c., may be moved towards the right or left, and the position of the folder thus adjusted to suit exactly the printing the margin, or any design, upon the paper to be folded.

Instead of the blade, or folder E, by which the first fold, or doubling of the paper, is produced, being acted upon by springs, so as to bring it into the upraised position ready for another sheet being put beneath it, a counter-balance weight may be attached to the short arm J, which will effect the same object as is obtained by the springs. The return of the second folding-blade may also be caused by means of a vulcanized India rubber, or any other spring, instead of the spiral one represented in the engraving.

THE PATENT-LAW AMENDMENT BILL.

The Government measure has now been pushed through both Houses of Parliament;* but with a distinct understanding, specially assented to by Lord John Russell,

* Contrary to all expectation the Bill has *been lost at the eleventh hour.* On the Bill going back to the House of Lords (on Thursday evening) to obtain their assent to the Schedule of charges, and to the different amendments made in the Commons, the

that "there shall be a searching inquiry into the whole matter before a Select Committee next year."

Why this understanding? Why, but that the inquiry which took place before the House of Lords is (as we have been for some time contending), altogether unsatisfactory? and that the Government measure would have been in danger from the opposition growing up against it, if this pledge of early reconsideration had not been given?

We now publish the entire Bill, with the exception only of a Schedule of forms.

The alterations made by the Commons in the measure since it came from the Lords have been considerable. Of the original clauses some have been altogether struck out and others greatly modified, while no less than twenty-nine new clauses have been added.

The first difference which we notice regards the point on which we were at issue with the Attorney-General. The Bill as it came from the Lords did, as we stated, propose to perpetuate the old practice of issuing three patents for the United Kingdom; and Sir Alexander Cockburn did *not* describe it correctly to the House when he said that it would, on the contrary, consolidate them. We charitably conjectured that it might have been Sir Alexander's intention to have the Bill altered in Committee to suit his speech; and this turns out, in fact, to have been the case. The three patents perpetuation clause (a pet crotchet of Mr. Webster's, by-the-bye,) was cancelled, and a clause of consolidation substituted; whereby it is enacted, that in future one patent shall suffice for *the whole* of the United Kingdom of Great Britain and Ireland, the Channel Islands, and Isle of Man, including also, if the Crown shall see fit, Her Majesty's Colonies and Plantations abroad. We consider this so great a benefit, that for the sake of it we should have been contented to put up (for a time) with whatever else there may be objectionable in the Bill.

The proposition to exclude the Colonies has, it will be observed, been abandoned; and matters left in regard to them in *statu quo*.

The preliminary examination clauses remain much as they were, but with two or three notable additions.

All applications for patents are to be advertised in the *Gazette*, and "any person having an interest in opposing" may oppose; lodging with the Commissioners "particulars in writing of their objections." Of course this puts an end to the *whole of the old caveat system*. We agree with the Patentees' Protection Association in regarding this system of advertising for opposition as likely to expose inventors "to great risk and difficulty." Once it is made known to all the world that a person is occupied with some improvement in any branch of art or manufactures, nothing will be easier than to learn all about it from his servants or intimates — enough, at least, to make unprincipled opposition easy, and a forced participation of profits practicable.

The strength of this objection would be lessened if the Commissioners would direct the subject of the application only to be advertised, *without the name of the applicant*.

According to the Bill, as it came from the Lords, the Examiners were to inquire into "the novelty" only of the invention; but the Attorney-General proposed that the words "and utility" should be introduced. On the motion, however, of Mr. Muntz, most ably seconded by Sir James Graham, the addition was struck out. We are not sure, however, that this will wholly remove the point of utility from out of consideration in the grant of patents, for we observe that there is a new form of declaration prescribed in the Schedule, which may

---

Lord Chancellor stated that he had not been consulted on the measure, though it materially affected the office of the Great Seal; and this, as Lord Monteagle said, "put an end to the whole thing." We wrote our present article in anticipation that the Bill would most certainly pass, and we do not now withdraw it because it contains much that will have an important bearing on the *proceedings of next session*.

practically have the very same effect as the cancelled addition.

Hitherto it has been considered sufficient that an inventor should state in his petition and declaration that he believed his invention to be of "public utility," and the Courts have held that the *amount* of that utility was immaterial (*Lund's Treatise*, iv. 17.) But by the new form of petition and declaration, an applicant is required to declare that he believes his invention will be "of *great* public utility." Of course this will offer but small obstacle to inventors, whose geese are proverbially all swans. But the Attorney or Solicitor-General may say: "Here is a petition referred to me, praying for a patent for an invention which is alleged to be of great public utility. Well, I must inquire whether it is so or not." And so the whole question, which was supposed to be got rid of, will be opened. We doubt not this discrepancy between the Bill and the Schedule has arisen from oversight, and will be among the things sure of amendment next session. Meanwhile, and as long as these objectionable words exist in the Schedule, an unheard-of difficulty will be thrown in the way of putting patent rights in force. The constant question will be—Is this an invention of *great* public utility? And what answer a jury may give to this, no person can beforehand pretend to say. What one jury may think of great value, another—better or worse informed—may be of opinion is of no value at all; and *vice versâ*. The pirate will then have but to calculate whether the thing pirated is likely to be considered by a jury of the "great" class or not, and to arrange his speculations accordingly. All small things will then be fair game; and he will be only taking advantage of the statutory bounty thus offered, for the first time in the history of our jurisprudence, on petty larceny.

The clause by which it was provided that the previous use or publication of an invention in any part of the world should invalidate a patent for it, had first the teeth taken out by the interpolation of the words, "to the *knowledge of the applicant*;" and ultimately *it was abandoned altogether.*

Of good things contained in the Bill—besides the consolidation of the three patents—there are several deserving of great praise. If we have not noticed them before, it has been from a desire to concentrate public attention on what we still consider to be defects of so vital a nature, as not to be compensated for by any other advantages whatever, and only to be submitted to for a time on such an assurance of early revision as the Premier has given.

Foremost among these additional good things deserves to be ranked the provision made for the *printing and publishing of specifications* (*See* clause xxix.) The merit of originating this belongs to the Patentees' Protection Association.* All the difference between their plan on this head and the Government's is, that the Association proposed that they should be printed and circulated at the expense of the patentees—being prepared to show how this could be done without exacting from patentees a shilling more than they have been in use to pay under the old parchment system; whereas the Bill throws the whole of this expense on the Consolidated Fund. We are not sure that the framers of the Bill knew what they were about in doing so; for the cost to the country will be something enormous. As it is, however, we feel grateful for the boon.

The next most valuable change to be noticed relates to the legal proceedings for the *enforcement of patent rights* after they have been obtained. The clauses on this head are particularly deserving of attentive consideration. They involve large and most salutary reforms. The Chancery jurisdiction in respect of patents is in effect abolished, power being given to any of the Common Law Courts in which an action for infringement is brought, "to make such order for injunction, inspection, or account," as to such Court may seem fit, and as could hitherto only be obtained by going into Chancery; while, the proceedings

* Suggestions to the same effect are to be met with elsewhere, but they are all mere repetitions, without acknowledgment, of the plan of the Association.

at common law are greatly simplified and improved.

We may point also as deserving of special commendation to the arrangements made with respect to the patronage arising out of the Bill. According to the measure as settled by the Lords, the Commissioners of Patents were to have the appointment of all the subordinate officers—Examiners, Registrars, Clerks, &c.; but by the Bill, the whole of this power is transferred to the Lords of the Treasury. The practical difference to the public will be this: that in the one case the appointments would have been left in the hands of legal functionaries not directly amenable to Parliament, and not much in the custom of consulting public opinion, in the disposal of their patronage; while in the other, they are entrusted to a Board which is immediately under the control of the House of Commons, and more so than any other public body whatever.

On the whole, we think we have occasion to congratulate the Invention Interest on the position it has now attained. The opposition made to the Government Bill, No. 3, (by the Patentees' Protection Association, more especially), has not plucked from it all the evil there is in it, but it has had the important effect of extracting from Government a pledge of a further and early "searching enquiry"—implying of course a fair consideration of whatever alterations and amendments may be necessary. If inventors would be only true to themselves, and act unitedly and resolutely,—they might soon achieve all that they want, which is no more than all they have a right to. Though now put under the harrow of a most inquisitorial system of preliminary examination, they may, if they will, heave it off. The very toad does the same when oppressed. The more "searching" future inquiries are the better; the deeper the enquirers go, the more surely will they find inventive talent, and the encouragement of inventive talent at the bottom of all the advances which the world has made in Knowledge, Virtue, and Power.

Continuation of Proceedings in Committee of the Commons.

[We give only the more notable portions of these proceedings, and more is unnecessary.]

4th August.

Mr. J. Greene considered that the present Bill was worse than the old patent laws. He complained that the patentee would still be compelled to approach the Crown as a petitioner and suppliant for what was his just right. He also considered that the arrangement for inquiry before an examiner, previously to provisional registration, was highly objectionable. A board of examiners was some time ago appointed in France with regard to patents, but the system was found so inefficient that it was very soon abandoned. In America an examiner was appointed; but, although he exercised very slight powers, his authority was considered dictatorial, and all parties were dissatisfied with the system. He considered that the present Bill was an inroad upon the industry of the community, and he therefore felt bound to oppose it.

Mr. Labouchere said, though this Bill varied considerably in form from the measure sent down by the House of Lords, it did not differ from it essentially in intent or substance. A petition, expressing the opinion of the Society of Arts on the matter, had been presented by the right honourable member for Manchester, and the petitioners stated that, though this measure was not in all respects such as they desired, they yet believed it would effect an immense improvement on the existing law, and they considered it most important that the Bill should be passed this session.

Mr. Ricardo admitted that this measure would be an improvement upon the present law, but he did not think the House and the country had had a fair opportunity of considering the provisions of the Bill.

M. W. Williams believed the provisions of this Bill would afford encouragement to miserable and paltry inventions, which would have the effect of trammelling our manufactures to an enormous extent.

Mr. Muntz, although he did not altogether approve of the Bill, had been requested by his constituents to give it his support, rather than allow the session to pass without any legislation on the subject.

Sir D. L. Evans said a numerous body of his constituents were anxious that this measure should be adopted at once, because they believed it would materially improve the present law.

Mr. J. Greene moved that the clause relating to the appointment of examiners be omitted.

The Committee divided:—

For Mr. Greene's motion	2
Against it	46
Majority against it	—44

5th August.

On clause 14, Mr. Muntz said this clause provided that the examiners might from time to time inquire into the novelty and utility of inventions. He moved that the word "utility" be struck out of the clause.

The Attorney-General said the object of the clause was to prevent absurd and useless inventions, however novel they might be, from obtaining provisional protection.

Sir J. Graham observed that the difficulty was as to the construction to be put on the word "utility." On the one hand it might be understood to mean public utility, while on the other it might be regarded in the sense of utility to the person who sought to obtain a patent for his private gain. He entertained objections to this measure, because it proposed to create a set of new officers, designated examiners, who, it appeared to him, were to perform the functions which had hitherto been performed under the responsibility of the law advisers of the Crown. It was impossible to believe that the law officers, before they formed a judgment upon an application, and advised the Crown to grant a patent, did not satisfy themselves that the invention was likely to be useful, and was of such a character as to justify them in recommending the Crown to allow an exclusive right to be exercised against the community for a limited period. He wished to ask what had been the principle upon which the law advisers of the Crown had hitherto recommended patents, with regard to the two points,—novelty and utility? He had no doubt, although the law officers of the Crown were accomplished gentlemen, and had an opportunity of informing themselves with respect to improvements in trade and commerce, that before advising the Crown to grant exclusive patents as against the public, they had called to their aid persons who had really performed the functions of the proposed examiners. The law officers had, however, covered all these inquiries, but now it was sought to relieve them altogether of the responsibility. It appeared from this Bill that they were to delegate the inquiry to examiners, on whose report—without exercising any judgment of their own, except within the limits of such report—they were to ground their recommendation to the Crown.

The ATTORNEY-GENERAL believed that since he had been in office no patent had been granted for any invention, the usefulness of which was not apparent. [An honourable Member: "To the public, or to the individual?"] He meant to the public (hear, hear); for, so far as the individual patentee was concerned, he must take his chance. He (the Attorney-General) did not think that a patent for an invention that would be absolutely useless to the public ought to be granted. He did not consider that under this Bill the legal advisers of the Crown would be bound by the report of the examiners, who, in his opinion, would act, as it were, as the pioneers of the law-officers, and direct their attention to the different points of each case. In many cases which had occurred under the existing law, the legal advisers of the Crown had called in the assistance of scientific gentlemen, who had made reports to them, or aided them by personal suggestions. The advice thus obtained, however, was a sort of private assistance, and he thought it would be better that such assistance should be rendered by persons before the public, who were clothed with a certain official character, and who were responsible to the public (hear, hear), which was the course proposed by this Bill. Still he must say, that he would never consider himself bound by the report of the examiners; and, indeed, the fifteenth section of the Bill provided that if any person deemed himself aggrieved by reason of any report of the examiners, such person might claim to be, and should be, heard before the law-officers themselves.

Mr. HENLEY thought it might be a grave question how far the duties of the law-officers of the Crown would be devolved upon these examiners. He did not object to scientific questions being referred to the examiners, but it appeared to him that all matters, whether relating to science or not, would be submitted to their judgment.

The ATTORNEY-GENERAL said, he had no objection to strike the word "utility" out of the clause, and he moved accordingly that that word be omitted.

Mr. RICARDO opposed the motion, because, although all the gentlemen examined before the Committee concurred in the opinion that no tribunal could decide upon the utility of inventions, still he thought they had no excuse for granting these monopolies, except on the ground that the inventions were beneficial and useful to the public.

The Committee divided, when there were—

For the motion	48
Against it	1
Majority	47

Mr. LABOUCHERE, in answer to a question from Mr. Spooner, explained that the Crown could now grant patents extending to the Colonies (though there had been some doubt as to India), and that it was also in the power of the Colonies, of those at least that had local Legislatures, to grant patents within their own limits; this concurrent power had led to difficulty, complication, and inconvenience, and it was thought that it would be better to confine the patents granted here to the United Kingdom, and to leave the Colonies to deal with the matter as they thought fit. Indeed, a regulation had already been made, about six months since, that the Attorney-General should not pass patents extending to the Colonies.*

Sir J. GRAHAM thought this discussion became more complicated as it proceeded. (Hear.) Here was another most important question incidentally raised. (Hear, hear.) There really was not time now to give due consideration to a measure involving such interests as this did. It came down but lately from the House of Lords, and already thirty new clauses had been introduced into it in the House of Commons. Besides the provisions respecting compensation, those with regard to fees were changed, but there were still great objections to be urged to them; there ought not to be a farthing paid for the establishment from the public purse; the fees should be so regulated as to pay the whole expense (Hear.) Clause by clause the Bill would be found to touch important principles not yet fully discussed; and the further the House proceeded, the more they would be aware of the importance of the subject, the insufficiency of the time they could now devote to it, and the inadequate attendance of members representing the great commercial and manufacturing interests. (Hear.) He did not see how justice was to be done to so important a subject within the time that could now be given to it. (Hear, hear.)

Mr. LABOUCHERE was very sorry to find the right honourable baronet casting the weight of his great authority into the scale of the objectors, hitherto very few in number, to the passing of this Bill in this session. The disappointment on the part of the public, and the inconvenience sustained, if it should not pass, would be so great that nothing on his (Mr. Labouchere's) part should be spared to prevent its postponement.

Mr. CARDWELL had presented a petition against this Bill from parties who conceived that their competitors in the Colonies were to be allowed to proceed without paying the duties paid here.

Sir J. GRAHAM understood now, that by prerogative the Secretary of State, having come to a decision adverse to patents, had virtually abolished the law of patents throughout the Colonies, and the House was asked to continue in England a system which some of the advisers of the Crown condemned in principle, and which the Colonial Secretary had put an end to throughout our possessions abroad, giving our fellow-subjects there the privilege of using inventions denied to the inhabitants of the United Kingdom. (Hear, hear.) The importance of the subject became more apparent as the discussion proceeded, and the complication was increased by the division of opinion among the advisers of the Crown. The Vice-President of the Board of Trade had declared that he was opposed to the policy of the law of patents; and now the House learned that the Secretary for the Colonies agreed with him.

Mr. LABOUCHERE assured the right honourable baronet that he was mistaken. The Vice-President of the Board of Trade (Earl Granville) expressed an

* There must be some mistake here. We have, within the last six months, passed many patents, with the Colonies included.—ED. M. M.

opinion, shared by the majority of the Lords' Committee, that the principle of patent laws was vicious, and ought to be discarded; but he and all the Committee held it most desirable that this Bill, remedying monstrous and glaring evils in the present system, should pass into a law at once, and they thought that the public opinion of this country was not ripe for so great a change as the abolition of patents.

Sir J. Graham observed that this clause altered the law materially, and yet did not appear necessary for the objects desired by the right honourable gentleman (Mr. Labouchere).

6th August.

Mr. W. Williams and Mr. J. Greene repeated their objections to the Bill.

The Attorney-General said that he had, on reconsideration, resolved to leave the law with respect to the Colonies as before.

Sir James Graham urged on behalf of the country the necessity for further discussion of the whole measure.

Lord Palmerston and Mr. Labouchere replied, insisting on the disappointment which the public generally would experience if the Bill did not pass.

On the Compensation Clause being read,

Sir James Graham inquired if the Government had formed any estimate of the gross amount of compensation?—and next, if they had ascertained how many would have claims under the clause? It was his belief that the days of the patent law were numbered, and that it could not long be maintained if the House were now to establish a statutable claim to compensation, it would at no distant period be found to have established an inconvenient precedent.

The Attorney-General was inclined to think that the number of persons who would be entitled to compensation would be small, and the amount of compensation not large.

Sir James Graham observed that the Attorney General had not answered his question; but having raised a difficulty, he would offer a suggestion by which it might be met. Let the Government consider whether, on bringing up the Report, they could not make provision for having compensation granted only out of the fees to be hereafter received. By this course, the danger of establishing a permanent claim on the public purse would be avoided.

Mr. Labouchere thanked the right honourable baronet for his suggestion, and said it should be taken into consideration by the Government.

The remaining clauses were then agreed to, and the House resumed.

THE PATENT LAW AMENDMENT ACT, 1851.*

Whereas it is expedient to amend the law concerning Letters Patents for inventions: Be it enacted by the Queen's most excellent Majesty, by and with the advice and consent of the Lords spiritual and temporal, and Commons, in this present Parliament assembled, and by the authority of the same, as follows:

I. It shall be lawful for Her Majesty to grant Letters Patent for inventions, and to cause the same to be made and issued, in the manner hereinafter mentioned, and to cause to be inserted therein all such restrictions, conditions, and provisoes as Her Majesty may think fit, or as may be requisite in pursuance of the provisions of this Act, and such Letters Patent shall (subject to the provisions herein contained) be of the same force and effect as grants of a like nature heretofore made.

II. The Lord Chancellor, the Master of the Rolls, Her Majesty's Attorney-General for England, Her Majesty's Solicitor-General for England, the Lord Advocate, Her Majesty's Solicitor-General for Scotland, Her Majesty's Attorney-General for Ireland, and Her Majesty's Solicitor-General for Ireland, for the time being respectively, together with such other person or persons as may be from time to time appointed by Her Majesty, as hereinafter mentioned, shall be Commissioners of Patents for Inventions; and it shall be lawful for Her Majesty from time to time, by warrant under her royal sign manual, to appoint such other person or persons as she may think fit to be a Commissioner or Commissioners as aforesaid; and every person so appointed shall continue such commission during Her Majesty's pleasure; and all the powers hereby vested in the Commissioners may be exercised by any three or more of them, the Lord Chancellor or Master of the Rolls being one.

III. It shall be lawful for the Commissioners to cause a seal to be made for the purposes of this Act, and from time to time to vary such seal, and to cause to be sealed therewith all Letters Patent issued under this Act, and all instruments and copies proceeding from the office of the Commissioners, and all courts, judges, and other persons whomsoever shall take notice of such seal and receive impressions thereof in evidence in like manner as impressions of the Great Seal are received in evidence, and shall also take notice of and receive in evidence, without further proof or production of the originals, all such copies, certified under the seal of the said office, of documents deposited in such office.

IV. It shall be lawful for the Commissioners from time to time to appoint a person or persons as an examiner or as examiners for the purposes hereinafter mentioned, and at their pleasure to revoke the appointment of and to remove any person or persons so appointed, and to appoint others in lieu thereof; and it shall be lawful for the Commissioners from time to time to make such rules and regulations (not inconsistent with the provisions of this Act) respecting the business of their office, and all matters and things which under the provisions herein contained are to be under their control and direction, as may appear to them necessary and expedient for the purposes of this Act; and all such rules shall be laid before both Houses of Parliament within fourteen days after the making thereof, if Parliament be sitting, and if

* As passed the Commons, but thrown out in the Lords.—*See Note p. 104.*

Parliament be not sitting, then within fourteen days after the next meeting of Parliament; and the Commissioners shall cause a report to be laid annually before Parliament of all the proceedings under and in pursuance of this Act.

V. It shall be lawful for the Commissioners of Her Majesty's Treasury to provide and appoint from time to time proper places or buildings for an office or offices for the purposes of this Act.

VI. It shall be lawful for the said Commissioners of the Treasury from time to time to appoint for the purposes of this Act such clerks and officers (other than the officers whom the Commissioners of Patents for Inventions are by this Act authorised to appoint) as such Commissioners of the Treasury may think proper; and it shall be lawful for the said Commissioners of Patents from time to time to remove any of the clerks and officers so appointed by such last-mentioned Commissioners.

VII. The petition and declaration for the grant of Letters Patent for an invention shall be left at the office of the Commissioners, and there shall be left therewith a statement in writing, hereinafter called the provisional specification, signed by or on behalf of the applicant for Letters Patent, describing the nature of the said invention; and the day of the delivery of every such petition, declaration, and provisional specification shall be recorded at the said office, and indorsed on such petition, declaration, and provisional specification, and a certificate thereof given to such applicant or his agent; and all such petitions, declarations, and provisional specifications shall be preserved in such manner as the Commissioners may direct, and a registry thereof and of all proceedings thereon kept at the office of the Commissioners.

VIII. Every application for Letters Patent made under this Act shall be referred by the Commissioners, or according to such regulations as they may think fit to make, to one of the law-officers.

IX. The provisional specification shall be referred by the law-officer to one of the examiners for the time being appointed under this Act, and the examiner to whom such reference is made, being satisfied that the provisional specification describes the nature of the invention, shall give a certificate to that effect, and such certificate shall be filed in the office of the Commissioners, and thereupon the invention therein referred to shall be provisionally protected for the term of six months from the date of the application for Letters Patent for the said invention; and during the term of such provisional protection such invention may be used and published without prejudice to any Letters Patent to be granted for the same; provided always, that in case the title of the invention or the provisional specification be too large or insufficient, it shall be lawful for the person to whom the same is referred to allow or require the same to be amended: Provided also, that on the filing at the said office of any certificate of the Registrar of Designs under "The protection of Inventions Act, 1851," the invention to which such certificate relates shall be provisionally protected under this Act from the filing of such certificate, without prejudice nevertheless to the provisions of such last-mentioned Act.

X. The applicant for Letters Patent for an invention, instead of leaving with the petition and declaration a provisional specification as aforesaid, may, if he think fit, deposit with the said petition and declaration a complete specification or instrument in writing under his hand and seal, particularly describing and ascertaining the nature of the said invention, and in what manner the same is to be performed, and the day of the delivery of every such petition, declaration, and complete specification shall be recorded at the office of the Commissioners, and indorsed on such petition, declaration, and specification, and a certificate thereof given to such applicant or his agent, and thereupon the invention shall be protected under this Act for the term of six months from the date of the application, and the applicant shall have during such term of six months the like powers, rights, and privileges as might have been conferred upon him by Letters Patent for such invention, issued under this Act, and duly sealed as of the day of the date of such application; and during the continuance of such powers, rights, and privileges under this provision, such invention may be used and published without prejudice to any Letters Patent to be granted for the same; and where Letters Patent are granted in respect of such invention, the complete specification shall be deemed the description of such invention, and such specification shall be deemed to be filed in pursuance of such Letters Patent, and no other specification shall be filed in respect thereof; provided that all the provisions of the Act of the session holden in the fifth and sixth years of King William the Fourth, chapter eighty-three, concerning disclaimers and memoranda of alterations, shall extend and be applicable to such complete specifications as aforesaid: Provided also, that a copy of every such complete specification shall be open to the inspection of the public, as hereinafter provided, from the time of depositing the same, subject to such regulations as the Commissioners may make.

n case of any application for Letters or any invention, and the obtaining uch application of provisional pro- or such invention, or of protection ame by reason of the deposit of a specification as aforesaid in fraud rue and first inventor, any Letters ranted to the true and first inventor invention shall not be invalidated m of such application, or of such nal or other protection as aforesaid, y use or publication of the inven- sequent to such application, and he expiration of the term of such nal or other protection.

Where any invention is provision- tected under this Act, or protected n of the deposit of such complete tion as aforesaid, the Commis- shall cause such provisional protec- uch other protection as aforesaid to tised in such manner as they may

The applicant for Letters Patent, as he may think fit, after the inven- ll have been provisionally protected his Act, or where a complete specifi- as been deposited with his petition laration then so soon as he may after such deposit, may give notice fice of the Commissioners of his in- of proceeding with his application ers Patent for the said invention, reupon the said Commissioners shall is said application to be advertised manner as they may see fit; and sons having an interest in opposing t of Letters Patent for the said in- shall be at liberty to leave at the the Commissioners particulars in of their objections to the said appli- within such time and subject to gulations as the Commissioners may

So soon as the time for the deli- such objections shall have expired, visional specification or complete tion (as the case may be), and par- of objection (if any), shall be refer- e law officer to whom the application s referred to one or more of the exa- appointed under the provisions of ; and such examiner or examiners quire (subject to such regulations Commissioners may from time to ike) into the novelty of the said n, and the objections (if any) to ting of the said Letters Patent, and port thereon to such law officer; and icant and opponents (if any) respec- hall be entitled to a copy of such subject to such regulations as the sioners may think fit.

XV. It shall be lawful for any person who may deem himself aggrieved by reason of any proceeding in respect of any provisional specification, or the giving or refusing of any certificate as aforesaid, or by reason of any report of the examiner or examiners, to be heard before the law officer to whom the application is referred; and it shall be lawful for such law officer, if he see fit, to rescind or give such certificate, and to allow or require any entry, title, or provisional specification to be amended, or to refer any such report back again to the examiner or examiners, or to refer it to any other examiner or examiners or person or persons he may think fit, and to order by and to whom the costs of and occasioned by such appeal and reference are to be paid, and in what manner and by whom the amount of such costs are to be ascertained, and generally to make such order upon such hearing as to such law officer may seem just; and such certificate of the law officer shall be filed in like manner and be of the like effect as if the same were the certificate of the examiner to whom the provisional specification was referred; and if any costs so ordered to be paid be not paid within four days after the amount thereof shall have been ascertained, it shall be lawful for such law officer to make an order for the payment of same, and every such order may be made a rule of one of Her Majesty's superior courts of law at Westminster.

XVI. It shall be lawful for the law officer, if he think fit, on the receipt of the said report, and after such hearing, if any, as aforesaid, to cause a warrant to be made for the sealing of Letters Patent for the said invention, and such warrant shall set forth the tenor and effect of the Letters Patent thereby authorized to be granted, and such law officer shall direct the insertion in such Letters Patent of all such restrictions, conditions, and provisoes as he may deem usual and expedient in such grants, or requisite and necessary in pursuance of the provisions of this Act; and the said warrant, signed by the law officer, shall be deposited in the office of the Commissioners, and shall be the warrant for the making and sealing of Letters Patent under this Act according to the tenor of the said warrant; and no Bill, Signet Bill, or Privy Seal Bill, Writ of Privy Seal, or other warrant or authority whatsoever, save as herein provided, shall be necessary for or preparatory to the passing of such Letters Patent as aforesaid, the Act of the twenty-seventh year of King Henry the Eighth, chapter eleven, or any other law, or any usage, to the contrary notwithstanding: provided always, that the Lord Chancellor shall and may have and exercise such powers, authority, and discretion in respect to the said warrant, and the Letters Patent therein directed to be made under this Act, as he has heretofore had and exercised with respect to the Privy Seal Bill, and the making and issuing of Letters Patent for inventions according to the manner heretofore used.

XVII. All Letters Patent for inventions granted under the provisions of this Act shall be made subject to the condition that the same shall be void, and that the powers and privileges thereby granted shall cease and determine, at the expiration of three years and seven years respectively from the date thereof, unless there be paid, before the expiration of the said three and seven years respectively, the sum or sums of money and stamp duties in the schedule to this Act annexed; and the payment of the said sums of money and stamp duties respectively shall be endorsed on the warrant for the said Letters Patent; and such officer of the Commissioners as may be appointed for this purpose shall issue under the Seal of the Commissioners a certificate of such payment, and shall endorse a receipt for the same on any Letters Patent issued under the authority of the said warrant; and such certificate, duly stamped, shall be evidence of the payment of the several sums respectively: provided also, that such Letters Patent may be made subject to such other conditions for rendering the same void or voidable as to Her Majesty may seem fit.

XVIII. The officer of the Commissioners appointed to seal Letters Patent under this Act, so soon after the receipt of the said warrant as required by the applicant for the Letters Patent, shall issue under the Seal of the Commissioners Letters Patent for the invention according to the tenor of the said warrant applicable to the whole of the United Kingdom of Great Britain and Ireland, the Channel Islands, Isle of Man, and the Colonies and Plantations abroad, and such Letters Patent shall be valid and effectual as to the whole of such United Kingdom, and the said Islands and Isle, and shall confer the like powers, rights, and privileges as might, in case this Act had not been passed, have been conferred by several Letters Patent of the like purport and effect passed under the Great Seal of the United Kingdom, under the Seal appointed to be used instead of the Great Seal of Scotland, and under the Great Seal of Ireland respectively, and made applicable to England, the dominion of Wales, the town of Berwick-upon-Tweed, the Channel Islands, and Isle of Man, to Scotland, and to Ireland respectively, save as herein otherwise provided.

XIX. Provided always, that no Letters Patent, save as hereinafter mentioned in the case of Letters Patent destroyed (shall issue on any warrant granted a said, unless application be made to s Letters Patent within three montl the date of the said warrant.

XX. Provided also, that no Letters (save Letters Patent issued in lieu o destroyed or lost) shall be issued any force or effect unless the same be during the continuance of the pro protection under this Act, or where plete specification has been deposite this Act, then unless such Letters P granted during the continuance of t tection conferred under this Act by of such deposit.

XXI. Provided also, that in case s Letters Patent shall be destroyed other Letters Patent of the like te effect, and sealed and dated as of t day, may, subject to such regulation Commissioners may direct, be issue the authority of the warrant in pu of which the original Letters Pate issued.

XXII. It shall be lawful for the C sioners (the Act of the eighteenth King Henry the Sixth, chapter one other Act, to the contrary notwithst to cause any Letters Patent to be i pursuance of this Act to be sealed s date as of the day of the application same, and in case of such Letters Pa any invention provisionally registere the "Protection of Inventions Act, as of the day of such provisional regis or, where the law officer to whom th cation was referred, or the Lord Cha thinks fit and directs, any such Patent as aforesaid may be sealed s date as of the day of the sealing Letters Patent, or of any other day the day of such application or pro registration and the day of such seal

XXIII. Any Letters Patent issue this Act bearing date as of any day the day of the actual sealing there be of the same force and validity as had been sealed on the day on wh same bear date: provided always, tl where such Letters Patent are gran any invention, in respect whereof a c specification has been deposited u application for the same under this proceeding at law or in equity shall upon such Letters Patent in respect infringement committed before th were actually granted.

XXIV. It shall be lawful for any to whom Letters Patent at the time passing of this Act may have been under the Great Seal in respect of vention, but to whom Letters Pate n been granted for Scotland or Ire- the said invention, in case the said n may not have been used or pub- ithin Scotland or Ireland at the time aking of the application hereinafter ed, to petition Her Majesty for the f Letters Patent for the United King- Great Britain and Ireland, the Chan- nds and Isle of Man, under the pro- of this Act, and thereupon it shall l for Her Majesty to cause Letters o be issued in the manner herein- ntioned, and to be sealed and bear of the same day on which the Letters under the Great Seal bear date, with- payment of any fees whatsoever, but to the payments on or before the on of the third and seventh years vely as hereinafter provided: Pro- ways, that on the issuing of the said tters Patent the original Letters under the Great Seal shall be thence- id and of no force or effect.

. All Letters Patent to be granted his Act shall require the specification der to be filed in the office of the sioners, instead of requiring the be enrolled, and no other filing and lment shall be requisite.

I. All specifications deposited at e of the Commissioners, or filed in ce of any condition contained in any Patent granted under this Act, shall rved in such manner and subject to gulations as the Commissioners may me to time direct; and in case re- shall be made to drawings in any nal specification, or specification ed or filed, an extra copy of such s shall be left with such specification office of the Commissioners.

VII. The Commissioners shall cause pies of all specifications deposited or aforesaid, and of all disclaimers, and anda of alterations in relation to etters Patent and specifications, and provisional specifications after the f the provisional protection of the vention shall have expired, to be the inspection of the public at the of the Commissioners, and at an Edinburgh and Dublin respectively, sonable times, subject to such regu- s the Commissioners may direct.

VIII. The Commissioners shall cause printed, published, and sold at such and in such manner as they may think provisional specifications and com- ecifications deposited, and all speci- s, disclaimers, and memoranda of ons filed in the office of the Commis- under this Act, and such complete ations, specifications, disclaimers, and memoranda respectively shall be so printed and published as soon as conveniently may be after the deposit and filing thereof respectively, and all such provisional specifications shall be so printed and published as soon as conveniently may be after the expiration of the provisional protection obtained in respect thereof; and it shall be lawful for the said Commissioners to present copies of all such publications to such public libraries and museums as they may think fit, and it shall be lawful for the said Commissioners to allow the person depositing or filing any such provisional or other specification, disclaimer, or memorandum of alteration, to have such number, not exceeding twenty-five, of copies thereof so printed and published, without any payment for the same, as they may think fit.

XXIX. It shall be lawful for the Lord Chancellor and the Master of the Rolls to direct the enrolment of specifications, disclaimers, and memoranda of alterations heretofore or hereafter enrolled or deposited at the Rolls Chapel-office, or at the Petty Bag-office, or at the Enrolment-office of the Court of Chancery, or in the custody of the Master of the Rolls as Keeper of the Public Records, to be transferred to the office of the Commissioners.

XXX. The Commissioners shall cause indices to all specifications, disclaimers, and memoranda of alterations heretofore enrolled to be prepared in such form as they may think fit, and shall cause copies of such specifications, disclaimers, memoranda of alterations, and indices to be deposited at the office of the Commissioners, and such indices shall be open to the inspection of the public, subject to the regulations to be made by the Commissioners, and the Commissioners may cause all or any of such indices, specifications, disclaimers, and memoranda of alterations to be printed, published, and sold in such manner and at such prices as the Commissioners may think fit.

XXXI. The copies printed under this Act of specifications, disclaimers, and memoranda of alterations shall be admissible in evidence, and deemed and taken to be *primâ facie* evidence of the existence and contents of the documents to which they purport to relate in all courts and in all proceedings relating to Letters Patent.

XXXII. There shall be kept at the office of the Commissioners a book or books, to be called "The Register of Patents," wherein shall be entered and recorded in chronological order all Letters Patent granted under this Act, the deposit or filing of specifications, disclaimers, or memoranda of alterations filed in respect of such Letters Patent and specifications, all amendments in such Letters Patent and specifications, all confirmations and extensions of such Letters Patent, the expiry, vacating, or cancelling such Letters Patent, with the dates thereof respectively, and all other matters and things affecting the validity of such Letters Patent as the Commissioners may direct, and such register or a copy thereof shall be open at all convenient times to the inspection of the public at the office of the Commissioners, subject to such regulations as the Commissioners may make.

XXXIII. There shall be kept at the office of the Commissioners a book or books, entitled "The Register of Proprietors," wherein may be entered, in such manner as the Commissioners shall direct, the assignment of any Letters Patent, or of any share or interest therein, any licence under Letters Patent, and the district to which such licence relates, with the name or names of any person having any share or interest in such Letters Patent or licence, the date of his or their acquiring such Letters Patent, share and interest, and any other matter or thing relating to or affecting the proprietorship in such Letters Patent or licence, and a copy of any entry in such book, certified under the Seal of the Commissioners, shall be given to any person requiring the same, on payment of the fee herein after provided; and such copies, so certified and impressed, shall be received in evidence in all courts and in all proceedings, and shall be *primâ facie* proof of the assignment of such Letters Patent, or share or interest therein, or of the licence or proprietorship, as therein expressed: Provided always, that until such entry shall have been made the grantee or grantees of the Letters Patent shall be deemed and taken to be the sole and exclusive proprietor or proprietors of such Letters Patent, and of all the licences and privileges thereby given and granted; and any writ or scire facias to repeal such Letters Patent may be issued to the Sheriff of the county or counties in which the grantee or grantees resided at the time when the said Letters Patent were granted; and in case such grantee or grantees do not reside in England or Wales, it shall be sufficient to file such writ in the Petty Bag-office, and serve notice thereof in writing at the last known residence or place of business of such grantee or grantees; and such register or a copy shall be open to the inspection of the public at the office of the Commissioners, subject to such regulations as the Commissioners may make.

XXXIV. If any person shall wilfully make or cause to be made any false entry in the said Register of Proprietors, or shall wilfully produce or tender or cause to be produced or tendered in evidence any paper or other writing falsely purporting to be a copy of any entry in the said book, he shall be guilty of a misdemeanor, and shall be punished by fine and imprisonment accordingly.

XXXV. If any person shall deem himself aggrieved by any entry made under colour of this Act in the said Register of Proprietors, it shall be lawful for such person to apply, by motion, to the Master of the Rolls, or to the judges of any of the courts of common law at Westminster in term time, or by summons to a judge of any of the said courts in vacation, for an order that such entry may be expunged, vacated, or varied; and upon any such application the Master of the Rolls, or such court or judge respectively, may make such order for expunging, vacating, or varying such entry, and as to the costs of such application, as to the said Master of the Rolls or to such court or judge may seem fit; and the officer of the Commissioners having the care and custody of such register, on the production to him of any such order for expunging, vacating, or varying any such entry, shall expunge, vacate, or vary the same, according to the requisitions of such order.

XXXVI. All the Provisions of the Acts of the Session holden in the fifth and sixth years of King William the Fourth, chapter eighty-three, and of the Session holden in the seventh and eighth years of Her Majesty, chapter sixty-nine, respectively, relating to disclaimers and memoranda of alterations in Letters Patent and specifications, except as herein-after provided, shall be applicable and apply to any Letters Patent granted to any specification filed under the provisions of this Act: Provided always, that all applications for leave to enter a disclaimer, or memorandum of alteration, shall be made, and all caveats relating thereto shall be lodged at the office of the Commissioners, and shall be referred to the respective law officers in the said first-recited Act mentioned: Provided also, that every such disclaimer, or memorandum of alteration, shall be filed in the office of the Commissioners, with the specification to which the same relates, in lieu of being entered or filed and enrolled as required by the said-recited Act, and the said Acts shall be construed accordingly: Provided also, that such filing of any disclaimer and memorandum of alteration, in pursuance of the leave of the law officer in the first-recited Act mentioned, certified as therein mentioned, shall, except in cases of fraud, be conclusive as to the right of the party to enter such disclaimer or memorandum of alteration under the said Acts and this Act; and no objection shall be allowed to be made in any proceeding upon or touching such Letters Patent, specification, disclaimer, or memorandum of alteration, on the ground that the party entering such disclaimer or memorandum of alteration had not sufficient authority in that behalf: Provided also, that no action shall be brought upon any Letters Patent in which or in the specification of which any disclaimer or memorandum of alteration shall have been filed in respect of any infringement committed prior to the filing of such disclaimer or memorandum of alteration, unless the law officer shall certify in his fiat that any such action may be brought notwithstanding the entry or filing of such disclaimer or memorandum of alteration.

XXXVII. All the provisions of the said Act of the fifth and sixth years of King William the Fourth, for the confirmation of any Letters Patent, and the grant of new Letters Patent, and all the provisions of the said Act, and of the Acts of the session holden in the second and third years of Her Majesty, chapter sixty-seven, and of the session holden in the seventh and eighth years of Her Majesty, chapter sixty-nine respectively, relating to the prolongation of the term of Letters Patent, and to the grant of new Letters Patent for a further term, shall extend and apply to any Letters Patent granted under the provisions of this Act, and it shall be lawful for Her Majesty to grant any new Letters Patent, as in the said Acts mentioned; and in the granting of any such new Letters Patent Her Majesty's order in council shall be a sufficient warrant and authority for the sealing of any new Letters Patent, and for the insertion in such new Letters Patent of any restrictions, conditions, and provisions in the said order mentioned; and the officer of the Commissioners appointed to seal Letters Patent under this Act, on the receipt of the said order in council, shall cause Letters Patent, according to the tenor and effect of such order, to be made and sealed in the manner herein directed for Letters Patent issued under the warrant of the law officer: provided always, that such new Letters Patent shall extend to and be available in and for such part or parts only of the realm as the original Letters Patent extended to and were available in: provided also, that such new Letters Patent shall be sealed and bear date as of the day after the expiration of the term of the original Letters Patent which may first expire.

XXXVIII. In any action pending in any of Her Majesty's superior courts of record at Westminster and in Dublin for the infringement of Letters Patent, the plaintiff shall deliver with his *declaration* particulars of the breaches complained of in the said action, and the defendant, on pleading thereto, shall deliver with his pleas, and the prosecutor in any proceedings by *scire facias* to repeal Letters Patent shall deliver with his declaration, particulars of any objections on which he means to rely at the trial in support of the pleas in the said action, or of the suggestions of the said declaration in the proceedings by *scire facias* respectively; and at the trial of such action or proceeding by *scire facias* no evidence shall be allowed to be given in support of any alleged infringement or of any objection impeaching the validity of such Letters Patent which shall not be contained in the particulars delivered as aforesaid: provided always, that the place or places at or in which, and in what manner the invention is alleged to have been used or published prior to the date of the Letters Patent shall be stated in such particulars: provided always, that it shall and may be lawful for any judge at chambers to allow such plaintiff or defendant or prosecutor respectively to amend the particulars delivered as aforesaid, upon such terms as to such judge shall seem fit: Provided also, that at the trial of any proceeding by *scire facias* to repeal Letters Patent the defendant shall be entitled to begin and to give evidence in support of such Letters Patent, and in case evidence shall be adduced on the part of the prosecutor impeaching the validity of such Letters Patent, the defendant shall be entitled to the reply.

XXXIX. In any action pending in any of Her Majesty's superior courts of record at Westminster and in Dublin for the infringement of Letters Patent, it shall be lawful for the court in which such action is pending, on the application of the plaintiff or defendant respectively, to make such order for an injunction, inspection, or account, and to give such direction respecting such action, injunction, inspection, and account, and the proceedings therein respectively, as to such court may seem fit.

XL. In taxing the costs in any action commenced after the passing of this Act for infringing Letters Patent regard shall be had to the particulars delivered in such action, and the plaintiff and defendant respectively shall not be allowed any costs in respect of any particular not certified by the judge before whom the trial was had to have been proved by such plaintiff or defendant respectively, without regard to the general costs of the cause; and it shall be lawful for the judge before whom any such action shall be tried to certify on the record that the validity of the Letters Patent in the declaration mentioned came in question;

and the record, with such certificate, being given in evidence in any suit or action for infringing the said Letters Patent, or in any proceeding by *scire facias* to repeal the Letters Patent, shall entitle the plaintiff in any such suit or action, or the defendant in such proceeding by *scire facias*, on obtaining a decree, decretal order, or final judgment, to his full costs, charges, and expenses, taxed as between attorney and client, unless the judge making such decree or order, or the judge trying such action or proceeding, shall certify that the plaintiff or defendant respectively ought not to have such full costs.

XLI. There shall be paid in respect of Letters Patent applied for or issued as herein mentioned, the filing of specifications and disclaimers, certificates, entries, and searches, and other matters and things mentioned in the schedule to this Act, such fees as are mentioned in the said schedule; and there shall be paid unto and for the use of Her Majesty, her heirs and successors, for or in respect of the warrants and certificates mentioned in the said schedule, or the vellum, parchment, or paper on which the same respectively are written, the stamp duties mentioned in the same schedule; and no other stamp duties shall be levied, or fees, except as hereinafter mentioned, taken in respect of such Letters Patent and specifications, and the matters and things in such schedule mentioned.

XLII. The stamp duties hereby granted shall be under the care and management of the Commissioners of Inland Revenue; and the several rules, regulations, provisions, penalties, clauses, and matters contained in any Act now or hereafter to be in force with reference to stamp duties shall be applicable thereto.

XLIII. The fees to be paid as aforesaid shall from time to time be paid into the receipt of the Exchequer, and be carried to and made part of the Consolidated Fund of the United Kingdom.

XLIV. Provided always, that nothing herein contained shall prevent the payment as heretofore to the law officers in cases of appeal, or of opposition to the granting of Letters Patent, and in cases of disclaimers and memoranda of alterations, of such fees as may be appointed by the Lord Chancellor and Master of the Rolls as the fees to be paid on the hearing of such appeals and oppositions, and in the case of disclaimers and memoranda of alterations respectively, or of such reasonable sums for office or other copies of documents in the office of the Commissioners, as the Commissioners may from time to time appoint to be paid for such copies, and the Lord Chancellor and Master of the Rolls, and the Commissioners, are hereby respectively authorised and empowered to appoint the fees to be so paid in respect of such appeals and oppositions, disclaimers and memoranda of alterations respectively, and for such office or other copies.

XLV. It shall be lawful for the Commissioners of Her Majesty's Treasury from time to time to allow such fees to the law-officers (for duties under this Act in respect of which fees may not be payable to them under the provision lastly hereinbefore contained), as the Lord Chancellor and Master of the Rolls may from time to time appoint, and to allow such fees to the examiners appointed under this Act, and such salaries and payments to any clerks and officers to be appointed under this Act, and such additional salaries and payments to any other clerks and officers in respect of any additional duties imposed on them by this Act, as the said Commissioners of the Treasury may think fit.

XLVI. It shall be lawful for the Commissioners of Her Majesty's Treasury to allow from time to time the necessary sums for providing offices under this Act, and for the fees, salaries, and payments allowed by them as aforesaid, and for defraying the current and incidental expenses of such office or offices; and the sums to be so allowed shall be paid out of such moneys as may be provided by Parliament for that purpose.

XLVII. Accounts shall be kept of the fees mentioned in the schedule to this Act which shall be received thereunder, and of the payments out of the moneys so received; and such accounts shall from time to time be examined, tried, and audited by the Commissioners for examining the public accounts of the kingdom.

XLVIII. And whereas divers persons by virtue of their offices or appointments are entitled to fees or charges payable in respect of Letters Patent as heretofore granted within the United Kingdom of Great Britain and Ireland, or have or derive in respect of such Letters Patent, or the procedure for the granting thereof, fees or other emoluments or advantages:

It shall be lawful for the said Commissioners of the Treasury to grant to any such persons who may sustain any loss of fees, emoluments, or advantages by reason of the passing of this Act, such compensation as, having regard to the tenure and nature of their respective offices and appointments, such Commissioners deem just and proper to be awarded; and all such compensation

id out of the Consolidated Fund
ited Kingdom of Great Britain
l: Provided always, that in case
to whom any yearly sum by way
sation shall be awarded shall,
assing of this Act, be appointed
e or place of emolument under
ons of this Act, or in the public
m and in every such case the
such yearly sum shall in every
iinished by so much as the emo-
such person for such year from
or place shall amount to, and
that behalf shall be made in
o him of such yearly sum.
An account of all salaries, fees,
sums, and compensations to be
allowed, or granted under this
within fourteen days next after
all be so appointed, allowed, or
pectively, be laid before both
Parliament if Parliament be then
f Parliament be not then sitting,
fourteen days after the next
Parliament.
ers Patent may be granted in
applications made before the
this Act, in like manner and sub-
same provisions as if this Act
n passed.
re Letters Patent for England
granted before the passing of this
s in respect of any application
e the passing of this Act here-
ed for any invention, Letters
Scotland or Ireland may be
such invention in like manner
ct had not been passed: Pro-
rs, that in lieu of all the fees of
nd stamp duties now payable in
such Letters Patent, or in or
ning a grant thereof, there shall
respect of such Letters Patent
d or Ireland, on the sealing of
tive Letters Patent, a sum equal
rd part of the fees and stamp
h would be payable according to
ule to this Act in respect of
ent issued for the United King-
this Act, on or previously to the
such Letters Patent; and at or
xpiration of the third year and
h year respectively of the term
such Letters Patent for Scotland
sums equal to one third part of
d stamp duties payable at the
of the third year and the seventh
tively of the term granted by
ent issued for the United King-
this Act; and the condition of
rs Patent for Scotland or Ireland
ried accordingly; and such fees
shall be paid to such persons as the Commissioners of Her Majesty's Treasury shall appoint, and shall be carried to and form part of the said Consolidated Fund.

LII. The several forms in the Schedule to this Act may be used for and in respect of the several matters therein mentioned, and the Commissioners may, when they think fit, vary such forms as occasion may require, and cause to be printed and circulated such other forms as they may think fit to be used for the purposes of this Act.

LIII. In the construction of this Act the following expressions shall have the meanings hereby assigned to them, unless such meanings be repugnant to or inconsistent with the context; (that is to say),

The expression "Lord Chancellor" shall mean the Lord Chancellor, or Lord Keeper of the Great Seal, or Lords Commissioners of the Great Seal.

The expression "the Commissioners" shall mean the Commissioners for the time being acting in execution of this Act.

The expression "Law-officer" shall mean Her Majesty's Attorney-General or Solicitor-General for the time being for England, or the Lord Advocate or Her Majesty's Solicitor-General for the time being for Scotland, or Her Majesty's Attorney-General or Solicitor-General for the time being for Ireland.

The expression "Inventions" shall mean any manner of new manufactures the subject of Letters Patent and grants of privilege within the meaning of the Act of the twenty-first year of the reign of King James the First, chapter three.

The expression "Petition," "Declaration," "Provisional Specification," "Warrant," and "Letters Patent," respectively, shall mean instruments in the form and to the effect in the schedule hereto annexed, subject to such alterations as may from time to time be made therein under the powers and provisions of this Act.

LIV. In citing this Act in other Acts of Parliament, instruments, and proceedings, it shall be sufficient to use the expression, "The Patent Law Amendment Act, 1851."

The SCHEDULE *of Fees to which this Act refers.*

Fees to be Paid.

	£	..	..
On leaving Petition for Grant of Letters Patent	5	0	0
On Notice of Intention to proceed with the Application ..	5	0	0
On sealing of Letters Patent ..	5	0	0
On filing Specification........	5	0	0

	£.	s.	d.
At or before the Expiration of the Third Year	40	0	0
At or before the Expiration of the Seventh Year	80	0	0
On Extension of Period of provisional Protection	10	0	0
On leaving Objections to granting of Letters Patent	2	0	0
Every Search and Inspection	0	1	0
Entry of Assignment or Licence	0	5	0
Certificate of Assignment or Licence	0	5	0
Filing Application for Disclaimer	5	0	0
Caveat against Disclaimer	2	0	0

Stamp Duties to be Paid.

	£	s.	d.
On Warrant of Law-Officer for Letters Patent	5	0	0
On Certificate of Payment of the Fee payable at or before the Expiration of the Third Year	10	0	0
On Certificate of Payment of the Fee payable at or before the Expiration of the Seventh Year	20	0	0

SPECIFICATIONS OF ENGLISH PATENTS ENROLLED DURING THE WEEK ENDING AUGUST 2, 1851.

JOSEPH CROSSLEY, of Halifax. *For improvements in the manufacture of carpets, rugs, and other fabrics.* Patent dated January 28, 1851.

Claims.—1. A mode of manufacturing Brussels and cut pile carpets and rugs with thick backs. [This is effected by the employment, according to the degree of thickness required, of one or more additional sets of weft threads, which are tied in by a corresponding number of warps.]

2. The manufacture of carpets, in which printed warps showing defined figures are used, by the employment of flat or oval wires when weaving by power.

3. The weaving of carpets made with printed or parti-coloured warp, by applying thick weft to produce two similar corded surfaces, showing the same pattern on both sides of the cloth.

4. The wrapping or binding of the selvages of rugs made from printed warps, and also of rug backs for mosaic rugs (in which the pile is cemented on) during the act of weaving.

CHARLES GOTTHELF KIND, of Paris, engineer, and CHARLES ALEXIS DE WENDEL, ironmaster, also of Paris. *For improvements in the process and instruments to be used for boring the earth and sinking shafts of any given diameter for mining and other purposes, and in the means of lining such shafts.* Patent dated January 30, 1851.

The "improvements in the process and instruments for boring the earth," comprehend a novel construction of boring tool, a method of withdrawing a clod of earth to ascertain the nature of the soil at any desired depth, a compound boring tool for shafts of large diameter, and a scraper for striking the soil to the centre of the tool.

The "improvements in the means of lining shafts" have relation to those of a cylindrical form, and consist in the employment of oak cylinders or iron; the former are constructed like barrels, and retained in their proper positions, one above the other, by broad rings or hoops of iron; and the latter have internal flanges by which they are bolted together.

BENNET WOODCROFT, of Furnival's-inn. *For improvements in machinery for propelling.* Patent dated January 30, 1851.

The improvements sought to be secured under this patent comprehend—

1. Certain improvements in the construction of the blades of screw propellers, in combination with apparatus by which such blades may be turned through any or all the degrees of a circle, whereby the screw may be worked at any pitch, and whereby an increasing pitch screw may be made to move the vessel forwards or backwards by the concave sides of the blades, when using screws of an increasing pitch (which is preferred), without the necessity for reversing the engine.

2 and 3. Certain other improvements in the construction of the blades of screw propellers, in combination with apparatus by which the vessel may be moved forwards or backwards, or sideways, or be caused to remain stationary, without stopping or reversing the engine. Also, the use of the propeller as a rudder when required.

4. Causing the blades of screw propellers to overlap part of the stern-post and the rudder-post, or either of them.

SAMUEL MORAND, of Manchester. *For improvements in apparatus used when stretching and drying fabrics.* Patent dated January 30, 1851.

The present improvements are, to a certain extent, based on Mr. Morand's previously-patented constructions of finishing machinery. The links of the endless chains by which the fabric is carried through the machine, are connected with each other by joints at right angles to the surface of the conducting guides, and the chains move horizontally and laterally, instead of vertically, in returning to the feeding end of the

This arrangement of the links admits of the guides being so as to give a tortuous or serpen- o the fabric, and thus produce finish;" but when the ordinary ired to be obtained, the guides rallel with each other, and in the ital plane.

The mode herein described of and combining machinery for fabrics through machinery or or stretching and drying the

URDOCH, of Staple's-inn. *For rovements in preserving animal le substances.* (A communica- nt dated January 30, 1851.

part of this invention consists of preserving animal and vege- nces by exposing them to the current of dry air of a slightly perature, by which means the reservation of the substances is effected in their normal

ratus employed consists of a h horizontal tubes or channels and drying the air; a chamber h shelves or ledges, on which may be placed, or rods or bars they may be suspended; and a r mechanical contrivance by rrent of heated air is forced chamber to act on the sub- ained in it. The temperature d the continuance of the opera- ecessarily with the state and e articles to be treated, but a of from 65° to 85° Fahr., and om twenty-four to forty-eight ommended as suitable for most getables, such as cabbage, cauli- ach, beet-root, carrots, potatoes, all kinds, poultry, game, and , may all be subjected to the oned desiccating process after in certain cases preliminary d other operations, and they sequently treated in any way pear suitable.

d part of the invention consists l of composing a saline solu- ble to the preservation of ani- ces, such solution being used reliminary to the above de- m of desiccation, or in conjunc- y other process of preservation. n is composed of $10\frac{1}{2}$ ounces of aluminum, $10\frac{1}{2}$ ounces chlo- m, and $3\frac{1}{2}$ ounces of nitrate of lved in $2\frac{1}{2}$ wine pints of water, ed by injection to the carcases of animals in quantities varying with the sizes of the animal operated on.

The process of desiccation first described is recommended as suitable for treating glue made during the summer months, in order to dry it for preservation.

Claims.—1. The exposure of animal and vegetable substances in suitable chambers to the action of a forced current of dry air, in order to dry such substances for the purpose of preserving them, and indepen- dently of any preliminary or subsidiary pro- cesses or methods of treatment.

2. The injection of the carcases of ani- mals with a saline liquid, for the purpose of preserving them, whether such injection be employed alone or in conjunction with the desiccating process above described, or any other preliminary or subsidiary process that may be deemed advisable.

ALFRED VINCENT NEWTON, of Chan- cery-lane, mechanical draughtsman. *For improvements in manufacturing looped and other woven fabrics.* Patent dated January 30, 1851.

RICHARD JOHNSON, of Manchester, wire- drawer. *For certain improvements in an- nealing articles of iron and other materials.* Patent dated January 31, 1851.

The present improvements have special relation to the pans or ovens employed for annealing wire, and consist in the applica- tion of a central flue, through which the products of combustion from the fire are caused to pass downwards, being reverbe- rated in that direction by the closed top of the oven, before escaping through the chimney. The oven employed is of a circular form, having a ring of fire-bars all round its inte- rior; and the coils of wire are disposed in an air-tight annular case, which occupies the centre of the oven—the flue being formed by the circular space in the annular case or box in which the wire is contained.

Claim.—The use of a flue or flues situate in the interior of annealing pans or ovens.

CHARLES MARSDEN, of Kingsland-road, engineer. *For certain improvements in boots and shoes.* Patent dated January 31, 1851.

Claim.—The constructing of boots and shoes with air passages through the same, kept open at certain parts by the aid of metallic plates, or strong substitutes of lea- ther.

WEEKLY LIST OF NEW ENGLISH PATENTS.

Edward de Mornay, of Mark-lane, London, gen- tleman, for improvements in machinery for crush ing sugar canes, and in apparatus for evaporating saccharine fluids. August 5; six months.

Levi Bissell, of the City, County, and State of

New York, in the United States of America, engineer, for certain new and useful improvements in the means of sustaining travelling carriages and other vehicles, which improvements are applicable to other like purposes. August 5; six months.

Edwin Deeley and Richard Mountford Deeley, of Andman Bank, Stafford, flint and bottle-glass manufacturer, for certain improvements in the construction of furnaces for the manufacture of glass. August 6; six months.

Robert Hyde Greg, of Manchester, manufacturer and merchant, and David Bowlas, of Beddish, Lancaster, manufacturer, for certain improvements in machinery or apparatus for manufacturing weavers' healds or harness. August 7; six months.

Lockington St. Lawrence Bunn, of Walbrook, London, merchant, for improvements in the manufacture of kamptulicon. August 7; six months.

Alphonse Rene le Moire de Normandy, of Judd-street, Middlesex, gentleman, and Richard Fell, of the City-road, in the same county, engineer, for improved methods of obtaining fresh water from salt water, and of concentrating sulphuric acid. August 7; six months.

WEEKLY LIST OF DESIGNS OF ARTICLES OF UTILITY REGISTERED.

Date of Registration.	No. in the Register.	Proprietors' Names.	Addresses.	Subjects of Design.
July 31	2895	Cox and Wilson	Oldbury, Oxfordshire	Travelling label.
Aug. 1	2896	C. H. Wagnor	Birmingham	"The Flexibility Regulator" (penholder.)
,,	2897	W. Steer	St. John's-wood	Manifold-bladed razor.
,,	2898	J. Griffiths	Liverpool	Apparatus for opening, closing, and fastening skylights.
,,	2899	G. Granger	Worcester	Steam saucepan.
2	2900	W. Card	Westminster	Card's "Melodias," or flute tuner.
,,	2901	T. Melling	Ramhill Iron-works, Lancaster	Moulding box.
,,	2902	J. Whitworth	Birmingham	Button.
5	2903	S. Wilson	Glasgow	Life-preserving travelling bag.
,,	2904	D. Adamson & Co.	Hyde, Chester	Multitubular boiler.

WEEKLY LIST OF PROVISIONAL REGISTRATIONS.

July 30	268	J. Coxon	Paddington	Diagonally-seated omnibus.
31	269	W. Green	Paddington	Safety cash drawer or model till.
Aug. 5	270	J. Taylor	Birmingham	Dress fastening.

NOTICE TO CORRESPONDENTS.

The large space which we are obliged to devote to the subject of the New Patent Law—so all-important to our mechanical readers—compels us again to postpone the appearance of the conclusion of "M. S. B.'s" article on "Industrial Schools," and other valuable articles, in type.

CONTENTS OF THIS NUMBER.

LONDON: Edited, Printed, and Published by Joseph Clinton Robertson, of No. 166, Fleet-street, in the City of London—Sold by A. and W. Galignani, Rue Vivienne, Paris; Machin and Co., Dublin; W. C. Campbell and Co., Hamburg.

Mechanics' Magazine,

UM, REGISTER, JOURNAL, AND GAZETTE.

2.] SATURDAY, AUGUST 16, 1851. [Price 3d., Stamped, 4d.

Edited by J. C. Robertson, 166, Fleet-street.

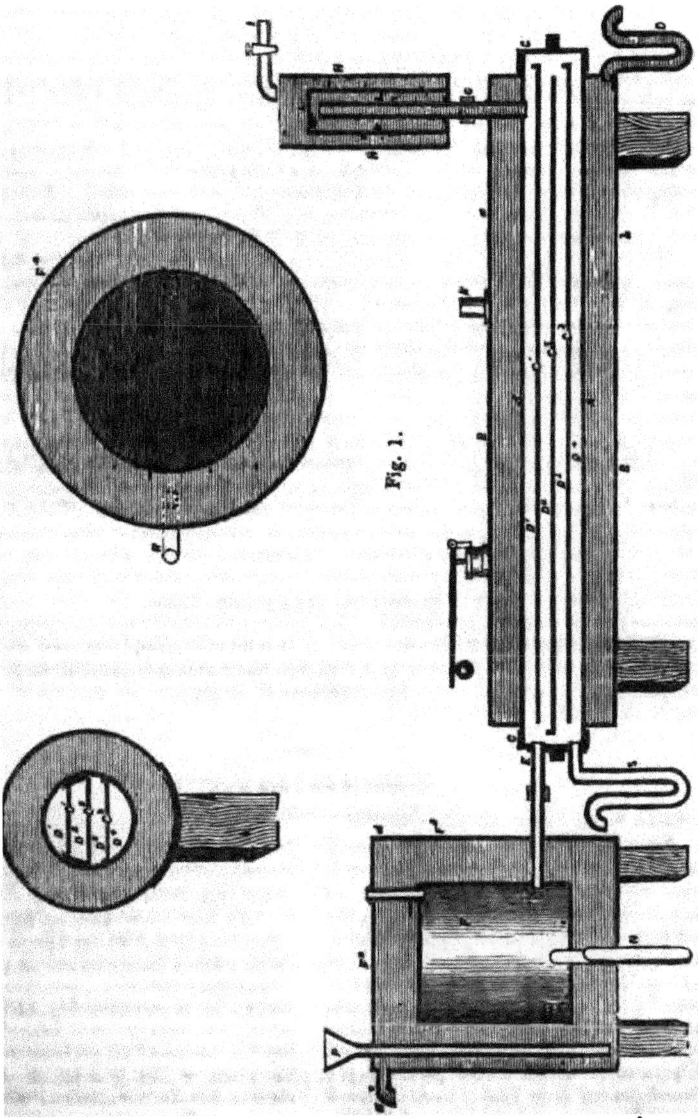

Fig. 1.

V.

MESSRS. BRIAND AND FELL'S PATENT WATER-FRESHENING APPARATUS.

(Patent dated Feb. 11, 1851. Specification enrolled Aug. 11, 1851.)

Specification.

Our invention has for its object the obtaining of fresh and pure water from salt water, and water containing mineral substances in solution, or containing other impurities which render it unfit for culinary or manufacturing purposes. The apparatus which we employ for this purpose is represented in the annexed engravings, of which fig. 1 is a longitudinal vertical section, fig. 2 a cross section on the line *a b* of fig. 1, and fig. 3 a horizontal section on the line *c d*. A is a cylinder, which is surrounded by or encased in a steam jacket B; C C are caps or covers to the cylinder. C^1, C^2, C^3 are a set of partitions, by which A is divided into four compartments or channels D^1, D^2, D^3, D^4; and each of these partitions is turned up at one end, so as to allow the water to flow along them towards the opposite end only. At one end the cylinder A is connected by a pipe E, with a condensing and aërating vessel F, which is contained within a closed tank F^2. At the other end the cylinder A communicates by a pipe G with a box H, into which the water to be purified is first introduced from some source of supply connected with the tap I. K is a pipe, by which steam from some steam-generator is admitted into the steam jacket B, so as to heat the cylinder A. (For steam, hot air may be substituted.)

While steam is thus being supplied to the steam-jacket, the water to be operated upon (whether salt or other impure water) is allowed to flow from the tap I into the box H, and down the pipe G into the cylinder A, where it falls upon the upper surface of the plate C^1; but to prevent any of the water returning in a state of vapour through the pipe G, the upper end of that pipe is sealed by a hood L, by which free communication from the the interior of the cylinder A with the external atmosphere is entirely cut off in that direction. The water which falls upon the plate C^1, after running along the channel D^1, falls down upon the next plate C^2, by which it is conveyed back again to the other end of the chamber, where it falls upon the plate C_3, along which it flows in a similar manner till it falls into the compartment D^4. While the water is thus conveyed successively from one end of the chamber to the other, it gets heated, and a portion becomes vaporised, and goes off by the pipe E into the condenser F, where it gets mixed with atmospheric air admitted by the pipe M, and is finally condensed by coming in contact with the sides of the condenser, when it is run off by the inverted syphon-pipe N and collected for use. O is another inverted syphon from which the condensed steam which has been employed in heating the cylinder A is collected. The water obtained from this pipe not being aërated may be employed for washing. P is a pipe for supplying cold water to the condenser, and R another pipe by which the water as it gets heated escapes. S is an inverted syphon by which the residuum of the impure or salt water operated upon escapes from the cylinder A.

INDUSTRIAL SCHOOLS.

(Concluded from page 66.)

Notwithstanding it has been already advanced in the *Mechanics' Magazine*, that *religion should be the basis on which education should be founded*, the subject cannot now be silently passed over, seeing the movements that have of late been made to render education merely secular. The advocates of this project profess themselves to be Christians: now it is acknowledged by all Christians that Holy Scripture is the foundation of their faith; that, to speak *humanly only — the* precepts of our Saviour are all of them such as the moralist would ardently desire to see observed; then, why should the lecture of Holy Writ be denied to schools? The different views that have been taken of some particular passages, in the New Testament especially, have, it is true, given rise to considerable variations in ceremony, and even as to some doctrines; but Christians of all persuasions refer to Scripture as the foundation of each of their particular ordinances and observances. There is one school, "The and Foreign," wherein the Bible read, but without any comment, to its application to the peculiar r practice of any particular church ; the children are very remark-ll-grounded in Scripture in this and it is applied therein to the ion of morality, but not to the of any particular doctrine or ies; hence parents of very dif-eligious credence do not hesitate their children to this school—of the Established Church of d, no less than Dissenters of all nations—certain that their off-vill therein be grounded in Chris-wledge as set forth in Scripture; it at the same time it will rest with themselves (the parents) to he application of this knowledge own interpretation of it. What objection that can be made to instruction when given in this ?

ebra, mathematics, carpentry, it is true, be taught without refer-Scripture or religion; but there y branches of purely secular edu-hat can *not* be imparted without g into the subject of religious his-example. In that of our own , to go no further, how uninstruc-w incomplete would it be, if di-of many events, the rise and s of which have been solely on of differences in our religion! continental history has been influenced by religion, and can-detailed without specifying the s motives that have given rise to able changes in policy, in man-d in freedom. Should the Cru-e omitted in a secular course of ? If so, how will the great changes rought in the habits, the wealth, lustrial progress of the people be ed for? So in the history of the orld, and in the ancient history Old one, Pagan rites and super-must of necessity be spoken of, how intimately they influenced ceedings of states. Voyages and could not be divested of their ces to religion; nor could the rise gress of the mundane art—archi-, be well taught without giving s of temples, both those of the ns and those for Christian wor-stronomy and natural history lead both the child and the man to reflect on the wonderful harmony and suitableness to purpose that pervade the heavens, and the earth, and all of its productions—is the schoolmaster to be debarred from replying to the child's inquiries of how all these prodigies originated? Impossible! It will indeed be found impossible to confine instruction to that which is purely secular.

Scripture should not, however, be made the *spelling-book* of schools; the portions of it read should be so with due reverence, consequently no boggler should be permitted in school to read aloud any portion of Holy Writ.

Another point which certainly is not sufficiently considered by advocates for the better education of the people is this, How far should it be *gratuitous*? We all admit that no one should be left by the nation to die of want—we willingly provide food, lodging, clothing, for the *pauper*, and education for the *pauper* child—well that it is so. This provision for the *pauper* affords, however, no reason for reducing in a mass the industrious classes to pauperism; yet this it is that is attempted by the project of providing education to all children indiscriminately, and gratuitously, by a rate on householders, that is, an additional *poors-rate* under another name. There seems no more reason for teaching gratuitously the independent honest workman's child either geography, history, or even reading, than there is for finding the father and his whole family in food, clothing, and lodging. If our population has had the effect of reducing the wages of labour below the amount that can provide for a family, it would seem more real benevolence to raise the rate of pay for work done, than to provide gratuitously for any of the necessities of the artificer or labourer, whether clothing, bread, or education for his child. Indeed a similar view of the question seems in some quarters slowly to be gaining ground, in the hope of keeping up the ancient independent spirit and self-respect of the industrious classes. The present Dean of Hereford, whose success in the school of his late parish, King's Somborne, is so well known, made that school nearly self-supporting, though in a very poor parish. The school fee was paid for *all* the children, even for those of farm-labourers, whose wages in some cases were as low as six shillings a week; moreover, all school books were paid for by the *parents*, remained their property, and were taken home by the children after school. This, the very reverend Dean says in his publication, not only maintained the self-respect of parents; but the taking home of books contributed materially to the home comforts of family evenings, to the good preservation of the books, and to the keeping up of school learning after children were taken away from school, that they might earn a pittance in the field. At King's Somborne, the school fee varied in proportion to the means of parents; at the British and Foreign School, Southwark, it is higher than in the National schools,—being 2d. and 3d. a week; an approximation to the self-supporting system, though insufficient to cover all expenses.

Here exists proof positive that artificers out of their usual earnings can afford excellent school education for their children; what one such man can do with ease, others might equally well accomplish if similarly circumstanced. Take, for example, an instance that can be vouched for; a sail-maker earning thirty shillings a week—not the highest pay given in his craft, but that for which this steady man has been content for many years to work for the same master. This artisan fearing for his children the contamination frequent in national schools, has all along placed his offspring in a private school, paying sixpence a week for the two eldest, girls, and for some time past, for a little boy, fourpence a week, leaving only a baby without schooling. The girls have acquired good proficiency in reading, writing, arithmetic, needlework, knitting. The eldest girl, at nine years old, knitted in school a set of anti-macassars, for which she was paid enough to purchase materials for a little winter cloak; the two girls do needlework for the family, of the average weekly value of their school fees. These children all attend school with pleasure; so much so, that when now and then they are with an aunt for a week's country air, they entreat as a favour to be allowed to pass the day in the village school. Neither father nor mother abstain from any of the comforts of their station; they and the *children are always respectably clad*, their habitation consists of three well-furnished rooms, the whole family have at least one meal of meat a day, they can afford an occasional country visit to a relative, and have withal money at all times in hand, ready for the purchase of edibles in large quantities when they can be so obtained at a less price than when procured by driblets.

Let it not, then, be credited that the prudent artificer needs to be *pauperised* by providing gratuitous education for his children.

National schools,—though a penny a week be paid for each child,—are, in fact, *pauper* establishments, since so little as a penny a piece cannot provide suitable masters, or even books, slates, and paper. Whatever the amount of education afforded, all schools that are not self-supporting are literally *pauper* schools; and *paupers*, it may be said, are all persons who forfeit their independence, so far as to accept such pauperising schooling for their children. This, though the true light, is not that in which such schools have been viewed; consequently, very many self-respecting parents avail themselves of the instruction national schools so cheaply afford to children. Far from intending to cast any slur on those who, under present circumstances, avail themselves of these institutions, the object of these observations is to rouse attention to the mischievous effects to be apprehended from a parish rate for general educational purposes.

For simplicity sake, the above has been in general stated as if in relation to boys only, but in every instance the observations and suggestions are equally applicable to girls, and should be considered as in relation to them no less than to boys.

In looking minutely into the actual state and practice of national schools, there seems one particular for which it seems extremely difficult to provide a remedy; namely, abandonment of the decidedly unruly pupils having more contaminating and prejudicial vices. A child that has proved particularly unmanageable or vicious, is usually expelled from school by the monthly committee; yet children, so excluded, are as the lost sheep for whose recovery there would be rejoicing; still the reclamation of an erring individual ought not to out-reign a probability that his depravity might lead to similar vices in scores of his companions. There does indeed exist an institution where depraved young girls are received, and most of them happily reclaimed,—the "School of Discipline" at Chelsea. This meritorious establishment is partly maintained by voluntary subscriptions,—partly by weekly payments for the children: the lowest sum at which a girl is received is half-a-crown a week, for which sum she is boarded, lodged, clothed, and instructed. When not admitted into some such institution, the child turned out of a school becomes a wanderer in the streets, acquiring there additional vices from bad companions, and is usually altogether lost to virtue, whilst the inmate of the school of discipline becomes moral, and so well instructed in domestic duties that the demand upon the school for servants is greater than it can supply.

Humanly speaking, that which at present appears to be the first essential step in giving better education to the rising generation of the industrial classes, is to keep children and youth out of the way of street contamination; and secondly, to inspire them with sentiments of self-respect, and with a reliance on their own industry and moral conduct for present comfort, future independence, and general estimation in this world. Here, as throughout the whole series of these Hints on Education,* it is left to the care of spiritual pastors to inculcate the importance of religious duties, and their bearing on the hopes of a future glorious immortality.

M. S. B.

THE SAFETY-LAMP. IMPROVEMENTS SUGGESTED BY MR. GOLDSWORTHY GURNEY.

A deputation from the coal-miners of Northumberland, Durham, and Lancashire recently applied to Mr. Gurney to know how the Davy Safety Lamp might be best improved. Mr. Gurney has made a report to them (published at length in the *Mining Journal*), from which we make the following extracts:

"From a series of experiments which, some years since, I carefully made with accurate photometers, it was shown that the aggregate amount of loss of light by the safety-lamp was upwards of three-fourths of the whole, compared to that quantity of light which the same flame gave out without being surrounded by wire gauze. The proportion lost horizontally is no measure of the whole or absolute quantity, and the loss is far greater upwards and downwards than horizontally, because the rays strike the gauze obliquely, and cannot so readily pass the meshes of the lamp. The obliquity of the wires in the gauze prevents the rays from passing through; these rays, thus intercepted, falling on a dark unreflecting surface, are lost, and practically so for ever. * *

"The angle of reflection of light, as you know, is always equal to the angle of incidence; and when these angles, on polished surfaces, are so made that all the rays of radiated light; viz., the first or incidental rays, shall be reflected parallel to each other, the loss by reflection is very little. It is scarcely appreciable even at considerable distances. Opaque matter floating in the atmosphere partially interrupts the passage of these rays, otherwise there would be little or no loss.

"The Bude Light, when placed in a true parabola (a reflector made to concentrate all the radiated rays), and converted into a bundle of parallel ones, was seen 94 miles on a clear night, in an experiment on the Pyrennees from a mountain where the curvature of the earth permitted; and the oxy-hydrogen lime light, in an experiment made some years since by the Trinity House, placed in a parabola at Purfleet, threw a visibly distinct shadow on a white screen at Blackwall; the distance is 11 miles. I mention these facts, because I mean to show you that it is possible, by taking advantage of some of the laws there in operation, to improve the illuminating power of the safety-lamp. Light which falls on polished bright surfaces, it is found, is reflected back again without much loss—that from a good mirror, for instance; but rays which fall o[n] dark surfaces are absorbed or destroyed.

"It is manifest in your lamp that all th[e] rays of light which fall on the surface of th[e] dark unreflecting wire of which your lamp [is] made are intercepted, and lost for ev[er]. Now, taking into consideration the kno[wn] laws of reflection, it is equally manifest t[hat] if the gauze presented a reflecting instead [of] an absorbing surface, the rays intercep[ted] by the wires would be reflected back, [and] turned to account. The first suggest[ion], therefore, that occurs for the improvem[ent] of your lamp, with a view of getting [more] light, is to substitute a burnished or pol[ished] wire, of which to make the gauze, inste[ad of] the black iron wire, of which it is now [made].

* *Mechanics' Magazine*, Numbers 1431, 1432, 1435, 1440, 1441, 1447.

Every wire would then have a reflecting surface; every ray of light falling on and intercepted by it would be reflected back, and find its way through the meshes or openings of the net-work—perhaps not at the first reflection, but on the second or third; for if the reflected ray, in the first reflection, chanced to fall on a wire on the opposite side instead of an aperture, it would be reflected back the second time, and most likely find its way out on the other side. From a series of experiments carefully made with a lamp so constructed, the loss of light does not appear to amount to one-eighth of the whole. Silver-plated wire, drawn through a burnisher, is highly reflective, and the increase of its expense over the common wire is so little, that the extra quantity of light gained by it will more than cover the difference of cost in a short time.

"Let us now consider the loss of light at the lower part of the lamp, downwards. This is occasioned by the light falling on, and being intercepted by, the oil-can. This vessel, in your lamp, has a dark absorbing surface, and as the flame is placed low down in the lamp, it cuts off a considerable quantity of light, otherwise available. To avoid this loss we must also take advantage of the laws of reflection. It is evident, if the oil vessel was also made with a reflecting surface, the rays of light which fall on it might be turned to account; its surface might easily be made to reflect the rays at such angles as would be most useful. * * * *

"We will now proceed to consider whether the quantity of light given out from the flame itself might not be increased. In all burning of oil the absolute quantity of light given out from its flame will always be in a direct proportion to the perfect state of its combustion. An open flame without a glass chimney does not give out so much light from the same flame as with one. An Argand lamp is an instance. The reason is that the glass chimney produces a current of air which strikes the flame of the burning wick, and renders the combustion of the oil more perfect. The glass chimney acts, in producing a current of air upwards, like the upcast shaft for ventilation in your mines. You are aware that the upcast shaft occasions a stream, or current of air, at and around to a considerable distance from its base. We may take advantage of this fact to obtain extra light in your lamp. If the flame be surrounded by a glass chimney, it would, possibly, break, and be subject to other objections; but if a small metallic chimney, smaller at its base, be placed a little above the level of the flame, the current produced by it will draw the air in sideways, *and strike at the point of greatest* intensity of combustion, and effect as perfect a combustion as if a glass chimney surrounded it. This metallic chimney should be made of very thin iron-plate, 4 inches long, and about three-fourths of an inch diameter. It may be secured in its situation, rather above than below the level of the apex, by iron stays. All smoke would be consumed, even if the lamp fell on one side.

"This chimney has no practical objection. It would not break nor be injured by the flame, even if the lamp be turned on one side; and in such case it would pass the flame through its centre, and prevent any smoking or injury of the polished wire-gauze. The obstruction of light by it upwards will not be so much as might at first appear. The side of this cylinder will be in a line with the direct rays of light.

"In confirmation of the action of these upcasts, I may state that I placed metallic chimneys on the burners in lighting the House of Commons ten years ago. They have been used ever since, and are still used there with practical success. They are placed above the apex of the flame, to pull in a strong current across the point of greatest intensity of the flame, in the modification of the Bude Light, with which the House is lighted. It may be observed, in passing, the globes of ground glass, which surround the burners in the House of Commons, to soften the light, have a net-work of wire weaved around them, of large mesh, to prevent any separation of the glass in case of breakage. The meshes are so large that they do not obstruct the light, and the wires so small that they are not visible. This net-work might be similarly used about the glass of the safety-lamp, as modified by Dr. Clanny and others, if you use it to prevent the glass falling out in case of breakage."

THE PATENT LAWS.—THE LATE DEFEAT.

The conduct of Government in regard to this measure has been altogether most extraordinary. Lord Granville, as their organ in the House of Lords, succeeded in passing through that House a Bill for the Amendment of the Patent Laws, which was represented to embody as large and complete a measure of reform as circumstances would admit of. The Bill is then taken down to the Commons; where, strange to say, the chief opposition it encounters, is from the Government's principal legal organ in that House,—the Attorney-general. Sir Alexander Cockburn takes charge of the Bill; but so little is he pleased with it as it has come out of the hands of the Lords, that he must needs have some thirty new clauses introduced, involving as many changes and alterations,—some of a formal character merely, but others directly affecting the fundamental principles of the measure. Then Sir Alexander pushes the Bill, with this vast heap of additions, through the Commons, in the teeth of a strong opposition, founded chiefly on the circumstance that the session was on the point of closing, and that time was wanting to give the measure that full and calm consideration to which it was entitled. Sir Alexander is deaf to all remonstrance, and, backed up by the forces ever at the command of the ministerial whipper-in, has the Bill read a third time and passed. The Bill is thereupon carried back to the House of Lords to have their assent to the numerous changes made in it, when it is at once met *and quashed* by an intimation from the Lord Chancellor that, though the Bill, as amended, dealt very freely with the functions of his high office—abolished them, indeed, *quoad hoc* altogether—he had never once been consulted about it! Was there ever the like of such want of concert and management in the members of one and the same administration? What was Lord Granville about that he did not take counsel of the Attorney-general in the first instance, and have the Bill adjusted to his liking before it left the Lords, so that it might be sure of *his* full support, at least, when it reached the Commons? What again was Mr. Attorney about that while literally forcing the Bill through the Commons, with as much loquacious pertinacity as if the existence of the Ministry depended upon it, he could not find a leisure moment to acquaint the Lord Chancellor with the sort of measure which was about to come before him, and to secure for it, if possible, his lordship's approbation and assent? And what were Ministers about that they could allow Mr. Attorney to drag the Government through such a week of angry wrangling and bickering as that with which the session closed in the Commons, only to be snubbed, and put down at last by their own Chancellor in the Lords?

Sufficient it was for the justification of the Lord Chancellor's part in this comedy of blunders, that he was asked in the last moments of the Session to concur in amendments of which he knew nothing, and which he could not possibly have then time to consider; but had reasons been wanting, his lordship would have found them in abundance on a critical examination of the Bill as returned to the Lords. Our readers are already acquainted with our own objections to the Bill on the general grounds of its injustice to inventors, and tendency to impair the industrial resources of the country; but there are other defects in it, which, though merely of a technical character, and therefore passed over by us, could not fail to have caught the attention of so acute a legal mind as Lord Truro's, and insured of themselves the rejection of the Bill. For these grounds, we beg leave to refer to the following paper of "Objections," which was drawn up by an eminent member of the common law bar of extensive practice in Patent cases, and placed in the hands of the Lord Chancellor just previous to the "closing scene" of this memorable affair.

Objections.

The Bill if passed will take away the right of appeal to the Lord Chancellor against the grant of a patent, and no other right of appeal is given in lieu of it.

The power to make patents is by the Bill transferred to a Board of Commissioners, who will in making a patent act in pursuance of a warrant directed *to them*.

The Lord Chancellor will not receive any Privy-seal Bill, or other warrant for making a patent, and he will not therefore as Lord Chancellor have any jurisdiction.

The proviso at the end of Section 16 seems to have been left in the Bill by mistake; it was no doubt a useful proviso in Section 10 of the Bill as originally framed, but it is difficult to see how the Lord Chancellor can exercise any power, authority, or discretion in respect of a warrant not directed to him, and which is not in his custody, power, or control.

The consequence seems to be, therefore, that all appeals against a decision of any law-officer will be virtually taken away, and the Bill makes no provision for any appeal from a law-officer to the Commissioners.

The Commissioners do not indeed appear to have any judicial powers vested in them, nor does the Bill in any way authorise them or the law-officers to take evidence upon oath.

Clause 37 of the Bill respecting disclaimers and memorandums of alteration will be inoperative, and therefore the power of entering disclaimers will, in effect, be taken away as to patents granted under the new law.

The clause enacts, That the provisions of the 5 and 6 Wm. IV., c. 83, respecting disclaimers, shall apply to patents granted under the new law, but those provisions apply to English, Irish, and Scotch patents solely, and they only authorise the issuing of a fiat, by an English law-officer "in case of an English patent;" by a Scotch law-officer, "in case of a Scotch patent;" and by an Irish law-officer, "in case of an Irish patent."

The patents granted under the provisions of the Bill will be for the whole of the United Kingdom, and not therefore within the meaning of the power contained in 5 and 6 Wm. IV., c. 83, s. 1.

And if it should be contended; that a patent for the United Kingdom is also a patent for each of the parts, it is to be observed that there is no provision in the Bill *as to which of the law-officers* of the Crown may authorise the entry of a disclaimer.

The provisions of Section 39 of the Bill, respecting "the place or places at or in which, and in what manner the invention is alleged to have been used or published," &c., will impose great hardship upon a defendant, for the evidence in support of his case as to the validity of the patent will always be known to the plaintiff without the defendant possessing any corresponding advantage.

The proviso, at the end of Sect respecting the right to begin, will, in render the proceedings by *scire fac* practicable, and there is no saving actions or patents prior to the pa the Bill.

If a defendant as well in a *scire fa* also in a patent action were to be all address the jury after the whole of t dence in the case has been given, m tice would be done than by the pro Section 39.

The Bill when passed will also, in entirely take away the writ of *scire* to repeal any patent granted under i visions, and there is no other pro substituted for the action of *scire fa*

The writ of *scire facias* it is well can only be founded upon a record court out of which it issues, and as p will cease to be made in Chancery af passing of the Bill, they will not be e in that court, and consequently the be no record upon which such a writ issue.

The Bill is, indeed, silent as to any ment or other record being made of p granted under it provisions.

COAL-WHIPPING BY STEAM.

Messrs. E. and A. Prior, of Thames-street, have patented a pl discharging colliers by steam power Monday last we were on board the brig *Crocus*, 365 tons burthen, and nessed the operation. The engine, whi designed by, and constructed und direction of Mr. F. H. Trevithick, up but little room on the deck of a v the action is direct, the piston-rod w the crank axle, on which is the drum winds the chain, drawing up the box below. The engine, when not at wo kept in a barge built for the purpose its mast, tackle, chains, &c., by wh can be slung on board with the gr facility, and taken away when the h the vessel has been emptied of its con The boxes used hold 5 cwt., and by plan more than twenty tons per hour c delivered into the barges. A 70-ton has been filled in three hours; and o occasion 209 tons were delivered from sel in nine hours and a half, with one consisting of four men; but the oper is yet young, and from the observations made on our visit, we have no doubt a quantity of 30 tons per hour may be accomplished. All parties connected with the coal trade are well aware how important is not only the mere cost of whipping coals, but the saving of time, and the prevention of that detention to which the shipping in the trade has hitherto been subject. A gang of men, by the aid of ropes and blocks, could barge about 50 tons a day.—*Mining Journal.*

ON THE ABUSE OF INFINITE SERIES.

Liebnitz having, as he thought, proved by a metaphysical argument that the Infinite Series

$$1-1+1-1+1-\&c.,$$

must be equal to $\frac{1}{2}$; confirmed his result by observing that the fraction

$$\frac{1}{1+x}$$

gives, when developed,

$$1-x+x^2-x^3+\&c;$$

and that when $x=1$, the fraction becomes $\frac{1}{2}$, and the series

$$1-1+1-1, \&c.$$

It does not appear to have been perceived how very fallacious this latter argument is. In fact, by developing the fraction

$$\frac{1+x+x^2 \ldots\ldots\ldots +x^{n-1}}{1+x+x^2 \ldots\ldots\ldots +x^n}$$

we arrive at a very different result. The numerator of this fraction

$$=\frac{1-x^n}{1-x};$$

the denominator

$$=\frac{1-x^{n+1}}{1-x}.$$

Therefore the fraction

$$=\frac{1-x^n}{1-x^{n+1}}=(1-x^n)\left\{1+x^{n+1}+x^{2n+2}+x^{3n+3}+x^{4n+4}+\&c.\right\}$$

$$=1-x^n+x^{n+1}-x^{2n+1}+x^{2n+2}-x^{2n+3}+\&c.$$

When $x=1$, this also degenerates into $1-1+1-1+1-1$, &c. And the generating fraction becomes

$$\frac{n}{n+1}.$$

So that, according to the doctrine of infinite series, that which Liebnitz thought he had proved to be equal to $\frac{1}{2}$, is $\frac{1}{2}$, $\frac{2}{3}$, $\frac{3}{4}$, $\frac{4}{5}$, &c.; any one of them which you please!

Again;

$$\frac{1}{1+x+x^2 \ldots\ldots +x^n}=\frac{1-x}{1-x^{n+1}}=(1-x)\left\{1+x^{n+1}+x^{2n+2}+x^{3n+3}+x^{4n+4}+\&c.\right\}$$

$$=1-x+x^{n+1}-x^{n+2}+x^{2n+2}-x^{2n+3}+\&c.$$

Make $x=1$, and this also becomes $1-1+1-1+\&c.$; while the generating fraction changes into

$$\frac{1}{1+n}.$$

So that the proposed series is now proved to be equal to $\frac{1}{2}$, $\frac{1}{3}$, $\frac{1}{4}$, $\frac{1}{5}$, &c; any one of them.

These observations are not intended to disparage this branch of learning, but only to show that the most plausible arguments into which the doctrine of infinity enters are to be received with caution, and that even the most illustrious writers err on those subjects with which they are most familiar. I hope, in a future Number, to set forth what I esteem the true theory of series, and to show that in using them we are not obliged to borrow anything from metaphysics, nor to introduce any principles which do not possess that simplicity and clearness which belong to the elements of science.

ALEXANDER Q. G. CRAUFURD, M.A.

MR. ROCK'S PATENT SPRINGS.

(From Patentee's Specification.)

I first take a blade of spring steel such as is generally used, and having prepared a mould (of cast iron) with a longitudinal hollow, of a tapering form similar to that shown in fig. 1 or in fig. 2, I heat the steel, and set it down into the said hollow by means of a "fuller" and hammers, so as to raise a hollow rib on the steel corresponding in shape to the hollow in the mould. This having been done, the spring is complete with the exception of the ends, which may be formed according to any of the usual methods. Or, instead of following the method just described, the springs may be made of the same form by means of rollers formed with appropriate ribs and hollows.

When a spring of greater strength is required than can be given in a single blade or leaf, I use two or more blades of similar form, placing them together in the ordinary manner of a laminated spring. For certain purposes—as where the spring may have to sustain irregular action—I place together two leaves formed in the manner above described, with their concave sides inwards; and I fill up the concavity with gutta percha or other soft substance, to prevent vibration and exclude wet. By this arrangement a kind of hollow tube, with side flanches, is formed, calculated to resist forces acting upwards as well as downwards; but as the force will in practice generally be in excess downwards, it is necessary to draw or thin down the ends of the auxiliary plate, in order that it may always keep close to the main plate, and not "gape" when under pressure. In springs of this kind, I call the plate or leaf upon which the eyes or other end fastenings are formed the "principal" plate, and the other I call the "auxiliary" plate. Fig. 3 represents one of these springs in its complete form; A is the principal plate or leaf, B the auxiliary. Fig. 4 is a cross section of the same on the line *a b*.

Fig. 1.

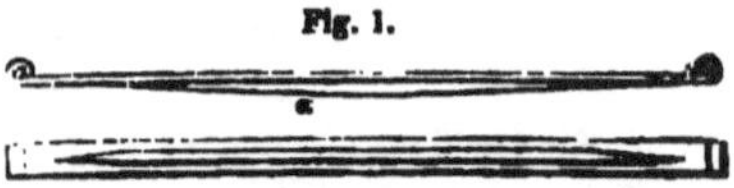

Fig. 2.

Fig. 3.

Fig. 4.

Fig. 5.

Fig. 6.

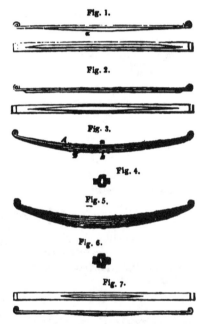

Fig. 7.

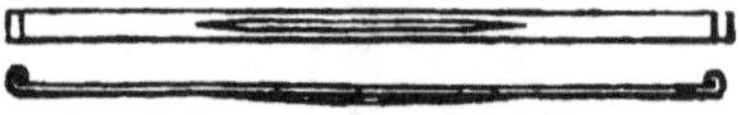

When very great strength is required, I double either the principal or auxiliary, or both, according to the strain to be resisted. Fig. 5 is a side view of a spring made with two principal plates and two auxiliaries, and fig. 6 a cross section of the same at or near the centre.

The form of rib most suitable for these improved springs is founded on the following considerations. In consequence of the nature of the work performed by springs rigidly fixed to bearings or blocks at a certain part or parts, and free to all other parts, it is requisite (to ı proper degree of strength) that uld be almost wholly inelastic at ; where they are fixed to their ı. Since, therefore, in the im- spring the strength is given by ring rib, and not by the number ı of varying length, I make the er, that is, with a greater versed he bearing-place than elsewhere; rder that the other part of the ay not be so stiff as to master the ere it leaves the bearings, and e spring to break at that point, I ı the height of the rib somewhat from the bearing towards the the blade through about one- the distance, then continue the her third at a less rapid rate of ion, and for the remaining third, : exception of a small space near it, which is quite flat, I diminish th of the rib so as to cause it to te in a point. I mention this of distance to give an approxi- ea of the form which I use, but intending to restrict myself to act limits. Before the spring bent or curved the rib appears, ewed in longitudinal section, as of double curvature, as seen at

.

ıer material point to be noticed springs is, that in cross section near the bearing, the rib has a thickness at its apex than at its here it joins the flat parts of the hus giving it greater power to mpression at the part where that s most useful. This difference ness diminishes towards the point spring, and is scarcely or not at rvable as the rib approaches the ıg point.

same objects which I effect by the height of the rib, I also hough not so certainly, by vary- width in the manner shown in

n a very easy spring is required, a larger portion of the spring to- he ends flat—that is, without any shown in fig. 7.

MICAL PRODUCTS OF PEAT.

ort on the chemical products of Irish s been presented by Sir Robert s Director of the Museum of Irish Industry. The inquiry was undertaken at the suggestion of Lord Clarendon, in consequence of the numerous applications for information by foreign authorities, and by individuals interested in the welfare of Ireland, consequent upon the announcement of the method patented by Mr. Rees Reece and the establishment of the Irish Peat Company. The two principal points to be arrived at were the actual nature and quantities of the products that could be obtained, and the economic conditions of the process so as to determine its importance as a commercial speculation. With regard to the first—namely, the products obtainable—the result of the experiments was to establish the general correctness of Mr. Reece's statements, the comparison of the materials gained by that gentleman with those gained by Sir Robert Kane being as follows:

From 100 parts of peat.	Statement in Mr. Reece's prospectus.	Average results of Museum trials.
Sulphate of ammonia	1·060	1·110
Acetate of lime	·700	·305
Wood naphtha	·185	·140
Paraffine	·104	·125
Fixed oils	·714 }	1·059
Volatile oils	·357 }	

With regard to this comparison, Sir Robert Kane remarks:

"It is evident that the quantity of ammonia obtained is rather greater than that expected by Mr. Reece; secondly, that the quantity of oils and paraffine may be considered the same; thirdly, that the quantity of wood naphtha expected by Mr. Reece is more than we obtained in average, but not more than was obtained in some Museum trials. That the quantity of acetate of lime expected by Mr. Reece is more than double that which was in average obtained in the Museum, unless the commercial acetate of lime calculated for by Mr. Reece shall contain such excess of lime, &c., as shall render its weight double that which the pure article, calculated in the result of the Museum trials, should have. This latter circumstance may possibly explain the difference.

"It may, therefore, be admitted that the statements made as to the quantities of those bodies obtainable from peat have not been exaggerated, and, indeed, are such as should immediately be inferred to be obtainable from a body of its constitution compared with coal and wood."

The amount of produce being thus determined, the Report touches upon the other question; namely, the cost of its production. This is described to be so difficult as to preclude the possibility of any absolute conclusions. The facts that the manufacture is of a novel character, involving numerous and complex collateral operations, and that it is to be established in localities and amongst populations unaccustomed to manufacturing industry, are referred to as calculated to have an important influence in augmenting the expense. At the same time also some strong doubts are expressed as to the practicability, on a large scale, of the mode upon which Mr. Reece relies for an inexpensive supply of the heat necessary for conducting his entire operations, which consists in employing as fuel the gases produced in the blast furnaces by the distillation of the peat after all their condensible vapours shall have been secured. On the whole, however, even the conclusions thus arrived at, and which have evidently been framed with all the caution that would be generated by a sense of the responsibility of encouraging extravagant views, are not unfavourable to the hope that the process on its introduction may prove sufficiently profitable, and that eventually, as the difficulties attending it shall gradually be lessened by experience, it may realize some of the sanguine expectations originally indulged in regarding its possible influence in the regeneration of Ireland. Subjoined is the final paragraph of the Report:—

"Although the excessive returns stated by the proposers of the manufacture may not be obtained, it is yet probable that, conducted with economy and the attention of individual interests, the difficulty connected with so great complexity of operations would be overcome, and the manufacture be found in practice profitable; and certainly it must be regarded as of very great interest and public utility that a branch of scientific manufacture should be established specially applicable to promote the industrial progress of Ireland by conferring a commercial value on a material which has hitherto been principally a reproach, and by affording employment of a remunerative and instructive character to our labouring population."

SPECIFICATIONS OF ENGLISH PATENTS ENROLLED DURING THE WEEK, ENDING AUGUST 12, 1851.

ALFRED VINCENT NEWTON, of Chancery-lane, mechanical draughtsman. *For improvements in manufacturing looped and other woven fabrics.* (Being a communication.) Patent dated January 30, 1851.

The *first* part of this invention has relation to looms for weaving Brussels and tapestry carpets, and other piled fabrics. The inventor of the present improvements obtained former letters patent for weaving Brussels and other carpets, in the specification of which was described a method of carrying and supporting the pile wires as they passed between the warp; and that part of the present improvements by which the same object is effected, consists in the employment of guides or supports, on or through which the pile wires slide as they pass between the warps. This part of the invention also consists—1. In a method of causing the moveable guides or supports, which have a vertical movement, to open and close so that, to ensure the introduction of the wire in a proper position between the warps, it may be introduced when the guides are closed, and may be liberated when in its place by the opening of the guides or supports. 2. In the employment of a stop motion, acted on by the wires as they are introduced, so that in the event of the wires failing to be introduced, the loom may be stopped.

The *second* part of the invention consists in a method of regulating the tension on the warps, and their delivery from the bobbins used in looms for weaving Brussels and tapestry carpets and other figured fabrics, and for other purposes where a uniform tension is required; and this is effected by suspending a weight on the yarn in front of the bobbin or spool, which is so arranged as to give off the yarn at intervals (when the slack is taken up) by means of a catch, which holds it from returning until raised by a weight.

The *third* part of the invention has relation to looms for weaving tapestry carpets. In this description of manufacture, three warps are generally employed—the ground warp, which controls the length of cloth produced by each beat-up of the lathe; the slack ground warp, for forming the back of the fabric; and the figuring warp, by which the figuring loops or face of the fabric is produced. The latter is previously printed with the desired pattern or figure so elongated that when the warp is drawn up into the figuring loops, the pattern shall be shown of a determined size. It is highly important that each of the figures in the carpet should be of a regular length, in order that, when the carpet is put together, the pattern may be accurately matched; and it is, moreover, desirable that the body of the fabric should be of a uniform texture throughout. When the figures are produced by the Jacquard, it is only necessary to measure out the cloth; but in tapestry carpets, the figuring warp must be also measured out, in order to prevent the figure produced from being imperfect. The improvements specified under this head of the invention, and by which the above objects are attained, consist—1. In regulating the delivery or giving-off of one parate warps or chains by the ach separate warp, in combina- ground or controlling warp, ates the length of cloth made at p of the lathe; the delivery of or controlling warp being also y its own tension, and fixed or y a brake when the lathe beats ; and the whole being combined ar and positive take-up motion up the finished cloth, whereby re produced of a regular mea- l, and the body of the cloth is, practicable, of uniform texture. ployment of an index measuring o indicate the amount of figur- en up in the process of weaving, e the tension to be varied when than sufficient figuring warp is in. (When this is combined x measuring-apparatus to indi- ount of cloth woven, the length on the figuring warp and on the s may be ascertained to deter- ire figure—the delivery of the self-adapting to any inequality w in the formation of the loops.) mployment of grids or fingers dependently of the lathe, for ward and exerting a tension ft before the lathe completes its novement, and also for retaining en the lathe commences to go vent it from being drawn back- retiring movement of the lathe. th head of the invention has a method of delivering out the ps in weaving tapestry carpets, f which is to prevent any imper- e pattern owing to irregularities s of figure on the figuring warp; effected by attaching to deter- s of the figuring warps a clamp worked in connection with end- ving a given and positive deli- , by which means the length of ed will be varied according to arities which may exist in the s elongated figures on the figur- ilst the length of the figure will r.

part of the invention relates to eaving cut pile fabrics. In car- improvements into effect, the ploys the longitudinally grooved s in ordinary use when weaving ms. The woven cloth passes having cogs which enter between ad is properly presented to the eciprocating knife, which works ve on the figuring wires without ming in contact therewith, and s the face of the fabric, and in succession cuts each range of loops as they are brought under its action by the intermittent motion of the carrying rollers. When the pile is cut, the wires are pushed out and delivered into an appropriate trough, from which they are removed by wings or projections on endless belts, and delivered into a second trough at the end of the lathe, in a proper position for entering between the warps. When the lathe is carried back, and the shed opened, the wires are pushed in under the looping warps, being guided and supported by slots formed in the front face of the dents of the reed, which then pass below, and leave the wires between the warps to be woven into the fabric.

The *sixth* part of the invention has relation to the weaving of figured piled carpets double, and consists—1. In combining with the loom employed for this purpose a Jacquard apparatus for operating the figuring warps of the two fabrics. 2. In dividing the figuring warp in such manner that half of the threads shall be in connection with each cloth, so that the colours required for the figures can be selected from the top and carried to and tied into the bottom cloth, or *vice versâ*, by which means fabrics may be produced either with a greater variety of colours in the figures for the number of warps woven into the body of the fabric, or with the same variety or number of colours a thicker pile may be produced in proportion to the number of figuring warps woven into the body of the fabric. 3. In arranging the Jacquard in two parts,—one to operate the figuring warps connected with the bottom cloth and arranged in the ordinary manner, and the other to operate the figuring warps of the upper cloth, and arranged in a reversed manner. 4. In a method of determining the space between the two cloths to measure the length of the pile, by the employment of two vibrating bars, with a space between them for the passage of the double fabric, and equal to the intended thickness of the two fabrics with the pile between them, the front faces of the said bars being curved so that the two fabrics, as they are separated by the cutting of the pile, shall be carried one up and the other down, and thence around rollers to the take-up rollers or beams. 5. In combining with the two curved bars a long vibrating knife or knives, with the cutting edge or edges in a line midway between the edges of the bars, the whole being arranged so as to vibrate freely to the beat of the lathe and the shedding of the warps.

The *last* part of the invention has also relation to looms for weaving figured piled carpets double, and consists: 1. In connecting the intermediate bars or plates which separate the two cloths, and pass between the slips of the reed, with the loom, in such manner that they shall be free to vibrate to the beat of the lathe and the shedding of the warps. 2. In the employment of a stop or stops, to arrest the motion of the bars or plates at a determined point, so that the beat-up of the reed on the cloth shall cause the woven cloth to advance upon them.

DAVID DAVIES, of Wigmore-street, Cavendish-square, coach-maker. *For certain improvements in the construction of wheel carriages, and in appendages thereto.* Patent dated January 31, 1851.

The patentee describes and claims,

1. An improved wheel-plate or wheels locking plate working on two centres for the purpose of enabling a carriage to be turned in a short space.

2. A square or rectangular umbrella for carriage use. [The frame of this umbrella is constructed and covered in the ordinary manner, but in order to compensate for the unequal lengths of the corner and side ribs, they are formed with eyes at the end, so as to slide freely on vertical wires attached to the runner between two flanges, one at the top and the other at the bottom, the runner being made longer than usual for this purpose. The ends of the corner ribs rest on the lower flange of the runner; and when the umbrella is to be expanded, and the runner is raised up the stick, they first come into action, the continued raising of the runner brings the ends of the side ribs to bear on the lower flange, and when the runner arrives is raised to its full extent, it is caught by a spring catch and prevented returning until released therefrom as in the ordinary construction of umbrella.]

3. A pedomotive or exercising carriage for one or two riders [which consists of two wheels, one free to turn within the other and carrying the seat for the rider who sits astride of it with his feet touching the ground on either side. The carriage is caused to progress by the action of the feet of the rider against the ground, and is guided by handles attached to the inner wheel. When the carriage is designed for two persons, the seat is made of sufficient length, and they sit one on each side of the carriage, which is then propelled by their joint exertions].

4. A retiring carriage or portable urinal having urinals and (in some cases) a water-closet fitted to the interior thereof.

5. A peculiar construction of railway break [which is an improvement on one registered by Mr. Davies in 1840, and in which the shoes (that act on the wheels are brought into play by means of levers, which are connected at their free ends to nuts on a shaft with a right and left-handed wheel cut thereon, and are caused to recede from or approach one another by the turning of this shaft].

GEORGE BRADSHAW, of Bishopsgate-street Within, hosier. *For certain improvements in fastenings for garments.* Patent dated January 31, 1851.

The improved fastening which forms the subject of this patent is intended to supersede the use of strings or other modes of securing garments hitherto practised, and to render unnecessary the employment of a separate fastening for each separate article, being capable of being shifted from one to another as occasion may require.

It consists of a band of any desired width composed of a number of strands of vulcanized India-rubber, covered with a filamentous material, and having attached to each end a piece of silk, cotton, linen, or leather, for the twofold purpose of preventing the elastic bands reeving or fraying out, and of serving as a stay to which to fix the buttons or studs employed for holding together the articles to be so secured. Each part or end of the garment or article held together is provided with a button-hole to receive the buttons or studs on the band by which the fastening is effected; but, in order to prevent the band becoming accidentally detached when in use, a transverse slit of the width of the band is formed in one part or end of the garment or article, through which slit the end of the band is passed before inserting the button on that end in its button-hole. Exemplifications are given of the application of this fastening to shirt-collars and fronts, drawers, trousers, belts for riding or walking, &c., &c.; but it is equally suited for other purposes where it may be convenient to unite two ends or articles together by an elastic and adjustable fastening.

Claim.—The fastening of garments by means of a moveable elastic fastening adapted and applied to garments in the manner before described.

JOHN DAVIE MORRIES STIRLING, of Black-grange, North Britain, Esq. *For improvements in the manufacture of metallic sheets and coating metals, in metallic compounds and in welding.* Patent dated January 31, 1851. No claims.

JUAN NEPOMUCENO ADORNO, of Golden-square, gentleman. *For improvements in the construction of maps and globes, and in apparatus for mounting the same.* Patent dated January 31, 1851.

Claims.—1. The dividing of the terrestrial and celestial spheres into any number of triangles of the two harmospherions, con-ng them of metal, wood, gutta percha, er suitable material with screws, or other means of junction, to be together as globes, or separated in to be kept, transported, or mounted aes, or on brackets to ornament or suspended by chains in cabins on ard.

he construction of maps representing t surfaces those triangles of the har- rions or any of their parts with rical shapes, suitable to be folded, en opened and joined together to with those maps solid polyhedrons ain maps.

he double maps superimposed the l on the terrestrial, or *vice versâ*, y kind of painting, engraving, or the convex, or in the concave sides said triangles and their geometrical ations.

he construction of frames of wood, ercha, or any other suitable material lded or separated into pieces for the portability.

he construction of frames for globes e zodiac inclined at an angle, to pted for the easier solution of the mical and geometrical problems of e, and with travelling spheres of the sun, moon and planets, or any of parately.

UEL ALLEN, junior, of Birmingham, cturer. *For certain improvements in nufacture of buttons.* Patent dated ry 1, 1851.

hen manufacturing sewn-through according to these improvements, a disc of covering material composed fabrics, one coarse, and the other mented together, is employed, on s placed a cupped ring of metal, hav- llar or neck raised around the central the edge of the covering material is thered up all round to the neck of pped ring, a plain ring to form the the button, passed over the neck of pped ring, the edges of the covering compressed between the two, and ole secured together by riveting or ng out the neck of the cupped ring plain ring, which forms the back of ton.

hen manufacturing florentine, or buttons, a plain disc of metal, to e a flat face to the button, is first on the back of the disc of covering l; the cupped ring is then placed on c of metal, and the edges of the g turned over and secured by riveting the neck of the cupped ring over a ing, between which and the cupped e covering is compressed, as above described. The cupped ring also serves to hold a dished piece of some strong fabric for attaching the button to the garment.

Claims.—1. The mode of constructing sewn-through buttons, as described by gathering the edges of the disc of the covering material to the neck of the cupped ring constituting the body of the button, and securing the same by the pressure of the ring which constitutes the back of the button, the said ring being secured by riveting or spreading out the neck of the cupped ring constituting the body of the button.

2. The application of this mode of construction with the modification described, to the manufacture of florentine or covered buttons.

NATHANIEL JONES ARMIES, of Manchester, manufacturer. *For certain improvements in the manufacture of braid, and in the machinery or apparatus connected therewith.* Patent dated February 1, 1851.

Claims.—1. The manufacturing or forming a firm flat braid, having an uniform selvage upon one side or edge thereof, and with portions only of a selvage upon the other side or edge of the same, between which portions of selvage the loops are formed.

2. The manufacturing or forming a firm flat braid in any braiding machine having more than thirteen spindles, such braid having portions only of a selvage on each side or edge thereof, between which portions of selvage the loops are formed.

3. The manufacturing or forming a firm flat braid, upon any machine having thirteen spindles, or less, with portions only of a selvage upon each side or edge thereof, between which portions of selvage loops are formed, such loops exceeding one-eighth of an inch in length, and being more than two in number between such portions of selvage, such two loops being formed from the threads that may be on two spindles.

4. The manufacturing of an elastic braid by plaiting or braiding in one or more threads of India rubber, or other suitable elastic substance, such elastic substance forming one of the threads of which such plait or braid is composed.

5. The combination of an ordinary flat braiding machine, having one or more tongues of iron, or other material, over which the loops are formed, with spindles having an unequal tension given by the application of suitable weights, which weights, in the process of manufacturing, are attached to the thread, or threads forming the braid, by the thread or threads passing through an eye of the weight in each spindle. Also, the application of a spring or other suitable elastic substance to the upper end of braid

spindles, for the purpose of pushing aside the finger when required, in order to produce the peculiar effects above mentioned, in the manufacture of braid by such machinery.

Alexander Alliott, of Lenton Works; Nottingham, engineer. *For certain improvements in cleaning, dyeing, and drying machines, and in machinery to be used in sugar, soap, metal and colour manufacturing.* Patent dated February 3, 1851.

Claims.—1. In reference to dyeing machines of the centrifugal class,—certain arrangements whereby yarns may be dyed therein of different colours in separate and distinct divisions.

2. In reference to centrifugal cleaning and drying machinery,—the employment of an inner case, flange, lids, and perforated channels, in the modes or manners respectively represented.

3. In reference to the manufacture of sugar,—certain improvements in the "forming machine," patented by Richard Archibald Brooman, whereby the liquid matters are expelled from the loaves sideways as well as endways; and also, a separate machine for the same purpose.

4. In reference to the manufacture of soap,—the employment of a machine of the peculiar construction represented.

5. In reference to the manufacture of metals,—the adaptation of centrifugal action to separating, dividing and mixing the same while in a fluid state.

Alfred Vincent Newton, of Chancery-lane, mechanical draughtsman. *For improvements in communicating intelligence by electricity.* (Being a communication.) Patent dated February 3, 1851.

Claim.—A method or methods of rendering available the conducting power of electric telegraphic wires, so that they may be employed to transmit one or more electric currents in the same or opposite directions, during the time that must necessarily elapse between the transmission of succeeding signals which have reference to one and the same communication.

Angier March Perkins, of Frances-street, Regent-square, engineer. *For improvements in railway axles and boxes.* Patent dated February 5, 1851.

This invention consists in constructing railway axles and boxes in such manner as to obtain a lengthened bearing between the axle and box. According to one method adopted by the patentee a cylinder of iron of the required size is first made and lined with a compound metal capable of melting at a temperature less than 500° Fahrenheit. (The manner in which the cylinder is lined is as follows: a core bar is made of a size such that when inserted into the co[illegible] the axle cylinder a space shall be [illegible] round of about $\frac{1}{16}$ths of an inch. T[illegible] bar is turned quite smooth and of [illegible] decreased diameter at one end and is [illegible] by caps in the centre of the cylinde[illegible] lined. The whole is then heated to b[illegible] 500° and 600° Fahr., and the com[illegible] metal poured in until the vacant sp[illegible] entirely filled; when the metal has [illegible] become quite cold the core bar is re[illegible] by raising the cylinder and allowing [illegible] fall down on the small end of the co[illegible] which should be made to project sli[illegible] for the purpose. The bar will be [illegible] forced out, and the lining will be smo[illegible] the interior, and very nearly cylindrica[illegible] order to form the axle, a square bar of wr[illegible] iron, having its corners rounded off [illegible] nearly to fill the interior of the cyl[illegible] is inserted therein, and boxes are se[illegible] on the ends of the square bar, from [illegible] the railway carriages may be suspend[illegible] the ordinary manner, the wheels of th[illegible] riage being attached to the cylindrica[illegible] and turning therewith on the bar. [illegible] ends of the axles are fitted with l[illegible] washers, for preventing the escape [illegible] material employed for lubrication.

Another method of constructing axl[illegible] boxes in which the principle of a lengt[illegible] bearing is carried out, consists in th[illegible] ployment of a cylindrical axle turning [illegible] of and having its bearings on the in[illegible] of a cylinder lined with soft metal, as [illegible] described, the wheels of the carriage [illegible] attached to the cylindrical axle, an[illegible] carriage itself hung from the outside [illegible] der.

Claim.—The manufacture of railwa[illegible] and boxes herein described.

William Onions, of Southwark, [illegible] neer. *For certain improvements i[illegible] manufacture of steel.* Patent dated [illegible] ruary 7, 1851.

These improvements have relation [illegible] manufacture of cast steel by melting m[illegible] together and annealing the product, [illegible] carrying them into effect the patentee [illegible] in a crucible two parts by weight of h[illegible] tite ore, or iron ore of like character t[illegible] known as Cumberland ore, reduced by [illegible] ing or grinding to a coarse powder, and [illegible] thereto four parts by weight of steel [illegible] ordinary make, and ninety-four par[illegible] weight of iron, preferring to use iron [illegible] of Cumberland or other ore of a si[illegible] nature, and which iron possesses the [illegible] perty of being rendered malleable b[illegible] nealing. He then melts these ingred[illegible] and as it has been found better to cas[illegible] articles at once instead of running the [illegible] into ingots, and then re-melting i[illegible] ne molten metal to be run off into ulds of a form calculated to pro- article desired to be obtained. The thus obtained are annealed accord- e method practised when annealing made of Cumberland or other iron, may be rendered malleable by an- and a quantity of pulverized ore is etween them in the ordinary man- e being taken to operate upon as nearly as possible of a similar s. The annealing oven should be y raised to a red heat, and the tem- maintained at that point for 120 hich has been found sufficient for n inch square, after which the oven be allowed to cool gradually, the as of heating the oven to a red heat ling it occupying each 24 hours. icles after being annealed in this are to be removed from the oven eed to shape by turning or other- nen they may be tempered in the usually adopted when tempering composed of cast steel.

.—The mode described of manufac- ast steel by melting certain matters , and annealing the products.

IAM ONIONS, of Southwark, engi- *For improvements in the manufac- certain parts of machinery used in* . Patent dated February 7, 1851.

' parts of spinning machinery" here to are fliers and pressers which the proposes to construct by casting om Cumberland or other like iron, n annealing them, instead of, by from iron or steel as is usually prac- The articles are cast in sand moulds, for the sake of convenience it is l to make in separate halves; when t, they are exceedingly brittle, but rendered malleable by annealing. r process is performed in the man- tomary when annealing articles of land or other iron, which may be l malleable by annealing, that is to articles are placed in boxes, with d ore between them, and are sub- a red heat for a time, which varies g to the size of the article operated ; for a full-sized flier for a roving , ninety-six hours has been found t. When the leg of the flier is to ed, or coiled for the passage of the it is preferred to cast the leg of suf- ngth, and to coil the end of it after ealing process; and when the leg is ollow, or tubular, it is also cast in anner as to admit of its being bent ubular form after annealing. The are subsequently finished by filing rdinary manner.

Claim.—The manufacture of fliers and pressers for spinning machinery by casting them, and then causing them to be annealed.

FRANÇOIS MARCELIN ARISTIDE DUMONT, of Paris, engineer. *For improved means and electric apparatus for transmitting intelligence.* Patent dated February 7, 1851.

Claims.—1. A system or mode of applying electricity to internal communication in large towns, and the interchange of communications between the inhabitants of such towns.

2. Certain improvements in electric bells or alarums, as patented by Mr. J. O. N. Rutter. [These improvements consist in the application of a peculiar system of knobs and keys in connection with the main wires, instead of the contact pieces used for completing the circuit and causing the sounding of the alarum.]

3. A system of transmitting telegraphic dispatches by a peculiar arrangement of keys and appendages thereto.

4. A method of effecting submarine telegraphic communication. [According to this system, the wires instead of being laid at the bottom are suspended from buoys at suitable distances apart, at such a depth as to be out of danger from vessels passing over them.]

CHARLES DE BERGUE, of Arthur street West, engineer. *For improvements in the construction of the permanent way of railways.* Patent dated February 7, 1851.

It is well known that in several of the different permanent ways of railways, as heretofore constructed, the ballasting or ground in or on which they are laid frequently subsides, and that very unequally; and, consequently, the bearers or supports for the rails sink in certain parts more than in others, and the upper surfaces of the rails do not maintain the level or uniformity with which they were originally laid. Now this invention consists:

Firstly. Of a mode or modes of constructing iron longitudinal bearers or sleepers for more effectually supporting the rails of permanent ways, for which purpose such longitudinal bearers are constructed on the principles which commonly govern the construction of cast-iron girders (in order to obtain as much strength as practicable without unnecessarily increasing the quantity of metal), by so disposing the metal, or disposing or combining the metals of which they are composed, as best to resist the strains to which they are subjected by reason of the inequalities of the ground or ballast, being chiefly tension at their lower portion or base, and compression at their upper portion. In accordance with this object, the patentee makes his bearers in such form or forms that their bases, bottoms, or feet shall be considerably stronger, or have a greater transverse sectional area than their upper portions, in every case making their upper parts of such forms as may be necessary for enabling them to receive and support the descriptions of rails to be placed upon them. [By making the base of his longitudinal bearers stronger than the upper parts, the patentee is enabled to construct them so that their feet or bottom flanges shall rest firmly on the ballasting. Their transverse sectional form may be varied in such manner as may be deemed best for the purpose, provided only they are constructed on the principle hereinbefore mentioned.]

Secondly. Of transverse iron ties or bearers to be used in conjunction with longitudinal bearers or sleepers. These ties are made to have a transverse sectional form similar to the letter T inverted, so that the ends of any longitudinal bearers or sleepers (having flat bottoms) may rest upon the tie, and assist in keeping the longitudinal bearers steady, and prevent them from canting. [The patentee makes these ties of wrought iron by rolling, or he uses ordinary T-iron for the purpose; and in laying down his longitudinal bearers, he places the ends of the bearers so that their feet may rest upon the horizontal flanges of the tie, and he bolts them together, making provision for expansion and contraction by an enlargement in the proper direction of the holes through which the bolts pass.]

Thirdly. Of certain forms of rails; namely, an angular rail, made so as to fit against one side and upon the top of the upper part of longitudinal bearers or sleepers, made according to the first part of this invention; and also a saddle-rail, so formed that the interior or under-side of the tread, or working surface of the rail, shall rest upon and be supported by the upper part of a longitudinal bearer or sleeper; the sides or flaps of the rail being bent either to a right angle with the bearing surface of the rail, or to such other angle as may be requisite to make the rail fit upon and be held more securely by the top of the bearer, upon which it shall be placed. And for this purpose, instead of bending each of the sides or flaps to the same angle, the one side may be bent to one angle, and the other to another different angle. The surface on one side and on the top of the bearer against which an angular rail fits should be planed true to receive it. So also the upper surface of the webs of the bearers, on which the improved saddle rails are intended to rest, should be planed, or otherwise made sufficiently true to receive them, or a strip of lead, or other metal, or suitable material, may be introduced as required. The rails are connected firmly to their respective bearers by bolts, which may be either square or round; but in either case the holes into which they fit should be a trifle wider than the bolts, in order to allow for expansion and contraction. Or they may be connected to the bearers intended to receive them by set screws.

In laying his improved permanent way, the patentee disposes the longitudinal bearers and rails relatively to each other, so that the junction of any two lengths of the latter may be midway, or at some intermediate distance between the junction of the former, in order the better to preserve the required uniformity of the rails. If it should be preferred to use ordinary transverse wooden sleepers instead of the improved ties in laying permanent way, the same may be constructed either with the ends of the longitudinal bearers made to abut square against each other, or to abut obliquely against each; which latter method will tend to prevent the transverse sleepers from rocking. In whichever of these ways the ends of the longitudinal bearers may be formed, they are connected to the transverse sleepers by bolts, using a washer above and a triangular nut below, as practised on the Great Western Railway.

No claims.

William Edward Newton, of Chancery-lane, civil engineer. *For improvements in apparatus for milking animals.* Patent dated February 10, 1851.

This improved apparatus consists of an air and water-tight bag composed of India-rubber, gutta percha, or other suitable material, encircled by an elastic strap or band at top, and provided with an aperture at bottom fitted with a silver tube of a size capable of entering the milk duct of a cow's teat; which tube is provided with a piston also of silver, and packed so as to be air-tight; or the piston may be made of gutta percha, in which case no packing would be required.

When using the apparatus, the bag is turned down, so as to expose the silver tube, which is inserted in the milk duct of the cow's teat, and the bag is raised all round, so as to enclose the teat, and prevent the air entering; the piston is then withdrawn and a free passage left for the milk through the tube into a can placed underneath. The flow of the milk is facilitated by the contraction of the elastic band round the mouth of the sack, and by the warmth caused by the exclusion of air between the sack and the cow's teat, which is about equal to that produced by a calf in the act of sucking.

Claim.—The sack made of India-rubber or other suitable water and air-proof mate-

mbination with the strap or band
ssing the teat and neck of the
he exhauster tube and piston, or
or more of the foregoing parts in
n with a sack for the purpose of
ws.

FAIRBAIRN, of Leeds, machinist,
HETHERINGTON, of Manchester,
*For certain improvements in
for casting pipes, railings, gates,
al implements, and other metal
nd also in preparing patterns or
r the same.* Patent dated Fe-
1851.

sent improvements have relation
of moulding, which, as applied
spinning and weaving machinery,
of the subjects of a patent granted
ne gentlemen, 18th July, 1850
ation of which was duly enrolled
r. p. 117.) The articles of which
to be taken are cut in half in the
f their widest plane, and the two
e attached to opposite sides of a
iaphragm, from which two half
s successively taken; these half
then fitted accurately together,
ect casting obtained. The paten-
se to economize time and labour
ng a greater number of articles
when so cut in half, to the oppo-
of the plate, forming suitable
or the passage of the metal, and
ining several castings simulta-
When moulding articles, of which
e "under cut," those parts of the
to be loosely attached, so as to be
the sand-mould, from which they
carefully removed before the two
ls are finally secured together for

D STUART NORRIS, of Warring-
ngineer. *For certain improve-
le construction of the permanent
ilways, bridges, locks, and other
wholly or in part constructed of
improvements in breaks for rail-
ges.* Patent dated Feb. 10, 1851.
—1. A method of joining, fixing,
ting the bars or other metallic
f railways, bridges, locks, and
llic erections, by pouring molten
her suitable metal on to or about

employment of a portable cupola
pose of melting the metal to be
in casting portions of railways,
cks, and other similar erections at
he situation where such castings
d to be used. [The barrel of the
omposed of metal lined with fire-
re clay, in the ordinary manner,
, worked by a winch-handle, is
mounted in the same frame, to facilitate the
melting of the metal employed.]

(The "improvements in breaks for rail-way carriages" are not specified.)

JOHN STEPHENS, of the Allyards, Astley Abbotts, gentleman. *For certain improvements in threshing machinery.* Patent dated February 10, 1851.

The machine employed for threshing wheat, beans, peas, &c., consists of a drum with a grooved, fluted, or corrugated surface, revolving almost in contact with two grooved surfaces placed at unequal distances therefrom, in order to effect a more perfect separation of the grain, to which a reciprocating motion at right angles to that of the drum is imparted by means of a crank-shaft actuated from the main spindle of the machine. A rubbing, instead of a beating action, is thus produced, by which the grain is effectually separated from the straw, which passes through the machine without being broken to such an extent as in threshing machines of ordinary construction.

The same principle of producing a rubbing action is also applied to machinery for threshing barley, oats, and seeds generally of a smaller size than wheat. In this case a card cylinder is substituted for the unyielding friction surfaces employed in the machine above described. This cylinder has a reciprocating motion communicated to it by a crank-shaft, as well as a motion round its axis; and, from its peculiar formation, acts in the manner of a brush, preventing injury to the grain or seed operated on, at the same time that it effects its perfect separation from the straw.

Claim.—The application to threshing machinery of friction surfaces (whether rotary or not) having a traverse motion whereby, in combination with a rotating friction surface, the ears of corn or other vegetable produce may be subjected to a rubbing action for the purpose of separating the grain or seed therefrom.

JOHN HARCOURT BROWN, of Fir Cottage, Putney, gentleman. *For certain improvements in the construction and building of ships, boats, buoys, rafts, and other vessels and appliances for preserving life and property at sea.* Patent dated February 10, 1851.

The first part of this invention consists in constructing ships, boats, and other vessels with double keels, and double or duplex rudders, and in adapting to their propulsion either the screw or paddle wheels. The two keels are made parallel with each other, and extend along the whole length of the vessel commencing at the bows at the water-line, where they unite with the cut-water, and terminating at the stern in two stern-posts, from which the duplex rudders are hung in the usual manner. The space between the two keels is arched for the sake of obtaining greater strength, and they may be also bound together at intervals by stays connecting them one with the other.

The second part of the invention consists in constructing the frames of life-boats, rafts, and buoys, of hollow cases or tubes composed of gutta percha in combination with metal, wood, or other suitable material. Under this head of the invention the patentee describes a life-boat with a double keel, and deck entirely covered in with the exception of apertures for the rowers and passengers. These apertures are covered with pieces of India rubber, having slits for the admission of the occupants, for wh or slings are placed immediately un them. The bottom of the boat is with lockers for containing bread &c. Rafts and buoys are constr the same principle, the air-tight cases being made of such sizes a nected together in such manner be found most convenient.

Claims.—1. The construction o boats, and other vessels with doubl

2. The construction of the fr ships and other vessels with hollow cases of gutta percha combined wi rubber and other suitable substance

3. The adaptation to ships' bo other vessels of a duplex rudder.

WEEKLY LIST OF NEW ENGLISH PATENTS.

Jonathan Grindrod, of Birkenhead, Chester, consulting engineer, for an improvement in the machinery for communicating motion from steam engines or other motive power, and in the construction of rudders for vessels. August 14; six months.

John Plant, of Beswick, Manchester, manufacturer, for certain improvements in the manufacture of textile fabrics. August 14; six months.

Thomas Skinner, of Sheffield, for improvements in producing ornamental surfaces on metal and other materials. August 14; six months.

Stephen Moulton, of Bradford, Wilts, India-rubber manufacturer, for certain improvements in the preparation of gutta percha and caoutchouc, and in the application thereof. August 14; six months.

Aime Nicolas Derode, of No. 37, Rue Paris, France, gentleman, for a certain p uniting cast iron to cast iron and other m for uniting other metals together. Augu months.

Joseph Birkbeck Blundell, of New-c Kent, gentleman, for improvements in for sweeping and cleansing roads and ways 14; six months.

Henry Glynn, of Bruton-street, Berkel gentleman, and Rudolph Appel, of Gerr Soho, annastatic printer, both in Midd improvements in the manufacture or tre paper or fabrics, to prevent copies or im being taken of any writing or printing August 14; six months.

WEEKLY LIST OF DESIGNS OF ARTICLES OF UTILITY REGISTERED.

Date of Registration.	No. in the Register.	Proprietors' Names.	Addresses.	Subjects of Desi
Aug. 6	2905	Fisher and Bramall	Sheffield	Crank-handled screw
7	2906	A. Rabett	Newgate-street	Military-sleeved shir
8	2907	J. Fuller and Co.	Southwark	Neoteric ventilating
9	2908	T. Porter	Manchester	Hooks and eyes for c articles of dress.
11	2909	Myers and Son	Birmingham	Peristaltic pen.
13	2910	J. H. Ferguson and Co.	Queen-street, Cheapside	Railway companion.

CONTENTS OF THIS NUMBER.

LONDON: Edited, Printed, and Published by Joseph Clinton Robertson, of No. 166, Flee in the City of London—Sold by A. and W. Galignani, Rue Vivienne, Paris; Machin Dublin; W. C. Campbell and Co., Hamburg.

Mechanics' Magazine,

MUSEUM, REGISTER, JOURNAL, AND GAZETTE.

No. 1463.] SATURDAY, AUGUST 23, 1851. [Price 3*d*., Stamped, 4*d*.

Edited by J. C. Robertson, 166, Fleet-street.

ULLMER'S PATENT SELF-INKING HAND-PRINTING PRESS.

Fig. 1.

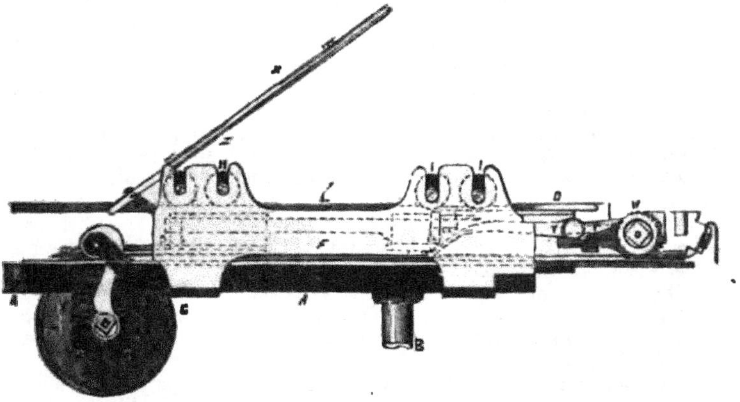

Fig. 2.

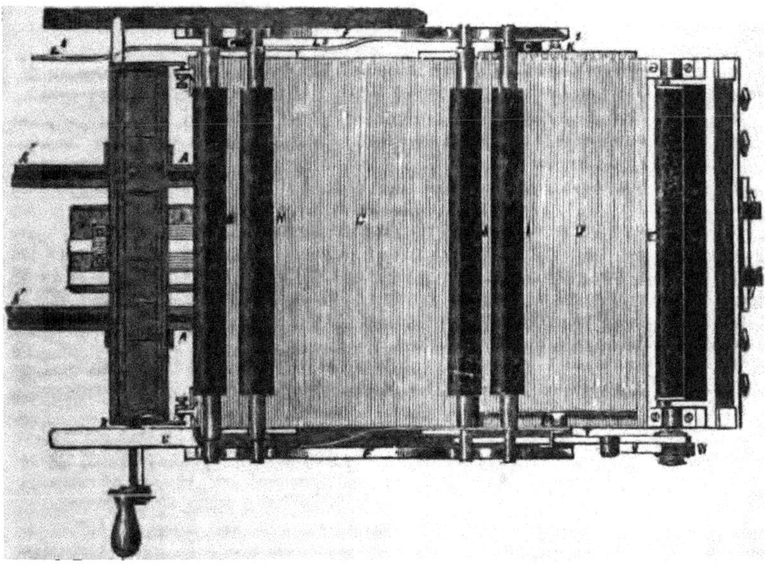

ULLMER'S PATENT SELF-INKING HAND-PRINTING PRESS.

(Patentee Mr. Edwin Ullmer, of the firm of E. and W. Ullmer, 110, Fetter-lane. Patent dated February 12, 1851. Specification enrolled, August 12, 1851.)

Specification.

My invention has for its object to expedite the process of printing, by causing the same movement by which the table or bed is pushed out and in beneath the platen of the press, to effect at the same time the inking of the types, whereby a considerable saving of labour and time is effected, and likewise a saving in the space occupied by the press and inking-table.

Fig. 1 is a side elevation, and fig. 2 a plan of the bed of a hand-printing press with its appendages constructed according to my invention. AA are the two rails, upon which the bed C, which carries the type-form, is moved out and in; they are attached to the press by the inner ends A^1 A^1, and are supported at their outer ends by a pillar B. D is the ink-table, and E the ink cylinder, both of which are attached to the bed C, and move along with it; FF are two cheek-pieces, or carriages which are connected to the rails AA by cross bars GG, and thus occupy a fixed position in relation to the table; HH are the inking rollers which have their bearings in the cheek-pieces FF; II are the distributing rollers, of which there may be one, two, or more; K is the rounce, with crank-handle L, by which the sliding in and out of the table is effected. When the table is pushed inwards the type-form passes beneath the two inking-rollers HH, which have their bearings in open upright slots, in which they are free to rise or fall, so that they adjust themselves readily to the surfaces of the type and of the ink table.

When the bed has been fairly brought beneath the platen and into the proper position for the impression being taken, then the inking rollers will have passed over the surface of the ink-table, and taken up a fresh supply of ink for re-inking the types for the next impression. The drawing out of the bed causes the inking and the distributing rollers to retraverse the ink-table, and the inking rollers to pass over the surface of the types; but in order to more effectually distribute the ink, the ink-table is made to partake of a movement from one side of the bed of the press towards the other, so that the inking rollers may not traverse continually back and forward over the same space. This shifting of the ink-table is effected by the following arrangements:—The ink-table D is fitted into V-shaped grooves in the bed of the press, whereby it is capable of being slid from one side of the press to the other. K^1 K^2 are two friction pulleys, which are attached to lugs upon the ends of the ink-table and bear against the cheek-pieces FF, or the guides L^1, L^2, L^3. As the bed is rolled into the press, the friction pulley K^1, bearing against the guide L^1, keeps the ink-table towards the further side of the bed of the press. When, however, the pulley K^2 comes against the guide L^2, then the ink-table is pushed towards the near side a certain distance. Again, when the same pulley comes upon the guide L^3, the ink-table receives a further impulse towards the near side of the bed of the press, but upon again running the bed out from the press, the friction-roller K^1, does not come against the guide L^1, so as to cause a movement of the ink-table in the reverse direction to take place, until the ink-table has completely passed from underneath the inking-rollers.

Thus, each time the bed is rolled out or into the press, the position of the inking-table is changed in relation to the surface of the inking-rollers, and thereby prevents any streakiness, or blotching, which would otherwise appear upon the printing. Fig. 3 is an elevation of part of a framing of a hand press, showing another method of effecting the change of position of the ink-table in reference to the inking-rollers. FF are the cheek-pieces as before; H is the inking-roller, and D, the inking-table, which is mounted in V-shaped grooves in the bed of the press, as represented in fig. 1. P is a screw spindle which works into a nut Q, formed upon the lower side of the ink table; this spindle carries at one end a friction-wheel R, which rests upon a ledge S, formed upon one of the cheek-pieces F. Every time the bed is made to move either out of, or into the press, the rolling of the wheel along the ledge causes the screw spindle to revolve, and draw the ink-table towards one side of the bed of

›ress or towards the other side, according to the direction in which the bed is : to travel.

he rotation of the ink cylinder E, by which a continuous supply of fresh ink is lied to the distributing rollers, and to the ink-table (by the distributing rollers ng in contact with it each time they pass over), is effected by means of the : T, which is jointed at one end to the spindle of the ink-cylinder, the other end sliding upon a ledge U, connected to one of the cheek-pieces. When the bed been fully brought into the press, the free end of the lever T, passing beyond end of the ledge U falls down by its own gravity, and causes the click V, which inted to it to pass over one or more of the teeth of the ratchet-wheel W, which eyed to the spindle of the ink cylinder. On the return of the bed in being d out of the press, the free end of the lever T comes upon the ledge U, by h it is raised up and carries with it the click V, ratchet-wheel W, and ink cylin- E, whereby a fresh portion of the surface of the cylinder is supplied with ink presented to the distributing roller. X is the tympan, and Y the frisket,

Fig. 3.

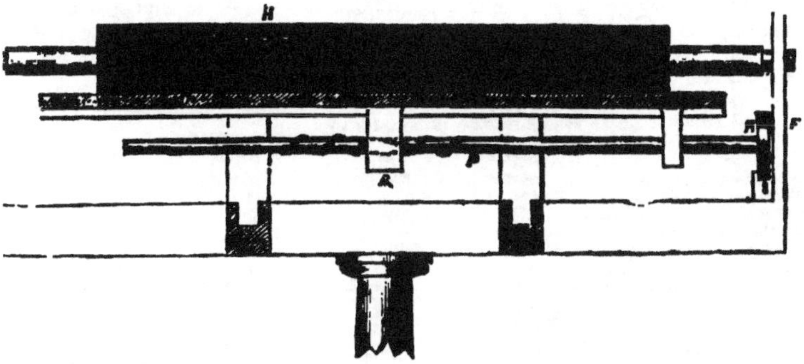

:h open the reverse way to the same parts in presses now in general use; they jointed together at the lower edge, and open like the leaves of a book. Z is a e by which, as the bed is drawn out of the press, the tympan and frisket are ght up into the position represented in fig. 1., upon which the workman opens e tympan, which falls towards the press and takes hold of the printed sheet; he puts on another clean sheet in the place of the one removed, throws over the pan and rolls in the bed into the press, upon which the tympan and frisket slide the guide Z, and place the paper ready to receive the impression upon the . In larger presses two guides may be employed for raising the tympan and et instead of one, as already explained; in which case there would be one ed on each side.

generally prefer, in constructing printing presses according to my invention, the press should be so disposed that the length of the sheet, instead of the dth, may lay across the press, so that the bed may have to travel as short a nce as may be. When the press is of a smaller size, and worked by a single l, then the rolling out or in is most conveniently effected by making the rounce ch size that a single turn of the crank handle performs the operation; but if press is to be of a large size, and such as two hands might be employed upon, the rounce may, with great advantage be increased in dimensions, so that r the rolling out or in of the bed may be performed by one-half, or any other on of a turn of the rounce. When two hands are employed, the pressman ls at the near side of the press, and the other attendant, or assistant, at the er side. The duties of the assistant are to roll the bed out and into the press, hich action the inking of the types is also effected, and he further removes printed sheets while the pressman supplies the clean ones.

THE CUNARD AND COLLINS' LINES OF ATLANTIC STEAMERS.

(From the *Journal of the Franklin Institute.*)

About a year has elapsed since this line commenced their trips across the Atlantic, and already enough has been demonstrated to prove them as unsurpassed in speed, as it has been universally conceded they are unequalled in model, and for the beauty and convenience of their internal arrangements. Enough at any rate is already known, to justify me in drawing your attention to the many unfair and injurious reports that have been brought to bear against them, and principally, I regret to say, on this side of the Atlantic; for, strange as it may appear, more open and concealed enemies to the line have existed among us than in England, taking the published statements as the standard by which to judge. It is my present purpose fairly to state the case, and to show, as I think conclusively, that the steamers of this line are a triumph to our skill, and a credit to those who designed and constructed them. These vessels were commenced just after the appearance of the *Europa*, *Canada*, *Niagara*, and *America*, and were intended to surpass them in size and speed. But as soon as these ships were fairly commenced, the *Asia* and *Africa*, larger and more powerful than the former steamers of the line, were put under contract, and, by great exertions, brought out nearly at the same time as the steamers of this line; so that, to achieve a name, we had not only to do all that our vessels were intended to do, but to excel the last two steamers also. How well they have accomplished their object, the public can determine; but they, looking only at the shortness of the passage, without regard to circumstances, are unable to understand the real value of our success, which is the point to which I wish particularly to call your attention.

The dimensions of the contending vessels are as follows:

	Atlantic.	Pacific.	Baltic.	Asia.
Length on Deck	285 feet.	284 feet.	287 feet.	280 feet.
Breadth of Beam	45½ ,,	45 ,,	45 ,,	40 ,,
Depth of Hold	32 ,,	32 ,,	32 ,,	27¾ ,,
Tonnage (Custom House)	2771 ,,	2686 ,,	2718 ,,	2072 ,,
,, (Carpenter's)	3040 ,,	2900 ,,	2920 ,,	2240 ,,
Load Draught	20 ,,	20 ,,	20 ,,	20 ,,
Diameter of Cylinders	95 inches.	95 inches.	95 inches.	96 inches.
Length of Stroke	9 feet.	9 feet.	10 feet.	9 feet.
Nominal Horse-power of both Engines	800 H. P.	800 H. P.	828 H. P.	816 H. P.
Diameter of Paddle-wheels	35 feet.	36 feet.	36 feet.	36 feet.
Length of Paddles	12½ ,,	11½ ,,	11½ ,,	9¾ ,,
Immersed Midship Section	725 ,,	720 ,,	720 ,,	640 ,,

The term nominal horse-power has become a mere conventional unit for expressing a certain size of cylinder, without reference to the power exerted; and the *actual* horse-power exerted by the engines of either American or English steamers greatly exceeds the nominal. This may be attributed to the increased pressure of steam used since the rules for calculating nominal horse-power were established by Watt. In England, they designate the size of their cylinders by horse-power, a term rarely alluded to in this country among engineers, who always speak of the diameter of the cylinder and stroke of the piston. From an examination of the above Table, it will be seen that the *Asia* has cylinders one inch larger than the *Atlantic* or *Pacific*, and the same length of stroke; estimating the nominal horse-power by rules established in England, the power of the *Asia*=816, and the *Atlantic* and *Pacific* 800, as will readily be seen from the following rules:

1st. The square of the diameter of the cylinder in inches, multiplied by the cube root of the stroke in feet, and divided by 47 = nominal horse-power, thus,

$$\frac{95^2 \times \sqrt[3]{9}}{47} = 800 \text{ horses-power.}$$

It was quite a mistake to suppose that the new steamers of the Cunard line carry but 7 lbs. of steam; the average pressure is about 13 lbs., while that of our steamers is about 15 lbs.; and when you take into consideration that our vessels cut off the admission of steam into the cylinder at four-ninths the stroke, and that *they* generally use it to a much longer point, it will be apparent that the actual horses-power developed will be the greatest on their steamers.

From the statistics given, we find that the immersed midship section of the *Asia* is 80 square feet less than the *Pacific* or *Baltic*, and 85 square feet less than the *Atlantic*, giving per square foot of immersed section to the

Atlantic	1·10	horses-power.
Baltic	1·16	,,
Pacific	1·12	,,
Asia	1·28	,,

having an important advantage er of the others.

g demonstrated that the *Asia* has ersed midship section and greater he question may properly be asked, the *Pacific* and *Atlantic* equal in e *Asia?* I answer, by superior hich, at high speed, gives us decided e over them in that particular. As of this, I would state that the new steamers now building (the *Arabia* sia) will resemble them in model, also have tubular boilers. In one however, they have always had l advantage; their firemen and en- have been long in the trade, are ed to its severe duties, and are per- their parts; while with us great have taken place, and the ship often ea with one-half at least new hands ngineer's department. Until we l crews well accustomed to their duties, we shall suffer materially. respect, I am happy to state that a provement has taken place within few months, and I trust that soon have the advantage of picked crews nd with our friends across the . For it is evident that they are beaten from the want on their part best exertions to prevent it. It be forgotten that the *Asia* is 600 burthen than the ships of this line, ntly, has less accommodation for sengers and freight.

g shown, as I think, that we have shed much, and that we have urpassed our opponents in speed as n the comfort and convenience of , I wish to show that we have had position to contend with at home oad; there they have done us the say that the ships of this line were o the country; while among us, no as the *Atlantic* finished, than the ggerated accounts were given of the ill-advised would-be friends; her stated to be 3,000 to 3,500 tons, ines of 1,000 horse-power each, h also having about the same rela- e truth.

r first passage out, she had the mis- break her air-pump bucket, which detention of thirty hours; when ly the same parties that had t brag shook their heads wisely, ated that proper attention had not l to their machinery; and so it has r since.

cific was laid up a few weeks last do some repairs to her boilers and y, and put a new saloon on deck, as *immediately announced* that a new bed plate was to be put under one engine at a cost of 20,000 dollars. To be sure, the story was contradicted the next day, but if the truth only had been uttered, no contradiction would have been needed. So when the *Atlantic* broke her shaft, not satisfied to wait a reasonable time for information, one party very wisely hints that possibly the boilers have exploded, or some other accident, *serious* in its nature, has occurred, and that it is strange we cannot build machinery with the English. If the Cunard steamers are behind time, the wind must be ahead; but if these ships are the same, some accident has happened to their machinery.

I very much question if these same parties can call to mind a single accident that has happened to the Cunard line. The chances are, they would *deny* that one ever did happen; and if you should state that one had broken a shaft, and only made the passage this way as far as Halifax, and sent her passengers to Boston by another steamer: that another had been lost on the coast of Maine; and that about a year since, on a passage this way, one of them broke a side lever, causing much injury to and rendering one engine entirely useless, and that she left on her return trip with but one engine, and took her usual number of passengers, he would deny the whole, although perfectly true. There is another class of friends, who think these steamers can be very much improved, and have published *their opinions* in relation thereto. One who must think himself a paragon of wisdom, states that the machinery of the *Baltic* weighs 1,800 tons (its real weight is 825); that to ensure a safe passage, she should carry 1,800 tons of coal (her bunkers hold but 1,050, which usually leaves two or three days' supply on hand); and that she should have 60-feet wheels. With these improvements, he promises increased speed, although all his suggestions have not a grain of common sense in them.

And here let me notice an article in *Appleton's Mechanics' Magazine* for March, by B. F. Isherwood, Chief Engineer, U. S. N., who compares the *Pacific* (the clipper of the sea), with the English war steamer, *Arrogant*, a propeller. You might as well compare the *American Eclipse with a Donkey*. There is great injury done by suggesting the possibility of comparison; one has crossed the Atlantic at a speed of 13 sea miles, while the very best speed of the other, with sails and steam, is about 9. But see how the comparison is made, and that, too, by an American. He states, (page 153), that the *Arrogant* is 200 feet long between perpendiculars, and that the

"*Pacific*" *has almost the identical dimensions of the* "*Arrogant*," *and is* 210 *feet between the same points*; he also says, he is *fortunately* in possession of the exact data of the two vessels, while he has hardly a single item right; but, leaving the small errors to take care of themselves, I will state what will considerably affect his *exact* data. The *Pacific* is 271 feet between perpendiculars, and 284 feet on deck; a small difference of 61 feet in length, or nearly 30 per cent. If the hulls are exact with this difference, we must yield the point. Mr. Isherwood says, the *Arrogant* made 8·35 miles per hour in the river, and that the *Pacific* made 12 miles. Now the latter can go, in the river, 15 miles per hour with the greatest ease, and has done it often. I have no disposition to follow this gentleman further, but I think I am not going too far, when I say it becomes all who write for scientific journals to pay some regard to facts—we do not look for romance here, or at least, not such a one as the comparison of the *Arrogant* and *Pacific*. As an American engineer, I feel proud of the success of the Cunard steamers; for about twelve years they have crossed the Ocean with great success and with a constantly increasing speed, as each addition was made to the line. There is no reason for envy, for the world is large enough for us both, and neither can make an improvement which does not in some degree benefit the other; and while I am pleased with their success, I may be still more so at our own, and may rejoice that we have at length become the *master* of the *mistress* of the seas.

PRACTICE.

INVENTORS AND WHAT THEY HAVE DONE.

A world without inventors would consist only of forest and swamp. Before they appeared it was, and where they are not it is, an Australian jungle, through which men affiliated with beasts roam in quest of miserable subsistence and shelter. The difference between the civilized and troglodytes is, one class contrives, the other does not. Nothing is clearer than that mechanical inventions are ordained to animate, clothe, and adorn a naked and torpescent world—to infuse into the species the elements of increasing vigour and felicity. Even as arts multiply and flourish, the chief labour of working out the great problems of existence continues to devolve upon inventors. Without them the prospects and hopes of the present had neither been seen nor felt. It is they who, by discovering new physical truths, are establishing the grandest of moral ones—*perpetual progress*—illimitable advancement in social, civil, and intellectual enjoyments.

The fact has scarcely, if ever, been glanced at, that nearly every marked advance of civilization began with, and is due to inventors. Without disturbing old records, it is enough to turn a leaf of modern history. The substitution of fire-arms for primitive weapons, has wrought an entire change on the face of society. Another and ever-memorable epoch was introduced by the revivers of printing and inventors of type-founding; another by steam as a motor; to say nothing of the revolutions brought about more recently by spinning-jennies, power-looms, ocean-steaming, gas lights, photography, railroads, telegraphs, &c., which so honourably distinguish our times from all that preceded them.

But for the artificer's skill, the sublimest of the sciences had not been attempted, nor the sublimest triumphs of human reason and research achieved. By means of two inventions, the extremes of creation are brought within the range of human observation, and the grandest of conceivable miracles demonstrated. With the microscope, the human eye discovers animated worlds in drops of liquid and grains of fecula, and may yet detect ultimate atoms in the most attenuated of the gases. By the telescope, the same eye penetrates and wanders at leisure through a space far beyond what was once thought the limits of an arch-spirit's flight. Leaving the satellites of remote planets behind, it resolves the infinitely more remote nebulæ, and, sweeping round the awful horizon, takes in what would seem half the universe.

At a more favourable time than Fitch lived in, Fulton rose, and steamers began to creep up rivers, next dashed over lakes and inland seas, and now are rushing in fleets over every ocean. Whitney appeared, and forests were swept away to make room for cotton-fields—thus turning the soil from harbouring beasts of prey to raising clothing for half mankind. Daguerre, and the sun turns portrait painter—exemplifying a classic myth. Stranger still, Morse and his compeers have bridled the most subtle, fitful and terrific of agents, taught it to wait, silent, and prompt as a page in a monarch's antechamber, and, when charged with a message, to assume the character of a courier whose speed rivals thought and approaches volition. From the beginning, means more or less rude and refined have been employed for the conveyance of material things, but not until now has the transportation of thought—of thought divested of aught visible or ponderable—been attained. Indian runners hasten with information through floods and forests, over hill and dale; but to carry it,

avey themselves as packages contain-
or as tablets on which it is impressed.
with the contents of our mails—
commune with distant minds through
ss medium of printed and written
whereas, by means of artificially-
lightning, a postal system is estab-
kin to the spiritual; for by it thoughts
le to dart through space unclogged
bols and envelopes, and consequently
ded by carriers and postmen.
wildest freaks of fancy have been
ly verified in the telegraph as *outré*
mps and more attractive fairies; giv-
our to the proposition that in Nature's
are germs of every popular supersti-
d that no prevalent delusion is with-
corresponding truth. Be this as it
he chiefs of modern Prosperos, by
of a few strips of metal, release from
acid, spirits so agile and obedient,
n the slightest tap of its master's
each one flies with messages over a
d leagues of latitude, delivers them,
, and is in waiting for others before
nals can be repeated, or the pulse beat
An ancient elf boasted of putting a
round the earth in forty minutes—
modern sprites can really do it within
one. If art and science allied have
uch things, what is it they cannot do?
achinery does not *think*, it does that
nothing but severe and prolonged
g can do, and it does it incomparably
In the composition of astronomi-
d nautical tables accuracy is every-
Many a ship has been wrecked
h wrong figures, in "Guides to Navi-
;" but absolute accuracy, continued
h abstruse calculations that occupy
s, and sometimes years, is too much
ect even from the most sagacious,
s, and careful. But suppose it
d, the next difficulty is to transfer
ults, untainted with error, to printed
; a source of mistakes which few
s authors and printers can appreciate.
er persons were told of the impossi-
of copying from manuscript millions
res without misplacing, leaving out,
erting more or less, they would hardly
heir assent. It is enough to say that
tion in elaborate and difficult calcula-
s unattainable with certainty by human
g; nor is it to be expected in the
sional labours of the most expert com-
rs.
r, automata have been made to work
rithmetical problems with positive
ty and admirable expedition, relieving
maticians and others of an incalcu-
mount of mental drudgery—drudgery
s worn out *the strongest constitu-*tions. Moreover, they carry the use of numbers further than the clearest intellects dare follow—to an extent that language lacks terms to express. In human computations, minute errors creep in and corrupt the whole, often requiring months of the closest ratiocination to find out; but calculating machines detect their own mistakes at once, correct them, and then shutting out interference of human fingers as well as heads, and with them the chance of marring a work, they print their tables as well as compose them, thus producing works to which entire confidence can safely be given.

The power inventors wield is not less manifest in the changes they have wrought in the habit, customs, and occupations of females, than it is obvious in the pursuits of the other sex, in the out-door world. They have not only broken up the time-honoured arrangements of the kitchen, wash-house, and dairy, but have invaded the parlour and even boudoir. A century ago the rock and spindle were common;—in Europe are women who still twist thread with their fingers. Fifty years since, a wheel had a place in every dwelling, and carding no less than spinning was a domestic duty. With thrifty housewives the shuttle, too, was not a stranger. Within twenty years knitting was indispensable, not a few of our farmers still wear home-made hose. Then straw plaiting, tambour working, lace making, plain and fancy embroidery, with other delicate operations of the needle, were, and are still taught as necessary accomplishments. Such they will hardly be held much longer, since these and various other performances are now done by automatic fingers, with a precision, regularity, dispatch, delicacy of touch and finish that no human organs can rival.

Most, if not all, the fine arts have been subdued by mechanism. The lathe is still to be met with in its primitive forms, in the potter's wheel, the spring-pole instrument, and also as used in the modern Egyptian's atelier—(seated on the ground, this artist employs one hand to revolve the object to be formed, holds the cutting tool in the other, and presses it on the rest with his toes). The lathe, so long confined to shape articles whose sections were circles, now produces oval, elliptical, epicycloidal and eccentric work; copies medallions, and even busts in equal, enlarged, or reduced proportions—performing the work of the engraver, die-sinker, and statuary or sculptor.

The richest figured tapestry and damask in relief, are now produced by magic mechanism. Looms rival the palette and burin; besides gorgeously-coloured carpets, they weave landscapes equal to oil paintings, and portraits after the finest line engravings. Then, from the increase in number of sewing machines,* the time would seem not distant, when the needle itself, and thimble will be exhibited in museums with distaffs, spinning-wheels, knitting-wires, tambour-frames, hand-looms, lace-making bobbins and pillows, and other antiquarian curiosities, as evidences of imperfect civilization. In chromolithography, automaton artists rival the finest touches of old masters, and shortly will multiply by millions their most esteemed productions.

Though not suspected, the power of inventors over human affairs, is already supreme; machinery even now governs the world, though the world does not acknowledge it.—*Mr. Commissioner Ewbank.*

THE AMERICAN REAPING MACHINE.—SEE VOL. LIV., P. 481.)

(Extract from a Letter by Mr. William Dray to the *Times*.)

I have taken some interest in the American reaping-machines exhibited at the Crystal Palace, and availed myself of the first opportunity of testing Mr. M'Cormick's invention. This I succeeded in doing on my farm at Farningham on Wednesday last, before the principal farmers and influential men of Kent, and yesterday at Mr. Ross Mangles's farm at Guildford. The result is so far satisfactory that the saving to the farmer by its use must be very great. I imagine the machine will reap, effectually, 12 acres per day; but, taking it at a lower rate, the result will show a clear saving of 40 per cent. For instance, say the machine, employing two horses and two men, will cut 10 acres per day, which is a moderate computation, this will be the result:

	£	s.	d.
Two men, per day, 2s. 6d.	0	5	0
Two horses, per day, 5s.	0	10	0
Hire of machine	0	10	0
Binding and shocking 10 acres, 3s. 6d.	1	15	0
Making a total of	£3	0	0

This is taking the outside expense.

By the old system, reaping, binding, and shocking, at 10s. per acre, would cost 5*l.*—showing a clear saving of 2*l.* on one day's work, no mean consideration, besides giving the farmer the opportunity of taking advantage of fine weather for clearing his crops with certainty, and at a time when good labourers are often difficult to obtain.

* Four patents have been issued from this office for such machines during the past year.

BATTLE OF THE RAILWAY GAUGES AMERICA.

It would be a good thing if all the rai tracks in our country were of the width; but what is the best gauge som say? Almost all our railroads hav narrow gauge—the New York and Railroad however has the broad gau splendid track, and we can have v gauges in this State. It would appea other States have strange notions such things. The *Cincinnati Gazette*

The laws of Ohio established the or width of the railway track at 5 f inches, while those of Indiana fix gauge at 5 feet 8½ inches—making ference of an inch and a half in the of the tracks. This difference is suf to prevent the use of the same rollin chinery on both tracks. It is vexatio detrimental to the interest of railway panies in both States. The legal ga both States was inconsiderately ad looking to no practical good. Roads i State have been built, and are now ru machinery adapted to each gauge, an ficulty is experienced in connectin lines of road of different gauges so secure the greatest advantage with th delay and cost of transportation.—*Sci American.*

ON MR. FAIRBAIRN'S PATENT TUB CRANE. BY SIR DAVID BREWSTE

(*From Reports of British Association.*)

"These structures indicate some addi examples of the extension of the tu system, and the many advantages may yet be derived from a judicious bination of wrought-iron plates, and a ful distribution of the material in all constructions which require security dity, and strength.

"The projection or radius of the these cranes is 32 feet 6 inches fro centre of the stem, and its height 3 above the ground. It is entirely com of wrought-iron plates, firmly riveted ther on the principle of the upper side calculated to resist tension, and the or concave side, which embodies the c construction, to resist compression. form is correctly that of the prolonge tebræ of the bird from which this ma for raising weights takes its name; truly the neck of the crane tapering the point of the jib, where it is 2 feet by 18 inches wide to the level of the gr

* For description of this crane, with engr see our last vol. p. 381.

it is 5 feet deep and 3 feet 6 inches From this point it again tapers to a of 18 feet under the surface, where it ates in a cast-iron shoe, which forms on which it revolves. The lower or e side, which is calculated to resist ssion, consists of plates forming ells, and varying in thickness in the f the strain; as also the convex top, is formed of long plates chain-riveted wers; but the sides are of uniform ss, riveted with T-iron, and covering 1½ inches wide over each joint. This ment of the parts and distribution materials constitute the principal ele- of strength in the crane. The form jib, and the point at which the load ended, is probably not the most ble for resisting pressure. It never- exhibits great powers of resistance; form, as well as the position, may e considered as a curved hollow beam one end immoveably fixed at A, and er end C, the part to which the force ied. Viewing it in this light, the hs are easily determined; and taking periments herein recorded, we have formula

$$W = \frac{A\,d\,C'}{e},$$

eight of argo in tons.	Deflection at the point of the jib in inches.
11	2·05
12	2·22
13	2·40
14	2·60
15	2·80

in turning the crane round with a 20 tons there was no perceptible on in the deflection, and the perma- t, after removing the load, was ·64

m the above experiments, it appears e ultimate strength of the crane is greater than is requisite either in r practice, and, although tested with double load, it is still far short of ate powers of resistance, which it bserved are five times greater than hat it is intended to bear.

a load of 63 tons, the weight it would re- quire to break the crane. With 20 tons the ultimate deflection was 3·97 — ·64 of a per- manent set = 3·33 inches, the deflection of the jib due to a load of 20 tons. The following constitute the experiments made at Keyham Docks:

"Experiments made to ascertain the re- sisting powers of a new wrought-iron tubu- lar crane, erected at Keyham Dockyard, Devonport, November 8, 1850:

Weight of cargo in tons.	Deflection at the point of the jib in inches.	
2	·32	
3	·50	
4	·65	With 5 tons suspended the crane was turned completely round, without any altera- tion in the deflection.
5	·90	
6	1·05	
7	1·20	
8	1·35	
9	1·50	
10	1·70	

With this weight the crane was again turned round; the deflection in eight minutes in- creasing to 1·85 inches, when it became per- manent after sustaining the load during the whole of the night, a period of about 16 hours.

"On 9th November the experiments were resumed as follows:

Weight of cargo in tons.	Deflection at the point of the jib in inches.
16	3·00
17	3·20
18	3·50
19	3·73
20	3·97

"The advantages claimed for this con- struction are its great security, and the facility with which bulky and heavy bodies can be raised to the very top of the jib without failure. It moreover exhibits, when heavily loaded, the same restorative princi- ple of elasticity strikingly exemplified in the wrought-iron tubular girder. These con- structions, although different in form, are nevertheless the same in principle, and un- doubtedly follow the same law as regards elasticity and their powers of resistance to fracture."

DOES THE MOON INFLUENCE THE WEATHER?

remote ages a traditionary opinion ailed among the rude—and civilized ople of all nations, that the moon d the weather. A few years ago, nch astronomers reported against tion as a fallacy, and the question ught to be settled; but in the July Number of the *American Journal of Science and Arts*, Mr. J. W. Alexander contributes a short article on meteorological coinci- dences, in which he states as the result of a long-continued series of observations, "that the third day before the new moon regulated the weather on each quarter-day of that lunation, and also characterized the general aspect of the whole period. Thus, if the new moon happened on the 26th of May, 1851, the term day was the 24th; the weather on which the 24th of May determined was to be on the 26th of May, and on the 3rd, 11th, and 19th of June, the quarter days respectively of that lunation." This is an important discovery, and shows that the influence of the moon is appreciable, contrary to the generally-received opinion among the learned.

CONSTABLE'S COMPENSATING FLY-WHEEL.

(Provisionally Registered.)

Sir,—Permit me to invite the attention of your readers to the model of a mechanical contrivance which I have placed in the Industrial Exhibition, under the title of "A Compensating Fly-wheel." Its purpose is to convert into an uniform force the varying force derived from reciprocating steam-engines. As to its mode of effecting this, I beg to present you the following explanations:

The fly-wheel, as ordinarily applied to steam-engines, effects two purposes—it carries the engine through the dead points of the crank revolution, and it does this effectually; it also, in a measure, corrects the variations inseparable from a power communicated through a crank; this latter service, however, it performs only approximately, and, being *fixed* upon the main-shaft, it transmits all its uncorrected irregularity through any train of machinery connected with it, in many cases to the great detriment of the manufacturing work it has to perform, and often occasioning the rapid destruction of the gear-work through which the power is transmitted. This defect is incurable under the existing method of using the fly-wheel; for although every augmentation of its weight will bring its oscillatory movements within a more limited range, yet no weight of metal will ever entirely correct them. In the scheme involved in the model to which I invite your attention, the hopeless task of compelling the fly-wheel to steadiness is abandoned, it being permitted to take up its oscillatory motion according to its pleasure or caprice, while all the subsequent machinery is secured from partaking, i
slightest degree, of these oscillatio

The means by which I propo
effect this, consists in releasing th
wheel from its *rigid* connection
the main-shaft, and substituting
fore a spring or springs, through
the force is conveyed, and then,
further device, correcting the
remaining irregularity that result
the re-action of the springs when
different degrees of enforcement.

It will be apparent to every me
cal person that constructive arrang
comprehending these principles of
may assume a great variety of
and combinations. I will proceed
scribe that which I have adopted
model in the Exhibition.

The model, the essential p
which are shown in the accomp
diagrams, consists of a fly-whee
feet diameter, having six arms, b
loose upon a cylindrical end of th
shaft; referring to the diagra
fig. 1 a portion of the fly-wheel is
marked A, and the end of the
shaft C; immediately behind t
wheel are three other arms, issui
a boss which is *fixed* on the sam
shaft (fig. 2) D; E is a pin for
tion with the crank-arm or conn
rod. On the face of each altern
of the fly-wheel arms there lies
spring, which is partially com
between two studs (fig. 1) F F
are fixed to the arms of the w
bolt G G passes freely throu
springs and studs, which bein
upwards (towards the rim of the
forces the spring, by means of
at the lower stud, into closer c
sion. The bolt is connected
roller H by a leather strap J; t
dle of this roller passes through
in the rim, and carries on the o
another roller (fig. 2) I, which
manner is fastened to a pin at t
end of the arm B by the strap I

The fly-wheel, during the pa
revolution in which the stean
power superior to the resistin
will advance by acceleration,
fixed arm not partaking of the i
speed, the space between the
and the arm B will be increased
the action of the rollers the sp
be compressed; that period bein
and the impelling power fallin

the resisting power, the latter will prevail, and bring back the wheel again to the place, with respect to the mean place, from which it started. Now, when the fly-wheel advances, by its oscillating movements, it will leave the arm B behind, and the strap connecting it with the roller, I, will occasion the roller to move on its axis, which will result in the drawing up of the bolt, and compressing the spring; and the reverse will happen when, through its oscillations, the fly-wheel loses speed; and thus the fly-wheel will go on oscillating twice in every revolution, the oscillations playing smoothly and harmlessly upon the spring.

Fig. 1.

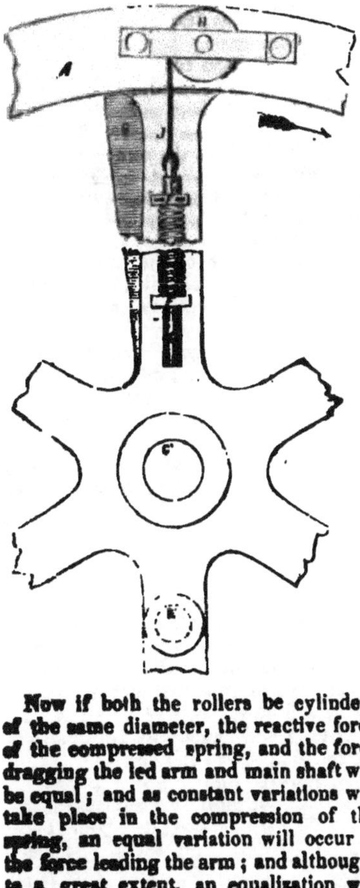

Fig. 2.

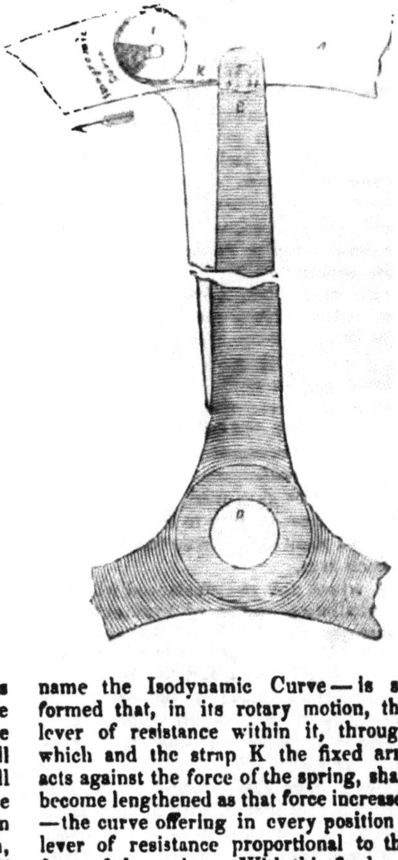

Now if both the rollers be cylinders of the same diameter, the reactive force of the compressed spring, and the force dragging the led arm and main shaft will be equal; and as constant variations will take place in the compression of the spring, an equal variation will occur in the force leading the arm; and although, to a great extent, an equalization will have been effected, the problem of the conversion of a fluctuating force into a constant force has, as yet, received no solution.

But the rollers are not both cylinders; the one marked I, is bounded by an irregular curve, a part of which—distinguished by red colour on the disc of both the model and diagram, and which I name the Isodynamic Curve—is so formed that, in its rotary motion, the lever of resistance within it, through which and the strap K the fixed arm acts against the force of the spring, shall become lengthened as that force increases—the curve offering in every position a lever of resistance proportional to the force of the spring. With this final appliance, a perfect uniformity of force is obtained, and the problem proposed is solved.

I propose now to inquire what will occur in the actual working of a specific steam engine with these appliances. In the case I propose, the model shall be extended in all its parts linearly in the ratio of 3 to 10; the fly-wheel will then be of 10 feet diameter, and will have 1,500 lbs. weight in its rim. The engine to be of 10-horse power, clear of all loss by friction or otherwise, in its own parts; strokes, 40 per minute. The steam may be applied to the piston at any amount of pressure, and be worked expansively or not. The springs, as in the model, to be three; coiled of steel wire, ·25 inches in diameter; the coils to measure 2¼ inches in diameter to the middle of the wires; the space between the coils, when the spring is relaxed, to be equal to the diameter of the wires; the number of coils, 70; its whole length, when uncompressed, consequently, will be 35 inches. Such a spring will suffer compression of 1·17 inch for every 10 lbs. of compressive force. I propose that each spring shall be confined between the studs with a force of 40 lbs., by which the length will be reduced to 30·32 ins.: this compression is not necessary to its proper action, but it keeps the length of the spring within convenient compass. The breadth of the rim I propose shall be 7 inches; therefore, the circumference of a circle at the middle of the rim will be 29·6 feet; and the speed, at this place, will be 1,184 feet per minute. The mean force communicated at any point in this circumference will be equal to 279 lbs. Let this force be transmitted through three springs to the three arms, then we shall have force 93 lbs. applied to each arm.

A wheel thus circumstanced will have an oscillating motion of 1·26 inch from its mean place, and the measure, between the extremes of its oscillations, will be 2·52 inches; and as the axis of the rollers is placed in this circle, the variations in the compression of each spring will be of the same measure—(2·52 inches.)

Now, since the compression of the spring is 1·17 inches, with a force of 10 pounds, a compression of 2·52 inches will be the exponent of a force of 21·54 pounds, and 21·54 pounds will be the variation of force communicated, from the fly-wheel, to the fixed arm; and since the mean force, applied to one arm, is 93 pounds, the maximum force, when the fly-wheel is at the extreme of its advance, will be 103·77 pounds; and the minimum, in the contrary position, 82·23 pounds; and this variation is so moderate, that, for most of the purposes to which steam engines are applied, it would have scarcely any appreciable disturbing effect.

For the final correction of this remainder, the Isodynamic Curve must have a length equal to the extent of the action of the springs, or the wheel's oscillations, 2·52 inches; and the radius of resistance of the minimum force must be that of the maximum, as the smallest force of the spring is to the greatest; that is, as 82·23 to 103·77.

With respect to the minimum weight of fly-wheel that may suffice in a given engine, under these arrangements, it must depend upon the range of action that can be obtained in practicable springs. In the case I have assumed, I have proposed a wheel with a rim of 1,500 pounds; if I had taken half that weight, the length of the arc of oscillation would have been doubled; and if we could employ springs in which the range of compression would be twice as great as in those proposed, when acted on by the same force, we should have a regulating power of the same efficiency; and if we desire to perfect the uniformity of force, we might do so by employing an Isodynamic Curve of twice the length, keeping the radii which constitute the levers of resistance, of the length proposed in the first case, and the action of the engine would be the same.

This invention claims not merely to improve, but to *perfect* the action of the Reciprocating Steam Engine; it is simple, of easy and inexpensive construction, and but little liable to go out of repair; therefore, I venture to believe that it will be appreciated by all persons practically or theoretically interested in the action of steam engines.

I remain, Sir,

Your most obedient Servant,

WILLIAM CONSTABLE.

Note.—A proper method of trying the action of the model is, to stand at the right-hand side, holding one of the arms of the fly-wheel in the left-hand; then applying the palm of the right-hand to the fixed arm in advance, and forcing the arms apart. It will require a force of about forty pounds to bring the springs into action. The two arms marked A and B, at the back, are in the right position to each other for handling.

THE MOON'S SURFACE.

number and magnitude of crater
l mountains with which every portion
moon's surface appears to be covered,
to lead to the conclusion that these
ally the craters of extinct lunar vol-
; the frequent occurrence of the
l cone being considered as the result
last eruptive efforts of an expiring
o, a feature of volcanic craters on the
surface. This central cone has been
to exist in the majority of the lunar
; and the conclusion consequently
s probable that they are the result of
ne kind of action which has produced
n the volcanos of the earth.

cause of the vast numbers of such
le mountains with which the lunar
is covered, has been assigned by
to the rapid consolidation and con-
n of the crust of the moon, whose
r bulk being only 1-64th of that of
rth, while its surface is the 1-16th,
consequence of these proportions, a
ng or heat-dispensing surface four
greater than that of the earth in rela-
its bulk. From this consideration,
been suggested by the exhibitor, that
rapid cooling and collapse of the
f the moon on its molten interior, the
atter under the solid crust has been
action forced to find an escape through
perincumbent solid crust, and come
n those vast volcanic actions which in
remote period of time have covered its
with the myriads of craters and vol-
features that give to its surface its
able character.

vast magnitude of the lunar craters,
also been suggested, are due to this
collapse of the moon's crust on its
interior,—the action as regards the
dispersion of the ejected matter being
ed by the lightness of the erupted
, seeing that the force of gravity
gives the quality of weight to matter
moon, as on the earth, is less on the
s of the moon than on the earth,—so
e collapse action had to operate on a
ght material.

causes of those vast ranges of moun-
seen on the moon's surface has been
ted to be produced by the continued
s of the collapse action of the solid
f the moon crushing down or follow-
e contracting molten interior, which,
gradual dispersion of its heat, would
from contact with the interior of the
rust, and permit the crust to crush
and so force a portion of the original
out of the way, and in consequence
action cause such to assume the form
rrangement of mountain ranges. In illustration of this important action, the familiar case of the wrinkling of the surface of an apple, by reason of the contraction of the interior and the inability of the surface to accommodate itself to the change otherwise, has been adduced.

The origin or cause of those bright lines which radiate from certain volcanic centres on the moon's surface (Tycho, for instance) has been illustrated by the experiment of causing the surface of a globe of glass filled with water to collapse on the fluid interior, by rapidly contracting the surface while the water had no means of escape. The result was the splitting or cracking up of the surface of the globe in a multitude of radiating cracks, which bear the most remarkable similarity to those on the moon. This subject is also illustrated by reference to the manner in which the surface of a frozen pond may be made to crack by pressure from underneath — so yielding radiating cracks from the centre of divergence, where the chief discharge of water will take place, while simultaneously all along the lines of radiating cracks the water will make its appearance :—thus explaining how it is that the molten material, which had in like manner been under the surface of the moon during that period of its history, appears to have come forth simultaneously through the cracks, and appeared on the surface as basaltic or igneous overflow, irrespective of surface inequalities.—*Illustrated Catalogue of the Great Exhibition.—Part II.*

CAVE'S IMPROVED DIVING BELL.

A new and happy modification of the diving-bell has been invented by Mr. Cave, the eminent engineer, for the purpose of descending to the bottoms of rivers, and carrying on works there with greater facility than by the ordinary diving-bell. On the front of a dredging vessel is placed a large chamber, made of sheet iron, having the form of an elongated hemispherical cup, 22 feet 9 inches in diameter, and 16 feet 3 ins. in height. In the centre of the bottom of the vessel there is a large opening which communicates with the river, and in it is placed vertically a large cylinder of sheet iron, open at either extremity, and which can, by means of grooves, be lowered to any depth that may be required. When it is desired to examine the bottom of the river, it suffices to lower the cylinder; and, by the aid of an air-pump, a large quantity of compressed air is forced into the chamber. The water by that means is expelled underneath the cylinder, until at length the bottom of the river is left dry. The workmen can then descend inside the cylinder, and proceed with the work without any difficulty.

For communicating from without with the chamber there is provided an ante-chamber, for persons to go in and out without allowing the compressed air to escape from the inner chamber. The doors of the inner chamber are hermetically closed, by which means the loss of compressed air is small, and is easily replaced. For the purpose of opening the inner door, it is necessary to open a valve to allow a small quantity of air in the inner chamber to escape into the ante-room, to restore a balance, and make the pressure the same on both sides of the door. And a similar contrivance is necessary in the outer door; but before the valve is opened in the outer door, care must be taken to close the inner door and valve. There is another contrivance for forming an air-tight connection between the vertical cylinder before described and the chamber. This is effected by a flexible joint or tube made of leather; one end is fastened to the bottom of the chamber, and the other to the top of the cylinder. This leather flexible tube allows some play in the cylinder, so as to adapt it to various depths of water, or variations in the depths of the river. The compression of the air is very easily accomplished by the steam engine which usually accompanies dredging vessels. The engine works two air-pumps, which communicate by a pipe to the chamber, and supply compressed air at discretion: of course, the density of the air must be in proportion to the depth of the water. It would appear that the workmen do not feel any particular difficulty in working in such an atmosphere; the only inconvenience in the augmentation of the density of the air is a slight pressure and noise in the ears. This vessel on the Seine is only an experimental one, to show that all descriptions of work can be performed under water with the greatest facility. M. Cave has already established two similar vessels for scouring the mud-banks of the Nile. The dimensions of them are much larger, the cylinders being 29 feet 3 inches by 19 feet 6 inches.—*The Architect and Civil Engineer.*

THE MAMMOTH CAVE OF AMERICA.

The last Number of *Silliman's Journal* contains an interesting account of the Mammoth Cave, in a letter addressed to Professor Guyot by Professor Silliman, Jun., who has recently made an exploration of its mysteries; and also, in connection with Mr. R. N. Mantell, made a collection of the animals found there. One atmospheric ph
attracted the attention of these
and taxed their ingenuity for a s
explanation; viz., The blast o
blowing outward from the mou
cave, which rendered it nearly
to enter with a lighted lamp. If t
air has a temperature of 90° Fahr
amounts to a gale; but if the a
has a temperature of 59—60°,
is observed, and the flame of a la
a favourable position indicates
immediately occurred to me (said
Silliman) that there must be tw
one above of warmer air, passi
and one below of colder air, pa
ward, and the reverse; but experi
satisfied me that this was not the
one current could be discovere
inquiry of our intelligent guide, I
that this phenomenon had attrac
tention, and that he was satisfied
observations that only one curre
and that this flared *out* when th
air was above 60° and *inward* wh
below 60°.

The phenomenon is accounted
fessor Silliman on scientific pri
follows:—The mouth of the cave
communication between the exter
the vast labyrinth of galleries a
which stretch away for many m
solid limestone. The air in th
ground excavations is pure and e
which may in part be accounted
nitre beds of incredible extent, a
gen which is consumed in the fo
the nitrate of lime must have its
of free oxygen disengaged, thus
this subterranean atmosphere wi
portion of the exhilarating princ
temperature of the cave is unif
summer and winter; and this i
very near to the annual mean of t
air. The expansion which acco
elevation of temperature in the
immediately felt by the denser
cave, and it flows out in ob
the law of motion in fluids, and t
current continues without inter
long as the outer air has a highe
ture than the cave.

The phenomena of life within t
comparatively few, but interesti
are several insects, the largest of
sort of cricket, with enormousl
tennæ. There are several species of
mostly burrowing in the nitre ear
are some small species of wat
supposed to be crustaceous. Of
are two species, one of which,
known, is entirely eyeless; the
external eyes, but is quite blind.

al, except the bats, is a rat, which is bundant. Professor Silliman is of a that the excavations of the Mam- Cave have been formed by water, and other cause.

ON BUILDINGS IN NEW YORK.

hin the past two years, a great num- cast-iron buildings have been erected w York. The designer and con- r of these buildings is Mr. Bogardus, ll-known inventor, who may well be "the man of Iron Genius." His ouses are manufactured in an iron in Centre-street, this city, where the ss is now conducted under the firm of lus and Hoppin; the latter a gentle- ' exquisite taste and genius also.

reat number of our new free-stone ck buildings have cast-iron fronts on t stories. The castings are beautiful; are glad to see this metal, so plastic lurable and strong, coming into more use.

32nd street, near North River, is an wer in course of erection, for the e of an Observatory and an alarm- The spot itself is rather elevated. undation is laid 14 feet below the of the ground, and is supported and with iron shafts sunk in solid rock. ilding is entirely open between the s, thus offering but little resistance wind; and such is the weight of ma- hat when done, a pressure of 14,000 uld be required to move either of its s or stories from the perpendicular, ey not fastened at all.

tower when done will be about 100 ft. o.mm'nding from that elevated point ct view of nearly the whole city and It is to be surmounted by an Ob- ry, where a watch is to be kept sub- direction of the Fire-department. cent is to be by a winding staircase inside, and it is to be completely fire- —*Scientific American.*

AMERICAN FLYING-SHIP.

r Hoboken village, on the other side river opposite our city, there is a enclosure 290 feet long, in which is t wonderful apparatus—or rather artificial dragon—*nearly ready for ing.* It is a huge cigar-shaped bal- 260 feet long, and 24 its greatest er. It has a car 64 feet in length, arp at either end, width 5 feet, height 4 inches, the whole composed of a , light wooden frame, covered with canvass, with doors and glass windows. It is to be propelled by two of the most beautiful engines ever constructed. They are made of gun-metal and cast-steel, are of 12 horse power, and are to work 20 inch stroke 66 times per minute, which will give 400 revolutions to two propelling fans. The entire weight of the car, float, and fixtures, is but about 4,000 pounds, leaving 2,500 pounds surplus. It is designed to run about 200 feet above the surface of the earth, at a rate of speed varying from 25 to 50 miles per hour. It is calculated that the gas will have an upward buoyant force sufficient to raise more than 6,000 lbs. above the ground. The engines only weigh 181 lbs. They are constructed by Mr. Robjohn, a most ingenious mechanician, one who can make a balloon go if neat and well-constructed machinery can do it.

It is designed to drive this vessel by steam; and to obviate the necessity of coal, Mr. Robjohn says he has discovered a plan of decomposing water, which is converted into steam, by the combustion, and this steam is again condensed and returned for decomposition.

The most skilful and best of men are oftentimes led away by enthusiasm, and it is a good thing for science perhaps that it is so.—*Scientific American.*

AN ELECTRICAL FACTORY.

The following remarkable phenomenon connected with frictional electricity developed at a thread-mill in Glasgow may prove interesting to many of your scientific readers, as it affords a strong connecting link in the chain of evidence establishing the identity of the electricities developed by friction and by chemical action.

For some time past the hands employed at the factory in which this occurrence takes place have been seriously annoyed by receiving smart sparks and shocks when approaching or handling the machinery. The construction of the mill is that of a number of flats or floors one above another, laid over with a coating of asphalte, on which the machines are placed, bolted to a sole-plate of iron. The ceilings are supported by a series of iron columns running down the centre of each floor, and having connec tion with the earth,—but owing to the circumstance of the asphaltum floor, in a state of tolerable insulation as regards the machines. The power is derived by drum-shafts, running parallel to the wall, and supported on hanging brackets attached to lateral iron beams in connection with the columns; motion being communicated to the machinery by leather and gutta percha belts. Each floor of the factory, therefore, assumes the condition of a vast electrical machine, the lathes representing the prime conductor, and the drums and belts the exciting medium. As may be supposed under these circumstances, the amount of fluid continuously generated is something considerable, and likely to have caused much discomfort to the workpeople at a time when their lathes were not in connection with the earth.

The result of my experiments may be stated as follows:—The electricity developed at shaft and drums—negative; that at the lathes—positive. When the current of electricity was connected so as to flow through a jar of solution into the earth, a feeble but continuous stream of gas was liberated at the electrodes. The most remarkable experiment, however, was its power of inducing continuous magnetism in a bar of soft iron surrounded with a helix in the manner of a voltaic magnet. The magnetism there developed deflected a magnet either way, and had a sensible effect on the suspension of a small bar of iron at the poles of the magnet, which exhibited in all respects the phenomena incidental to the voltaic current. Unfortunately, the coil of my galvanometer was of a different construction from that which I required for the experiment,—and I have in consequence been unable for the present to complete the interesting and valuable fact of a permanent deflection being obtained upon that instrument.

N. J. Holmes.

Athenæum.

SPECIFICATIONS OF ENGLISH PATENTS ENROLLED DURING THE WEEK ENDING AUGUST 21, 1851.

Benjamin Ledger Shaw, of Huddersfield. *For improvements in cleaning and preparing wool and fibrous or textile materials, and in the manufacture of coloured yarns of wool and other fibres, and in weaving.* Patent dated February 5, 1851.

Claims.—1. In machinery or apparatus for "cleaning and preparing wool,"—the application of a series of chains composed of plates or links of thin steel or other suitable material, notched or indented on their edges, and so arranged as to cover the entire exterior surface of the receiving drum or roller, or disposed in grooves cut for that purpose in the surface of the said roller. Also, the general arrangement of machinery shown, for the purpose of cleaning and opening wool, &c., [the principal feature of which, in addition to the employment of a drum constructed as above described, appears to be, that the feed rollers are caused to move at slightly-increasing velocities, and thus to reduce the sliver of wool in thickness in its passage through them, and consequently facilitate its being cleansed.]

2. A method of producing party-coloured yarns. [For this purpose, a number of slivers of different colours are wound on a large bobbin, from which they are simultaneously fed into a carding engine, and subjected to the operation of carding, from which they are delivered in short lengths, which are pieced up end to end, and subsequently drawn and spun in the ordinary manner, the colours being preserved throughout these processes in the same order which they occupied when the slivers were pieced together.]

3. In "looms for weaving,"—combining and connecting together the apparatus for actuating the harness or heddles and the sliding shuttle-boxes so that they may go correctly together. Also, a peculiar arrangement of apparatus for suspending and actuating the sliding shuttle-boxes; and a method of regulating and governing the action of the picking-sticks or levers by means of a suitable pattern surface.

Charles Xavier Thomas (de Colmar), Chevalier de la Legion d'Honneur, of Paris. *For an improved calculating machine, which he calls* "Arithmometer." Patent dated February 10, 1851.

This machine is intended to perform, not only the simple arithmetical operations of addition, subtraction, multiplication and division, but also the more complicated ones, such as the extraction of the square and cube roots, and other operations wherein fractions and decimals are involved. The complexity of the apparatus is necessarily such as to render futile any attempt to convey, in less than a very lengthy description, and without the aid of drawings, an accurate idea of its construction.

Benjamin Heywood, of Manchester, coach-builder. *For certain improvements in railway and other carriages.* Patent dated February 11, 1851.

These improvements have relation to the the windows of railway and other carriages, and consists—

1. In making such windows with a single pane of plate glass without a frame, and having its edges rounded off to prevent injury to the cloth or other material with which the grooves for the pane to slide in are lined.

2. In lining the grooves in which the pane of glass slides with India-rubber, velvet, or other similar material, for the purpose of making a good joint.

3. In attaching the glass strings to such

windows of carriages, by means of which the string is fastened, and is secured to the glass by rivets passing through holes bored therein, f some flexible material being in-tween the clip and the pane of prevent fracture of the latter when re inserted.

. HAYTHORNE REED, late of the ers, of the Harrow-road, gentle-r *improvements in saddlery and* Patent dated February 11, 1851. sent improvements have relation saddles and consist in an arrange-pparatus whereby the girths may ed or loosened at the pleasure of without the inconvenience of dis- or removing the feet from the

ls of the girths have straps at-them, which are inserted in an on the saddle-tree, and having a eel and handle at one end, by ay be turned and the straps thus nd it, and the girths consequently In addition to the lever or re is also a stop for preventing t-wheel running back, and it is g this stop from the teeth of the eel that the girths are loosened ired. The apparatus may be one or both sides of a saddle of ption.

WEBSTER, of Leicester, engineer. *vements in the construction and applying carriage and certain ags.* Patent dated Feb. 11, 1851. vention consists in constructing ing springs in such manner that hall be applied to the ends of their means of inclined planes or sur-

tentee prefers generally that the rfaces should be arranged so as to he letter V inverted, and that the e springs should act on the interior out this disposition of parts he does himself to. Exemplifications are prings constructed on this princi-d as bearing springs for railway and trucks, and wagons, draw uffer springs, and springs for apparatuses. In some cases the e made to act on two inclined sur-pposite directions.

—The construction and application in such manner that the arms or eof shall be acted on by inclined surfaces.

A MARCH PERKINS, of Francis-egent-square. *For improvements ating and heating ovens.* Patent ruary 11, 1851.

This invention consists in applying to the heating of ovens wrought-iron tubular apparatus, through which hot water circulates, thus obtaining an equable degree of heat throughout all parts of the oven. The interior of the oven is lined with tubes which communicate with main water-heating pipes, which again are connected by branches with a series of coils heated by the fire in the furnace. In order to prevent the temperature rising to an injurious degree, the flue of the oven is provided with a damper, to which is attached a rod having its end bent into a hook and resting on a cup containing a compound metal, fusible at a given temperature, say 600° Fahr.; when the tubes have attained the melting point of the fusible metal in the cup, the metal becomes melted, and the end of the rod sinks and opens the damper in the flue, admitting cold air, and thus lessening the draught and reducing the temperature to the required degree.

Claim. —The improvements in ovens whereby the same are heated by water in several circulations proceeding from main circulating tubes in connection with coils of tubes heated by the fire.

CHARLES HOWLAND, of New York, engineer. *For improvements in bell telegraphs.* Patent dated February 11, 1851.

The patentee describes and claims the application of telegraphs constructed on this principle for effecting communication in hotels or dwelling-houses from one room to another, and on board steam vessels from the pilot or captain to the person in attendance on the engines. The arrangements are in both cases extremely simple. In hotels the telegraph is employed, not so much to convey messages, as to indicate that attendance is required in a particular room or portion of the establishment. The bells, hammers, and indicating plates, with the apparatus by which they are more immediately actuated, are contained in a case or box, the front of which constitutes the dial-plate, and is perforated with a number of holes, in tiers, corresponding to that of the number of rooms with which communication is to be established. Behind the dial are formed grooves, in which slide number-plates communicating each by a bell-wire with one of the rooms, and bearing a corresponding nnmber to it. The pulling of any one of these bell-wires actuates a tumbling lever, which causes one of the hammers to rise and strike its bell, and at the same time releases the number-plate, which slides down its groove, and exhibits through one of the perforations in the dial the number of the room requiring attendance. The office of the bell, it will be seen, is that of an alarum to indicate that a signal is being made. As soon as a signal has been made, the number plate is raised to its former position by the attendant pulling a bell-wire, —the hammer falling back to a state of rest by its own weight.

On board steam vessels, the sliding plates bear, instead of numbers, the words of command generally employed in giving directions to the engineer: the working of the telegraph is in other respects precisely similar to that above described.

HENRY FRANÇOIS MARIE DE PONS, of 24, Boulevard Poissonnière, Paris, gentleman. *For improvements in constructing roads and ways, and pavements of streets, and the ballast of railways.* Patent dated February 17, 1851.

These improvements consist in constructing the surfaces of roads, streets, &c., with a concrete composed of iron ore, in the state of stone iron reduced to small pieces, granulated cast iron, broken, or in chippings or shavings, other metals, reduced to a similar state, volcanic schist, otherwise called volcanic gluten, previously calcined and pulverised so as to convert it to the state of cement, all kinds of cement and hydraulic mortars, lime, plaster, sand, stone, iron dross, slag, bitumens, asphalts, sulphur, sulphate of alumina, or sulphate of alumina and iron. The whole, or any number or combination of these materials are spread on the surface of the road, which should be Macadamised or otherwise rendered smooth and even, and subjected to rolling in order to bind them well together, and with this view the materials whilst undergoing the rolling are to be watered with water or with a solution of alumina. The patentee prefers in all cases to have a substratum of iron dross and to lay the other materials employed on that stratum.

Claim.—The combination of certain substances, and the employment of certain processes applied to the construction of roads, streets, and ways, and the covering of ways, with or without rails for the running of locomotives.

GUSTAV ADOLPH BUCHHOLZ, of Agar-street, civil engineer. *For improvements in motive power, and in propulsion.* Patent dated February 17, 1851.

This invention comprehends—

1. An improved method of constructing toothed or cogged wheels with a plain surface equal in diameter to the pitch circle of the wheel, on one or both sides of the teeth or cogs, the wheels being so geared together that while the teeth work into one another the plain surfaces may roll evenly one over the other, thus diminishing friction to a great extent.

2. The application of cogged wheels constructed on this principle to the bearings of shafting of every description, for the purpose of reducing friction and wear.

3. Certain simple and compound engines, in which a rotary motion is communicated to an axis or shaft by constant pressure exerted on springs actuating eccentrics.

4. A method of propelling vessels in any required direction, without the aid of a rudder, by the employment of three screws (one at the stern and one under each quarter of the vessel) geared together in such manner as to produce the above-mentioned effect.

CHARLES COWPER, of Southampton-buildings, Chancery-lane. *For improvements in moulds for electro-metallurgy.* (A communication.) Patent dated February 17, 1851.

This invention consists in constructing moulds for electro-metallurgical purposes by the employment of gelatine, glue, or other glutinious or elastic material in combination with metallic wires, or strips or pieces of metal imbedded therein, by which the electric current may be quickly and simultaneously conveyed to all parts of the surface of the mould. When making moulds of gelatine or glue, the surface of the model is first oiled all over, and covered with a number of fine metallic wires, or strips or pieces of metal, the ends of which should be allowed to project; melted glue or gelatine is then poured on so as to imbed the wires, and allowed to cool, when a mould will be obtained suitable for use. The interior of the mould is rubbed over with black lead or covered with foil, and the exterior varnished or oiled before immersing it in the bath. When immersed, the mould is connected to one wire of the battery, and a piece of metal to the opposite wire; or, the mould may be connected directly to a piece of metallic zinc when using the apparatus known as the single cell battery. When the moulds are made of gutta percha, it is preferred to mix with it naphtha or some other solvent, and to roll it into sheets, which are warmed and applied to the surface of the model—previously covered with fine wires, as above described. India-rubber moulds are obtained in the same manner, the India-rubber being previously treated with bisulphuret of carbon or other solvent, and rolled into sheets; or, the India-rubber may be dissolved, and a mould obtained by applying successive coats of the solution to the model —one coat being in all cases allowed to dry before another is laid on. Moulds constructed as above described are particularly suitable for operating on metallic articles.

Claim.—The forming of moulds for electro-metallurgy by covering the models with metallic wires, strips, or pieces of metal,

; the gelatine or other elastic or
aterial to form the mould in
that the wires or other metal-
s may remain imbedded in the
be adapted for conveying
current to the surface of the

BEADON, junior, of Taunton.
*ments applicable to the roofing
uildings and other structures.*
February 18, 1851.
vements sought to be secured
atent comprehend—
naking of the lowest range of
with the ends turned up, so as
itter.
bined gutter and coping, the
h forms an ornamental cornice.
ing, crease or ridge tile for
; the ends turned up, so as to
r on each side, thus preventing
running down the sides of the
abling it to be collected and
ilable for use.
king of flat titles with rebated
o overlap each other at the sides,
flush on the upper surface.
aking of valleytiles with their
res so formed as to admit of
g cement to form a watertight
h a knuckle joint for the same

s tile with an ornamental face,
table at the back to carry rain
er.
tee does not confine himself to
ar material of which to form
d copings, tiles, &c., as glass,
other substances are alike suit-
purpose, nor does he confine his
particular forms which he has
ey may be varied without de-
the principle of his invention.
DICKASON ROTCH, of Furni-
ntleman. *For improvements
al apparatus for separating
other matters.* (A communica-
nt dated February 18, 1851.
tion consists—
ending the axes of centrifugal
which the drums are attached,
and socket bearing, and in
driving strap to act on a hemi-
m forming the exterior of the
uch manner as always to main-
in a proper position for driv-
ine, even when the oscillations
shall have caused the axis to
blique position instead of the
in which it is suspended when
rest.
ing the drum of such machines
d *metal, by which means* a
greater degree of strength is obtained in proportion to the thickness of the metal than when uncorrugated metal is employed; whereby, also, the wire-gauze lining or perforated cylinder is kept from contact with all parts of the internal periphery of the drum. When the drum is made of uncorrugated metal, the wire gauze or perforated cylinder may be kept from contact with all parts of it by the interposition between it and the drum of a coil of wire.

Claims.—1. The mode herein described of suspending and driving the axes of centrifugal machines for separating fluid from other matters.

2. Forming the drum of corrugated metal, and keeping the wire gauze or perforated cylinder from contact with the periphery of the drum.

PETER WOOD, of the firm of Bury and Co., dyers, finishers, and calenderers, Salford. *For improvements in printing, staining, figuring, and ornamenting woven and textile fabrics, wood, leather, or any other material, substance, or composition, and in machinery and apparatus employed therein.* Patent dated February 24, 1851.

Claims.—1. The printing, staining, figuring, and ornamenting woven or textile fabrics, and any other material, substance, or composition, by the use of metal in a finely-divided or precipitate state, mixed with starch or other thickening, and the finishing of these surfaces by calendering or pressure. (For details of this process, see last vol., p. 497.)

2. An arrangement of apparatus for the purpose of applying the printing mixture, or powdering the surface of fabrics previously prepared with glair or other adhesive composition. [The fabric under operation is first submitted to the printing roller, which revolves in contact with and is supplied by an endless band dipping into a trough containing the printing paste. After receiving an impression, it is passed over rollers through a box, where it is sprinkled with powdered metal (when required) by means of a brush revolving in a perforated cylinder, supplied with the metal by a hopper in the top. The fabric then passes under a rotating brush, also contained in the box, by which the superfluous metal is removed, and is finally carried over and under steam-heated cylinders, by which it is dried and prepared for subsequent calendering or pressure with rollers, in order to develop the brilliancy of the metal.]

Specifications due, but not Enrolled.

CHARLES WILLIAM TUPPER, of Oxford-terrace, Middlesex, gentleman, and ALPHONSE RENE LE MIRE NORMANDY, of

Dalston, in the same county, gentleman. *For improvements in the manufacture of iron coated with other metal, commonly called "galvanised iron."* Patent dated February 12, 1851.

DAVID FERDINAND MASIRATA, of Golden-square, Regent-street, Middlesex. *For a new mechanical system with compressed air, adapted to obtain a new motive* Patent dated February 18, 1851.

HUGH LEE PATTINSON, of Scot's Gateshead, manufacturing chemist. *improvements in the manufacture tinson's oxichloride of lead.* Paten February 18, 1851.

WEEKLY LIST OF NEW ENGLISH PATENTS.

Lot Faulkner, of Cheadle, Chester, machinist, for certain improvements in the method of obtaining and applying motive power. August 21; six months.

James Roberton, of Oxford-street, Manchester, chemist, for improved methods of producing or obtaining printing dyes and other substances used in printing; which improvements, in whole or in part, are applicable to other like useful purposes. August 21; six months.

John Walters, of Sheffield, York, manu for certain improvements in knives an August 21; six months.

John Treasahar Jeffree, of Blackwall, for an improved apparatus for facilitating perfect combustion of fuel, whereby fu steam vessels and chimnies, or shafts tories, may be dispensed with. August months.

WEEKLY LIST OF DESIGNS OF ARTICLES OF UTILITY REGISTERED.

Date of Registration.	No. in the Register.	Proprietors' Names.	Addresses.	Subjects of Desi
Aug. 18	2911	Thurston and Co.	Catherine-street, Strand	Pool-marking board.
,,	2912	Joseph Page	Birmingham	Universal portable w screw.
19	2913	J. Whitehouse & Son	Birmingham	Lock knobs.
,,	2914	R. Brightman & Son	Bristol	Sportsman's boot.
,,	2915	James Park	Bury, Lancashire	Steam boiler or gene
,,	2916	Jacob Bonallack	Whitechapel and Holloway	Staves and stays fo cart bodies.

WEEKLY LIST OF PROVISIONAL REGISTRATIONS.

Aug. 15	271	J. C. Nesbit	Kennington-lane	Guanometer, or in for testing the guano.
,,	272	James Cook	Knightsbridge	Portable mangle.
16	273	Thomas Newcomb	East-lane, Walworth	Economic shadowl lamp.
,,	274	Oswald Deitz	Gt. Pulteney-street	Mechanical blood ex

CONTENTS OF THIS NUMBER.

LONDON: Edited, Printed, and Published by Joseph Clinton Robertson, of No. 166, Flee in the City of London—Sold by A. and W. Galignani, Rue Vivienne, Paris; Machin *Dublin; W. C. Campbell and Co., Hamburg.*

Mechanics' Magazine,

[...]EUM, REGISTER, JOURNAL, AND GAZETTE.

[...]464.] SATURDAY, AUGUST 30, 1851. [Price 3*d.*, Stamped, 4*d.*

Edited by J. C. Robertson, 166, Fleet-street.

[...]IOTT'S PATENT IMPROVEMENTS IN CENTRIFUGAL MACHINERY.

Fig. 1.

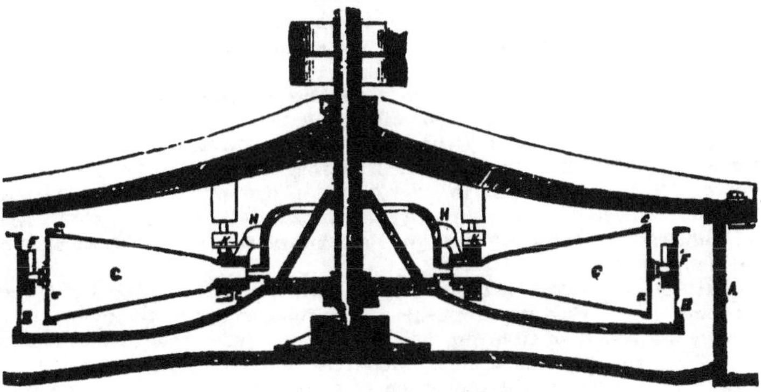

Fig. 2.

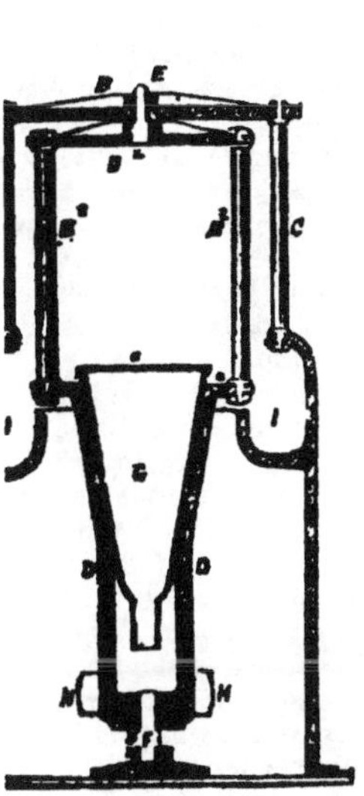

Fig. 3.

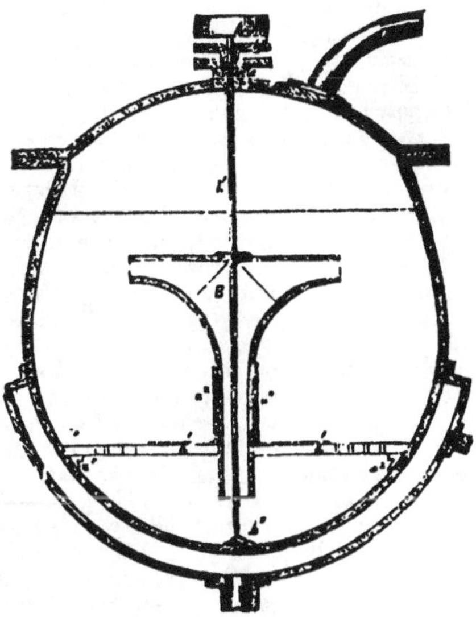

ALLIOTT'S PATENT IMPROVEMENTS IN CENTRIFUGAL MACHINERY.

(Patent dated Feb. 3, 1851. Specification enrolled Aug. 3, 1851.)

Specification.

Firstly.—My invention in so far as it relates to the manufacture of sugar, has reference to the machine for moulding sugar into loaves, called the "Forming Machine," which was the subject of letters patent granted to Richard Archibald Brooman, of the date August 16, 1849, and consists in certain new and improved arrangements of the parts of the same, whereby the liquid is driven from the loaves laterally as well as longitudinally, that is, sideways as well as endwise.

Secondly.—My invention in its relation to the manufacture of soap, consists in subjecting the fatty and alkaline materials whilst boiling, to the action of centrifugal force, in a machine suitably constructed for the purpose, whereby their intermixture is accelerated and a great saving of both time and cost in the process of saponification effected.

Thirdly.—My invention in so far as it relates to dyeing machines, has reference to those of the centrifugal class, and consists in making such additions to and modifications in the same, that yarns can be dyed therein of different colours in separate and distinct divisions, without having recourse to the usual practice of knotting for the purpose, or, in other words, so that any determinate length of the yarn may be dyed of one colour only, another length of a second colour, and so on, and without there being the least intermixture of any one of the colours with the others or any of them.

Fourthly.—My invention has relation to centrifugal cleaning and drying machines, and consists in certain improvements in the construction of the cylinders thereof, whereby the labour of emptying and working the same from time to time is much diminished, and the escape of the liquid so much facilitated that the difficulty hitherto experienced in the application of these machines to heavy and sticky substances, as some classes of sugars, colours, pigments, &c., is greatly diminished.

Fifthly.—My invention in so far as it relates to the manufacture of metals, consists in the adaptation of centrifugal force to the separating, dividing and mixing of the same while in a fluid state.

My improvements in the machine for moulding sugar into loaves, called the "Forming Machine," are represented in fig. 1 (a vertical section), A is the outer case or framework of the machine; B, the cylinder; CC, two of the forms, of which the number in the circle arranged around the spindle D, will be more or less according to the size of the machine; the forms are fitted into bearings in the cylinder, so that they may each revolve upon its own axis; FF are the outer bearings, which are fixed upon the side of the cylinder B; the inner bearings are formed in a ring G, which is attached to the cone of the cylinder, by the joints HH; II are pulleys fixed to the neck of each form; K is a ring covered with leather on the underside, which occupies a position immediately over the whole set of pulleys II, and upon which it can be made to bear more or less as required, by means of screws or springs attached to the top of the casing. The lids *a a* have suitable perforations, and are fastened by the catches *c c*. When the sugar is set in the forms, and a syrup or other liquid is to be passed through the loaves for the purpose of purifying the same, then the cylinder, being in motion, the ring K is caused to bear upon the pulleys II, by which the forms are made to revolve in their bearings F and G, so that while the centrifugal action of the cylinder is causing the syrup to pass longitudinally or endwise, the revolution of the forms upon their own axis causes also a dispersion of the fluid outwards towards the sides of the loaves. In some cases the forms are put each separately into a machine of the description represented in section in fig. 2, and there subjected to the lateral dispersing process. A is a pan, which carries at top a cross head or bearing B, which is supported by the pillars CC; D is a cylindrical case or frame, which is connected to a second cross-head, D^2, by pillars, B^2 B^2, and revolves in top and bottom bearings at E and F; G is one of the sugar-loaf forms, and H a pulley, which is keyed to the cylindrical case D. When the form has been put into its place, as represented in the engraving, a quick rotary motion is communicated to the machine through the pulley H, when the centrifugal tendency imparted to the syrup causes it to be driven to the sides of the moulds, when it is discharged through the perforated lid into the channel II, formed in the top of the pan A. The sugar mould, or pan, I, may have perforations in it (but small enough to prevent the escape of the crystals of sugar) to facilitate the discharge of the liquid.

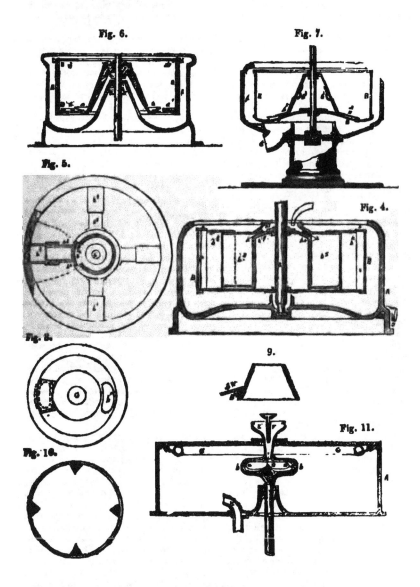

Fig. 6.

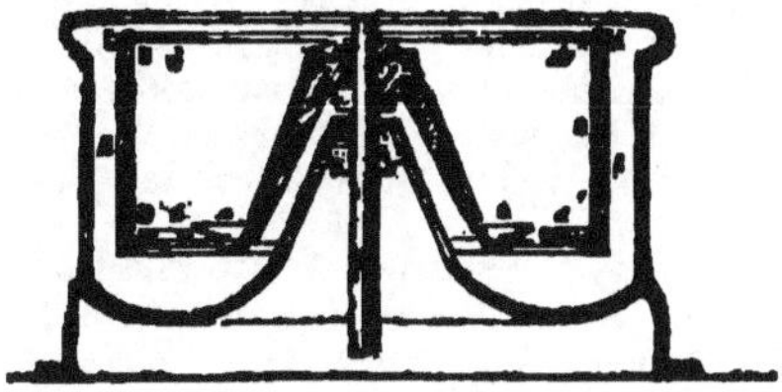

Fig. 7.

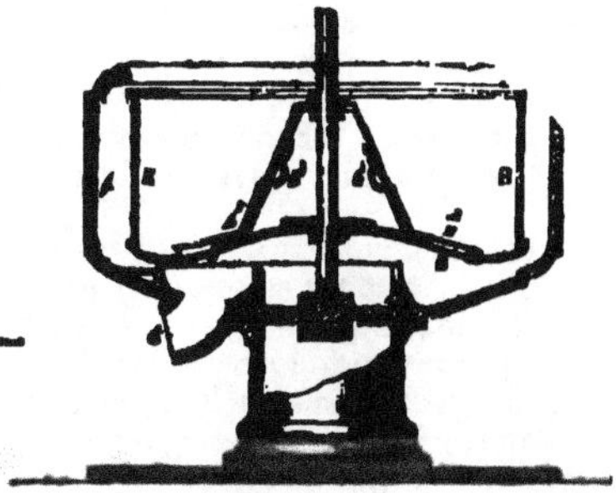

Fig. 5.

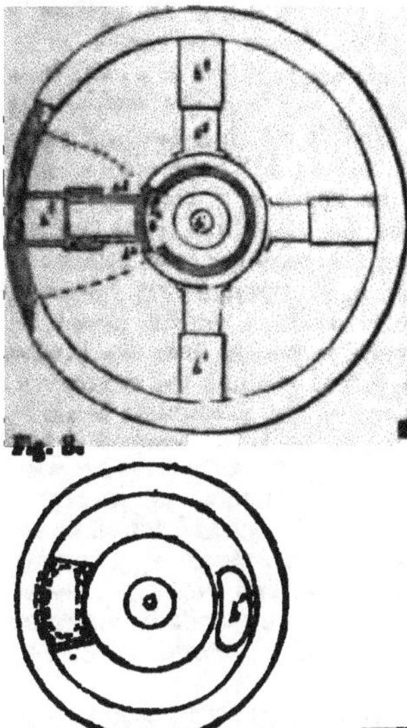

Fig. 4.

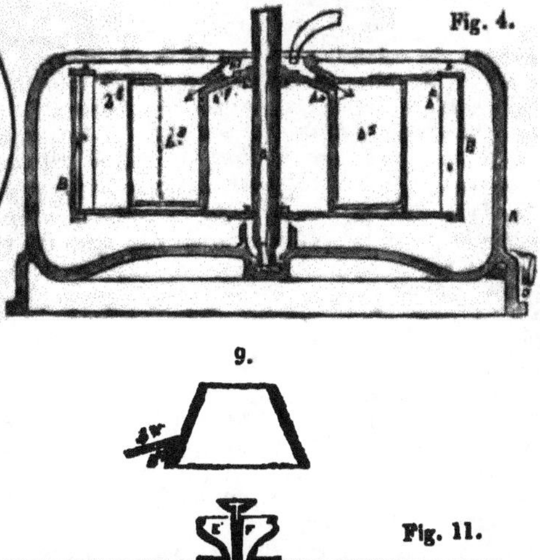

Fig. 8.

9.

Fig. 11.

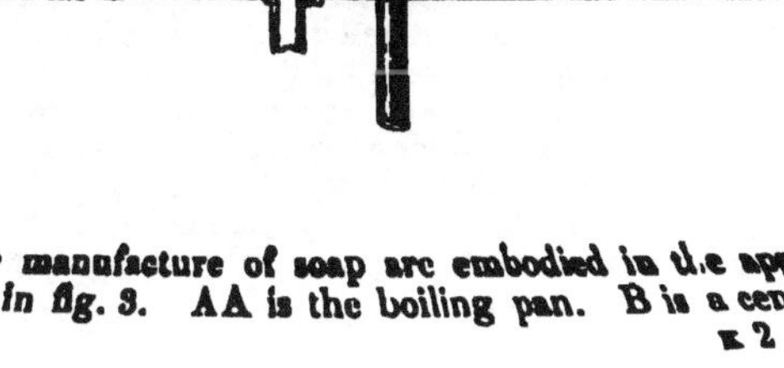

Fig. 10.

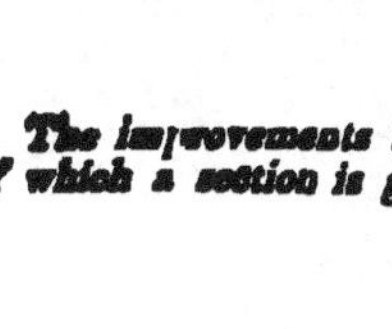

The improvements in the manufacture of soap are embodied in the apparatus of which a section is given in fig. 3. AA is the boiling pan. B is a centrifugal pump, the lower end of which reaches below a false bottom a^1, and is fixed on to the spindle b^1, which has its bearings b^2, b^3, at the top and bottom of the pan. The top bearing b^2 has a stuffing-box b^4, through which the spindle b^1 passes.

The false bottom a^1 rests on a flange a^2 in the pan, and has a pipe a^3 fastened on it, fitting loosely over the lower part of the pump; the false bottom has an opening in the centre of the same size as the pipe a^3, and is perforated beyond the flange of the pipe a^3, for the return of the liquor into the bottom. D is a pipe for connecting the pan with an air-pump, when it is desired to draw off the air and gases that may be formed in boiling.

The action of the apparatus will be as follows:—The materials being placed in the pan, and heat applied to them, the pump is made to revolve by turning the pulley b^5, when the liquid will be thrown from the bottom, by means of which a thorough mixing of the materials will take place, thus assisting the saponification of the fatty matters. When it is desired to treat the mass with sulphites, the pump must be raised a little above the level of the mass and sulphurous gas introduced above.

Fig. 4 is a sectional elevation of the improved dyeing machine before mentioned; and fig. 5 a plan of the same partly in section. A is a pan which forms the framework of the machine, and also answers for catching the liquids used. B is a perforated cylinder with interiorly projecting rings b b, which turns on the spindle b^1 in bearings at top and bottom, and has upright wires b^2, for the purpose of preventing the spreading of the liquid, soldered inside. The interior of the cylinder is further covered with wire gauze b^3; b^4 is a dome which forms the centre of the cylinder; $b^5 b^5$ are boxes perforated at the ends, which are fastened to a cylindrical ring b^6, fitting loosely and turning round the dome b^4, and form the channels for delivering the colouring liquids; b^7 is an inverted tundish which is fastened round the top of the ring b^6, and has openings into the boxes; b^8 b^8 are covers open at both ends, which fit loosely over the boxes b^5; the space c is occupied by the yarn to be operated upon. D is a pipe for conveying the liquid into some receptacle to be used over again. The method of working the machine is as follows:—The yarns are placed round the cylinder in the space c, the boxes b^8 are moved up to the yarn, and the machine set in motion; when it attains a suitable speed, the liquid with which it is desired to operate upon the yarn is poured into the tundish b^7, and is by the centrifugal tendency imparted to it, forced through the channels b^5 and that part of the yarn which comes opposite the end of the perforated boxes b^5, the other parts of the yarn remaining untouched; the liquid being pumped up and made to pass through the yarn as often as desired. When sufficient liquid has been passed through to produce the desired effect on the yarn, the machine is kept in motion until the yarn is partly dry, when the boxes b^5 can be turned round on the dome b^4 to act on a different part of the yarn either with the same or any other colouring matter. Thus yarns may, by moving the boxes and working as before described, be dyed, as shown in plan fig. 2. By perforating the ends of the boxes b^5 with more holes in the centre of the ends than at the sides a shaded effect may be imparted to the yarns.

Fig. 6 is a sectional elevation of part of my improvements in centrifugal cleaning and drying machines. A is a cylinder fastened on the spindle B (driven from below); the periphery of this cylinder may be made of sheet metal or of framework, but if of sheet metal, suitable arrangements, namely, perforations or other suitable openings must be made to carry off the liquid. a' is a flange projecting upwards. D is an inner case or cylinder which fits loosely in the cylinder A, and is constructed of wire-work or other suitable reticulate material and has rings d' projecting inwards at top and bottom. This cylinder may be made very light, as it will have only to bear its own weight with the weight of the material operated upon; the force imparted to it in work being resisted by the cylinder A. The advantage of this improvement consists in the ease with which the cylinder D can be lifted out and emptied when used for operating upon substances of a cohesive nature.

Fig. 7 is a sectional elevation of a further part of this branch of my improvements; and fig. 8 is a plan. A is the pan, as before, but constructed so as to enable a vessel to be brought under it, having also an opening a' for carrying off the mate-s after being operated on in the machine. B is the cylinder, the improvement which has relation to the bottom openings $b^1 b^1$, and consists in a method adopted preventing the escape of the materials through these openings without the neces- of having recourse to water-tight joints. For this purpose a flange b^2 fitting ely on the dome of the cylinder is used, and a couple of spring bolts b^3, for venting this flange rising; or lids b^4 fitting in recesses b^5 on the dome may be ployed, as shown separately in fig. 9,

he advantage of this improvement consists in the ease with which the material rated upon can be introduced and removed from the machine, no time being lost making any joint to secure the opening.

n the machines represented in figs. 6 and 7, the materials are poured into the nder whilst revolving, and are immediately thrown by centrifugal force over the ge or the lids, on to the internal periphery of the cylinder, and when sufficiently ned and dried, the loose flange or loose lids are removed. The opening in the nder is then brought opposite the opening in the pan, and the materials are dis- ged through the opening in the cylinder and the opening in the pan into a vessel, therwise, as desired. When used for washing substances such as corn, or other erials requiring washing, the lids b^4 are the most convenient; and a perforated e should be employed in addition to the fast dome, into which the water is intro- ed in manner well known. Crystallized and granular substances, and substances naturally drainable nature, can be dried, and, when desirable, also can have ids passed through them in these machines.

ig. 10 shows an arrangement in plan for assisting the drying of substances in e machines. The improvement consists in making perforated channels $a\,a$ in the nder (which is shown with the top removed), which project inwards or towards interior of the cylinder. This arrangement is found advantageous in separating liquid parts from such descriptions of substances as starch, raw sugars, pigments, other materials which have a tendency to force the liquid inward when operated n these machines.

Iy improvements in relation to the manufacture of metals consist, as before ed, in the adaptation of centrifugal force to the separating, dividing, and mixing he same while in a fluid state. Fig. 11 is a vertical section of a machine suitable this purpose; A is the outer casing; B is a centrifugal cylinder or machine fixed n the upright spindle C; D is one of the bearings for the spindle C, the lower of which is stepped in another bearing not shown in the engraving. The inner ace of the cylinder B is covered with a lining $a\,a$ of fire-lumps or other suitable erial being a bad conductor of heat; $b\,b$ are two of a series of slots made in the phery of the cylinder at the point of its largest diameter; E and F are two fun- or supply pipes; the outer and larger one (E) is fixed to the top of the casing and its lower end terminates within an opening formed in the cover of the cen- igal cylinder A, but is not connected to it. The smaller funnel F is supported he larger one, and occupies a central position in relation to it. When any two ore metals are to be mixed by the aid of this machine, they are poured in a ten state into the funnels E and F, in the proper proportions in which it is red to mix them, taking care that the cylinder is in motion while the pouring in he metals is taking place. The centrifugal action causes the metals to be pro- d through the holes or slots $b\,b$, and thereby an intimate mixing of the metals ke place. By the same action metals may also be divided into minute particles, speed of the machine and the size of the slots regulating the fineness of the divi- . G is a pipe, which is fixed round the interior of the casting, and is perforated numerous holes upon the under side, so that water or other liquid, or steam, as may be showered upon the metal while it is being dispersed by the machine. the steam or gases may be introduced into the casing A through the pipe H. this means sulphurous or other gases may be made to act quickly upon the als operated on.

IMPORTANCE OF INCLUDING IN ANY SYSTEM OF METROPOLITAN WATER-S ARRANGEMENTS FOR THE SPEEDIER EXTINCTION OF FIRES.

The General Board of Health, having laid considerable stress on the importance of connecting fire-extinguishing works with those for the supply of water to the metropolis, and that measure having been first devised by Sir Samuel Bentham, and proposed by him to the Prime-Minister in the year 1830, his said proposal on the subject, as it enters into many details of execution, seems of sufficient interest to justify its publication, although the general tenor of it has been already given in No. 1268 of the *Mech. Mag.*

The water and fire-extinguishing works in Portsmouth Dockyard, first proposed by Sir Samuel in the year 1797, and executed shortly afterwards according to his plans, have served—it can hardly be doubted—as examples for the several works that have since been executed on the same system in many towns—Hamburgh, for example. In that town the works in question appear to be precisely similar to those of Sir Samuel at Portsmouth, and in both cases they have been attended with unquestioned success in the speedy extinguishment of fire.—M. S. B.

(Copy.)

Brig.-Gen. Sir Samuel Bentham, K.S.G., to the Right. Hon. Robert Peel.

Lower Connaught-place,
19th Feb., 1830.

Sir,—The sensation produced by the late destructive conflagrations has at last induced me to communicate a plan which occurred to me some time ago for the better security of the metropolis against the ravages of fire. It cannot be considered in the light of an untried project, for it is grounded on the efficacy which has been experienced of the water and fire-extinguishing works which, according to my proposals, in the years 1797 and 1801, were introduced first in Portsmouth Dockyard, and similar works have since been executed in the other principal royal dockyards.

As in regard to Portsmouth Dockyard the efficiency no less than the economy of the works in question depended on their being combined with the water-works for other purposes, and with the steam engines for manufacturing purposes, so in regard to the metropolis it appears to me that works already existing might be made the basis of
the plan I have to sugg[illegible]
pense to be incurred compared to th
in view would be light, and the effici
the fire-extinguishing apparatus w
secured in a manner much more cert
it could ever be expected to be, if
this purpose alone. To give an ide
plan in view, it may be best to tra
part of one of my official communic
the subject, and then to state bri
modifications of that plan as would b
in adapting it to the metropolis.

Inspector-General of Naval Work
Navy Board, 9th April, 180

[A considerable part of that comm
was here transcribed, stating that th
of fresh water was from a deep well la
according to Sir Sam. B.'s propo
that water was pure and soft, and r
pump, "worked at pleasure by eith
two 30-horse steam engines which
pen to be at work for the use of t
mills." "The water so raised is
either into the cisterns near the
mills, or into the two large cister
form the roof of the two building
wood-mills, thus covering them wit
of water." "For the supply of t
engine boilers, of the blacksmith's
the metal mills, of the shipping,
other ordinary purposes where a hi
is not required, the water, in orde
any unnecessary expenditure of
generally taken from the lower cist
the mortar-mills, but for servin
levels," * * "the water is drawn fr
upper cisterns."

"To convey the water so rais
parts of the dockyard, a system
pipes is laid, beginning with an
main, which extends from the ne
the cisterns at the wood-mills; from
system of six-inch mains branches
the eight-inch main," * * "and
about the principal parts of the y
of the Camber-dock, and returns
the eight-inch main at the new we
intermediate communications by
four-inch pipes. A six-inch branc
carried to the extremity of the
island, and another along each si
Great Basin," &c.

"Another system of six-inch pipe
off from the eight-inch main, is carr
the Camber-dock, and, after c
round the northern parts of the
turns" * * "into the eight-inch m
new well."

"These mains, with their inferior

of them be supplied at pleasure with ater either from the cisterns near the mills, or from the cisterns at the the wood-mills; and in case of fire y, on simply turning a cock, be supp- ith salt water either from the reser- the harbour, by means of a forcing rovided expressly for this purpose, the salt-water well, and worked by of the steam engines at the wood- ising water to the amount of a ton te, and with a force to throw the ver the highest buildings."

the contrivance of making the six- ins return into the eight-inch main, s supplied by *two* ways to any one , the circulating course *all over the* "Several large stop-cocks to the s of the pipes are placed in proper es in different parts of the yard, for pose of facilitating any additions or ons; and upon all these pipes, whe- ins or branches, at intervals of about dred feet, fire cocks are fixed of two ers, which fire cocks are made suita- the screwing on of the largest hose ng to the extinguishing engines, so turning the cocks the fresh water e cisterns over the wood-mills may lied by means of these hose to the ishing engines, or may at the first be thrown immediately by the force sed into the ships, or into the first- f the buildings."

regard to the wood-mills themselves, he two 30-horse steam engines, the , and the pumps are fixed; as their in the case of fire, was looked upon the greatest importance to the secu- ll other parts of the yard, so extra- y precautions are taken for the ex- ing of any fire that might happen in ildings. From each of the cisterns er which cover the two principal s, a three-inch pipe is brought down h of the floors below, and into this each floor a fire-cock is fixed, with always screwed upon it in readiness ving water on turning the cock; and cock is fixed at the same place for g water into buckets; so that within ildings themselves there are no less ; fire-cocks, with their hose, and six or delivering water, and by which the water might be brought to act o or three of the nearest hose on any thin the buildings in the course of a nds."]

ngst additional works Sir Samuel ended was, "as a farther safeguard vent of fire, a forcing pump adapted ng the water through all the pipes ms sufficient to throw the water over the highest buildings, conformably to the drawings sent herewith." "The handles of the pump are adjusted to the applying the force of forty men, which would be sufficient to throw three-quarters of a ton of water per minute through the pipes in a jet over the highest building in the yard." "The pump itself to be fixed in an arch under- ground."

In a note to the letter to Sir Robert Peel Sir Samuel said—"I have understood that a fire-extinguishing apparatus of the same kind as that in the wood-mills has lately been introduced, and extremely well adapted for the protection of Windsor Castle." The let- ter to Sir Robert then continued as follows:

"In regard to the *efficiency* of the fire- extinguishing works at Portsmouth, several alarming fires have broken out there since they were established; but the vast quantity of water promptly thrown upon the flames by means of those works, have always extin- guished them before they could make any progress.

"Such having been the arrangements in Portsmouth Yard, it remains to exhibit how the same ideas are applicable to the metro- polis.

"The supply of water for the use of the inhabitants, it is well known, is by means of mains proceeding from the reservoirs of the several waterworks. Most of the water- works possess means of raising water to a reservoir sufficiently high to supply water at the top of the highest dwelling-houses. Here, then, as in Portsmouth Yard, there already exists a system of water-pipes ex- tending, it may be said, into every street, and supplied, as at Portsmouth, with water under a head sufficient to force water from them in a jet as high as the first-floors of all houses, to the second, or even the third of many in the lower parts of the town.

"There seem two ways by which this ex- isting apparatus might be adapted to the ex- tinguishing fire:—1st. By such an alteration of the plugs as should enable hose to be screwed upon them, by which water would be thrown as high as the first-floors at least, in as large a jet as the large diameter of these plugs would furnish. 2nd. By adapting in front of every house where it might be thought desirable, lesser screw plugs ready for screw- ing on a hose to the service-pipes propor- tioned to their size. Further, supposing the mains to be strong enough to bear an extra pressure, or into only such of them as are so, water might be forced by the steam en- gines already existing at the several water- works in such manner as, by these several hose, to throw the water over the tops of houses without the intervention of any im- moveable fire-extinguishing engine.

"The whole expense attendant on these alterations would consist in the alteration of the plugs and the provision of suitable hose. The larger ones, of course, would be so far a public concern that the expense would be to be paid, either by subscription amongst insurance-offices, or as included in the rate paid by inhabitants for the supply of water. The smaller plugs and hose being peculiarly suitable for application to each particular house at the moment fire might be discovered, the provision of this small apparatus might be optional with each householder, and a suitable hose might be provided and kept, if desired, on the premises themselves.

"In addition to these means of converting the existing water-works into fire-extinguishing works, it may be observed that there already exist in town, many deep wells capable of affording a great supply of water, and steam-engines often on the spot with pumps for raising it; these whenever thought desirable, and where permission might be obtainable from the proprietors, might be connected with the general system of pipes to act in addition to the apparatus at the water-works.

"In particular buildings of great value or of peculiarly hazardous nature,—such as libraries, manufactories, theatres,—apparatus might be introduced into the interior, such as is above described as existing in the wood-mills. So also at any time forcing-pumps might be applied in any of the several public wells in the streets.

"But however well contrived and efficient, such a system of fire-extinguishing works might be in itself, there could be no dependence placed upon them unless suitable regulations were adopted for ensuring their *prompt* and *certain* application to use in time of need.

"Referring no farther back than to the two last fires of importance (the Argyle-rooms and the English Opera-house), it is notorious that want of immediate arrival of engines, want of water after they did arrive, and in the latter case want of presence of mind and experience in the persons who might at first have made the fire-apparatus within the house available, even perhaps to the extinguishing the fire before it burst into the body of the theatre, precluded the possibility of saving these structures; these, therefore, are the most important mischiefs to be guarded against.

"At the time of the Argyle fire it was a severe frost, the fire-plugs were frozen. The fire-plugs, it must be observed, are at present down in holes in the carriage-way of the streets, without any means of closing those holes, so that the mud and water of the *streets runs into* them, and being exposed to the atmosphere is frozen. The ex
for obviating this mischief is to
wooden plugs (as bad conductors of h
fill up the holes. The mains the
are usually laid at sufficient depth
freeze even in the severest frosts.

"Another frequent, and indeed acco
present arrangements, irremediable so
delay in the procurement of water,
the turncock of the division must be
for, before the plugs can be opened-
to be sent for at some perhaps cons
distance; in the daytime he may be
home, at night asleep; he has then
roused, to dress, before he can set
that often half an hour or more is th
To remedy this, means of turning o
should be in the hands of persons
and it may be said on the spot, what
part of the town—such persons exis
police. It may at first sight bear an
ance of impropriety or impracticabil
every police-constable should carry w
a turnkey, but in point of fact a poli
staff, without being either too cumber
heavy, might be made to answer th
pose; and if it be called to mind h
frequent the occasions are when su
has occurred, it may well be though
the while not to reject a suggestion
nature because it professes to effect
portant purpose by very simple mea

"Suppose householders to prepare
small hose as above-mentioned, it m
kept in a receptacle outside the
either under lock and common k
policeman having one, or in a re
easily broken open in case of nee
larger hose and jets might be kept i
priate receptacles outside of build
distances not too far apart to be
manner openable by the police-const

"Should any forcing-pumps be intr
they, as in Portsmouth and othe
Dockyards, should be *below* the pave
as to be secure from frost; the m
working them might be a capstan-p
bars to be put in in time of need, w
sufficient to produce an immense for
easily be collected.

"In order to ensure the readiness
in case of fire of this whole syst
keeping it as much as possible in
use would be most effectual. Th
engines at water-works, for exampl
work night as well as day with adva
the proprietors; but even were the
work in the night the fire might
damped up ready to make the engi
at any time; so, for insuring the rea
of hose, instead of sending carts a
water streets, this business might be
a much less expense by making us

hose in question with a suitable orifice for delivery, and every householder might thus have the space before his house watered by his own hose; as, indeed, I have already seen practised at Knightsbridge.

"Besides the system of works above indicated, there is a variety of minor apparatus used in foreign countries for saving life and property, that it would be desirable to introduce and improve upon here. At St. Petersburgh, for example, the fire-guard are provided with exceedingly light ladders, with hooks of various kinds on long light poles for pulling down inflamed portions of buildings, and for hooking out articles from the interior of houses; to which might be advantageously added (as has been proposed) strong elastic webs extended upon frames for inhabitants in case of need to jump into.

"It would be well that the several persons whose duty it would be to make use of the whole apparatus should be kept in the exercise of the duties assigned to them by signals calling upon them occasionally to the performance of those duties—not precisely the same signals as for real fire, yet it should be by signal, and *unexpectedly*, for it is the unexpectedness of the occasion that is so frequently found to excite a panic where coolness and presence of mind are of so much importance.

"Those late conflagrations indicate likewise the need of furnishing the police with instructions for their guidance on occasion of fire. The zeal, the usefulness, of the new police on such occasions has been observed and well appreciated; but it was also observed at the English Opera-house that the police by breaking open the pit-door let a burst of flame into the audience-part of the house, and all was from that moment lost. This same error of opening doors has often been attended with the same mischievous results; but it cannot be expected that this class of persons should be acquainted with such physical effects as that without air fire will cease of itself; nor, indeed, do persons aware of this call it sufficiently to mind in case of accident—how often has fire in a chimney been extinguished by throwing a wet blanket over it. I have known washing curtains in a drawing-room, with other furniture, burnt, yet the fire self-extinguished because the chimney happened to be closed and the doors shut. Not only, therefore, might a suitable set of instructions be furnished to policemen, but were they inserted in the paper of parochial information periodically sent round to householders it might prevent the spreading of many a fire.

"In conclusion; I must add that the cost of carrying such a plan into execution would materially depend on the judiciousness of the means employed, and the manufacturers applied to. The engineer and mechanist of Portsmouth Dockyard afforded important assistance in the arrangement of the above-mentioned plan for Portsmouth, he has contrived the details, and has carried into execution the fire-extinguishing works in other dockyards; and as he is still in the public service, the Admiralty, it may be supposed, would not object to his giving such information on the subject as his experience enables him so well to afford.

"I have the honour to be, Sir,

"Your very obedient Servant,

"Samuel Bentham."

The original of the above letter was left at the Horse-guards for Sir Herbert Taylor, with a note from Sir Samuel, commencing as follows:

"Horse-guards, Friday, Feb. 19, 1830.

"My dear Friend,—According to your advice, that what I had to propose relative to security against fire should be addressed to Mr. Peel, the inclosed letter was written yesterday, and copied this morning, when I immediately brought it here.... Not finding you, I added my signature.... and have to request that you would be so kind as to cause it to find its way to Mr. Peel.

"I am, Sir, &c.,

"Samuel Bentham."

Sir Herbert accordingly presented the letter himself to Mr. Peel,—who feared the public were not yet ripe for such a measure; but what most effectually deterred Sir Samuel from farther urging its adoption was that, as he understood, strong opposition would be made to it by a powerful body of persons who conceived that their private interests would be injured by its introduction.

In the year 1844, when Lord Lincoln was at the head of the department of the Woods and Forests, a copy of the proposal to Sir Robert Peel was inclosed to his Lordship, with the following note:

"Queen Square-place, 1844.

"My Lord,—In consequence of a Bill being now pending in Parliament for the protection of the metropolis from the ravages of fire, I venture to take the liberty of inclosing a proposal on the subject which was presented through Sir Herbert Taylor to the Right Honourable Sir Robert Peel in February, 1830, by my husband, the late Sir Samuel Bentham, as it appears to offer some advantages which may be worth your Lordship's consideration.

"I have the honour to be, my Lord,

"Your Lordship's obedient Servant,

"MARY SOPHIA BENTHAM."

The above correspondence affords an instance, amongst the many that might be adduced, of inattention to salutary public measures when private interest is not concerned in their adoption. In this case the plan was gratuitously offered, and the services of the mechanist of Portsmouth Dockyard would have been obtained without additional expense to the public; but it was not till private engineers for their personal interest adopted Sir Samuel's plan that it was carried into execution in various towns.

MARSDEN'S PATENT VENTILATING BOOTS AND SHOES.

(Patent dated January 31, 1851; Specification enrolled July 31, 1851.)

My improvements in boots and shoes have for their object the constructing them in such manner that they may admit of a free circulation of air through them, and thereby prevent that over-heating of the feet which takes place with boots and shoes of the ordinary kind, especially in walking, to the great discomfort, and not unfrequently serious injury of the wearer.

Fig. 2.

Fig. 10.

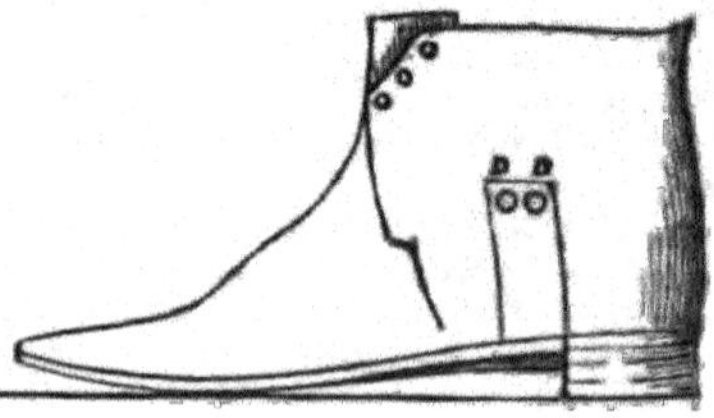

Fig. 1.

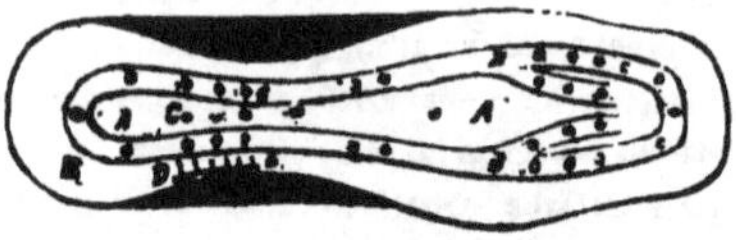

Fig. 11.

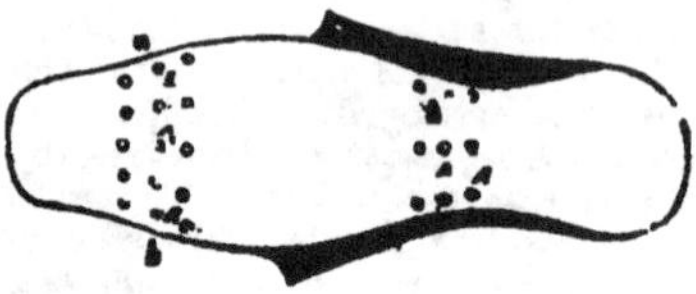

Fig. 5.

Fig. 12.

Fig. 15.

Fig. 3.

Fig. 7.

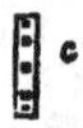

Fig. 16.

8.

Fig. 6.

Fig. 13.

Fig. 14.

4.

9.

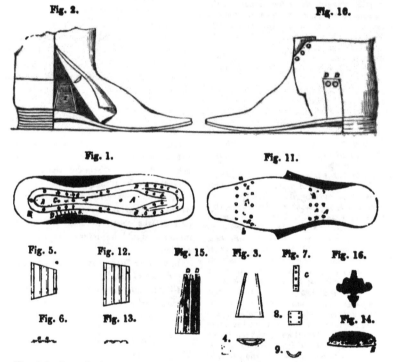

Fig. 1 is plan of a boot fitted with arrangements for this purpose. In this figure the outer sole is removed, to show more clearly the passages through which the air is made to circulate. A is the inner sole; BB, a series of grooves cut out of the lower side of A, and which when the outer sole is put on, form so pipes or channels. (These may be fitted, if deemed neces- h metallic tubing, or quills). e holes which communicate he inner sole A with the inte- he boot; DD are other holes, an into a conical vertical pas- formed on the inside of the boot the outer covering or leather, lining, as represented in fig. 2, e upper of the boot is partly to show the position of the pas- l also in separate explanatory s. 3 and 4. Instead of only one age E, there may be a similar n each side, or more if desired. pper end of the passage E is the external atmosphere, and at end it terminates in the holes before mentioned), which again nmunication with the passages ed in the inner sole by means of thin corrugated metal-plate is inserted between the outer r sole previous to their being otherwise joined together. In- the metallic-plate F, a piece of ather or other stout pliable sub- ith grooves cut in it, may be sub- Fig. 5 is a plan, and fig. 6 an w of one of these pieces of plate; ows *a a* are placed accurately holes DD, so that there may be assage for the upward escape of ir from beneath the soles of the l the passage is carried to such a s to exclude, under all ordinary ances, the entrance of wet or dirt.

G, fig. 7, is a thin strip of brass which is laid along the outer edges of the holes DD, to prevent the stitching from closing them up. Figs. 8 and 9 are represen- tations of another bent strip of metal, which encircles the mouth of the pas- sage E.

Another method of effecting the ven- tilation of boots and shoes is shown at fig. 10 (a side view) and fig. 11 (a plan with the outer sole removed); AAA are the air-holes; BB other holes, which communicate with a passage formed be- tween the lining and the upper-leather of the boot. These two sets of holes are connected by thin pieces of corrugated brass plate, represented in figs. 5, 6, 12 and 13, and sewed in the manner before explained between the two soles. Fig. 14 is a cross section of the boot taken on the line *a b*, of fig. 11. The passage formed between the lining and upper-leather is carried some distance up the leg of the boot upon the inside by means of a grooved piece of leather C, shown sepa- rately in fig. 15, which terminates at top in two holes DD, which are open to the external atmosphere, and may be at any height.

The boot or shoe may be also venti- lated from the top by making a few holes in the vamp or upper, and covering them over with a few ornamental pieces of leather, such as represented in fig. 16; or by using a light under vamp with cut holes in it, and making these communi- cate with a common aperture of escape in the external vamp.

ON THE THEORY OF ALGEBRAIC EQUATIONS. BY JAMES COCKLE, ESQ., M.A., BARRISTER-AT-LAW.

(Continued from vol. liii., p. 331.)

Third and Concluding Series.

VI.—CUBIC AND QUINTIC EQUATIONS.

s term *quintic equation* I mean an equation of the fifth degree. The present es of the Theory of Equations require, or at all events its promised pro- ggests, the necessity of some such nomenclature. I propose to call equations xth, seventh, and the respective higher degrees, by the respective names of *ptic, octic, nonic, decic,* &c., *equations.* Or, if an English should be pre- a Latin derivation, we might term equations *fifth-ic, sixth-ic, seventh-ic,* ording to their degrees. As we adopted the one or the other method we ave the terms *eleventhic, twelfthic, twentiethic* (or *twentythic*) *thirtiethic, thic,* &c., or the respective synonyms *undecic, duodecic, vigintic, trigintic,* tc. Either system would be preferable to that of which I gave a specimen in g "On Mathematical Nomenclature," published at pp. 509-510 of vol. xlv. *Mechanics' Magazine.*

Before proceeding, let me beg your readers to consider my paper "On Mr. Bills's Rule for Cubics" (*Mech. Mag.*, vol. liii., pp. 586-8), and also that "On two remarkable Cubics" (*Ibid* pp. 448-9), as incorporated with this series of *Notes*. With respect to the latter paper, I have to express my thanks to Mr. Lockhart and Mr. Bills for pointing out an oversight which I had committed, and for which I there attempted to compensate by communicating portions of their letters to this Magazine. Let me also add that, since the publication of the last of these *Notes*, the Second and Concluding Part of my *Analysis of the Theory of Equations* has appeared in the *Philosophical Magazine* (in the Supplementary Number for December, 1850.).

The equations which constitute the sixth line of the last of these Notes ought to have been numbered (4). In the *last* step of the division, by which the final expression for x is there obtained, the divisor is written

$$-\rho+\alpha\,\sqrt[3]{n\rho}\,;\ \text{and in the } \textit{first}\ -3z+\alpha\,\sqrt[3]{n\rho}-\alpha.$$

The venerable and distinguished mathematician, Mr. Lockhart, has made an interesting and important observation on equations of the fifth degree. Now, by slightly changing the form of his results, we obtain

$$(a^2+b)^2+a^2b=p,\ \text{and}\ ab\left\{(a^2+b)+b\right\}=q \ldots\ldots\ldots\ldots\ldots\ldots (6.)\,.$$

And, since every quintic equation is capable of being reduced to the trinomial form given by Mr. Lockhart, we see that if we were able to obtain a finite solution of the system (6), treating a and b as the unknowns, and p and q as known, we should be able to solve the general quintic; for we should be in possession of quadratic factors of the trinomial quintic.

While upon the subject of equations of the fifth degree, it may interest some of your readers if I give one of the unsuccessful expedients by which I sought to solve the general quintic. This is its outline. Since every quintic can be reduced to any trinomial form, and the coefficients of that form may be modified by well-known artifices, we may start with the assumption

$$x^5+2\,x^3+b=0 \ldots\ldots\ldots\ldots\ldots\ldots (7.).$$

Multiply (7) into x, and we have

$$x^6+2\,x^4+bx=0=\left\{x\,(x^2+1)\right\}^2-(x^2-bx) \ldots\ldots\ldots\ldots\ldots\ldots (8.)$$

Now, if x^2-bx were a perfect square, the right-hand side of (8.) would be decomposible into cubic factors, and the given equation would consequently be capable of solution. Not being so, can we make it a square by substituting for x a linear function $(\alpha y+\beta)$ of a new unknown y, and then determining α and β so as to satisfy the requisite conditions? The answer is—*No*. For we shall find ourselves conducted to the nugatory result $a^2b^2=0$, and are thus stopped. Without particularizing other attempts, let me state that the failure of my Method of Vanishing Groups (and the same remark applies to Mr. Jerrard's process) to accomplish the solution of the general equation of the fifth degree does not arise from any want of logical coherence in the processes themselves, but from those processes giving a final result which, in the case of a quintic, takes the form of a vanishing fraction. Mr. Jerrard's processes, however, enable us to reduce the general equation of the fifth degree to any one of the four following forms at pleasure, viz.:

$$x^5+x^4=a \ldots (1);\ x^5+x^3=a \ldots (2);$$

$$x^5+x^2=a \ldots (3);\ x^5+x\ =a \ldots (4).$$

These wonderful results may also be obtained, and with greater facility, by my own processes. The solution of any one of the above four forms would comprise that of the general equation. I do not mention this for the purpose of encouraging any one to attempt such solution—for I would rather dissuade from such attempts—but merely to show that the labour spent upon the subject of quintic equations has not been wholly unrequited. I cannot conclude this paper better than by presenting your readers with the following portion of a letter with which Mr. LOCKHART has recently favoured me.

2, Pump-Court, Temple, April 25, 1851.

[*Extract from a Letter from James Lockhart, Esq., to Mr. Cockle.*]

"My dear Sir,

* * * * * * *

I was very sorry to learn of the death of Mr. Davies—he was a first-rate mathematician, as great, I think as any of his cotemporaries. Mr. Potts of Cambridge loses a great assistance in the forthcoming work on Porisms.

"Mr. Davies understood the subject particularly well, and was anxious to further the study of the doctrine. I take the liberty to send you a few remarks of mine on the roots of the cubic $x^3-3x-1=0$ which apply to all cubic equations.

"I published the forms of the roots in the *Mathematician*, but not the inferences I now give.

"It is most desirable to connect these, if possible, with equations of the 5th degree. These forms cannot be lost, however they may be combined. It is the extrication of them from such combination that is required.

* * * * * *

"I am, my Dear Sir, truly yours,
"JAMES LOCKHART.

"Parsons-green, Fulham, April 9, 1851."

[*Mr. Lockhart's 'remarks' referred to in his Letter.*]

"let $x^3-3x-1=0$

$$\text{numerical roots}\left\{\begin{array}{l} +\ 1{\cdot}879385241\ldots \\ -\ 1{\cdot}53208888\ \ldots \\ -\ 0{\cdot}34729635\ \ldots \end{array}\right.$$

$$\text{algebraical roots } +x,\ -\frac{1}{x+1},\ -\frac{x+1}{x}$$

give to any one of these last any one of the preceding numerical values; for instance,

$$\text{call } +x=+1{\cdot}87938\ldots \text{ and apply it to } -\frac{x+1}{x},\ \text{the result is } -\frac{2{\cdot}879..}{1{\cdot}879..}=-1{\cdot}53208..$$

$$\text{apply it to } -\frac{1}{x+1}\ldots\ldots-\frac{1}{2{\cdot}879}..=-0{\cdot}347296..$$

$$\text{call } -\frac{x+1}{x}=-1{\cdot}53208..\ \text{apply it to itself}\ldots-\frac{{\cdot}53208..}{-1{\cdot}53208}=-0{\cdot}347296..$$

$$\text{apply to } -\frac{1}{x+1}\ldots\ldots-\frac{1}{-1{\cdot}53208}=+1{\cdot}8793852..$$

$$\text{call } -\frac{1}{x+1}=-{\cdot}347296 \text{ and apply it to itself } -\frac{1}{{\cdot}652703}=-1{\cdot}53208$$

$$\text{apply it to } -\frac{x+1}{x}\ldots\ldots-\frac{{\cdot}652703..}{-{\cdot}347296..}=+1{\cdot}8793852."$$

PHANTASMAGORIA LANTERNS.

The phantasmagoria lanterns exhibited,* are a scientific form of magic lantern, differing from it in no essential principle. The images they produce are variously exhibited, either on opaque or transparent screens. The light is an improved kind of solar lamp. The manner in which the beautiful melting pictures called dissolving views are produced, as respects the mechanism employed, deserves to be explained. The arrangement adopted in the instruments exhibited is the following:—Two lanterns of the same size and power, and in all respects exactly agreeing, are arranged together upon a little tray or platform. They are held fast to this stand by screws, which admit of a certain degree of half-revolving

* By Messrs. Carpenter and Westley, of Regent-street.

motion from side to side, in order to adjust the foci. This being done in such a manner that the circle of light of each lantern falls precisely upon the same spot upon the screen, the screws are tightened to the utmost extent, so as to remove all probability of further movement. The dissolving apparatus consists of a circular tin plate, japanned in black, along three parts of the circumference of which a crescentic aperture runs, the interval between the horns of the crescent being occupied by a circular opening, covered by a screwed plate, removeable at pleasure. This plate is fixed to a horizontal wooden axis, at the other end of which is a handle, by which the plate can be caused to rotate. The axis of wood is supported by two pillars, connected with a flat piece which is secured to the tray. This apparatus is placed between the lanterns in such a manner that the circular plate is in front of the tubes of both, while the handle projects behind the lanterns at the back. The plate can, therefore, be turned round by means of the handle, without difficulty, from behind. A peg of wood is fixed into the axis, so as to prevent its effecting more than half a revolution. The widest part of the crescentic opening in the plate, is sufficiently so to admit all the rays of the lantern before which it happens to be placed. On the plate being slowly turned half round, by means of the handle behind, the opening narrows until it is altogether lost in one of the horns of the crescent. The light of that lantern is gradually cut off as the aperture diminishes, until it is at length wholly shaded under the moveable cover occupying the interval between the horns of this crescentic opening. In proportion as the light is cut off from one, it is let on from the other tube, in consequence of the gradually increasing size of the crescent revolving before it, until at length the widest part of this opening in the plate is presented before the tube of the second lantern, the first being, as we have seen, shaded. This movement being reversed, the light is cut off from the second lantern, and again let on from the first, and so on alternately. Thus while the screen always presents the same circle of light, yet it is derived first from one lantern, then from the next.

When in use, a slider is introduced into each lantern. The lantern before the mouth of which the widest part of the opening in the plate is placed, exhibits the painting on the screen, the light of the other lantern being then hid behind the cover. On turning the handle, this picture gradually becomes shaded, while the light from the second lantern streams through the widening opening. The effect on the screen is the melting away of the first picture, and the brilliant development of the second, the screen being at no instant left unoccupied by a picture.

The principle involved in this apparently complex, but in reality simple mechanism, is, merely the obscuration of one picture and the throwing of a second in the same place on the screen. And it may be accomplished in a great variety of ways. Thus, by simply placing a flat piece of wood, somewhat like the letter Z, on a point in the centre, so that alternately one or the other of the pieces at the end should be raised or depressed before the lanterns, a dissolving scene is produced. Or, by fixing a moveable upright shade, which can be pushed alternately before one or the other of the lanterns, the same effect is produced.

Individuals exist in this metropolis whose sole occupation consists in painting the minute scenes or slides used for the phantasmagoria lanterns. The perfection to which these paintings are brought is surprising. There are two methods by which the sliders now employed are produced. In one of these, the outline and detail are entirely the work of the artist's pencil. For pictures representing landscapes, or wherever a spirited painting is required, this is the exclusive method employed. The colours are rendered transparent by being ground in Canada balsam and mixed with varnish. The other method is a transfer process. The outlines of the subjects are engraved on copper plates, and the impression is received from these on thin sheets of glue, and is then transferred to a plate of glass, the impression being burnt in the same manner as is effected in earthenware. Sliders produced in this way receive the distinctive name of copper-plate sliders. The subject is merely represented in outline, it being left to the artist to fill up with the necessary tints, &c. The advantages of this method for the production of paintings of a limited kind are obvious. Latterly photography on glass has been employed to obtain pictures for the magic lantern.—R. E.—*Official Illustrated Catalogue of the Great Exhibition.*

ECCENTRIC MOVEMENT OF THE FIXED STARS.

At a meeting of the Berlin Academy of Sciences, held on May 31 last, the venerable Alexander Von Humboldt made an interesting communication upon some observations of singular movements of fixed stars. It seems that at Trieste, January 17, 1851, between 7 and 8 o'clock P.M., before the rising of the moon, when the star Sirius was not the horizon, it was seen to perform kable series of eccentric movements. and sank, moved left and right, and ies seemed to move in a curved line. ervers were Mr. Keune, a student in ar class of the Gymnasium, and Mr. , a saddler, both certified to be persons. The family of the latter eld the phenomena. Mr. Keune, with leaned immovably against the wall, us rise in a right line above the roof ighbouring house, and immediately ak out of sight behind it, and then ppear. Its motions were so con- that for some time the beholders it was a lantern suspended by a t also varied in brilliancy, growing ly brighter and fainter, and now being for moments quite invisible, the sky was perfectly clear. As far known, this phenomena has been d but twice before, once in 1799 e Peak of Teneriffe by Von Hum- imself, and again nearly fifty years y a well-informed and very careful , Prince Adalbert of Prussia.

LDSWORTHY GURNEY'S DIFFEREN- TIAL BAROMETER.

r. Gurney's Report to the Coal Miners of umberland, Durham, and Lancashire.)

well known that slight changes of eric pressure will sometimes occa- h discharges of fire-damp in a coal to endanger, and cause explosion. of the barometer indicates a reduced of the atmosphere, and it is known -damp is relieved thereby from its situation in air-pressed recesses he workings, or in the goaves, or aces; and under diminished pressure es by its reaction and its elasticity galleries.

s a mechanical explanation, and so goes satisfactory; but as explosions ings of the mine oftentimes occur at, etimes before the change of balance, to mechanical pressure taking place t to account for it, we must look for auses beyond mechanical disturb- Meteoric changes, and some of the portant, sometimes occur entirely ected with any sensible change of eric pressure. Hail and snow, wind , wet and dry, heat and cold, often ithout the slightest change in baro- pressure.

ame cause which produces wind on ay probably disarrange fire-damp. e cause which brings about meteoric ena may do so, and be the same which produces a change in the weight of the atmosphere; so that the fall of the barometer is an effect, and not a cause, of some hidden disturbing influence, which simultaneously or previously occasions fire-damp escape, and the danger of explosion in a coal mine. Nature operates by insensible agencies, by powers we have no conception of, by laws far beyond our reach. The same agent probably which produces a difference even in the solvent condition of the atmosphere may at the same instant act to the very centre of the earth. We cannot tell; it surely may act near the surface. The hygrometric state of the atmosphere, for ought we know, may be more important for your consideration than its barometrical condition.

I may hereafter suggest for your consideration some known conditions on this head, and its connection with fire-damp. We will at present confine ourselves to the barometer. The moment of change, or break of balance, is the most important time to adopt precautionary measures, for the disturbing cause may have been acting for hours previous to any perceptible change in the barometer. It is highly desirable to be able to detect the first and smallest reaction. It has, I believe, been noticed at sea—where sailors, from experience of its importance, pay great attention to the rise and fall of the barometer—that the period of reaction is the most important period to enable them to anticipate a change in the direction of the winds, or the coming of an impending storm. It is equally, if not more important, for the pitmen to pay attention to the same fact.

The ordinary barometer is too sluggish. The changes of the surface of mercury from the concave to the convex, which it assumes on the break of balance, cannot possibly be read off. This fact and its importance was shown by the Honorary Secretary of the South Shields Committee, in his report "On Accidents in Coal Mines." You may observe the allusion he there made to the sympiesometer as a more delicate instrument for the purpose. There are, however, objections to this instrument, and another, of greater delicacy than the common barometer, is required. On this account it was that the one referred to was devised. This instrument was founded on the aneroid, substituting a moving fluid as a pointing range or index, instead of the complicated movement of hands and levers. The same principle can be applied to the ordinary barometer, with which you are more familiar. I will explain this arrangement as simply as I can.

If a tube of glass, closed at one end, be filled with mercury, and while so filled we dip its open end in a cup of mercury, holding the sealed end vertical, the mercury will fall down the tube, leaving a vacuum in the upper part, until the height of its column exactly balances that of the atmosphere. This will vary between 27 and 31 in., according to the weight at the time. This, you know, is the common barometer. You will observe that the quantity of mercury which rises or falls in the tube, to balance the changing weight of the atmosphere, will always be equal to that which is necessary as an equivalent to make up the required height of the column, irrespective of its solid contents. Thus, in a barometer tube of large diameter, a larger quantity of mercury will be required to go up the tube, to obtain a given level, than one of a smaller diameter; yet the variation of the level, the space moved through up and down, in both cases, will be exactly the same. You will observe, therefore, we can gain nothing by increasing the size of the barometer tube, to enable us the better to read off minute changes.

Let us suppose the barometer to be 1 inch diameter, or (say) equal to 1 inch sectional area, the rise of the column of 1 inch in height will require 1 cubic inch of mercury to come up the tube into the vacuum. If the tube be one-quarter inch sectional area, then only one-sixteenth of a cubic inch will have to go up to raise the level. The disturbance, or space moved through, in both cases, will be an inch—no more or less in one than the other; and both will be an exact balance of the atmosphere, representing the change of pressure. I wish to call your particular attention to this, because the instrument about to be described is founded upon it.

In order that you may better understand me, we will, in the first instance, suppose we have a barometer tube about 1 inch diameter, and 4 or 5 feet in length, closed at one end. We fill this with mercury to within 3 or 4 inches of the top, and then fill it up with water till it runs over the open end. On placing the finger air-tight on the open end, and inverting the tube, keeping the end still closed, the mercury and water will change places; the water will rise to the surface of the inverted tube, and the mercury will fall to the bottom. While the finger is still on the end of the tube, place the end under the surface of mercury in a cup or basin. The mercury will now fall to balance the weight of the atmosphere, and the water will fall on it, a vacuum being left above the surface of the water. The water, it is manifest, swimming on the mercury will rise and fall with it. An inch of mercury will displace a cubic inch of water, and, of course, drive it up or draw it down the tube, higher or lower in the vacuum, as the case may be. The rise of level of the mercury and water in this case will be the same, because the diameters of the portions of the tube through which they move are equal; but if they are unequal, then their range would be unequal. Let us suppose the upper part of the tube, where the water works, to be less than that where the mercury works. It is evident the displacement of the water, depending on the rise of the mercury, will be equal to its volume; but having less room to move in will extend to a greater distance than the mercury, proportionate to the difference of the respective sizes of their tubes. Suppose the mercurial tube to be an inch sectional area, and the tube above the water connected with it to be one-tenth of an inch; 1 inch rise of the mercury would then displace a cubic inch of water, which, being driven up the tube into the vacuum, would extend to 1000 tenths, or 8 feet 4 inches in height. We have here a great range, an index by which we can sensibly read off the most minute disturbance or change of level. The change in the surface figure of the mercury, however minute and otherwise insensible, may be accurately ascertained. The water range, or small tube, is called the "differential tube," and the larger one the "barometer tube." An instrument of such extensive range or length of the differential tube as the above can hardly be wanted. It would require correction for the height of the water, and be inconveniently long; but if wanted to save height and room, the differential tube may be turned zig-zag or spiral. It would then come within a convenient size, and the columnial pressure of the water would be of little consequence, and hardly require correction. If you wish, you may have it of such sensibility; but an instrument of less range will be sufficiently sensitive for your purpose. Let us take such areas in the tube as we know will not interfere by capiliary attraction, or by the attraction of cohesion, to disturb the certain and uniform working of the differential fluid, yet at the same time be sufficiently delicate for practical purposes. A differential tube of less than one-fourth of an inch in diameter will not work well. The fluid, whether water, spirit, or oil, will hang and stick to the sides of the tube, so as to interfere with its perfect freedom of action. The size of the mercurial one is of little consequence; so far as respects the ratio of proportion, it is only necessary to have the part of it where the surface of mercury ranges up and down to be of large dimensions.

...FICATIONS OF ENGLISH PATENTS ENROLLED DURING THE WEEK ENDING AUGUST 24, 1851.

...Y RICHARDSON, of Aber Hirnant, ...rth Wales, Esq. *For certain im...ts in life-boats.* Patent dated ... 22, 1851.

...culiar construction of this life-boat ...ndered sufficiently apparent by the ...ich is, for

...raming or constructing a life-boat ...r more tubes or pontoons of iron, ...r other suitable material, divided ...al partitions into water-tight com...s, and connected or braced toge...rnally with iron or other staunch...tays, upon which is laid an open or ...deck, with the necessary seats for ...pants.

...AM STONES, of Queenhithe, sta... *For improvements in the manufac...safety-paper for bankers' cheques, ...xchange, and other like purposes.* ...ated February 24, 1851.

...ject of this invention is to produce ...which shall by the discoloration of ...ce indicate when an attempt has ...le to obliterate or remove writings ...s. For this purpose the patentee ...bromine, or iodine combined with ...ide or ferricyanide of potassium ...h. Any combinations of bromine ..., with bases of various descriptions, ...sed, but the patentee prefers iodide ...ium, both on account of its being ...dily obtainable in the market than ...ds of bromine, and because the ... the paper is in no degree affected ...s. This substance is mixed with ... or with the size, or the finished ...ay be prepared by being saturated ...olution of it; the ferrocyanide or ...ide of potassium is mixed with the ...may like the metallic iodide be ...t a subsequent stage of the manu...and the starch is added to the pulp ...size.

...uantities required for the prepara...manufacture of safety paper may be ...ut the following proportions are ...be suitable for the treatment of one ...post, weighing about 18 lbs.; 1 oz. ... potassium, one-quarter ounce fer...e or ferricyanide of potassium, and ...d of starch.

...ffect of chlorine, or of a mineral ...m employed to obliterate or remove ...from this prepared paper, is to break ...f the salts, when the iodine will be ..., and an iodide of starch formed, ... insoluble, and of a dark colour: ...*getable or mineral acids* are em...*he small quantity of iron* contained ...*uld be affected by them*, and the solution would combine with the ferrocyanide of potassium, and give rise to a compound of the natu... of Prussian blue, by which all the parts of the paper adjacent to that operated on would be deeply and indelibly tinged, and the fraud attempted to be practised thus made apparent.

Claim.—The application or use of iodine or bromine, in their combined state, in combination with ferrocyanide, or ferricyanide of potassium and starch, for the purpose above set forth.

ROBERT HAWTHORN and WILLIAM HAWTHORN, of Newcastle-on-Tyne, engineers and partners. *For improvements in locomotive engines, parts of which are applicable to other steam engines.* Patent dated February 24, 1851.

Claims.—1. The employment in locomotives of double compensating beams and springs for distributing the weight of the engine equally on all the axle bearings, whatever may be the relative positions of such beams and springs with reference to the axle bearings: Also, a method of varying the spring from the centre of the beam towards either end.

2. The construction of the slide valves of locomotives, as worked in connection with each other, back to back in one steam chest, either with or without a passage formed through the plate (which separates the valves), and the piston of the valve into the exhaust ports. Also, the construction of the slide valves of stationary and marine engines, with an opening through the valve for the passage of steam either into or from the cylinders.

3. The construction of the link motion as hinged or connected by an eye-joint to the ends or other parts of the slide-rod or rods of locomotive and other steam engines, and the connection of the ends of the eccentric rods with the long sliding block.

ROBERT ADAMS, of King William-street, gun-maker. *For improvements in rifles and other fire-arms.* (Partly communicated.) Patent dated February 24, 1851.

The improvements claimed under this patent are—

1. The forming of the interior of rifle-barrels with projections or ridges in lieu o. the grooves ordinarily employed, the number of such ridges being varied at pleasure, although three, placed at equal distances apart, will be found a convenient arrangement.

2. The application to the interior of gun-barrels of a projecting tubular piece, to receive the charge, and for the end of the bullet (which is to be suitably hollow

to rest against when rammed home. Guns constructed in this manner will require to be loaded with a tubular ramrod, or by means of a ramrod having its end hollowed out to receive the charge, over which the barrel is inverted, and the whole then turned over so as to throw the charge into the tube —the ball being then rammed down as customary.

3. Certain improvements in gun and pistol locks, the object of which is to combine simplicity and certainty of action.

4. An improved safety catch or lock to prevent the hammer being accidentally struck down when the gun is at half or whole cock.

5. A peculiar construction of barrel for guns loaded at the breech. The end of the barrel is mortised out to receive the moveable chamber in which the charge is placed; the chamber in this case being centred on a vertical pivot and turning horizontally.

6. A peculiar construction of revolver pistol in which one barrel only is employed in conjunction with a series of chambers which successively come into a correct line with the barrel by rotating on an axis.

Edward Lloyd, of Dee Valley, near Corwen, North Wales, engineer. *For certain improvements in steam engines, which improvements are in part or in the whole applicable to other motive engines.* Patent dated February 24, 1851.

Claims. — 1. Certain constructions of steam valves for single and double cylinder engines, the peculiar feature of which is that the valves have a semi-rotary or vibratory motion in their seats, and thus admit the steam alternately on both sides of the piston or pistons, the passages in the valves being made to correspond with the number of cylinders employed.

2. The arrangement of two cylinders end to end, in double cylinder expansion engines, the steam acting alternately on both sides of the two pistons, which are fixed to the same piston-rod. [Another arrangement is also shown in which the cylinders are placed vertically side by side, and have their piston-rods connected to a cross-head, to which the crank-connecting rod is attached at such a point of its length as to balance the unequal pressure of the piston-rods of the two cylinders on each side thereof. The steam-valve is worked by the crank-connecting rod taking into a fork formed on its side, and imparting to it the necessary vibratory motion in its seat, and the force-pump is worked by the guide-rod of the crosshead which connects the pistons].

3. The application of these improvements to other motive engines where the moving power is air, water, or other fluid.

4. The application of a single vibratory valve for regulating the action of the steam in double cylinder expansion engines.

Henry Dircks, of Moorgate-street, engineer. *For improvements in the manufacture of gas, in gas-burners, and in apparatus for heating by gas.* Patent dated Feb. 24, 1851.

This invention consists; firstly, of improvements in the method of charging gas retorts, of diminishing the quantity of coal tar produced in the manufacture of gas, and in obtaining gas from the residual coal tar resulting from the destructive distillation of coal when manufacturing gas for illuminating purposes; secondly, of improvements in the construction of burners, whereby a film of gas of greater or less thickness may be obtained at pleasure; and thirdly, of improvements in heating stoves, ovens, and furnaces, by burning gas intermixed with atmospheric air.

1. According to the ordinary process of manufacture, the gas, on issuing from the retorts, is conducted to the hydraulic main, and thence by a suitable pipe to the condenser. Now it is at this last pipe that the patentee proposes to intercept its progress, and to cause it to pass through a boiler containing coal tar in the proportion of ten to fourteen gallons for every ton of coal undergoing distillation, the vapour from which rises and becomes mixed with the crude gas, which then, in this combined state, passes through retorts heated by a suitable furnace, and containing boghead cannel coal, coke, and other suitable substances, by which the gas tar vapour is decomposed, whilst the gas is finally conducted to the condenser for purification. The patentee also subjects the gas, when combined with tar vapour, to the action of heat, by passing it over incandescent surfaces of suitable material. The quantity of tar produced may be diminished, by introducing into the retorts with the coal employed for charging them, coke, lime, sawdust, peat, turf, or other substance capable of absorbing tar. When manufacturing gas from tar, oil, resin, or fatty matters, the patentee causes them to be projected in drops on the surface of heated boghead cannel-coal coke, instead of the ordinary coke or tiles employed for this purpose.

For charging the retorts, the patentee makes use of shoots with double bottoms and ends, capable of holding half or the whole of a charge, into which the necessary quantity of coal is placed, the shoot is then raised by means of a travelling crane running on a suspended railway in front of the range of retorts, and the end of it introduced into the retort; and when it has been forced in to its greatest extent, the shoot is withdrawn, and the charge left in the retort, being prevented from being drawn out with the shoot by the end piece.

burners described are constructed
forms, but in all, the same prin-
reserved, viz., that the size of the
through which the gas passes for
ay be increased or diminished at
The burners are constructed
two flat plates of metal, for which
latina is preferred, or one concave
er convex plate are employed, the
g forth between them. Sometimes
heated before issuing from the
nd this is effected by the flame from
r acting on the interior of tubular
th double sides placed above it.
en using gas for the purpose of
toves, the interior of the stove is
with a flue of earthenware, and a
in it, the end of which is either
th wire gauze or perforated with
apertures. The burner consists
of smaller diameter than that first
rforated at its end, and placed con-
y with it, an annular space being
ound for the admission of air to
the gas and support combustion.
ng furnaces, ovens, &c., the burner
milar construction.

.—Intercepting the progress of the
in its passage from the hydraulic
the condenser, and causing it to
ith the vapour of coal tar, and, when
ned, to pass over through or among
ent matters in suitable closed
.

method of passing the crude coal
gh heated apparatus on its passage
retort to the hydraulic main.

employment in the same retort
th or adjoining the coal of the
scribed of diminishing the produce
r.

employment of the light coke of
cannel coal when used along with
hen heated alone in close vessels
production of gas from coal tar,
, or fatty matters.

method of charging retorts.

peculiar construction of burners
.

methods described of burning gas
for the purposes of heating.

AS WICKSTEED, of Old Ford, Mid-
ivil engineer. *For improvements
manufacture of manure and in
y to be used therein.* Patent dated
24, 1851.

.—1. The manufacture of manure
ge water and other liquids contain-
izing salts by subjecting the same
tion of lime, and subsequently to
al force.

construction of centrifugal drying
with one or more moveable rings
s of wire gauze or other perforated material, so as to admit of the substances under operation being transferred
from one cylinder to the other.

3. The construction of centrifugal drying machinery in such manner that the drying of manure may be commenced at one velocity or distance from the centre of rotation, and completed at another velocity or distance from the centre of rotation.

4. A peculiar construction of centrifugal drying machinery in which the manure is supplied, dried, and discharged, during the continuous rotation of the machine.

5 The general arrangement and combination of apparatus for manufacturing artificial sewage manure, as described.

Specifications Due, but not Enrolled.

FRANCIS CLARK MONATIS, of Earlston, Berwick, builder. *For an improved hydraulic syphon.* Patent dated Feb 24, 1851.

ISAAC LOWTHIAN BELL, of Washington Chemical Works, near Newcastle-on-Tyne, chemical manufacturer. *For improvements in the manufacture of sulphuric acid.* Patent dated February 24, 1851.

WEEKLY LIST OF NEW ENGLISH PATENTS.

James Palmer, of Paddington, Middlesex, artist, for improvements in delineating objects, and in apparatus and materials for that purpose. August 23; six months.

Edward Clarence Shepard, of Duke-street, Westminster, gentleman, for improvements in obtaining and applying motive power. (Being a communication.) August 28; six months.

Thomas Brown Jordan, of Lambeth, Surrey, engineer, for improvements in machinery or apparatus for cutting, dressing, planing, and otherwise working slate, and also for framing and setting the same. August 28; six months.

James Edward M'Connell, of Wolverton, Buckingham, engineer, for certain improvements in locomotive steam engines and railways axles, parts of which are applicable to stationary and marine steam engines. August 28, six months.

William Johnson, of Millbank, Westminster, gentleman, for improvements in ascertaining the weight of goods. August 28; six months.

Pierre Armand Lecomte de Fontainemoreau, of South-street, Finsbury, Middlesex, and Boulevart Poissonnière, Paris, France, for certain improvements in apparatus for gas lighting. (Being a communication.) August 28; six months.

LIST OF SCOTCH PATENTS FROM 22ND OF JULY TO THE 22ND OF AUGUST, 1851.

William Johnson, of Lincoln's-inn Fields, Middlesex, civil engineer, for improvements in machinery, or apparatus for the manufacture of envelopes. (Being a communication.) July 23; six months.

Daniel Towers Shears, of Bankside, Southwark, copper merchant, for certain improvements in the manufacture and refining of sugar. July 25; six months.

Alexander Alliott, of Lenton, Nottingham, engineer, for improvements in cleaning, dyeing, and drying machines, and in machinery to be used in sugar, soap, metal, and colour manufacturing. July 31; six months.

John Davie Morries Stirling, of Black-grange North Britain, Esq., for improvements in the manufacture of metallic sheets, and in coating metals and alloys of metals in metallic compounds, and in welding. July 31; six months.

James Whitelaw, of Johnstone, Renfrew, North Britain, engineer, for certain improvements in steam engines. August 1; six months.

Charles Cowper, of 20, Southampton-buildings, Chancery-lane, Middlesex, for certain improvements in piling, fagotting, and forging iron and steel for plates, bars, shaft axles, tyres, cannons, anchors, and other similar purposes. (Being a communication.) August 6; six months.

Peter Robert Drummond, of Perth, for improvements in churns.) August 6; six months.

Robert Oxland, and John Oxland, both of Plymouth, chemists, for improvements in the manufacture and refining of sugar. August 6; six months.

James Buchanan Mirrlees, of Glasgow, Lanark, North Britain, engineer, for certain improvements in machinery, apparatus, or means for the manufacture or production of sugar. August 8; six months.

Joseph Mansell, of Red Lion-square, Middlesex, manufacturing fancy stationer, for improvements in ornamenting paper and other fabrics. August 8; six months.

William Onions, of Southwark, Surrey, engineer, for improvements in the manufacture of certain parts of machinery used in spinning. August 11; four months.

Alphonse René le Mire de Normandy, of Judd-street, Middlesex, gentleman, and Richard Fell, of the City-road, in the same county, engineer, for improved methods of obtaining fresh water from salt water, and of concentrating sulphuric acid. August 13; six months.

David Farrer Bower, of Hunslet, Leeds, manufacturing chemist, for certain improvements in preparing, rating, otherwise called rotting and fermenting flax, line, grasses, and other fibrous vegetable substances. August 20; six months.

LIST OF IRISH PATENTS FROM 21ST OF JULY TO THE 19TH OF AUGUST, 1851.

Thomas Allan, of Edinburgh, gentleman, for improvements in electric telegraphs, and in apparatus connected therewith. July 23.

William Beadon, jun., for improvements applicable to the roofing of houses, buildings, and other structures. August 2.

David Ferdinand Masirata, of Golden-square, Regent-street, Middlesex, gentleman, for a new mechanical system, with compressed air, adapted to obtain a new moving power. August 9.

Hugh Barclay of Regent-street, Middlesex, for improvements in the means of extracting or separating fatty and oily matters, in refining and bleaching fatty matters and oils, animal and vegetable wax, resins, and in the manufacture of candles and soap. August 11.

WEEKLY LIST OF DESIGNS OF ARTICLES OF UTILITY REGISTERED.

Date of Registration.	No. in the Register.	Proprietors' Names.	Addresses.	Subjects of Design.
Aug. 21	2917	W. Price	Manchester	Imperial copying press.
22	2918	J. Britten	Birmingham	A grate.
23	2919	J. Carter	Delabole, Cornwall	Filtering apparatus.
25	2920	J. Young	Wolverhampton	Flooring cramp.
,,	2921	J. A. Drake	Mells, Somerset	An instrument to be used in cases of prolapsus uteri.
26	2922	J. T. Moss	Bayswater	Crank spit.
,,	2923	J. and J. Holmes	Regent-street	Cloak shawl.
27	2924	H. J. and D. Nicholl	Regent-street	A garment.
,,	2925	E. McMorland and Co.	St. Paul's-churchyard	The pelia or hooded shawl.

WEEKLY LIST OF PROVISIONAL REGISTRATIONS.

Aug. 20	275	T. Forbes	Ely-place, Holborn	Parallel vice.
21	276	J. Roberts	Portsmouth	Paper clamp.
22	277	F. J. Earl	Bermondsey	Perpetual calendar.
,,	278	P. Warren	Longton, Staffordshire	Danger signal for railways and carriages.
26	279	H. Studdy	Torquay	"Stoker's Ventilator."
27	280	J. Boydell	Regent's-park Terrace	Iron support.

CONTENTS OF THIS NUMBER.

LONDON: Edited, Printed, and Published by Joseph Clinton Robertson, of No. 166, Fleet-street, *in the City of London*—Sold by A. and W. Galignani, Rue Vivienne, Paris; Machin and Co., *Dublin*; *W. C. Campbell* and Co., Hamburg.

Mechanics' Magazine,

MUSEUM, REGISTER, JOURNAL, AND GAZETTE.

No. 1465.] SATURDAY, SEPTEMBER 6, 1851. [Price 3*d*., Stamped, 4*d*.

Edited by J. C. Robertson, 166, Fleet-street.

THE NORTHUMBERLAND MODEL LIFE-BOAT.

Fig. 1.

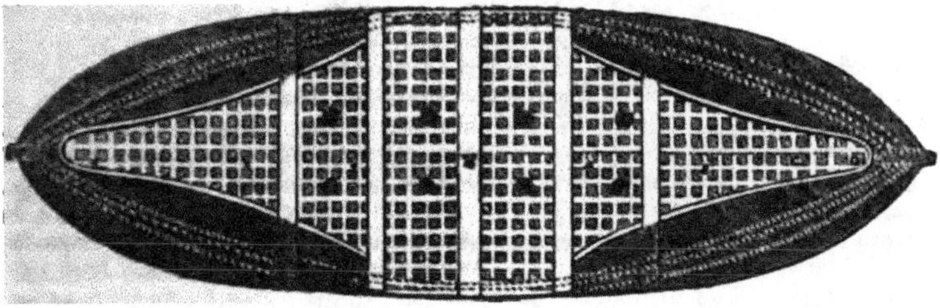

Fig. 2.

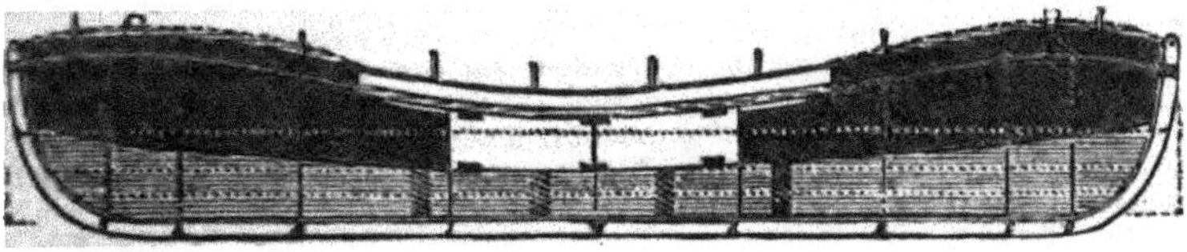

Fig. 3.

2

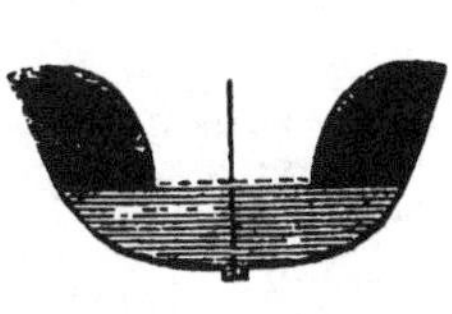

D

H

B

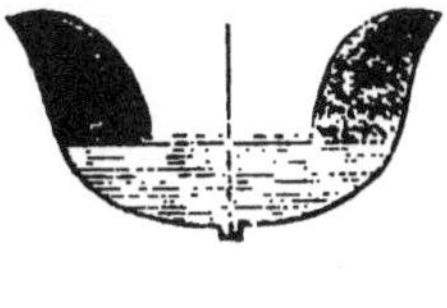

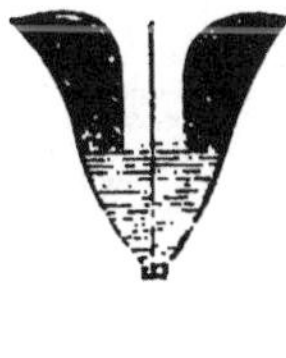

THE MODEL LIFE-BOAT DESIGNED BY MR. PEAKE, AND RECOMMENDED BY THE CO MITTEE APPOINTED TO EXAMINE THE PLANS WHICH COMPETED FOR THE NORTHU BERLAND PREMIUM.

In October last the Duke of Northumberland made a public offer of a reward one hundred guineas for the best model of a life-boat. His Grace, who is him a seaman, pointed out to intending competitors that the chief defects in the existi life-boats requiring to be remedied, were—that they had not the power of self-righti in case of being upset,—that they did not free themselves quickly enough of wat —that they were mostly too heavy for transporting elsewhere, &c.; but while reco mending special attention to these points, his Grace left the form, construction, a all other particulars entirely to the skill and judgment of the respective designe The prize was not a large one (hardly sufficient to cover the cost of the commo model which could be constructed), but it appears to have been sufficient for t purpose. No less than two hundred and eighty competitors, from all parts of t United Kingdom, responded to the Duke's offer. The models and plans sent in w referred for examination to a Committee composed of Sir Baldwin Walker, S veyor of the Navy; Captain Washington; Mr. Watts, Assistant-Surveyor R.N Mr. Fincham, Master Shipwright, Portsmouth; Commander Jerningham; a Mr. Peake, Assistant-Master, Woolwich; with power to the Committee to aw the prize to the most deserving. The Report of this Committee has now b "printed" in a very handsome style, at the expense of his Grace, and copies of privately distributed, with a degree of liberality which leaves small occasion regret that it has not been placed, by the usual mode of publication, more wit the reach of the public at large. The Report is a very able, impartial, and instru tive document. The Committee have been at infinite pains to deal justly and fai by the competitors, and show such good grounds for the award which they h pronounced, that we anticipate a more general concurrence in it than is ordina the fate of decisions on prize competitions. Thus it was they proceeded in inquiry:—

A general review of the models soon pointed out that they might be advantageo grouped according to their characteristic features; thus there were several models in form of pontoons; catamarans or rafts formed a second group; a third group may described as having for its type a troop-boat or steamer's paddle-box boat; a fourth partaking chiefly of the north-country cobble; and lastly, a group composed of the or nary boats in every-day use, slightly modified according to the nature of the coast they w intended for.

After examining the models separately, so as to ascertain their form for pulling or saili their dimensions, capacity for holding water, area of delivering valves, weight, nature a amount of extra buoyancy, and trying experiments in the Thames on their relative stabil power of self-righting, and readiness in freeing themselves; having also prepared a desc tion of several, and added a few remarks, each model was brought forward in turn bef the General Committee, the description and remarks read over, discussed, corrected, a agreed upon.

The difficulty then arose, where so many boats were nearly alike, of deciding on the r tive merits of each. In order to insure that no good quality should be overlooked, and obviate the possibility of bias, the Committee agreed upon those points which they co dered the essential qualities of a life-boat and their order of precedence. A certain num was then given to each of these qualities according to its importance, so that the wh numbers should make up 100. It may be satisfactory to the several competitors to kn what, in the opinion of the Committee, are those qualities and the degree of importa they attach to them. They are as follows:—

	Nos.		Nos.
Qualities as a rowing boat in all weathers	20	Suitableness for beaching	4
Qualities as a sailing boat	18	Room for, and power of, carrying passengers	3
Qualities as a sea boat; as stability, safety, buoyancy forward for launching through a surf, &c.	10	Moderate weight for transport along shore	3
Small internal capacity for water up to the level of the thwarts	9	Protection from injury to the bottom	3
Means of freeing boat of water readily	8	Ballast, as iron 1, water 2, cork 3	3
Extra buoyancy: its nature, amount, distribution, and mode of application	7	Access to stem or stern	3
Power of self-righting	6	Timber heads for securing warps to	2
		Fenders, life-lines, &c.	1
			100

be seen by the above formula that the Committee consider it an essential requisite boat that she should be a good rowing boat, able to get off the beach in any weather a boat can live at sea, as without the power of doing this, other good qualities are il. To this, then, is awarded the highest number. As on the coasts of Norfolk and where the wrecks generally occur on outlying sands, all the life-boats go off under , as it was evident some of the best models were prepared with this view, it was ed that these also were entitled to be placed on a par with boats built chiefly for but as rowing is the general rule around the coasts, and sailing the exception, a ference was made in favour of the former.

ther qualities speak for themselves and call for no further notice beyond the gene-rk that the numbers are so distributed that a boat that could neither pull nor sail n if she combined all the other essentials of a life-boat, could not by this scheme igh rank; it being considered that if a builder could construct a boat to combine and safety in all weathers at sea, with speed in rowing, he would find no difficulty ing the internal fittings requisite to render his structure a good life-boat. On the and, pulling or sailing well alone would not enable a model to take a high rank; fact, most of the best models combine the qualifications of speed and stability, d internal fittings.

bove details will appear tedious; but the Committee felt that they had a delicate ponsible task to perform, and they would rather incur the reproach of being weari-an that the humblest competitor among the number should not be fully aware of the s that guided them in awarding the premium.

preliminary formula arranged, all difficulties vanished. Each model was again forward in its turn, each quality of it was named and examined in order, the num-proportion of the whole number, according to its merits, was proposed, agreed nd set down in a column. After some days, when many models had been examined, umbers were added up, and the relative order of merit in the several boats established. boats that stood first on the list were then, for the third time, brought forward and ogether side by side, their several points again examined, and the models carefully d with each other; the result was a confirmation of the former numbers. They und to stand as follows:—

eeching, of Great Yarmouth	84	J. and E. Pellew Plenty, Newbury	77
inks, Appledore, Devon	78	Harvey and Son, Halifax, Ipswich	74
Teasdel, Great Yarmouth	75	Semmens and Thomas, Penzance	72
arrow, South Shields	72	Willem van Houten, Rotterdam	70
almer, Nazing-park, Essex	70	J. and J. Harding, Whitby...	70
r Robinson, Hartlepool	70	Thomas Gaze, Mundesley	70
nd Laurie, Commercial-road	70	George Lee, Tweedmouth, Berwick	68
Greener, Aston, Birmingham	70	William Falkingbridge, Whitby...	66
John and Son, Spittal, Berwick... ...	67	Thomas Costain, Liverpool	65
mond, Scarborough	67	Joseph Hodson, Blythe	65
Cambridge, Filey, Yorkshire	65	W. Goodridge, Swansea	65
r, Newcastle	65	John Cockey, Portsmouth	64
and Haines, Brighton...	65	Benjamin Birch, South Shields	63
Wake and Sons, Sunderland	63	J. Bertram, East-street, Manchester square ...	63
mpson, Rotherhithe	63	Robert Blair, South Shields...	63
ven, Padstow	63	T. and J. White, Cowes	62
wer, Ipswich	63	Charles Gurr, Portsea	62
rowsmith, Gosport	62	John Lister, Sunderland	60
anes, Liverpool	61		

Committee therefore declare James Beeching, of Great Yarmouth, to be cessful candidate for the premium offered by the Duke of Northumberland.

ever, though James Beeching's was, on the whole, the best plan which the tition had produced, the Committee thought that something still better might rived, embodying all the best features of all the plans submitted to them. herefore requested Mr. Peake, one of their number, to prepare a drawing g a plan and sections of a 30-feet boat, in which, profiting by the experience in the examination of the models, all the best qualities of a life-boat should bined.

views which the Committee desired to see carried out by Mr. Peake are thus th at large in the Report:—

1. With regard to *form*:—

form best adapted for the general purposes of a life-boat is that usually given to a oat, that is, both ends alike, but with more breadth of beam; fine lines to enable t to pull well, but *sufficient fulness forward* to give buoyancy for launching through a surf; good sheer of gunwale, say an inch for each foot of length, but rounded off t the extremes; a long flat floor; sides straight in the fore-and-aft direction; the g strake in the midships to tumble home to protect the thole-pins, and the bow strake out to throw the sea off; as much camber or curvature of keel as can be combin steady steering and safe launching from a beach, in order that the boat may be quickly to meet a heavy roller when about to break on her broadside.

2nd. As regards dimensions:—

In point of length, life-boats may be conveniently divided into three classes—fron 25 feet, from 25 to 30 feet, and from 30 to 36 feet—which last may be conside maximum, and a length rarely required. The smaller-sized boat is handy on those p the coast where it is difficult to find a crew, a difficulty that would be found to exte great part of the shores of this kingdom. Such a boat would be easily transported shore, easily launched, and readily manned, and, except in some special cases, woul rally bring on shore the whole of the crew of a stranded vessel; and as the boat's cre not consist of more than six men, there would, in case of an accident occurring, b lives perilled. The two boats already alluded to as built by Plenty, one on the c Devon, and the other on the coast of Lincolnshire, are respectively 18 and 24-feet and they have saved 120 lives within the last few years.

The medium, or 30-feet boat, to pull ten oars double banked, is probably the best s for the general purposes of a life-boat at all places where a sufficient crew can be found to man her. Such boats are in use at Liverpool, Shields, Dundee, and othe ports where no difficulty is experienced in finding a crew; and on a special occas Liverpool, one is said to have brought on shore 60 persons. At less populous place the coast a 27-feet boat would be found more easily manageable.

The maximum, or 36-feet boat, is adapted for such places as Yarmouth, Lowestoft &c., where it is the invariable custom to go off under sail, and where there is n difficulty in finding beachmen to launch or man the boats, however large. The wre Yarmouth and Deal occur generally on outlying sands, and the boat that happens to windward on the coast, according to the direction of the wind, goes off under canvas wreck. Thus should a wreck occur on the Yarmouth sands in a south-east gale, the field or Lowestoft boat would push off while, in a north-east gale, the Caistor or (boat would put to sea. The boats actually in use at these places are from 40 feet to long; they weigh from four to five tons, and cost from 200*l.* to 250*l.* each. They fore form the exception to the general rule; but they are powerful boats, are adn manned and handled, and have been the means of saving some 300 lives within t 30 years.

With respect to breadth of beam, in a rapid tideway, as the Tay, the Humber, the l Channel, the shores of the Isle of Man, the Shannon, &c., a boat somewhat of the form, but with ends like a whale-boat, would be more suitable than a wider boat. In exceptional cases the breadth of beam might be one-fourth the length; but for a lif where the requirements are, roominess for passengers, width to pull double-banked, st to resist people moving about and occasionally pressing down on one side in rescuing from the water, it should never be less than one-fourth. The Tyne boats have a b of fully one-third the length, and some more, but such would not seem to be the bes portions; probably as 1 to 3·3, or 9 feet of beam to a length of 30 feet would best s the purposes of a life-boat.

As to depth, it seems only necessary to observe, that a boat that has to be lau through the surf on a beach should not be too shallow in the waist. The well-l masulah or surf-boats at Madras have sides 8 feet deep. This height, however, wou suit a boat that has to pull off a lee shore against a gale of wind, where the less s exposed the better. As a general rule, the inside depth to the gunwale amidships sho one-third the breadth; this would give a 30-feet boat a whole depth of 3½ feet amid and a depth not exceeding 6 feet might be taken at the stem and stern-posts.

The weight suitable to a life-boat does not seem to have received much conside from our builders, to judge from the difference in existing boats. Those at Holy I at Yarmouth, and Southwold, as before mentioned with their gear, weigh about five whereas many of the models sent in are said to weigh less than half a ton. The between these two extremes will be near the truth. For however desirable lightness transport along a beach, a certain weight of boat is necessary to resist the force (waves and to retain momentum, so as not to risk being driven back by the sea; which consideration 1 cwt. or 1½ cwt. for each foot of length would be a fair genera The weight of gear would vary from 5 cwt. to 15 cwt., according as it comprises oars, r sails anchor, cable, warps, &c.

ever be the length of the boat, care should be taken that the space between the should not be less than from 28 to 30 inches, as in pulling in a seaway it is imprac- always to keep stroke, and if the thwarts are too close, the loom of one man's oar to strike the back of the man abaft him. This is a common complaint in life-boats. s should be short to pull double banked, and of fir, as being lighter, more buoyant, ler than ash, which is too pliant. They should pull with iron thole-pins having rope ts secured to them, and the pins should be so placed that the boat may be pulled ay, by the men merely turning round on the thwarts.

As regards extra buoyancy:—

rto all our boats have been of wood, but the testimonials in favour of metal boats strong. Galvanized iron (if that process prevents oxidization, which does not yet be established) would be the most economical, and the corrugated form of it would ngth. But if metal boats be adopted, copper might be preferred as more durable e tractable. The boats in which Lieutenant Lynch, of the United States navy, ed the rapids of the river Jordan in 1848 were of copper, and that officer reports vourably of them. It is said that a copper boat is now supplied to every vessel in ted States revenue service, if not to the navy at large. The first cost of such boats e heavy, but the material would always be of value. In metal boats it is affirmed air-tight cases could be more easily built into the boat (if in any case such were ble), and kept from leaking. About one-tenth of the whole of the models sub- are in favour of the use of iron in boats. The Committee are far from advocating ption of metal boats as life-boats, but they would recommend a fair and full trial of any convenient opportunity.

e construction of wood boats, well-seasoned Scotch larch, from its durability and s (its specific gravity being little more than double that of cork), would be found t material, but neither Polish nor Italian larch, should be trusted to. American edar is both light and durable. One advantage in having wood boats is that we have the benefit of the skill of the numerous boat-builders around the coasts, the building of metal boats is confined to a few hands; and there is an advantage g a boat built by an experienced man, who designs and executes his own work.

utta percha, caoutchouc, kamptulicon, and other similar materials, the Committee experience that can be relied on. A gutta percha boat was taken out to the Arctic last spring, but the time of trial was too short for any decisive opinion to be formed erits. It is stated that the material shrinks, and it certainly will not bear a continued nor do we know the effect of heat and cold upon it. It is, however, quite possible me of these materials may prove useful in the internal fittings of a life-boat. A ation of gutta percha and cork, by CLARKSON of the Strand, and another consisting a percha, between two layers of thin wood, or a lamina of wood, coated with gutta and caoutchouc, by Mr. FORSTER, R.N., seem likely to be well adapted for air- A notion seems prevalent that gutta percha is very light, but its specific gravity is ss than that of water, or, in other words, it will hardly float. JEFFERY's marine y also be found useful in the internal fittings of a boat, in joining cork, &c.

As regards extra buoyancy:—

a buoyancy, or that required beyond what the materials used in the construction of at will afford, is the characteristic feature of a life-boat, and as such its nature, t, and distribution, deserve the most deliberate consideration. If sufficient buoyancy obtained by cork, it is far preferable to air-cases, as not being liable to accident. ore mentioned, the Committee have reason to believe that cork may be used entirely the flat or floor of the boat, so as to reduce the internal capacity, and enable the free herself from water. The only doubt is as to its weight; but it appears that ries considerably in weight as well as in price, the commonest description of cork, used by fishermen as floats for their in-shore nets, does not exceed 12 lbs. weight per foot, and costs about 12*s.* a cwt.; a heavier sort, also used by fishermen, weighs 5 lbs. per cubic foot, and costs 15*s.* a cwt. These two might be advantageously dis- in the bottom of a boat, formed into a solid mass by marine glue, and the boat then bid defiance to accidents, as thus armed, even if bilged against a rock, she would

e is one point connected with the use of cork for buoyancy that deserves a passing . From the number of models that have cork outside on the bottom, and from the cork thus placed being reckoned as buoyancy in the written descriptions that accompany there seems reason to believe that some of the builders have considered that cork so l is of more avail as buoyancy than if it were in the bottom inside the planking. Now a *fallacy; for if the same form for* the external body of the boat be taken in each case, and the whole weight of the boat be the same, the same weight of cork being employed, whether applied externally or internally, the boat would float at the same water line. There are those who argue that the cork when placed to form the exterior shape of the boat gives additional buoyancy, or, in other words, causes the boat to swim higher out of the water, but with the knowledge that the water displaced in each case must be equal in weight to the weight of the boat, and that those weights are equal, it follows that the results on the immersion must be the same; or that the position of the cork could have no effect on the depth of immersion, or on the buoyancy of the boat.

If, however, a doubling of cork be brought on outside, in addition to the complete boat, then, of course, the external form of the boat being altered, the buoyancy will be increased by the difference of the specific gravities of cork and water, less the fastenings. Cork is useful outside to protect the bottom of the boat against a rocky beach, but, from its usually unfair surface, it has a tendency to retard the speed of the boat. Extra cork should not be applied to the bottom of a boat without due consideration, for if, as in several of the models which have large air-cases under the flat and no ballast, cork were applied to the bottom outside, its effect must be to raise the boat out of the water and thereby lessen the boat's stability, and allow her the more easily to upset.

With respect to air; the great difficulty of rendering vessels permanently air-tight makes it unfit for general use, unless great care and watchfulness be exercised. In those instances in which the air-cases are built into and form part of the boats, it seems doubtful whether any of them can be depended upon for a year, and from various inquiries that have been instituted the Committee have reason to believe that there does not exist at this moment a complete air-tight case (undetached) in any life-boat that has been six months in use around the coasts of Great Britain. As to air cases that are detached, they may be better, but unless in the form of small casks, as in the Liverpool boats and in those of the Shipwreck Institution, there seems sufficient reason to suspect them all. Metal air-cases offer rather a more reasonable prospect of security; but when a life-boat was laid open in Woolwich Dockyard a few years since, it was found that from corrosion there were several holes half an inch in diameter in the copper tubes, supposed to have been air-tight; in fact copper, like other metals, is liable to corrode, and the more so when placed in conjunction with sea-water. The weight, too, of copper tubes makes them objectionable. It has been the practice of TEASDEL, an experienced life-boat builder at Great Yarmouth, to build his detached air-cases of thin boards of willow wood, which is both tough and light, and to cover them with painted canvas; but it might be an improvement to interpose a sheet of gutta percha between two thin boards, according to FORSTER's process.

Before quitting the subject of the nature of extra buoyancy, the Committee consider themselves called upon to warn sailors, and the public generally, against one of the schemes submitted to them, and termed by the inventor "Patent Buoyancy,"—otherwise, dried rushes done up in linen or canvas bags. This material, however promising it may be at first sight, sooner or later is sure to imbibe moisture when used, and then, instead of having floating properties, it must act as ballast. Mr. CHATFIELD, Assistant Master-shipwright at Woolwich Dockyard, has officially reported on the unsuitableness of these rushes for giving buoyancy. At Bude, on the north coast or Cornwall, in October last, might be seen a boat, called a life-boat, thus fitted; it had been lying in a canal for a few months, and was completely waterlogged, so much so that the water was awash with the gunwale. It is to be feared that some of the so-called "life-boats," which all sea-going passenger steamers are required to carry by a recent Act of Parliament, are fitted with these dried rushes. They will assuredly fail them in time of need. What are termed life-belts and life-buoys, similarly fitted, are also extensively distributed on board our river steamers. For this latter purpose, neither air nor rushes should be used; cork only should be trusted to, and from the great external resemblance of the spurious articles to the genuine cork life-buoy in form, size, and colour, it may be feared that many persons are deceived into purchasing what may in the hour of danger cost them their lives.

The amount of extra buoyancy may be much less than it has hitherto been customary to give in a life-boat The cubical contents of the air-cases of many existing life-boats, and of a great part of the models submitted for competition, measure from 200 to 300 feet, equivalent to the support of from 6 to 9 tons of dead weight. Now, if only intended for buoyancy to balance the extra weight likely to be put into a life-boat, this amount is unnecessary. The Liverpool life-boat, already alluded to as having on one occasion brought on shore 60 persons from a wreck, had not above 60 cubic feet of extra buoyancy; this is too little, but in a 30-feet boat, provided with ample delivering valves, 100 cubic feet, or the equivalent of 3 tons, is sufficient extra buoyancy for all general purposes.

The distribution of extra buoyancy requires great care. As a general rule, it should be placed high in the boat, so as not to affect her stability; but circumstances require this rule to be slightly modified. To reduce the internal capacity of the boat that she may rise under the weight of a heavy sea that may fall on board, and to enable the delivering-valves to act freely, a certain amount of space should be occupied under the flat or floor of the boat, so as to exclude the water; and the question is, so to fill this space with a material of less specific gravity than water, yet sufficiently heavy to ensure the boat's stability when the flat or flooring is laid at from 12 to 14 inches above the keelson, or about the water-line of intended immersion; thus acting generally as ballast, but on emergency as extra buoyancy. From the various plans submitted, this would seem the most difficult problem to solve in the whole arrangements of a life-boat. In some existing life-boats, and in many of the models sent in, reduction of internal capacity is attempted by placing a tight deck fore and aft at from 16 to 18 inches, and even in some at 24 inches above the keelson, with only air beneath; the result is, that all the weights in the boat are raised above her centre of gravity, and there is a risk of her upsetting when a sea is shipped. Some of the models thus fitted on being tested as to their stability, by having a bucket of water thrown into them when afloat in the river, went over directly. Other competitors, foreseeing this result, added an iron keel to their boats; while some inserted a well or tank amidships for water ballast, which, as long as it remains in its place, compensates for the amount of air, and restores the equilibrium. Others have tried a combination of cork and air, alternately distributed, so as to preserve the requisite stability of the boat. But although conceding full merit to water-ballast, which has the advantage of being taken in only when the boat is afloat, and thus leaving her light for transport along shore, the Committee have come to the conclusion before stated, that cork is the safer material, and that it may be placed under the flat, up to about 12 inches in height above the keelson, and combine the properties of ballast generally, with extra buoyancy in case of need; if upon the cork a light water-tight deck be placed, with a grating over it, the cork will be preserved and very little water will remain to inconvenience the crew or passengers.

The next point to attend to in the distribution of the extra buoyancy is to place the requisite amount of air vessels in the head and stern-sheets of the boat from the floor up to gunwale height, always taking care to leave a passage of 18 inches wide up to within 2 feet of the stem and stern-post, to enable a man to stand there and receive people from the wreck, as it commonly occurs that a boat cannot go alongside a stranded vessel, but has to receive the men either over the head or stern of the boat. Air-cases may also be placed on the bows and quarters without occupying space that should be reserved for passengers; but they should not be fitted within a few feet each way of the middle of the length of the boat, as, if the forces are nearly equally balanced, it will be difficult for a boat to right herself against the resistance offered by the flat tops or projections of midship side air-cases. This appears to be a point that has been overlooked by most of our boat-builders. In all cases in which air-cases are used, it is recommended that they be not built into the boat, but be detached so that they may be examined to test if they are air-tight, there being great reason to fear that such is not the case in general. Besides, air-cases built into the sides are liable to open with the working of the boat, or to be stove in going alongside a wreck, as in a recent instance, and thus a boat would be disabled. If they be used under the flat, and it is considered necessary that they should form part of the boat, they should always be divided into compartments, so that if one failed, the others would not be affected by it.

5. As to internal capacity for holding water:—

The more the internal capacity is reduced consistently with leaving space for a rescued crew, the better the life-boat. If practicable, the internal capacity for holding water up to the level of the thwarts of a boat 30 feet long should not exceed three tons. It may be diminished by side air-cases from the thwarts to the floor, or by air-cases under the thwarts. On this latter mode of reducing capacity there is a difference of opinion, some contending that it is an advantage to break up a sea, and prevent the water rushing fore and aft the boat, while others think that it is better to let the sea have a fair range, and that then much of the water that comes in over the bows would go out over the stern. The balance seems rather in favour of filling up under the thwarts; it has the certain advantage of reducing capacity.

6. As to the means of freeing the boat of water:—

In order to efficiency, every life-boat should be provided with the means of freeing herself rapidly of any water she may ship. This would seem a self-evident proposition; but it appears not to be admitted as such by the designers of many of the models, as in them no provision is made for it beyond a bucket for baling. Not to multiply proofs of the necessity for such an arrangement, it is sufficient to cite, as decisive on the point, the recent instance of the Liverpool life-boat, in October last, having been obliged to cut her tow-rope and bear up for the Mersey, in consequence of having shipped a succession of seas. If a boat has large internal capacity, say from six to seven tons, which is not unusual in the Yarmouth boats, and she ships a heavy sea, or a succession of seas, or if, as is commonly the case, while under storm sails the crew pull out their plugs and let the boat fill up to her water-line for ballast, should a sudden squall carry away her masts, how is that weight, in addition to the weight of the boat, to be propelled by 12 oars against a heavy sea? It would be impracticable, and the relief of the wrecked vessel must, in such a case, be abandoned.

By means of sufficient delivering-valves or tubes, led through a platform or flat laid about the level of the water-line, there seems no reason why the water when shipped should not be carried off rapidly. The area of the valves or tubes should be ample, not less than one square inch for each cubic foot of capacity; more would be better. A question may arise whether it is better that the boat should free herself by tubes through the bottom, or by scuppers in the sides, as shown in several of the models; the former is the more direct and quickest action, but the tubes are liable to be choked in the possible case of a boat grounding on an outlying sand-bank, or on the bar of a river harbour; it will be better, therefore, to be provided with both to meet such an accident. The tubes and scuppers might be closed by self-acting valves if thought necessary; a simple modification of an apparatus known as Kingston's Valve would answer the purpose effectually.

7. Provision for self-righting :—

The power of self-righting is a contested point among the best boat-builders; but they seem hardly to have given the subject full consideration. The accidents that have happened to life-boats have not been carefully investigated, and the necessity for meeting these accidents with a remedy has not forced itself upon their minds. But a remedy is necessary. Recent and sad experience has shown that a life-boat may be upset and may drown the crew from want of being able to right herself. Had the South Shields life-boat that upset last year possessed the means of self-righting, there is reason to believe that many of the crew might have been saved, whereas twenty of the best pilots out of the Tyne were drowned. This, however, is not the only instance of a boat upsetting and remaining bottom up, as will be seen hereafter; but it is sufficient to prove the absolute necessity of grappling with the difficulty, if difficulty it be, and of overcoming it. Most life-boats have good sheer of gunwale, and, consequently, raised extremities, in which air-cases (or light cork) should be placed, in order that when the boat is bottom upwards, their buoyancy may co-operate most effectually with the weights in the bottom of the boat (now raised, it may be, considerably out of the water) to restore her to her originally upright position. The higher the centre of gravity of a vessel or boat is above the centre of buoyancy, *cæteris paribus*, the less is her stability; and by the separation of these two centres, a condition of instability will ensue, the effect of which will be, that with the slightest motion, the boat will reverse her position, or right herself. To determine the necessary extent of separation of these centres in each case involves careful calculation. The best mode of applying this principle will readily occur to most boat-builders. The objections to the raised air-cases at each end are the wind they hold in pulling off a lee-shore, and the difficulty of approaching the stem and stern of the boat; the latter may be modified, the former must be tolerated for the greater benefit in another respect that arises from their adoption. If air-cases be used in the extremes to obtain the buoyancy, a thin layer of cork on the top will afford great protection to them, and better footing for the crew when necessity requires them to stand on them.

8. As to ballast :—

If the requisite stability, and righting power, can be obtained without ballast, it is very desirable to avoid the incumbrance it causes, in case of having to transport a boat along shore. In this respect water-ballast has a great advantage, as it is not taken in until the boat is fairly afloat, and may be discharged directly she again touches the beach on landing. Water-ballast, if used in immediate connection with air-tight cases, as it always must be, requires very good workmanship in the bulk-heads or partitions of the well, in order that they may not become leaky by straining when at sea, or by shrinking when the boat stands ashore, which she sometimes does for a year together. A doubt may arise, too, whether a boat does not require her ballast as much or more at the moment of launching than at any other time; lightness has its advantages, but in launching through a surf a boat requires a certain weight so as not to be readily thrown aside by a breaking sea. All these circumstances considered, the Committee incline to the opinion that ballast given by cork inside in the bottom of the boat is best adapted to meet the varied contingencies to which a life-boat is subjected.

Although a minor point, it may be as well to add, that a moderate-sized cork fender, say

4 inches in diameter, should be carried round the sides and both ends of the boat at about 6 inches under the gunwale; but there is no occasion for the unwieldy fender, occasionally 24 inches deep, that may be seen in some life-boats. Holes in the bilge pieces, to enable a man to lay hold of them, should the boat be upset; timber heads, to make warps fast to at each bow and quarter; long sweep oars for steering at each end; a stout roller in the stem and stern-post to receive the cable; spare oars, one for each two that the boat pulls; life-belts, life-buoys, and life-lines; hand-rockets; heaving lines, and such minor fittings are indispensable in every life-boat.

The model boat which Mr. Peake has designed in accordance with these views is represented in the engravings prefixed to this paper. Fig. 1 is a plan; fig 2 a sheer plan; fig. 3 a body plan; and the other figures are cross sections at different points. The parts engraved in dark lines represent air-chambers—those in light lines, cork. The following are the principal dimensions, with some other explanatory particulars:—

Extreme length	30 feet.
Length of keel	24 ,,
Breadth of beam	8 f. 9 in.
Depth to under side of keel	3 f. 6 in.
Sheer of gunwale	2 f. 4 in.
Extra buoyancy, cork and air, equal to	3 tons.
Internal capacity up to the level of the thwarts	4 ,,
Area of delivering valves	300 sq. in.
Proportion of delivering area to capacity	1 to 3
Weight of boat and fittings, or displacement	38 cwt.
Ballast, iron band, equal to	3 ,,
Draught of water with thirty men on board	16 in.
Number of oars, double-handed	10
Diagonal built, of rock elm, and copper fastened.	
Rig of boat—a fore and mizen-lug sail.	
Cost, complete	100*l*.

MR. GOLDSWORTHY GURNEY'S DIFFERENTIAL BAROMETER.

(Concluded from page 176.)

The instrument which I have now before me, and which works well, was thus constructed:—A small glass cylinder, 4 inches long, and 1 inch in diameter, about the size and thickness of a small phial bottle, was blown with two mouths and necks, one at each end. A glass tube, ¼ of an inch in diameter, and 3 feet long, sealed at one end, had its open end cemented air-tight into one neck of this vessel. Sealing-wax, gutta percha, plaster of Paris, or any other air-tight luting, will answer for this purpose. Into the other end or neck, a tube, 2 feet 6 inches long, open at both ends, ⅛ inch in diameter, was cemented. The tubes and the connecting cylinder, the joints all air-tight, were now laid along a wooden staff, and secured to it by wire ligatures into the open end of the compound. Distilled water was poured, until it rose up to about 2 inches in the working cylinders. Mercury was now poured in, so as to raise the water to the top of the tube. The finger was then placed tight on its surface, and the staff inverted, when the mercury and water changed places. The open end, the finger still tightly on, was placed under the surface of a cup of mercury. The finger was removed, and the mercury fell in the ordinary way; the water followed, and a vacuum was formed above its surface. The water was now in vacuo, and a quantity of air was seen *escaping from it in little* bubbles rising to the surface, as is usual when water is placed under the exhausted receiver of an air-pump. This escapage of air would interfere with the vacuum, therefore the finger was again placed tight on the open end of the tube under mercury; the tube withdrawn, and the staff inverted; the mercury and water again changed places. The water, deprived of the air which it previously held in solution, came to the top, the finger was withdrawn, and a further portion of mercury poured in, so as to raise it again to the surface, as at first. No air bubbles will now escape. Some plan is essentially necessary for graduating the instrument to correct inequalities, because it is more than probable the proportion of areas you may seemingly obtain by measurement may not be mathematically true, for it is difficult to measure them exactly. The tubes may be taper; the supposed cylinder may really be a cone, or any thing else.

After having obtained the proper quantities of mercury and water to fill the tubes, and stand at their proper and respective levels, the water about half-way up the differential tube, and the mercury about half-way up the bulbous cylinder (which, if it does not approach in the first, we must add to, or take from, in the second filling), mark off accurately where the surface of the mercury stands, and also where the surface of water stands. Now pour mercury in the under cup or vessel until the surface of the mercury in it rises in the tube exactly an inch above this mark: make another mark here, and scribe off very exactly the height of which this inch of mercury had driven up the water. Again raise the mercury an inch, or any other known measurement, in the same way. By pouring in more mercury in the bottom vessel again mark off, very accurately, the proportional rise of the water on the differential scale, and so on, up or down, until the whole working ranges are obtained; thus the height of mercury will make its own scale or graduation of the water level. The division thus obtained may be subdivided at will.

The instrument I have is graduated on the differential scale into tenths of an inch. The diameter of the working part of the mercurial column being 1 inch, and the differential tube being ¼th of an inch, their proportional rise will be as 16 to 1, so that every tenth rise of the barometer will move up the water column 16 tenths. This, I think, will be sufficiently sensible, and the graduations, perhaps, small enough for your purpose. You may, of course, make it more or less so.

Another way of graduating this instrument is to turn the staff out of the perpendicular, until the arc described shall occasion such an incline plane of the mercury as will rise it a given distance up its scale; the corresponding displacement up the differential tube may then be marked off.

In the former arrangement the differential tube is sealed, and the water in vacuo. Here no error can arise from evaporation, or any other atmospheric interference. It should be observed that, if spirit be used instead of water, an atmosphere of its vapour will pervade the vacuum. If this amount of vapour was a constant quantity it would be of little consequence, seeing we have only to deal with differential changes; but its amount will vary, and be sensibly increased or diminished by change of temperature. This vapour will interfere with the working of the instrument, and require troublesome correction to be made for it. If you think the vapour of the water likely to interfere with the vacuum, and require correction, you had better use oil; in such case no possibility of error could arise from temperature or evaporation.

For measuring the heights of mountains, or obtaining absolute weights, a graduated correction may be made; but water without any correction seems to be sufficiently accurate for the purposes of pitmen or sailors. Perhaps I hardly need observe that the absolute weight of the atmosphere can always *be read off from the mercurial* column, *making correction by allowing* about a foot of water to be equal to an inch of mercury.

This instrument, so far as I know, is perfectly new; but the principle is so manifest, and the construction so easy, that it seems more than probable it must have suggested itself to many. There are no objections to it that I can see; if there be, I think they may be removed. I shall be happy to confer with you on the subject, and assist so far as I can.

WATER PASSAGE BETWEEN THE ATLANTIC AND PACIFIC ACCOMPLISHED.

In February, 1849, the *Times* directed public attention to the desirableness of a transit route from the Atlantic to the Pacific, by way of Lake Nicaragua. At that time we exercised an inoffensive control over the Mosquito territory, which commanded the entrance to the passage; our West Indian Colonies gave us an interest, as they still do, in everything connected with the region; we possessed already lines of steamers both on the Atlantic and Pacific side: our traffic from ocean to ocean, in the shape of specie and passengers, was so considerable as to call for our fullest activity with regard to everything that might facilitate it; and, above all, the privilege if necessary for the construction of the route in which Nicaragua, Costa Rica, and the Mosquito territory were interested, was instantly obtainable on such moderate terms as would have been consequent on the absence of competition. Nothing was done, however. A scheme based upon commercial considerations, and which to be appreciated required a comprehensive faith in the inevitable march of western civilization, had few attractions after a period of excitement when fortunes were to be made by anything rather than the exercise of judgment and a knowledge of the laws of progress; the enterprise, moreover, according to some, was not sufficiently exclusive, and would merely result in the loss of our capital for the benefit of the Americans, who were "too shrewd to undertake it themselves." A few months afterwards the announcement came, that the Americans, aroused to the value of the route, had lost no time in opening negotiations in Nicaragua, and had actually succeeded in concluding a contract. The cry was then raised that they were seeking to grasp Central America; that they would construct the canal and exclude our vessels from entering it upon equal terms, and that we must forthwith prevent their proceedings by contesting, in the name of the King of Mosquito, the right of Nicaragua to concede the route.

o this was the arrival of two
's from the Canal Company to
sh capitalists, upon little more
terms, a clear half of all the
ch they had acquired by their
and which were now held
guarantee of protection from
ents of the United States and
. The desire for participation,
again cooled as soon as it was
could be gratified. The pru-
ishmen in 1850 was enough to
or the entire history of 1846,
00,000,000*l.* had once been
moderate outlay to be incurred
r three years for local traffic,
xpenditure by England and the
combined, of 300,000*l.* per
elve years, to revolutionize the
globe, was something almost
for a congress of capitalists
e. A direct purchase, thereof the question, and it was
ly from a sense of patriotism
ame that would be incurred if
accepting the offer should be
us, that at the eleventh hour
t was entered into with the
ng its renewal at a fair price,
ny should ultimately succeed
ing the feasibility of their un-
October last this agreement
d in the *Times*, and Messrs.
l White, the Commissioners of
, returned to New York with
e to their colleagues that they
in their work without any
from this side until the results
would be no longer doubtful.
ate there have been almost
rtions of impending failure.
itish capital" their attempts,
would end in hopeless insol-
Rica, moreover, would con-
ry through which they would
ite, also, might still be made
quarrel. Other competing
would extinguish them. Their
t sent out (one of which is at
Lake) were "totally wreck-
ertion that they were forming
twelve miles of land from the
Pacific was alleged, on the au-
travellers, to be a delusion;
within the last fortnight we
tatement that their privileges,
n guaranteed by England and
been suddenly revoked by the
gua. A more remarkable his-
ngements was perhaps never
was certainly never terminated
ive or interesting way than by
one of the simple paragraphs in the *American News* of Monday morning:—"The passengers by the *Pacific* steamer arrived at San Juan del Sur (Nicaragua) with the California mails on the 29th of July, crossed the Isthmus in thirty-two hours, and arrived at New York after a passage from San Francisco of about twenty-nine days."

Thus, within thirty months of the time when the question of Nicaragua first became generally discussed, the American Company have entered into and secured a contract with that State, have gained the advantage of a protective treaty between Great Britain and the United States, have established a line of the fastest steamers both on the Atlantic and Pacific, have completed a survey which shows that the difficulties which for 300 years have frightened the world from attempting a junction of the oceans were absolutely fabulous, have carried 200 passengers in a few hours down a river which was represented as almost impracticable from shoals and rapids even for Indian canoes, have removed all the uncertainties and terrors that rendered the Isthmus the great stumbling-block of a safe and cheap passage to Australia, have brought California a week or ten days nearer to New York, and have secured for themselves the monopoly of a traffic which is the most marvellous that has ever been known, and the disposal of fertile lands and trading stations and natural docks that promise ultimately to receive the commerce of the world.—*Times.*

CANAL NAVIGATION BY STEAM.

Last year, on Captain G. F. Morrice, R.N., being appointed traffic manager to the Kennet and Avon Canal Company, his attention was drawn to the present expensive, tedious, and sluggish operation of towing the canal barges, which are usually 70 feet in length, 13½ feet wide, and when laden with 50 or 60 tons, draw from 3 feet 8 inches to 4 feet of water. Captain Morrice saw the advantages which would arise by the application of steam instead of horses to their propulsion, and after several experiments, which he made on the river Avon between Bristol and Bath, and upon a portion of the canal, he was satisfied that a speed double that of horses might be arrived at provided a steamer could be constructed of sufficient power which would not require too great a draught of water, and which would not by the commotion produced in the water destroy the banks of the canal. These matters were represented to the Kennet and Avon Committee, and they ordered a boat to be procured under the direction of Mr. Thomas E. Blackwell their engineer; the building, construction, design, and fitting of which were intrusted to Mr. John Jones, engineer, who is well known as the inventor and patentee of the "Cambrian" steam engine, and who was for years manager of the Great Western steam-yard, Bristol. Mr Jones at once set about building an iron steamer for the purpose, 47 feet long, 9 feet beam, and 5 feet depth of hold, fitted with the "Cambrian" engine and Griffiths' patent propeller. By their combined action all the difficulties hitherto experienced were expected to be overcome, and, as tested by the vessel's working, have been successfully so. The advantages obtained by the use, in combination of these patents, are—*First:* That by Mr. Jones's engine, the "Cambrian," (for description of this see *Mech. Mag.*, vol. xxxvii., p. 1,) any required speed is communicated to the propeller by direct action, without gearing or other secondary movement, and also, that while the propeller is revolving at the highest rate of speed, the steam piston is only required to travel at the ordinary rate, and that any sized crank can be applied, obviating the very great evil resulting from the use of short-stroke engines, in which a great portion of the power is exhausted in overcoming the friction and centre pressure which necessarily take place in them. *Second:* Griffiths' screw propeller is also quite a novelty, and entirely opposite in principle to the screw propellers hitherto in use; for although the screw, as an instrument for the propulsion of vessels, has occupied the attention of engineers for some years, yet comparatively little improvement has been made in it for the last ten years, the almost invariable opinion having been held that the best form of screw was that which offered the least resistance to the water in its centre, and with blades wide at the extreme ends and gradually diminishing towards the centre. Griffiths' propeller, in direct opposition to these principles, has its centre formed into a sphere, one-third or more of the entire diameter of the propeller, with the blades narrower at their extremities, gradually growing wider up to their junction with the sphere. With the ordinary screw the water is drawn through the central portion, and driven outwards with great velocity, at right angles, by the centrifugal action of the blades, consuming about 25 per cent. of the total power in destroying the effective action of the screw blades upon the water. In Griffiths' patent, on the contrary, the sphere of about one-third the total diameter causes the water to come in the right direction on the widest and most effective portion of the bl where they lay hold of and drive it aw a direct line with the vessel's cours which means all commotion of the is prevented—an invaluable result for navigation. Captain Morris, Naval S tary, reports that, upon trial, the perf ance of the boat in power and speed answered every expectation, having quently towed from Bristol to Bath (18 m in three hours forty-five minutes, but n cluding the time lost in passing throug locks—about eight to twelve minutes; in towing a barge from Bath to Bristo distance has been accomplished in two h forty-five minutes, not including the pa through the locks. The usual time for h to perform the journey up being from si twelve hours, and down about seven h The boat was then tried from below Ra to London, and, where free from weeds, t upwards of four miles an hour; and passing Richmond, where the water wa 7 feet deep, a speed of six miles per hou obtained, and, on getting past Kew-b she went over the ground at a rate of miles an hour, with a little ebb tide i favour. A barge was afterwards towe her from London up the Thames to St bridge, with the tide as far as Richm and a speed of eight miles an hour wa tained, and from Richmond to Stain distance of twenty-two miles, at a sp three and three-quarters to four mil hour—and here the superiority of t by steam over horses was apparent, barge with four horses hardly made than one mile and a half per hour. A of her speed without towing was made, although she is only a barge, her spee equal to the Westminster-bridge river st the *Blue Bell*, running along side of her Westminster to London-bridge at the of 10 miles an hour. Captain Morris that the expense of working these tug be much less than that of the horses req to perform the same work, while much than double the speed will be obtained that, too, without the very heavy exp attendant upon the wear and tear of and ropes; they will also be worked perfect ease to the bargemen.

THE INTRODUCTION OF STEAM NAVIGATION.

Sir,—I felt somewhat surprised at American contemporary setting forth (i *Mech. Mag.*, August 23) Mr. Fulton originator of steam navigation. Now you have aided not a little in endeavo to establish the right of the late Mr.

n, of Falkirk, to that invention; ık I am right in saying, a model mington's first steam-boat brought vas submitted to your inspection weeks ago, and is now to be seen oyal Polytechnic Institution, Re-t. This boat was constructed and at Grangemouth, in 1801. It sals on the Forth and Clyde Canal, he Forth and Carron rivers, in 2, and 1803. It also carried the engineer, Mr. Robt. Fulton, eight g the canal in an hour and twenty luring which trip he was allowed etches of the boat and her machi-l received answers to his various respecting these. Indeed, I have n what Mr. Fulton really *invented* benefited the world;—certainly, entitled to the honour of having he steam-boat.

ertion of this will much oblige

Your obedient Servant,

THOS. HOBSON.

rch-street, Bermondsey,
ugust 29, 1851.

orrespondent ascribes a meaning mmissioner Ewbank which our ex-s him (p. 146) hardly warrants. rely says, that "at a more favour-than Fitch lived in, Fulton rose, ers began to creep up rivers," &c. gh Mr. Symington was undeniably demonstrate the practicability of vessels by steam—as we have, wer, been at some pains to show less certain, that Fulton did even Symington to bring steam navi-general use; and there is nothing vbank's language but what is quite with this state of things.—ED.

—•—

AL-MATCH BETWEEN THE NEW YACHT "AMERICA," AND THE H YACHT "TITANIA."

victory achieved in this case by Jonathan has been represented *pectator* and other respectable as furnishing a practical refuta-he wave-line principle of ship-but, according to the statement elligent eye-witness, which ap-he *Times* of the 2nd instant, it ing of the sort—BOTH *vessels fact, built upon that principle.* t of the theory of construction that of best carrying it out is it belongs to our transatlantic We subjoin the greater part ter referred to.—ED. M. M.]

The object proposed by the Yacht Club was understood to have been the determination of the relative values of the English and American forms of yachts in strong winds and a rough sea. For this purpose it was stated that the yachts would not be started unless there should be a strong breeze.

There was not, however, so much wind as could have been desired on Thursday morning; nevertheless, at half-past eight, the orders of the Earl of Wilton, the commodore, were sent on board, "that the vessels should be alongside the *Xarifa* at ten, near the Nab Light," and should start before the wind, to run twenty miles out to sea, rounding a steamer sent out for the purpose, and then return, beating up against the wind, to the Nab Light.

Both vessels had been put into dock previous to the race, and many and curious were the examinations made of their bottoms. They resemble each other under water much more than would be supposed from their different aspects above. They are both built on the wave-line principle; they both have great depth of keel; they have the same difference of draught fore and aft, and they both have the gripe entirely rounded away, or, in other words, the keel rounds up in a quarter circle, nearly into the stem at the water-line. At the surface of the water their water-lines are similar, but the transverse sections are very different—the *America* having a wedge bottom, like the vessels of Sir William Symonds, which project out above the water beyond the water-line, while the *Titania* has the straight up and down side, such as is seen in the vessels of the School of Naval Architecture. Above water the vessels are entirely different. The *Titania* has the raking stern of English yachts; the *America* is cut off close to the sternpost. At the bow both show the hollow wave-line, but the *America* carries the hollow up into the harping, while the *Titania* presents the usual aspect on deck of a sharp-bowed yacht.

The main and essential differences, however, between the vessels are their dimensions, tonnage, and cut of the sails. The *America* is 22½ feet beam, the *Titania* 18; the *America* has a keel over 90 feet long, and the *Titania* under 70; hence, the *America* is some 210 tons by our measurement, and the *Titania* 100 tons. To estimate the value of such difference, we have only to refer to the scale of time for tonnage, to see that in an English yacht race the time allowed for difference of tonnage would be 55 minutes.

It was, however, in the cut of the sails that the difference between the vessels was most striking to the eye as they got under weigh to start. The general arrangement of the sails is not very different, nor the rake of the masts. The *Titania's* sails, cut by Eversfield, of Gravesend, are the best English cut, beautifully gored and curved; the *America's* are simply "flat as a board;" and, during the race, they remained so! The value of this fact, however the object is effected, is most important, and will appear from the sequel of the match. Both yachts had arrived at the Nab Light by a quarter past 11 o'clock, at 11h. 19m. 15s. the signal-gun was fired from the *Xarifa*, the Blue Peter was hauled down, and both vessels started, the *America* taking the lead. The course was S.E., and right before the wind, which was a fresh breeze. There was a rolling sea, but by no means heavy. As the wind freshened the *Titania* gained upon the *America*, and as it fell the *America* gained upon her. For half an hour they alternately gained and lost upon each other, and the issue appeared very doubtful until the wind fell a good deal more, when the *America* gradually gained upon her rival, and at last took her station permanently ahead. They rounded the steamer as follows:

1. America	at	2h. 3m. 0s.
2. Titania	at	2h. 7m. 6s.
Difference in favour of America		0h. 4m. 6s.

Thus the American gained the first half of the match, running free, by a very small difference; and this fact is very important to naval architecture, as showing that in point of resistance to going through the water, or in adaptation of the forms of the vessels to speed simply, the advantages were nearly equal. The second part of the match, in beating up against the wind, was of a totally different character. It was evident to every qualified spectator that the American must win. The advantage obtained by the American cut of the sails was at once evident to the eye. The American was able to steer one point of the compass nearer to the wind than the *Titania* without the least flutter of her canvas. Although, therefore, the *Titania* appeared to hold her own very fairly in point of distance run, and although when the wind freshened she appeared to be gaining in speed over her competitor, yet the distance the American would be able to make good to windward over her adversary on each tack by laying her course 11¼ degrees nearer the wind was such as to put competition beyond a doubt. During the race the wind freshened and the sea rose, but both yachts carried all their lower sails, without taking in a reef, and both proved themselves wholesome and easy sea vessels. They passed the Nab to the S.E. as follows:—

1. America at	5h. 31m.
2. Titania at	6h. 22m.
Difference in favour of America	0h. 51m.

If from this difference on the whole match we subtract the difference on the first half we find the America to have gained 46 minutes 54 seconds in beating to windward—a most important advantage, and one which shows clearly how valuable an art the better cutting and management of the sails is for the purpose of giving windwardly qualities to ships.

If any thing could add strength to the importance of the practical conclusions which this important experiment suggests, it is the circumstance that in beating home to Cowes from the Nab Light both vessels either passed or gained upon every other yacht under sail by a long distance. This fact shows how much may be gained by the study of the forms of two vessels in some respects so much alike, and in others so dissimilar. The public are certainly indebted to the owners of both yachts for giving them the advantage of so interesting an experiment. It is much to be desired that advantage should be taken of the experiment to advance the interests of scientific shipbuilding and of practical seamanship; and it is for the advancement of those objects that these notes are offered for your acceptance.

THE WETTEST PLACE IN GREAT BRITAIN.

Hitherto the hamlet of Seathwaite, in Borrowdale, has been generally considered the wettest spot in Great Britain; but there is a place at a little distance from it, where the average fall of rain is found to be greater by nearly one-third. We extract the following account of this discovery from the remarks appended to the "Synopsis of the Fall of Rain in the English Lake District and on the surrounding Mountains, in the Year 1850," supplied by Mr. J. F. Miller, of the Whitehaven Observatory, for the information of his meteorological friends:

"The new station is about a mile and a half distant from Seathwaite in a south-westerly direction, and 580 feet above it, or 948 feet above the sea level, at the extreme southern termination of the valley; it is on the shoulder of Sprinkling Fell, or the Stye, about 100 yards south of the road leading over Stye Head to Wastdale. The actual quantity of water measured on Sprinkling Fell in 11 months of 1850, is 174·33 inches; but the receiver was found running over on four different occasions, by which I calculate 5 or 6 inches, at least, must have been lost to the instrument; hence, if we add

5·67 inches for overflow, and 9·49 inches for the computed fall in January (7·34 inches at Seathwaite), the result is 189·49 inches for the fall on the Stye in 1850, with 143·96 inches at Seathwaite.*

"The wettest year since the commencement of the experiments in 1848, when 160·89 inches fell at Seathwaite, and computing the fall at the new station for that year in the same proportion which the two localities bear to each other in 1850, we have 211·62 inches for the depth of rain on the Stye in 1848. An inspection of the tables kept at the coast during the last eighteen years, shows that the period (1844—1850) over which the lake district gauges have been in operation, has been far from a wet one. At Whitehaven, the average fall from 1844-50 inclusive, is 43·543 inches; but in the eleven years preceding 1844, the average is 48·53 inches; and the average of the last eighteen years is 46 58 inches. And if we analyse the period of seven years comprehended between 1844-50, we find that only three of those years have exceeded the full average; whilst of the remaining four, one year has been characterised by drought, and the other three by unusual dryness. Even in the year 1848, when 161 inches fell at Seathwaite, the depth at Whitehaven was only 47·34 inches, or three-quarters of an inch above the average of eighteen years; but in 1835, the fall was 54·13; in 1846, 58·97 inches; and in 1841, 55·97 inches. Hence, the maximum annual depth in the mountain district of Cumberland may far exceed the computed fall of 211 inches at the Stye in 1848, enormous — almost incredible — as is the quantity for a climate situated in the heart of the temperate zone."

* *A new and capacious gauge* was planted on *the Stye early in January, 1851.*

SPECIFICATIONS OF ENGLISH PATENTS ENROLLED DURING THE WEEK, ENDING SEPTEMBER 4, 1851.

CHARLES FREDERICK BIELEFIELD, of Wellington-street North, Strand, papier-maché maker. *For improvements in the manufacture of sheets of papier-maché or substances in the nature thereof.* Patent dated February 24, 1851.

The methods usually adopted when manufacturing sheets of papier-maché have been to paste together a number of sheets of paper until the requisite thickness is obtained, or to take a number of sheets of pulp from the frame, and laying them one on the other, to compress the whole between felt, for the purpose of expressing the water; and sometimes, when the sheets have not been required of very great thickness, the pulp has been run direct into frames of the required depth. Now the present improvements consist in making such sheets by rolling or compressing fibrous and other materials, ground to a putty-like consistence, into frames of the required depth, in a manner similar to what has been hitherto practised when forming moulded articles of papier-maché, and then treating such sheets with heat and oil, as has been formerly practised in treatment of articles made of papier-maché. The sheets of papier-maché, when so made, may be painted and varnished, or japanned in the ordinary manner; or they may have colouring matter introduced among the materials employed in producing them. The patentee describes a method of making a paste with fibrous and other materials, to be employed in manufacturing sheets according to his invention, which, however, he does not claim, as others may be adopted if preferred. To make this paste, he takes 80 lbs. water, 32 lbs. flour, 9 lbs. alum; 1 lb. copper; and to these, when mixed, he adds 15 lbs. resin, dissolved in 10 lbs. boiled oil, and 1 lb. of litharge; he subsequently grinds the mixture up with 55 lbs. to 60 lbs. rag-dust (for which paper-makers' half-stuff may be substituted), and thus obtains a paste of the consistence of putty, suitable for rolling or pressing into sheets as above mentioned.

Claim.—The mode described of manufacturing sheets of papier-maché and substances in the nature thereof.

JOHN HINKS, of Birmingham, manufacturer, and JAMES VERO, of Burbage, manufacturer. *For certain improvements in the manufacture of hats, caps, bonnets, and other coverings for the head.* Patent dated February 24, 1851.

Claims.—1. The waterproofing of felt hats, caps, bonnets, and other coverings for the head, without altering the appearance or diminishing the flexibility of the same, by applying thereto a solution either of caoutchouc or India-rubber, or of gutta percha, or both.

2. The waterproofing of the bodies of hats, caps, bonnets, and other coverings for the head made of felt, to be covered with velvet, silk, or other plushes or fabrics, so as to preserve the flexibility of the same, by treating them with a solution of caoutchouc or gutta percha, or both in combination.

3. The waterproofing of japanned and painted felt hats, caps, bonnets, and other coverings for the head, without destroying the flexibility of the same, by treating them with a composition of whiting and lamp-black mixed in water, with the addition of boiled linseed oil and solution of India-rub-

ber in oil of turpentine, and subsequently varnishing them with a composition of asphalte and resin, or pitch, dissolved in boiled linseed oil, and mixed with a solution of India-rubber in oil of turpentine.

4. The covering of hats, caps, bonnets, and other similar articles, by attaching thereto knitted coverings of silk or other suitable material, stretched so as to envelope the article without requiring a seam to be made.

5. The making of the bodies of hats, caps, bonnets, and other coverings for the head, of knitted fabrics treated with a flexible waterproof composition.

Gabriel Didier Fevre, of Paris, gentleman. *For certain improvements in apparatus for manufacturing and containing soda water and other gaseous liquids, and also in preserving other substances from evaporation.* Patent dated February 24, 1851.

The apparatus here specified possesses, to all appearance, no particular feature of novelty; the patentee, however, lays great stress on the fact that the aërating powders which are contained in a compartment of the vessel used for holding the liquid operated on are kept from contact with that liquid, and wetted only sufficiently to produce the gas necessary for impregnating the liquid contained in the vessel. He also particularly claims the use of water for wetting the aërating powders, even though the liquid to be aërated, should happen to be wine or any other liquor. As for the "improvements in preserving substances from evaporation," they may be perhaps intended to be exemplified by some peculiar construction of stoppers which M. Fevre specifies, but they are certainly by no means so "particularly described or ascertained" as to fix their identity.

Samuel Cunliffe Lister, of Manningham, near Bradford. *For improvements in preparing and combing wool, and other fibrous materials.* Patent dated February 24, 1851.

1. The improvements in preparing have relation to the preparing of long wool, and consist in placing a comb in front of a screw gill, porcupine roller or ordinary rollers, so that as the wool or hair is brought forward by the gill or rollers the points or ends of the wool may be combed, worked, and opened out by the comb placed in front of the gill, the wool being at the same time held fast and drawn into a sliver by a pair of rollers placed about three inches from the front of the gill for coarse wool and one inch for fine wool, but this distance may be varied. The gill should have porcupine feed-rollers, and also be heated, which may be conveniently done by a fire-box placed under the gill-comb in such a position that the heat may ascend among the teeth of the comb. The comb which works the wool between the gill-comb and the drawing-rollers has motion imparted to it by a crank, which is preferred to other means which might be adopted on account of the high speed at which it can be worked. Supposing the gill to run at the rate of 125 fallers (gill-combs) a minute, the crank motion should cause that comb to make 700 strokes during the same time, but this number may be varied, fine wool requiring a greater number of strokes than coarse.

2. When combing wool by hand, the workman lashes on the wool to the comb, and it is retained thereon by the fibres of the wool looping around the teeth of the combs, and this is generally also the case in machine combing, although the method of lashing on the wool is sometimes reversed, being in this case effected by the comb approaching to take the wool from the feed-roller, and then retiring to cause the wool to be retained on the teeth. According to this system, when the wool is detached from the combs, the loops prevent its being readily removed, and in consequence more noil is produced owing to the breakage of the fibres. Mr. Lister's improvement consists of a peculiar arrangement of screw-gill or other suitable means by which he is enabled to lash in the wool so as to preserve the fibres straight and without looping, by which means they may be readily detached from the comb without producing noil in so doing.

This part of the invention also refers to the machine for combing wool patented by Mr. Josué Heilmann in 1845, and consists in applying heat to such machine during the operation of combing and in making the teeth which first comb the wool as fine as the back rows, which has not been hitherto done. "The inventor of the machine," says Mr. Lister, "appeared to think that heat was objectionable, but, on the contrary, I have found that it is very beneficial in preserving both the wool and the teeth of the comb from injury, for if the teeth are broken, it is evident that the fibres must be injured, and it is well known to all experienced combers that heat causes the teeth to pass through the wool with greater facility, especially if the wool be worked in a moist or damp state." The fire-box employed for the purpose is placed in such a position that the heat ascends and warms the teeth as soon as the noil is removed, and before they begin to comb the wool. Mr. Lister further observes that in Heilmann's machine the combs are subject to be broken and injured, and in consequence they have been usually made with the teeth stronger and coarser in front; that is, those which first begin to comb the wool. The injury to the combs arises from two causes, the want of heat and the wool not being properly prepared. Now, when the combs are heated, and the wool previously well carded, the teeth which first begin to comb the wool may be made as fine, or finer than the back rows, especially for operating on carded wool on account of the knots being small and difficult to extract.

Another improvement consists in carding on a common carding engine the noil or short silk produced in combing that which is long; and, also the tow or noil made in heckling or combing long flax, and afterwards combing it either on the combing machine described, or on any other description of machine which may be found suitable for the purpose. The chief thing to be observed being that the fibres are preserved in the carding, and then that the combs shall be sufficiently fine to take out the knots; say, for instance, not less in number than 24 to the lineal inch, which number may be increased to 40 when operating on silk noil. Mr. Lister does not claim for combing, or carding and combing, long silk or flax, but only the refuse or noil which has not been hitherto so treated. Cotton, also, after being carded may be combed with great advantage upon the machine before-mentioned, especially for spinning to high numbers.

The other improvements specified, which have relation to Preller's and other similar machines in which moveable combs are employed; and to a machine of Ramsbotham's, in which the wool is fed on to a circular comb and worked by a screw gill, consist in the application of a greater number of teeth to the combs employed therein than is usually the case, by which better top and more work is produced, and the knots are effectually removed, more especially so when the wool has been previously opened by carding.

Alpaca, mohair, goats' wool, and course Russian wool, previously carded, Mr. Lister combs in machines having a greater number of teeth than usual, whether the machines used are Preller's or those of any other description, and he claims their application to these purposes.

Amedée François Remond, of Birmingham, gentleman. *For improvements in the manufacture of metallic tubes or pipes, and the machinery or apparatus connected therewith, which improvements are applicable to other like purposes.* (A communication.) Patent dated February 26, 1851.

These improvements consist in manufacturing tubes from sheets of metal, which are first stamped by successive operations into the form of a tube closed at one end, and finally passed between dies or rollers until sufficiently elongated. The main features of the improvements will be evident from the claims, which are for

1. Converting metals into a tubular or other similar form, with one end closed, by means of stampers and dies, and subsequently elongating the same into pipes or tubes.

2. The elongating of hollow pieces of metal, closed at one end, into tubes or pipes, by steam or hydraulic machinery acting directly on such closed end, and forcing the metal between dies or rollers—the mandril on which the tube is formed being connected to the piston of the said machinery.

3. The manufacture of metal tubes by means of grooved rollers, traversed and caused to move tangentially at a speed corresponding to that of the mandril.

Thomas Ellis the elder, of Tredegar Iron-works, Monmouth, engineer. *For certain improvements in machinery or apparatus to be employed in the manufacture of blooms or piles for railway and other bars or plates of iron.* Patent dated February 27, 1851.

According to the methods of rolling blooms at present practised, the bloom, after having passed through the machine, has to be raised to be introduced between the rolls a second time, in order that it may be again operated on. Mr. Ellis's improvements consist in causing the rolls to have a motion first in one direction and then in the other, so that the bloom of iron, after being once drawn through the rolls may be returned through them in the opposite direction; and this operation may be repeated until the bloom shall have been sufficiently rolled. For this purpose, the ends of the rolls are furnished with pinions, which are geared into and actuated by racks, which are connected with a crank by a suitable connecting rod in such manner that the revolution of the crank may cause the racks to be moved backwards and forwards, and, through them, the rollers to revolve alternately in opposite directions. The blooms, after having been sufficiently rolled, may be manufactured in the ordinary manner into bars or plates for railway and other purposes. The same principle of causing the rolls to move alternately in opposite directions may be also applied to the rolling of bars and plates of iron, care being taken that the throw of the crank, the length of the rack, and the size of the pinion shall be so regulated as to cause the roll to pass over a space of greater length than that of the metal operated on.

Claims.—1. The application of a crank and rack for giving motion to rolls first in one direction and then in another, when rolling blooms or piles of iron.

2. A rack and pinion working in the manner described, for the purpose of giving motion to rolls for manufacturing blooms or piles of iron.

HENRY WILLIS, of Manchester-street, organ-builder. *For improvements in the construction of organs.* Patent dated February 28, 1851.

The improvements claimed under this patent comprehend—

1. The construction of valves (especially for the large pipes in pedal organs) in such manner that the resistance of the air, when opening such valves, may be overcome gradually, instead of by a sudden impulse, which is the action of ordinary valves. [For this purpose, Mr. Smith employs elastic valves composed of leather or India-rubber, attached to the seat at one side, and opened by a rolling motion. Or, he makes use of a valve of a thin curved plate of brass, which fits on its seat in the same manner as the valve of a trumpet fits the reed—in both cases adapting suitable springs to cause the return of the valve on the removal of the pressure from the keys or pedals.]

2. A method of supplying the bellows of the swell organ exclusively with air drawn from the swell-box. [The bellows which supply the swell organ with wind are placed in the swell-box itself, so that the air, as it is expelled from the pipes, is drawn into the bellows and again used. The effect produced by this novel application of the bellows, is that the compression of air in the swell box, when the front is closed and the pipes of the swell organ are sounding, and consequently emitting air from the external bellows, is entirely prevented, and the liability of the pipes to sound out of tune, on account of the vibration being checked, is thus avoided. In addition, a more subdued tone is obtainable than by the plan ordinarily in use, there being no tendency of the air to force a passage through the crevices of the swell box.]

3. The employment of threaded centering for the bellows levers, or for any analogous purpose in organs. [Instead of having a cylindrical fulcrum-pin for the bellows-levers, and affixing the same in the levers by means of collars against the sides of the lever, Mr. Willis employs, as above-mentioned, a fulcrum-pin with a coarse thread cut on it, which he screws into the lever, thus preventing the liability of the lever to shift on its fulcrum, and causes it to work steadily and without noise or creaking.]

4. A method of bringing the stop movements within reach of the performer's hands, so that by the use of manual power he may, without raising his hands from the key-board, be enabled to draw or shut off any stop or "composition" of stops that may be required. [This is effected by the employment of what are known in the trade as "pneumatic levers," the motive-power for actuating the stops being obtained from the compressed air contained in the bellows.]

5. The application to pneumatic levers of an improved construction of escape valve whereby the repeating power of the lever is increased, and the touch of the organ improved. Also, a method of arranging and combining the pneumatic levers for giving motion to the stop-action.

CHARLES FELTON KIRKMAN, of Argyle-street, gentleman. *For certain improvements in machinery for spinning and twisting cotton, wool, or other fibrous substances.* Patent dated February 28, 1851.

Claims.—1. A peculiar combination of arrangements whereby a permanent twist may be put into rovings or yarns, and the same wound into bobbins without the twist being removed, both ends of the yarn or roving being fastened, and neither end being allowed to turn round.

2. The employment in drawing, spinning, and twisting machinery of rotary drawing and twisting frames, consisting of sets of rollers which, in addition to their revolution on their axes, have a second motion communicated to them by causing the frames in which they are mounted to revolve independently of the motion of the rollers, by which means the roving is simultaneously drawn out, and spun or twisted by the rotary motion of the drawing frame.

3. A method of driving or actuating spindles by means of a screw or worm acted on by a wheel. Also, the application of this method of actuating spindles or shafts for driving any parts of drawing, spinning, twisting, and winding machinery where a great speed combined with an even and regular motion is required. [In place of bands and pulleys or ordinary toothed gearing, Mr. Kirkman mounts on the spindle a worm with a sharp short pitch, and drives this worm by means of a wheel with teeth made to correspond with the thread of the worm. When several spindles are arranged side by side in a frame, the toothed wheels employed to act on the worms of the spindles should be mounted on a common shaft at the back of the range of spindles.]

4. A method of drawing, spinning, and twisting yarns and converting them in the same machine into cords or threads.

WILLIAM MILLWARD, of Birmingham, plater. *For certain improvements in electro-magnetic and magneto-electric apparatus.* Patent dated February 28, 1851.

This invention consists; *firstly*, of an improved method of charging or magnetising iron and steel bars to be used as permanent magnets or electro-magnets; and, *secondly*, of certain new *forms* of electro-magnetic machines.

The first branch of the improvements is carried into effect by the employment of an electro-magnet formed by a current of electricity produced from a magneto-electric machine, instead of that generated in a voltaic battery; and such an electro-magnet may be very advantageously used for magnetizing large bars of steel, or for producing very powerful magnets. Any of the known forms of magneto-electric machines will serve thus to convert a bar of steel to an electro-magnet, but the patentee prefers to use one composed of four, eight, or any other number of permanent magnets, having double the number of armatures, and coiled with strong wire of about 60 feet in length. The machine about to be described has been found to answer well in practice. In this machine, the steel magnets are composed of eight plates of a U form, weighing about 30 lbs. each plate, and there are eight such compound magnets, all the north poles of which are arranged on one side of the machine and the south poles on the other side, although this precise arrangement is not essential, and may be varied. The armatures are of soft iron, weighing about 15 lbs., and are coiled with about 60 feet of copper wire of No. 4 gauge, and insulated in the usual manner. The armatures revolve in a brass wheel, and are caused to pass as near to the poles of the magnets as practicable, the commutator or break acting on the whole eight magnets at the same instant, so that the current of electricity shall always pass in one direction, and the surfaces of the whole of the 64 plates be in combination at the same time. The bar of soft iron used as the electro-magnet with this machine weighs about 500 lbs., and is coiled with bundles of about 30 copper wires of No. 16 gauge, and about 60 feet in length (the bundles are formed by binding a series of uncovered wires together into one covered strand or bundle), and the power of the electro-magnet will depend upon the power of the permanent magnets used in the machine, both as to the weight it will support from a keeper, and as to its capability of rendering bars of steel permanently magnetic by contact therewith. It will therefore be evident that by having two sets of the permanent magnets, and changing them in such machine, their supporting power may be increased by continued charges or passes from the electro-magnet thus produced.

In one form of electro-magnetic machine represented and described under the second head of the invention, the steel bars or permanent magnets are eight in number (these bars may be of cast or soft iron, but when soft iron is employed, bars of steel permanently magnetised will have to be used in conjunction with them) of a U form, and arranged around a circle with their poles pointing towards the centre. Each arm of each of the magnets has attached to it straight bars of steel, also rendered permanently magnetic (of which any desired number, and of any length or size, may be employed according to the strength of magnet required), which are so placed as to be out of the influence of the armatures when the latter are revolving. The poles of the U-shaped magnets are, on the contrary, as nearly as possible in contact with the armatures which revolve within the circle formed by them, either between the poles or in front of them. Instead of the bars which form the circle being of steel and magnetised, they may be made of soft iron, and depend for their magnetism upon the magnetic bars before named placed around them.

In another form of machine both the magnets and armatures are stationary, and the commutator alone has motion between the poles of the horseshoe magnets and the armatures being mounted on a spindle and caused to revolve by a band from some driving machinery. The commutator or break-piece, is composed of a brass centre with four radial arms of soft iron either solid or formed of two or more plates.

LOCK CONTROVERSY.—REPORT.

"Whereas for many years past a padlock has been exhibited in the window of the Messrs. Bramah's shop in Piccadilly, to which was appended a label with these words—'The artist who can make an instrument that will pick or open this lock will receive 200 guineas the moment it is produced;' and Mr. Hobbs, of America, having obtained permission from the Messrs. Bramah to make a trial of his skill in opening the said lock, Messrs. Bramah and Mr. Hobbs severally agreed that Mr. George Rennie, F R.S., London, and Professor Cowper, of King's College, London, and Dr. Black, of Kentucky, should be the arbitrators between the said parties, that the trial should be conducted according to the rules laid down by the arbitrators, and the award of 200 guineas decided by them on undertaking that they should see fair play between the parties. On the 23rd of July it was agreed that the lock should be enclosed in a block of wood and screwed to a door, and the screws sealed, the key-hole and hasp only being accessible to Mr. Hobbs; and when he was not operating, the key-hole to be covered with a band of iron and sealed by Mr. Hobbs; that no other person should have access to the key-hole. The key was also sealed up, and not to be used till Mr Hobbs had finished his operations. If Mr. Hobbs succeeded in picking or opening the lock, the key was to be tried, and if it locked and unlocked the padlock it should be considered a proof that Mr. Hobbs had not injured the lock, but picked and opened it, and was entitled to the 200 guineas. On the same day, July 23, Messrs. Bramah gave notice to Mr. Hobbs that the lock was ready for his operations. On July 24, Mr. Hobbs commenced his operations; and on August 23, Mr. Hobbs exhibited the lock open to Dr. Black and Professor Cowper. Mr. Rennie being out of town, Dr. Black and Professor Cowper then called in Mr. Edward Bramah and Mr. Bazalgette, and showed them the lock open. They then withdrew, and Mr. Hobbs locked and unlocked the padlock in the presence of Dr. Black and Professor Cowper. Between July 24 and August 23, Mr. Hobbs's operations were for a time suspended, so that the number of days occupied by him were sixteen, and the number of hours spent by him in the room with the lock was fifty-one. On Friday, August 29, Mr. Hobbs again locked and unlocked the padlock in the presence of Mr. George Rennie, Professor Cowper, Dr. Black, Mr. Edward Bramah, Mr. Bazalgette, and Mr. Abrahart. On Saturday, Aug key was tried, and the padlock was lock locked with the key by Professor Cowper nie, and Mr. Gilbertson, thus proving Hobbs had fairly opened the lock withou it. Mr. Hobbs then formally produced ments with which he had opened the loc

"We are, therefore, unanimously of o Messrs. Bramah have given Mr. Hobbs portunity of trying his skill, and that has fairly picked or opened the lock, and that Messrs. Bramah and Co. do now Hobbs the 200 guineas.

"Sept. 2, 1851. "GEORGE RENNIE, Ch
"EDWARD COWPER.
"G. R. BLACK."

[Mr. Hobbs claims to have picked a Chubb's unpickable, but this Mr. Chubb

WEEKLY LIST OF NEW ENGLISH PATENTS.

John Wallace Duncan, of Grove-end-road, St. John's-wood, gentleman, for improvements in engines for applying the power of steam or other fluids for impelling purposes, and in the manufacture of appliances for transmitting motion. September 4; six months.

Henry Alfred Jowett, of Sawley, Derby, engineer, and John Kirkham, of Peckham, Surrey, engineer, for improvements in hydraulic telegraphs and in making signals. September 4; six months.

John Poad Drake, of St. Austell, Cornwall, for improvements in constructing ships and other vessels, and in propelling ships or other vessels. September 4; six months.

Dominique Julian, of Sorques, France, for improvements in extracting the colouring properties of madder, and in rendering useful the water employed in such processes. September 4; six months.

Baron Charles Wetterstedt, of Grosve Commercial-road, for improvements in animal and vegetable substances. Sep six months.

William Imray, of Milton-road, Live improvements in the manufacture of bri tember 4; six months.

Timothy Kenrick, of Edgbaston, War founder, for improvements in the manu wrought-iron tubes. September 4; six

Benjamin Hallewell, of Leeds, wine for improvements in drying malt. Sep six months.

Pierre Armand Lecomte Fontainen 4, South-street, Finsbury, for certain imp in preserving animal substances from means of a composition applicable to th certain diseases. September 4; six mon

WEEKLY LIST OF DESIGNS FOR ARTICLES OF UTILITY REGISTERED.

Date of Registration.		No. in the Register.	Proprietors' Names.	Addresses.	Subjects of Des
Aug.	29	2926	William Dray	Arthur-street and Swan-lane	Turn-rest plough.
	,,	2927	W. Hibbert	Manchester	Hat.
	,,	2928	H. Bowser	Finsbury-pavement	Collar.
Sep.	1	2929	G. Beattie	Edinburgh	Brick.
	,,	2930	Somervell, Brothers	Kendal	Improved spring fo
	,,	2931	T. T. Read	Hull	Improved capstan.
	2	2932	S. White	Manchester	Improved gas-retor
	3	2933	I. G. Reynolds	Bristol	"The Februa," or f
	,,	2934	G. Boswell	Rickmansworth	Ventilating chimne

CONTENTS OF THIS NUMBER.

LONDON: Edited, Printed, and Published by Joseph Clinton Robertson, of No. 166, Fle in the City of London—Sold by A. and W. Galignani, Rue Vivienne, Paris; Machin Dublin; W. C. Campbell and Co., Hamburg.

Mechanics' Magazine,

MUSEUM, REGISTER, JOURNAL, AND GAZETTE.

No. 1466.] SATURDAY, SEPTEMBER 13, 1851. [Price 3*d*., Stamped, 4*d*.

Edited by J. C. Robertson, 166, Fleet-street.

IMPROVED TUBULAR STOVE.

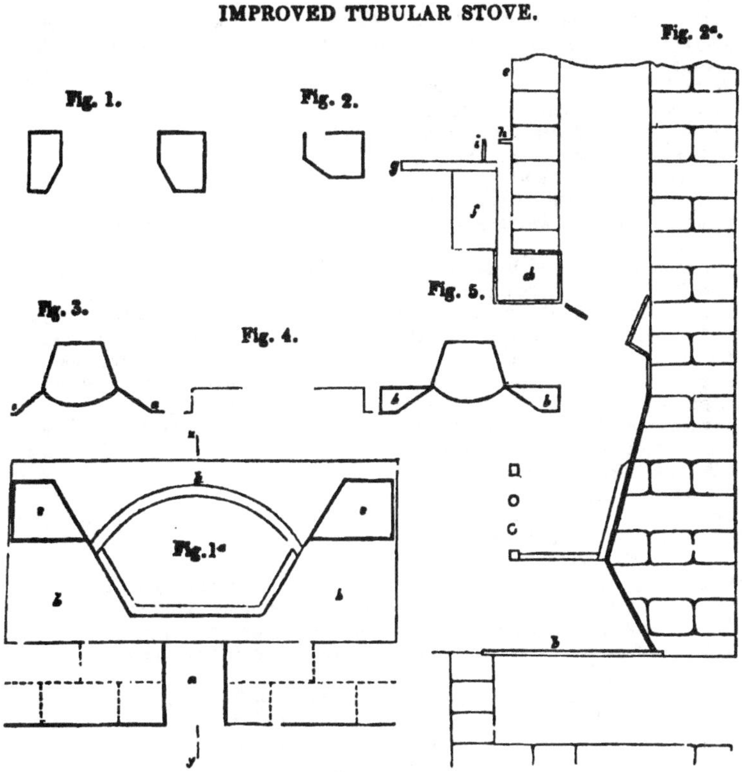

Fig. 3ᵃ.

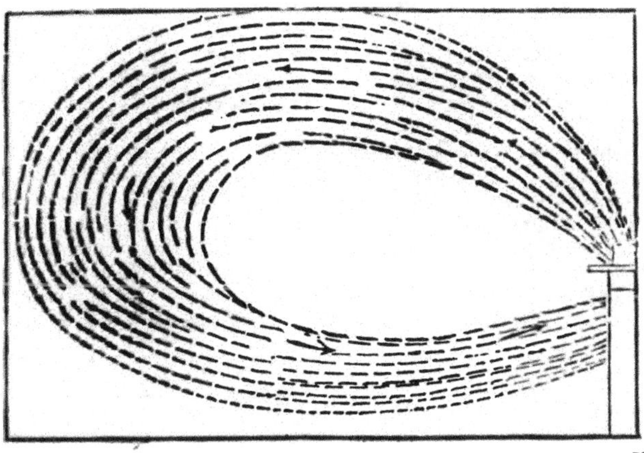

ON WARMING AND VENTILATING.—WITH A DESCRIPTION OF AN IMPROVED TUBULAR STOVE.

The present age is honourably distinguished by the zealous efforts which are being made for the promotion of measures likely to improve the health and welfare of the masses.

Parliament is at present considering the means of insuring an abundant and cheap supply of water to our large towns; and the sewerage of the metropolis, though a few years ago the boast of this country, is proved to be very defective; while the principles which should govern the construction of sewers are only now being determined with a view to an improved system.

While previous deficiencies are thus supplied, and previous errors corrected, so that we have reason to hope for uncontaminated air around our houses, and pure water in abundance, we have yet to secure WITHIN our homes a copious supply of pure air, especially during that large portion of the year, when a low temperature necessitates artificial warmth, and precludes an unregulated admission of the external air.

It has been remarked by a high authority, that "the standard for pure air is generally low, in consequence of the comparatively indifferent condition of the state of the atmosphere in the greater number of habitations." The wealthy are accustomed so to vary the occupation of their rooms, as to experience but little inconvenience and disturbance of health from want of purity in this prime necessary of life, but very different is the case of those forming that numerous class whose limited means confine them to the use of one or two rooms.

Invalids, unable to take walking or riding exercise, derive great benefit from being out in the air; that they do so is attributable to their breathing a pure atmosphere. On the other hand, there are numbers of persons with wealth to purchase, and *general* health to enjoy those pleasures which require the congregation of numbers under one roof, who are obliged to deny themselves such pleasures in consequence of the suffering induced by the contaminated air which is their invariable accompaniment. The *heat* of crowded rooms is by many thought to be the cause of their painful sensations; but it is known that a temperature much higher than that prevailing in such assemblages can be borne without prejudice to the health.

The statements of various writers on the subject, and the constant succession of new forms of stoves,* abundantly evidence the interest felt on the subject of ventilation, and the difficulties to be overcome.

Amidst all the improvements which have been made in our modern buildings, there would seem to be no advances in this direction; the most splendid rows of modern town mansions are deformed by grotesque groups of zinc piping and cowls, which are found necessary to remedy the defectiveness of the chimneys, as constructed. There have been of late years numerous works published, which treat at length on the difficulties, and suggest improved plans, but it may be said of all that they do not meet the requirements of the case. The scientific elaborateness in most of them is such as almost to prevent their application to our public buildings, and quite precludes their adoption in ordinary dwelling-houses.

No plan which is not perfectly simple and reasonably inexpensive will be generally adopted.

The Arnott stove (which is admirably adapted for some purposes, though not conducive to ventilation) has grown into general disfavour, from the difficulty attending its management and the annoyance consequent on its mismanagement. It is doubtful if the climate of this country is such as to admit of the general use of close stoves; few persons being able to endure the oppressiveness which is invariably attendant upon them, possibly because our atmosphere is both milder and more moist than that of the more northern parts of Europe. There is also reason for thinking that the light from an open fire is as congenial to the health as it is to the

* The word stove is sometimes applied indifferently to the open fire place and the close stove; in the present paper it is to be understood as meaning the former.

the cheering effect of a fire newly lighted in a cold room can hardly be d either to the slight warmth it gives or to its influence on the imagination. r plans may be found suitable for large buildings, it may be confidently at for ordinary rooms an open fire-place will be universally required, and mproved method of warming and ventilating our houses will be generally unless it be simple, self-acting, and inexpensive in construction and use. lso be affirmed that no considerations of health will induce the adoption of which infringes on personal comfort.

dmission of cold air at the skirting, or through the floor or window of a evitably produces a cold current, which is at least unpleasant, and there persons who will submit to the annoyance when they do not *see* the mis- iich will follow from closing such supplies. At present smoky chimneys, y warmed and imperfectly ventilated rooms, are common to every house; persons have deliberately considered why! We see doors and windows fit as closely as possible, while the smoke-flue is constructed to secure an draught. Now, supposing a fire to burn well, a very large quantity of air om the room up the chimney; so a corresponding quantity must enter the some means, and either air must enter through crevices around the doors lows, or there must be a current down the chimney.

lain fact would appear to be disregarded both by those who build and those ıpy houses, as the builder exercises all his skill to make the fittings as close le while the occupiers not unfrequently stop up the small openings the for- left. Any person who attempts to empty a bottle-full of water, quickly, ting it, finds that if the water flows at all, it does so by sudden gushes; on ; it a little, it flows more freely, but still with no regularity; but on adjust- ngle at which it is held, so that only a due proportion of the orifice is occu- the water coming out, leaving sufficient space for the air to enter freely, a ontinuous stream is obtained, and the bottle is rapidly emptied. Everyone ır with this fact. Now, is not the action of a chimney strictly analogous? ht we not to provide an inlet for air approximating in size to the outlet, if d secure a steady continuous flow through the latter?

ıeasurement around a door may be taken at about 20 feet; and if we assume en closed, the space that admits air is one-eighth of an inch, we have alto- ı area of opening of 30 inches, or something more than one-third of the n ordinary chimney-pot—the most contracted part of the smoke-flue: con- , supposing a room to have only one door, it will be necessary that the , floor, or skirting, admit twice as much as the door, in order to supply the of the chimney. As, however, all such inlets are narrow slits, the air will impeded by friction; and in the best-constructed modern houses, the space the closed door and its frame is not nearly so much as that assumed—while ows are expected to be almost air-tight. It must be admitted that the com- ll of the stove-maker and the cowl-maker do, in many cases, overcome the left by the architect and builder, that the imperative demands of the chimney and that the chimney does not smoke. Modern fire-places are generally so ed that the fire is within about six inches of the hearth; the heat is, there- ected back to the fire so strongly as to cause a very rapid combustion:—the of highly rarefied air thus produced has so great an ascending power, that rapidly into the chimney, and compels the entrance of air into the room by ink that may have escaped the skill of the joiner. From those there must proceed rapid streams of cold air, chilling every one within their range, occupants are warmed only on the side they present to the fire, by means rect radiated heat. In severe frosty weather, when in order to maintain the warmth a very large fire is kept up, a room may be observed to be unusually , although the external air may be comparatively still: this effect arises increased draught of the chimney, consequent on the rapid combustion. It efore follow that, if the fire be increased, the warmth obtained will not be tion to the increased combustion—more cold air being thereby drawn into .

now consider how the air of the room is changed, or rather the movements it undergoes—for it is questionable if it be changed when the windows and doors are closed. Let us take the case of a room, on a winter's evening, occupied by several persons. It must be borne in mind that, while a temperature of from 60° to 65° is as high as is generally desirable, the temperature of the human body is 98°; consequently, all exhalations from occupants, as well as the respired air from their lungs, will ascend. The heated air from the lamps or candles will also rise to the upper part of the room. The air immediately in front of the fire, at a distance of from two to three feet above the hearth, being highly heated, will also be found to be constantly ascending.

All the air which enters by the doors and windows to supply the chimney draught, being of a much lower temperature than that of the room, makes its way down to the floor, however high it may enter; and it creeps along the lower part to the fire-place without any great admixture with the general air of the chamber. In the mean time, the upper part of the room is constantly supplied with contaminated air from the occupants and the candles or lamps. How is this to escape? It probably moves to the several sides of the room, where it is partially cooled and descends; but as its temperature is not reduced so low as the fresh air entering the room, it will rise again to the breathing level, while the fresh air feeds the fire.

The ordinary open fire-place unquestionably carries off a considerable quantity of air above as well as through the fire; but observation shows that this draught is from a low level. If a line be drawn from the mantel, to the same distance in front of the fire-place that the mantel is above the floor, it will be seen, on burning a piece of paper, that the smoke will enter the chimney only when the paper is held *within* the line. If held *without* the line (except a few inches from the floor), the smoke will be observed to ascend to the top of the room, thereby proving that, before the chimney can be instrumental in carrying off contaminated air, the temperature of the air must be so reduced as to make it float at a very low level.

Arnott's ventilators (which are sometimes fixed in the chimney-shaft, near the ceiling) are of advantage in carrying off the contaminated air; but many chimneys have so sluggish a draught, that the ventilators will not act unless they are balanced towards the chimney,—in which case, puffs of smoke will occasionally enter the room. They require nice adjustment, and are frequently objected to on account of their unsightliness.

The writer of this essay has been led to reflect much on all these particulars, in consequence of his having had many opportunities of observing the effect of warming by steam, hot water, close stoves, and open fire-places—in the latter case, accompanied with at least an average number of smoky chimneys; his observation was, perhaps, also quickened by a lively sensitiveness to stagnant air. He had also seen numerous instances of pertinacious exclusion of fresh air where the place of supply was so situated as to cause temporary discomfort. The annoyance attending the draught from a door, left ajar to prevent the chimney from smoking, led to the idea of carrying an air-tube from the external walls to the hearth-plate to supply the consumption of the fire. There seemed to be no great difficulty in this, and there appeared to be every probability of thus remedying a material evil. The question of where the openings should be made then arose. Supposing the air to have egress at or about the hearth-plate, it would at once flow to the chimney, thereby hindering the withdrawal of the air of the room. On examination, an ordinary register stove showed that but a small proportion of the whole space allotted to it was occupied by the fire-grate, the frame around and above the grate forming a mere ornamental facing. There seemed to be no obstacle to substituting for this mere shell of cast-iron a horizontal and two upright tubes; and if they were connected with that beneath the hearth-plate, a continuous tube would be formed around the front of the fire. Air passing through these tubes would, it was imagined, acquire some warmth. The next consideration was, how to provide for the admission of the fresh warmed air into the room? and here it was important to select a place where the inflowing current should not be felt. The upper part of the mantel, from its proximity to the stove, and its position relatively to the occupants of the room, seemed a suitable place for the ingress of the air. Supposing the upper horizontal tube to have an opening or slit extending along the top throughout its whole length, the stove to be set sufficiently forward to leave this slit beyond the line of the chimney-breast, and the upper mantel and mantel-shelf to be so formed as to continue the passage-way from the tube, the air would be discharged into the room at about four feet above the level of the floor. This idea appeared so promising, that it was determined to test it in a new stove, which was arranged thus:

First.—A cast iron hearth-plate, having two openings in it of this form (see fig. 1), was laid down, leaving a hollow between it and the brick trimmer.

Then two hollow jambs, forming tubes of cast iron, were fitted to the openings in the hearth-plate, and these were connected by a head-piece of the same form and dimensions as the jambs, but having along the top side an opening extending the whole length of the head-piece; a section through the head-piece showed thus. (See fig. 2.)

Within these there was fitted an ordinary office fire-grate, consisting of a skeleton frame with front and bottom bars, the sides and back being of fire brick. Care was taken that none of the burning fuel should be in contact with the side tubes. In fixing the stove, it was arranged that the head-piece, and consequently the side tubes, should stand forward into the room sufficiently to secure the whole of the air opening in the former, being beyond the chimney breast. In fixing the mantel-piece the upper slab was kept forward from the chimney breast, leaving a space to correspond with the air-opening in the head-piece. The external air being then conducted by a zinc tube to the space between the trimmer and the hearth-plate, was found to pass up the two hollow jambs, thence into the hollow head-piece, whence it passed up the back of the mantel through openings in the mantel-shelf into the room. The extremely disagreeable, if not unhealthy, effect produced by highly heated metal, in contact with air which is afterwards diffused about a room, suggested great caution in subjecting the air-tubes to the direct action of the fire. It has been stated, that so long as the heating surface does not attain a temperature exceeding 212°, the air will not be vitiated; though it is difficult to see the peculiar charm of this temperature, except that it is the boiling point of water under an atmospheric pressure of 30 inches of mercury. Eider down, it is said, will not bear a temperature exceeding 200°, without emitting an offensive odour: and when we consider the various and delicate particles of organised matter which float in the air, it may be questioned if any surface which is to warm the air by contact should be heated much beyond 120°.

On a trial of the stove, arranged as before described, the air was observed to rush rapidly through the tubes into the room as was expected, and very soon after the fire was lighted the air stream exhibited a temperature of from 90° to 100°.

No very accurate observations were made during four or five weeks after the stove was first in operation, beyond the fact that the result aimed at, and expected, was attained; viz., a supply of fresh-warmed air,—which was however but limited in quantity, from the delivery opening having been made too small.

About this time another case presented itself, which offered a good opportunity for again trying the same arrangement. The dining-room of a house, a few miles from town, was rendered scarcely occupiable in consequence of the chimney smoking. The room—about 16 feet square and 8 feet high—was on the ground-floor, with no basement story beneath. It was a few inches below the level of the garden in front, and had a provision for the circulation of air beneath the floor. The house was old, with doors and windows fitting but indifferently. The stove was of modern make, but the fire invariably burnt dull, and the room was scarcely ever well warmed. As the room was required for daily use by the family, alterations to the old stove would have caused inconvenience; a new register-stove, of the ordinary dining-room kind, was therefore selected, with the view of trying how such an one could be fitted with air-tubes.

The *plan* of the stove was thus (fig. 3): To the cheeks and the back of the face, *a a*, of the stove were riveted two strips of sheet-iron bent in this form (fig. 4). The plan then showed thus (fig. 5). Two side-chambers, or tubes, *b b*, were thereby obtained, the back of the upper face of the stove was then fitted with a rectangular tube, also of sheet iron, which was connected with the two side tubes, making *uninterrupted communication* between them. The old stove was set very far back, in the chimney shaft, and was fitted with a large marble mantel, the displacement or altering of which was, on many accounts, objectionable; it was therefore determined in this instance to admit the air into the room immediately under the mantel-piece, and afterwards to adopt means of preventing any inconvenient rush of air in a horizontal line into the room. The area of a section of each of the side tubes exceeded 18 inches, while that of the horizontal tube was about 40 inches. The width of the face of the stove was 3 feet; and to furnish the means for the admission of the air into the room, an opening of 1 inch in width was left at the top of the horizontal tube, in its vertical face, extending its whole length. A hearth-plate of cast iron was then provided with openings corresponding to the horizontal section of the side tubes, as shown on the plan. The hearth-plate was then set with a chamber beneath, extending in the middle as far back as the line of the front fire-bars. From this chamber a zinc tube, 6 inches square, was carried under the marble hearth, beneath the floor, through the front wall into the garden, where it was carried up to the height of 14 inches, and was furnished with a cap having sides of finely perforated zinc. A passage-way of 36 inches area was thus obtained beneath, around, and above the *front* part of the stove, while there was a corresponding ingress for the air from the garden, and egress from the upper part of the stove into the room.*

The stove was set in the ordinary manner; and within twenty-four hours of the removal of the old stove a fire was lighted in the new one, and the room was in occupation. It was at once evident that the tendency to smoke was remedied, and the fire burnt freely and cheerfully. As it was felt to be an important point to introduce the air into the room in such a direction that its entrance should not be perceptible, the air-opening was furnished with a metal plate bent in such a form as would direct the air-stream towards the ceiling, and also admit of the supply being diminished or stopped entirely as might be found desirable.

The complete and agreeable change in the character of the air of the room was at once apparent to every one, and instead of the room being barely habitable in cold weather it was found to be the most comfortable in the house.

(*To be continued.*)

THE DUKE OF SOMERSET'S GEOMETRICAL TREATISES.†

Up to the era of Herschel, Babbage, Airy, &c., we were undoubtedly behind the mathematicians of the Continent in the application of mathematics to physical science. Their subtle and refined modes of analysis which, until the period mentioned, were almost unknown in this country, enabled them to penetrate into the nature of subjects which could not be approached through any other medium. Whilst, however, Laplace, Legendre, Poisson, and many others, were suns shining with an effulgency peculiarly their own, we had a number of ingenious mathematicians spread over the country as village school-masters, or as labourers at the loom and anvil. In this respect England stood alone; in no other country was mathematical knowledge so widely diffused, or so generally and ardently cultivated. The time-honoured *Ladies' Diary*, and other publications of that description, which were truly indigenous to this country, not only excited an ardour for mathematical pursuits by the curious problems which they proposed, but their aim was to encourage the study of the mathematics, and to make mathematicians. The result has been that those unassuming publications have been the means of forming more mathematicians in this country than our most richly-endowed establishments have produced. Whilst the Continent had its celebrated analysts who, for their scientific achievements, had the highest honours and the choicest dignities

* A complete form of the stove on this principle is represented in fig. 1*a*, fig. 2*a*, and fig. 3*a*, and afterwards described.

† "A Treatise in which the Elementary Properties of the Ellipse are Deduced from the Properties of the Circle, and Geometrically Demonstrated. By the Duke of Somerset."
"*Alternate Circles, and their Connection with the Ellipse.*" By the same Author.

ighty nation could bestow, con- n them, we had a truly noble iathematicians, having only their il attainments for their reward; le encouraged were they by the even their favourite *Diary* was undred per cent. upon its cost! taught sons of science, however, the spirit of pure geometry. ations which their labours sup- ; a storehouse of beautiful pro- l theorems—from which book- en of titled orders, have habitually mselves without leave and without gment. Dr. Simson, Dr. Stewart, others, have done much for geo- eir labours have rendered their erable to every cultivator of the metry in all climes and countries. ss, the unpretending publications e been adverted to have made far netricians. Let any one who is doubt the assertion turn to the he *Mathematical Companion*— id new series of the *Repository*, ee what Hinton, Wood, Mabbott, rke, Cunliff, Campbell, Fletcher, wry, Swale, Wright, Butterworth, amented Davies, the accomplished ind the venerable Gompertz, have m so many gems of pure geo- found elsewhere? Is there any tant so suggestive of principles, ting such a familiar inlet into the re of geometrical analysis, as Geometrical Amusements?" It opinion of Professor Davies — a judge on the subject—that we have omparable to it in the language. ithstanding the utility of pure geo- a mental exercise, and its varied n to the arts and sciences—it has been considered rugged and for- at all events not fashionable. its most ardent cultivators have admiration by their labours within narrow circle; but, alas! they thing else. When contemplating al productions of Butterworth, nhabited *of the country that suf*fered him to toil at the loom for his daily sustenance; and at last, when he could no longer eke out an existence by that labour, generous friends had to support the honourable old man by an eleemosynary subscription. If we go through the geometricians of this order, who conferred so much credit on England as a nation, we shall find that England has done nothing for them in return. The State, whilst it pensions off in comfort and competency its foreign dancers and outlandish mountebanks of various kinds, suffers its own sons of genius to pine in obscurity, and to end a life, devoted to the cause of real science, in all the wretchedness of chilling want.

The cold indifference with which abstract mathematics are regarded by what is popularly termed "the men *in power*," in this country is proverbial. Their devotees are looked upon by the tricksters who get highest up in the tripos of cajoling, and become Senior Wranglers in political duplicity, as mere dreamers in useless metaphysics and propounders of paradoxes. The fact is,—pure geometry cannot be successfully applied to bamboozle one party, or to hoodwink another; and therefore its cultivation is condemned as being unserviceable. The geometer, in the opinion of the political adventurer, is in the same category with the visionary theorist or crazy speculator. Since, then, the promoters of abstract science,—especially geometers,—are treated with so little respect or consideration, we have deemed it a duty to notice the works named at the head of this article—the performance of one of our highest aristocracy—one of the most venerable chiefs of our nobility; namely, His Grace the Duke of Somerset. The work was published some time ago, but it has only recently come under our notice. Quite apart from the intrinsic merits of the works, we have much pleasure in introducing them to the notice of our readers.

It is refreshing to find such a cultivator of pure science. Kindred spirits have an affinity for each other. No doubt the noble author of this work would have been delighted with the beautiful problems of Butterworth, and their neat and unique demonstrations, had they been brought to his notice; and had the age and the circumstances of the life of the venerable geometer been made known to his Grace, what a delicious pleasure the noble Duke would have experienced in befriending a man so distinguished, so honoured and so deserving it, in the decline of his valuable life! We happen to know enough of the unostentatious Duke to give an opinion on the point. We had named the honoured Butterworth as one of a class. Unfortunately, however, the publications containing the scientific performances of our self-taught men, although they are filled with the choicest gems of abstract science, are comparatively little known. Titled authors help themselves most liberally from them, but say nothing as to whose brainwork it is, or where it may be found. We have no association for the protection and advancement of self-taught scientific men, — we mean of the humble, but meritorious classes. Were such a society in operation, the merits and privations of such men could be made known to fellow-labourers like our noble author, who have the means, as well as the inclination, to encourage and assist. Our talented correspondent, Mr. Wilkinson, has rendered essential justice to this class of students in his interesting historical sketches of our mathematical literature; he possesses just the ability to form a scheme, and the perseverance to carry it out for their protection and advancement; and we trust he will some day turn his active mind to the congenial subject.

We now proceed to give the reader some account of the works, the titles of which are cited above. Perhaps the best mode of effecting this object is, to let the author speak for himself. In the *advertisement* to the first-mentioned Book, he says—"The following work consists of three books. The first book consists of concentric circles, and especially such as are now for the first time called alternate circles.

"The second book is chiefly upon the three classes of parallels, which a[re em]ployed in deducing the ellipse fr[om the] circle.

"The third book applies the pri[nciples] developed in the two preceding, to [explain] and demonstrate some of the simple[r pro]perties of the ellipse."

We think the definition of each bo[ok will] explain to some extent the nature an[d scope] of the work.

"*Book First.*

Definitions.—1. Concentric circ[les are] those which are described on the sam[e plane] and from the same centre.

2. When of three concentric cir[cles the] area of one is equal to the areas of [the two] others, these others are called a[lternate] circles.

3. The greatest of three concentric [circles,] of which two are alternate, is cal[led the] centred circle.

4. The least of three concentric ci[rcles, of] which two are alternate, is called th[e inter]nal circle.

5. When of two alternate circles [one has] been mentioned in a proposition, th[e alter]nate circle means the other.

6. An ordinate to the diameter of [a circle] is a straight line, placed in the circle [at right] angles to the diameter.

7. An abscis of the diameter of [a circle] is a part of the diameter intercepted [between] the centre and the ordinate.

"*Book Second.*

Definitions.—1. A class of par[allels is] any number of straight lines paralle[l to one] another, and meeting the circumfe[rence or] the produced diameter of a circle.

2. Parallels of the first class ar[e termi]nated by the circumference of a ci[rcle, and] met by other parallels in its dia[meter or] produced diameter, figs 2, 3.

3. Parallels of the second class ar[e drawn] from the points in which those of [the first] class meet the circumference of th[e circle,] and are terminated by those of the thi[rd class,] fig. 1.

4. Parallels of the third class a[re those] which meet in the diameter or p[roduced]

the parallels of the first class, and inated by those of the second, fig. 1.

arallel of the first class is said to nd with one of the third when they the diameter of the circle.

arallel of the second class is said to nd with one of the third when each corresponds with the same parallel rst class.

"*Book Third.*

tions.—1. The points in which pa- the second class meet their corre- parallels of the third class are in a as curved line, which is the circum- f an ellipse, fig. 1.

traight line is said to be in an ellipse e straight line is terminated both the circumference of an ellipse,

ch a line bisecting two other such t are parallel, is a diameter of the fig. 2.

ich parallel lines so bisected are s to the diameter that bisects them.

e diameter being bisected, the point ion is the centre of the ellipse.

line drawn parallel to the ordinate the centre, and terminated both the circumference, is the conjugate , fig. 2.

e former diameter is also said to be e with the latter.

axis is a diameter that makes equal ith its conjugate, fig. 3.

oblique diameter is one that makes angles with its conjugate, fig. 2.

. diameter common to an ellipse and is a straight line drawn through the f each of these curves, and meeting umference of the ellipse in the points it is met by the circumference of e. (See fig. 3.)

When two lines, or a line and its part, circle and the other in an ellipse, nates to the same diameter, and cut e same points, these ordinates are correspond.

Of the diameter the part that is in- d between the centre and the ordi- alled the abscies.

When the external circle has for its diameter a diameter of the ellipse, and the diameter conjugate with this is that of another circle, and an alternate circle is described cutting the first diameter, the points of section are the principal points.

14. When the diameters are axes, the principal points are focuses.

15. The third proportional to the two conjugate diameters is the parameter of that which is just in the proportion."

Probably the reader will perceive, from these definitions, that the aim of the author has been to simplify the geometry of the ellipse by making it depend exclusively upon the properties of the circle and straight lines. The noble author has certainly given to a matter usually considered somewhat abstruse a different aspect, which, if we mistake not, will prove *practically* useful. In order, however, that the work may be rendered as generally serviceable as possible, or, at all events, that its real usefulness may be thoroughly tested, the author should have published it at a less cost, so that it may easily get into the hands of that class of geometers whose heads are well stored, but their pockets scantily. If there be anything original and useful in the work—and we think there is much—the Duke may be assured that the humble and self-taught description of mathematicians just mentioned are more likely than any others to carry out his suggestions, and to extend the simplicity and utility of his work.

To make the author's treatment of the subject as apparent as our limits will permit, we give a proposition, and its demonstration from each book :—

Proposition 8.—*Book* I.

If two circles within and concentric with a third, have placed in that third a tangent of one equal to a diameter of the other, these two circles are alternate.

Let EFG and HIK be two circles within and concentric with ABD, and having C the common centre ; and let AB, a tangent of HIK, be equal to EF, a diameter of EFG : then EFG and HIK are alternate circles. For if they be not alternate, then the two circles ABD and EFG remaining the same, let EFG and some other circle be alternate. And let OPQ be the other circle. From the centre C draw CK perpendicular to AB, and produce CK till it meet the circum-

Fig. 1.

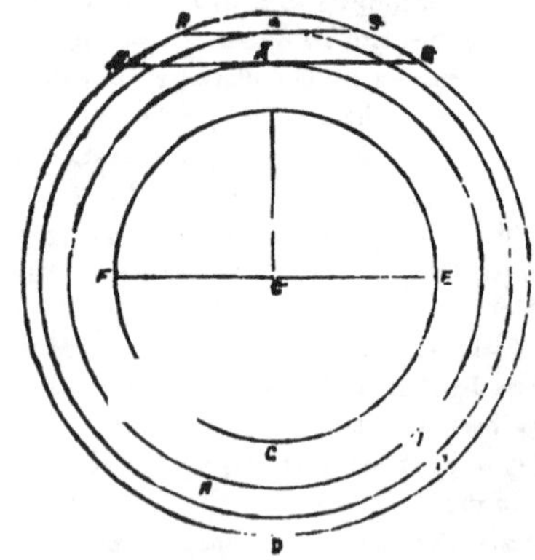

ference of OPQ in A; from which draw RS touching OPQ in Q, and meeting in the points R and S, the circumference of ABD. Then because EFG and OPQ are alternate circles, and that in the external circle ABD there is placed RS a tangent to OPQ, and that EF is a diameter of the other alternate circle, by the sixth proposition of this book RS is equal to EF.

Now, by construction, AB is equal to EF, therefore RS is equal to AB.

Wherefore by the 14th proposition of Book III., "Elements," RS and AB are equally distant from the centre C. Therefore by the third definition of that book, CQ is equal to CK, the greater to the less; which is absurd.

Proposition 4.—*Book* II.

Theorem.—Every parallel of the third class is bisected by the diameter in which it meets its corresponding parallel of the first class.

Fig. 2.

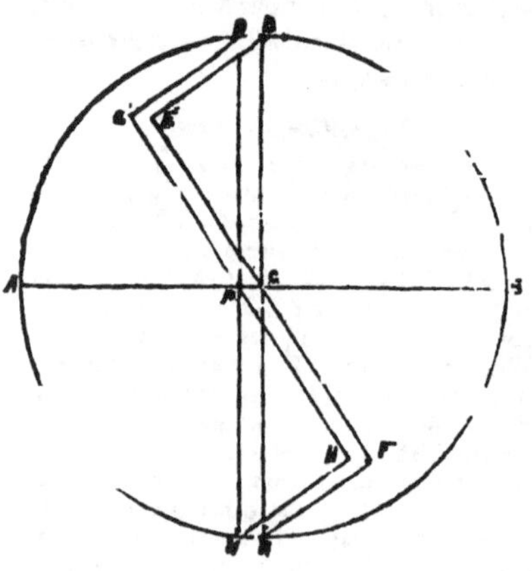

Let EF and GH be parallels of the third class, and let DR and ON be parallels of the first class corresponding with EF and GH respectively. Then as EF is to DR, so is GH to ON.

Let EF and DR cut the diameter AB in the point C, and let GH and ON cut it in P. Draw DE, OG, RF, and NH parallels of the second class. Then by the 3rd proposition of this book, the triangles DEC, OGP, RFC, and NHP are similar.

Wherefore as EC is to CD, so is GP to PO, and so is FC to CR; and so is HP to PN.

Therefore, by the 12th proposition of Book v. of "Elements," as CE and FC together are to CD and CR together, so are GP and HP together to PO and PN together. That is, as EF to DR, so is GH to ON.

Proposition 21.—*Book* III.

Problem.—The two axes being given to find the focuses of an ellipse.

Let AB be the greater and EF the lesser axis of the ellipse AEBD. The focuses are required.

Fig. 3.

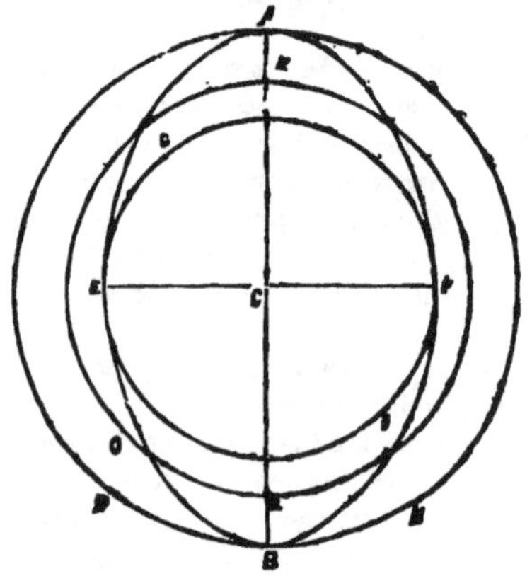

Let C be the point in which the line AB is cut by the line EF.

From the centre C, at the distances CA and CE, describe the circles AHD and EIG, and the alternate circle MNO. And let M and N be the points in which AB is cut by the circumference of MNO.

Then, because every diameter of an ellipse passes through the centre, and that C is the only point through which AB and EF both pass C in the centre.

And because the centre of an ellipse is the point of bisection, C is the point of bisection.

Therefore the lines AB and EF are bisected in C, and are diameters of the concentric circles AHD and EIG.

Wherefore by the 13th Definition of this

d N are principal points. And
, they are focuses.
y.—The focuses are equally dis-
he centre of an ellipse.
ior says, in an advertisement to
book, "that it is a sequel to the
consists of two sections.
st contains some geometrical pro-
monstrated by an algebraic pre-

cond section employs alternate
simplify or to render to lower
such equations as express the
of the ellipse."
ior is a mathematician quite of
ool: he usually employs the old-
otation *aa*, *xxx*, &c., instead of

in section the second—
ling to the method hitherto used
aticians, the equations to the
quadratic. The reason of this is,
ists have been used to express
ties of the ellipse by signs and
ting the proportions that certain
d parellelograms bear to each
t in the method now proposed,
upon the theory exhibited in the
Treatise, the properties of the ellipse are to be expressed by signs and latters denoting the proportions that certain straight lines bear to each other. These will at first produce only simple equations."

The author gives many illustrations in proof of this doctrine; this is one of them:—Calling the diameter of the alternate circle $2u$, the ordinate $2y$, the diameter of the ellipse $2t$, and its conjugate $2c$, then (prop. xv. 3) $2u : 2y :: 2t : 2c$, or $ty = cu$, which is a simple equation.

Having extended this article to a considerable length, we are prevented from giving more extracts from this part: we would, however, recommend our mathematical friends to give the book an attentive consideration. We think "the *vein* of it is good," and that "the world may find it after many days." In the mean time we congratulate the noble Duke on the result of his studies, and Geometers on their having in their list so distinguished a fellow-labourer.

(*To be concluded in our next.*)

DRAY'S TURN REST PLOUGH.

[Registered under the Act for the Protection of Articles of Utility. William Dray, of the firm of Dean, Dray, and Co., of Arthur-street, and Swan-lane, London Bridge, Proprietor.]

Fig. 1.

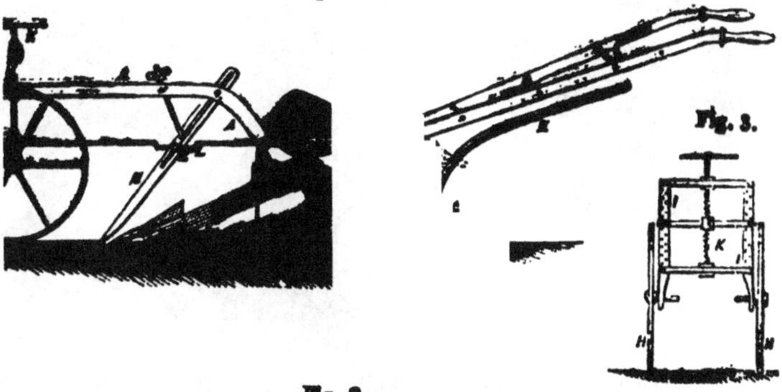

Fig. 3.

Fig. 2.

a side elevation, and fig. 2 a
he plough. AA is the beam;
BB the two mould-boards, which are jointed at front to a breast mould-board C,

which is common to either of the pieces BB, and with one or other of them forms the entire mould-board; D is the share which abuts against the breast of the mould-board C, and with it is capable of being turned from one side of the beam towards the other; the amount of travel allowed to them being about 45°, or one-eighth of a whole revolution; E is a handle which is connected by links FF to the mould-boards BB, and by means of the spindle G to the share D, so that when the handle is turned towards the one side or other, it puts the share and mould-board on that side in the proper position for turning over the furrow on that side; HH are the two wheels, which adjust themselves so as to run in the furrow and surface of the ground (upon the return of the plough), by means of the parallel motion II, to which the wheels are connected. Fig. 3 is a front elevation of the wheels and parallel motion; K is a screw for regulating the height of the wheels in relation to the share of the plough, so as to cause it to enter more or less into the ground; L is a crank spindle, to the fore-end of which there is attached the draught link M; the crank L[1], takes into a socket formed upon the side of the coulter N, and brings the coulter to its proper position when the wheel takes into the furrow; or the crank may be acted upon by the lever P, indicated by the dotted lines in fig. 1. O is a link by which the two mould-boards are connected and made to move together.

READ'S IMPROVED CAPSTAN.

Registered under the Act for the Protection of Articles of Utility. Thomas Travis Read, of Hull, Engineer, Proprietor.]

Fig. 1.

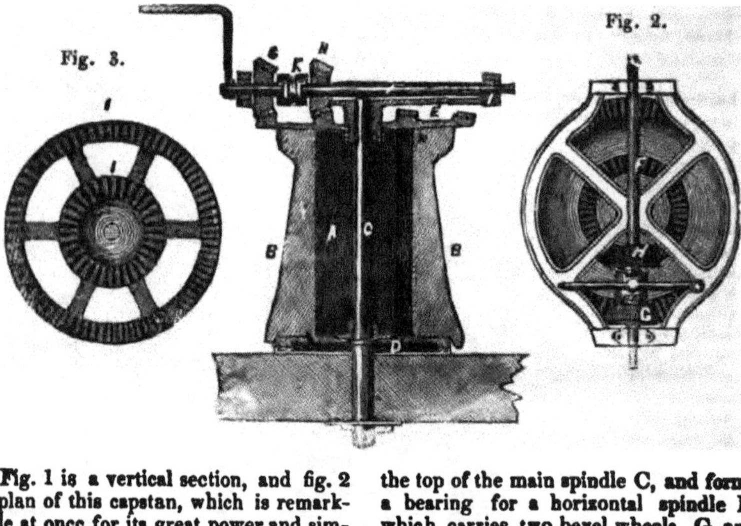

Fig. 1 is a vertical section, and fig. 2 a plan of this capstan, which is remarkable at once for its great power and simplicity. AA is the barrel; BB the whelps; C the spindle, on which the barrel revolves: this spindle is securely fixed at its lower end, so that it cannot itself turn. D is the base-plate, which is furnished with a paul plate, as usual, to prevent the return of the capstan. EE is a frame, which is affixed to the top of the main spindle C, and forms a bearing for a horizontal spindle F, which carries two bevel wheels, G and H. These bevel wheels gear into toothed circles, or wheels, I I (shown separately in fig. 3), affixed on the head of the capstan, and they are perfectly free to turn upon the spindle F, unless when carried round with it by means of the double clutch K, which can gear into one of them only at a time. When a quick speed can be obtained with the power applied, then the wheel H is made to act upon the inner circle of the teeth; but if the power so applied is not sufficient, then the wheel G is made to act upon the outer circle of teeth.

PREVENTION OF EXPLOSIONS IN BOILERS. BY DR. J. L. SMITH.*

The explosion of boilers are devisable into two classes; namely, the simple bursting of the boiler when the plates are ruptured and opened by a steady pressure from within, and these explosions, the effects of which seem to exceed any thing that could be imagined would result from an accident of this character; this last has been properly called by the French *explosions fulminantes.* Not only is the boiler burst under these circumstances, but the whole or a part of it is projected with the rapidity and force of a rocket, and although weighing six or eight tons, may be thrown to a distance of many hundred feet from its original position, even when the rupture of the boiler is of small dimensions.

It will be remarked, that in these cases the opening made in the boiler is always in the end opposite to the direction in which the boiler is thrown; in other words, it is projected by the reaction of the steam, which escapes against the air; and in fact, is subjected to precisely the same action as a rocket. No gradually-increased pressure could cause a result of this description, and it is not so much an immense pressure that is required to produce this as a very rapid development of that pressure. An iron bomb-shell, several inches in thickness, may be burst by two or three pounds of powder, although we might condense with safety into the same shell fifty times the amount of gas formed by the combustion of the powder; the fact is, that although the powder does not form gas enough to explode even a much thinner shell, if the gas were gradually formed, still, in the actual state of things, its rapid development more than compensates for the want of quantity, and the effects produced are far more terrific than those which could be produced by a steady pressure, however great.

The same fact is still more remarkable with that class of bodies called fulminating powders, which, when exploded in small quantities in close vessels of great strength, will burst them with prodigious violence, even when the gas formed from the combustion of the powder would not more than fill the capacity of the vessel; this is owing entirely to the instantaneous production of the explosive force.

These remarks are merely to show the necessity of separating the study of the two classes of boiler explosions; the first of which arises from either deficiency in the strength of the boiler or from carelessly overburdening a boiler of proper strength (a too frequent occurrence). The engines of most steamboats, when in proper motion, are said to exhaust the boiler three times in the course of a minute; so then, when the engine is stopped, this large amount of steam will accumulate in the boiler if proper care be not used to prevent it. The water in the boilers, when the boat is not in motion, and with the steady fires, will acquire fifty degrees of heat in two and a half minutes,—and let us suppose that the pressure of the boiler is one hundred pounds per square inch; now, at the end of two and a half minutes, stoppage, the pressure will have increased two hundred pounds,—for it has been ascertained that an increase of fifty degrees of heat will double the pressure of the steam; and in two and a half minutes more, or five minutes from the stoppage, the steam will have accumulated to four hundred pounds pressure to the inch. Now, there is doubtless much carelessness in observing the boilers when a boat stops, and consequently the rapid accumulation of steam becomes a frequent cause of the first class of explosions.

The consideration of the more immediate causes of the second class of explosions (*explosions fulminantes*), are those which mostly excite our attention; and for their explanation we have to search for some condition of the interior of boilers that will give rise to a rapid evolution of steam.

A cause very commonly admitted, and one in which there is, doubtless, great truth, is, that of the water getting below the fire line of the boilers, a considerable portion of the boiler becoming exposed to the fire without water being in contact with it, thereby getting greatly overheated and diminished in strength.

While the engine is at rest and the safety valve closed, the water remains quiescent under the accumulated pressure on its surface; the moment this pressure is diminished by working the engine or opening the safety valve, a frothing or boiling in the water takes place, its surface is raised and brought against the overheated sides of the boiler, an immense amount of steam is rapidly formed, and neither the safety valve nor the cylinders will allow a sufficiently rapid escape of steam, to prevent, in many instances, a terrific explosion; particularly

* From the New Orleans *Daily Crescent.*

when the boiler is already overcharged by a steady pressure, weak in its natural construction, and the iron diminished in strength by the heat to which it is subjected. In the minds of most persons this is considered a frequent cause of this class of explosions.

It is insisted on by many, that accidents occur when the proper quantity of water is in the boiler, the boiler strong, and the pressure not beyond its normal condition. Now, it is asked how is it possible, under these circumstances, for one of these explosions to take place? In vain have explanations been sought after, and we either quit the subject in despair of finding out the cause, or in disbelieving that it can happen under those circumstances. I must, however, state, that after a very close examination of the subject, I see reason to believe that a rapid formation of steam may take place under the circumstances alluded to above, and explosions result.

In studying the causes of the explosion of boilers, it is usual to proceed by taking into consideration the ordinary properties of water, and the laws of liquids with reference to heat, presuming that we understand all the laws that water obeys in being heated from the freezing point to 600° or more. But the fact is, that as far as our knowledge goes, water does not obey a regular set of laws with reference to the action of heat upon it—in this respect it has many peculiarities.

Commencing with water at the ordinary temperature, and gradually abstracting heat from it, the water contracts, and continues to do so until it reaches 40°, when any diminution of temperature will cause it to expand until it reaches freezing point. Here, then, is a circumstance occurring in water contrary to what we knew of as belonging to other liquids, which contract by the abstraction of heat down to their freezing point. Again; it is usually considered that if any thing is fixed, it is, that water becomes solid when reduced to 32° (its freezing point); yet, singular to say, this is not the case, for, under certain circumstances (as that of perfect repose), water may be reduced to a temperature 10° or 12° below the freezing point without becoming solid; this, then, is another peculiarity belonging to water that no other liquid we know of possesses.

If we consider the temperature at which water may be heated in the open air, without boiling, a third still more remarkable peculiarity will be seen. Nothing appears better settled than that water boils at 212° in open vessels, and that it cannot be heated beyond that; now it has been clearly proved, and the experiment lately repeated by Professor Faraday, at the Royal Institution, that water completely deprived of air may be heated in an open vessel to 250° or 260° without commencing to boil, but once the ebullition begins, steam is given off in an explosive manner, and all the heat that has accumulated between 212° and 260°, enters instantly into the formation of steam, the water falls to 212°, and the regular ebullition continues. Seeing then the existence of this latter peculiarity in water, what then is there irrational in supposing that water in a boiler may acquire a temperature beyond that due to the pressure of steam on the surface?

When the boiler is perfectly close, and the water heated, it does not boil, but gradually increases in temperature, a pressure of steam accumulating above the surface of the water, just in proportion to the temperature of the water; and according to experiments made, water in a boiler at 300° ought to furnish steam exerting a pressure of a little more than 60 lbs. to the square inch. Suppose then a boiler to be in that condition, and the water should continue to acquire heat, without that heat converting any additional quantity of the water into steam; (a very rational supposition when we see that water in an open vessel can, under certain conditions, be heated 40° or 50° above 212°, without the formation of steam;) and suppose the water arriving to 320°, and the pressure of steam be at 60 lbs., which is 30 lbs. lower than what is due to water at 320°. Should now any disturbance of this state of things take place, either by the opening of the safety-valve, or by the working of the engine, the water would evolve with prodigious rapidity an amount of steam due to this excess of 20°, and we cannot properly estimate the rapidity with which this will take place, as every particle of water gives out almost at the same instant a portion of steam, and even this rapidly-increased pressure of 30 lbs. might be sufficient to endanger the boiler.

Now, if it be possible for the water to acquire 40° or 50° more temperature than it ought to have, in consideration of the pressure of the steam upon the surface we should be liable at any moment to have the pressure on the boiler doubled almost instantaneously.

These, then, are indications that I advance, thinking them worthy the attention of those engaged in investigating the subject, and may be, seeking for the cause of those explosions, when every thing connected with the boiler is apparently as it should be. This is one of those explanations that can only be properly determined by experiment.

There are many supposed causes of the

of boilers that are sometimes in-
on with a great deal of tenacity by
ocates. It is occasionally stated,
other explanations fail to account
ticular explosion, that electricity
the cause, without telling how this
y could be formed, and when
how it could burst a boiler.
it is possible to render a boiler,
m escaping, electric, it must be in-
r this purpose either on glass rods
other non-conductor of electricity.
e case of boilers as they are con-
this insulation does not exist;
y any hazard a prodigious amount
icity was accumulated in a boiler,
known conductibility of metals,
ricity could be discharged from
r without disturbing the most deli-
of its structure. And, could we
ite the electricity of a flash of
on a steam boiler, from the known
of metals to electricity, it would be
d with the same harmless result to
r.

er cause, occasionally advanced, is,
xplosion of hydrogen gas has burst
r. This, however, is no more ten-
n the cause of electricity; for we
t learn how this gas is formed, and
ned, how it can produce an explo-
irect experiments have been made
rating water to dryness in a boiler,
ining from time to time the steam,
finding any hydrogen gas mixed
In order to form hydrogen
e water, or vapour of water in
r, it is necessary that the sides
boiler should be at a red heat,
perature at which the iron is di-
to one-sixth its original strength;
therefore presumable, before this
things could occur, the integrity of
r would be injured by the ordinary
of steam. But suppose that the
be formed by the decomposition
pour of water, against the red-hot
the boiler,—that the iron combines
oxygen of the water, and liberates
ogen, how is this to explode? for
gas is inexplosive; it is a mixture
gen and oxygen gas that is explo-
pure hydrogen gas; and as there is
that we can possibly divine, by
s hydrogen could be furnished with
this hydrogen explanation for the
of boilers becomes groundless.

is yet another explanation, more
y regarded as a cause of explosion
er of the last two; it is, the injec-
old water into an over-heated boiler
been *exhausted of its water*. It is
on *of those who have investigated* this experimentally, that no danger can arise from letting cold water into a hot boiler; and among those whose investigations have led them to adopt this conclusion, are Mr. C. Evans, of Pittsburgh, and a committee of gentleman at the north, now engaged in investigating the subject of the explosion of boilers.

Mr. Stillman (of the firm of Stillman and Allen, engine-makers, New York) on one occasion, seeing the engine of the shop stop without any apparent cause, on examination found that all the water was out of the boiler, and the flues very hot; he immediately turned the cold water on, and as the water entered the boiler, he could trace its elevation by looking into the flues, and see them gradually cooled down. This experiment might appear dangerous to some, but upon a little reflection, I have no doubt it would seem a harmless one. Boiling water admitted under the same circumstances might be dangerous.

An explanation has been attempted to account for the explosion of boilers by M. Boutigny, of France, which involves a peculiar condition of water on hot surface; it has, however, not sufficient practical bearing to be insisted upon here.

So, from what has been stated in this lecture, the immediate causes of the explosion of boilers are reducible to two. An immense steadily-increased pressure, producing the bursting of boilers properly speaking—and a rapid formation of a large amount of steam, productive of explosions; and the manner in which these conditions were probably brought about, have been mentioned in full.

It is not sufficient that we should know the cause of these accidents, but it is equally important to seek after the manner of remedying them.

First of all, the frequency of these accidents would be most wonderfully diminished could the use of high pressure steam be done away with, and low-pressure boats be substituted. Doubtless, the ingenuity of our mechanics could devise some system of low-pressure boat adapted to these waters, if steam-boat proprietors would be willing to enter into so great a reform.

Again; high-pressure engines could be made to work with a lower presure than is now used, which might be done, and their power rated lower; as it is, the pressure used in steam-boats on the western waters surpasses everything that is dreamed of in other parts of the world, except on locomotives, whose boilers far exceed in strength every other description.

As regards the construction of the boilers, I would merely state that the ties used to strengthen the boiler can not be too many, and that they ought not to be fastened directly to the plates of the boiler, but wherever they are riveted on, an iron band should go around the boiler, and the rivets pass through the boiler plate and the bands; this would doubtless give great additional strength to the boiler.

The safety-valves should be larger or more numerous, to give sufficient vent to any sudden formation of steam. Some contrivance should be used to prevent the boiler from being overheated, and there is none better than the plugs of fusible metal placed in that part of the boiler immediately over the fire, which, on the water arriving at too high a temperature, would melt, and let the water out upon the fire. There is yet another method that has been devised for this purpose, that was proposed some years ago by Dr. Riddell, and has since been made the subject of a patent by some one else. It is based upon the difference of the expansion of copper and iron by heat. A copper rod, nearly the length of the boiler, is fastened in one end of it, and the other end is attached to a valve which slides over an opening on the bottom of the boiler: now this rod lies immersed in the water, and as the boiler and water become heated, the boiler and the copper rod both expand, but the copper rod expands more for the same amount of heat than the iron does, the consequence is, that the valve on the bottom of the boiler is moved; there is nothing easier than to so arrange the position of that valve with reference to the opening in the boiler, that when the water, and consequently the rod, are heated to a certain temperature, the valve will uncover the opening, and let out a portion of water upon the fire, which being partly extinguished, will permit the boiler to cool, the rod to contract, and the opening be again covered up. This method could no doubt be rendered practical, although I am not aware of its having had any application.

The next most important step to take in this matter is that by legislation; for, let the ingenious invent what remedies they please, to prevent the explosion of boilers, there are always a number of the reckless and indifferent who will not adopt them, and even neglect the ordinary precautions of safety. In this place a boiler explosion takes place—forty or fifty lives are destroyed—a kind of investigation is made, which generally results in finding out nothing as to the cause of the explosion, and exonerating all those responsible in the matter. This is not owing to any deficiency of will or zeal on the part of those whose business *it is to investigate it*, but from want of a proper method as well as judicious laws to enforce the results of their investigation.

Why not, as in other countries, make it criminal to carry over a certain pressure of steam? for most surely it is criminal to jeopardize the lives of those who entrust themselves to your care. Certainly there is no more wholesale species of manslaughter than that which results from many of the explosions occurring on these waters, and there is no body of men more responsible for it than the Legislature of the various States, for it is in their power to diminish the frequency of them.

Let them, as I before said, enact laws making it criminal to carry over a certain pressure of steam; make it necessary to have a safety valve enclosed beyond the reach of any one on board of the boat; make it punishable to insert any thing but fusible metal in the holes of the boiler meant for that purpose. The Legislatures of the various Western States should also make a joint appropriation of some twenty or thirty thousand dollars, to investigate the explosion of high-pressure boilers, not by experiments made on small apparatus, but on boilers of the dimensions ordinarily in use. And if these means be resorted to, the columns of our papers will not be daily filled with those lamentable boiler explosions which have become a reproach to the steam engine on these waters, and are considered in other parts of the world as discreditable to a nation that has such just cause to boast of its rapid strides in civilization, arts, and science.

The *Franklin Journal* makes the following remarks on the preceding lecture:—As regards the bearing of Faraday's experiments, to which Dr. Smith alludes, we think that if he will read them again attentively, he will find them as inapplicable to the case of steam boilers, as the spheroidal state of the water proposed by Boutigny. For the phenomenon depends on the perfect stillness of the water, requires that the surface of the vessels should be very smooth, the water free from air (which is never the case in practice), and the vessels comparatively deep and not too large (a common glass-test tube will show the experiment very satisfactorily, as many an unfortunate chemist who has been manipulating in too much of a hurry can tell). The shaking of a steamboat or locomotive boiler, the deposit of insoluble matters, or the rough surface of the inside of a boiler in use, by giving rise to an evolution of steam, would always in practice prevent the phenomenon, though we suppose that in stationary boilers with hanging tubes the case might occur.

ERIMENTS WITH FIRE-ARMS.

experiments have recently taken
Woolwich to test the capabilities
Colt's and Mr. Adams's revolvers,
Sears's needle guns and carbine;
wing report of which we extract
daily papers:

xperiments were commenced with
t's revolving pistols at 50 yards
d the practice was very good, in
stances the whole of the six balls
the target, which was about six feet
Mr. Adams then tried his revolv-
l, which differs from Mr. Colt's in
o ramrod, the balls, which were all
ical shape, being merely placed in
lving receptacles with the thumb,
percussion caps placed in their
sition for firing by a spring feeder.
r. Colt's six charges are fired, the
the pistol is drawn back by the
n each firing, but the action is dif-
Mr. Adams's invention; the works
scharge the percussion-caps do not
outside projection, and are yet
of causing the firing to be very
succession, and the reports very
if the balls had been projected
at velocity. Mr. Colt used both
and conical shot, but all those
Mr. Adams were conical, and wad-
an elastic material glued to the
he firing of the revolving pistols
d, the rifle-stand was removed from
0 yards range, and the experiments
ed with Mr. Sears's needle-gun.
, 1½ ounce in weight, used in Mr.
gun is of a conical shape, and the
1½ drams of powder, with a small
of igniting composition, is made up
lar manner to a ball-cartridge, and is
its firing position with the greatest
loading being at the breech of the
d consequently no ramrod is re-
When the charge is placed in the
, a sliding cover is pushed forward,
effectually prevents the smallest
of the powder escaping, that a
range is obtained with a compara-
aller charge than can be obtained
ther means yet tried, and the com-
of the projective materials is so
that, apparently, hundreds of
may be fired without materially
e interior of the barrel, or render-
essary to cleanse it. The perfect
f the combustion was repeatedly
placing clean white paper over the
re the charge was inserted, and it
riably exhibited as clean and free
in as it was when applied. The
experimented with was one of the
service gun-barrels, fitted with Mr.
Sears's invention, in which the charge and ball are inserted in one cartridge under the gun, and the firing took place at 400 yards range, to show its capabilities when fitted with Mr. Sears's invention. The firing with both guns, and with a carbine carrying a ball of one ounce in weight, with a charge of 1½ drams of powder (about half the usual service charge) was remarkably good. There were 135 rounds fired from the two muskets and 80 from the carbine, several times in rapid succession,—the pieces being capable of firing 20 rounds in two minutes; and yet there was no appearance of any of the parts heating or getting the least out of order during the whole time of trial.

THE SUBMARINE TELEGRAPH.

The preparations for making another attempt to lay down a line of electrical communication at the bottom of the sea between the English and French coasts are rapidly progressing to completion. That it is possible to transmit a current of galvanic electricity from one shore to the other by means of an insulated wire deposited in the bed of the Channel was fully demonstrated by the experiment of last year, unsuccessful as it was in permanently establishing the communication. The only question now remaining to be solved is, whether sufficient measures can be adopted to insure the connecting chain against the casualties which proved fatal to the project as originally designed, and according to the plan now finally determined upon there is every prospect of attaining the desired end. The essential conditions upon which the success of the undertaking have been shown to depend are, in the first place, that the conducting medium should be so securely isolated as to resist the pressure of the immense mass of water to which it will be subjected, and should unite the flexibility requisite to allow of its being easily coiled and uncoiled with a strength and weight sufficient to enable it to retain its position at the bottom of the sea, and to resist any ordinary violence from the natural action of the waves; and, secondly, that the points on either coast to be connected should be so chosen that the line between them present no other causes of accident than have been foreseen and guarded against in the preparation of the connecting line, and that the effect of these be reduced to the least possible amount. The line of communication which is now being manufactured at Wapping will, it is expected, entirely fulfil the required conditions. It consists of four copper wires of the thickness of an ordinary bell-wire, cased in gutta percha, and twined with a corresponding number of hempen strands steeped in a mixture of tar and tallow into a rope of about an inch diameter. Another strand, similarly prepared, is wound transversely round this, and finally ten wires of galvanized iron, about a third of an inch thick, are twined round this central core and form a solid and at the same time flexible casing. The whole, when thus completed, has the appearance of an ordinary 4½-inch metallic cable. The machinery by which this is effected is extremely simple, and the work proceeds, night and day, with the utmost regularity. A huge coil is thus being formed in one continuous piece at the rate of about 1½ mile a day, and will finally attain the length of 24 miles. The weight of the entire rope when finished, it is estimated, will be from 170 to 180 tons. From time to time, as the work proceeds, a galvanic current is passed through the wires, and their conducting power is tested by a galvanometer. The results thus obtained have been hitherto highly satisfactory, and as the whole length is now nearly completed there is every reason to anticipate that the same success will attend the work to the end.

Messrs. Crampton and Wollaston, the engineers who have undertaken to carry out the necessary measures for establishing the Submarine Telegraph, have recently been engaged in exploring the coast in the neighbourhood of Calais, for the purpose of selecting the most advantageous point towards which to direct the line of communication. In performing this task, they have had the advantage of the experience of Captain Bullock, who is engaged in a survey of the Channel, and whose steam-vessel, the *Fearless*, has been placed at their disposal. The starting-point on the English coast which has been fixed upon as the most advantageous, on account of the nature of the beach and the comparative freedom from the danger of ships anchoring in the immediate vicinity, is the South Foreland. The line will be conducted down a shaft practised perpendicularly in the cliff and along a short tunnel, communicating with it at right angles, to the beach, where it is proposed to bury it at some considerable depth beneath the shingle, to the lowest level of ebb-tide. The spot slected on the French coast for similar reasons is situated about four miles to the south of Calais, near the village of Sanngate. The beach at this point consists entirely of sand, and a ridge of sandhills, called *dunes*, extends for a considerable distance along the coast, protecting the adjacent country from the inroads of the sea. The line will here be buried again to some depth beneath the sand, and on reaching the *dunes* the communicating wires will be con[...] underground to the telegraphic stat[...] Calais, and brought into immediate [...]munication with the network of teleg[...] lines radiating from that point, an[...] bracing the principal towns of the con[...] The line from the South Foreland to [...]gate, although not the shortest that c[...] drawn between the two coasts, is that [...] presents the fewest chances of acciden[...] points at the two extremes are those [...] the low-water mark is nearest the hig[...] the character of the bed of the Channe[...] between, as far as can be ascertain[...] careful soundings, offers no pecul[...] likely to affect the integrity of th[...]ducting line.

Every facility that could be giv[...] establishing the connection of th[...] marine wire with the telegraphic sta[...] Calais has been granted with the [...] alacrity by the French Government, [...] as by those persons whose local righ[...] to be consulted before the line could [...] down; and should the experiment [...] successful, as far as regards the tr[...]sion of the galvanic fluid across the C[...] the communication with Paris will be [...] immediately established.—*Times*.

SPECIFICATIONS OF ENGLISH PATEN[...]ROLLED DURING THE WEEK, [...] SEPTEMBER 10, 1851.

WILLIAM MILNER, of Liverpool, [...] box manufacturer. *For certain im[...]ments in boxes, safes or other depo[...] for the protection of papers or other [...]rials from fire.* Patent dated M[...] 1851.

The improvements claimed as consti[...] this invention, are

1. The closing or securing of the [...] or lids of boxes, safes or other depo[...] for the protection of papers or other [...]rials from fire, by means of a cont[...] bolt or bolts on all sides of the sa[...]tending or expanding into suitable [...]nuous grooves, rebates or recesses, [...] within the front of such safes or [...] whether the simultaneous action o[...] continuous bolt or bolts be effected by [...] of grooved plates acting against pins [...] or by other equivalent mechanical [...]trivances; and whether such cont[...] bolt or bolts be expanded or extende[...] the door or lid into suitable grooves, [...] or recesses formed within the front [...] safes or boxes, or *vice versâ*, that is [...] the front of the safes or boxes into g[...] formed in the door or lid.

2. The coating of the interior surf[...] the outer part of safes (constructed as

d, or according to the methods by Thomas Milner, in 1840,) with formed of pulverized quartz, or ilar infusible material mixed with roportion of hard wood-dust, for the of more effectually resisting the intense fire on the interior of such

AM EDWARD NEWTON, of Chan-, civil engineer. *For improvements ble bedsteads, and in sacking bot-* Being a communication.) Patent arch 3, 1851.

ame of the portable bedsteads which of the subjects of this patent is d of four telescopic corner-posts mber of other vertical rods, which ected to the corner-posts and to er by levers arranged on the prin-the lazy tongs—one of the ends of r being attached to a ring sliding own on the rod, so as to admit of frame being folded up into small . With the same view, the corner-s composed of three tubes, each ithin the other, and being held by spring catches: these tubes d, in order to prevent their turning ne within the other. The sacking ded by rings and hooks from the osts and rods, and arrangements for raising the head end of it to a ufficient to dispense with the em-t of a bolster.

nprovements in "sacking bottoms" n making the same with gores or f elastic or other material, inserted ent parts to suit the swell of the the shoulders, hips, &c.

.—1. Making the frame of a port-stead of a series of vertical rods, d or combined together with levers rinciple of the lazy tongs, for con-of folding into small space when .

spending the sacking bed of canvas suitable material or fabric by means , rings, or other equivalents there-hed to the sides or edges thereof; rting into the sacking, gores, gus-pieces to fit the swell of the body at lders, knees, hips, &c.

ED VINCENT NEWTON, of Chancery-chanical draughtsman. *For im-nts in the preparation of materials roduction of a composition or com-applicable to the manufacture of knife and razor-handles, inkstands, bs, and other articles where hard-ngth, and durability are required.* a communication.) Patent dated , 1851.

nufacturing the compositions which form the subjects of this patent, the inventor employs caoutchouc or gutta percha, or both combined, which he mixes with equal parts of sulphur, and submits to a temperature of 250° to 300° Fahr., for a time varying from two to six hours, and thus obtains a substance possessing characteristics analagous to those of bone, horn, or jet. Carbonate or sulphate of magnesia or lime, or carbonate or sulphate of lime, or calcined French chalk and magnesian earths, together with salt of lead or zinc of any colour, shellac and rosin, or other vegetable and mineral substances, are also recommended to be added to the composition in proportions varying from four to eight ounces to every pound of gutta percha or caoutchouc. The composition may be moulded into articles of the required forms, and then subjected to heat, and hardened by being enclosed in fine sand or other suitable material. Or, after having been rolled into thin sheets, it may be applied to wood or iron, and caused to adhere thereto by heat. Ornaments composed of this composition may be also caused to adhere to elastic bands by submitting the bands to the vulcanizing process while in contact therewith.

Claims.—1. The treatment of caoutchouc or of gutta percha, or of caoutchouc and gutta percha combined, in the manner described, for the production of a composition having the qualities above mentioned.

2. The completion of the manufacture of the composition or compositions described, after their application to the manufacture of the articles mentioned in the title.

Specification Due, but not Enrolled.

JAMES LEACH, of Littleborough, Lancaster, cotton spinner. *For certain improvements in machinery or apparatus for carding, spinning, doubling, and twisting cotton and other fibrous substances.* Patent dated March 3, 1851.

RECENT AMERICAN PATENTS.

APPARATUS FOR ROLLING TAPERED METALLIC RODS. *William Clay.*

My invention consists in the adaptation to rolling machinery of pistons bearing against confined columns of water, or other non-elastic fluid, the ends of the piston-rods maintaining or affording the means of keeping the bearings of the rollers from shifting their positions, excepting as the columns of water are allowed to relax their resistance, by a slow and gradual escape of the fluid from the cylinder or chamber through an adjustable valve.

Claim.—What I claim as my invention,

is permitting the rollers to recede from each other by means of the hydraulic apparatus, constructed and arranged substantially as described.

And, secondly, the adjustable screw in conjunction with the apparatus claimed above, whereby bars of metal are enabled to be rolled tapered for a portion of their length, and parallel for the remaining part thereof.

IMPROVEMENT IN SMUT MACHINES. *John Hollingsworth.*

Claim.—The scouring and freeing wheat of smut and other impurities, by throwing up the grain on to the inclined face of a chimney, fitted to an opening along the top of the concave in combination with the inclined aprons U, for transferring the grain from end to end of the cylinder, that it may be discharged, as set forth.

IMPROVEMENT IN WINNOWING MACHINES. *Oliver Elnier.*

The object of my improvement is to effect a more perfect separation of the wheat from the pieces of straw, chaff, cheat, small grains, dirt, &c., than can be performed by machines of the ordinary construction, unless the grain is run through them two or more times.

Claim.—Placing the screw in an inclined position above the fan, and extending the whole length of the machine, by which the wheat is thoroughly sifted before being acted upon by the blast, in combination with the direction of the blast at right angles to the screw, as above set forth.

WEEKLY LIST OF NEW ENGLISH PATENTS.

Gail Borden, Jun., of Galveston, Texas, in the United States of America, manufacturer, for improvements in the treatment of certain animal and vegetable substances, to render them more convenient for use as articles of food, and for their better preservation. September 5; six months.

John Blair, of Irvine, Ayr, North Britain, gentleman, for certain improvements in beds or couches and other articles of furniture. September 11; six months.

John Rowland Crook, of Birmingham, hatter, for improvements in hats, caps, and bonnets. September 11; six months.

David Main, of Beaumont-square, Middlesex, engineer, for improvements in steam engines and in furnaces. September 11; six months.

William Jean Jules Varillat, of Rouen, France, manufacturing chemist, for improvements in the extraction and preparation of colouring, tanning, and saccharine matters from various vegetable substances, and in the apparatus to be employed therein. September 11; six months.

WEEKLY LIST OF DESIGNS FOR ARTICLES OF UTILITY REGISTERED.

Date of Registration.	No. in the Register.	Proprietors' Names.	Addresses.	Subjects of Design.
Sept. 5	2935	A. Marion and Co.	Regent-street	Pencil-cutter and sharpener.
,,	2936	John Classon	Dublin	Royal Victoria ink-holder.
,,	2937	William Healy	Dorset-street, Salisbury-square.	Portable bath-heating apparatus.
,,	2938	Benedict Barnard, Alfred Rosenthal, and George Burton	Cheapside	Pearl edge braiding-machine.
6	2939	Barnard Riege	Finsbury-square	German air-gun.
,,	2940	John Ranson	Bury	Graver-holder for engraving print-rollers.
,,	2941	Edward Davis	Leeds	Pressure gauge.
10	2942	William Heslop Barnes	Coningsby, Boston	Portable mangle and linen press.

WEEKLY LIST OF PROVISIONAL REGISTRATIONS.

Sept. 8	281	Thomas Lewis	Lymington	Multiplex coat.
10	282	Julius Roberts	Portsmouth, Lieut., R.M.A.	Spur.

CONTENTS OF THIS NUMBER.

LONDON: Edited, Printed, and Published by Joseph Clinton Robertson, of No. 166, Fleet-street, *in the City of London*— Sold by A. and W. Galignani, Rue Vivienne, Paris; Machin and Co., *Dublin; W. C. Campbell* and Co., Hamburg.

Mechanics' Magazine,

MUSEUM, REGISTER, JOURNAL, AND GAZETTE.

No. 1467.] SATURDAY, SEPTEMBER 20, 1851. [Price 3*d*., Stamped, 4*d*.

Edited by J. C. Robertson, 166, Fleet-street.

HORN'S PATENT CARPET-BEATING AND CLEANSING MACHINERY.

Fig. 1.

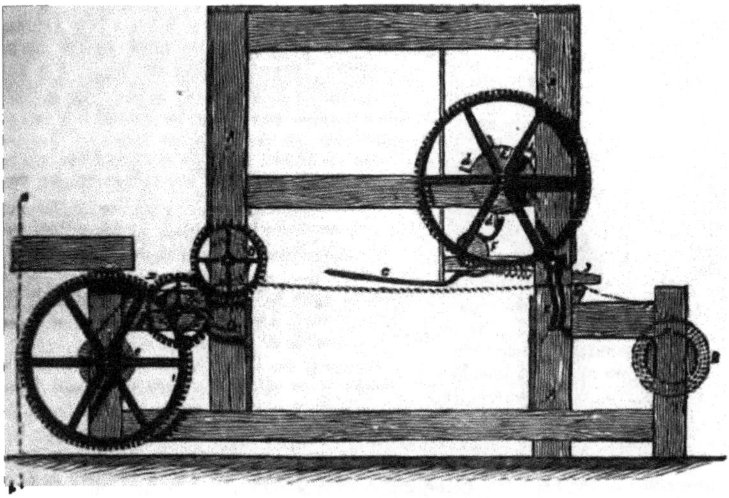

Fig. 2.

HORN'S PATENT CARPET-BEATING AND CLEANSING MACHINERY.

(Patentee, Thomas Horn, of 13, Little Stanhope-street, May Fair. Patent dated March 10, 1851. Specification enrolled September 10, 1851.)

Specification.

My apparatus or machinery consists of two parts, both, or either, of which may be used for cleansing carpets, matting, and other similar fabrics, according to my invention. One part of this machinery is intended to be employed to beat and brush out the dust from carpets, matting, and other similar fabrics—an operation now performed by hand; the other part of the machinery is intended to be used for further cleansing such articles by scouring them, and then drying them, so as to make them fit for use.

Figs. 1 and 1ᵃ represent a side elevation of an arrangement of machinery, embracing both the said objects; the parts on the right of the central line *a b* (fig. 1) being those appropriated to the beating and brushing processes, and those on the left of that line (fig. 1ᵃ) those required for scouring, dressing, and drying. A A is a strong framework; B is a wooden roller, mounted in standards at the rear or front end of the apparatus, on which the carpet (or other article) to be cleansed is rolled; and C is a similar roller, mounted in standards on the opposite side (of the first division of the apparatus), by which the carpet is drawn through the machine, and on which it is wound after being beaten and brushed. Each of the rollers B and C is furnished with a canvas web, which is sufficiently long to extend from the one roller to the other. One end of each of those webs is attached to its roller, and the other end of it is furnished with metal eyelet-holes, by means of which one end of a carpet to be cleansed may be secured to it.

The rollers B, C, and E, must be made of such a length, and consequently the machine and the canvas webs must be made of such a width, as to receive the widest carpets which the machine is intended to be capable of cleansing.

E is a large wooden cylinder, which is mounted on the framework a little way in advance of B, and at a sufficient height above it to allow the carpet to pass freely between them. *d d* are several tappets, which project at right angles from the cylinder E, and each of which catches against an eccentric F as the cylinder E is made to revolve. G is one of several beaters or flails, which are fixed parallel to one another across the whole width of the machine, each of which beaters is fixed upon an eccentric, the eccentric being mounted upon an axle in the frame. The tappets *d d* alternately raise and liberate each beater; and in this way they are made, by means of their weight, and by springs attached to them, as shown in the figure, to give a succession of smart strokes to the carpet passing beneath them. D D′ are two cylinders, upon which are mounted radial brushes, to brush the carpet on both sides as it passes between them. One end of a carpet to be cleansed is to be secured by means of twine to the end of the web attached to the roller B, the other end of the carpet being in like manner secured to the end of the canvas attached to the roller C. The web attached to the roller B, and the carpet secured to it, is then to be wound upon that roller, the canvas web attached to the roller C being, therefore, drawn through the machine with one end of it near to the roller B, and in such a position that it may draw the carpet through the machine whilst it is undergoing the operations of beating and brushing. A bar *x*, which extends across the whole width of the machine, is placed in such a position that the carpet, and also the web to which it is secured, may be drawn over it. Above this bar is placed a similar bar *y*, mounted upon hinges, and weighted so as to cause it to press slightly upon the carpet and canvas web when passing between the bars, and thus form a drag for the purpose of causing the carpet to be stretched between the two rollers B and C sufficiently tight to sustain the action of the beaters.

When the carpet leaves the brushes, it passes over a bar or roller Z; the object of which is to prevent the carpet pressing too much upon the brushes mounted upon the cylinder D′. Motion is given to the different parts by wheel gearing, in manner represented in the engraving; that is to say, the carpet is, by means of the canvas web attached to the roller C, drawn through the machine, and wound on that roller by ning a crank-handle H, which turns two wheels of equal numbers fixed on the is of the brush cylinders D and D′. On the axis of the wheel D′ there is a ion, which takes into a wheel I on the roller C. The cylinder E, through which beater is actuated, is turned by a crank handle K, which turns a pinion *f* which tes into a wheel L on the cylinder E. The whole of the apparatus just described mely, that on the right side of the line *a b*, is boxed in by closely fitting boarding, order to prevent the dust from flying about, and also to deaden the noise caused the beaters. Should the carpet be so charged with dust as not to be sufficiently ed from it by passing it once through the machine, it may be passed through a cond time, or oftener if necessary, to effect the object in view. If it is further quired to cleanse the carpet from stains or grease or to freshen it in appearance, it next transferred from the beating and brushing division of the machinery to the uring and drying part, which is on the left of the line *ab*, and may be left unin-med. M is a damping cylinder, which may either revolve along with the carpet, may have a reciprocating motion given to it by being connected to some of the volving parts of the machinery, as shown in the engraving, for the purpose of isting in the scouring the surface of the carpet or other fabric. This cylinder is ated interiorly by steam, hot air, or water, and has several layers of felt, blanket-g, or other fabric wrapped round it, which are kept constantly in a moist state by jet of water *m* directed upon the cylinder from some convenient source of supply, by having the cylinder perforated with numerous small holes, and introducing e moistening liquid by a pipe into the interior of the cylinder. N is a brushing linder, on which are fitted several brushes of a considerably finer quality than those ounted upon the brush cylinders D and D′; and upon the end of the cylinder N mounted a pinion which takes into the wheel R, from which it receives motion.

Fig. 1a.

When the roller M is intended to have a reciprocating motion, that motion is oduced by the means shown in the engraving, the motion being received from the nion upon the end of the roller N, as shown. When this reciprocating motion not required, the connection of the parts with the pinion at the end of the roller must be broken, leaving the roller M to be turned by the carpet as it passes under . O is a drying cylinder, kept at a moderate heat by steam or hot air. P is a ceiving roller, to which is attached a canvas web (similar to those attached to the llers B and C) for the purpose of drawing the carpet or other fabric through the achine, motion being given to this roller and its web by means of the crank-han-e Q, which works a pinion which takes into the wheel R on the end of the ller P.

In order to pass a carpet through this part of the machinery, it is first discon-ected from the canvas web attached to the roller B, and the end of the carpet then

N 2

disconnected from that web, is then secured to the web attached to the roller P, in like manner as already mentioned.

The carpet or other fabric as it is drawn from the roller C, will pass beneath and be subjected to the action of each of the three cylinders M, N, and O, and then be finally wound upon the receiving roller P. In case the passing of the carpet once through the machine should not produce the required effect, it may be passed through it a second time, or oftener if necessary.

A modification of the preceding arrangement of machinery, in so far as regards the first part or division of it, is represented in fig. 2. The chief points of difference in this modification are; first, that the beaters are made of gutta percha or some other similar flexible material, and are hinged to a rotating cylinder, and they are thus made to strike or act upon the carpet or other article. And, second, that the brushes have a to-and-fro or reciprocating instead of a rotary motion given to them. S is a rotating cylinder with its beaters (which I call centrifugal beaters), and T the to-and-fro movement by which the brushes are actuated. The wheel and other gearing by which these parts are worked is clearly shown in the engraving, and needs no further description.

Either division of the machinery may, as before mentioned, be used independently of the other. For example, a carpet may be beaten and brushed by hand in the ordinary manner, and then passed through the scouring and drying part of the machinery only; or a carpet may be passed through the beating and brushing part of the machinery only, without passing through the other part. In using that part of my machinery which is for scouring carpets and other fabrics, any of the agents or materials ordinarily applied in scouring such carpets or other fabrics, may be used to effect or complete the desired scouring or cleansing operation.

ON WARMING AND VENTILATING.—WITH A DESCRIPTION OF AN IMPROVED TUBULAR STOVE.—(CONCLUDED FROM P. 206.)

This stove was fixed at the latter end of December, and has been in daily use for four months without the slightest difficulty of management, and with entire satisfaction.

During this time careful observations have been made upon it; and the results are in many respects remarkable. Within an hour after the fire is lighted, the air issuing from the air-passages is warmed to a comfortable temperature, and soon attains a heat of 80°; at which it will be maintained during the day with a moderate fire. The highest temperature that has been attained has been 95°, whilst the lowest on cold days, with only a small fire, has been 70°. The result of twenty observations was: two instances where the temperature was 95°, in which the fire was large and the door of the room open, whereby the draught through the air-tubes was diminished; five instances below 80°, averaging 75°; the remaining fourteen gave an average of 80°. The mean temperature of the room at the level of respiration was 61°, while the uniformity was so perfect that thermometers hanging on the three sides of the room rarely exhibited more than one degree difference, although two of the sides were external walls. As might be expected, there was no sensible draught from the door and window. On observing the relative temperatures of the inflowing and general air of the room, it appeared that there must be a regular current from the ceiling down to the lower part of the room, and thence to the fire. The inflowing current being of a temperature nearly approximating to that of the body was not easily detectable by the hand, but on being tried by the flame of a candle, it was observed to be very rapid, and to pursue a course nearly perpendicular towards the top of the room, widening as it ascended. It was also noticed that the odour of dinner was imperceptible in a remarkably short time after the meal was concluded. In order to trace the course of the air with some exactitude various expedients were made use of. It was felt to be a matter of great interest to ascertain, if possible, the direction of air respired by the lungs. The smoke of a cigar as discharged from the mouth has probably a temperature about the same as respired air, higher rather than lower, and was therefore assumed to be a satisfactory indicator; on its being repeatedly tried, it was observed that the smoke did not ascend any material height in the room, but tended to form itself into a filmy cloud, at about three feet above the floor; at which level it maintained itself steadily, while it was gently wafted along the room to the fireplace. In order to get an abundant supply of visible smoke of a moderate temperature, a fumigator, charged with cut brown paper, was used. By this means a dense volume of smoke was obtained in a few seconds; and it conducted itself as in the last-mentioned experiment. On discharging smoke into the *inflowing air-current*, it was diffused so rapidly that its course could not be traced, but in a short time no smoke was observable in the room. Another experiment was made with a small balloon charged with carburetted hydrogen gas, and balanced to the specific gravity of the air. On setting it at liberty, near the air-opening, it was borne rapidly to the ceiling, near which it floated to one of the sides of the room, according to the part of the current in which it was set free; it then invariably descended slowly, and made its way with a gentle motion towards the fire. The air has always felt fresh and agreeable, however many continuous hours the room may have been occupied, or however numerous the occupants.

It is difficult to estimate the velocity of the inflowing current; but if it be assumed to be 10 feet per second, there would pass through the air-tubes in twelve minutes as much air as will equal the contents of the room.* And as it appears that the air so admitted passes from the room in a continuous horizontal stream, carrying with it up the chimney the rarefied air, the exhalations from the persons present, the vitiated air from the lamps or candles, and all vapours rising from the table, it is by no means surprising that the air should always be refreshing and healthful. Since this stove has been fixed, two others have elsewhere been fitted up on the same principle, and have been found to exhibit similar satisfactory results.

That this form of stove should conduce to economy of heat must be apparent, when it is considered that the air which is of necessity drawn into the room by the fire has free admission in this case through appointed passages, in which it must take up more or less warmth; while in an ordinary stove the air is drawn in at the doors and windows, much of it at the temperature of the external air. It is, however, not at first sight so evident how there is at the same time an economy of fuel. It may be asked, where does the wasted heat of an ordinary stove go to? The intense heat to which the hearth, the cheeks, and jambs of an ordinary stove are raised, produces through the fire a very rapid draught, which carries up the chimney a large proportion of the heat which should be diffused Now, by converting the hearth-plate and jambs into air-tubes, part of the heat is absorbed as given out by the fire, and the intensity of the draught is diminished without hindrance to a steady and continuous current up the chimney.

At the same time there is the cheerfulness and comfort of radiated heat from an open stove, and facility for the enjoyment of the fireside without being subject to an inconvenient heat, whilst in most modern stoves the draught is so great that we have either an intense fire scorching all within its range, or a small fire which burns out so rapidly as to require constant attention. Dr. Reid is of opinion that a fire burns most advantageously for warming a room when air enters only in front, and that an ash-pit should be provided with means for preventing the ingress of air below the fuel. The cooling of the hearth-plate by passing air beneath it tends to check combustion, while a direct horizontal current of air to the whole surface of the stove promotes a lively combustion in front of the fire; the only part which radiates effectively into the room. The results obtained by this construction of stove may be thus summed up:

I. The prevention of a smoky chimney.

As provision is made for a regular inflow of air to the room, the outflow of the smoke may be deemed tolerably certain, as an open door or window almost invariably secures that end. There is surely no reason why chimney-shafts should not be

---

* It has been estimated that in a chimney of 50 feet in height, having a mean temperature of 20° above the external air, the velocity of the ascending current will be 11½ feet per second, and a gentle pleasant wind is computed to move at the rate of 10 feet per second, or nearly seven miles per hour.

made to contribute to the picturesque character of a building. Some of our most beautiful modern buildings are quite deformed by the chimneys and their adjuncts.

II. Equal distribution of warmth and freedom from draughts.

While in an ordinary stove the heat is obtained by radiation only, in this we have also a copious supply derived from the warmed air, which is dispersed over the room insensibly.

III. Complete ventilation, unaccompanied by any sensible current.

There being a continuous flow of air of a temperature exceeding that of the room distributed over the upper part, while there is a continuous drawing off of air at the lower part by means of the fire, it follows that there will be a regular current in the upper half of the room *from* and in the lower half *to* the fire-place. The movement of the air is so different from what usually takes place within the walls of a house, that it is probable some persons will have doubts about the desirableness of it. Man's labours are most successful when he follows Nature as a guide. The air around the globe we inhabit is constantly in motion, mainly in a horizontal plane, which secures to the productions of the earth a more complete change than they could otherwise have. So with ourselves, by a horizontal movement of the air the lungs obtain a supply which has not been influenced by contact with other parts of our frame.

The means by which these very important results are obtained are essentially of the most simple and inexpensive nature. All that is, in fact, necessary is, to have around the fire-grate four tubes with open ends, connected together; one placed horizontally beneath the hearth, and furnished with a duct from the outer air, while another, placed horizontally beneath the mantel, has an opening left in it for the egress of the air which has passed from the lower one. The whole being set somewhat in advance of the fire-grate, it is almost impossible that the air can be overheated. The arrangement of the joists around the hearth presents a difficulty in the carrying of the supply tube to the hearth in some houses; but where the chimneys are against an external wall, the expense of such an adaptation is insignificant while in the construction of new houses it will be shown to be actually economical, independently of the saving of fuel which results.

There can be no doubt that spacious and lofty rooms are both elegant and convenient. They are also found to be more comfortable and healthful, even when occupied, by a very limited number of persons. This preference for large rooms, is based on a consciousness that a longer time is required to contaminate the air to an unbearable extent than in small ones, which implies a practical admission of the imperfectness of ordinary arrangements in regard to ventilation. As it is found difficult to obtain fresh air, we naturally wish for it as little poisoned as possible, and therefore select large rooms; which nevertheless only mitigate the mischief. It must be admitted that it is desirable that air that has been in contact with us, should not return again even in a diluted condition. If we can establish in a room a continuous flow in one direction, in an abundance proportioned to the number of persons present, there can be no reason why even a small room should not be perfectly healthful. In the stove already described, this continuous change of air is secured; hence it is stated to be economical, as it admits of the rooms of a house being constructed of a lower elevation than is usual, without prejudice to their healthfulness.

The true principle for the construction of our houses in reference to ventilation would appear to be this. First, that every stove should be an instrument for the introduction, as well as the withdrawal of air; which should, before it enters, be warmed to a suitable temperature; secondly, given the quantity of duly-warmed air that the stove may be instrumental in introducing, you have a measure of the capacity of the chamber for occupants; thirdly, that when a room is intended for the occupation of a large number of persons, the number of stoves should be increased. Supposing it practicable to make arrangements for impelling air into a room, with an increased rapidity when the number of persons in it is great (a matter extremely difficult to accomplish), it will be found to be objectionable. In certain conditions of the atmosphere a rapid movement of the air around the body is distressing from the absorption of moisture from the skin. Hence we may assume a certain rate of movement as a maximum; if this does not change the air in a sufficiently short time, we ought to increase the inflow and outflow. If this be done, it will be immaterial whether the room be occupied by one or by fifty persons. In a river, of which the current flows at the rate of two miles an hour, the quantity of water which passes a bather is precisely the same whether the width of the stream be 10 yards or 100 yards; but in the one case five persons may swim abreast, each having a continuous supply of fresh water; in the other fifty persons. The amount of animal heat evolved in densely-crowded rooms is such as probably defies any methodical ventilation; but there do not appear to be insurmountable difficulties in the case of moderately filled rooms. The quantity of fuel-heat distributed must of necessity be diminished where a large number of persons are in one room, but the supply of air should of course be increased. The tubular stove warms a room, partly by direct radiation and partly by communicating heat to the air passing beneath the hearth and through the tubes, which are warmed by the radiated heat of the fire. Now it must be borne in mind, that the smallest increment of heat will cause air to move upwards; if therefore the air in the tubes be raised in temperature only 1°, an inflow will be obtained if no obstruction be offered. Where we have a fire burning beneath a chimney with a good draught, we have in effect a manageable, economical and powerful pump. If a bright metallic screen be placed before the fire, it will prevent direct radiation; and by reflecting to the grate the heat which would otherwise have radiated into the room we increase the pumping power of the fire. By forming the screen so as to make it interpose between the tubes formed in the hearth and the sides and the fire, we cut off the supply of radiant heat to the air-passages.

It is however probable that the air would still take up some heat in its passage over those parts of the tubes which are in contact with the brick-work around the fire-grate. Suppose, however, that instead of making the fire in the grate it be made in a fire basket placed within the grate, then all the heat would be intercepted by the screen, except the insignificant quantity which might be radiated or reflected to the upper tube. When we consider the high temperature at which air is agreeable when it is in motion,—75° having been found not too high with rapid movement; it seems not improbable that with such an arrangement a fire might be made to impart a cooling effect, even in warm weather. In the case of diseases the importance of an abundant supply of fresh air is now generally admitted; there is however considerable difficulty in ventilating a sick room, even in the warmest weather, without hazard to the patient; open doors and windows are attended with draughts, not under control. By admitting at the mantel-shelf air which is at once distributed over the upper part of the room in a diffused state, and at the same time withdrawing air at the fireplace, the risk of cold from a draught would be avoided. But a still more important point appears to be attainable. When we consider the number of lives annually sacrificed by attendance on the sick bed,—not only in contagious diseases, but also in numerous others, where continued occupation of one room of necessity so contaminates the air, as to make it poisonous to delicate organizations, worn with fasting, watching, and tender anxiety,—we must be impressed with the immense importance of mitigating, if we cannot subdue, so serious a drawback on human health and happiness. Hospitals, barracks, and other such rooms, require peculiar attention in the ventilating arrangements, or they become nurseries for disease.

Supposing the beds in a sick room or hospital to be ranged about 4 feet from the wall which faces the fireplace, and that we can produce a continuous flow of the air in the lower half of the room to the fire, the medical attendant, nurse, or visitor might approach the bed-side of the patient with but little if any risk of inhaling the air which has passed the patient, while the latter is constantly supplied with fresh air from the upper part of the room. They who are called by duty to visit the houses of the labouring classes, find their power of endurance more taxed, and their health more hazarded than in the wards of a hospital rife with disease; the bulk of the population living in small rooms, frequently occupied day and night continuously, by parents and children in sickness and in health, at times even by the dead as well as the living. In no case is there more urgent necessity for the application of a cheap and simple method of warming and ventilating; while our houses are constructed with such a disregard to the natural laws, we cannot say that we profit by the means placed at our command by a bounteous Creator, for the mitigation of human ills. We fly by thousands to our coasts for change of air to restore enfeebled health, while we passively submit to have our houses constructed in a mode irreconcileable with common intelligence.

The *principle* on which the tubular stove is constructed is not new. It is nearly a century and a half since the Cardinal Polignac published a description of a stove which was furnished with very complete arrangements for supplying an apartment with fresh air, warmed by its being made to circulate through passages at the back of the stove; and within these few years, some very excellent stoves have been introduced on the same plan. They are, however, all of such a construction as to make them necessarily expensive, and therefore to place them out of the reach of the poor.

There is also this difference between them and the plan here adopted. In the former, the air is made to circulate at the back of the stove; while in the latter, it is carried through passages in front of the fire-grate, thereby preventing the possibility of the air being overheated, and admitting of a modification of the heat supplied. As this is accomplished by simply forming the face of the stove (where a register stove is used), as a continuous tube instead of a mere shell, the only additional cost would be from the increased weight of metal, which would probably not exceed one-fourth. For the dwellings of the poor, tubes of wrought iron, built around the chimney jambs would, without doubt, be found quite effective and very economical.

It is not evident that there is any obstacle to the construction of air flues in the walls of a house adjoining the chimney stacks. Supposing a general air flue to be carried up at the side of the smoke-flues, with branches to the several fire places, we have at once a ready means of connecting the latter with the external air.

Assuming an opening of 48 inches to be sufficient for the supply of ordinary rooms, the only requisite would be to construct such air flue at the lowest front of the house of a width of 18 inches, and a depth of 9 inches, for the supply of three rooms on different stories, diminishing the width 6 inches as each floor is passed. The flue might be carried horizontally under the floor of the basement, with an external upward branch to any level that might be desired, to insure a pure supply. If this external duct were placed on the north side of the house, it seems likely that there would be an upward current day and night without the assistance of a fire,— the upper part of the air-flue would probably be warmer than the external air on the north side of a house, a few feet above the ground; and if the lowest part of the external duct were formed in the same way as a drain trap, there could scarcely be a probability of the air current being reversed. The opening for entrance of air to the external duct might be carried up to any height where the air could be relied on as being uncontaminated, provided the temperature of the height selected was not higher than that of the internal air flues. The temperature of the air at 6 feet above the earth is not usually more than 2° higher than at the earth's level; and if the air supply be drawn from the north side of the house, a relatively low temperature may be insured at all seasons of the year. The importance of this consideration in reference to districts subject to malaria can scarcely be overrated. When *close* stoves are used, it is found necessary to adopt means for moistening the air. It is not evident why the same necessity does not exist in the case of an open fire place, except on the supposition that the general air of rooms so warmed is not raised to the same temperature as is obtained from close stoves.

Whenever it is found necessary to add moisture to the air it is easily accomplished when there is an inflow of warmed air, as in the stove described. Supposing the system of warming and ventilating to be adaptable for large rooms used for public meetings, there are grounds for thinking it might act favourably in the transmission of sound. If the stoves were placed on one side of the room, and the speaker on the same side, the air stream would tend to carry his voice directly to his auditors.

Description of a complete form of the Tubular Stove.

Fig. 1 (p. 201) represents a horizontal section.

Fig. 2 a vertical section on the line X Y; *a* is a flue 6 in. × 9 to conduct the external air from the outer wall to the under side of the hearth plate *b*. It is here shown as formed in the back of the chimney shaft,—but it might enter at the side, or front, where convenient for it to do so; *c c* are openings in the hearth plate *b*, communicating with two upright tubes of a corresponding form, which conduct the air entering at *a*, upward, to the horizontal tube *d*. This tube is fitted to the two upright tubes, and has an opening extending along its whole length. Supposing the width of the stove to be 3 feet, this opening should be 1¼ inch wide. The stove should be set 1¾ inches forward, from the chimney breast; *f* is the upper mantel, which stands forward from the chimney breast *e* 1½ inch; *g* is the mantel-shelf, which has part of the back next the chimney breast cut away to continue the air-passage; *h* is a thin slab of marble, 1½ inch deep, built into the chimney breast, and extending to the width of the mantel; this serves as a support for a chimney-glass, and also to divert the current of air from flowing directly up the chimney breast; *i* is a strip of metal (or marble), which may be made ornamental, and which serves to guide the air stream upwards. By moving *i* to *h*, the supply of air may at once be diminished or stopped at pleasure, or *i* may be fixed, and the same end be attained by having between *i* and *h* a thin strip of metal fixed on centres at the extremities, and made to act like a throttle valve. As the mantel shelf is commonly more than a foot wider than the stove, the opening between *i* and *h* need not be greater than 1 inch, in order to obtain a total opening of 48 inches; which should be maintained throughout all the passages.

Fig. 3 exhibits the circulation of the air in a room fitted with the tubular stove.

Ll.

STEEL-PEN GRINDING.

Sir,—Every one is aware of the evil results of dry-grinding. In Sheffield and elsewhere, generation after generation of fork-grinders follow each other to the grave, not by the natural course of human decay, but owing to the tubercle-smitten lungs, induced by the mineral particles inhaled during respiration. A new branch of industry has sprung up which threatens to consign other victims to the tomb. I advert to steel pen-making, now conducted by thousands of girls in Birmingham, and in every note of which, dry-grinding is employed. Steel pens are about the most abominable invention for writing purposes that ever was made,—inflexible, lasting but a short time, worthless. They bear no sort of comparison with goose-quill pens, which by splitting up the quill into two equal halves, then subsequently quartering it, can be formed four from each quill, cheaper, better, and every way preferable, without, so far as mineral dusts are concerned, a shadow of prejudice to the health. The French, also, make good pens from the legs of fowl; of which I enclose one. I have long thought that a writing-machine might be invented for general use, by which every one, so far as penmanship was concerned, might clothe his or her thoughts in decent attire, to the infinite comfort of printers and all others. I know there is an ingenious machine for this purpose in the Exhibition, but it is at once much too cumbrous and expensive for common use. I should like to see a writing-machine as cheap and as portable as a common letter-copying-machine,—a writing-machine by which one might write a letter and copy it at the same time. Writing is an art in which few excel. I hope to see the time when elegant writing-machines shall be seen in every sitting-room, so that every miss and master, and mammas and papas likewise, shall be able to produce a fair legible line, in place of the horrible scrawl in which most of us indulge. Add to this advantage, that we should be probably able to write quicker with a copying-machine, as well as devote the time now thrown away in learning caligraphy, to drawing and the fine arts generally.

H. McCormac, M. D.

Belfast, September 8, 1851.

THE "AMERICA."

Sir,—Lord Bacon observes, "That we may not reason with certainty from consequences, if the results forming our premises depend on several antecedents, and if we are ignorant how many of them are *elements* and how many are only accidental, or mere *circumstances*." The lesson which we are bound to learn from our defeat by the yacht *America* is affected by a difficulty of this sort; for, supposing that we have resolutely set aside the opposition offered by our national prejudices to instruction in such matters, and from such masters, it is soon discovered that, with every desire to profit by our humbling, we cannot easily determine what exactly are the improvements, the novelties even, or the distinctive features of our victorious rival.

A short time ago I ventured timidly to declare in your pages my convictions regarding several of the fallacies adopted by our yachtsmen, and obstinately adhered to, propagated, improved upon, and refined by our sail-cutters; and I found that not a few men, well qualified to judge, altogether agreed with my remarks and wished for a change. Precisely at this point in so many branches of mechanical improvement do these sharp-witted Yankees step in, boldly putting into practice what the Englishman so long thinks about; and then we quietly lag after them, congratulating our country upon the *steadiness* of its progress, and smiling at the hasty ingenuity of our friends.

Here, however, we have a clear case, in which we can ill afford to laugh, but must put up with defeat, and prepare for a renewal of the struggle. I have no fear but that the hull of the *America* will be reproduced often enough to give her lines a fair trial during the next season. Her long entrance, drooping heel, beam amidships, and full run, will be seized upon and adopted without acknowledgment, and, perhaps, from being under water, without detection.

But it will be, indeed, a wonderful piece of enlightenment if our sail-makers give up at once the bagging foot, bellying leach, and reduplication of jib over the fore-sail, which have grown to such preposterous defiance of all theory. Leaving, then, the "below-water" of the *America*, I desire to make a few observations upon her sails; as I am convinced that her superiority lies chiefly there, and will be longer her peculiar property than the excellences of her build.

If we resolve the action of the wind blowing upon a surface into pressures along the normal and tangent at any point, it is found that, practically, the tangential force may be neglected, and, but for the existence of this principle (either thus generally or in some modified degree), no vessel could sail obliquely against the wind. Another property, indeed, contributes to the facility of beating to windward; namely, the difference which may be created between the resistance offered to motion in the directions of the shorter and longer axes of the vessel: but as this is wholly dependent on the form of the hull, I do not further allude to it.

Now, so far as I see, the Americans have surpassed us in the department of sail-making just in proportion to their faithful application of the principle I have mentioned. The *America's* sails have been cut so as to present a surface to the wind as nearly a plane as possible. But let us observe how very differently the sails of Rattray or any first-rate maker are cut. First, there is the ridiculous inclination in azimuth of the gaff to the main boom. This would have been long ago evident, and soon corrected, but for the fact that, if any part of a taut sail holds wind, it will prevent the other parts from shivering, and may induce the belief that the whole sail is drawing well.

A gaff mainsail, with the boom hauled well aboard, has a certain superficies of an elliptical shape (and which I figured in a former communication), properly inclined to the wind, and, singularly enough, straining pretty equally every fastening of the canvas. Not only does the sail other than this particular surface contribute nothing to the propulsion of the ship, but it is a serious impediment to its progress; and it is so, because of the very principle thus violated.

For on the windward side, from the peak to the boom, there is a ledge of canvas directly acted upon by the current generated in the direction of the boom; and besides this, there is the

iangular surface of sail, the apex
ı is the peak, and which is un-
ly at least in a neutral state, if
tively aback.
area is kept full only by sailing
tain extent off the wind. Next,
peak lies so much more to lee-
ours than in the *America's* rig,
n of the normal forces acting on
will be a pressure to heel over
el, applied in a direction so much
re inclined to the horizon, and
e the more depressing and less
ive. Surely this is enough, but
all; for the rib of taut canvas
poken of, and the whole sail thus
to the horizon, direct a down-
rrent of air along the canvas and
on the boom, which has almost
ated the roomy space given to
of the sail; and it being, I sup-
and that a dozen yards of pendent
canvas would actually hold wind,
urdity is perfected by this, the
ition of our sail-makers. A very
xamination of each of these faults
hat they encourage each other,
heir combination, produce effects
onceal their existence.
rring to the principle before
it is clear that when the wind has
the sail, the sooner we get rid of
so employed the better, whereas
ppears to be a latent fallacy which
ossesses the mind, and by which
y curved surface seems better
to *catch* the wind than one abso-
lain. Long since have our wind-
eased to display those swelling
hich the Continental and eastern
still use in theirs, and our yachts-
ust even be taught by our millers
en the pressure is oblique to the
n of motion, and when the sup-
the sail itself is free to move,
rigidly accurate plane surface is
on which the wind will act most
lly.
e years ago I made a large num-
experiments upon the effect of
the foot of a cutter's mainsail to
m, but I confess that, although
nvinced that prejudice alone pre-
ur constant adoption of the prac-
t I have been like other Britishers
ded by the universality of the
at I have continued to withstand
uth. *Had the America* been
by our cutters we should, no
doubt, have ridiculed a sail thus lashed to the boom, but now we feel constrained to explain away its effects; or if this is impossible, as a last resource, and most reluctantly, to examine thoroughly whether this novelty has been an ingredient in the causes of success or not.

For my own part I should have felt chagrined if so decided a superiority over her rivals as exists in the foreign yacht had been called forth by her constructors from the same lines, and the same system of rigging and cut of canvas, as have been used so long by ourselves, for the particular improvements on these are almost exhausted, or at least very fanciful and minute. But to be distanced by those who have bravely opened up a new path and broached and practised new theories in defiance of the prejudices which bound *us* to the former modes, is only to show more clearly than before, that *originality* is not our forte in ship-building, and to be prepared for the concession to the Americans of excellence in sail planning; as to the French we have been forced to accord the palm in hull building, reserving to ourselves the first place in sailing the ship when launched, rigged, and manned. As in a new dye for cotton, or a new construction of steam boiler, or a new quarry of undiscovered stone, so in a new principle of rig and build we have surer hopes of progress than in continual and supplementary improvements on ancient materials or processes, the very principles of which have been overlaid in perfecting their application.

One word as to the experiments on the resistance offered to various forms in passing through fluids. Many costly experiments have been fallacious on account of their having been performed upon linseed, turnip seed, small shot, and other substitutes for a fluid where there is free motion of the particles *inter se*.

It is surely unnecessary to say with what distrust results so arrived at should be reasoned from; but another scarcely less unfaithful means has been employed for applying the force to the moving body, which force is supposed to represent the pressure of the wind upon the sails.

To apply this force at a point near the water-line is to do precisely what will never be done by the natural agents of which the action is sought to be investigated. Putting this aside as preposterous, let me ask what endeavours have been made to ascertain the comparative speed of models where the force is applied at that point where in reality the resultant of all the pressures on the sails may be found to act? How many models have been made to move not in the direction of the keel, but in a line at a considerable angle to it? Lastly; what models have been moved when well "heeled over"— the actual position in which the ship itself will have to do service? These experiments on models, always unsatisfactory, have been deprived of all chances of accuracy in their results, by using imperfect substitutes for the resisting fluid, wrong points for the application of the forces, and wrong directions for their actions, as well as fallacious conditions in minuter details which it is unnecessary to refer to.

The discarding of a foresail and the elongation of the jib by the Americans is merely an extension of the principles which have for ten years been rendered apparent as correct by the increased length of foot given to that sail in every new yacht; but the raking of the foremast rather than the mainmast is an alteration which experience alone can assign its proper importance to as an element of success.

Fig. 1.

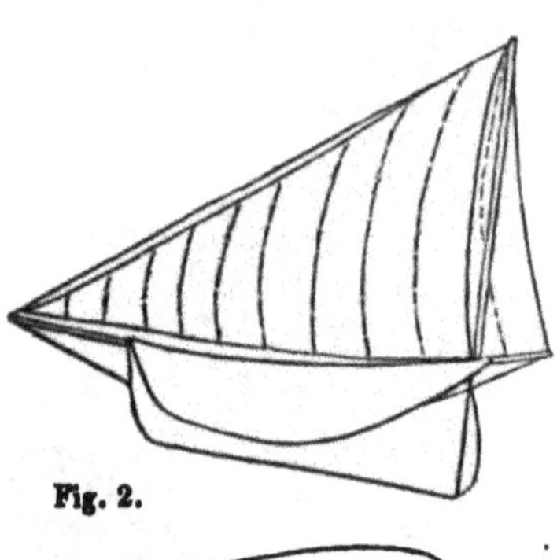

Fig. 2.

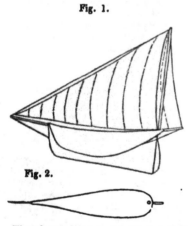

The above sketch represents a mould and rig which I think are new, and which possess several of the peculiarities, *if not the* excellencies, of the *American rig. The* mast is stepped along the stern-post, and one long jib is the only sail, with a strong spar as a bowsprit, a "drooping heel," and a "fine entrance" with a very full run. Fig. 1 is a vertical section, and fig. 2 the water-line, the sheet being made fast to a "jigger pole," which should be stout enough to carry a backstay and a sort of stern-bob-stay.

I am, Sir, yours, &c.,

JOHN MACGREGOR.

Temple, September.

THE GOVERNMENT REPORT ON PEAT.

To Sir Robert Kane, Queen's College, Cork.

Sir,—In a Report to the Chief Commissioner of Woods, professing to be on the nature and products of the destructive distillation of peat, by Mr. Rees Reece's patent for treating of peat, and bearing your signature, I find the following passage, printed as extracted from a communication of that person to you:

"To get sixteen tons of peat on the Bog of Allen, in its present state, requires 100 tons of bog and 84 tons of water to be evaporated. I think these *facts* fully demonstrate that no artificial means can ever be employed to harvest peat; and in reflecting on the various processes that have been proposed for that purpose, how little must the parties have studied the subject. Doubtless, the best of these is that proposed by Mr. Rogers, which may be said to be a brick-machine, with compressing power given from an oscillating beam; and probably the most absurd, the turf-kiln of Mr. R. Mallett, Ph. D. I have seen some of them on the Bog of Allen—one about three miles from Monasterevan. To suppose that any artificial means of applying heat to evaporate 80 or 84 tons of water to get 20 or 16 tons of peat, must suppose an arrangement of physical chemistry unknown at present."

Blue-books are supposed, by the uninitiated, to consist of positive and ascertained truths, and not to admit mere unsupported personal dicta, and never when given in words and with the obvious purpose of personal discourtesy. As the above passage was addressed to yourself, and had nothing to do with your subject in hand—being, as the lawyers say, in every sense, "irrelevant and impertinent," and, therefore, might have been well omitted—you will pardon my addressing you as responsible for this petty attempt at literary assassination. The word is justified, in that mass of Blue-book buyers take their "ex cathedra" contents for granted, and that no equally legiti-

are afforded for repelling attacks
ements published under such
sanction.

r two, however, may right mat-
aterim. There never has been,
discover, a single one of the
bove adverted to yet erected in
have made diligent inquiries
resident at Monasterevan, and
cquainted with peat-preparing,
now of none. When Mr. Rees
efore says, "he has seen some
the Bog of Allen," the assertion
at one reply.

ointed this out, I might leave
eece's opinions to be appraised
hen, in some year or two, the
he better able to estimate its
lue.

a fact that ordinary peat bogs in
drained, contain from 80 to 90
weight of water—not even in
winter; but passing this, kiln-
rf always presumes the previous
rainage of the bog whence it is
pper parts of the bog then con-
5 to 80 per cent. of dry turf;
20 to 35 per cent. of water at

like to know "what arrangement
chemistry, unknown at present"
happy lucidity of thought Mr.
expresses it—" prevents this
water from being dried out by
l heat of as much fuel as may be
o evaporate it?"

withstanding the impossibility
by Mr. Rees Reece, whose dog-
pt in your pages without doubt
kilns for the artificial drying of
cal in principle with mine (as
the Transactions of the Institu-
l Engineers of Ireland for 1845)
ve long been valued, and in con-
Holland, Wirtemburg, Olden-
ay, and Prussia; and these not
rf for domestic uses, but enable
be made with it; and if there
of the industry that abounds in
, and less talk with us, they
been in use in Ireland long

s such a mode of drying turf is
absurd," "doubtless the best,"
Mr. Rees Reece, "is a brick
ith compressing power, given
illating beam by Mr. Rogers."
, I suppose, are ignorant that
W. Williams, Lord Willoughby
nd a score of other fully compe-
, have spent fortunes in trying
the water mechanically out of
vain. This nature does forbid
so as to pay!

No man likes to be gibbetted as a fool, even in · Blue-book, or by such a person as Mr. Rees Reece; but, lest any one may fancy I have also some interest to serve in connection with these turf-kilns, I beg to add that I have none; nor have I ever seen or spoken to Mr. Rees Reece, to my knowledge.

Six years since I took pains to bring before my own profession these and other methods of preparing peat long in use, and of acknowledged value on the Continent. There has not been energy enough to try them in Ireland; yet no one has attempted to show that what succeeds abroad should not (the cases being similar) succeed here—and success in this instance would be of no small national value, as will hereafter be admitted.

I cannot but think, then, it would have been well that either you had not brought my name most discourteously into print, and on a matter quite beside your subject, or, as an unbiassed reporter, had added—to the very safe oracle you have uttered as to Mr. Rees Reece's project, which sums itself into—"it may pay, or it may not pay"—some consideration of the accuracy of the dicta which he has hazarded, and you have put forth, in connection with my name.

I am, Sir, your obedient Servant,

ROBT. MALLET.

Dublin, Sept. 1, 1851.

THE WATER-SUPPLY AND DRAINAGE OF TOWNS.—REPORT OF THE GENERAL BOARD OF HEALTH RESPECTING THE METROPOLIS.

The public voice often leads to a decision on questions requiring considerable knowledge of details, but with which few can be acquainted, considering the time that must necessarily be consumed where voluminous publications have to be consulted. On this account abstracts of them can hardly fail of being useful. The question of the water supply of the metropolis and its efficient drainage affords a case in point; public opinion will doubtless materially influence the arrangements that will be made on the subject, yet the public have not before them any concise statement of the facts on which a sound judgment must be based. With the view, therefore, of contributing to the supply of this want, the following abstract has been drawn up of the "Report of the General Board of Health on the Supply of Water to the Metropolis."

This Report includes information of a high order—geographical, geological, chemical, medical, engineering—as also matters of great importance in legislation, and in political economy. The investigations of the Board are professedly confined to what has a relation to the metropolis; yet the principal features of the Report are applicable to all considerable towns, so that it may be made extensively useful far beyond the sphere of London.

The Report is preceded by a Table of "Contents," so ample and explanatory as to afford easy means of reference to any particular on which full information may be desired.

Where, in the following abstract and observations, the expressions of the Board are employed, they are distinguished by inverted commas; as are also those of persons whose evidence is quoted.

Some of the information afforded is such as individuals may profit by; for instance, the means indicated of softening water are easily practicable on a small scale, and every private family might render water salubrious by boiling it, although it were much contaminated with organic matters in a state of decomposition.

The Board have inquired fully into that essential point—What are the qualities of water the most conducive to health, and to economical employment for domestic and manufacturing purposes?

On this question the Board refer to a great mass of evidence; and first, as to the effect of lime contained in water.

It cannot be gathered from the evidence that, excepting in calculous disorders, a small quantity of lime in water has been *proved* to be injurious to health, though it has always been suspected of being insalubrious, and is thought to be so by many of the witnesses.

The Report states that Professor Clark has invented the term of *degrees of hardness;* this will doubtless be found to be of great practical convenience. One degree of his lime scale, as it may be called, denotes that one grain of carbonate of lime is contained in a gallon of water; the several *higher degrees of hardness* denote that for *as many as are the degrees* marked, so many are the grains of carbonate of lime contained in each gallon of water.

That lime injures water for all economical purposes is fully proved, as also for most manufacturing processes, "but the natives of London are very little aware of it." We find no attention paid to it in any of the "schemes for water-supply which have obtained the aid of parochial boards."

Hardness is unfavourable to all culinary operations; thus, according to M. Soyer's evidence, hard water gives a yellow tinge to vegetables boiled in it, and gives them a shrivelled appearance. Hard water "does not open the pores of meat so freely as soft water does." Infusions of all kinds are stronger when made with soft water than with hard; so that tea made with water of 6½ degrees of hardness requires nearly a third less of the tea than when made with water of 16 degrees of hardness, being about that of Thames water. This evidence was corroborated by that of Mr. Phillip Holland and of Professor Clark. Hard water requires more fuel than soft to raise it to a boiling heat. Hard water occasions greater expense than soft in washing, on several accounts; as the greater quantity of soap or of soda required, the extra labour in washing, the greater wear-and-tear of the clothes themselves. "As far as the home-market is concerned," "more money is expended in washing clothes than in the manufacture of the fabric, or of the clothes themselves." In London, "before a shirt is worn out, five times as much money as it originally cost will have been expended on it in washing." The alkalies and mineral ingredients used in washing with hard water never fully leave the clothes.

The data on which the Board calculate the expense of mere washing are the only ones throughout the whole Report that appear to be erroneous; in this instance they seem to be so, because the prices stated for washing—whether by the piece, or from the weekly average expenditure of individuals or families—are not for washing alone, but include the charge for expenses incurred in collecting linen, drying, and folding it, mangling some of the articles, ironing several of them, starch, blue, &c.; which items constitute together at least half the expense of laundry-work. The oversight in this respect does not, however, invalidate the strong testimony given of the increased cost on several accounts occasioned by washing in hard water.

From Mr. Donaldson's evidence, it appears that for every 100 gallons of water used 2 oz. of hard soap are required to soften it for each degree of hardness; thus water of 5° of hardness requires 10 oz. of soap; if the water be of 15° of hardness, 30 oz.

ount of the money expended for in the metropolis was another According to one estimate the ills in London rise to the enor- n of 5,000,000*l.* a year, at the ite of little more than a shilling a ndividual for the whole popula- is the Board seem to think ex- and, indeed, it appears to be so; rs in some families may, indeed, n more than five shillings a week -work, as specified in the calcula- it the laundry expenses of the rtion of the industrious classes do nt to more than sixpence a head —and this description of persons great bulk of the metropolitan n.

ter is more agreeable and effective for baths, and for all purposes of e. Several witnesses affirm that, age, soft water is more agreeable ; though those accustomed to hard ome reconciled to its use.

re many manufactures that can- vantageously carried on with hard nce several of the great London have, at a heavy outlay, caused to be dug on their premises for se of obtaining water that is soft. ater of the Thames supplied to s stated to be of from 14 to 16 ' hardness, and that of the other l streams which contribute to the London is about equally hard. d have had 424 different specimens ested; they were taken from dif- tant places. Water from wells gs averaged nearly 26 degrees of , that from rivers and brooks 13 that from land and surface drain- nite 5 degrees. About 26 tons of delivered in the metropolis *daily*, dissolved in the water with which lied.

sor Clark has indicated an econo- ns of depriving water of its carbo- me by means of quick lime; and olland describes a mode of farther it by the addition of "a little f ammonia, or of soda."

ure of other inorganic impurities such as iron, or clay, is prejudicial onomical purposes; but, excepting ds, being seldom found in consi- quantity in any of the waters with wns are supplied, little notice is such impurities.

oard's inquiries respecting organic s are extensive and important. By mation they have elicited, it does ar that animals and vegetables nhabiting water are, whilst in a *living* state, insalubrious]; but that, when in a state of decomposition, they render water containing them highly injurious to health. The evidence on this head proves the fact incontestibly. Organic matters, when in that state, produce disease taken internally as a beverage, and also when the gases aris- ing from them are inhaled. A remarkable instance of the former is given by Dr. Gavin:—"A thirsty navigator drank of the Hackney brook, and was almost imme- diately attacked with cholera, and subse- quently speedily died." The instances adduced are innumerable of the deleterious effects of the emanations from water ren- dered putrid by the decomposition in it of organic matter; but "chemistry has not to the present moment succeeded in isolating those substances, or in characterizing them by particular reactions." Water got at Hun- gerford-market contained in a gallon above 13 grains of volatile and organic matter in suspension, besides 43 grains of inorganic matter. When that water was boiled down, it emitted a strong acid smell, and when heated a smell like that of burning wood. Exclusively of epidemic periods, "in ordi- nary times, it is known that troops who have drank water polluted with animal or vegetable matter in a state of decomposition are peculiarly subject to dysentery."

The *boiling* of water appears to greatly diminish the deleterious effects of such water. "There was the case of a man who lived in the Coburg-road, Camberwell parish, in a semi-detached house, in a healthy situation, and with a garden behind the premises. His wife had noticed that the water supplied to them was exceedingly bad; and having been informed that it was likely to affect the health of her family, she invariably boiled and filtered it. All kept in perfect health except the father, who objected to drink this water from its being flat and unaërated: he would still drink it as it came from the water-butt, and the consequence was that he was attacked with choleraic diarrhœa: he afterwards drank no more of it, and got well."

It is shown that lead is more corroded by pure than by impure water; but it has been unexpectedly found that filtration through sand separates the lead.

These preliminary inquiries having been gone through, as to the economical and the sanitary qualities of different waters, the next question to be considered is, From what source can London be supplied with water that shall be the most free from vitiating matters?

The Board report, that "The quali- ties for the water-supply of the population appear to range themselves in the following order:

"1st. Freedom from all animal and vegetable matter.

"2nd. Pure aëration.

"3rd. Softness.

"4th. Freedom from earthy, mineral, or other foreign matters.

"5th. Coolness in delivery at a minimum temperature,—neither warm in summer nor excessively cold in winter.

6th. Limpidity, or clearness."

To the above are added, as popular tests, that all special flavour or taste in water is objectionable.

A great mass of evidence is brought forward in proof that the streams from which London is supplied with water, *before* they come to be charged with sewer-water, already contain a vast quantity of animal and vegetable matter; that it is not any one of those streams, but all of them, that are polluted with it,—the Thames itself to an excessive degree. That though the Thames and its tributaries be largely derived from land-springs, through chalk strata, their water is in a turbid state when delivered, much of this turbidity being occasioned by animal and vegetable matter so completely in chemical solution that the common filters will not remove it; and it appears that the water of the New River, and of the water companies generally, is also charged with impurities of the same nature.

"We must state, as our conclusions upon this topic of inquiry, that if the water of the Thames could be early protected from the sewerage of all the towns draining into it, and from the sewerage of the metropolis, —if it could be purified from animal and vegetable matter as completely as deep-well water, or as some of the surface water from the chalk districts, as proposed by Captain Vetch,—we should nevertheless feel compelled, upon the evidence recited, to pronounce water of such degrees of hardness to be ineligible for the supply of the metropolis, and to recommend, as we now do,—

"That the water of the Thames, the Lea, the New River, the Colne, and the Wandle, as well as that of the other tributaries and sources of the same degrees of hardness, should be as early as practicable abandoned.

"Deep-well water is free from surface animal and vegetable impurities, but it has generally more of mineral impurity, and is usually unobtainable in sufficient quantity at a moderate expense."

Since the number that has been made of deep wells in the metropolis of late years, it seems certain that the supply from this source would be inadequate to the wants of London. Already great brewers have arranged amongst themselves to brew respectively on different days, so as to equalize the demands on the water-bed: it is further stated that water is higher in the wells on Mondays than on any other days, by reason of there being no brewing on the Sunday. This difference in the level of the water-bed is felt as far from town as Tottenham.

"Seeing the disadvantages inseparable from river and well-water, attention has been directed to other sources of supply." "Professor Clark states, that nowhere has there been made such important improvements in the collection and purification of water-supplies as in Lancashire."

"The improvement in the collection is due to the application of the principle we have above stated; that is to say, that the nearer to the actual rain-fall the water is collected, the freer it will be from adventitious impurities. The new practice in Lancashire has been to take some elevated ground, —generally sterile moorland, or sandy heath; and to run a catch-water trench or conduit round the hill, midway, or as high up as may be convenient for the sake of fall, regard being had to the space of the gathering-ground. An embankment is thrown across some natural gorge, at the nearest point at which a reservoir may be formed without the expense of excavation. Into this the rain-water is led, and stored, to be used in dye or print-works, or for other manufacturing purposes, having in many instances been previously filtered. The economy and efficiency of these filters, which merely act as strainers, are much praised by Professor Clark. They serve to show, however, how much more economically filtration may be conducted on a large than on a small scale; and how sordid and erroneous is the administration, whether of water companies, or of local boards, which neglects or refuses filtration of the supplies used for the general population. But until recently, with the exception of a very small proportion, the supply of towns was delivered without any previous filtration whatsoever, and more than half of the supply of the metropolis is still so delivered.

"The new process of land-drainage furnishes a means for the filtration and depuration of impure water on a large scale, with considerable advantages over the larger sand-strainers or common filters." "Where the drains have been tolerably well adjusted, the water from this deep drainage is seen running away perfectly pellucid. Where there happen to be two branch outfalls into one main,—the one a branch outfall from mere surface-drained land, the other an outfall from thorough-drained land,—the water from the thorough-drained land may be seen running perfectly limpid, whilst the water surface-drained land runs away d of the colour and consistency of from the inorganic or organic which it contains."

ard caused to be tested 424 specimens of water from different parts of the The results were as follows:

and springs, average hardness... 25·86.
and brooks, average hardness... 13·05.
and surface-drainage, average dness .... 4·94.

ard were early desirous of invesrhat matters were taken up by ing through different sorts of soils, ollege of Chemistry declined the ly, however, "the examination made by Professor Way," "with rtant results. From this examiwill be seen, that clays and loams rs of chemical action for the reganic and inorganic matters from a extent never before suspected, t will be practicable to use agriainage arrangements on gatherls, as means of filtration and more purification of water on a larger is at present accomplished."

(*To be continued.*)

ION OF ELECTRIC TELEGRAPH WIRES.

k, of Ayr, has invented a proing for the wires of the Electric especially meeting and obviating of to which the wires are liable ating of the chafing sea on a rocky will be recollected that the telemmunication opened last year as ent between Dover and Calais was l a few hours after it was made, tting in two of the telegraphic the rocks forming the French rolling sea chafed it asunder in one place, on the sharp edges of A cheap and effective protection has been a desideratum for which have much cudgelled their brains. opposite qualities must be comreme hardness, with perfect flexa the invention of Mr. Dick these es are completely blended. The cast-iron—at least *hard* enough: rm adopted to secure flexibility is Nature herself has selected for promost delicate and powerful teleparatus yet known to man—the rves which radiates from the brain remities of the higher animals, eir spine or vertebral column, ene of a man, or that of a more ature, *the snake or eel*, might *taken as a pattern*; but in that *ould have been a complication* of "processes" and interlocking projections to imitate: Mr. Dick has taken a simpler form, and has thus been unconsciously hit on the form selected by Nature for the backbone of the shark,—an apparatus at once powerful and more almost than any other flexible. A large bead of iron is threaded on to the cord of electric wires (which is previously encased, as at present, in a thick tube of gutta percha); then a perforated cylinder, like a "bugle," is threaded on to the string next to the ball; then another ball is threaded, and then another cylinder, and so on. The two ends of each cylinder are made concave, so as to receive the convex surface of the two balls on each side of it. Thus the whole string of iron "beads and bugles" makes an iron tube, which protects the electric cord on which they are threaded, and is at the same time so flexible that a rope of it, massive enough to weigh thirty or forty pounds to the lineal yard (without the telegraphic cord), will double up in a loop that will lie round the rim of your hat. The merits of the contrivance are its perfect simplicity and effectiveness; it consists of balls and cylinders, the chief cost of which must be only that of their cheap material, castiron. With such a protection, one would think that the wires of the submarine telegraph would be safe against the beating of any sea, on any coast.

The invention would also be useful in protecting wires under our street thoroughfares, where the vibration and crushing pressure caused by heavy vehicles rapidly passing might be of evil effect to the cord of message-wires.—*Spectator*.

BISHOPP'S DISC ENGINE.

We had the pleasure the other day of inspecting a disc engine which is in the course of exhibition at the works of Messrs. G. and J. Rennie (by which eminent firm it has been built), for the purpose of enabling shipbuilders and nautical men to judge of its suitableness for screw propulsion. The engine is a considerably improved edition of the one in use at the *Times'* office (see *Mech. Mag.*, vol. li., p. 241.) The first of the improvements now introduced by Mr. Bishopp consists in dispensing with the slips and springs employed in packing the cones, by which arrangement the easiness and smoothness of the motion of the disc is much increased. Secondly, the cylinder is entirely enclosed by a steam jacket. Thirdly, the groove on the top of the cylinder, hitherto employed as a guide for the disc, is dispensed with, and the bow of the diagonal shaft connected to a second bow crossing it at, or almost at, right angles, the ends of which last are pinned to opposite sides of the cylinder, in such manner as to admit of the bow rocking on those pins as fulcra when the disc is in motion, and at the same time preserving the disc in steam-tight contact with the cones. A fourth improvement consists in connecting the end of the diagonal shaft with the driving crank by a ball-and-socket joint, which is capable of sliding in a radial slot, but kept up to the periphery of the crank by a block of vulcanized India-rubber, so as to prevent jarring.

The working of the engine is beautifully smooth and easy, and the high speed obtained renders it peculiarly adapted for driving the screw, without employing any such expensive and easily-deranged gearing as is necessary with the engines ordinarily applied to that purpose. The weight of the engine also is trifling, and it occupies but small space, and, in consequence, can be placed at a low level in the vessel.

COLOUR-BOXES AND DRAWING INSTRUMENTS.

The following special prizes have been just offered by the Council of the Society of Arts—urged thereto by a consideration of "the present high price of the necessary materials" for acquiring a knowledge of drawing.

"*The Society's large medal* to the person who shall produce the box having the greatest number of the best colours for general use, and brushes, which may be sold retail to the public for *one shilling*. The maker must agree to make not less than 1000 of such boxes, and keep the same on sale.

"The Council will be prepared to purchase not less than 1000 of the successful boxes, provided they agree in all respects with the specimen submitted in competition.

"*The Society's large medal* for the best and cheapest set of the following instruments:

"One pair of 6-inch compasses, with shifting pencil leg.

"One wooden 4-inch triangle (equilateral).

"One ruler with one fiducial edge, 1 foot long, divided into inches and tenths, and marked with a scale of chords.

"To be contained in a case.

"The Council will be prepared to purchase not less than 100 sets of the successful case, provided they agree in all respects with the specimens submitted in competition.

"Specimens of the colour-boxes and cases of instruments, accompanied by the name of the manufacturer, and by a written engagement to produce not less than the above numbers respectively, to be sent to George *Grove, Esq., Secretary* to the Society, 18, *John-street, Adelphi*, on or before *December 1, 1851.*"

SPECIFICATIONS OF ENGLISH PATENTS ENROLLED DURING THE WEEK, ENDING SEPTEMBER 17, 1851.

VICTOR HYACINTHE LIBERT GUILLOUET, of Condé sur Noirot Calvados, France, chemist. *For certain processes for increasing on manufactured fabrics the several shades of indigo.* Patent dated March 16, 1851.

This invention consists in subjecting textile fabrics dyed with indigo to the action of an elevated temperature under steam pressure.

Indigo is a substance naturally insoluble, and in order to fix it when used as a dye for manufactured fabrics, it must be rendered soluble, which is effected by means of deoxygenating agents, such as sulphate of iron and oxide of calcium. Then, by immersing the articles in such solutions, different shades of colour may be obtained, varying in intensity, according to the duration of their immersion. Indigo is also of a volatile nature, and this is one of its distinctive peculiarities. The patentee proposes to avail himself of this volatile principle by subjecting the dyed fabrics to the action of an elevated temperature, under a certain pressure, in metallic vessels closed air-tight, and sufficiently strong to resist the internal pressure necessary for acting on the atoms of the indigo, and causing them to be fixed and amalgamated with the fibres of the manufactured articles, and to produce a change in their physical constitution. The form of the vessel employed is immaterial, but it is necessary that it should be provided with a safety-valve, and a blow-off cock for expelling the atmospheric air when the steam is first admitted. The fabrics to be operated on, having been previously dyed of any required pattern, are placed in layers in the vessel on a wooden framework, and enveloped in a blanket, for the purpose of preventing contact with the sides of the vessel, and to absorb the vapours produced by the first application of steam, which is then admitted at a pressure varying from two to six atmospheres. At the expiration of from twenty to thirty minutes, the cover of the vessel is raised, and the fabrics are withdrawn; and, after allowing them to cool, they will be in a fit condition to be folded and packed. The effect of this operation on the indigo colour will be to impart to it a violet tinge; nor will any detrimental effect be produced on other colours with which the goods may have been dyed; on the contrary, their richness and brightness will be rather increased. The cloth will be found to have undergone a material diminution in length, but the shrinkage in width will be hardly perceptible: it will at the same time have assumed a closer and finer texture, and will have become thicker and softer.

—The subjection of textile fabrics, indigo, to the action of an elevated ure under steam pressure, for the of increasing the shades of colour d fabrics.

Murray, of Canterbury, barter and captain. *For improve- saddlery and harness.* Patent rch 10, 1851.

ject of the present improvements duce a saddle which shall be less the back of the animal than those in use, and easier and safer for , especially when obliged to remain ngth of time on horseback. With , Captain Murray provides his sad- springs, which he attaches longitu- the saddle-tree, and on which he padding for the seat. The springs in number, of double curvature, entre one is forked at one end, to date it more readily to the form of One end of each spring is at- to the cantle end of the saddle by at around a plate thereon, and the l is secured by pins on the pommel to slots formed in the end of the the sides of the springs are left r free, and their full extent of play itudinal direction is thus secured. to prevent the padding falling into s between the springs, they are to ected together by an elastic lace. ngement confines the whole of the to the edges of the saddle, and thus he back of the animal very consi- at the same time that the elasticity rings renders the saddle much easier comfortable for the rider.

— The combination of springs, ns, and plates, for the purpose of saddles safer and easier both to and the horse.

h Galloway, of Southampton- , Chancery-lane, civil engineer. *ovements in steam engines.* Patent rch 10, 1851.

improvements consist in arranging y between the piston of a double- eam engine and its load, in such that the varying effort of the steam, rking expansively, may be caused to with a comparatively uniform effect ad, and in such manner that, both he up and down stroke, the load moved through a greater space for xtent of motion of the piston, when by the greatest pressure of steam, moved through for a similar extent n of the piston when acted on by asing effort of the expanding steam. atentee does not confine himself to icular arrangement of mechanism this object, but he claims—The employment of mechanical means interposed between the piston and the load of a double-acting steam engine, to compensate for the decreasing effort of expanding steam, and thus, as nearly as may be, to obtain a practical uniformity of effort in respect to the load.

Peter Armand Lecomte de Fontaine-moreau, of South-street, Finsbury. *For improvements in compressing air and gases for the purpose of obtaining motive power.* (A communication.) Patent dated March 10, 1851.

The invention here sought to be secured consists in compressing air and gases, by introducing into suitable vessels containing them, water or other liquid, by means of hydraulic pressure, and employing the air or gases, in their compressed state, to produce motive power.

Claim.—The use of one or more recipients, into which a liquid is introduced by one or more hydraulic presses, by which the air or gases are expelled or compressed.

George Roberts, of Selkirk, manufacturer. *For an improved manufacture of certain yarns of linen, wool, silk, cotton, or other fibrous substances.* Patent dated March 10, 1851.

The particular sorts or descriptions of yarns to which this invention has reference are those known as mottled, marled, clouded or spotted yarns, of whatever material they may be composed, whether linen, wool, silk, cotton, or other fibrous substances, and the invention consists in manufacturing such yarns from rowans or cardings, which are produced by combining slivers of different colours, or mixtures of colours, in such manner that each rowan shall contain a portion of each such sliver.

In carrying his invention into effect, the patentee takes portions of wool or other material, which having been cleaned (when necessary), and dyed of different colours according to the pattern of yarn to be produced, are submitted to the action of the machine called a "teaser," (and in the case of wool oiled in the usual manner.) The separate portions are then successively fed into the machine known as the "scribbler," having cylinders and rollers covered with cards, from which they issue in continuous streams or slivers, which are to be wound on bobbins or spools. A number of these bobbins or spools, with the coloured slivers thereon, are then placed in a suitable frame near the feeding end of a carding or finishing engine, and, having been arranged in order according to the required pattern, the ends of the slivers are introduced between the feed rollers, and the machine put in motion, when rowans or short cardings will be produced, each containing a portion of each of the slivers on the bobbins. A heck is placed in front of the feed rollers to keep the slivers apart, and the divisions of the heck are made to correspond with the length of spot of any particular colour which is to be produced on the carding or rowan. The cardings produced by this machine are joined together end to end, in the usual manner, passed through a "billy," to bring them to the state of slubbings, and finally spun to the required size of yarn in the "mule." It is evident that, by following this mode of manufacture, and arranging the bobbins in different order in the frames, before carding the slivers thereon, an almost endless variety of patterns may be produced on t —No claims.

Specification Due, but not Enr

JEAN BAPTISTE ALPHONSE B Paris, gentleman. *For improveme manufacture of coverings for roo partitions, furniture and other sin cles, and in boxes, tubes, and oth articles, and in the preparation facture of materials to be emp such purposes, and also in machi employed in such, or similar man* (Being a communication). Pat March 10, 1851.

WEEKLY LIST OF NEW ENGLISH PATENTS.

Alexander Parkes, of Birmingham, Warwick, chemist, for certain improvements in the manufacture of copper, and in the separation of some other metal therefrom, and in the production of alloys of certain metals. September 11; six months.

George Phillips, of Upper Park-street, Islington, Middlesex, chemist, for preventing the injurious effects arising from the smoking of tobacco. September 18; six months.

John Wormald, of Manchester, Lancaster, maker-up and packer, for improvements in machinery or apparatus for spinning and doubling cotton, wool, silk, flax, or other fibrous substances. 18; six months.

John Simpson Leake, of Whitehall S Chester, manfacturer, for certain imp in the processes and machinery or app ployed in the manufacture of Salt. Sep six months.

John Livesey, of New Lenton, N draughtsman, for improvements in th ture of textile fabrics, and in machine ducing the same. September 18; six m

WEEKLY LIST OF DESIGNS FOR ARTICLES OF UTILITY REGISTERED.

Date of Registration.	No. in the Register.	Proprietors' Names.	Addresses.	Subjects of De
Sept. 12	2943	William Healey	St. Martin's, Leicester	Resilient straps fo vests, &c.
15	2944	Joshua Jackson	Wolverhampton	Ink bottle.
17	2945	Jesse Shaw	Bishop's-place, Fulham	Machine for clea rants.

WEEKLY LIST OF PROVISIONAL REGISTRATIONS.

Sept. 12	283	Thomas Humphrey Roberts	Plymouth	Drag apparatus.
13	284	Joseph William Lea	Birmingham	Match box.
15	285	Peter Warren	Longton	Danger signal for r railway carriage
17	286	James Cockings	Birmingham	Soirée union back
,,	287	William S. Adams	Haymarket	Hinged lid for spe and boxes.

CONTENTS OF THIS NUMBER.

LONDON: Edited, Printed, and Published by Joseph Clinton Robertson, of No. 166, Fl *in the City of London*— Sold by A. and W. Galignani, Rue Vivienne, Paris; Machin *Dublin*; *W. C.* Campbell and Co., Hamburg.

Mechanics' Magazine,

SEUM, REGISTER, JOURNAL, AND GAZETTE.

1468.] SATURDAY, SEPTEMBER 27, 1851. [Price 3*d.*, Stamped, 4*d.*

Edited by J. C. Robertson, 166, Fleet-street.

BROOMAN'S PATENT SCREW-MANUFACTURING MACHINERY.

Fig. 3a. Fig. 2.

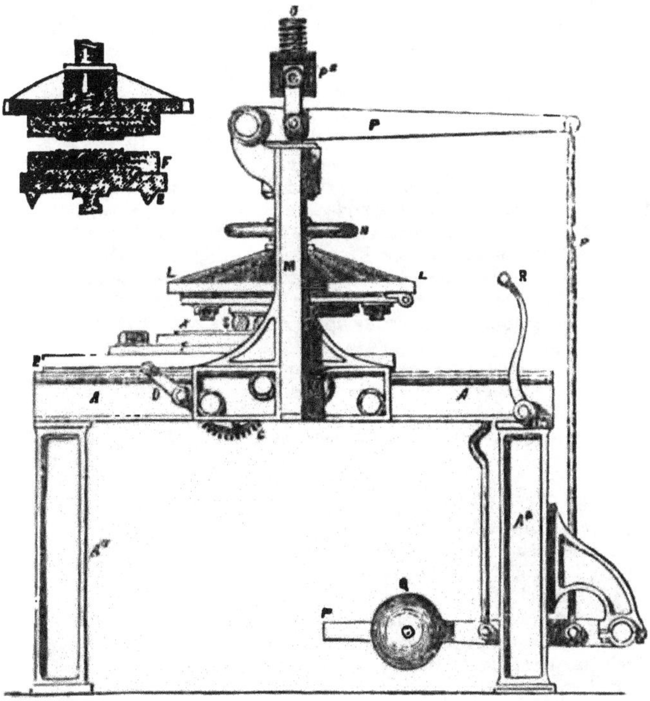

Fig. 3.

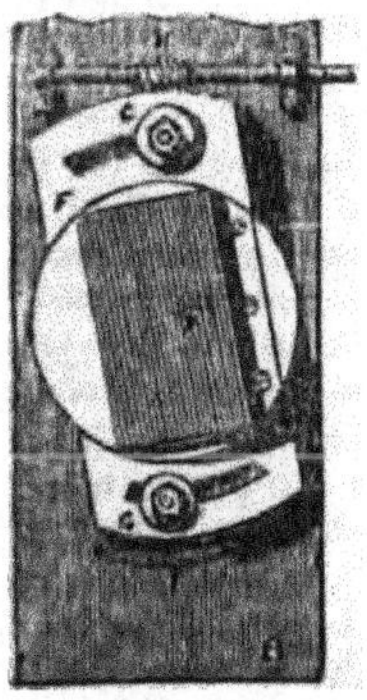

Fig. 4.

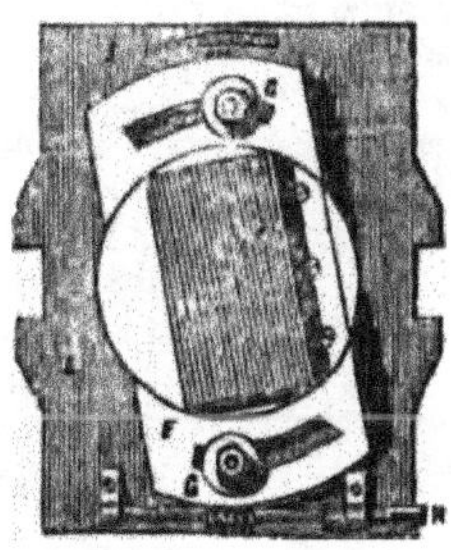

BROOMAN'S PATENT SCREW-MANUFACTURING MACHINERY.

[Patent dated March 15, 1851. Communication from Abroad. Specification enrolled, September

Specification.

FIRSTLY,—This invention consists in manufacturing screws of all sizes, more than one thread, from one and the same set of dies, and in produc threads, whether square or angular, by rolling compression. The mechanic binations and arrangements by which these improvements are effected are sh the engravings annexed, and in the following description thereof. Fig. end elevation of the screw-rolling machine which is employed, with the di in their places, and fig. 2 a side elevation of it. A is a strong rectangul supported by standards $A^a A^a$, and surmounted at about the middle of its le a bridge-like piece M, which is securely bolted by its two sides or cheek sides of the table. The upper surface of the table is made with two V-groo ning in the direction of its length, in which grooves a bed-plate B, move fro, according as it is actuated by a crank-handle D, which turns a pinion th into a rack C, affixed to the under part of the bed-plate. To the top travelling bed-plate, one of the dies K', is secured in the manner shown se in plan and section in figs. 3, and 3^a. F is the die-carrier, which fits on t cular part E, raised on the centre of the bed-plate B, yet not so tightly but carrier may turn freely on E as a pivot, either to the right or to the left. K' is dove-tail jointed into the carrier F, as shown in fig. 3^a. The carri slot at each end, and there are screw bolts GG, fitting into these slots, by wh carrier and die can be made fast in any position to the right or left, which be proper to assign to them. The other, or upper die, K^2, is fixed in the bri piece M, by arrangements precisely similar to those just described (see fig fig. 3^a), only that the positions are reversed; that is to say, the die K^2 is do into the under side of a top-carrier F^2, and that carrier is attached to the un of a top plate L, which slides up and down the bridge-piece M. The two dies are constructed precisely alike: and consist each of a flat rectangular plate or other suitable metal, having on one side or face a series of straight grooves (formed either by hand or machinery), of like width and depth threads of the screws intended to be produced. The grooving is effected w plates are in a (comparatively) soft state, and the plates are afterwards h sufficiently to enable them to serve as dies. A screw-spindle O, passes rig through the cross head of the bridge-piece M, and into the boss of the top by the turning of which the one way or the other the two dies can be approach to, or recede from one another as required. N is a wheel by w screw-spindle O is turned round: PPP are a series of levers, with weig attached, which are connected with the top-plate L, by a crank cross-head lowered or raised as required, by means of a handle R, so as to bring down die with any degree of pressure, and at any rate of velocity on any objec between it and the under die. S represents the plain cylindrical metal rod, o the screw thread is intended to be produced. The principle on which this done falls now to be explained. If two grooved plates, such as has been de were fixed with the grooves of the one fitting exactly on the ridges of th and coinciding with a horizontal line drawn through the centre of the mach pressed against a plain metal rod introduced between them at right angl said horizontal line, the effect would simply be to impress the metal rod with of straight parallel grooves and ridges, but if the two die-plates, instead of b parallel, are moved horizontally on their centres, one a little to the right, t exactly the same extent to the left, then on the rod being rolled between the pressure, and at right angles to the horizontal line drawn through the cent machine as aforesaid, the plates will indent the rod in continuous spiral li responding exactly in pitch to the angle at which the dies have been set. sure of uniformity throughout the thread of the screws, it is of course essen the two dies (at whatever angle they are set) should be kept with the utmost *steadiness in that* relative position. The fixing of the die-plates at their *angles of inclination* is effected by means of the screw bolts GG, as before e

Fig. 1.

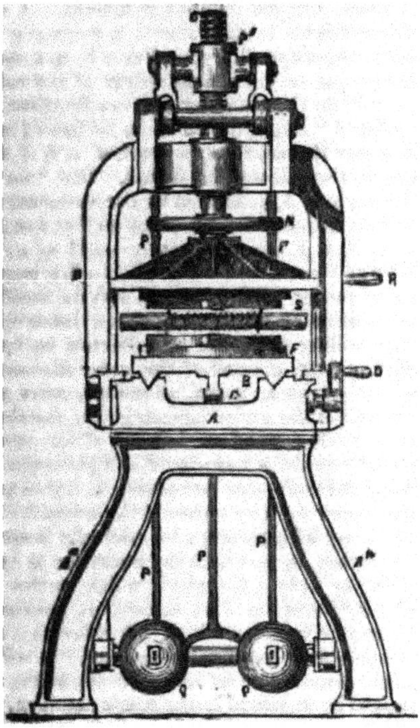

Fig. 8.

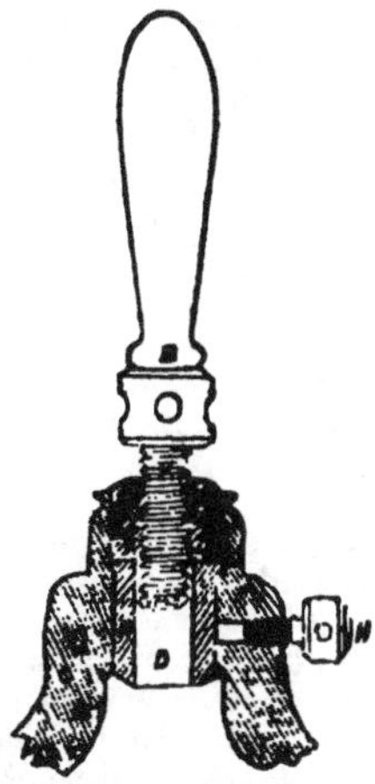

Fig. 3b

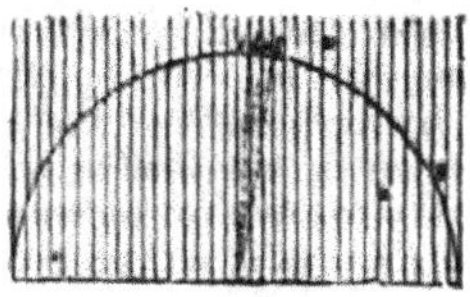

Fig. 9. Fig. 5. Fig. 6.

Fig. 7.

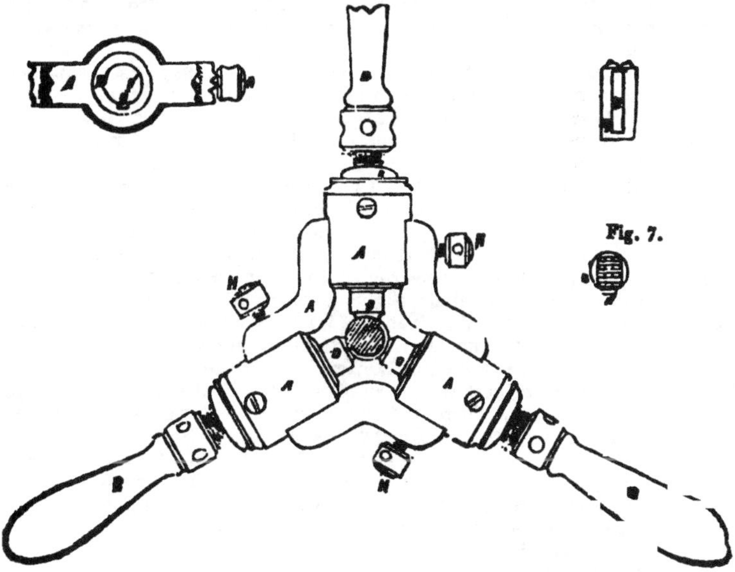

The angle at which the dies are set, must of course depend upon the diameter of the rod to be screwed, the pitch of the thread, and the number of threads. A suitable index for this purpose may be constructed in the manner shown in the engraving, fig. 3[b]. Let *a*, *a*, *a*, represent the parallel ridges upon the die. From *b*, as a centre describe the arc *c*, *d*, *e*, *f*, *g*, with a radius equal to the circumference of the rod or bolt to be screwed, then the angle *c*, *b*, *d*, will be that at which the two die-plates require to be set in regard to each other, when a "single thread" is to be formed upon the rod. If the screw is to have more than one thread, then the arc *c*, *d*, *e*, *f*, is to be divided into equal portions corresponding to the pitch of the thread. If a "double threaded" screw is to be formed, then the angle *c*, *b*, *e*, formed by the extremities of two such divisions with the centre *b*, is that required to be given to the dies; if a treble thread, then three such divisions, *c*, *d*, *e*, *f*, are to be taken,—and so on for any number of threads which the diameter of the rod will admit of. In this manner the angle corresponding to any diameter of rod or pitch of thread may be obtained, and a number of those angles may be marked or tabulated upon an index plate which may be attached to the face of the sliding bed-plate B, as shown in fig. 3, and traversed by a needle-pointer projecting from the end of the under die-carrier. The better to insure perfect accuracy in using such an index, an endless screw may be affixed to the bed-plate at the opposite end of the under die-carrier, as also shown in fig. 4, and made to work into a series of teeth formed on that end of the carrier. In operating with the machine, the rod to be screwed is introduced at right angles to a horizontal line drawn through the centre of the moveable bed-plate. It is then gripped fast between the dies by lowering the upper die (by means of the wheel N) into contact with the rod, and then causing it to act with pressure by gradually lowering the weighted levers PPP. The under bed-plate B, in which the under die is fixed, is then moved to and fro on the V grooves, as before described, which carries the rod along with it. The length of rod which can be thus screwed is necessarily limited by the breadth of the grooved surfaces between which it is rolled, but that breadth may be increased to any extent likely to be required in practice. It will be further obvious that two or more pairs of dies, each pair set at a different angle, and producing screws of different patterns, might be mounted in the single machine, and worked simultaneously by one and the same moving power. And so also the die-plates might be made with two sets of grooves crossing each other at any required angles, in opposite directions; in which case, when they are rolled on a plain rod, as before described, they will produce at one and the same time both a right and a left-hand thread.

Instead of the dies being flat, as before described, they may be made of the form of segments of circles, or even of entire circles; and instead also of the under die being moved to and fro, it may have a continuous rotary motion given to it. But the modes of combination and operation embodied in the method or machine before described are preferred, as being in the whole the most practically convenient.

The principle of using dies moveable on their centres into different positions in relation to the rod to be screwed, which forms the distinguishing feature of the method or machine just described, may also be applied to the improvement of ordinary stock dies in the manner represented in the engravings annexed. Fig 5 is a side elevation of a triple stock constructed on this principle; fig. 6 and 7[a] longitudinal and end views of the dies. Fig. 8, a section of one division of fig. 5, and fig. 9, an end view of the same. The dies DDD, are made of hardened steel and perfectly cylindrical, and one or both extremities are made with sharp parallel cutting edges, which are formed in the first instance by a tap, the same practice being followed in this respect as in the ordinary dies. They fit freely into sockets, D^1 D^1 D^1, bored out for them in the circular bosses EEE, of the stock A, and so that they may be readily turned round on their centres, so as to vary at pleasure the angular bearing of their cutting edges. When turned into their assigned positions, they are then made fast by set screws HH. Each die has a feather *f*, which fits into a slot in the boss E. BBB are screw handles, by the turning of which the tool is worked, the screwed parts causing the dies to collapse or recede as required. *The angles at which* the dies in this description of stock should be set may be *ascertained by the diagram* fig. 3[b], before described; and from this diagram the ngles may be transferred by means of a bevel stock to the dies. The k should be applied first to the diagram, and being set at the proper angle, of it should be applied to the face of the screw stock, while the line of the ed upon the die is made to correspond with the other leg of the die stock.

PROFESSOR YOUNG'S "INTRODUCTION TO ALGEBRA."*

: glance at the title of this work r Young's latest—should cause a feeling of regret that his great mind should have been employed bject which must have been far nial to them than the more ad- ies and higher departments of nal science, yet this feeling will away under the reflection that it er-mind which is best capable of h even the most elementary por- cience, and that it is the most te proficient in a subject, who est appreciates the difficulties t the path of a beginner.

our readers are probably aware h opinion which we, in common rest of the mathematical world, of the scientific talents and attain- Professor Young, and of the high sh we set upon his researches, which have enriched the pages and the readers of this Magazine. rofessor Young's numerous lite- rs, none perhaps have met with a ral circulation, or filled a larger usefu'ness than that portion of prised in his "Elementary Trea- lgebra."† It would be no easy point out a work better adapted to ich it has in view than the treatise re have just alluded. The person s to its perusal a steady purpose ful and thoughtful mind, cannot ive from it not only extensive in- , but he would also have before st pleasing model of mathematical style. The object of the new Introduction differs somewhat from that of the Elementary Treatise. The Intrduction is intended more particularly for schools and the junior class of students.

Irrespective of its smaller size and compass, and the more simple character of the work, the "Introduction" has for one of its objects the placing before the student a greater proportion of examples than it was necessary or proper to introduce into the larger work. And we think there can be but little doubt that Professor Young has acted judiciously in following the advice of his friends, and giving this little work to the public. In our opinion, which is grounded on the same principle as one which we have before expressed (see vol. xlv., p. 66,) in reviewing Professor Young's "Three Lectures on Mathematical Study," there is no more effectual way of impressing the elements of science upon the mind than by the judicious use of examples; and those who take this view of the subject will do well to glance at the Professor's new work. It is to be observed that the answers to the examples are given in a separate cover, a form which has great obvious convenience to recommend it.

Of those who peruse the "Introduction" there are probably many who will proceed to the study of the larger work. Of those who do not, many will gain from the smaller one, a somewhat comprehensive view of the general scope of algebraic science. For it is a very striking feature of Professor Young's "Introduction" that, without swelling into undue detail or disproportionate development, it touches on most of the points comprehended within the sphere of algebra, and affords to the student indications by means of which he may judge of the direction in which the further fields of knowledge lie, and from which he may learn to shape

---

* roduction to Algebra, &c. By J. R. , Professor of Mathematics in the Royal Institution, Belfast. London: Simoms re, 13, Paternoster-row, and 26, Done- Belfast. 1851.

† ly edition of this work upon which we moment lay our hands, is the fourth, at the foot of its title-page the words Souter and Law, 131, Fleet-street.

his course accordingly. We repeat that there is no more distinguishing feature of the Introduction than the symmetry of its proportions. Even the subject of logarithms is (of course briefly) treated of; and we need scarcely say that Horner's process is not forgotten. In the case of Horner, as indeed of every other mathematician of whom mention is made, the fairness and candour of the Professor are most remarkable and praiseworthy.

The work is intended "for the use of schools and private students," and we do not doubt that it will prove of great benefit to both.

THE WATER-SUPPLY AND DRAINAGE OF TOWNS. — REPORT OF THE GENERAL BOARD OF HEALTH RESPECTING THE METROPOLIS.

(Concluded from page 237.)

Professor Way was asked: "For the collection of the rain-fall of a given district, what soil would you prefer?" "I should say, decidedly, a sand. Rain-water, when collected at a distance from towns, is fit for every purpose. All that is required from the collecting surface in this case is, that it shall perform its office without imparting to the water anything to render it impure Sands which have been washed by rain for ages are most likely to fulfil this condition, and would possess the further advantage of allowing the ready escape and collection of the water."

Specimens of surface-water collected at various points in districts surrounding London were examined,—amongst these was a specimen from the Ruislip Reservoir, the water from which is used for feeding the Grand Junction Canal; this water is from the surface-drainage of a tract of clay land, and "gives 8 degrees of hardness,—or just one-half the impurity from lime of all the rest of the Companies' specimens put together."

The searches made south-west of the metropolis appear, however, from the nature of the strata, to be attended with the greatest promise of success. The evidence of Mr. Donaldson is given at length respecting Richmond Park as a gathering-ground, as "this one plot of land is illustrative of the principle of improved supplies." The soil of that park "is a sandy or gravelly loam, incumbent on a clay subsoil." The water drained from it was "perfectly clear, soft to the feeling, well aërated, and pleasant to drink." Mr. Donaldson said, in h
dence, "I am, from long observation
that water passing through a bed of
tion does leave behind, not only the
in mechanical suspension, but much
matter in chemical solution. This is
which has not hitherto received the a
which its importance deserves. I a
sure that a bed of vegetation will de
its food saline and other matter in s
which no sand or other artificial fi
separate from the water. I have see
containing a considerable quantity
age from a farm-yard, which has pass
well-drained pasture land, and the
which has drained through it has c
perfectly clear from the manure in sol

"It is to be presumed, howev
there might be an extent of manu
shallowness of the filter bed of earth
would *not* detain the matter in s
No doubt of it." Animalculæ hav
been met with by Mr. Donaldson
low spring-water as it came from the
or drain.

The ordinary rain-fall at Richmo
is estimated at 25 inches: the usual
lation of the quantity of water c
from rain-fall is the third of the am
fall.

The Board has in review various
the vicinity of the metropolis which
be applicable as gathering-grounds, b
a preference to the tract "commenci
Bagshot and Woking sands, and ex
to Hampshire." Farnham has b
some time supplied with soft-water
drainage of less than two acres of t
mon land. The water "is delivered a
limpid at all times of the year."
is no reason why the same quantity
should not be obtained from the who
tract of waste land there, ten miles l
five broad." "The improvement
of these tracts has hitherto been give
despair, and the growth of fir is
mended as the only agricultural pur
which they are fitted." The Bagsho
are estimated as covering an area
square miles. "Beneath these san
retentive stratum of marl and clay,
from five to fifteen or twenty feet in
ness."

"The portion of this district to
our attention has been more imme
directed, comprises an area of no le
100 square miles, lying east and we
line from Bagshot to Farnham." "
collected in this district at the surfa
mediately after rain-fall, are of the
degree of purity, being in large qu
not exceeding one degree of hardness

"The chief practical result deduci

ervations is, that by arrangements
ting the water before it has tra-
y great extent of surface, a quan-
ient, as it appears, for the domestic
f the whole metropolis will be ob-
at a very high degree of purity,
equal to the present supply of
."

nature of the source requires a
ion of the rain during periods of
num fall, for a regulation of the
uring periods of a minimum fall.
age-room must therefore be very
. The primary engineering disad-
of this district are, that it presents
natural hollows, such as are avail-
many of the northern towns, for
ge of water without extensive ex-
. Here the excavations for storage
must be very large and extensive.
he modern engineering practice of
nd open reservoirs, we would rather
he custom of the Roman engineers,
mmend covering the service reser-
l aqueducts to the utmost extent
le."

foundation for proximate esti-
lans have been got out by our en-
inspectors for extensive covered
, and for the conveyance of the
deep conduits, also covered. They
the total expense of storing and
to the metropolis this new and im-
pply, inclusive of reasonable com-
for waste land for reservoirs at
re than one million sterling. We
eve that two years' saving from the
ne purer water, would fully repay
on of the outlay."

oard evidently give a preference
btainment of water from suitable
-grounds, which, on account of its
nd purity, they consider as supe-
hat from the sources of supply in-
y Captain Veitch. This gentleman's
is, however, given at length in
, No. 2: it appears by that evi-
t he indicated several different avail-
ces of supply, and he recommended
account of their perennial abund-
onsidering that the population of
polis has nearly doubled itself in
" "and that great solicitude is en-
lest the same rate of increase may
to 1890, I consider it a most im-
easure to secure all the best sup-
water that can be obtained near
efore they be appropriated to other
minor importance." The sources
ly adverted to are as follows:
Hertford, where "there is a sin-
ting of four copious streams of
ceeding from chalk valleys; viz., the Lea, the Verulam, the Beane, and the Rib, which jointly have a discharge of 9,360,000 cubic feet per diem;" "the Ash, the Stort, and springs which join the river Lea below Ware, about 4,320 cubic feet of water per day. Such are the resources of the river Lea and its tributaries, and which for the paramount object of supplying the increasing population of the metropolis with so needful an element of health and consumption ought to be held sacred for that purpose alone." The conjoint waters of the river Lea, at Field's Weir, "amount to 14 millions and a half of cubic feet, or ninety-four and a quarter millions of gallons per day." In respect to other sources, Captain Veitch said, "I conceive, in the first place, that the water of the river Verulam is the first to be secured, and rendered available for the public good at London; the water of this river taken a little above Watford, is a never-failing stream, derived from springs, and yielding three millions of cubic feet of water per diem, at an altitude of 158 feet above high water in the Thames." "Similar to the supplies of water on the north-east of London, which unite to constitute the river Lea, those on the north-west of London unite to constitute the river Colne, and consist of the following streams: The Colne proper, an insignificant brook in dry weather:" The "Verulam, a fine stream," is chiefly fed from springs, and is clear and constant, with an average yield of about 3,000,000 cubic feet per diem: The Gade, chiefly fed by springs, yields a supply of about 4,000,000 cubic feet per diem: The Chess, a lime stream fed by springs, 2,000,000 feet per diem. The above streams have their water united a little way above Rickmansworth." The river Mole is only noticed as a probable source that may be available.

For the supply of water to the south side of the Thames, Captain Veitch had "especially directed his attention to the waters of the river Darenth," which yield about 2,600,000 cube feet per diem." "The waters above specified," may all be delivered to reservoirs, 140 feet above high-water mark. Captain Veitch was asked, "Do you not consider the above quantity of water as unnecessarily great? If water can be brought to London from such short distances, and at such an altitude on the gravitation system alone, cool and clear in quality, I do not consider that any quantity of such water, and under such conditions, can be deemed over abundant for the health of the population." The supply of water to Rome, under the Empire, "by the Roman aqueducts, amounted to about 50,000,000 cubic feet per day, for the use of a population presumed to have consisted of about 1,000,000." The popula''on of London is now about two millions and a quarter; Captain Veitch conceives that forty years' hence it may be increased to 4,000,000 of souls.

A statement of *all* the schemes that have been proposed for the supply of water to the metropolis has been drawn up by Mr. Henry Austin, and is published in the Appendix, No. 2, of the Board's Report; but as they are all of them variations only of the principle schemes, following up the Board's recommendation of gathering-grounds, or deep wells, or Captain Veitch's, to take advantage of existing springs and streams, it seems superfluous in this abstract to enter into particulars of these minor schemes.

As to deep wells, it is abundantly proved that they could not afford a sufficient supply. Whether to prefer the collecting of water from gathering-grounds, preserving it pure in covered reservoirs, or whether the bringing water impregnated with lime from streams affording a perennial supply, and providing for the purification from lime of the water before delivery, seems ma'nly a question of *£. s. d.* The greatest expense attendant on the former mode seems to arise from the immensity of reservoir required,—that reservoir necessarily a *covered* one; for it is admitted, that in an *open* reservoir the water would become even more contaminated with organic impurities, animal and vegetable, than it is found to be, even in the Thames. These reservoirs would have to provide not only for a sufficient supply during the dry weather of ordinary summers, but also for seasons of extraordinary drought; the leaving London so much as a single day without water is too horrible a chance to be risked. There are some seasons when, for perhaps two months, there is no material rain-fall. How many acres of covered reservoir would suffice for the supply of London during such a drought? It is easy of calculation; so is the expense of depriving water of lime, which would be to be compared with the cost of covered reservoirs. The depriving water of carbonate of lime, by Professor Clark's process, might perhaps suffice—further purifying it in Mr. Holland's mode would add to the expense; but even that is compensated for, according to his evidence, by enabling thus the water to save soap in washing, and t› require less tea, &c., in making extracts.

Should the gathering-ground system be adopted, Professor Way's experiments, in addition to experience at Farnham, and to general agricultural and other observations, indicate that the gathering-grounds should be under the complete control of whatever person or persons may have to govern the *supply of water* to London; this would be essential, since those grounds shoul
times be kept at a certain degree of
ble production: enough of that gr
assimilate the whole of whatever
tious manure might fall upon the g
as also any moderate quantity that
permitted for the purpose of profita
tivation of the area chosen as a ga
ground.

The properties desirable in wat
having been thus exhibited, and the
from which a supply may be obtai
next consideration is that of the m
cient, and the most economical
bringing water to the metropolis,
distributing it therein with the great
venience and at the smallest cost to
habitants.

The General Board of Health
opinion, that "even if the same so
supply as those taken for the New
were eligible, and if those works b
to the public, they ought to be aba
and the Roman principle of cover
nels reverted to, as Captain Veit
poses."

Captain Veitch, in his evidenc
many particulars, showing that th
originally adopted for the conductio
New River was defective, on accoun
being an open canal, following
sinuosities of the ground, as on a
line; that it had an inclination of on
inches per mile, but that "within t
sent century great ingenuity and g
pense have been applied by the Ne
Company to correct the evils of t
and vicious mode of conduit first ad
He observed, that "a great obje
the conveyance of water for domes
poses, in an open earthen channel,
the water must have a very slow mo
exceeding half a mile per hour, to
the current wearing the channel-b
bringing in turbid water; the slow
is again attended with serious evils,
tions of silt and decayed vegetable
take place, which require to be cle
from time to time; in the warm se
long and broad a surface exposed
atmosphere gets heated to a degree
able to the production of vegeta
animal life of the lower forms, and
giving rise to a considerable qua
waste from evaporation; the hig
perature of the water rather facilit
decoction of leaves and other v
matters, which get blown into th
River, to the manifest injury of the
but there are other pollutions of
character, to which all open ca
subject. It is true the New Rive
pany have five acres of settling-gro

,* and thirty-eight at Newing-
s deposit of solid matters."
the objections to all open water
aducted in earthen channels,—
sies of which will, however, be
appreciated by a contrast with
tions that may be obtained for
water, if conveyed in covered
structed of stone or brickwork,
ted in straight lines with an
l efficient descent, crossing val-
ankments or arcades, and pierc-
tunnels or adits; for example,
f the river Lea might be con-
London in such a channel from
distance of twenty miles, instead
d with a speed of one mile per
d of half a mile; that is, the
ld be accomplished in twenty
d of eighty."
linson, in giving an account of
queduct, instanced it as serving
warning and an example. The
elf has cost upwards of 40,000*l.*
sively of reservoirs; it is sup-
a solid foundation wall, 17
the "true aqueduct or water-
ructed with a brick lining upon
foundation." The whole struc-
ed up with earth on each side.
be difficult to devise a more
ork. "Nevertheless, in several
t has been found necessary to
r-way with iron."
linson is of opinion that "iron,
d cast, may be much more ex-
ployed in water-works than has
in the practice." "Where it
ht advisable to cross a valley or
nverted syphon-pipe, an elevated
a tube aqueduct may be con-
ht, elegant, nay, even graceful
." "Telford set an example in
ed Pont-y-Cypoylte aqueduct,
6 feet in height, 1,007 feet in
has a cast-iron water-way as
perfect now as the day it was

linson himself, in 1846, pro-
n to the Corporation of Liver-
g in a supply of water from the
"The several intermediate val-
have been crossed by inverted syphons, or by means of elevated aqueducts of wrought iron." His proposal was submitted to Mr. Fairbairn, who in his report to the Chairman of the Liverpool Waterworks Committee, October, 1846, said that "tubes 6 feet deep and 2 feet wide, with close tops, can be made of sufficient strength to carry 33 tons of water on 100 feet span." "The weight of 100 feet of such a tube will be about twelve and a half tons."

In respect to durability "care must be taken to prevent oxidation, and in order to do this effectually, it will be necessary to make the top of the tubes, as all the other parts, perfectly water-tight, and the tube being always full of water, it will be a great security against corrosive action in the interior. On the outside the usual preservatives must be applied; with these precautions the tube might last for an almost indefinite period of time."

"The effects of winter, or the change of temperature, will not be severely felt on a long and somewhat flexible tube. Internally the temperature will not vary considerably, as it never can be above 60°, and never lower than 32.°"

Thus it seems that Mr. Fairbairn is in favour of iron tubes.

Mr. Rawlinson appears to advocate economy in public works.—He said that "If modern science has taught us how to make a steam engine, it has not yet fully inculcated the necessity there is that rigid economy should be studied in all engineering works." He gives the aqueduct of Spoletto as an example of the small quantity of masonry that suffices for piers, "The middle arch of that structure is 328 feet in clear height, supported on piers 10 feet 6 inches thick." The Pont du Gard, near Nismes, France, might also have been noticed for the small quantity of masonry in its piers; in this instance they have occasionally to resist the immense force with which the waters of the Gardon come down suddenly upon the piers after storms in the mountains, converting a shallow brook into an impetuous broad river.

The mode of execution the Board prefers is not indicated in their Report; but it is said that, from proximate plans and estimates "got out by our engineering inspectors for extensive covered reservoirs, and for the conveyance of the water in deep conduits, also covered," they estimate the "total expense of storing and bringing to the metropolis this new and improved supply, inclusive of reasonable compensation for waste land for the reservoirs, at little more than one million sterling." The Board, in their twenty-fifth question to Mr. Stirrat, of Paisley, said—"We find at present we

gs and cats are strained off by a
the water enters the settling-
New River Head, but other im-
st be so separated. A panic has
en occasioned by a report that the
was poisoned, as it happened during
at occasioned by Lord George Gor-
all the water was then for a short
, on examination, was found to have
quantity of refuse madder, thrown in
nse.

can cover a reservoir at about 1,000*l.* an acre." Mr. Stirrat had previously observed that, "as to covering or roofing the *storage* reservoirs, that is altogether unnecessary, as nothing of the kind" (the growth of vegetation and production of animal life) "affects us in so *deep* water." So, in other evidence, it is intimated that where waters are *deep*, the evils in question do not exist. As to loss by evaporation in uncovered reservoirs, Mr. Stirrat said—"It is a great mistake to imagine that evaporation takes place to any extent, even in the height of summer, from the surface of a reservoir where the water is of any considerable depth. The deposit of dew, I think, counterbalances it. I have one pond 10 feet deep, on which I made the experiment, and found, in the heat of summer, that in two months it did not go down one-sixteenth part of an inch; and there might have been a small escape to account for even that diminution."

Whether on *political* grounds the water-supply of London should be from one single source, or from many different sources, does not seem to have been adverted to by the Board itself or by any of their witnesses; nor, though the water were derived all of it from one and the same source, whether it might not be expedient to convey it to town in two or more conduits rather than by a single aqueduct. That an army of foreign invaders would easily find its way to London, and burn it, is not to be feared; but that attempts to so destroy it, should be made by its own populace, experience has proved, and that in furtherance of this scheme cutting off the water was designed. Had Lord George Gordon's riots lasted another day, the mains at the New River-head would, it was dreaded, have been cut off, though such troops as could be spared were ordered for the protection of those works. It should not be lost sight of that, in all metropolitan disturbances, a great portion of the rioters is made up of those whose object is plunder, and that plunder is facilitated by extensive conflagration. In that fearful night, when his Lordship's mob had the upper hand, no less than fourteen separate fires were counted from one house-top; and then it was that the abundant supply of water saved the town. Of late, too, when setting fire to houses was one of the projects in an intended insurrection, the water would doubtless have been cut off. On such accounts this political question in regard to water-supply seems well worth consideration; and it will be hereafter shown by what arrangements water might be conveyed to every part of the town, though all but a single conduit were destroyed.

M. S. B.

MR. HOLLAND'S DOMESTIC APPARA[TUS FOR] THE PURIFICATION OF WATE[R ON] CLARK'S SYSTEM.

Considering the facility wit[h] water may be greatly purified f[rom car]bonate of lime, and from many o[ther con]taminating matters, and that so[me years] are likely to elapse ere the m[etropolis] will be supplied with pure water[, it may] be useful to extract from th[e Report] of the General Board of He[alth the] means by which, on a small s[cale, the] present supply to London may [be freed] from its most injurious contents, [as prac]tised by Mr. Holland, and to giv[e a copy] of the drawing he furnished, s[howing the] simple apparatus of his contrivan[ce for the] application of Professor Clark's [process].

Mr. Philip Henry Holland[, in his] evidence, said, that "the Lambe[th Com]pany's water is very hard, an[d so foul] as to be unfit for drinking. It [is, I] may generally, offensive, and [in hot] weather swarms with visible [animals.] Even when filtered it is unfit for d[rinking,] and is, as I believe, unwholeso[me. To] render it fit to drink, I am oblig[ed to add] to it lime-water, as recommende[d by Dr.] Clark, which precipitates the c[halk and] much of the organic matter, r[enders it] less hard, and removes its unpleas[ant taste] and smell."

"This water has about 17° [of hard]ness, and requires very nea[rly one-]twelfth of lime-water to be adde[d to pre]cipitate the dissolved chalk. [This is] rather more than the Londo[n water] generally requires."

"How do you mix the lim[e-water?] I have two jugs, one holding [eleven] times as much as the other. [I fill the] smaller one with a saturated sol[ution of] quick-lime (lime-water), put thi[s quan]tity into the larger one, and f[ill it up] with the water to be purified, [and mix] the two together; chalk is forme[d by the] union of the bi-carbonate of lim[e in the] water with the quick-lime add[ed, and] falls slowly to the bottom of the j[ug, car]rying with it most of the organic [impuri]ties; the clear water is filtered fo[r use."]

"To insure the regular additi[on of] lime, I have constructed a simpl[e instru]ment which mixes the lime-w[ater for] use, and requires no attention, [beyond] adding fresh lime once or twice a [week.] It is very simple, as may be se[en from] the engraving.* J is the lime-w[ater]

* These engravings we shall give in o[ur next.]

which is placed in the cistern, with its top just above the water line; in this I put a few pounds of quick-lime, which requires renewing from time to time. On the top of the jar is a zinc-dish D, into which the water from the supply-pipe runs, and is stopped when the cistern is filled. This water entering is divided into two portions; one thirteenth part runs through the narrow notch in the pipe B, into the jar, the rest runs direct into the cistern through a notch in the side of the dish, the width of which is regulated by the slide S. The water in the jar dissolves the lime, and entering at the top displaces an equal quantity of this lime-water from the bottom, which overflows through the pipe E, and mixes with the fresh water entering at S.

The proportion of the lime-water is regulated "by the slide at S, by which the width of the notch may be increased or diminished, so that twelve times as much water may enter the cistern directly as it enters the jar, which of course displaces an equal quantity of the lime-water it contains. In order that the flow through the two notches may continue proportionate, whatever be the rapidity with which the water enters, the width of the larger notch diminishes upwards, by which compensation is made for the increasing proportionate resistance to the flow through the narrower notch as the dish fills."

As to the quantity of lime that would be expended in this process, according to Professor Clark, nine ounces of lime that is dry, pure, and freed perfectly from carbonic acid, "will require not less than 40 gallons of water for its solution, it is then called lime-water and is perfectly clear and colourless." This quantity will purify 560 gallons of hard water in addition to the 40 gallons in which the lime has been dissolved: together 600 gallons of water purified at the expense of nine ounces of lime. The chalk precipitated may be burnt again for lime.

Professor Clark's experience has led him to conclude that about sixteen hours are required for the settling of chalk in the water after it has been so treated with lime-water; and he considers it important that the *due proportion* of lime-water should be used; "no more, nor no less, but certainly no more," than what is requisite to combine with the bi-carbonate of lime in the water.

After water has been so limed, still Mr. Holland does not consider that the Lambeth water is a *first class* water. "Nevertheless, it is not much to be complained of, and I have good reason to be grateful to Dr. Clark for showing so easy a way of improving it so much."

Mr. Holland "improves the water still more," by "adding to it a little oxalate of ammonia, or of soda;" which is well worth while, for washing for instance, as the purification of the water would save half the expenditure for soap; and for making tea with water that has been treated with the oxalate, "10 parts of tea go as far as 18 without it;" and the saving pays the expense, "over and over again:" saving in his family nearly as much as the amount of his water-rate.

The oxalate of ammonia is prepared for convenient use, "by dissolving one troy ounce of oxalic acid in a quart of water, and adding as much carbonate of ammonia as will saturate it; if somewhat more be added, it will be advantageous rather than otherwise This quantity ought not to cost more than *3d.*, and would soften above 30 gallons of water."

The above-mentioned two jugs, or one jug of the requisite proportion to the usual receptacle for water, are habitually to be found in the humblest dwellings; so that with little trouble, and at very little expense, Professor Clark's invention might very generally be availed of: but Mr. Holland's apparatus seems more certain of producing the desired effect. It is simple, but, like all other new contrivances, a single one would not probably be either well or cheaply made. Would not manufacturers find their account in bringing into the market apparatus of this kind of various sizes, and at a moderate price? Now that pottery is prepared in such a degree of perfection for pipes, &c., it may be that earthenware—*Hollands* should they be called? might be afforded cheaper than metal ones, and would on one account be preferable,—that of non-liability to oxidation, especially in water that dissolves zinc.

The greatest obstacle to the use of Professor Clark's process on a small scale, is the difficulty of obtaining small quantities of lime that is fresh burnt; but a lump of it that had been once well burnt, if made red hot in a common fire, parts with the carbonic acid the lime had combined, with since taken from the kiln. Should Dr. Clark's process and Mr. Holland's apparatus come into use, it would be advisable, with each one sold, to furnish a paper of popular instructions for its use, as also some general information as to the *rationale* of the process, the saving obtainable by its means, and its sanitary importance.

Throughout the Report, and the evidence on which it is grounded, it is exhibited that the humbler classes of the London population suffer most severely from the impurity of the water with which the inhabitants are supplied; and that its foulness is often the chief cause of a general resort to strong drinks amongst the poor: hence philanthropy could hardly exert itself in a more hopeful cause than that of supplying the people with enough to drink of water sufficiently purified. Little as is the trouble or cost of Professor Clark's process, it cannot be hoped that it would be extensively employed by the poor: some plan, then, should be devised for furnishing them with enough to drink of water that had been purified. Many a scheme, at first considered fanciful—as that about to be suggested may at present be thought—has afterwards proved practicable and useful; as perhaps the present proposal, if realized, might be: it is, that some amongst the traders resorted to by the humbler classes should be induced to furnish customers with small quantities of potable water. All who purchase food of any kind buy bread; bakers, then, would be the most appropriate tradesmen to supply purified water—seeing that to their shops the industrious poor habitually resort; so that their time, their capital, would not thus be encroached upon. A baker might well afford the time required to serve gratis a pint or a quart of purified water with every two pounds loaf; but he could hardly be expected to incur the expense of providing the purifying apparatus itself. Doubtless, in even the poorest parishes, some benevolent and sufficiently wealthy person might be found who would furnish a *Holland* to some baker of his district, who, on his part, would engage to supply customers with enough to drink of water really potable. So parish distributors of bread to out-door recipients, might at each delivery of a loaf give also a small measure of good water; and the landlords who have already so judiciously and benevolently supplied their tenants "in Rose-court, Dockhead, those of ten houses on a small part of Jacob's Island, Bermondsey," with water laid on to the houses from a tank, instead of there being only a common tap, would add greatly to the value of the boon were the water in these tanks to be in a body softened and purified by Clark's process.

It is not to be supposed that confirmed drunkards would be reclaimed by furnishing them with potable water; but many a young person might by such means be preserved from tippling, and very many of the most worthy and industrious of the poor would drink good water at their meals rather than purchase beer or spirits: money thus saved would go far in the acquisition of meat or other nourishing food.

It must be observed that as Professor Clark has found that after lime has been added to water, it requires on an average about sixteen hours to settle, the *Holland* apparatus does not appear to have provided for the due subsidence of the chalk produced. This would seem to require the use of *two* cisterns, each of them containing about a day's supply; the same *Holland* might serve for both, with the slight addition to it of a double set of pipes, slides, and balls to the cock, whereby water from the same service-pipe might be made to flow at pleasure into the one or the other of these cisterns, a due portion of lime-water having passed through one and the same vessel containing lime. This arrangement would do away with all need for filtering as practised by Mr. Holland.

FRESHENING SEA-WATER.

Many plans have been brought forward at different times for converting salt water into fresh, and some have been even tried with perfect success (Mr. Charles Clark's, for example); but notwithstanding, and though it has been found that a ship going a long voyage can much more conveniently carry the quantity of coals necessary for obtaining (by distillation) the fresh water required for the use of the crew and passengers, than appropriate an equal amount of its tonnage to the storage of fresh-water casks, yet we still daily see ships taking their departure for distant shores, bur-

ith a supply of the very element which they have to plough their ly in an impure, but not unpuri-te. The causes of this, no doubt, able to something else than an nce to the importance of an t and never-failing supply on ip of fresh water; it may be plexity of the best apparatuses employed for the purpose, and ble attending the use of them; y be the extra expenditure which olve, at a time when the exces-sure of competition makes pound-uperior to all other considera-If either be the real cause of ll carrying water to sea—which as absurd a thing in its way as ng coals to Newcastle"—there prospect, we are glad to say, of speedily removed. A plan, we son to believe, has been devised ill render the obtaining of fresh t sea nearly as simple a process of heaving the lead, and at no at all. We allude to an inven-Mr. Robert Bowie, M.R.C.S., he has provisionally registered iew to patenting it, and is on the f subjecting to the test of a ned trial on board ship, under n immediate superintendence. he name of Robert Bowie our are familiar. He became first o them, as to us, by his spirited vindication of the claims of his -law, the ingenious but ill-re-WILLIAM SYMINGTON, to the able honour of having first re-team navigation to practice; and ently he figured as an inventor of several patented improve-nore remarkable for their clever-n (we fear) for their money-pay-erties. At a later day he devoted with a rare enthusiasm and to the stay of that dreadful the cholera, in a quarter of the lis where it raged most fearfully, h such signal success that the ment Board of Health were in-solicit him to visit other cities ns, to aid them with his advice, vestigate generally their sanitary n. Mr. Bowie at once consented; ll of humanity and state neces-instantly abandoned a large practice, and placed himself and unreservedly at the com-mand of the Official Guardians of the Public Health. His reward—what has it been? Only the usual reward of those who reckon upon public gratitude, as represented by the public's head ser-vants of the present times, as good for anything. His "reward" is a passage for himself and family to Australia—paid for out of his own pocket—exile from the land, to one of the most truly noble of which he is nearly allied, and to which he has been himself an ornament—exile from the land to the welfare of whose people he has literally *sacrificed himself*.

Mr. Bowie goes out to Australia in the *Athenian*, one of the emigrant ships, fitted out by the Association identified with the name of that admirable lady, Mrs. Chisholm; and arrangements have been made for enabling him to give his plan a thorough trial during the voyage.

Our best wishes, and the best wishes of many indebted friends and patients go with him. He may fare better (it will be odd if he don't!) than he has done at home; but he will nowhere find truer and sincerer (albeit, more influential,) friends than he leaves behind.

THE DOVER AND CALAIS SUBMARINE TELEGRAPH.

Our readers cannot fail to remember that the successful completion of this important line of communication was last year prevented by the insufficiency of the wire or wires provided for the purpose. Since then a cable has been manufactured for the projectors by Messrs. Newall and Co., of Gateshead (at their works at Wapping), by which every difficulty of the case is supposed to have been overcome. It consists of one continuous copper wire insulated by gutta percha, then countered with yarn, and the whole incased in wire rope. On Thursday last, this cable was laid down from Dover to within three miles of the French coast, and by this time has, no doubt, been laid the whole way.

Friday, September 26.

PLANT'S PATENT PROCESS OF PUDDLING IRON.

The following statement, showing the successful working of Mr. Plant's process of puddling iron, described in our vol. lii., p. 61, has been furnished to us by the manager of the Llyavi Iron-works:

" *Week Ending* 23rd *August*, 1851.

	Puddle Bars produced.				Coal used		Coal used per Ton of Bar-iron.		
	tons.	cwt.	qrs.	lbs.	tons.	cwt.	cwt.	qrs.	lbs.
No. 1 furnace, working with Mr. Plant's patent......	22	0	3	3	11	4	10	0	22
No. 9 furnace, worked by the ordinary mode......	14	4	0	0	11	5	15	3	10

Both No. 1 and No. 9 furnaces were boiling pig-iron of the same description; the yield of iron at both furnaces was about the same, but the quantity of puddled bar produced from No. 1 furnace was rather superior to that from No. 9. Last week's work is a fair average."

IMPROVEMENT IN SAW-MILLS.—BY LEMUEL HEDGE.

The nature of my invention consists in so arranging and combining the machinery as to prevent the necessity of straightening and straining longitudinally any other part of the saw, except that portions of it which is at the time at work.

Claim.—What I claim as my invention, is the driving belt-saws by the friction surface of two cylindrical pulleys or drums, which gripe the saw-plate below the wood which is being cut, but at some part of its tangent line, so that the strain to which it must be subjected in cutting, to keep it in the line of the tangent, shall not be at any part of its curved path; but this I only claim, in combination with straining-rollers, which gripe the saw above the lumber on which it acts; the said rollers being controlled by a brake, or the equivalent thereof, substantially as described, whereby the saw during its action is kept in a strained condition along its entire line of action, that it may cut in a straight line, and to avoid its being under tension where the flexions take place along the curved portions of its track, as specified.

I also claim, in combination with the mode of driving a belt-saw by means of cylindrical rollers or pulleys, the employment of a belt passing around the outer one of the said driving rollers, and applied to the outer surface of the saw, when it passes around the lower deflecting or guide pulley, substantially as herein described, by means of which the saw is bent by the pressure of the belt applied to its outer surface, instead of being communicated through the metal itself—thus avoiding, in a great measure, the tendency to break the metal.

And, finally, I claim in combination with the mode of driving a belt-saw, the employment of fenders or scrapers interposed between the driving rollers and the wood to be sawed, and placed each side of the saw, as described, to catch the sawdust, and conducting it away from the bight of the driving rollers or the saw, and thus avoid clogging."—*Franklin Journal.*

SPECIFICATIONS OF ENGLISH PATENTS ENROLLED DURING THE WEEK ENDING SEPTEMBER 25, 1851.

HENRY ALFRED JOWETT, of Sawley, near Derby, engineer. *For improvements in railway breaks and carriages.* Patent dated March 10, 1851.

This invention consists in the application of hydraulic power for working the breaks of railway carriages. For this purpose, the rods which actuate the break-blocks are connected to a plunger working in the horizontal portion of an elbow-shaped hydraulic barrel—the piston of the vertical portion of the barrel being actuated by a lever-handle placed within reach of the guard, by depressing which handle the plunger of the horizontal portion of the barrel is forced out, the blocks brought into contact with the wheels, and the stoppage or retardation of the carriage effected. A modification of this arrangement is also described, in which the vertical barrel has two plungers, one concentric to the other, and actuated by separate lever handles—the object being to admit of an increased degree of pressure being applied to the breaks.

Claims.—1. The application of hydraulic pressure for locking the wheels of railway breaks and carriages under every modification and arrangement of construction.

2. The combination of two pistons or plungers working in one hydraulic cylinder, for the purpose of forcing the breaks against the wheels of railway carriages, by which they may be locked, and finally stopped.

3. The combination of a hollow plunger

and plunger mounted and working for the purpose of effecting the making of railway carriages.

..m Galloway and John Gal-of Manchester, engineers. *For ...nts in steam engines and boilers.* ..ed March 10, 1851.

..provements claimed under this ..prehend,

..construction of steam boilers with ..ambers or flues, containing within ..al water pipes or tubes, open at to the water spaces of the boiler, admit of the water circulating ..ough them; the tubes being ..om end to end in such manner ..ecrease in diameter shall be not three-fourths of an inch for each ..gth.

..construction of steam boilers with ..ambers, containing tapering water ..ing flanges at their ends of such ..s to allow of the flange at the of the tube passing through the in the flue, which receives the upper end of such tube.

..construction of steam boilers with ..ambers, containing conical water flanges at the ends of which are that one shall be inside and the ..e outside of the said chamber.

..construction of steam boilers with ..ambers, containing tapering water ..inishing in diameter and distance ..hey approach the chimney end of

..construction of steam boilers with ..aining conical water tubes com-.. with the water spaces of the placed diagonally or in an in-..ition, and secured by flanges to chamber.

..lacing of the conical water spaces ..eam boilers within one or more ..lues, such flues being above or .. the sides of the boiler furnace.

..construction of steam boilers with ..ambers, containing conical water ..h flues or chambers having flat top and bottom to receive and ..s flanges of the water tubes.

..construction of steam boilers with ..es placed side by side, and sup-..fuel by a fan or feeder, rotating in opposite directions, so as to .. one furnace and then the other.

..application of a damper or .. the furnaces of steam boilers, ..as to fold up to the under-side ..e bars, and thus regulate the ..of air to the burning fuel.

..construction of steam boilers, ..liment collector placed underneath, and communicating freely with the water spaces of the boiler, and provided also with an agitator or agitators, and a blow off cock for the removal of the sedimentary impurities.

11. The construction of the slide-valves of steam engines, with notched or Vandyked edges, for the purpose of admitting the steam gradually.

12. Constructing the slide-valves of steam engines and the faces on which they work, with two or more passages on the escape side of the valve, so as to facilitate the escape of the exhaust steam by affording a large amount of passage or opening for a slight movement of the valve.

13. The construction of the slide valves, of steam engines with a vandyked edge on the steam side, in combination with double passages in the valve and seat on the exhaust side, for facilitating the admission and eduction of steam into and from the cylinder.

14. The construction of steam engines with three parallel pipes, in connection with the valve apparatus, the centre one of which forms the steam pipe to the valve-boxes, and the others the eduction pipes to the condenser.

15. The construction of steam engines with four slide valves placed horizontally, and actuated by four excentrics on a revolving shaft placed between the valves.

16. The construction of steam engines with square, angular, or diamond-shaped slide valves having seats of a corresponding form, and being applied without a valve-box, and provided internally with longitudinal partitions, dividing them into two portions, one of which forms the steam pipe and the other the eduction pipe for the exhaust steam.

17. The application to steam engines of a throttle valve, adapted to the passage between the slide valve and cylinder for the regulation of the supply of steam thereto.

George Robins Booth, of Portland-place, Wandsworth-road. *For improvements in generating and applying heat.* Patent dated March 10, 1851.

These improvements are exemplified by certain peculiar constructions of fire-places or furnaces, adapted generally to heating and evaporating purposes, and the generation of steam for locomotive, marine, and stationary boilers. The furnaces are divided internally into compartments, in the first of which, called the retort, and constructed with or without fire-bars, according to the uses to which the furnace is to be applied, the fuel is ignited, and from which the smoke and products of combustion pass through a neck or flue into a chamber or receiver, where they become mixed with a certain proportion of atmospheric air, and thence through tubes to an expansion chamber, which serves as a reservoir of heat, and from which it is applied as may be required. The patentee describes also an improved drying-room, and claims the application generally of his improvements, or any modification thereof.

THOMAS DAWSON, of Milton-street, Euston-square, machinist. *For an improved method of constructing umbrellas and parasols.* Patent dated March 13, 1851.

Claim.—1. A method of constructing walking-stick umbrellas, and several modifications thereof, that is to say, in so far as regards the making the case serve for the stick or central rod of the umbrella, the mode of joining the ribs with the stretchers, the screwing of the umbrella-stick to the runner, and the constructing of the ribs (in some cases), external to the cloth covering, as also the application of the said method and modifications in so far as applicable to the construction of parasols.

2. An improved method of constructing parasols, that is to say, in so far as regards the employment of expanding ribs, and an elastic band to gather up the extra portion of covering like a curtain, and also the application of the said method in so far as applicable to the construction of umbrellas.

JESSE ROSS, of Victoria-terrace, Keighley, York, gentleman. *For certain improvements in machinery and other apparatus for combing wool and other suitable fibrous substances, and in applying or working the same.*—Patent dated March 13, 1851.

Claims.—1. An improved sheeting machine, in so far as regards the employment of two preparatory porcupine-combing rollers in combination with the ordinary sheeting drum or cylinder, the said rollers having their teeth set in rows, and the rows of one roller gearing, or taking into the spaces between the rows of the other.

2. A modification of the sheeting machine, in so far as regards the feeding from both ends by means of fluted rollers, each taking or gearing into a porcupine-combing roller, having its teeth pointing in a direction opposite to those of the sheeting cylinder; also, the application generally of the same combination to sheeting machines whether fed from both ends, or from any number of different points.

3. A peculiar method of drawing the sheet from off the sheeting drum.

4. An improved comb-filling machine, in so far as regards the employment of metal porcupine-rollers and revolving brushes; also certain modifications thereof.

5. Certain improvements in the same patentee's formerly patented combing machine, that is to say, a compound to and fro and up and down movement of the lashing fan, the attaching of the comb-gate springs to the lower arms of the gates, the brush, and the working of the saddle-combs by cams, as described.

6. An improved method of, or apparatus for heating combs.

GEORGE LITTLE, of New Peckham, electro-telegraphic engineer. *For improvements in electric telegraphs, and in various apparatus to be used in connection therewith—part of which improvements are also applicable to other similar purposes.* Patent dated March 14, 1851.

The present improvements have for their object the improvement of the electric telegraph, and consist—1. In suspending the indicators, by means of magnetic attraction, within tubes containing spirits or other suitable fluids, which act as lubricants to prevent the needles or indicators continuing in contact with the sides of the tube when moved to the right or left by the passage of a current of electricity through coils placed in juxtaposition with the tubes containing such indicators; and 2. In supporting the indicators from floats of blown glass or other buoyant material, enclosed within similar tubes containing spirits or other suitable fluids, the dipping or sinking of such indicator or float in the fluid, when acted on by a current of electricity passing through a coil surrounding the tube, being applied to indicate conventional signals.

Claims.—1. Any arrangement in accordance with the mode described for the suspending of indicators, by means of magnetic attraction, for electric telegraph purposes.

2. The supporting of indicators, by means of floats, in tubes of spirits or other suitable fluids, for electric telegraph purposes.

HERBERT TAYLOR, of Cross-street, Finsbury, merchant. *For certain improvements in the manufacture of carbonates and oxides of barytes and strontia, sulphur, or sulphuric acid from the sulphates of barytes and strontia, and for consequent improvements in the manufacture of carbonates and oxides of soda and potassa.* Patent dated March 15, 1851.

1. To produce the carbonates of barytes and strontia, and sulphur or sulphuric acid, according to this invention, sulphate of barytes or strontia is reduced to the condition of a sulphuret by calcination in a suitable vessel or furnace, when mixed with the requisite quantity of carbon or carbonaceous substance, or by exposure to an incandescent current of carbonic oxide. The sulphuret, in conjunction with water, or, what is still

irely dissolved, and decanted from ing insoluble matters, is placed le close vessels, and a current of cid gas, generated by any suitable made to pass into the liquid. A of the base of the sulphuret emcipitates, and sulphuretted hydrolved, which, on being conducted ppropriate apparatus, is either contact with just sufficient atmor to convert its hydrogen into oper precautions being taken to xplosion), or the sulphuretted s brought into contact with nitrous rs and an excess of atmospheric never of these methods be adopted, ill be set free, and deposited by ; the gases or vapours through ranged chambers. When it is obtain sulphuric acid, the sulphurogen is ignited in an excess of le air, and is thus converted to acid vapours, which are then into apparatus such as usually in the manufacture of sulphuric treated therein in the manner adopted in such cases. When position is completed (which is by the evolution of sulphuretted gas ceasing, or by testing the he decomposing vessel with lead ch should not change colour), the rawn off, and the carbonate with-

in the oxides of barytes and stronrbonates of these bases are submtense heat in a suitable furnace ntire evolution of their carbonic place, which will be materially r the injection of steam; by which l oxide will be produced. Or if, ing the solution of the sulphuret or barytes, the same be in a boilnd highly concentrated, and then cool, almost half of the base will ized in the form of hydrated oxide, rosulphated sulphuret will remain n; this, on being evaporated in ls after separation from the cryspart with its sulphuretted hydro- small portion of sulphur will be The escaping gas (the watery ing first condensed) can be treated rted into sulphuric acid, as before . A simple sulphuret of the base will remain in the vessel, which, re-dissolved, may be passed rehrough the same process as above, whole of the oxide and sulphate of have been obtained; the hydrated m then *be purified from* the por*phuret adhering to them* by wash*crystallization*, and can then, if thought proper, be converted to carbonate by exposure, in a moist state, to carbonic acid, or by passing a current of that gas into a solution made of the crystals in water.

To produce carbonate of soda or potassa, the sulphates of these bases in solution at the ordinary temperature, are brought into contact with the carbonate of barytes, when mutual decomposition takes place, which is greatly accelerated and perfected by passing a current of carbonic acid into the mixture. By this process, sulphate of barytes is formed, and carbonate of soda and potassa remain in solution, and may be obtained by crystallization, or by evaporating the solution to dryness. To obtain hydrated oxides, or caustic soda, or potassa, the oxides of barytes or strontia (the latter being much more readily reducible from the carbonate, is preferable) are slacked with water into a thin paste, to which a proportionate solution of sulphate of soda or potassa is added, when sulphate of barytes or strontia precipitates, and the caustic soda or potassa remains in solution, and may be decanted and used in that state, or evaporated to dryness, and converted to a hydrated oxide by melting it at a lowered temperature. The carbonate of strontia may be also used to decompose the sulphate of soda or potassa, in conjunction with carbonic acid; but this material is very inferior to baryta for the purpose, as, without the aid of carbonic acid, no decomposing effect would be produced.

Claims.—1. The combined process of manufacturing sulphur or sulphuric acid, carbonates of barytes and strontia, and their oxides, from the sulphates of these bases, by the before-described methods.

2. The making of carbonates of strontia and barytes and their oxides from the sulphates of these bases by the methods described.

3. The producing of sulphur or sulphuric acid from sulphates of barytes or strontia by the method above mentioned.

4. The manufacture or improvement in the process of producing carbonates of soda and potassa, by treating or decomposing the sulphates of these bases by the carbonates of barytes or strontia, employing in aid thereof a current or sufficient quantity of carbonic acid, as described; the use of carbonic acid being optional with baryta, but imperative with strontia.

The patentee does not claim the production of the oxides of barytes or strontia by the calcination of the carbonates of these bases, when the carbonates are produced or obtained in any other than the method or manner before described.

HERBERT MINTON, of Hart's-hill,

tleman, and AUGUSTUS JOHN HOFFSTAEDT, of Bridge-street, Blackfriars, gentleman. *For improvements in the manufacture of faces or dials for clocks, watches, barometers, gas meters, and mariners' compasses, or other articles requiring such faces or dials.* (Partly a communication.) Patent dated March 17, 1851.

This invention consists in the substitution of porcelain or earthenware for the wood or metal and enamel, employed in the manufacture of faces or dials for watches, clocks, and other articles requiring such faces or dials. In carrying their invention into effect, the patentees either press clay or other suitable plastic material into moulds of the required shape, or they throw the faces on the wheel, and afterwards turn them down to the proper form, after which they are passed through the usual firing processes, and subsequently the devices or figures are printed on the biscuit in the manner adopted when ornamenting articles of earthenware, or painted by hand after the glazing, and passed through the enamelling kiln in order to cause the colours to adhere. Or the faces may be manufactured from powdered clay by pressure, and subsequently have the required devices or figures printed or painted thereon as above described. These devices may be also obtained on the faces or dials according to the methods practised when manufacturing encaustic tiles, is which hollows corresponding to the figures are formed and subsequently filled in with coloured clay in the state of paste or in strips. Faces manufactured in this manner do not require glazing, but they may be subjected to that operation, by which the brilliancy of the colours employed in producing the [figures or devices thereon will be much increased.

Claims.—1. Manufacturing faces or dials of plastic clay or other analogous plastic material, having devices or figures thereon, suitable for the purposes to which they may be applied.

2. Manufacturing faces or dials for watches, clocks, and other articles requiring such faces or dials, from powdered clay, or other analogous compound, by pressure, and applying the requisite devices thereto by any of the well-known printing or painting processes

3. The application of the process now employed in the manufacture of encaustic tiles to the manufacture of faces or dials for watches, clocks, and other articles to which faces or dials may be applied.

JAMES HART, of Seymour-place. *For improvements in the manufacture of bricks, tiles, and other articles made from plastic materials, and in the means of making parts of the machinery used therein.* Patent dated March 17, 1851.

Under the first head of his invention, which consists mainly of improvements in a machine which formed the subject of a patent granted to Mr. Hart in 1848, the patentee claims:

1. The making of the moulds used in this machine larger at the bottom than at the top, or part where the clay or plastic material is forced in, so as to insure uniformity of substance and form after the shrinkage caused by baking.

2. The application of a flange projecting from the top sides of the ends of the moulds to facilitate the removal of the mould and brick contained therein from the machine.

3. The application of additional stops to the endless chains (which form the bottoms of the moulds) to prevent the moulds from slipping during the moulding; and the employment of suitable means to retain the chain of plates in position during the same operation.

4. The arranging of the hexagonal or chain wheels, so that one or more of them shall revolve freely on their axes.

5. The application of projections to pug-mills for guiding the moulds as they enter the machine.

6. The supporting of the road or way over which the moulds travel on a seating resting on springs adjustable to suit the thickness of the moulds and the pressure to be sustained by them.

7. The employment of clutch-boxes to throw the chain into and out of action, and for starting and stopping the machine.

8. An arrangement of wheels for giving motion to the chain of plates which form the bottom of the moulds.

9. The employment of revolving scrapers and apparatus working therewith to remove the superfluous plastic material.

10. The application of stuffing-boxes to the shafts which pass through the sides of the pug-mill, and the application of a dividing-plate, or partition, in pug-mills between the two rows of moulds, and the flanges on the moulds, in order to keep the material in its proper position.

11. An improved door to the sand barrel, and an additional riddle or sieve for distributing the sand more evenly over the surface of the moulds.

12. The application of leather or other packing to the road wheel-shaft to prevent injury to the machinery when travelling.

13. A mode of applying or connecting shafts to the machine, so that they may be readily applied or removed.

14. The application of a screw or screws

piston or pistons for the purpose
clay or other plastic material
oulding orifices.
using the crushing rollers to move
ent surface speeds.
employment of ground coal or
ead of sifted ashes as now usually
in the manufacture of bricks.
ond head of the invention relates
g machine for boring holes in the
e moulds and plates which form
ing chain, in which two drills,
ultaneously on the opposite ends
te or mould, and caused to ap-
h gradually as the operation of
gresses, are employed.
NDER ROBERTSON, of Holloway,
nd JAMES GLOVER, of the same
r *improvements in the rolling and*
of metals, and in the manufac-
tallic cases and coverings. Patent
ch 20, 1851.
ect of the improvements specified
first head of this invention is to
may be required a uniform elas-
e, or a uniform dead pressure on
of the rolls employed for rolling
ating metals. For this purpose
l of the same length as the rolls,
e of sliding up and down imme-
r them, is mounted in the same
to each end of this crosshead is
y keys a rod or plunger, which
st the brasses fixed at the ends of
The centre of the crosshead is
out of a cylindrical form, and a
ng is inserted and inclosed by a
f metal. A nut is tapped through
of the bridge-piece at the top of
in which works a screw turned
wheel, and bearing upon the disc
ring in the centre of the cross-
urning the hand-wheel, the screw
vered, and the crosshead and rods
ereto caused to descend and act
s so as to produce a uniform elas-
re at both ends, which may be
o any desired extent by lowering
When a dead pressure is re-
spring in the crosshead is dis-
th, and the screw made to act
olid abutment. Another method
ng an elastic pressure on the rolls
application of hydraulic power,
case the cylinder in the crosshead
nnected with a force-pump, and
ure produced in the ordinary

nd part of the invention has rela-
ethod of making boxes, cases, or
rom William Betts' patent metal,
foil, or Betts' patent metal pasted
paper, and variously ornamented.

The method of making boxes or coverings from Betts' metal is as follows:—The patentees take a slightly-tapering block of iron of any desired form, and wrap round it a sheet of metal of about the 300th part of an inch thick, and of such a size that the edges just overlap; a small portion also must be allowed to project over the smallest end of the block; they then run a slightly-heated soldering bit or iron along the overlapping edge (using a copper straight-edge as a guide), so as to cause the two parts to adhere. They then fold in the projecting part of the sheet of metal over the small end of the block, place upon it a piece of slightly less size than the end of the block, and run a soldering-bit along the edges of that piece, and thus finish the covering, which is then to be drawn off from the block. The use of solder is thus entirely dispensed with. In order to prepare tinfoil or Betts' metal for use, it is glued or pasted to paper, passed through a pair of flatting rollers, and then embossed or ornamented with transparent colours, mixed in turpentine or copal varnish; or it is coated with varnish, and then ornamented by sprinkling it with flock. The metal paper, thus prepared, may be used for paper-hangings and various other purposes, as well as for the manufacture of boxes, cases, or coverings.

Claims.—1. The producing a uniform elastic pressure on each end of the rolls in the rolling and laminating of metal.

2. The producing a uniform dead pressure on each end of the rolls in the rolling and laminating of metals, by means of the arrangements described.

3. The manufacture of William Betts's patent metal in the manner described.

4. The manufacture of metallic coverings made wholly or partly of tinfoil, or of William Betts's patent metal, pasted or glued on paper, and embossed, painted, or otherwise ornamented, as described.

WEEKLY LIST OF NEW ENGLISH PATENTS.

Frederick Hale Thomson, of Berner's-street, Middlesex, gentleman, and George Foord, of Wardour-street, in the same county, chemist, for improvements in bending and annealing glass. September 25; six months.

Charles Green, of Birmingham, Warwick, for improvements in the manufacture of brass tubes. September 25; six months.

Richard Archibald Brooman, of the firm of J. C. Robertson and Co., of Fleet-street, London, patent agents, for improvements in presses and in pressing. (Being a communication.) September 25; six months.

Robert Roberts, of Dolgelly, Merioneth, mine agent, for an improved method of quarrying certain substances. September 25, six months.

Charles Watt, of Kennington, Surrey, chemist, for improvements in the decomposing of saline and other substances, and in separating their component parts, or some of them, from each other; also in the forming of certain compounds or combinations of substances, and also in the separating of metals from each other, and in freeing them from impurities. September 25; six months.

James Garforth, of Dukinfield, Chester, engineer, for certain improvements in locomotive steam engines. September 25; six months.

David Stephens Brown, of the Old Kent-road, Surrey, gentleman, for an improved agricultural implement. September 25; six months.

Ernst Kaemmerer, of Blomberg, in the kingdom of Prussia, iron-founder, for his invention of improvements in sowing, depositing, or distributing seeds over land. September 25; six months.

WEEKLY LIST OF DESIGNS FOR ARTICLES OF UTILITY REGISTERED.

Date of Registration.	No. in the Register.	Proprietors' Names.	Addresses.	Subjects of Design.
Sept. 18	2946	Henry S. Rogers	New Oxford-street	Eye renovator.
"	2947	R. and W. Wilson	Wardour-street, Soho	Bath heater.
"	2948	R. Sorby, R. Sorby, jun., and T. A. Sorby	Sheffield	Point for scythes.
"	2949	R. Sorby, R. Sorby, jun., and T. A. Sorby	Sheffield	Point for a reaping-hook.
19	2950	W. Scott	Exeter	Air-regulator, with air-strainer for the admission of pure air into apartments without a draught.
"	2951	T. Cook and W. J. Corsan	Plumstead / Shadwell	Alarum for houses.
"	2952	Robert Hammond	Kirkgate, York	General two-horse reaping machine.
"	2953	James Guest	Birmingham	Penholder.
"	2954	The Grangemouth Coal Company	Grangemouth, Falkirk	Drain-pipe, chair, and sleeper.
22	2955	The Rev. E. H. Johnson	Lindfield, Sussex	Skim plough.
23	2956	S. A. Bell & J. Black	Bow-lane, Cheapside	Matchless match-box.
"	2957	Joseph Taylor	Wolverhampton	Tittley's protection segmental slide-cap for locks.
"	2958	Henry McEvoy	Birmingham	Hooks for dress fastenings.
"	2959	Beach and Minte	Birmingham	Inkstand

WEEKLY LIST OF PROVISIONAL REGISTRATIONS.

Sep. 18	288	Henry Maling	Home-office, Whitehall	Elevation sight for rifles.
"	289	Francis Evans	Deptford	Music stand.
"	290	John N. Gibbs	Wendling, Norfolk	Economic heating apparatus for forcing-houses, greenhouses, conservatories, hothouses, &c.
19	291	Ebenezer Poulson, sen.	Monkswearmouth, Sunderland	Life-boat.
20	292	George Lomas	Camberwell	Spring-lever ventilator for shop plate-glass or other windows.
23	293	Joshua Rhodes	Camberwell	Transit indicator and universal almanack.
24	294	Alfred Ford	Ebury-street, Eton-square	Safety-spring for railway carriages.

CONTENTS OF THIS NUMBER.

LONDON: Edited, Printed, and Published by Joseph Clinton Robertson, of No. 166, Fleet-street, *in the City of London*— Sold by A. and W. Galignani, Rue Vivienne, Paris; Machin and Co., *Dublin*; *W. C. Campbell* and Co., Hamburg.

Mechanics' Magazine,

MUSEUM, REGISTER, JOURNAL, AND GAZETTE.

No. 1469.] SATURDAY, OCTOBER 4, 1851. [Price 3d., Stamped, 4d.

Edited by J. C. Robertson, 166, Fleet street.

MESSRS. RIDLEY AND EDSER'S PATENT SAFETY-HINGE AND BURGLARY-ALARM.

Fig. 1. Fig. 3. Fig. 4.

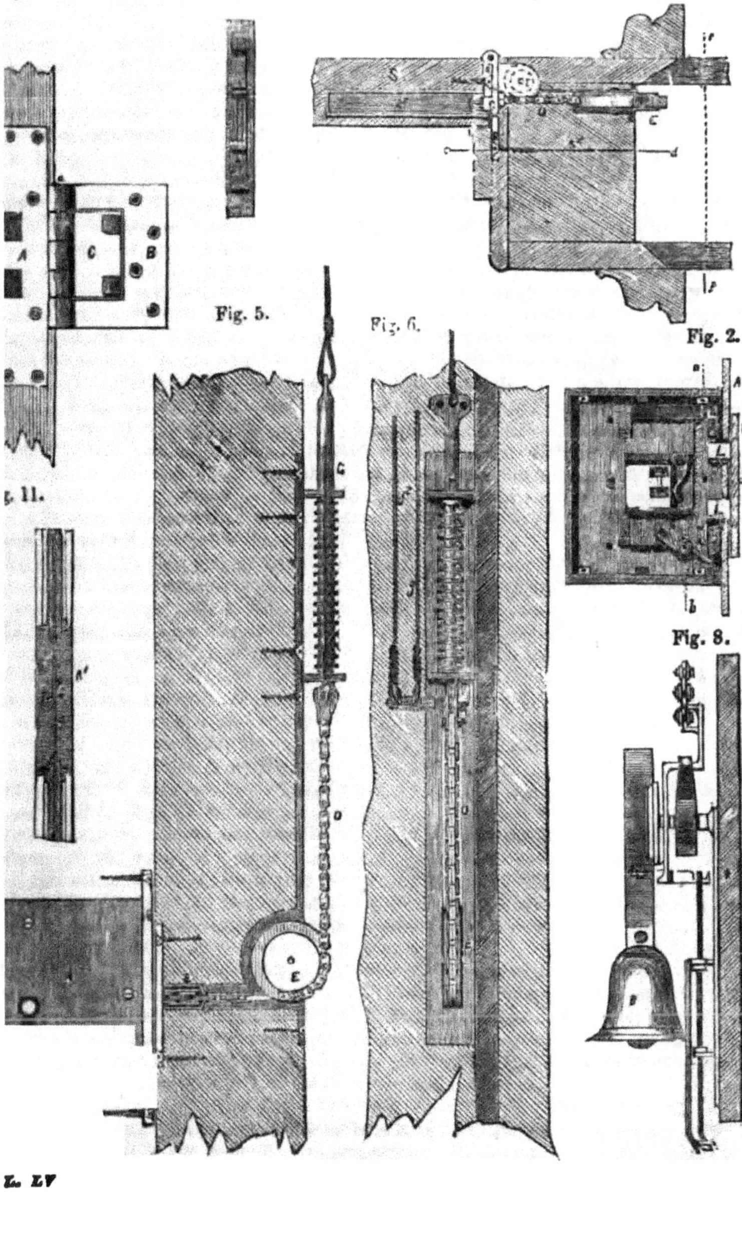

MESSRS. RIDLEY AND EDSER'S PATENT SAFETY-HINGE AND BURGLARY-ALARM.

[Patent dated March 24, 1851. Specification enrolled, September 24, 1851.]

Specification.

Firstly. Our invention for the detection of burglars and prevention of burglaries, consists of a safety hinge to be attached to doors, shutters, or windows, an of an alarm apparatus to be actuated by the movements of said hinge. Fig. 1 is front view of the same; A and B are the two wings or flaps of the hinge; A^2 is tumbler-lock, which is attached to the inner face of the wing A; C is a supplementary flap, which falls into a recess in the inner face of the wing B, and turns on pin *a*, common to all the three pieces. The supplementary flap C, serves a keeper to the lock A^2, having two staples LL, into which the bolts L^1 L^1 of the lock take, when actuated by a key in the manner of the ordinary chest door locks, or by the movements of the safety-hinge in the manner to be presentl explained. Fig. 2 is section of the safety-hinge through the lock attached to th wing A, and fig. 3 a cross section on the line *a b* of fig. 2.

Fig. 4 is a plan showing how this safety-hinge is attached to a door, and th apparatus in connection therewith, whereby its movements are made to give notic of any burglarious entrance being attempted. Fig. 5 is a side elevation on the lin *c, d*, of fig. 4, and fig. 6 a front elevation on the line *e, f*, of fig. 4. The wing the safety-hinge, which carries the lock, is sunk into the style S of the door, and th other wing, which carries the keeper, sunk into the door-post S^2. To the back the keeper there is attached a chain D, which after passing over the guide-pulley EE, is connected to the lower end of a spring-rod G, which puts in motion th alarm apparatus; the construction of which is represented separately in figs. 7 and 8 the former being a front view, and the latter a section on the line *g, h*. In thes figures the apparatus shown is supposed to be connected with three differen entrances to a house,—the front-door, area-door, and back-door; but it may, by mere multiplication of the same parts, be connected to any number of doors, or t any number of window shutters or sashes, as well as doors. O is a bearing-plate which is affixed to one of the walls of the apartment to which it is desired to communicate the alarm; T a bracket, which carries a set of vertical levers R^1, R^2, R^3 To the lower ends of these levers there are hooked on, by rods running in guides a set of labels, indicating the doors with which they are respectively connected—a "Front-door," "Back-door," "Area-door;" and at top, each lever has tw chains, or wires, branching off from it, one connecting it with the spring-rod G o the safety-hinge of the particular door with which it is in communication, and th other connecting it with a fourth lever R^4, on the bracket T, to which last leve there is attached by a spiral spring an alarm-bell, B, W.

Supposing, therefore, each door to have its safety hinge properly set, that is, wit the one wing let into the door-post and the other into the door style, and both con nected by a lock and keeper, as before described; then it follows that, on the doo being forced open, it will pull the chain D, attached to the back of the keeper-fla C, and that will draw down the spring-rod G, which again pulling back the cor responding lever R of the alarm apparatus, will set the bell ringing, and simulta neously therewith unhook the rod of the label bearing the name of the particula door which is attempted to be opened, and thereby allow it to fall below the line the other labels. In fig. 7 an exemplification is given of the relative positions whic the labels would, in such a case of attempted burglary, assume: the label belongin to the front door (which is that supposed to be forced open) is shown as unhooke and dropped down, while the others remain in their original places. When it desired to have an alarm-bell on the outside of a house, which may be put in motio by the movements of a safety-hinge such as has been described, and rouse the polic or others to the detection of the burglars, independently of any person inside, w either connect the alarm-bell with the safety-hinge in the manner before describe or by means of an apparatus such as represented in front and end elevations in fig 9 and 10.

This apparatus consists of a train of wheelwork acted upon by a spring or weigh and which, when liberated by the movement or pulling of any of the wires J^3, give ction to the outside alarm-bell. A is the framework; B the spring-barrel; C C e wheelwork; D an escapement wheel which actuates the lever E, to the upper d of which there is attached a connecting rod F, which again, through the lever , imparts to the bell H an oscillating motion; I is the lever to which the wire I² is nnected, and by which the wheelwork is liberated upon pulling that wire.

Fig. 7.

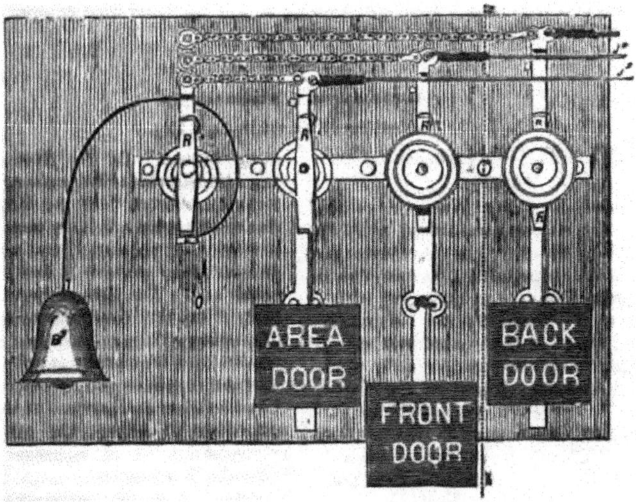

Fig. 10.

Fig. 9.

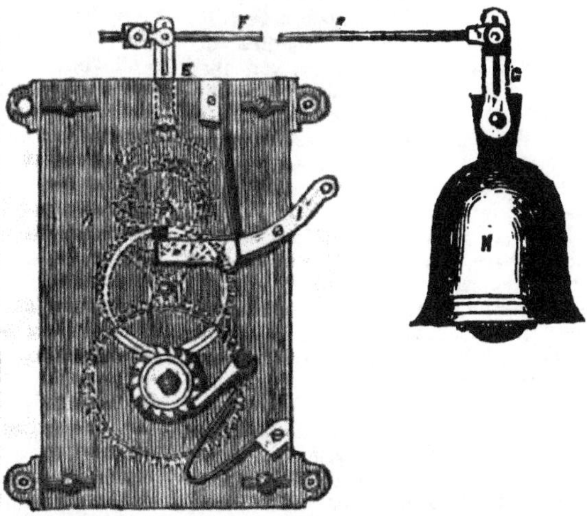

Secondly. For further security against burglars and burglaries, we propose to fill and strengthen the panels and styles of doors, shutters, and other framed car-nters' work exposed to be broken through for burglarious purposes, in the manner own in fig. 11, which is a vertical section of a door-rail and part of two panels th those additions. A A is the style, which is formed in two thicknesses, a¹ and *Previous to these thicknesses* being put together, the iron plate b is sunk into one of the halves or thicknesses, and imbedded in marine glue. The other the style is also glued to its place over the iron plate, after which the whole fast by screws *c*, the heads of which are sunk in the rail, and afterwards co with wood, so as to conceal completely the fastenings. The panel is for cementing, by means of marine glue, to each side of a plate of metal, a canvas, and gluing over the canvas a veneer of wood. Doors and shutters so are put together in the usual manner, and exhibit externally all the appear doors and shutters formed entirely of wood, and at the same time present gre culties to their being forcibly broken open.

MATHEMATICAL PERIODICALS.

(Continued from vol. liv., p 494.)

XXVIII.—*The Mathematical Repository.*

Origin.—The first Number of this extensive serial was published in 1795, under the title of "The *Mathematical Repository:* containing many ingenious and useful Essays and Extracts, with a collection of Problems and Solutions, selected from the correspondence of several able Mathematicians, and the works of those who are eminent in the Mathematics." The work consists of two portions under the respective designations of the "Old" and the "New Series," of which the former comprises *three* volumes of the Mathematical, and *two* volumes of the Philosophical Department: the "New Series" is almost entirely devoted to Mathematics, and is comprised in *six* volumes; the first of which was completed in 1806, and the last in 1835. In the "Address to Correspondents," issued with the first Number, "the Editor of the *Mathematical Repository* ...... begs they will accept his warmest expressions of gratitude for their present favors ...... and assures them that nothing but an ardent desire of promoting the study of the Mathematics could have engaged him in this undertaking." He announces, that "the Repository will be open for all, particularly where a superior knowledge or merit leads the way: nor shall the gleanings of diffident merit or conscious knowledge be overlooked: he therefore hopes that every lover of the Mathematics will feel it their duty to exert themselves in favor of a work undertaken solely for their use and amusement." The remaining portion of the "Address" is the same as that quoted from the *Yorkshire Repository*, (*Mech. Mag.*, vol. liii., p. 504,) and hence the *probability that both* were written by the same person. In the *first* vol the *Philosophical Repository*, t face was written "for the inform those who may not have seen th bers as they appeared, and who, account, may require a few expl remarks." The Editor therein d that "on the first appearance *Repository*, it was designed for tension and improvement of the matical Sciences only; but, on consideration and advice, it was expedient to enlarge the origina by including in it whatever re Natural Philosophy ...... The Number of the *Repository* was th divided into *two* departments, whi regularly been continued in every quent Number .... A free sion," he observes, "has been to all which appeared to deserve and a collection of useful questio their solutions, has been inser appearing to be well calculated excite and to satisfy inquiry."

The simultaneous publication Mathematical and Philosophical ments continued until the appear "No. 14, May 1, 1804," whi nounced that, "the present Numb cludes the *third* volume of the matical department and the *secon* Philosophical, and terminates t series of the *Repository*. The or 'New Series' followed in due the size being changed from s large octavo, and was issued at ir intervals until the completion sixth volume, in 1835; when th was discontinued."

Editor. — Mr. Thomas Le F.R.S., Professor of Mathemati Royal Military College, Author Synopsis of Data for the Solu Triangles," &c., &c.

tents.—Part 1. *Philosophical* *tment.* The first volume of this nent contains viii + 368 pages of losely-printed matter, embracing almost every topic of Philosophi- erest, and is dedicated "To Wil- Ierschell, L.L.D., and Fellow of yal Society, in testimony of sin- espect for the numerous advan- which Astronomy and Philosophy eceived from his judicious labours ir improvement .... by the ." In the course of the volume rer than 91 Questions in various ments of Natural Philosophy are ed, amongst which are several for prize medals were awarded. The of Newton Bosworth, Olinthus ry, Sir John Byerley, Richard William Marrat, J. H. Swale, Dr. , Benjamin Gompertz, Henry , &c., &c., frequently occur in this ment, and the subjects treated both that interest and originality might be expected from the efforts of so many distinguished s. Amongst the miscellaneous s we have, "A Comprehensive nt of the Newtonian Method of ophising. By Robert Thorp, M.A., nicated by Olinthus Gregory;" ecture on Optics. By Dr. Saun- ,with notes by Olinthus Gregory;" the Management of Heat. By Rumford;" "On the Progress of oderns in Natural Philosophy. By on) B(osworth);" "On the Origin ings and Rivers. By Mr. Gregory usticus;" "On the Origin of Ne- By the Rev. James Jogglebelly, rk;" "On the Dimensions of the By Newton Bosworth;" "In- ng Extracts;" "Description of an ved German Key for extracting . By Mr. Charles Brown;" "On t Light with some remarks on ustion. By Mr. Kay;" "Obser- s on Negroes;" "On the Inven- f Clocks. By Mr. C. Brown;" nentary Books on the Mathematical es recommended, with directions e Student. By Cantatrigos," "An on Putrefaction. By Dr. Crane;" Phosphorescent Bodies. By Charles , Surgeon;" "On the Circulation ids in Vegetables. By Hortus;" Excitability. By Dr. Crane;" "On . By Benjamin Gompertz;" "On Sensation. By Mr. J. T. McDonald;" "On Keeping Interest Accounts. By Mr. Henry Boiley;" &c., &c., &c.

The second volume of this department contains iv + 124 + 102 pages, the *first* portion of which consists of a continuation of the Philosophical Department, and the *second* is devoted to "A Review of Mathematical and Philosophical Books." The Queries are continued up to No. 124, the Answers to which are mostly supplied by Dr. Crane; J. B.; N. Y.; Thomas Crosby; Thomas Bower; J. F.; Richard Shillitoe; Thomas Boole: John Dawes; &c., &c.; and are equally as interesting and useful as those in the first volume. Amongst the Miscellaneous papers are found "some Interesting Extracts. From the *Annales de Chemie;*" "Essay on a New Division of the Year, Universal System of Standard Measures, Scales of the Barometer, Gamut in Music, &c. By J. K.;" "On Teaching Geography, and the most proper Books for that purpose. By W. Marrat;" "On the preparation of Mephitic Alkaline Water, for curing the Stone or Gravel. By Mr. Thomas Crosby;" "Fragments of Science;" "On the Planets discovered by Piazzi and Olbers;" "Improvement of the Method of Impregnating Water with fixed Air. By Richard Shillitoe;" "On the Building of the Pyramids. By S. B. G., Esq.;" "*On the Identity of Light, Fire and Electricity.* By Mr. Thomas Squire;" "On Purifying Corrupt Water by Charcoal. By Mr. Boole;" "On the Periodical Rising and Falling of the Barometer. By Mr. T. Squire;" &c., &c. That portion of the volume devoted to Reviews is paged separately; and has a distinct title-page: 37 Mathematical and other works are reviewed; but as most of these are merely signed by the letters Y, S, H, N, D, T, W, it is impossible at this distance of time to assign each review to its real author.

Playfair's *Elements of Geometry* is the first work noticed, and its peculiarities are very clearly stated. The author of the paper seems disposed to agree with Playfair's alterations of the Fifth Book, and considers "his edition of the *Elements of Euclid* superior to every other which has yet been published." The next review is devoted to an examination of Hutton's *Course of Ma-*

thematics, several extracts from which are given "as we conceive them to be of great utility." The work itself is spoken of in the highest terms, although exception is taken to the manner "in which the writer has treated proportion." Manning's *Algebra*, 2 *vols.*; Dix's *Surveying*, and Bryan's *Astronomy*, follow in order; after which Gregory's *Lessons, Astronomical and Philosophical*, are noticed in a manner which shows that the writer had *read* the book upon which he offers an opinion. The *Transactions of the Royal Society of Edinburgh* (1799) are examined at considerable length; Mr. (afterwards Sir) James Ivory's *New Method of Resolving Cubic Equations* is extracted from his paper on this subject. We may here remark, that Mr. Ivory here *first* gave the *criterion* for the roots of the complete cubic

$$x^3+ax^2+bx+c=0,$$

which has since been published by Mr. Lockhart, as Question 167 in the *Mathematician*. It is as follows: In the general equation

$$x^3+ax^2+bx+c=0,$$

if $$a^2b^2+18abc,$$

is greater than

$$4\,a^3c+4\,b^3+27c^2,$$

the roots are *real*; if *less*, two roots are *imaginary*; and if *equal*, two roots are equal to each other," (*Mech. Mag.*, vol. xlviii., p. 226). The *Transactions of the Royal Society of London*, and Vince's *Trigonometry*, are also noticed at some length; but the latter work by no means reaches the standard required by the writer of the critique, who remarks, that "a *complete* treatise on Trigonometry has long been a *desideratum*." The leading features in *The Principles of Algebra*, by William Frend, are very clearly stated, but the writer carefully avoids the expression of any opinion on the questions in dispute between the *positive* and the *negative* algebraists. Several short comments on Dr. Hutton's *Principles of Bridges*, Horsley's *Elementary Treatises*, and Atwood's *Dissertation on the Construction and Properties of Arches*; in which the last-named work is rather severely criticised; the writer being of opinion that "Mr. Atwood's Dissertation can never be of any use in the Art of Bridge-building, since the skilful Architect is not likely to practice and confide in an *old and imaginary theory*, which has been rejected as fallacious." Donna Agnesi's *Analytical Institutions* are reviewed at a greater length than most of the other mathematical publications, and several instances of the excellence of this course of instruction are offered to the reader's notice. Exceptions, however, are taken to some parts of the work, especially to the proof that "like signs give +; unlike −;" which is based upon proportions by Agnesi. A mode of proof is offered which the reviewer "trusts will be found unobjectionable;" it appears to be based upon Frend's principles, and argues that since "$+a\times-m$ has *no meaning*, for *m* must be an abstract number;" in order to avoid there being "no proof" he takes $+a\times(m-n)$ and so arrives at a proof "free from metaphysical subtleties." Dr. Gregory's *Astronomy*, and Fenwick's *Four Essays on Practical Mechanics*, next come under review; and the work concludes with a well-deserved encomium on Mr. Thomas Keith's *Introduction to the Theory and Practice of Plane and Spherical Trigonometry*.

T. T. W.

Burnley, Lancashire, Sep. 25, 1851.

(*To be continued.*)

M. DE COLMAR'S ARITHMOMETER.

This instrument was invented thirty years past, by M. Thomas de Colmar, and has the incontestable advantage of being the first apparatus produced for operating simultaneously on several figures at the same time. The mechanism is simple and ingenious, and performs the four rules of arithmetic with facility and precision without the possibility of any mistake. Thus, powers of figures, the extraction of their roots, calculations of interest and discount are obtained with the utmost facility.

The whole mechanism is contained in a box, 14 inches long and 10 inches broad, for machines working with ten figures combined; and for machines producing the result of sixteen figures, the dimensions are 22 inches by 6 inches. The upper part of the machine is divided lengthwise into two parts, one of which manent, the other moveable. The able plate is pierced with ten small perforations, in each of which ten s, from 1 to 0 successively appear. it plate be raised and brought to- the right or left, units will appear same order, and in a column, the of the operation will be written ie same plate. The permanent is provided with notches, and the slide in them when pressed by the , and point out on the scales con- g the ten figures, the number to be ted upon.

ltiplication is produced by means andle set on the right of the ma- , which carries at each rotation ie holes the number indicated by iobs of the grooves.

th respect to the internal me- im, it consists of a cylinder set ntally on a square axle; this cylin- s provided with nine teeth or es, the first of which runs through hole length of the cylinder, and rest diminish progressively and pond with the primitive figures 8.

e knobs of the grooves are con- l with a wheel, provided with ten mounted on a square axle, which cogs with one, two, or with nine according as the knobs are placed. transmission of motive power, as med by the handle and the dial ning the figures in the holes, will in the hole the figure pointed out knob of the groove. All the me- m consists in this cylinder, which, eated as many times as the number ures desired, will produce the ma- .

mall lever is adapted to each cylin- or performing the functions of re- .

ien one unit is written in one hole, is desired to add to it nine other —for instance, the knob over the e is placed on the figure 9, one m of the handle is performed, g which time the figures 0, 1, 2, 3, 6, 7, 8, 9, 0 pass successively in the when arrived there, the lever of linder will meet with a ten-toothed corresponding to the following towards the left, and will advance istance of one tooth of the wheel rting the deal of that hole. The fig. 1 will appear in that space, and the fig. 10 will be produced as the sum of 9 and 1.

With regard to the mechanism for setting the apparatus to the particular operation to be performed, it is very simple and ingenious, and consists of a key with ears, putting in motion a lever, which causes to approach or recede an iron band serving as bearing for the angle wheels, which put the hole dials in motion. The key has only to be turned to subtraction or addition to cause one wheel or the other to cog with the dial-wheel, which causes it to turn from right to left or from left to right.

Thus, when the key is at addition, the figures on the dial appear in the following order: 1, 2, 3, 4, 5, &c.; and when the key is at subtraction, they will be in the contrary order, 9, 8, 7, 6, 5, &c.—(*From a Correspondent.*)

THE GUTTA PERCHA WORKS.

The extensive and highly-interesting establishment of the Gutta Percha Company, situated near the City-road Basin of the Regent's Canal, is worthy of attention even beyond the general average of such centres of industry, for the peculiar character of the substance operated upon necessitates the employment of new processes, new machines, and new tools. An incessant course of invention has marked the manufacturing history of this material during the brief period of its existence. If the gutta percha is to be applied to some new useful purpose, tools and processes of novel character have to be employed; if an ornamental application is determined on, methods are adopted for developing any natural beauty which the grain of the substance may present; if an attempt be made to supersede leather, or wood, or papier-maché, or metal, by this singular gum, great pains are bestowed on a study of the special qualities to be imitated, and the process of imitation often requires operations and tools differing considerably from those before employed.

A pervading odour is sensible throughout the buildings in which the gutta percha is stored and manufactured. If it were necessary to characterize this odour, we might, perhaps, liken it to a hybrid between tan-bark and old cheese—an odour to which one is not, at first, easily reconciled. But this becomes dissipated after a time.

When we direct our attention from scent to sight, and look around the establishment, we see the very history of the manufacture pictured in the buildings themselves. Every separate block of building speaks of a particular application of the gutta percha, or some particular mode of preparing it for use. If we see a building somewhat more fresh and modern than its neighbours, we may infer that some new, or comparatively new, process is there carried on; and the area is thus becoming dotted about with workshops and ware-rooms, which will not much longer yield each other sufficient elbow space. It is only when we bear in mind the very recent introduction of this remarkable substance, that the extent to which the manufacturing arrangements have grown can be duly appreciated. Store-rooms for the newly-imported gum; steam engines and boilers for supplying the agency whereby the manufacturing processes are conducted; large buildings filled with the machines and tools for working; workshops in which the finishing processes are conducted; a canal quay for unshipping the raw material, and shipping the finished goods;—all speak of a busy series of operations. It is also proper to remark, that another extensive establishment of a similar character is carried on at West Ham, and that minor manufactories are now scattered over London and other towns.

In the store-room the blocks and lumps, of slightly-varying colour and texture, generally present a fair outside, and it is not till the first process has been gone through that the fraud can be detected. This process consists in cutting the block into slices. There is a vertical wheel, on the face of which are fixed three knives or blades; and while this wheel is rotating with a speed of two hundred turns a minute, a block of gutta percha is supplied to it, and speedily cut into thin slices—much on the same principle as a turnip-cutter performs its work. Woe to the steel edges if a stone be imbedded in the block! all alike, the soft and the hard, are cut through, but not with impunity.

These slices show that the gutta percha is by no means uniform in different parts, either in colour or texture. To bring about a uniformity is the object of the shredding or tearing process. The slices are thrown into a tank of water, which is heated by steam to such a temperature as to soften the mass; the dirt and heavy impurities fall to the bottom, leaving a pasty mass of gum; and the mass being thrown into another rotating machine, is there so torn, and rent, and dragged asunder by jagged teeth as to be reduced to fragments. The fragments fall into water, upon the surface of which (owing to the small specific gravity of the material) they float, while any remaining dirt or impurity falls to the bottom. These fragments are next converted into a dough-like substance by another softening with hot water, and the dough undergoes a thorough kneading; it is placed in heated iron cylinders, in which revolving drums so completely turn, and squeeze, and mix it, that all parts become alike, and every particle presents a family likeness to its neighbour.

The kneaded state may be considered the dividing line between the preparatory processes and those which relate to the fashioning of the material. The soft ductile mass may be formed either into sheets or tubes. In forming sheets the mass is passed between steel rollers, placed at a distance apart corresponding with the thickness of the sheet to be made—whether for the heels of a rough booted pedestrian, or for the delicate "gutta-percha tissue," now so much employed by surgeons. By the time that the substance has passed through the rollers, it has cooled sufficiently to assume a solid firm consistency. By the adjustment of a few knife edges the sheet may be cut into bands, or strips of any width, before leaving the machine. In making tubes and pipes the soft mass of kneaded gutta percha is passed through heated iron cylinders, where a singular modification of the wire-drawing process reduces it to the desired form and dimensions.

From the sheets and tubes thus made, numberless articles are produced by cutting and pressing. Machines, somewhat like those used in cutting paper, are employed to cut the gutta percha into pieces. If for shoe-soles, a cutting press produces a dozen or so at one movement; if for string, or thread, narrow parallel strips are cut, which are then rounded or finished by hand; if for producing stamped decorative articles, the sheets are cut into pieces, and each piece is warmed and softened to enable it to take the impress of a mould or die. But the mode of casing copper wire for electro-telegraphic purposes is, perhaps, one of the most singular applications of the material in the form of sheet. Several wires are laid parallel, a strip of gutta percha is placed beneath them, another strip is placed above them, and the whole are passed between two polished grooved rollers; the pressure binds the gutta percha firmly to the wires, while the edges between the grooves indent the gutta percha so deeply, that it may easily be separated into wires, each one containing its own core of copper.—*Knight's Curiosities of Industry.—Part III.*

MR. HOLLAND'S APPARATUS FOR FILTERING WATER.

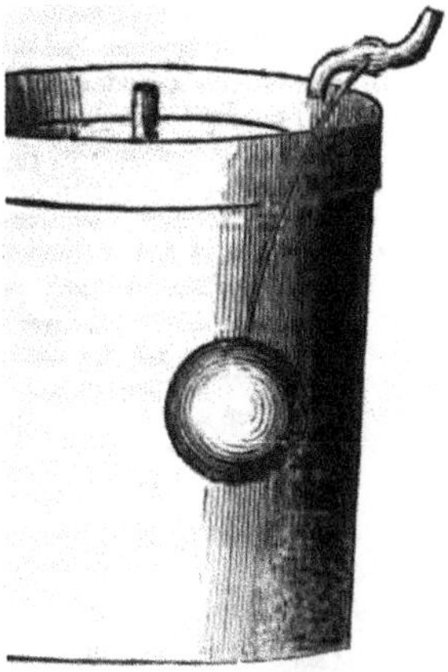

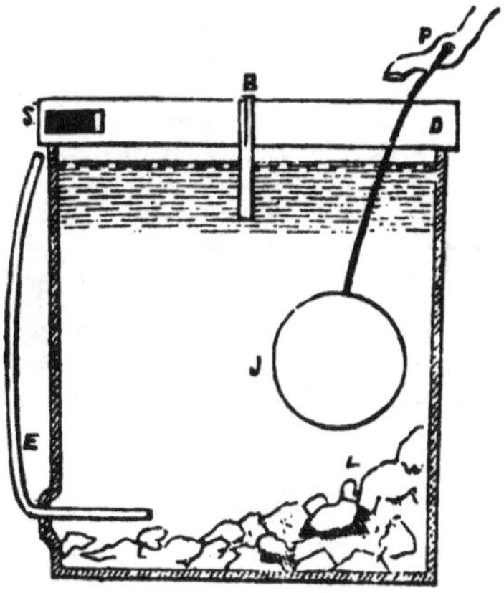

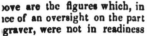

›ove are the figures which, in ıce of an oversight on the part ıgraver, were not in readiness to accompany the article in our last week's Number on Mr. Holland's Filtering Apparatus.

ICATIONS OF ENGLISH PATENTS ENROLLED DURING THE WEEK ENDING OCTOBER 2, 1851.

BESSEMER, of Baxter-house, Old ›-road, engineer. *For improve- the manufacture and refining of ' in machinery or apparatus used ing a vacuum in such manufac- ch last improvements are also applicable for exhausting and uids.* Patent dated March 20,

'ention comprises,

ode of clarifying, evaporating and ing saccharine fluids in open pans , of a construction and form :cording to the quality of the fuel oyed, and the nature of the previous quent processes which the fluids rgone, or are intended to undergo, f which forms of apparatus the heat ing evaporation of the saccharine pplied thereto by the agency of ierated in a separate compartment sel containing the fluid to be eva-

ain apparatus for producing a par- n in the pans or boilers employed centration of saccharine fluids, by e steam produced by their evapo- ıct on a piston in connection with ıpparatus, by which the motive the steam thus produced is caused s own exhaustion. Also, a method ıg a rotary motion suitable for the ›ose, by the emission of steam and water from the condenser, such steam having been first employed in the vacuum pan. Also, a method of heating the vacuum pans employed in the manufacture of sugar, by means of a steam generator in a separate compartment of the same vessel.

3. Certain apparatus for the production of a partial vacuum in the pans or boilers employed in the manufacture of sugar, where the motive power applied for producing such partial vacuum is obtained from a fall of water or other similar source of power, and not from the steam produced by the evaporation of saccharine fluids as described under the second part of the invention. The apparatus adapted to this purpose (which is a modification of the disc pump, the action of which for exhausting aëriform fluids is dependent on a well-known physical law, that water when so forced carries with it a certain portion of air), is intended to produce a low degree of exhaustion suitable for such purposes as the curing and filtration of sugars, and is also applicable otherwise to the forcing, exhausting, and impelling of fluids.

4. A method of curing sugars and separating the crystals from fluid.

5. A method of improving the quality of muscovado and other crude sugars, by separating the larger from the smaller-sized crystals, and removing the impurities which are generally found in crude sugars; the

larger crystals being subsequently treated with liquor or syrup, and again passed through a similar operation, whilst the small crystals are treated in the blow-up pan, and subjected to the operation of filtering before being finally manufactured for use.

6. A method of, or apparatus for, converting crystals of sugar into loaves or lumps, by subjecting the loose crystals to heat and pressure in suitable moulds.

Hector Ledru, of 28, Faubourg Poissonnière, Paris, civil engineer. *For improvements in heating.* Patent dated March 24, 1851.

The present invention consists in certain new combinations of apparatus for heating, by which a circulation of air is produced to and fro against the fire chamber or furnace, and the conduct pipes of the flame and smoke, but in an inverse direction, by which means all the caloric of the combustible is utilised.

Fig. 1 is a vertical section passing through the centre of the apparatus; fig. 2 is a horizontal section through the line A of fig. 1; fig. 3 is a horizontal section drawn through the line B of fig. 1; and fig. 4 is a front elevation of the apparatus.

Fig. 1.

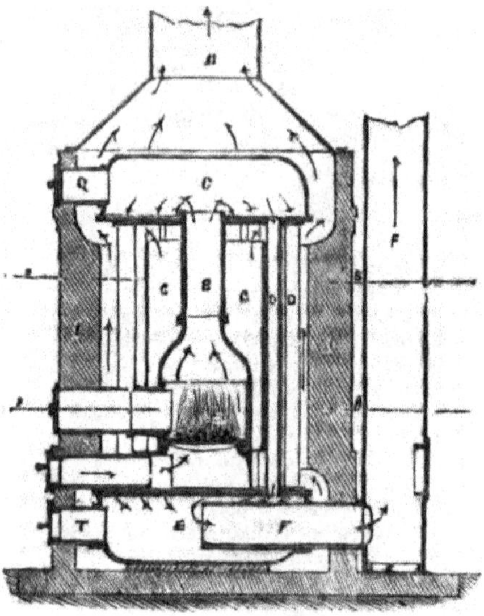

Fig. 4.

Fig. 2.

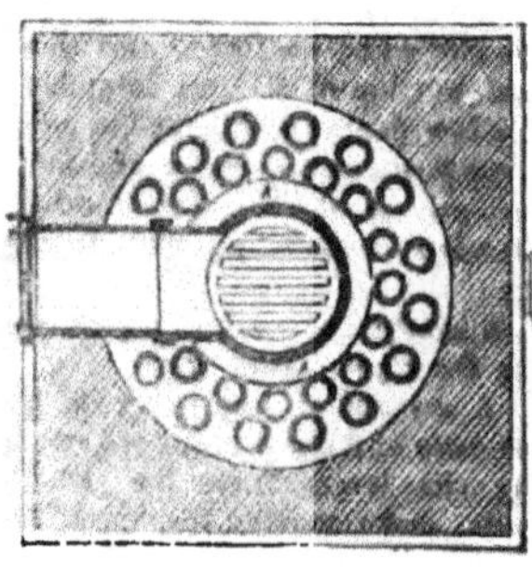

Fig. 3.

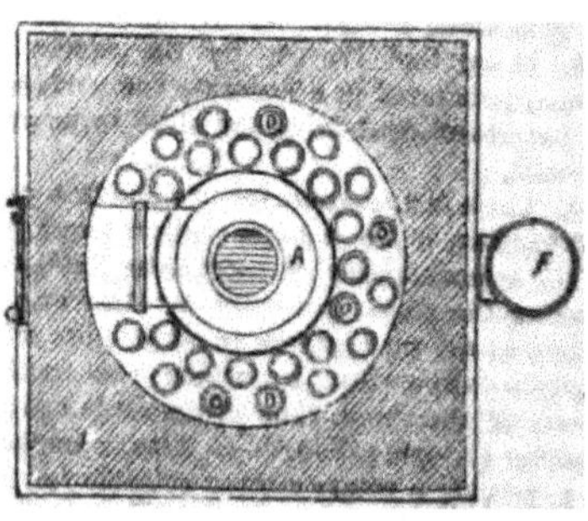

. 1, 2, 3, and 4, A represents the ... re the fire is made—it is of cast- ... is a discharge pipe for the smoke ... d gases; C cast iron or sheet iron ... s, within which the smoke and ... ses expand; DDD iron pipes com- ... g from the top recess C to the ... ss E. The smoke circulates from ... ttom within the pipes DDD, and ... recess E it escapes into the atmo- ... rough the smoke pipe F; the num- ... e pipes DDD is unlimited, and is ... ned to the diameter of the appara- ... an internal iron cylinder or casing ... ing the part A, and serving to neu- ... e effect of the radiation of the fire ... ipes DDD; the cylinder G besides ... a current at the central part, and ... e equilibriating of the temperature ... ole of the air—which escapes by the ... I. I masonry envelope surrounding ... atus; it is constructed with bricks, ... paration from the apparatus trans- ... he heat is not any impediment to ... nsion of the several metallic parts. ... table holes are made at the lower ... e masonry work, for the introduc- ... he air to be heated. When the ... ric air is introduced through an ... ound channel, having its ingress ... r recess E, the above described ... suppressed. The recesses C and E ... ded with stoppers, which can be ... when the cleansing is to take place. ... quired for the combustion is taken ... basis of the ash-pit, and a slide or ... regulates the necessary quantity. ... of the furnace is provided with a ... facilitate the combustion of the un- ... ses. The arrows represent in fig. 1 ... e of the smoke and heated gases. ... d lines in the said fig. 1 represent ... lation of the air to be heated, ... discharged through the opening H ... n temperature of 70° Centigrade; ... erature may be varied according to ... s to be produced. One of the ... parts of this new apparatus consists ... ting the products of combustion ... cinder pan, so as to permit an ... ntity of burnt air to be carried off ... rtical pipes DDD. This result is ... by the pipe F, through which the ... proceeds into the chimney. This ... s up the air from the very centre of ... recess E, to insure uniformity of ... the burnt air through the pipe F; ... ld not be obtained if the pipe took ... r from one side of the said recess ... that case the pipe DDD, nearest to ... h of the pipe F, would permit the ... a larger quantity of burnt air ... ment remote pipes from the said centre E—in that case the result would be that the surfaces presented by the pipes DDD would not produce all the necessary calorific effect; that is, they would heat a smaller quantity of air.

An important feature of the invention consists in having a small grooved opening, the size of which may be regulated at discretion; it is made in the door of the furnace for introducing fresh air into the current of flame and smoke which have escaped combustion; by that means much caloric is saved which otherwise would be lost in gas and smoke; the internal cylinder or casing G stands about eight or ten inches above the lower recess E, in order to permit a part of the cold air to enter the bottom of the apparatus in the direction shown by the black arrows, and rub against the furnace in order to cool it and prevent its becoming red-hot; the air thus heated escapes at the top of the apparatus through the opening H, which conducts it by means of pipes into the places to be heated. The apparatus being elongated to the upper recess *e* enables coke to be burned, which is introduced through an opening in the top of the apparatus, a little below the recess C. All the pipes DDD may be suppressed; but in that case an opening is made above the recess *e*, through which the smoke proceeds direct into the chimney.

It will be easy to perceive that this new combination of apparatus, the essential part of which consists in obtaining a very large surface and a good circulation in order to heat a considerable quantity of air, effects the saving of all the caloric of the combustible, and produces all possible effect from the caloric. The apparatus is so combined as to permit free dilatation of all its parts; by which the iron parts composing it are protected from breaking, and a long duration insured to the apparatus without necessitating repairs.

Another important improvement, relating

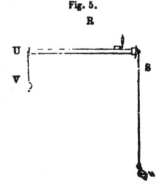

Fig. 5.

to the mode of cleaning the heating apparatus, is represented by fig. 5. It consists of a tube R, through the interior of which a cord S is passed; to the extremity of the latter a ball V or round brush is suspended, having a diameter of a dimension similar to that of the interior of the pipes DDD (figs. 1, 2, and 3,) which is to be cleaned; if a brush be employed it should be loaded with lead or other heavy metal, to precipitate its descent into the pipes DDD. The sweeping tube is provided with a small socket into which a candle or torch is fixed, for the purpose of giving light whilst cleaning; the cleaning tube is introduced through the opening into the upper recess C, and the ball is made to slide down alternately each of the pipes DDD, by means of the weight x and cord U, as far as the lower recess E; the passage of the said ball into the pipes DDD detaches the soot clogged to the interior of the pipes, and it falls into the lower part of the said recess E to be withdrawn through the opening T.

The patentee observes, in conclusion, that he does not confine himself to the precise details laid down, provided the general features be preserved; nor does he claim the several parts of the apparatus taken separately, but what he does claim is,—

1. The combination of mechanical apparatus for causing air to travel against the furnace and conduct pipes of flame and smoke for obtaining and supplying all the useful effect of the caloric, as described.

2. The apparatus for detaching the soot collected in the flues of the furnace.

HENRI and ALEXANDRE SIX, of Wazemmes les Lille, gentlemen. *For improvements in bleaching flax and hemp.* Patent dated March 24, 1851.

This invention consists in improved means of bleaching,—firstly, the fibrous parts of flax and hemp (but principally flax) in the straw—that is to say, in all the original length of the plants, before being peeled; secondly, flax which has been only partially peeled, so as to leave a certain portion of straw which is peeled off after the bleaching; and thirdly, flax after being completely peeled and reduced to the state of fibre, as it is generally found in commerce, before being spun, or combed into flax before being spun.

Firstly, the patentees employ a continuous system, which consists in subjecting flax and hemp in the straw, when once placed in the vats, to all the chemical operations which are usually employed for obtaining the different degrees of whiteness, without changing their position until the bleaching is entirely completed. The flax is placed in the vat upon a false bottom, disposed in a suitable manner for being raised by means of a crane, pulley, or other analogous means. The flax may be agitated in the vats, or withdrawn therefrom, according as may be deemed expedient.

Secondly, they apply the bleaching processes to flax which is partially divested of its straw, and in that case they allow the straw to remain on to the height of eight inches, more or less, at the foot of the plant, and complete the peeling operation after the matters have been bleached. This mode of operating has the advantage of preserving the flax in a more natural and more advantageous condition. In that state the matters form much less tow in combing than those that have been bleached after being peeled completely.

Thirdly, the patentees make use of hurdles for the bleaching of flax, either partially or completely peeled, or after being combed, as before described in the second division.

The form and dimensions of the hurdles vary according to the vats in which they are to operate. They may be made of wood, metal, or any other material, and are composed of an assemblage of laths made in the form of combs, and placed at a distance of one half-inch to two inches apart. The teeth of the combs composing the said hurdles are about one inch to two inches in height, are pointed, and at their base are about half an inch in diameter, and they serve to produce apertures through the layers of flax, thereby preventing the matters being clogged together, and also facilitating the filtration of the liquors employed, and with which the flax should be well impregnated, in order to secure as much as possible an uniform bleaching. These frames serve to spread out the flax in layers varying in thickness, and are placed one upon the other in the bleaching vats until the latter are full. The pressure arising from the superposition is entirely obviated, as the teeth or points of the hurdles at the bottom support those above them, and thus successively until the last. By that means the flax is left free between the various hurdles in such wise that the pressure of those at the top do not prevent the bleaching agents from penetrating with as much facility through the matters at the bottom of the vat as those placed above. The same result may be obtained in employing simple hurdles without teeth, made of wicker-work, wood, or in any other suitable manner, and placing them in the same position as above stated, care being taken to place a small cross piece between them, to prevent the pressure of the upper hurdles on the flax placed on those below. If the latter kind of hurdle is selected, care must be taken to intro-

rough the flax laid upon it rods ne-fourth of an inch in diameter, cilitate the filtration of the liquids the matters to be bleached.

of the great advantages claimed different modes of bleaching before d consists in obtaining the flax after g in such a natural and perfect con- to permit it to be spun as easily as l only been submitted to a steeping n. The flax bleached by this pro- l give a stronger and more beautiful than that which is bleached after un in the raw state.

Mode of Operating.

ats are provided with a false bottom, with holes, and set at a distance of eight inches from the real bottom, ween the latter and the false bottom pipe is set, in order to raise the ture of the solution contained in the required. When operating upon heaves, that is to say, in the straw, dles are placed upright upon the false ; if the flax is partly or completely r combed, it is spread and worked urdles, as before described. Then chlorine, acids, or water are alter- ntroduced into the bleaching vat in t quantities to immerse completely ters to be bleached. All the above are drawn off after they have pro- heir effect by a tap at the bottom of

patentees consider it superfluous to the chemical processes employed hing, because all the known means applied to obtain the object of their n with equal advantage; they wish understood also that they do not themselves to the precise details aid down with regard to dimensions ns, nor do they claim the different f the apparatus herein referred to eparately; but what they do claim invention is,

e application of a continuous series tions for bleaching the fibrous parts and hemp direct and simultaneously ame vat, and without changing their 's position, either in the straw, that y, before peeling, after they have rtially or entirely divested of the or after being combed, as before d.

e application of the bleaching pro- o flax which has been peeled at the art of the plant, the straw being left rtain extent at the lower part, as described.

e *application of hurdles*, when ope- rating upon flax that has undergone the operation of peeling and combing, for spreading the matters, as before described.

4. The application of rods to facilitate the filtration of the bleaching agents through the matters to be bleached, as before described.

David Farrar Bower, of Hunslet, York, manufacturing chemist. *For certain improvements in preparing, retting (otherwise called rotting), and fermenting flax, line, grasses and other fibrous vegetable substances.* Patent dated March 24, 1851.

(*Specification.*)—My invention of certain improvements in preparing, retting, otherwise called rotting, and fermenting flax, line, grasses and other fibrous vegetable substances, relates to certain means whereby the time required for effecting the operation will be considerably reduced, and the fibre of the rough plant will be less subject to deterioration than when submitted to the ordinary rating process. It is well known that after flax has been pulled and dried in the field, and the seeds removed therefrom, it is usually steeped for some weeks in water in order to detach the glutinous and other foreign matters contained in the plant; the stalks after being allowed to remain some weeks in the water are removed therefrom, and when dried are subjected to breaking and scutching to remove the woody matters from the useful fibre. I find, however, that the glutinous and other matters, although they may be partially detached by the water, are not wholly removed from the fibre, but when the latter is dried a portion of the glutinous and other matters still adhere to the same and render it unfit for the breaking and scutching operation. According to this part of my invention which has reference to the retting process now in common use, I steep the flax or other fibrous substance in the ordinary way, either in cold or warm water; if cold water is used, the flax must remain immersed for six days, but a much shorter time will answer if warm water is employed. I then remove the flax from the water, and pass it between rollers for the purpose of expressing the glutinous matters from the interior of the plant. After this, I again steep the flax in cold or warm water for another six days, and a second time submit it to the squeezing operation by passing the stalks between rollers; after which the flax is dried, broken, and scutched in the ordinary manner, and will be found to be much clearer of the glutinous and other foreign matters than if merely steeped for a much longer time, and operated upon in the ordinary way. For the finer descriptions of flax, and when a good coloured fibre is required, I steep the plant in a solution of caustic ammonia, or the neutral salt of one of the alkalies. The salts I prefer to employ are chloride of sodium (common salt) or sulphate of soda. The quantity of the alkali or salt required to be dissolved in a given portion of water will depend upon the temperature at which the process is to be carried on and the quality of the water employed; in other words, the quantity of alkali or salt required will depend upon whether the impurities consist of the salts of iron, or of lime, or other matters contained in the water. It is, however, impossible to give definite proportions for every case; I shall therefore merely say that if common rain-water is employed, I add one pound of caustic ammonia, or one pound of any of the neutral salts to every 150 gallons of water, and with this solution the process may be carried on at any temperature from 90° to 120° of Fahrenheit's thermometer, and the operation will be completed in about thirty hours. If, however, cold water is employed, the quantity of the alkali or salt must be somewhat increased, and the operation will be completed in about four days. The admixture of the above-named chemical ingredients with the water in which the plant is steeped will very much facilitate the retting process; but if, in addition, the fibres after being submitted to the action of the alkaline solution for a time, are passed between rollers or otherwise subjected to pressure to express or squeeze out the dissolved glutinous matters, the process will be still further facilitated and otherwise greatly improved.

Another improvement in the preparation of this kind of vegetable fibre consists in operating upon the plants or stalks in an exhausted vessel. In carrying out this part of my invention, the flax or other fibrous substance is pulled and dried in the field, and the seed removed in the ordinary manner. The flax is then placed in a cylindrical or other shaped air-tight vessel, and the air is exhausted therefrom by means of an air-pump or otherwise, after which I let into the exhausted vessel a solution of one of the following materials,—either caustic ammonia or chloride of sodium or sulphate of soda, or any neutral salt of any of the alkalies of ammonia, soda, or potash. These materials I add in or about the proportion of one part of ammonia, or one pound of one of the neutral salts to 150 gallons of water, which should be kept at a temperature of from 90° to 120° of Fahrenheit. By exhausting the vessel and drawing out the air contained in the cellular tissues of the plants, the latter will be brought into a condition to be more easily and rapidly acted on by the chemical *agents employed*. The alkaline solution being now let into the exhausted ve[ssel, the] flax will readily aborb it. In this s[...] state it should be allowed to rem[ain] from two to four hours, more or l[ess] may be required, after which the [...] should be run off from the vessel a[nd the] air again exhausted therefrom. Th[is will] have the effect of extracting the d[...] glutinous and other matters from th[e inte]rior of the plant. After this second e[xhaus]tion, the fibrous materials are remove[d from] the vessel and laid up in a heap, so th[at they] may cool down gradually without fe[rment]ing, after which they should be spr[ead] in a field or drying shed, and when d[ry] will be ready for breaking and scu[tching.] Or, instead of using an alkaline solu[tion I] sometimes employ hot water, merely [letting] it into the air-tight vessel after the [air has] been exhausted therefrom.

Instead of subjecting to mechanica[l pres]sure flax or other similar plants tha[t have] been steeped in open vessels, as de[scribed] under the first part of my invent[ion, I] sometimes take the plant after it h[as been] steeped long enough, and place it in a [cylin]drical or other shaped vessel, and su[bject it] to the exhausting operation for the p[urpose] both of extracting the glutinous and [other] matters from the interior of the fib[res and] removing the superfluous moisture.

Having now described my inventio[n, and] the best means with which I am at p[resent] acquainted for carrying the same into [effect,] I would observe, in conclusion, that al[though] I have stated certain proportions in w[hich I] propose to use the chemical ingr[edients,] having by experience found them to a[nswer in a] satisfactory manner with the before [named] proportions, yet I do not mean to c[onfine] myself thereto, nor do I intend to [confine] myself to the use of the alkaline or [neutral] salts named, as other alkaline neutra[l salts] will answer; nor do I intend to clai[m as] my invention steeping flax or other [similar] or analogous fibrous plants in warm [or cold] water or other chemical agents than [those] before named, for the purpose of det[aching] and removing the glutinous matter[s con]tained in the plant, except when suc[h pro]cesses are carried on in conjunctio[n with] my improvements; but that which I c[laim] to be new in the processes above des[cribed,] and therefore claim as the invention [secured] to me by the hereinbefore in part [recited] Letters Patent is,

1. Subjecting flax, line, grasses, an[d other] fibrous vegetable substances which [have] been steeped in water or other liq[uids to] mechanical pressure, for the purp[ose of] expressing or squeezing out the de[...] glutinous and other matters contained [in the] plant.

a steeping flax, line, grasses, or
gous vegetable fibrous substances
n of ammonia, or of chloride of
of sulphate of soda, or of any
il salt of the alkalies ammonia,
ash.
m the employment in the pre-
flax, line, grasses, and other
table substances, of the process
on as before exemplified and
o accelerate the detaching from
e glutinous and other matters
erein.

ck William Mowbray, of
gentleman. *For improvements
ry for weaving.* Patent dated
1851.

ation consists of improvements
y employed for producing terry
fabrics by what is called cross-
e loom employed for this pur-
he power loom, and the arrange-
as are adopted for producing
weavings by a single warp; which,
ay be varied, and the loom ar-
producing other classes of weav-

eculiarity of the invention con-
ying a series of guides to carry
threads which are to produce the
-pile on the face of the fabric;
ides are caused to work the pile
h they carry, in such manner
cross them over wires, which,
t new in themselves, are never-
roved in their action by moving
and by being combined with a
ide, through which the ordinary
ps pass. The pile-warp guides
leads affixed to a bar extending
whole width of the loom, and
vertical rods connected to the
ers on each side of the loom.
s turn on pins projecting from
, and carry at their free ends
llers, which run against cams by
are acted on so as to give to the
and-down motion, in order to
guides to take the pile threads
the shed, to be tied in by the
e again raised to produce the
r the next rows of pile. And
t the pile threads passing through
may be laid successively on one
n on the other side of the loop-
produce the crossing, the bar
y movement equal to one space
alternately to the right and then
on the ascent of the guides above
y warp threads. For the pur-
cting the more regular spacing
nary warps, and thus facilitating
e of the pile warp-guides and
threads between them, other combs or guides are, as above mentioned, employed, through which the ordinary warps are caused to pass. These combs or guides are also the means of sustaining, as on an axis, one end of the wires, whilst the other end of each of such wires is in the work produced, and resting on the breast beam. A rising-and-falling motion is also given to these guides by establishing a communication between the bar on which they are mounted and the cam-shaft giving motion to the pile warp-guides, by means of rods and levers, as above described, with reference to those guides. The object of giving this movement—which the patentee considers an important feature of his invention—to the body warp-guide, is to save as much as possible the time occupied in the movement of the pile-guides, by lowering the end of each of the wires when the shed is closed for the change, and whilst the sideway movement is being given to the pile-warp guide-bar. When it is desired to produce a cut-pile fabric, the patentee employs, in conjunction with the arrangement just described, a series of rotatory cutters, consisting of thin discs or circular plates of steel, with knife edges, mounted, and kept at suitable distances apart by washers, on a spindle, which is supported in bearings over the breast beam, and caused to revolve with the cutters taking into the slits at the ends of the wires, by which means the rows of loops are cut as the work is finished, and drawn forward over the breast beam by the action of the taking-up motion.

A modification of the arrangements first described is specified, in which two sets of guides are employed for the pile-warps, and by which a fabric with a much denser pile is produced. When working according to this arrangement, the patentee prefers to give to each set of guides a sideway motion in a direction opposite to that in which the other set of guides is for the time being moving, so as to effect the crossing of the loops produced by the individual guides. The means above described for cutting the pile are equally applicable in this case.

Another modification is also described, in which a Jacquard or other suitable pattern-surface is employed for operating the guides, and this arrangement is particularly adapted for producing terry fabrics with patterns thereon.

Claims.—1. The manufacture of terry and cut-pile fabrics, by employing suitable guides or instruments, and apparatus combined therewith, to control the movement of the pile-threads when producing pile fabrics by cross weaving.

2. The giving to the wires or instruments

used, when producing pile fabrics by cross weaving, an up-and-down movement, as described; also, the mounting of such wires or instruments on axes.

3. The so arranging apparatus in a loom that the pile of fabrics woven therein, when the pile is produced from the warp by cross weaving, shall be cut by rotatory cutters acting in the ends of the wires or instruments on which the pile is formed.

SAMUEL HOLT, of Stockport, manager. *For certain improvements in the manufacture of textile fabrics.* Patent dated March 24, 1851.

The improvements sought to be protected under this patent have relation to the production of certain peculiar descriptions of fabrics (having terry loops or pile on one or both sides or faces thereof) by the employment of a ground warp, and one or more terry or pile warps without the use or intervention of wires in the process of weaving.

According to the methods hitherto known and practised in manufacturing fabrics of this description, and also those classes of cloths known as "tucked cloths," the weft has been shot through the pile warp by itself, that is, separately from the ground warp; and this is also the case in weaving certain descriptions of terry carpets in which the pile warp threads are partially woven into a cloth by themselves or separately from the ground warp, which partially formed cloth is subsequently woven into the ground warp, and the "tucks" thereby produced. It will be seen from this, that as it were two cloths are employed, one being laid on the face of the other, and according to this arrangement also the tension of the warp threads requires to be varied during the weaving process. The peculiar feature of the present improvements is that the fabric is produced in one cloth, no shoot of weft being made through the pile warp separately, but every shoot being caused to pass simultaneously through a shed opened in the ground warp. The ordinary amount of tension is given to the ground warp threads, but the pile warp threads are slack, and the loops are formed by the weft acting on the slack threads, and drawing them up when the weft is beaten up by the lathe after every three shoots, at the same time that the weft passes freely over the ground warp, which is retained in a stretched condition by the tension of its weights.

Claim.—The weaving of either a double or a single faced terry or piled cloth, wherein no part of the terry or pile warp is woven n separately from the ground warp threads.

JAMES CHEETHAM, junior, of Ch ton, near Oldham, cotton-manufac *For certain improvements in the facture of bleached, coloured, or coloured threads or yarn.* Patent March 24, 1851.

This invention has relation to the p tion of bleached and coloured or coloured yarns of cotton, by subjecti material to the bleaching or colourin rations in a prepared state and subs to the completion of the carding p In order to obviate the imperfection dental to the ordinary practice of ble or colouring yarns after being spu tempts have at various times been m perform those operations on the raw rial, or previous to the completion process of carding, and thus to dispen some of the usual processes; but in l these cases the result obtained has be satisfactory and attended with a w material. Now according to the pres provements the perfect action of the l ing or colouring agents is insured, an ral of the operations usually adopted cially that of doubling in the manu of party-coloured yarns, may be dis with, the material being bleached o in the state of a continuous sliver, t dition of which, both as regards its ness and disposition of the fibre, p great facilities for the thorough pern through it, and consequent perfect a the chemical agents employed.

The ordinary course of manufac adopted when operating on cotton ac to this invention, until the completio last drawing, when the material is co through a slubbing frame, the a ments of which are so modified as to to it a rather greater amount of tw would be required for the ordinary quent processes; the object now b give to the slubbings or slivers s hardness and solidity to enable t undergo the bleaching or dyeing op without risk of fracture, or destroy continuity of the fibres composin These slivers are then wound on bobbins in the manner adopted when yarns for dyeing or bleaching, only is preferred to wind two or more sl the same bobbin, but without twist, then less liable to fracture, and when enabling the ends to be more readil and offering increased facility for unw As the ordinary operation of wring after the slivers have been bleached would, unless very carefully perform their breakage, the patentee prefers pel the moisture contained in them

xtractor." The slivers at this too much consolidated to permit ndergoing the operation of drawing, l be necessary therefore to remove rabundant twist, which may be y conducting them through the last ollers only of a slubbing machine, ng action being now required, or machine for the purpose provided one pair of rollers, the flyer of caused to revolve in a direction to that in which the twist had been y put in, and with velocity proportha amount of twist required to be to reduce the sliver to its original or the flyer may be caused to t a velocity sufficiently high to he existing twist entirely and resliver to the required extent in the direction. The slivers, or slub- then drawn in a roving frame, and tly spun into yarns by any of the ordinarily employed for that pur- obtain party-coloured yarns, the takes two or more coloured slivers to the required pattern to be pro- the yarn, which he draws down to g, and subsequently spins direct; dispenses with the process of doub- h according to this method is ren- necessary. Grey, or unbleached cot- lk, or wool, may be combined with er coloured slubbings or slivers, by this method of procedure, and n into threads or yarns.

tentee does not confine himself to details given, as the cotton may be or dyed after being carded, with- going the slubbing operation above ; but he claims, the bleaching, the dyeing, colouring or printing between the operation of carding, of spinning.

L Walker, junior, Birmingham, urer. *For a certain improvement improvements in the manufac- metallic tubes.* Patent dated , 1851.

rocess of manufacture adopted by er, is as follows:—He takes a skelp f metal, bends it to a U form, and es it through a pair of suitably olls, using at the same time a man- ternal support, by which the skelp p to a tubular form, and one edge overlap the other; he then either he partially-formed tube to a second peration, using also a mandril or upport as before, so as to compress apping portion of the metal, and a bevil of both the meeting edges, cts the same object by means of a draw-bench. The edges of the tube are then soldered, and the operation is completed.

The furnace employed for the latter purpose is one of a peculiar construction, adapted for conducting the flame and heated air over those portions only of the tube to be soldered. For this purpose it is furnished with a chamber behind the bridge, into which the tubes (either singly or several at a time) are introduced through apertures in each side, at such a level as to cause the flame to pass either under or over them on its way to the chimney, according as the junction of the edges of the tube is to be effected by solder applied interiorly and melted down between the abutting edges, or otherwise.

Claims.—1. The bending of skelps or plates of metal into a tube-like form, and the bevelling of the edges of the same by rolling.

2. The construction of a soldering furnace in which the flame and heated air are caused to pass in immediate contact with those parts only to which the solder is to be applied.

Peter Armand Lecomte de Fontaine-moreau, of Paris, and South-street, Finsbury. *For certain improvements in mills for grinding wheat and other grain.* (A communication.) Patent dated March 24, 1851.

This invention comprehends; firstly, certain mechanical arrangements for placing the motive shaft of mills in a parallel position with respect to the shaft of the mill-stones, and for working the bolting apparatus conjointly with the mill for grinding; secondly, a system of dressing the surfaces of mill-stones with channels or cuts according to a geometric form and shape, the dimensions of which may be varied; and of constructing the eye, for producing, when combined with the preceding improvement, a continuous circulation and supply of cold air over the said surfaces, and thereby grinding corn in a cold state; and thirdly, a new mode of cleansing and desiccating wheat and other grain, by the action of certain apparatus employed in conjunction with the mill for grinding; of which improvements, however, without the aid of drawings, a general idea only can be very readily communicated.

Claims.—1. The arrangements of mechanical parts for placing the motive shaft parallel with the shaft of the mill-stone.

2. The mechanical arrangements for setting conjointly in motion the bolting apparatus with the mill for grinding.

3. The radiating or dressing of mill-stone with channels or cuts according to a geome-

tric scale and form, and the mechanical arrangements of the eye for producing a continuous current of cold air on the said grinding surfaces.

4. The constructing and arrangement of mechanical parts for setting in motion with one axis only the cleansing apparatus.

THOMAS HILL, of Langside Cottage, Glasgow, Esq. *For improvements in wrought iron or malleable iron railway chairs.* Patent dated March 24, 1851.

The invention sought to be protected under this patent consists in the construction of railway chairs made from plates of wrought or malleable iron by machinery, by which tongues or lips are punched up of a proper form to embrace and hold securely the rail placed upon them. Chairs of this description may be divided into three principal classes, the form, size, and strength of which may be varied to any extent to suit the requirements of railway engineering: these are; first, chairs in which the lips are presented sideways to the rail; second, those in which the rail is held by the edges of the chair; and third, chairs in which one or more of the lips are presented sideways, and the remainder edgeways to the rail: the lips of each of these several varieties may either come close up to the sides of the rails laid upon them, or sufficient space may be left between the rail and the lips of the chair for the introduction of a wedge or key, as may be preferred.

When manufacturing chairs according to this invention, a flat bar of wrought iron should be selected, of a width equal to the length of the chair to be made, and of a thickness of about half an inch, more or less, according to the strength required in the chair. This bar is to be submitted to the action of a pair of dies provided with knives, which clip off a sufficient length of the bar to form the chair and punch up the lips, and also make the requisite holes for the admission of spikes to secure the chair to the sleepers of the line of railway. One of the dies is stationary, whilst the other is caused by suitable mechanical arrangements to move alternately up to and away from the stationary one. The moving die is also provided with a knife by which, in conjunction with the edge of the stationary die, that portion of the bar which is under the action of the dies is severed from the bar. The dies are exact counterparts of the opposite faces of the chair, so that when brought near to each other a space is left between them equal to the thickness of the chair to be formed, and by bringing the dies together the projections of the one enter the corresponding hollows of the other, and at the same force into them the port the plate at that time between the dis thus produce a perfect chair. In ing such chairs it is in all cases pr that the bar or plate of iron shou heated preparatory to its entering be the dies, as there is then less risk of ture, and less power is required to the machinery. The patentee descr machine suitable for making cha wrought iron according to his inv but he does not confine himself ther the dies may be attached to any pu machine of the ordinary constructi sufficient size and strength.

Claims.—1. A wrought iron r chair, the lips of which are forme that portion of the plate on which t is ordinarily supported.

2. A wrought iron railway chair, t of which are presented edgewise direction in which the rail lays wi place.

GEORGE GUTHRIE, of Appleby, berlain to the Earl of Stair, and resi Rephad by Stranraer. *For improv in machinery for digging, tilling, or ing land.* Patent dated March 24, 1

The machinery which forms the of this patent is intended to facilit operations of digging and working la the application thereto of steam powe machine is composed of a strong recta wooden framework mounted on four which are made very wide in the t prevent their sinking into the grou to facilitate the travelling of the m The motive power for working the movement or digging action is derive a pair of steam cylinders, supported framing, which serve also to supp power for propelling the machine. diggers are formed with three prong and they may be of such a number, a side by side, as to extend across th width of the machine. In addition downward motion imparted to the by a crank, for the purpose of causin to enter the ground, they are also acte after they have passed through the fi of the down stroke, by a second cr which they are raised to a horizonta tion, carrying with them the portion immediately in front and within the of their action. As soon as the grap arrived in this horizontal position, th on which they are mounted is caus means of a slide working on a sp screw of coarse pitch cut on that sl make a semi-revolution on its axis, throw off the earth which had been sup on the grapes, which are then retu

-iginal position, and carried back or operating on a fresh piece of over which the machine will by this re been moved.

-tentee does not confine himself to the nent of cranks for effecting the move- f the grapes, as other means capable lucing a similar motion approxi- -iosely to that of the spade in hand- may be adopted for the purpose; in- e of the above-mentioned arrange- the grapes, they may be fixed to ring shaft, and the earth may be l from them by causing them in their on to come in contact with suitable It may be necessary in some cases, to facilitate the action of the grapes, a series of cuts in the land previous grapes coming into operation; and be effected by attaching to the front nachine a series of circular cutters a revolving shaft, and brought into the machine is propelled forward.

-s.—1. The general arrangement of -es, forks, or spades, for digging and land.

-e system or mode of turning the r digging apparatus by a screw and g slide.

-e application and use of guides for irection to the grapes.

-e system or mode of digging or land by causing the grape or dig- -aratus to turn, or partially revolve, aded with earth.

-HEW HERRING, of Tonbridge-place, anter. *For improvements in the -iture of sugar and rum, part of -nprovements are applicable to eva- generally.* Patent dated March 1.

-e *Evaporator.*—This apparatus con- casing of any suitable shape, within re placed a number of vertical metal oined together so as to leave hollow between them, in which spaces hot m, hot water, or other heating me- rculates. The saccharine juice or be evaporated is supplied from nd caused to trickle down the sides et plates and pass into a receiver, iich it is returned, and repeatedly d to the evaporative action until rated, when (in the case of cane- is conducted away for crystalliza- n order to effect the purification of simultaneously with its evaporation, lver may be provided with a filtering , through which the fluid may be o pass before being returned through -erating chamber.

-e *Cleanser.*—The saccharine matter, after crystallization, requires to undergo the operation of cleansing, to fit it for subse- quent processes. With this view, the crys- tals are placed in a vessel of any convenient size, provided with a perforated false bot- tom, and liquor is pumped on them through a pipe furnished with a rose or jet, so as to cause it to be well dispersed over the crys- tals, and thus effect their purification by attrition. The liquor employed, which may be either cane-juice diluted to about 32° Baumé, or rum, passes off through the false bottom, and is pumped up for further use.

3. *The Continuous Still.*—This apparatus is constructed on the same principle as the evaporator above described. The wash is caused to traverse heated surfaces, by which the greater portion of the spirit is vola- tilized; the aqueous particles and the non- volatilized portion of spirit are subjected to a similar operation until all the spirit has been obtained. The vapours arising from this distillation are condensed in a worm, in the usual manner. The patentee has disco- vered that the quality of the rum thus pro- duced, and that of rum generally, may be improved by subjection to the action of air, and with this view he causes a current of air to traverse and pass through the rum, and thus produces an improvement in its flavour.

THE SUBMARINE TELEGRAPH BETWEEN DOVER AND CALAIS.

The length of wire-cable provided consi- derably exceeded the distance in a straight line between Dover and Calais, but, in con- sequence of a strong gale of wind blowing down Channel while it was being laid down, it took so large a bend to leeward that, when the whole was reeled out, it was found to fall short of the French shore by more than a quarter of a mile. The completion of the line is therefore suspended until an addi- tional piece can be spliced to the cable.

WEEKLY LIST OF NEW ENGLISH PATENTS.

William Hodge, of St. Austell, Cornwall, for im- provements in the manufacture of glass, china, porcelain, earthenware, and artificial stone. Octo- ber 2; six months.

William Henry Ritchie, of Kennington, Surrey, gentleman, for improvements in ornamenting glass. (Being a communication.) October 2; six months.

Thomas Cussons, of Bunhill-row, for improve- ments in ornamenting woven fabrics for bookbind- ing. October 2; six months.

James Warren, of Montague place, Mile-end- road, for improvements applicable to railways and railway carriages, and improvements in paving. October 2; six months.

Leman Baker Pitcher, of Syracuse, New York, America, gentleman, for improvements in apparatus for regulating motive-power engines. October 2; six months.

LIST OF SCOTCH PATENTS FROM 22ND OF AUGUST TO THE 22ND OF SEPTE

Dominique Julian, of Sorgues, France, for improvements in extracting the colouring properties of madder, and in rendering useful the water employed in such processes. August 25; six months.

George Jordan Firmin, of Lambeth-street, Goodman's-fields, manufacturing chemist, for improvements in the manufacture of oxalate of potass. August 25; six months.

Thomas Wilks Lord, of Leeds, York, flax and tow machine-maker, and George Wilson, director of the flax-works of John Fergus, Esq., M.P., Fife, North Britain, for a machine to open and clean tow and tow waste from flax and hemp, and other similar fibrous substances, and an improved mode of piecing straps and belts for driving machinery, and a machine for effecting the a communication.) August 27; six

Richard Fletcher, of Blackdowns ton, Gloucester, farmer, for an imp taining motive power. August 29;

Henry Dircks, of Moorgate-street neer, for improvements in the manu in gas burners, and in apparatus for September 1; six months.

Richard Archibald Brooman, of Robertson and Co., of 166, Fleet- Patent Agents, for an improved m facturing screws. (Being a commu tember 8; six months.

LIST OF IRISH PATENTS FROM 21ST OF AUGUST TO THE 19TH OF SEPTE

James Whitelaw, of Johnstone, Renfrew, North Britain, engineer, for improvements in steam engines. August 22.

William Mather, and Colin Mather, of Salford, engineers, and Ferdinand Kaselowsky, of Berlin, Prussia, engineer, for improvements for washing, steaming, drying, and finishing cotton, linen, and woollen fabrics. September 5.

David Farrar Bower, of Hunsle manufacturing chemist, for certain in preparing, rating, otherwise call fermenting flax, line, grasses, an vegetable substances. September 8

William Johnson, of Millbank gentleman, for improvements in weighing goods. September 9.

WEEKLY LIST OF DESIGNS FOR ARTICLES OF UTILITY REGISTER

Date of Registration.	No. in the Register.	Proprietors' Names.	Addresses.	Subjects o
Sep. 26	2960	James Lysander Hale, C. E.	Canton-place, Lambeth	Firewood.
,,	2961	George Pate Cooper	Suffolk-street, Pall-mall	Gorget shirt.
,,	2962	Richard Clayton	Gresham-street	Swimming gl
27	2963	Thomas Humphreys	Bridge-wharf, Deptford	American fire-
29	2964	George Howe	Gt. Guildford-street, Southwark.	Pressure gaug
30	2965	John Johnson Broadbent and Fieldhouse Fieldhouse	Bradford	Tappet-lever c
Oct. 1	2966	Samuel Brown	Marlborough-place, Kennington-cross	Economic filte

WEEKLY LIST OF PROVISIONAL REGISTRATIONS.

Sep. 25	295	A. A. De Reginald Hely	Manchester-buildings, Westminster	Pedestrian va
,,	296	Richard Clayton	Cheapside	Sylphide wate
,,	297	John E. Grisdale	Bloomsbury-street	Ventilating w
,,	298	W. E. Kirkman	Knightsbridge	Portfolio bracl
27	299	George Gotch	Islington	Window flowe
29	300	William Rowden	Northampton	Thumb-screw
,,	301	Joseph William Lea	Birmingham	Knife-cleaner.
,,	302	Robert Watson Savage.	St. James's-square	Invisible door

Erratum.

In Abstract of Mr. Ross's Specification, No. 1468, p. 256, top of second column, *for* " and-fro and up and-down movement of the lashing fan," &c., *read* "the compound to-and down movement, the lashing fan," &c.

CONTENTS OF THIS NUMBER.

LONDON: Edited, Printed, and Published by Joseph Clinton Robertson, of No. 166 in the City of London— Sold by A. and W. Galignani, Rue Vivienne, Paris; Ma *Dublin*; *W. C. Campbell and Co., Hamburg.*

Mechanics' Magazine,

[M]USEUM, REGISTER, JOURNAL, AND GAZETTE.

[No]. 1470.] SATURDAY, OCTOBER 11, 1851. [Price 3*d*., Stamped, 4*d*.

Edited by J. C. Robertson, 166, Fleet-street.

WIMSHURST'S PATENT SCREW-PROPELLING MACHINERY.

Fig. 1.

Fig. 2.

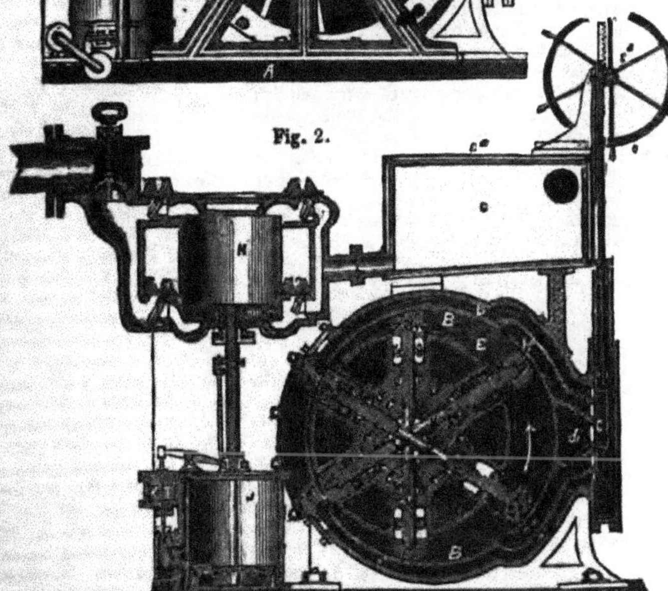

WIMSHURST'S PATENT SCREW-PROPELLING MACHINERY.

(Patent dated November 12, 1850. Specification enrolled May 12, 1851.)

Mr. Wimshurst, the shipbuilder, who was the first person to apply the s sailing vessels as an auxiliary propelling power, namely, in the *Novelty*, built ab years ago, and who has ever since continued to exert himself with a most praise perseverance to bring into general mercantile use this combination of the po wind and steam, has now patented a number of improvements in relation t which may be considered as exhibiting the matured results of his long p experience in this branch of naval combination, and of the great ingenuity a he has brought to bear upon it.

The first object with Mr. Wimshurst has been to improve the method of ap steam power to the working of the screw. He would still, as in the *N* apply the power directly to the screw-shaft, without the intervention of gear any other kind of multiplying power; but he would prefer, it seems, before other description of engine, one on the rotary principle, constructed in the m represented in figs. 1 to 8 inclusive of the accompanying engravings. Fig 1 end elevation of Mr. Wimshurst's proposed engine; fig. 2, a transverse section; a longitudinal section; and figs. 4, 5, 6, 7, and 8, separate views of differen detached from the others.

Specification.

In its general structure this engine so far resembles others of its class, that the is obtained by means of a drum revolving eccentrically within an external cylind the pressure of steam against a series of pistons successively protruded from the c ference of the drum as it revolves; but it differs from others in the following amon important particulars:

First. The external cylinder is bored, not of a perfectly circular form as usual, unequal diameters at certain points, by which a nearer correspondence to the ec path described by the pistons carried by the inner drum is obtained.

Second. The pistons have their chief bearing points on anti-friction rollers (wit recesses of the drum) whereby the frictional resistance to their movements is reduc minimum, or as nearly so as may be.

And *third*, the steam is admitted (during the forward and ordinary course of the from below into the external cylinder, and by its upward pressure on the pistons se a great measure to relieve the bearings and external cylinder from the weight and fri the moving parts of the engine.

In the figures, A represents the bed or foundation plate of the engine. B the e cylinder which has the peculiar form above-mentioned given to it by boring it fro centres in manner following:—Assuming the diameter of the cylinder to be 60 in is in the first instance bored out of a true circle from *a* as the centre in the diagr 1*; the centre or axial line of the boring tool is then shifted in a vertical line towards *b*, about one-eighth of an inch, or about a fifty-third part of the cylinde meter; which being done, it is again shifted in a horizontal direction towards d three-eighths of an inch, or about 160th part of the diameter, which will bring t line of the cutter bar to the point *f*. The tool is then placed at such a distance fr last centre that it shall just touch the circumferential point *g*, about 2 inches above while it is in this position, a cut is made through the cylinder from end to end, wh lune-shaped piece is cut away as indicated by the dotted lines. To make the lower the cylinder of a corresponding shape, a similar series of operations is gone through is to say, the cutter bar is first shifted from the centre *a* towards *c*, one-eighth of a and next horizontally towards *d*, about three-eighths of an inch, when a third cut i through the cylinder from end to end as before (beginning at the point *h*, about two from the under side of *d*). The slight ridges left by boring from different centres be then worked off, and the internal surface of the cylinder to be rendered quite s The result of the peculiar form thus given to the interior of the external cylinder although the inner drum is revolving eccentrically to the outer cylinder, yet the p which are of one continuous length, passing through the inner drum, are kept ne constant contact with the inner surface of the external cylinder, and with as little ment of the necessary packings as may be. C is the inner drum, which may be cast *solid or partially hollow* (as shown in fig. 4), and is divided by six slots or recesses, *many pistons to work in*, into six segments, $c^1, c^1, c^1, c^1, c^1, c^1$. C^1C^1 are bosses

ed to the ends of the drum (revolving with it) for the double purpose of strengthen- drum and affording a ready means of packing it at the ends. DD are two shafts keyed on one line with the two bosses, and carried by suitable standards FF, or y be supported by bearings affixed to the ends EE of the outer cylinder; these two orm the main shaft of the engine, which may be coupled to any other machinery.
details of the mode of packing the drum at the end are shown in fig. 3, and in nd 8. A metallic hoop H (figs. 7 and 8) is slipped loosely over each of the bosses nnected with the hoop there is a metallic ring y^2; y^3y^3 are segmental pieces of etween which and the ring y^2 and the bosses there are inserted packings (of cork uitable yielding material). y^4y^4 are pinching screws, by which the pressure given ork or other yielding packing is regulated; for regulating the pressure of the ring st the cylinder cover, there are pinching screws passed through the hoop H. egments, packing rings, and hoops are secured to the boss by screw bolts y^1y^1, so y all revolve along with the inner drum. When the drum revolves, the inner edge metallic ring packings runs against the inner edge of the cylinder cover (rendering t steam tight), but to prevent them or the bearings from being injured by the heat, le are inclosed in cases SS, which may be filled with water or other liquid, and made am-tight round the shafts DD by means of packings, such as represented in the . 4. TT are the packings, which are formed of cork or some other like yielding l. T^1T^1 are gaskins, and T^2T^2 metallic blocks by which the packings are pressed the shaft by means of the pinching screws T^3T^3.

Fig. 3. Fig. 7. Fig. 1ᵃ. Fig. 4.

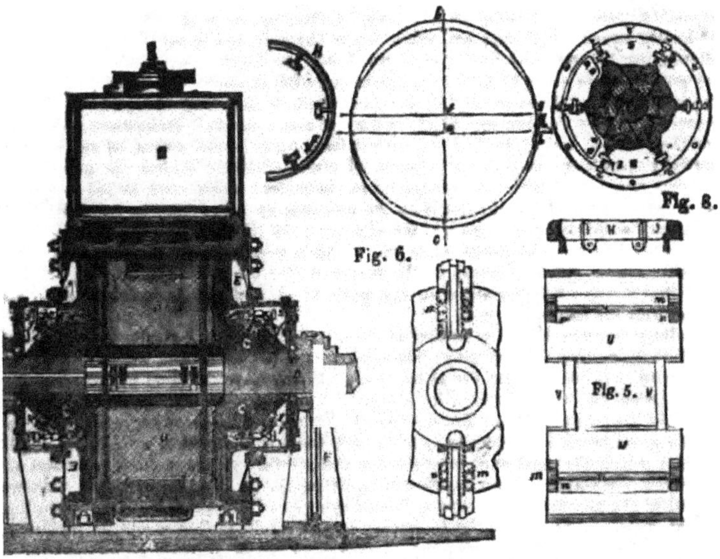

Fig. 8. Fig. 6. Fig. 5.

are the pistons, which are of a rectangular form, and connected together in pairs ns of bars V V, which are passed through the pistons, and secured at the ends by ; for which screws or nuts might be substituted. Each pair of pistons works to and ough the drum in the recesses made for them, and they bear at the sides against anti- rollers X X, placed at different distances, but near to the extremities of these s (when the pistons are fully protruded from the drum.) To keep these anti-friction true in every change of position, they have at their ends small pinion guides $m\,m$. 5), which gear into fixed racks $n\,n$ within the recesses in the body of the drum. stons are rendered steam-tight by means of packings, which are inserted at y^5y^5 in , running parallel to the recesses, and pressed by springs against the sides of the by which means the steam is prevented from passing through into the chamber side. W W are caps or head-pieces, which are affixed on the outer ends of the pistons, and made on their rubbing surfaces of a convex form; and x x are metallic packings, which are inserted into these caps at the extreme points where they come in contact with the exterior cylinder. To keep these packings pressed out against the inner face of the external cylinder, they are acted on by spiral or other suitable springs inserted in recesses made in the pistons and caps. The springs are each covered by a plate, between which and the metallic packing a gaskin is inserted, if required, or any other suitable packing. a^1 a^2 are the induction or steam ports (supposing the engine to be revolving in the direction indicated by the arrows), and b^1 b^2 the eduction steam ports; c is a reversing slide, which is worked when necessary by means of the hand-wheel O, and the pinion O^2 and rack O^3, the last of which is affixed to the spindle of the reversing slide; d is the exhaust port. The steam ports and passages a^1 a^2 and b^1 b^2 open into the cylinder at different parts of its circumference, which enables the steam to be worked more or less expansively at pleasure, without the necessity of having a separate expansion valve for the purpose. If it is desired to work the steam expansively, then the reversing slide valve is so placed over the ports that it admits the steam only into the lower one (a^1), in which case the quantity of steam admitted behind the piston during one-sixth of a revolution, after passing the port, has to expand in the compartment contained between that and the next succeeding piston, till, after having traversed over about one-half the entire circumference of the cylinder, it ultimately escapes through the exhaust port, the compartment having progressively increased in capacity during nearly the whole time, whereby the full expansive force of the steam is brought into action upon the piston, and transmitted to the main or driving shaft of the engine. If the steam employed in giving motion to the engine is not to be so much expanded, then the reversing slide is pushed further up, so as to open both the ports a^1 and a^2 for the supply of steam; and when this is the case, the steam entering by the port a^1, after acting through one-fourth of a revolution, is further augmented by steam through the port a^2, so as again to fill the compartment with steam at the same pressure as that in the steam pipes. After producing its usual effect, or impelling the pistons and shaft, the steam escapes by the passages b^1 b^2, and so on continuously. Sometimes, instead of opening the parts a^1 and b^1 to their full extent for supplying the steam to the engine, I only partially open them, so that the pressure of steam admitted behind the piston in the first instance may be less than that contained in the boiler; when such is the case, a separate set of passages (either cast along with the cylinder, or pipes may be substituted for these passages), and valves are employed for admitting the full pressure of steam behind the pistons, after they have advanced a portion of their course under the influence of the low-pressure steam. The reversing of the engine is effected by simply moving down the slide c, so that it may admit the steam to the ports b^1 b^2, and put the ports a^1 a^2 in communication with the exhaust port d.

In large engines, it will be further necessary to shut off the steam by the throttle valve previous to moving the reversing slide valve. G is the condenser, which I have placed over the engine, considering that the nearer the condenser is to the level of the water-line of the vessel (when the engine is used for marine propelling), the less will be the back pressure upon the air-pump piston. The condenser is enclosed in an outer casing G^2, the intervening space being kept constantly filled with the injection water, which is allowed to flow through it into the condenser, whereby the temperature of the condenser is kept at a comparatively low state. H is a double-acting air pump, the plunger of which is directly connected to the piston of an auxiliary "donkey engine" J, which in most cases I prefer to having the air pump worked by the rotary engine itself, as that would be revolving at too high a velocity for conveniently connecting the air pump directly to it. When, however, the rotary engine is of great power, and the cylinder of considerable length, then an auxiliary shaft may be employed to give strength, or prevent any risk of fracture from torsion; the auxiliary shaft in such case would be placed parallel to the main shaft, and be connected to it by wheel gearing placed at both ends of the cylinder. By adopting an arrangement of this sort, the auxiliary shaft may be made to revolve with much less velocity than the main shaft, and, under such circumstances, might be employed for working the air pump. K is a steam cylinder, the piston rod of which forms the slide-valve spindle of the cylinder J; the slide valve of the cylinder K is worked by tappets affixed to the piston rod of the air pump.

Although this engine is presented to us as more especially applicable to marine propulsion, yet the reader will understand that Mr. Wimshurst by no means confines himself to this, or, indeed, any particular application of it. The first trial made of it

[f]act, been made on land, and the first proofs which Mr. Winshurst has been [able to p]roduce of its capabilities have been derived from its application as a sta-[tionary] prime mover. We refer to an engine which has been constructed on this [plan for] the Butterley Iron Company, and is now in constant use there. Of the [dimensio]ns and performances of this engine, the following is an authentic account: [The] diameter of cylinder is 60 inches, length of piston 40 inches, width 12 [inches;] effective area of piston 450 square inches, equal to 5-feet stroke; pressure [of stea]m in the steam-box 20 lbs., pressure on the piston 13 lbs. The engine was [tried b]y a break applied to a 6-feet friction wheel keyed on the shaft, with a groove [on its ed]ge to receive a band 5 inches wide; lever applied 5 feet on a 4-inch ful-[crum;] weight on the end of lever 672 lbs.; speed of engine 65 revolutions per [minute.] If, therefore, we multiply 672 by 15 = 9580 lbs. on the circumference, [which i]s 18 feet 9 inches × 65 per minute = 1228 feet × 9580 lbs. = 356 horses power [(at 33,]000 lbs.) as indicated by the break. But if we take the pressure on the pis-[ton and] its velocity, we shall find it much less; that is to say 450 area of piston, [× 13 p]ressure = 5850 lbs., 65 revolutions; rate of piston 1040 feet per minute = [184 hor]ses power. The difference is to be thus accounted for: The friction wheel [worked] in oil, and in all engines some allowance must be made for friction, but in [this inst]ance the indicator was applied to the piston, and at the minimum pressure of [the] steam the engine made near thirty revolutions per minute. After realising [these sa]tisfactory results, it was determined to test this engine at the Great Britain [Colliery], belonging to the Butterley Company. All we had to do, was to bury two [wooden] sleepers in the ground, and to place our engine on them, securing it with [scre]w-bolts. At this trial we had no balance of any kind, no break or gear-[ing, b]ut a single rope brought direct to 10 feet, drew up the weight of rope, [26 cwt.], basket 13 cwt., chain 3 cwt., and 18 cwt. of coal, together 60 cwt., [190 yard]s in 32 seconds = 1015 per minute = 205 horses power. At this time the [pressure] in the steam-box was 28 lbs. × 450 area of piston = 12400, rate of piston [... revol]utions, or 912 feet per minute = 331 horses power (at 33·000 per horses [power).] This would indicate greater power than the weight raised by 126 horses [power,] which is no doubt to be imputed to friction on the pulleys, &c. The whole [work w]as done under the greatest command without the break; when the steam [was shu]t off, the engine stopped almost instantaneously, and when the weight was [raised], and the steam shut off, the whole weight would dance or spring up with [elasticity]."

(*To be continued in our next.*)

[INEFFICI]ENCY OF THE EXISTING ARRANGEMENTS FOR THE EXTINCTION OF FIRES.

[The] frequent occurrence of destruc-[tive con]flagrations in the metropolis, the [acknowl]edged efficacy of Sir Samuel [Bentham]'s arrangements for the prompt [extingui]shment of fire (see *ante* p. 166, [this] vol.), together with its applica-[tion in] a partial manner in all situations [where a]ny of the water companies have [mains,] indicate that it would be well [worth th]e while of many private persons [to adopt] his invention immediately, with-[out waiti]ng for a general introduction of [the mea]sure, as recommended by the [General] Board of Health.

[Sir Sa]muel grounded his proposal to [the Adm]iralty of 13th February, 1797, [on a co]nsideration of "The immense [losses th]at have been sustained in con-sequence of the devastations of repeated fires in His Majesty's Dockyards," and recommended in general terms the fire-extinguishing works, which, early in the present century, were carried into execu-tion at Portsmouth according to his plans; they provided a great supply of fresh water, and also for the raising sea-water for the extinguishment of fire, as already shown.

An ample supply of water for this pur-pose is already furnished throughout Lon-don by the several water companies; their mains, with fire-plugs upon them, are laid along all of our principal streets; the adaptation of those plugs to the recep-tion of hose is all that is wanting to carry into effect the proposal to Sir

Robert Peel, which appeared in No. 1464 of the *Mechanics' Magazine*.*

In the present state of the water-supply question, it cannot be supposed that any of the water companies would be at the expense of altering their fire-plugs, but they would probably give permission to private persons to adapt at their own cost the fire-plugs on the mains to the reception of fire-hose. This might be done by a joint subscription amongst neighbours for the protection of private dwellings; and in the case of many manufacturers, the small cost of a screw-plug and suitable hose would be but a trifling premium paid for the security thus afforded against extensive conflagration; now that Hamburgh is in proof that the introduction of Sir Samuel's plan is to the interest of insurance companies, they might possibly diminish the rate of insurance on property whenever it might be protected by efficient apparatus for throwing water direct from the mains.

Sir Samuel's views extended from the first far beyond the boundaries of a dockyard; in his first proposal of 1797, it was specified that the water mains "might be extended without the limits of the dockyard to the gun-wharf, victualling premises, barracks, and other public establishments;" he observed that such an undertaking on the part of Government would afford an example to the *public*, and concluded his proposal thus, "The reasons for affording the security above-mentioned to the naval establishments near Portsmouth seem to apply equally to *all other establishments elsewhere* in proportion to their importance. Where the establishment is less extensive the expense would be proportionably less. In the instance of Plymouth Dockyard, the water is already sufficiently elevated to be laid in according to the mode here proposed without the necessity of raising it by machinery."

To what public establishments other than the Royal Dockyards has his plan to this moment been adopted? Ships have been on fire in Portsmouth Yard, and the flames have there been speedily extinguished by Sir Samuel's apparatus; where no such works exist, the other day at Havre (France), two costly ships were burnt on the stocks. We a House of Commons consumed Exchange, important building Tower of London, &c., all lo the efficacy of the Portsmoutl had been experienced: may it be presumed that had the exan followed in regard to public ments generally, many of those been destroyed by fire would be ing to us at the present day?

There is a steam-engine fo water near Trafalgar - squar should it not force water in ca for the protection of the public in the vicinity? Why should hose be immediately fitted on t that supply water around th Exchange, the Bank of Engl Mansion-house, that beautiful St. Stephen's, Walbrook, the hall, &c., &c.?

The mode of altering fire-plu not be so well contrived if left ent private persons as if som plan were followed; but is it waiting for perfection to forego good attainable at small expens chief inferiority of separate bined arrangements would ar the want in the former case o organised and numerous set o always at hand to work fire-ex ing hose. This was a difficulty mouth; the police of the yard to be interfered with; but Si placed all the fire-extinguishi under the responsibility of the millwright, who was entirely (Sir Samuel's) control. By article of his instructions, officia 1805, he made the master-m responsible for the good repai steam engine, and for its bei in full working order." The w panies may be depended on in spect; but another article of th tions, the eighth, would be to be to in private establishments person appointed by the master important article, namely, "th apparatus prepared for extingui be kept in good repair, and in *readiness to work*, night as well Farther, the master-millwrigh cause all the engine-keepers bourers "To be properly instr the management of the pensto of the pipes and cocks for a

* *Erratum in that Number*, page 167, last line *but one, for* "immoveable" *read* "moveable."

o the different parts of the yard, they may be enabled to render eatest possible service in case of The master-millwright was far-require "four persons, taking it s weekly, to remain every night engine-house of the wood-mills, that they may be in readiness in fire happening in the night," &c. manager of a public establish- r a private manufacturer, would assign to some amongst his own the duty of remaining on the s at night, but it would be little were such men relieved at about ock in the morning, so that each would have a portion of the or sleep. So in a manufactory some hundreds of men might be ed, it would be only some few of at need to be instructed in the g on the hose to the fire-plug and nagement of the hose when so .

prompt application of water on t outbreak of fire is, by all who nsidered the subject, affirmed to nfinite importance; it is on this that persons on the spot are pro- o have immediate means of bring- -extinguishing apparatus into use with the most distant view of dis- with the eminent services of the rigade, and other firemen; for it valuable bodies of brave and nced men that are most to be d on for the extinguishment of t has gained any considerable o they also on arriving at the spot e best capable of employing with atest effect the powerful means l by hose and jets supplied with nmediately from mains.

M. S. B.

TS FOR THE IMPROVEMENT OF HOSPITALS.

le XXII. of Sir Samuel Bent- "Desiderata in a Naval Arse- pecified essential requirements in hospital, and as they seem equally le in every hospital for the recep- the sick and wounded, they may th insertion in the *Mechanics ine.* The article ran as follows:—

"XXII.

at *in regard to a* naval and marine hospital, it should be contrived no less with a view to the prevention of the spreading of contagion than to the recovery of the sick and wounded; that there should be an appropriate room for the immediate reception of the sick, for the changing their apparel, and other operations necessary in regard to cleanliness at their first entrance; that there should be the easiest means of removing patients from place to place; that apartments should be provided for naval officers of different ranks, as well as for seamen; that not only separate apartments should be provided for contagious disorders, but that the communication with the parts of the building containing patients afflicted with such disorders should be capable of being completely cut off from other parts of the hospital, and that the portion of the building so cut off should be in a great degree variable at pleasure; that the separation should nevertheless be such as, whilst it prevents communication of contagion, admits of the inspection of those who, without being in immediate attendance, have superintendence of the sick; that means of giving supplies should be provided without personal intercourse with the contagious quarters; that the whole building should be warmed, ventilated, lighted, and fumigated in the manner the most comfortable and healthy, and so as to prevent the spreading of contagion; and that it should be furnished with complete medical, surgical, chemical, culinary, and washing apparatus; and that it should have the appendage of an airing-ground, and a chapel for this establishment exclusively."

The above, like all the other articles of the "Desiderata," were solely intended to indicate what was desirable, without regard to whether it already existed in practice or whether it was new; but in this, as in most articles, there are many items not theretofore nor even now attended to,—such as easy means of removing patients from place to place, complete separation of contagious quarters of an hospital from its other parts, &c. These desiderata were not set down till after the obtainment of precise information in regard to hospitals in general for the sick and wounded, particularly through Sir Gilbert Blane, the governors, physicians, surgeons, and other officers of Haslar and Devonport hospitals, the physician of one of the greatest London hospitals; and, not content with investigation through superiors only, Sir Samuel inquired of patients themselves, of different classes, by what further accommodations in an hospital their sufferings might be alleviated.

A few of his contrivances as to detail in hospitals are noted in a paper headed "Hospital Projects," as follows:

"The ventilation and heating of each ward, though arising from the same source of air and heat, wholly independent of those of any other apartment; this not only to enable the ventilation and temperature of each ward to be at command within it, but also for the avoidance of contagion and unpleasant smells.

"However well heating might be effected by general apparatus, still an open fire in each ward, on account of its cheering influence on the sick; consequently, the front of the fire-grate to expose the greatest possible surface of radiant heat in proportion to the quantity of fuel consumed.

"A system of pipes for hot water in every ward, so arranged as to pass by each bed, its patient having a tap within convenient reach, so that he might always have hot water at command to rinse the mouth or mix with drink, without trouble to the nurse; connected with this pipe, a little apparatus to keep drinks warm, as in many disorders frequent sips are more grateful than distant copious draughts. In winter, water for this purpose might be heated in a boiler behind the fire. In a large ward the cost of such an apparatus is estimated at sixpence per annum for each bed.

"Bedsteads, especially in surgery wards, instead of the usual fixed ones, to be a kind of cot frame formed of metal tubes, and having loops from them for the insertion of poles, in the manner of those of a sedan-chair, with sockets also for fitting firmly on to legs fixed to the floor. On such a frame a patient might be removed without disturbance in a horizontal position, whether to a place the most convenient for changing bedding or clothing, for washing and dressing wounds, or to or from the surgery. After severe operations, such a bed would be particularly desirable.

"Lifts from floor to floor, large enough to receive such beds, or arm-chairs for patients suffering great prostr strength.

"Where, as in towns, extens ing-grounds are not attached to pital, the roof of the building that it may be appropriated as a place for convalescents at least.

"For accuracy, uniformity, save trouble, physicians' pres papers printed in columns. The headed perhaps respectively w number of the bed, day of mo of patient, pulse, tongue, medicin special directions, nurses' obse house surgeons' ditto, physicians

"Trays for medicines, with corresponding with those of th whereby the chance of admi wrong medicines would be mu nished.

"Communication with infect ters cut off by glazed partition of them in small hospitals rem at pleasure. In large building panopticon principle, the wings ble from the centre by means bridges."

M.

BRITISH MUSEUM MSS. ON SHIPB AND NAUTICAL AFFAIRS.

[The following memoranda were fou the unpublished papers of Sir Samuel His own observations are given in Ital tinguish them from the extracted matter

"2903. 2. Sir Will. Petty. of a ship."

"The desiderata in a ship are

"1. To cut the water easily

"2. To draw no dead water

"3. To feel her rudder, and t short and quickly.

"4. To beat off head seas, no live on growne seas, and not rov

"5. To bear a stout sayle.

"6. To lye well agrounde.

"7. To draw little water.

"8. To be roomy abaft.

"9. To lie near the wind.

"10. To sale well both by an

"Sir Will. Penn's Papers.

"3232. 1.—*This paper re Dockyard management, very int after the first 25 or 30 pages.*"

"Dialogical discourses of Ma fairs."

"2449. 5.—*Interesting, worth publishing.*

"1. Naval force of Great Br an Officer of Rank. 1791.

"*Good about arming and manning.*

"2. Treatise on the Science of Shipbuilding, by Isaac Blackburn, Shipbuilder, Plymouth, 1817. *Good.*"

"Nautical Tracts.

"A practical plan for manning the Royal Navy without impressment, by T. Trotter, M.D., in a letter to Lord Exmouth, 1819.

"*A fleet of* 60 *sail of the line, with the usual proportion of small vessels, might at all times be ready for sea in six weeks. Ten thousand troops might be embarked for duty in a month.*"

He appears to be an invalid, resident at Newcastle-on-Tyne.

"Series of Essays describing new systems and inventions relative to the Navy, by A. Bosquet, 1818.

"*Good expedients for preventing foundering from leakage; and recommending pigments for preserving ships from decay.*"

THE "AMERICA" AND SLIDING KEELS.

The success of the *America* seems likely to arouse investigation as to the various peculiarities by means of which that success was obtained. It turns out that, among other novelties, she had *sliding keels;* and probably this contributed not a little to her triumph.

Sliding keels, however, are well known to have been the invention of the late British admiral, Sir John Schank, though some one under the signature of "Breeze" spoke of them, in the *United Service Gazette* of the 20th ult., as having originated in America. In the same journal of the 4th inst., there was a letter on this subject from "John M. S. G. Schank," of Boston-house, who stated truly that "the invention did not belong to the Americans, but to Admiral John Schank, who fitted out with 'sliding keels' the *Trial* cutter, the *Lady Nelson* brig, and the *Cynthia* frigate." . . . "With these 'sliding keels' the *Trial* cutter was never known to have been beaten by any king's cutter, and had frequent trials upon a wind with the *Nimble*, *Sprightly*, *Spider*, *Ranger*, and *Revolution* cutters—the *Salisbury*, *Nautilus*, and *Hyena* ships and *Kingfisher* brig, and was never beaten by any of them. She would sail upon a wind nine and a half knots, and before the wind ten knots; and would tack and wear surprisingly quick." There are some very valuable sailing qualities attending these 'sliding keels,' and only recorded by the deceased inventor; therefore, as 'Breeze' suggests, 'as good a wrinkle may perchance be taken from the *America's* keel as her canvas.'"

Sir Samuel Bentham, impressed with the advantages afforded by sliding keels, also adopted them in his experimental sloops *Arrow* and *Dart*, and in the *Netley* schooner; but they were afterwards removed from these vessels (April, 1798), from an apprehension entertained by the other naval authorities of the day that the liability to leakage of the sliding-keel cases was attended with danger. The cases were therefore caulked up, the sliding keels abandoned, and eight inches of false keel added in their lieu.

The inveterate opposition manifested by the dockyard officers of that day to all of Sir Samuel's experimental vessels, caused a strong prejudice in naval men against them; so that trivial leaks, that would not have been even noticed in vessels of the usual construction, wer looked upon as highly dangerous when they occurred in the keel-cases of his sloops. In a letter to the Admiralty, 14th December, 1797, in speaking of the *Dart* and *Arrow*, he stated that, "All vessels hitherto fitted with sliding keels have leaked at the cases enclosing the sliders; the ordinary expedients for preventing leaks in the bottom of a ship are not applicable to the sliding-keel cases; in these vessels, however, expedients have been introduced which have diminished these leaks so much as that, notwithstanding the sliding keels are of a much greater extent than those of any other vessel, nevertheless the leaks of these vessels are not so great as are to be found in the general run of vessels, built in the ordinary mode, without sliding keels."

The difficulty of rendering such keel-cases water-tight has probably caused Captain Schank's invention to have been neglected; but surely the mechanical skill now applied to naval architecture will surmount this difficulty, as General Bentham would have done but for the worry his naval improvements kept him in; he gave up sliding keels, and many other improvements, rather than waste his time in confuting objections that were made to his own invention, or to the introduction of those of others.

The *Mechanics' Magazine* for March 30, 1850, gives, amongst other fragments from Sir Samuel's papers, his statement of the advantages sliding keels afford, and which must evidently have had considerable influence in the obtainment of the *America's* victory. Rarely as it is well to re-produce what has already been published in the Magazine, yet on account of the excitement the late contest has occasioned, it seems desirable again to bring forward the manner in which sliding keels contributed to the success of the *America:*

"*Sliding Keels.**

"The change of form of the vessel by the protrusion of sliding keels, and the power they afford of producing quickly this change at pleasure is of very great importance; a considerable acquisition of power is obtained by their means to influence the motion of the vessel through the water, and the action of the water on the vessel.

"Besides the giving the power of increasing the lateral resistance of the vessel by the direct impulse on the whole area of the extent of these keels, an augmentation of progressive motion is obtained, whilst no other disadvantage is incurred than that of the direct impulse of the edge which divides the water, together with the friction only of the sides.

"Perhaps a still greater use of sliding keels is that of the power they afford to increase the lateral resistance of *either end* of the vessel *at pleasure.*

"The imperfection of sliding keels is, that they require a depth of water beyond what would otherwise be sufficient for the vessel, and thereby it is deprived when in shallow water of the advantages which are acquired in deep water by means of such keels."

In regard, however, to that imperfection it is no more than a foregoing of the advantages obtainable by the use of false keels if a vessel has to navigate shallow water. Indeed, when the sliding keels were removed from the *Arrow* and the *Dart*, their permanent draught of water was necessarily increased by the addition of several inches of false keel.

From an existing longitudinal section of the *Arrow* and the *Dart*, it appears that there were no l four separate sliding keels in th of those sloops,—thus affordin of exercising their draught a end, or in midship, at pleasur out varying their draught at otl of their length.

The great advantages derival Admiral Schank's invention fully justify the construction vessels for the purpose of ex different modes of constructing k with the view of rendering the tight. Probably by forming metal, the desired purpose migh fected. No danger to these cas shot need to be apprehended, s sliding keel would not require to b up so high as the water-line; a the weight of the metal, it bei the middle line of the vesse might be so contrived as to ballast.

In the *United Service Gazett* 4th inst., there is also a letter fro tenant Shuldham, saying that back as the year 1829, Command neaux Shuldham, R.N., exhibi model of a boat with a *revolvin* keel, which the Lieutenant possessing several advantages; description he gives of it does sufficient to enable any observ be made respecting this invent is much to be regretted that no repository of inventions exists; of a record easily consulted man improvements are altogether lost others are often brought for persons who have no claim to t vention, but who profit by them credit and in purse, as if they h nated with themselves.

MR. VINT'S UNDER-WATER PA WHEEL.

We witnessed recently on the tine some very successful exp made with a model steambo with under-water wheels, on the Mr. Henry Vint, of Colchest worked by means of a pair of Boggett and Smith's rotary engi

Mr. Vint's improvements, as d by himself in the specification patent, consist "in inclosing sub propellers in cylindrical casings

* Mech. Mag., No. 1390, p. 245.

here) at the sides of a vessel; the of the propelling wheels and the pal part of the casings, being in- l to be submerged, or placed below vel of the water-line."

. 1 of the prefixed engravings is a side-elevation of a vessel as thus constructed; and fig. 2 a stern view. The wheels (one on each side) with the cases, are placed in recesses made for them under the quarters, or rather at about the beginning of the run, which Mr.

Fig. 1.

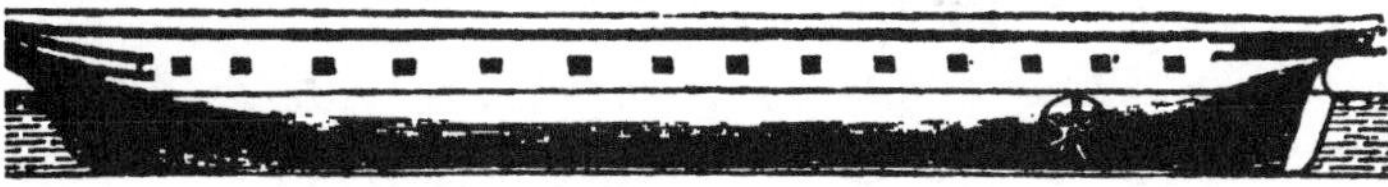

. 3.

Fig. 4.

Fig. 5.

Fig. 6.

Fig. 2.

inds by experiment to be the best n for them. Fig. 3 is an end- and fig. 5 a side-view of one of eels so inclosed in its case; fig. 4 of the wheel detached; and fig. 6 -view of the wheel and case with ter side of the case removed.

will be seen," says Mr. Vint, that lower part of the propelling wheel ely on the current (body) of water ch the vessel floats, whilst the re- g parts of the wheel are revolv- the cylindrical chamber occupied ead water (or partly with air, for ape of which there is a vent-hole ed), the chamber being open to e access of the water only at the part." Each wheel is composed radial blades, and when revolv- re are always three of these arms working efficiently in the water through which the vessel is moving.

The experiments which we witnessed were on the whole extremely satisfactory. There was great speed very quickly attained, and little if any obstruction perceptible from the tail water, which passed off freely from the successive strokes of the paddles, without noise or ripple.

The following Table has been kindly supplied to us by Mr. Vint, showing the results of experiments made in order to ascertain the comparative efficiency of the ordinary paddle-wheels, the screw, and Vint's wheel. To guard against fallacious results, the results were made in each instance with the same model, the same weight of ballast, the same motive apparatus, and on the same water:

TABLE.

Propellers.	Time in Seconds.	Distance in Feet.	Power Expended.
w	31⅓	84	63⅓
inary Paddle Wheels	39⅓	84	55
t's	31	84	56

h the clever rotary engine of t. Boggett and Smith, which was used in propelling Mr. Vint's model, our readers are already acquainted (see *ante*, vol. liv., p. 281). The exceeding simplicity and compactness of this engine, and the smoothness and equability of its motion, make it especially adapted for the production of high speeds; and in screw vessels it would, therefore, probably answer even better than in paddle steamers such as Mr. Vint's. The small space, too, which it occupies, and its consequent lightness and cheapness are other circumstances in its favour, which may be expected to contribute materially to its introduction into general use.

AEROSTATION.—MR. LUNTLEY'S PAMPHLET.

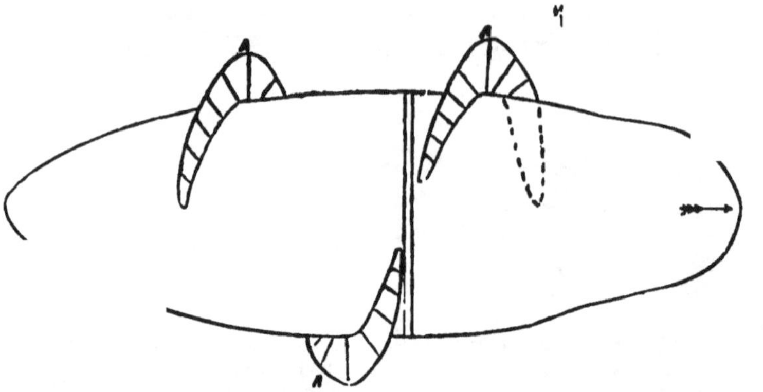

Sir,—As you have kindly permitted me to occupy many of your pages formerly with the subject of aërial navigation; allow me to obtrude once more. Mr. John Luntley has just published a pamphlet on that subject, in which he alludes to those papers, and states his obligation for "many valuable suggestions" obtained from them, without specifying the several parts of his pamphlet in which they are contained; and I feel that I may thus be considered as agreeing with that gentleman in his plan of a rotary balloon, as exhibited in the Crystal Palace, and in the plate printed with this pamphlet. Ever anxious to acknowledge and encourage any steps towards the advancement of this art, I am sorry to differ in opinion from Mr Luntley as to the practicability, or efficiency, if practicable, of such a construction for locomotive purposes. Could a balloon be constructed altogether in the helix form, and put in rotation round its axis, it would follow its lead, like a cork-screw, minus the retarding power of the car ropes *x x*; but a cylindrical balloon, with a bulky round screw excrescence at each end, if rotated, could not follow its own lead, and there would be a tendency, not to rotate round the axis of the cylindrical part, but round an intermediate point between that axis and the centre of resistance of the eccentric screw portion; thus creating a considerable loss of power. As the screw portion cannot follow its own lead because that lead is stopped by the cylindrical portion, it becomes merely a *circular* in lieu of a *flat* propeller; a form manifestly disadvantageous, as resistance from the air, in a given direction, is what is required to propel the balloon; and common surfaces are found to give the least resistance: hence the thin fins and flattened tails of fishes, and the extended thin surfaces in the wings of birds, &c., &c.

It is not my wish to criticise any further Mr. Luntley's ingenious pamphlet. I feel thankful to any man who has the moral courage to aid the progress of an art which has ultimately to lead to the most perfect civilisation of the human race; but which at present is considered as futile and hopeless by the same class of persons, who will be as glad to avail themselves of its advantages when perfected, as those who denied the pos-f steam navigation, and railroad me years ago, are at present.

. Mr. Luntley or others, in con- of his publication, try the ex- of a rotary screw-propelling let me recommend them to try form shown in the accompany- e, with a flat screw fin A, wind- d it; if the belt can really ch a bulky, and probably irre- indle efficiently to revolve, a he middle of the pin would not propelling power to any serious nd the belt could be applied, as tley proposes in his construc- form of the balloon is here made ncave in its outline, which pro- give less resistance than if en- nvex, if we may trust to the of most birds.

Sir, your obliged,

And obedient Servant,

GEORGE CAYLEY.

n, Oct. 4, 1851.

F OF SCIENCE IN BRITISH NAVAL ARCHITECTURE.

vestigation of the French Commis- was conducted on a sounder and ful plan than has yet formed the y such inquiry into the state of our ; it was much more analytical, s elements of comparison were with great exactness. Sufficient is essential to fair and satisfactory n; for generalization on such a this amounts to nothing without although many detached facts may t to bear on the subject in con- designs, yet they cannot bear so ue in their separate and isolated hey would bear in forming the ele- comprehensive generalization, be- inferences deduced from them can- own so certainly in one case as in . But this certainty is necessary, also attainable, since the ships of are sufficiently numerous for the d may be brought into any analy- parison which the claims of science Why, then, has this not been improving the British navy, as well mproving that of France? It is that a reply might be found in the the affairs of this branch of Go- t have not been conducted on a ly enlightened policy. Knowledge which could have been obtained by such means would have been useful only to ex- pose existing error—not to correct—before the restriction of tonnage was removed; for that was a bar to the improvements which such knowledge would have sought to introduce. Since 1832 a short period only has elapsed in which the supremacy of a system of construction resting rather upon opinion than upon science, made that class of information which might be gathered from the source now under notice, almost as useless as it was before. Under both of these preceding conditions of our naval administration the way to general improve- ment was not open, although it might be in- dicated; this way has, however, been gradu- ally opening since 1841; and some results which have been admitted go far enough to prove that liberty and encouragement would surely bring the improvement which has been wanted for ages, but which has not until now been attainable. It is yet with us, as it was described in the French report to be "an epoch of transition." With a sensible movement towards perfection there is an obvious uncertainty as to the counte- nance or check which may be given to this progress. To a great extent the navy requires to be formed on better models; and there is talent enough in the service to collect and apply information so far as to realize the improvement, which would leave room for little doubt as to the general cha- racter of ships of all classes. But will the country avail itself of this talent?

Mere practical qualifications in designing ships have been tried in our service, and found wanting. The surveyors of the navy were good practical shipbuilders, without scientific knowledge. Even their modes of calculating the contents of a ship's body, &c., were formerly rude and clumsy, in comparison of those which science has now shown to be better suited to the purposes. But yet, with all their disadvantages, a sound judgment and solid practical know- ledge of their profession kept them from falling into serious mistakes in forming their designs for ships; and although this sort of qualification is not sufficient of itself to obtain the highest degree of excellence, yet it can never be safely dispensed with, whatever may be the theoretical talents of a constructor. If the surveyors formerly adopted the forms of foreign ships, in many cases, instead of aiming at originality of design, the choice of models generally ex- pressed the soundness of their judgment in this respect; and they paid only a rational deference to the superior genius and accom- plishments of foreign constructors which it has always been deemed creditable to evince.

nted, 1832, by the French Government, into the causes of the inferiority exhi- he French vessels in the experimental s under Admirals Malcolm and Ducrest.

If practical knowledge were sufficient to carry on improvement in ship-building, then this improvement might have advanced under the administrations whose operations they had to control; for, so far as practical knowledge would realize a progress, the opportunities were presented from the beginning of the revolutionary war, except that they were bound by the restriction of limited tonnage. The ships they built were commonly what they were intended to be in their capabilities for service: they were not, it is true, excellent sailers, but they could always be depended on, with but few exceptions, for performing their evolutions; were able to keep the sea during long periods of arduous service in blockading; and they were efficient ships in regard to the qualities which were necessary in going into and sustaining an action.

The period in which the strictly partial qualifications of the old system could be thought sufficient in filling the responsible post of surveyor of the navy has passed away. The nation looks now with a more piercing eye into the several departments of Government; and whilst progress is exhibited almost everywhere else, it would be unreasonable to expect that the eye of cold indifference should rest with apathetic satisfaction on the proceedings of a department which has to do with the proper defensive power of the nation and its dependencies.—*Fincham's History of Naval Architecture.*

SISCO'S IMPROVEMENTS IN CHAIN-MAKING.

Experiments have been recently made at Woolwich dockyard to test a new description of chain cable, the invention of a French gentleman, Mons. Sisco, for the purpose of ascertaining their quality and value as compared with the chains now in use in the Royal Navy. The new chains are made of common hoop-iron of the breadth required, and wound on a reel by machinery into an oval shape and to the same breadth as the outer surface, which is rounded off after the whole has been brazed in passing through a furnace of molten metal. The usual test of an iron chain for naval service of two inches in diameter is 72 tons strain, but many links break with the application of far less power, and yet the other parts of the chain are found qualified to pass the required ordeal. M. Sisco's chain of two inches broad and two inches thick, with stays in the centre of each of the two links, was placed in the testing-frame, attached to a testing-chain of 2¼ inches in diameter, and on the hydraulic power being applied one of the links was lengthened five-eighths of an inch other one-eighth of an inch when it a strain of 110 tons, and the 2¼ inch chain broke off in two places when th reached 114 tons. The hoop-iron cl some openings in one of the links had been imperfectly brazed, but it appear to have been made otherwis tive. One link of the same dimensi inches thick and two inches bre afterwards placed in the testing-fra when a strain of 70 tons was applie lengthened one-twelfth of an inch; tons, one-eighth of an inch; with 1 three-sixteenths; with 110 tons, on of an inch; with 115 tons, five six and when it resisted 120 tons strai considered advisable not to conti strain, as it was so great as to lo stone frame on which the machine and liable to damage other parts powerful iron-frame of the machin strain applied on this occasion was more than had ever been previously and the hoop chain was only slightl on one side. When inquiries were to the price at which the hoop chai be supplied, it was stated they would more per cwt. than the common although their holding powers were greater in proportion.

FURNACES COMPELLED TO CONSUM OWN SMOKE.

By an Act of Parliament pa wards the close of last session (1 Vict., c. 91), it is enacted—

"That from and after the first January, one thousand eight hund fifty-two, every furnace employed employed in the working of en steam, and every furnace employed employed in any mill, factory, p house, dyehouse, iron-foundry, gla distillery, brewhouse, bakehouse, g waterworks, or other buildings use purpose of trade or manufacture wi city (although a steam engine be r or employed therein), shall in all constructed or altered so as to cons smoke arising from such furnace; an person shall, after the first day of J one thousand eight hundred and fi use any such furnace which shall constructed so as to consume or l own smoke, or shall so negligently such furnace as that the smoke arisi from shall not be effectually cons burnt, or shall carry on any trade ness which shall occasion any no offensive effluvia, or otherwise an ourhood or inhabitants, without using, satisfaction of the Commissioners, the racticable means for preventing or racting such annoyance, every person nding shall forfeit and pay a sum of ore than five pounds nor less than shillings for and in respect of every ring which or any part of which such e or annoyance shall be so used or sed."

SUBMARINE TELEGRAPH BETWEEN ER AND CALAIS.—REAL CAUSE OF NON-COMPLETION.

—The reasons assigned in your Jour- well as in all your contemporaries t exception, for the Telegraphic Cable is to connect England and France fallen short of the required length, ceedingly erroneous; and, no doubt, nay be all traced to one common . The cable is stated to have been l out of the direct line across, by the rs of a strong gale of wind which owing down Channel at the time of it down. A grosser untruth never tempted to be palmed upon the pub- There was *no gale*, but on the con- a perfect calm all the time. The sea not possibly have been in a more able state for the operation. I was pier at Dover when the *Blazer* started e cable on board, and in or about the r till nightfall, and never observed in a quieter state. A gale there was, dreadful one too, but it came the *day* and had nothing whatever to do with rtcoming of the cable. What, then, e real cause? Simply this, Sir,—the ungling manner in which the opera- as managed. While the *Blazer* was ng at the rate of only two miles r, the cable was allowed (in the first f the morning) to run out at the rate ! As a necessary consequence, the must have taken the ground in a series s or kinks, which sufficiently accounts h mile of cable not covering its mile und. The cable, as put on board, was ds of twenty-four miles in length, is considerably more than the direct e between shore and shore.

s only fair to the manufacturers of pe, Messrs. Newall and Co., to add, hey had nothing whatever to do with ying of it down. The engineers of legraph Company, Messrs. Wollaston rampton, having got the cable made, at they could do all the rest themselves, nce the ridiculous result, the shame h they or their friends have attempted by a *deliberate falsehood*. Roguery has been said to require only rope enough at any time to hang itself; and so, it seems, we may say of ignorance.

I enclose for your private satisfaction my real address, and am, Sir,

Your constant reader,

POSTCAPTAIN.

Dover, Oct. 7, 1851.

LIEBIG'S OPINION OF THE VALUE OF ELECTRO-MAGNETISM AS A MOVING POWER.

At the present moment electro-magnetism, as a motive power, is engaging great attention and study; wonders are expected from its application to this purpose. According to the sanguine expectations of many persons, it will shortly be employed to put into motion every kind of machinery; and, amongst other things, it will be applied to impel the locomotive engines on railroads, and this at so small a cost that expense will no longer be matter of consideration. England is to lose her superiority as a manufacturing country, inasmuch as her vast store of coals will no longer avail her as an economical source of motive power. "We," say the German cultivators of this science, "have cheap zinc, and how small a quantity of this metal is required to turn a lathe, and consequently to give motion to any kind of machinery!"

Such expectations may be very attractive; indeed they must be so, otherwise no one would occupy himself with them; and yet they are altogether fallacious; they are illusions, depending on the fact that those who entertain them have not made the necessary comparisons and calculations.

With a single flame of spirits of wine, under a proper vessel containing boiling water, a small carriage, of 200 to 300 pounds weight, can be put into motion; or a weight of 80 to 100 pounds may be raised to a height of 20 feet. The same effects may be produced by dissolving zinc in dilute sulphuric acid, in a certain apparatus. This is certainly an astonishing and highly interesting discovery; but the question to be determined is, which of the two processes is the least expensive?

In order to answer this question, and to judge correctly of the hopes entertained from this discovery, let me remind you of what chemists denominate "equivalents." These are certain unalterable values of effect which are proportionate to each other, and may therefore be expressed in numbers. Thus, if we require 8 pounds of oxygen to produce a certain effect, and we wish to employ chlorine for the same effect, we must employ neither more nor less than 35½ pounds weight. In the same manner, 6 pounds weight of carbon (in the form of coal), are equivalent to 32 pounds weight of zinc. The numbers representing chemical equivalents express, in the most general sense, the relative values or amounts of effect, and are applicable to every kind of effect which bodies can produce.

If zinc be combined in a certain manner with another metal, and submitted to the action of dilute sulphuric acid, it is dissolved in the form of an oxide; it is in fact burned (oxidised) at the expense of the oxygen contained in the conducting liquid. A consequence of this action is the production of an electric current, which, if conducted through a wire, renders it magnetic. In thus effecting the solution of a pound weight, for example, of zinc, we obtain a definite amount of force, adequate to raise a given weight one inch, and to keep it suspended; and the amount of weight it will be capable of suspending, will be the greater the more rapidly the zinc is dissolved.

By alternately interrupting and renewing the contact of the zinc with the acid, and by very simple mechanical arrangements, we can give to the iron an upward and downward or a horizontal motion; thus producing the conditions essential to the motion of any machinery.

Out of nothing, no kind of force can arise. We know that, in this case, the moving force is produced by the oxidation of the zinc; and, setting aside the name given to the force in this case, we know that its effect can be produced in another manner. If we were to burn the zinc under the boiler of a steam engine, consequently in the oxygen of the air instead of in the galvanic pile, we should produce steam, and by it a certain amount of force. If we should assume (which, however, is not proved,) that the quantity of force is unequal in these cases,—that, for instance, we had obtained double or triple the amount in the galvanic pile, or that in this mode of generating force less loss is sustained,—we must still recollect, that zinc can be represented by an equivalent weight of carbon (as coal). According to the experiments of Despretz, six pounds weight of zinc, in combining with oxygen, develope no more heat than one pound of coal; consequently, under equal conditions, we can produce six times the amount of force with a pound of coal as with a pound of zinc. It is therefore obvious, that it would be more advantageous to employ coal instead of zinc, even if the latter produced four times as much force in a galvanic pile, as an equal weight of coal by its combustion under a boiler. Indeed, it is highly probable, that if we were to burn, under the boiler of a steam engine, the quantity of coal required for smelting the zinc from its ores, we should produce far more force than the whole of the zinc so obtained, could originate in any form of apparatus whatever.—*Familiar Lectures.*

COLOURED DAGUERREOTYPES—HILLOTYPE—HELIOCHROMY.

It has been announced more than once, that Mr. Hill, of Westkill, New York, had discovered a simple method of making coloured daguerreotypes. (See vol. liv., pp. 273, 298, 332.) It is now stated in the *Daguerrean* (American) *Journal*, that one will be exhibited here this month. A great interest has been manifested in the discovery, but hitherto all has been very quiet and secret about the process. It has long been a desideratum with artists to discover such a process. Bequerrel, in France, has produced coloured pictures, but he was never able to fasten the colours; still many kept working in hopes of discovering this important secret, and before Mr. Hill has seen fit to present the public with any account of his process, M. Niepce St. Victor (nephew of the celebrated discoverer of photography, in France), has made the grand discovery, and showed his pictures to the world. Three of his pictures are now before the public in London, and the new art is called Heliochromy, or sun-colouring. The glory of the discovery will belong to him who first gives it to the world,—a fact not so well appreciated here as in Europe.

The three heliochromes now in London are copies of coloured engravings, representing the one a female dancer, the others male figures in fancy costumes; "and every colour of the original," says the *Athenæum*, the editor of which has examined them, "is most faithfully impressed on the prepared silver tablet." According to the *Athenæum's* account, which appears to be a hasty one, and is therefore not so precise and minute in the description as could have been wished, "the plate, when prepared, presents evidently a dark brown, or nearly a black surface—and the image is *eaten out* in colours." "We have endeavoured," it adds, "by close examination, to ascertain something of the laws producing this most remarkable effect; but it is not easy at present to perceive the relations between the colorific action of light and the associated chemical influence." The following is the *Athenæum's* description of the pictures:—

"The femal figure," it says, "has a red silk dress, with purple trimming and white lace. The flesh tints, the red, the purple, and the white are well preserved in the copy. One of the male figures is remarkable for the delicacy of its delineation:—here blue, red, white, and pink are perfectly impressed.

is injured in some parts; e number of colours which ost remarkable of all. Red, en and white are distinctly s intensity of the yellow is

here about yellow is the , as the colour has always be the one with which Mr. st difficulty, and which he able to fix on his plates in than that of buff.

e relative merit and artistic ew sun pictures, the *Athe-* speak very distinctly; and fer that they are not alto- ighest grade of excellence. owever, that "these results ose which were given to the ography was first announced; um adds, that "we may ex- see the heliochromes pre- scenes and chosen friends beauty of native colour."

een any printed account in ges giving anything like an r's process, but happily we n idea of it from recent pro- Paris Academy of Sciences. wn that a plate of silver im- tion of sulphate of copper le of sodium, and rendered by means of the battery, ible of receiving colour, awal from the solution, it is ction of light. Becquerel, plate to the coloured rays ctrum, obtained an image of such a manner that the red the plate an image of a red let, a violet colour; and so ays. The idea struck young there were some relation our which a body communi- , and the colour which the on a plate of silver which had d, and he therefore com- of experiments to test its le knew that strontium gave r to alcoholic flame. He red a plate of silver by pass- water saturated with chlorine de of strontium. He then ck of a drawing, containing lours, against the plate, and ole to the light of the sun tes, when the colours of the oduced on the plate, but the ter defined than the others. six other rays of the solar *me method used to* produce *is followed with* other sub- *loride of calcium* for an orange, the chlorides of soda or potassium, or pure chlorine, for yellow; and beautiful yellows have been produced by a solution of hydrochloric acid and a salt of copper. The green ray was produced by boric acid and the chloride of nickel; the blue ray was obtained by a double chloride of ammonia and copper, and a white ray with the chloride of strontium and sulphate of copper.

A silver plate prepared with water acidulated with hydrochloric acid and the battery, gives all the colours by the action of light, but the ground of the plate is always black. St. Victor found that all the substances which produced coloured flames produced coloured images by means of light. This is truly an important discovery. Oxygen has nothing to do with the colours, the same results were produced in vacuo; but water was necessary to the process, as dry chlorine produced no effect. The plate to produce these effects upon must be prepared with metallic silver, and that very pure. The baths are made of one-fourth by weight of chloride and three-fourths of water. After the plate is well polished by tripoli and ammonia, it is immersed in the bath at one stroke and allowed to remain for some minutes. It is then removed from the bath, rinsed in clear water and held over a spirit lamp till the plate becomes a cherry colour, at which point it is exposed to the light in the camera. It takes about two hours of exposure, but the process will yet be shortened. It has been exceedingly difficult to fix the proofs, and the process for this has not been made public.

The whole process to be successful must be nicely managed. The idea about the relation between the substances producing coloured flame and colours on the silver plate is a very important addition to the treasury of science.—*Scientific American.*

SPECIFICATIONS OF ENGLISH PATENTS ENROLLED DURING THE WEEK, ENDING OCTOBER 9, 1851.

THOMAS WOODS, of Portsea, upholsterer, and ROBERT WALTER WINFIELD, of Birmingham, manufacturer. *For certain improvements in bedsteads and couches, or articles for sitting, lying, and reclining upon.* Patent dated March 24, 1851.

The improvements claimed under this patent comprehend,

1. A mode of constructing cots for ship use, which may also be converted to sofa or standing bedsteads. [This is effected by lowering the frame which supports the sacking and resting it on fixed points inst- allowing it to swing on pivots, whi case when employed as a ship's cot

2. A method of stretching the sacking of camp and other similar bedsteads and articles, [by means of an arched rod placed at the head of the bedstead and provided with a rack at one end, into the teeth of which takes a click attached to the framing of the bedstead.]

3. A mode of making the pillar or corner blocks of metal bedsteads, and other similar articles, from tubing or shafting [by planing or otherwise removing portions therefrom, so as to leave the necessary dovetails for securing the parts together.]

4. Another mode of making such pillar blocks by rolling metal tubing into suitable forms, and then welding several of such rolled pieces of metal together and cutting off the compound tube thus formed, a length sufficient to form the pillar block.

5. A mode of securing ornaments to the head, foot or side frames of metal bedsteads, couches and other similar articles by forming the frame with grooves of such a depth that the ornaments, when laid therein, and secured by pins or rivets, shall come flush with the side of the framing.

EDWARD DUNN, of New York, residing at Brompton, master mariner. *For improvements in reciprocating and rotary fluid meters.* (Being a communication.) Patent dated March 24, 1851.

A description of the reciprocating meter (the invention of Captain Ericsson), sought to be protected under this patent, has already appeared in our present vol. (see p. 61), we now therefore give the claims only, which are made in respect of it. These are,

1. The operating, by means of cranks placed at right angles, the two plungers of different sizes, by the hydrostatic pressure of the fluid to be measured, when the operation is combined with definite stops for the purpose of regulating the stroke of the said plungers.

2. The forked ends of the piston rods, or other similar means by which the crank pin is allowed to move through a considerable arc in order to keep the plunger stationary while the position of the slide valve is reversed.

3. The operating of the crank shaft, valve gear, and valves within the fluid to be measured.

In respect of the "rotary meter," the claims are,

1. The employment of a uniform circular channel in combination with a contracted channel situated between the ingress and egress apertures of the meter.

2. The rotary paddle wheel having paddles projecting into and working in the said *uniform and contracted* channels.

3. The ingress and egress apertures proportioned and formed in a peculiar manner described.

4. The employment of a pipe and jet for admitting fluid, and giving motion to the wheel before the fluid enters through the ingress aperture proper.

5. The employment of a valve leading to the ingress aperture of the meter, by which any desired power of jet may be obtained before the fluid is permitted to enter the meter.

THOMAS HAWKINS, of Inverness-terrace, Bayswater, oilman. *For improvements in brushes.* Patent dated March 24, 1851.

The improvements which form the subject of this patent have reference principally to certain methods of securing the bristles, and attaching the handles of brushes in a more effectual manner than hitherto practised. The principal feature of novelty is the employment of a conical ferule, through which the bristles, previously cemented together, are passed point first, and drawn down until they are sufficiently tightly compressed. The handle is also held by a conical ferule, which is soldered to that containing the bristles, the space between the cemented ends of the bristles and the but end of the handle being filled in with cement or cuttings of cork, which serve to prevent the bristles from shifting, and at the same time do not materially increase the weight of the brush. The patentee describes several other methods of effecting the same object, which however are modifications only of that above mentioned; the bristles being secured by compression within a conical ferule, or between a ring or strips of metal and the but end of the handle.

JOHN PETER BOOTH, of Cork, feather purifier. *For an improved manufacture of fabric applicable to the construction of muffs, boas, tippets and other like articles, and also to the ornamenting of articles of dress and furniture and other similar uses.* Patent dated March 31, 1851.

The "improved manufacture of fabric" which forms the subject of this patent, is composed of parts or strippings of feathers, which are sown on to woven material of any suitable description, close together, so as to present, when made up, a continuous feathery or downy appearance. The feathers preferred are those of the turkey, from which the filamentous or downy portions are stripped off in such a manner as to bring with them a slip of the skin, or cuticle of the quill, by which the several strips are secured to the fabric, by sewing them on with a needle and thread, passing the thread through the woven material forming the foundation, and over or through the cuticle adhering to the portion of feathers.

.—The application of parts or strips ings of feathers, in the manner, and urpose described.

Gwynne of Lansdowne-lodge, and hill, merchant. *For improvements inery for pumping, forcing, and g of steam, fluids and gases, and laptation thereof to producing mo- the saturation, separation and de- ion of substances.* Patent dated 1, 1851.

stails of this invention we reserve ure Number.

Richardson, of Halifax, dyer. *ovements in dyeing and cleansing ds.* Patent dated March 31, 1851. tentee describes and claims:—

method of subjecting woollen and piece goods, or piece goods con- oollen or worsted, to the action of rs in the dye vat, by causing the hen sewn together in a continuous pass over and under a series of bmerged in the liquors, alternately te directions, until sufficiently satu-

arrangement of apparatus, on the ciple as the above, for cleansing ds. The cleansing-trough is divided e compartments containing water, yed cloth is passed through each of accession over and under a series , and is then conducted through a pressing-rollers for removing the us moisture to the folding appa- ich is of the ordinary construction.

Brunier, of Paris, civil engineer. *rovements in obtaining power by of "steam or" compressed air.* ated March 31, 1851.

atentee has disclaimed the words r" from the title of his patent. provements specified have relation employment of a vacuum vessel ection with air-compressing ma- for the purpose of ensuring a regular f air to supply the place of that l in working.

—The mode described of arranging y for obtaining motive power by f compressed air.

s Huckvale, of Choice Hill, *For improvements in treating rzel, and in making drinks and parations therefrom.* Patent dated 1851.

patentee takes mangelwurzel, clean, and cuts it into small pieces he size of beans or peas, which are sted and ground in the same man- ffee, for which the product thus may be employed as a substitute, *hich it may be combined as desired.*

2. The leaves of mangelwurzel are cut into shreds or strips, dried, slightly charred on a hot plate, and may then be used in the production of a beverage in the same manner as tea is now ordinarily employed.

3. The roots of mangelwurzel after being cleaned are cut into pieces, thrown into vats, and allowed to heat and ferment for two or three days, when the temperature is about 70°, when water is added and a fermented liquor obtained similar to cyder or perry.

4. The roots of mangelwurzel after having been washed and cut into small pieces, are slightly charred in suitable kilns or ovens, such as employed in the manufacture of malt, and may then be employed in the preparation of wort, which may be applied to any of the purposes for which wort made from malt or other substance is ordinarily used.

Claim.—The four several modes of treating mangelwurzel and making drinks to the preparations therefrom.

Richard Archibald Brooman, of the firm of J. C. Robertson and Co., of Fleet-street, patent agents. *For improvements in machinery for the manufacture of rope and cordage.* (A Communication.) Patent dated April 2, 1851.

Claims.—1. The causing of the fibres or threads taken from the bobbins in the manufacture of rope and cordage to revolve around said bobbins in a peculiar manner described, that is to say, under the operation of certain primary rings for the purpose of giving a double twist to said fibres.

2. A mode of forming the strand by collecting the fibres or threads in the tubes, so that said strand shall be composed of successive layers wound or twisted one upon the other.

3. A mode of laying the strands thus formed into rope by giving to said rope at one operation a double or compound twist.

4. The combination of revolving rings or carriers, producing like mechanical effects, with stands for holding the stock, &c.

5. The said ring or carrier, when combined with and revolving about several primary rings and their frames or stands, as shown in the drawings.

Auguste Motte, of Southwark, manufacturer. *For certain improvements in portmanteaus.* Patent dated April 2, 1851.

The improvements specified by Mr. Motte consist—

1. In forming portmanteaus from a single piece of leather or other material, the leather being so cut as to leave parts for overlapping at the edges, and the whole being secured by rivets or other suitable means.

2. In the employment of a deeper frame

than usual for the lower half of the portmanteau—the said frame extending upwards to near the top of the interior when the lid is closed down.

3. In the application of a washer or plate of metal when securing the handles of portmanteaus by means of rivets.

WEEKLY LIST OF NEW ENGLISH PATENTS.

Thomas Taylor, of the Patent Saw-mills, Manchester, Lancaster, for improvements in apparatus for measuring water and other fluids. October 9; six months.

Joseph Pimlott Oates, of Lichfield, Stafford, surgeon, for certain improvements in machinery for manufacturing bricks, tiles, quarries, drain-pipes, and such other articles as are or may be made of clay or other plastic substance. October 9; six months.

Sir John Scott Lillie, Knight Companion of the Order of the Bath, of Pall Mall, Middlesex, for improvements in forming or covering roads, floors, doors, and other surfaces. October 9; six months.

Henry Curzon, of Kidderminster, Worcester, civil engineer, for improvements in the manufacture of carpets and rugs. October 9; six months.

Henry Briggs, of Primrose-street, Bishopsgate-street, seed-crusher, for improvements in oil lamps and in apparatus for lubricating machinery. October 9; six months.

James Frederick Lackerstein, of Kensington-square, Middlesex, gentleman, for improvements in obtaining motive power. October 9; six months.

WEEKLY LIST OF DESIGNS FOR ARTICLES OF UTILITY REGISTERED.

Date of Registration.		No. in the Register.	Proprietors' Names.	Addresses.	Subjects of Design.
Oct.	3	2967	Parr, Curtis, and Madeley	Manchester	Machine for straightening bars or rods of iron.
	,,	2968	William Dray	London-bridge	Reaping machine.
	6	2969	George Lomas	Addington-square, Clerkenwell.	Spring-lever ventilator.
	,,	2970	J. Levilly	George-street, Hanover-square.	Parts of corsets or stays.
	,,	2971	Edmund Youldon	Torquay	Portable bathing machine.
	7	2972	Welch & Margetson	Cheapside	Hat reviver.
	,,	2973	Eliezer Edwards	Birmingham	Inkstand.
	8	2974	Joseph Stevenson and John Stevenson	Cripplegate Buildings	The soirée union back comb.

WEEKLY LIST OF PROVISIONAL REGISTRATIONS.

Oct.	4	303	John W. Hiart	St. John's Wood	The couvert (chimney.)
	6	304	Sidney Hall	Northampton	Safety contraction expanding joint for railways.
	8	305	J. F. R. de Francileu	South-street, Finsbury	Envelope letter.
	,,	306	J. F. R. de Francileu	South-street, Finsbury	Envelope letter.
	9	307	John Browne	Great Portland-street	Shifting paddle-wheel.

CONTENTS OF THIS NUMBER.

LONDON: Edited, Printed, and Published by Joseph Clinton Robertson, of No. 166, Fleet-street, *in the City of London*— Sold by A. and W. Galignani, Rue Vivienne, Paris; Machin and Co., *Dublin*; *W. C. Campbell* and Co., Hamburg.

Mechanics' Magazine,

EUM, REGISTER, JOURNAL, AND GAZETTE.

71.] SATURDAY, OCTOBER 18, 1851. [Price 3*d.*, Stamped, 4*d.*

Edited by J. C. Robertson, 166, Fleet-street.

TMSHURST'S PATENT SCREW-PROPELLING MACHINERY.

Fig. 2ª.

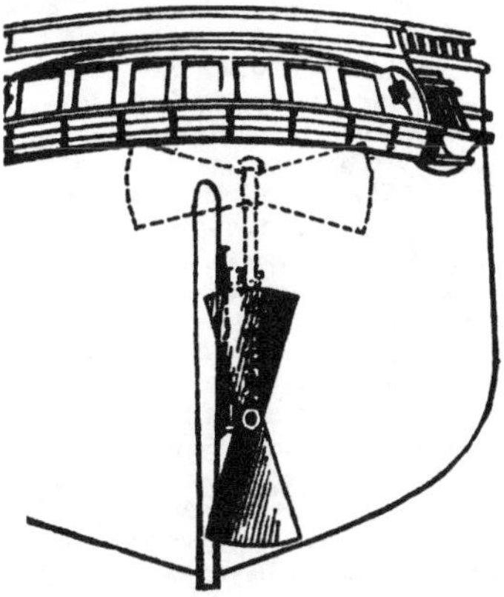

Fig. 1ª.

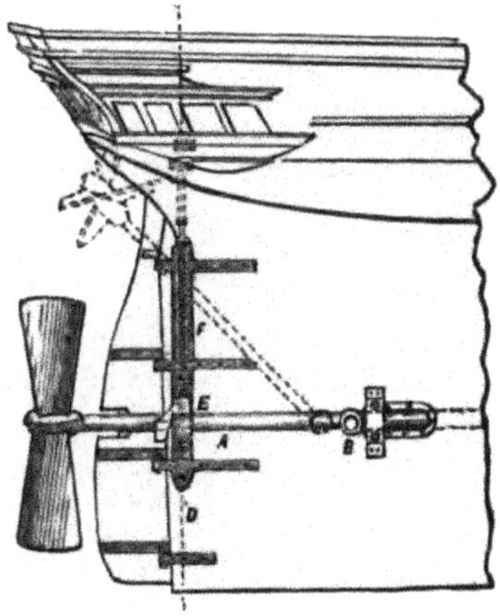

Fig. 3ª.

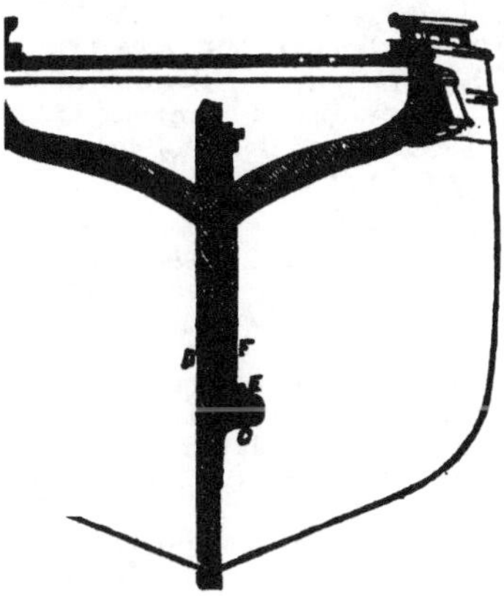

Fig. 5ª.

Fig. 7ª.

Fig. 6ª.

Fig. 8ª.

Fig. 4ª.

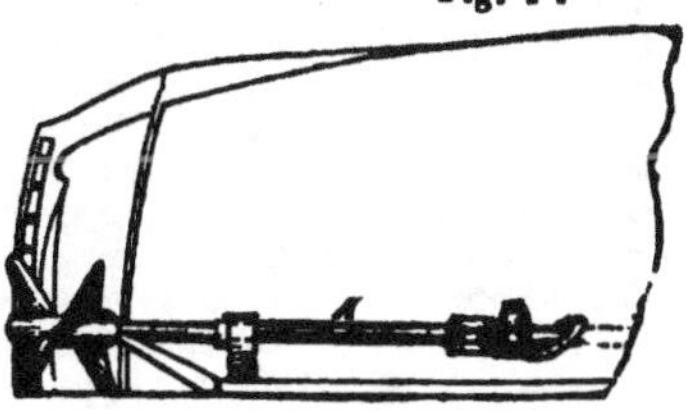

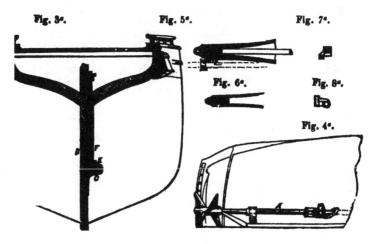

MR. WIMSHURST'S PATENT SCREW-PROPELLING MACHINERY.

(Concluded from p. 285.)

Mr. Wimshurst next proceeds to describe a new method of disposing screw in auxiliary steam-propelled ships, whereby "the propeller may be nientlу raised up out of the water when required, and retained there at plea so that "when the vessel is under sail, and the steam power not required, the will be free from the obstruction which would otherwise result from the pr being dragged through the water." Fig. 1ᵃ is a side elevation of the stern pa vessel fitted according to this improvement; fig. 2ᵃ is an end elevation of the fig. 3ᵃ is another elevation of the stern partly in section; and fig. 4ᵃ is a half the vessel viewed from below.

A is the propeller shaft, which may run parallel with the keel of the vessel, bu out at such an angle with it, that the point where the shaft passes through the hull situated in the run of the vessel. By this angular position the propeller shaft is to be extended so far as to admit of the propeller being placed abaft the rudder, an same time of the rudder being brought sufficiently round towards that side on wh propeller shaft is placed, as is requisite for the steering of the vessel. B is a join main-shaft A. C the lower half of the outer bearing of the shaft, which is securely the stern-post D. E is the outer bearing, which loosely encircles the shaft, but fi rately to the upper side of it when the propeller is lowered down into its place. Th end of the bearing E is of a T-form, and is free to slide up and down in a groov same shape formed in a metal plate affixed to the side of the stern-post, and by wh stern-post is greatly strengthened. F is a long screwed rod, the upper end of whic in a bearing, and the lower end in a socket. The rod is free to turn within the but cannot move either up or down; the lower part of the rod is screwed, and take corresponding screwed recess, or hole, through the bearing E, by which means, w propeller is desired to be raised out of the water into the position indicated by dott in fig. 1ᵃ, it is only necessary to turn the screwed rod F by means of a crank or mounted on its upper end, and which will raise the bearing E, and along with it th end of the propeller shaft and the propeller. By turning the screwed rod in the direction, the propeller is again lowered into its place when required. Figs. 5ᵃ, 6ᵃ, 8ᵃ are respectively separate plans of the under bearing, the T-shaped groove, and upper bearing. The groove may be of any other form capable of retaining the end sliding bearing E. In some cases, where the propeller is of great weight, I susp outer end of the propeller shaft by an outrigger attached to the stern of the vessel, i case a much smaller bearing would be formed on the outrigger than is required u stern-post to carry the main-shaft.

A description follows of an apparatus for steering from the bow, but this v over in order to lay before our readers Mr. Wimshurst's account of one wh consider of rather greater value: namely, "an improved steering apparatus, may be fitted into the bows or near to the stern of the vessel, with the shaft a angles to the keel, so that when the propeller is put in motion, the head of th sel may be at once carried round to either quarter, according to the direction in the apparatus is made to revolve." This arrangement is stated to be "parti applicable to tug-boats;" such a boat, so steered, would be enabled to turn ro a space about equal to her own length. The details of these improvements are in fig. 1ᵇ (longitudinal elevation), and fig. 2ᵇ (half plan).

A is a trunk which passes right through the bow, open at the top and bottom, bu perfectly water-tight from the hold or interior of the vessel. B is the steering app C, the main-shaft of the steering apparatus, which is placed "thwartships." It is ca two bearings KK, which are secured to cross beams affixed to the vessel. The appa put in motion by means of a steam-engine connected to the shaft F, through the in tion of the bevel gearing JJ. Fig. 3ᵇ is a longitudinal elevation on an enlarged scale in section, of the apparatus, and the shafts and gearing by which it is connected v steam-engine. B is the steering apparatus. C its shaft. JJ bevel gear, by which is communicated to the apparatus from the intermediate shaft F. GG are two spur gearing into each other, one is keyed to the shaft F, the other to the shaft E; to whic the power of the steam-engine is applied. When it is desired to reverse the dire *the revolution of* the steering apparatus for the purpose of putting the head of th n opposite course to that in which it moves, while the wheels GG are in gear, the follow- changes are requisite:—The shaft E is made to travel on end, from left to right, by ns of the pinion R, which gears into teeth turned on the shaft E, until the wheels GG disengaged from each other; but while the shaft E is being pushed on end, it causes ther intermediate shaft and pinion H to travel in the opposite direction through the inter- tion of the connecting beam link L, so that both of the wheels GG may gear into it, it then becomes an intermediate wheel between GG, communicating motion from one he other, and thereby causes the shaft of the apparatus to revolve in the opposite direc- . P is a clutch, and T a lever by which the shaft E may be engaged and disengaged n the engine shaft O.

Fig. 3^b.

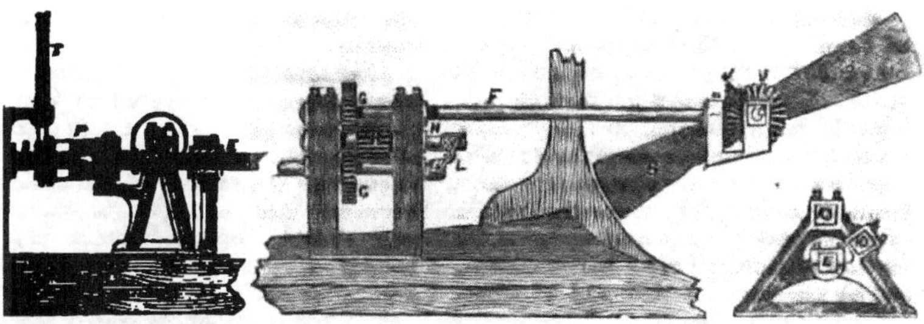

Fig. 1^b. Fig 2^b.

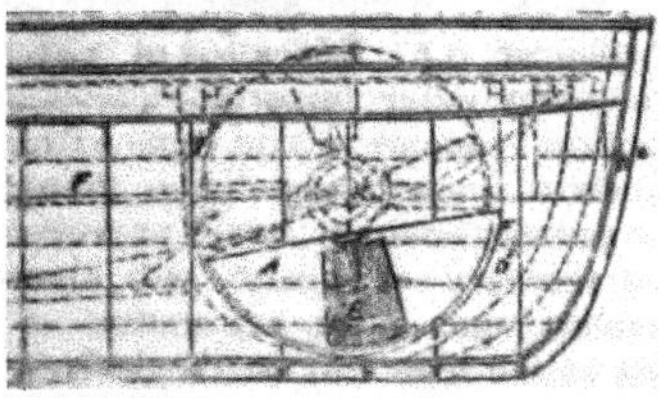

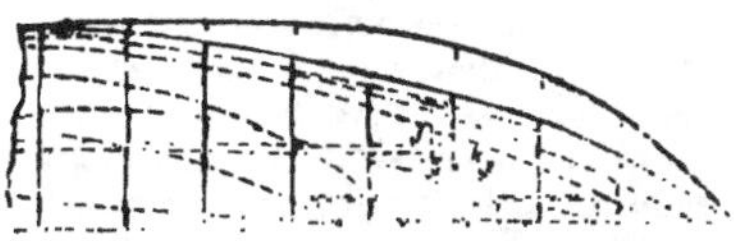

Ir. Wimshurst has also a very ingenious arrangement "for measuring and indi- ng the effective duty of propellers, for disconnecting them from the steam- ine, and turning round the engines and machinery connected therewith when *steam is not up;* as for instance when effecting repairs upon or examining the ine."

Another of his improvements consists in forming the run or stern-post of vessels er of wood or iron, with a metal plate or plates upon it, for the purpose of carry- the propeller-shaft bearing; when the propeller itself is placed abaft the rudder, at additional strength is thus given to the stern-post of the vessel. The metal es should be laid flush with the bottom of the vessel, to prevent unnecessary stance to the progress of the vessel through the water.

Ir. Wimshurst's claims are as follows:

irst. I claim a rotary steam-engine with broad-headed pistons, moving on anti-friction ers and guides in combination with an internal drum and external cylinder, constructed arranged as above described.

econd. I claim as an improvement in steam-engines the combining with a rotary engine, structed as above described, a separate cylinder or engine to work the air-pump.

hird. I claim as an improvement in propelling the raising and lowering the propellers screw vessels by means of the T-shaped groove, the carriage or bearing E and vertical w F, as represented in figs. 1^a, 2^a, and 3^a, and before described.

ourth. I claim as an improvement in propelling, the altering of the angles of the blades steering apparatus for bows of vessels by means of a travelling plate *f*, and screw *k*, combination with the arrangements for throwing the screw in and out of gear, as repre- ted and described.

ifth. I claim the constructing of vessels with apertures in the bows for the reception of *rtship propellers or steering* apparatus, as above described.

Sixth. I claim the constructing apparatus combined and arranged as above described, and represented in figs. 1, 2, and 3, whereby the propeller pressure may be indicated, the engine and propeller-shaft disconnected, and the engine moved round when the steam is not up.

Errata.

In the first portion of this article, page 285, line 25, *for* "a single rope brought direct to 10 feet drew up the weight," *read* "a single rope brought direct to a 10 feet *drum* drew up the weight."

Page 283, 12th line from the bottom, *for* "one-eighth of an inch," *read* "1⅛ inch."

HINTS FOR THE IMPROVEMENT OF HOSPITALS.

(Concluded from p. 288.)

Why should not provision be made in *all* hospitals for the reception of "*different ranks*" of persons as well as for the poor? This is a question touching persons of many descriptions, particularly operatives, seeing how much their sufferings from disease are augmented by their position in life, by the usually limited extent of their house accommodation, by the cost of medical advice and medicines, and by the want of skill and of experience in their attendants. The services of a stranger, it is true, are rarely supposed to equal in tender solicitude those of a wife, a child, or a mother: but in great public establishments for the sick there are advantages that fully compensate for the absence of relatives; so much so, that experienced physicians have affirmed that the proportion of cures in a hospital for the poor is greater than in private practice amongst the gentry. This has been attributed partly to the inexperience of private persons acting as nurses, but principally to disregard of injunctions respecting the administration of medicines, and by indulging the patient with improper diet. Independently of these considerations, here are many others indicating the need of *sanitariums*,* wherein the diseased of *all* ranks might at *moderate cost* obtain superior medical advice, the best unsophisticated medicines, experienced attendance, and suitable food and accommodations of every description. Attempts have been made to institute private establishments of the kind in question, but their charges are necessarily beyond the means of the humbler independent classes, and are not in sufficient esteem to be resorted to by superior classes: there has lately been established, too, a semi-gratuitous refuge for gentlewomen during illness, which promises the success such a valuable institution merits.

Mechanics of every rank, from wealthy proprietors to the operatives who serve them, now all take part in questions affecting the welfare of the masses; we have seen the success of associations for bettering the condition of the poor—why should not mechanics have the honour of associating, together with others, or without extraneous aid—to better the condition of the *independent self-supporting* classes of the community—that very honourable and honoured portion of it? Education of children, and many other important questions in political economy, would come within the sphere of such an association, though on the present occasion better provision for the recovery of health will alone be touched upon—the restoration of the sick at the least cost to themselves, and with the least amount of avoidable danger to the sane.

In contagious disorders there is much danger of infecting the sane; for example—scarlet fever, a disease now frequent and often fatal, is highly contagious; if one member in a family of operatives be attacked with it, it generally spreads to most of the children, perhaps to the parents themselves, and throughout the several families lodged under the same roof; but if on the attack of the first child it could be removed to a sanitarium, probably the other inmates of the dwelling would be spared the infliction.*

Sick quarters, at a moderate charge, are greatly needed for another numerous body of persons in the same rank of life as operatives—domestic servants. If a diseased domestic be tended in the master's house, it involves him in heavy ex-

* *This word is used to avoid complex terms, and the degrading notions which hospital conveys; it may stand till some better word be proposed.*

* In a family, last year, one of seven children was seized with scarlet fever; the physician advised immediate removal of the other children. A furnished house was taken for them all; six escaped the disease, but the precaution was attended with expenses that $100 did not cover.

ıd still heavier inconvenience: ·stablishments the unfortunate too frequently discharged withıe to go to, and in the dread that savings will be absorbed in doc- Were it possible to obtain adare in such a case, at a mode-, most masters would consider and a pleasure to pay a sick serpenses for a reasonable period.

ying an economical sanitarium ation, many circumstances indiadvantages that would result necting it with a general hospiach an establishment medical cal advice is provided of firstlence, besides which a compeical resident is always on the dy day and night to be called ıout delay in case of need; the are all of the best and of unsod quality; no useless addition to is ever made in a prescription ke of giving an apparent equiquantity for the price likely to d for a draught; the nurses are ed to their duty, they have beniliar with the symptoms that ich changes in disorders as call ation of treatment, and they serve medical orders; the food is the best of its several kinds; rd is had to temperature and n, and to general sanitary pro- the building itself and its ac-; and, in addition to these great s must be added, that requi-ll kinds can be afforded to at a much lower price, being l wholesale, than they can be for, in any private house.

ıospitals are not unfrequently in these it would be easy to requisite arrangements for the dation of *paying* patients; and l buildings, often added to, there hat might not be conveniently for the reception of paying

ot only the humbler classes of at might be benefited by pro- "different ranks" in a sanita-any are the men of superior no have neither wife nor child, relative to be had recourse to s—military and naval officers, nen, small capitalists, whose omes afford no more than the l attendance of the servant where they lodge; many are the single gentlewomen whose whole year's income would be consumed in payment of a nurse and the apothecary's bill, run up to a great amount during a short illness, though no unreasonable charges had been made. The general practitioner must be remunerated for his time wasted in visiting at distant houses, no less than for his skill acquired by a costly education.

Some prejudice would have to be overcome, particularly in the higher grades of society; but as nothing *gratuitous*, nothing that could be denominated *charity*, is proposed, there ought not to exist aversion to the profiting by such a sanitarium. There can be no humiliation where a person pays a fair price for commodities and services received.

"Different ranks" having to be provided for, the accommodations, the attendance, and the quality of diet should in each grade be suited to the habits of those different ranks.

Operatives and servants are accustomed to sleep often several in the same room, therefore for this description of patients a general ward might be provided for men, another for females,—some slight partition being made between bed-place and bed-place: greater privacy would be required for the next higher-paying class, for whose accommodation a separate cabin for each person might be enclosed within one large room, a portion of the cabin being of glass to enable the general nurse to see whether her services were requisite or not without disturbing the patient. At a still higher charge a separate room might be assigned to each patient.

As to diet, the very poorest persons in our great hospitals are provided with all that is requisite as to the essential properties of food,—even costly wines are not withheld when prescribed by the physician or surgeon: but as the higher ranks are accustomed to delicacies, certain of them might enter into the dietary of the high-paying class, divided, however, under the usual hospital regulations of spare-diet, half-diet, and full-diet. But for all ranks it should be a regulation strictly enforced that, without express permission of the physician, no extra article of food or beverage should be allowed of, either by purchase, or introduced by friends.

Though in a sanitarium, the expenses incurred would vary considerably in the case of different patients; yet the establishment would, on the principle of all insurances, be enumerated by a certain fixed charge to each and all patients of the same rank. These charges should not exceed the average of real cost to the institution, including proportion of rent, attendance, management, no less than the disbursement for food and medicines. The charges, if regulated on this principle, would be much below what similar necessaries could be obtained for in a private house, seeing that all articles would be obtained at wholesale prices, that the requisite staff of officers and attendants are already provided in hospitals, so that little addition to it would be needed other than nurses; and of these a single one could better attend ten or twenty patients when in contiguity than she could administer to the wants of two or three dispersed in different parts of a private-house. What the charges to be remunerated should amount to would be matter of calculation from data obtainable in existing hospitals; such as the total cost of medicines and of food, divided by the number of in-door patients, &c. Charges should be reduced to so much in each instance per diem, in order that patients should have to pay for all the time they had been inmates, and not a day beyond it.

Very many of the industrial classes are now members of some benefit club, from which they receive a certain allowance during illness. As a precaution against fraud, a usual condition is, that no sick-allowance should be granted to members doing any kind of work. However necessary this condition may be, it has sometimes a mischievous effect on individuals, by inducing a habit of idleness;—were a member during illness to have recourse to a sanitarium, he might be allowed to do what little work he might be able to perform whilst in the establishment, since the medical officer's testimony would be a better security against feigned disease than any other that a benefit society could obtain.

Relevant to the subject in question are Self-supporting Dispensaries. An establishment so-called, instituted six years ago in a suburban parish, has been attended with most salutary consequences; the members of it pay a very small weekly sum, more or less according to the number in family: for this they are entitled to medical advice at the Dispensary, if able to attend it; if not, at their homes,—and to medicines in all cases. The most prominent advantage of this Dispensary has been that of its checking disease at its first attack,—its subscribers, instead of waiting for advice till confined to bed, repair to the Dispensary on the first symptom of illness; and in most cases it is arrested immediately. The Dispensary Report for the year ending November, 1850, affords proof that disease in most instances yields to an early application of proper remedies; for of 2,052 cases treated, no less than 1,911 were cured, 39 other patients were relieved,—there were in the year no more that 17 deaths; the remaining cases were still under treatment. The happy results of early treatment were particularly manifested during the prevalence of Asiatic cholera, for during that scourge, in a population of many thousands, there were at the most but 11 deaths from it, including doubtful cases,—and some of them were imported from the metropolis in a dying state; at that sad time the clergy of the parish had seconded the medical practitioners in urging the inhabitants to seek advice on the slightest intimation of stomach derangement. This Dispensary cannot, however, be considered as a model of perfection; for though called self-supporting, the voluntary contributions of the gentry suffice for the purchase of medicines, the weekly subscriptions of the benefited members being appropriated as remuneration to the medical men. Similar dispensaries, but wholly self-supporting, might be easily established by the joint subscriptions of operatives living in a quarter of a town where a general practitioner would give his attendance on moderate terms.

M. S. B.

MATHEMATICAL PERIODICALS.

(Continued from p. 226.)

XXVIII.—*The Mathematical Repository.*

Part 2.—*Mathematical Department.—Old Series.*

The miscellaneous papers in this department are of a peculiarly interesting and valuable character, embracing many of the most elegant discussions of the *then* numerous staff of English geo-

In this respect the work is per-
qualled, both for extent and
y any other of our mathemati-
icals; and although the editor
igly biassed in favour of the
of the Greeks, yet he did not
occupy his pages to the exclu-
her branches of mathematics.
have reprints of whatever was
l valuable in more expensive
ate works; and numerous ori-
rs on various parts of algebra,
trigonometry, series, me-
pherics, and the fluxional cal-
inserted at intervals, several of
ve led the way to some of the
arkable discoveries of recent
The first volume of the *Reposi-*
ts completion, was dedicated to
Hutton, Esq., LL.D., F.R.S.,
ll testimony of esteem for his
l abilities . . . by the Edi-
ie second volume, in like man-
edicated to "The Rev. Nevil
e, D.D., F.R.S., Astronomer
io, by a long and indefatigable
n of his eminent talents, has
al honour and service to his
and to science." The third
oes not appear to have been
rtunate in obtaining a patron;
my copy must be imperfect,
cks a dedication.

Miscellaneous Papers.

A Dissertation on the Geome-
lysis of the Ancients. By the
n Lawson.

his is a reprint of the well-
act on Geometry which was
published by Mr. Lawson in
he "Collection of Theorems"
accompany the reprint, but is
itly proposed and answered in
portions.

. Eight Propositions with their
ations. From Dr. Stewart's
Theorems.

ese propositions are a reprint
which were demonstrated by
art himself, in his "General
" (1746), and are here given
from the original work. In
it articles of the *Repository* a
ble number of the remaining
are proposed for solution, to
monstrations were furnished,
Stewart's method, by Messrs.
d Swale. Dr. Small's demonstrations, from the *Edinburgh Transactions* (vol. ii., pp. 112-134), were also incorporated with those corresponding to them in the *Repository;* and it is here where the objections are urged with respect to the *un-geometrical* character of these investigations, which were afterwards satisfactorily removed by the late Professor Davies in his elaborate paper on "Stewart's General Theorems," printed in vol. xv., pp. 573-608 of the same Transactions.

Art. III., XI., XXIV. Lucubrations in Spheres. By John Lowry.

*** These lucubrations appeared just previously to the publication of Howard's "Elements of Spherical Geometry," and occasioned a warm controversy between Mr. Lowry and his former tutor. In the preface to the treatise Mr. Howard complains that he had "*in confidence,* communicated many of his discoveries to his friends and pupils, which confidence *has been abused,* and several of these have been already published in various periodical publications, in particular the article "Lucubrations" in the *Mathematical Repository;* and as the charge evidently pointed to Mr. Lowry as the offending pupil and friend, he replied to it in a short paper which forms Art. LX. in the *Repository;* which was again responded to by Mr. Howard in Art. XI., vol. ii., of the same work. Mr. Lowry corrected several oversights in the treatise on Spherical Geometry, which were readily admitted by Mr. Howard; but he avoids *denying the charge of a breach of confidence,* by pointing out "Euclid's Elements," "Emerson's Geometry," the *Diaries,* "Lawson's Tracts," "Simpson's Geometry," &c., &c., &c., as the *original* sources whence Mr. Howard had derived all his "*new discoveries:*" besides he *misquotes* the remarks from the preface, and thus throws himself open to receive the full force of the castigation which was afterwards administered by Mr. Howard with no sparing hand (*Repository,* Art. XI., vol. ii). On the whole, it appears that Mr. Lowry was more indebted to Mr. Howard for many of his earlier propositions in spherics than he desired at this time to admit; but the subject was afterwards pushed by the former gentleman to an extent which the latter never anticipated. Indeed, so valuable are some of Mr. Lowry's later investigations in this branch of mathematics that Professor Davies did not hesitate to say "that to *him alone* we owe every important spherical theorem that can be set down to the credit of Englishmen during at least a century past, or more." ("Supplement to Young's Trigonometry," p. 263.)

Art. IV., XV. New Tables for Finding the Contents of Casks. By John Lowry.

Art. V., XIV., XXVI., XLIV., XLIX. Tables of Theorems for the Calculation of Fluents. From "Landen's Memoirs."

Art. VII. A Curious Problem, with its Investigation. By Thomas Todd.

*** In a former article (*Mech. Mag.*, vol. liii., p. 453,) we noticed the avidity with which Mr. Todd seized upon matters for dispute; nor does his "ruling passion" forsake him on this occasion. During the course of a short "scholium," he falls foul of Dr. Hutton, the "author of *a* book of Mensuration;"—he informs "the readers of the *Repository*, and others, that two of his solutions in the *British Diary* for 1792 . . . are both spoiled by the Editor and Printer of that *Diary*;" whilst another question, "one would think by design, is made nonsense of;" and lastly, he calls upon "Mr. Thomas Keith to give his proof against [an] *equatement* solution given in the *Ladies' Diary* for 1789," when he expects "Mr. Keith will do his business effectually, or, like a man who loves truth, when he finds himself in errors, acknowledge them." Such language, of course, precluded all reply, nor do we find that any of the parties concerned ever publicly noticed Mr. Todd's requests.

Art. VIII. A New Property of the Cycloid. From "Landen's Memoirs."

Art IX. Observations on the Fundamental Property of the Lever. By the Rev. Samuel Vince, M.A., F.R.S.

*** This paper was also reprinted in the *Mathematical Companion*, No. I., p. 13, and has since been reproduced as Ques. 1648 in the *Ladies' Diary* for 1840.

Art. X., XXIII., XLV., L., &c. Investigations for determining the Times of Vibration of Watch Balances. By *George Atwood*, Esq., F.R.S.

Art. XII., XXV. On the Ellipse and *Hyperbolæ*. From "Landen's Memoirs."

Art. XIII., XXVII., XXXVII
Finding the Sums of certain Seri
Stirling's Differential Method. B
Jonathan Mabbott, of Manchester.

*** Mr. Mabbott was an able a
tensive contributor to the Mathe
periodicals of his time, and posse
more *general* knowledge of Math
cal science than many of his cont
raries. Since his death, all his boo
manuscripts have been disposed of
son, who was compelled to take th
in consequence of ill-health and in
circumstances. Several of his
have supplied me with the means
nishing a portion of two articles i
series of papers; and some man
collections, now in the possession
Henry Buckley, fully prove him t
been well versed in every thing re
to the summation of series.

Art. XXII., XLVI., LI., &c. (
Resolution of Indeterminate Pro
By John Leslie, M.A. From the
burgh Transactions, vol. ii., pp
212.

Art. XXVIII. General Pro
By Mr. John Lowry.

*** This paper contains the
analysis and synthesis of the two
ing Problems, subdivided into thei
ral cases after the manner of the
geometers; on this account it is c
siderable value to the young geo
and is well worth re-printing in
elementary treatise.

Problem I.

"Let there be any number as of
points A, B, C, &c., and let th
any other given point P: throug
is required to draw a right line
that the sum of the perpendicular
BY, CZ, &c., falling thereon fr
points A, B, C, &c., may be equ
given magnitude."

Problem II.

"Let there be any number of
points A, B, C, &c., and likewise
cle given in magnitude and posit
is required to draw a right line to
the circle, such that the sum of th
pendicular A×, BY, CZ, &c.,
thereon from the given points
C, &c., may be equal to a given

Art. XXIX., XXXV.,

strations to Lawson's Proposi-

These elegant demonstrations of ,'s Theorems were furnished by . Campbell, Lowry, Nicholson, Peletarius, and several others; ur at intervals through the whole hree volumes of the Old Series. rere subsequently demonstrated ntinued series from general Dia- by the Rev. Charles Wildbore, investigations have since been ed in the "Manchester Memoirs," p.p. 414—452, and again re- in the New Series of the *Reposi-* ol. vi., Part 3, pp. 9—26. Mr. Campbell has recently collected ionstrations from the *Repository* s included them in his neat ntitled "Lucubrations in various es of Mathematics. Liverpool, under the usual *erroneous* title iwson's Geometrical Theorems." Mr. Lawson was the first to pub- se propositions in the form of a *ion* is undoubtedly true, but that the *Author* of them cannot for a t be maintained in any sense of n. In the introductory remarks l to the first edition of the Tract ı they first appeared he disclaims *uthorship* in general terms, by ng that *many* of the Theorems . "*appeared before in English*," is entirely silent as to the *names* e Geometers to whom he had debted. Indeed so little were the urces of his information known, allusions to Dr. Stewart's writings ı any of the communications from ıy able geometers, whilst their s of demonstration preclude the ity of their ever having seen his An exception, however, must le in the case of "Peletarius;" r this may be, for he *faithfully tes* both the Analysis and Syn- of Dr. Stewart's "*Propositiones trica*," and forwards them to the *as his own productions!!!!* The ıg Table will show how much Mr. ı was indebted to Dr. Stewart's e works; several of the remain- eorems are to be found in Pappus, re the writings of Hugo D'Ome- Vieta, Fermat, &c., to be carefully *ed, no doubt the whole of the heorems in this Collection* might *ed to their proper owners.*

Lawson.		*Stewart.*
1	Corresponds to . . .	1 and 2
2	,, ,, ...	3 ,, 4
3	,, ,, ...	5
4	,, ,, ...	6
5	,, ,, ...	7
6	,, ,, ...	8
7	,, ,, ...	9 ,, 10
8	,, ,, ...	11
9	,, ,, ...	12
10	,, ,, ...	13
11	,, ,, ...	14
12	,, ,, ...	15
13	,, ,, ...	16 ,, 17
14	,, ,, ...	18
15	,, ,, ...	19, 20 21
16	,, ,, ...	22
17	,, ,, ...	23
18	,, ,, ...	24
19	,, ,, ...	25
20	,, ,, ...	26
21	,, ,, ...	27
22	,, ,, ...	28
23	,, ,, ...	29
24	,, ,, ...	30
25	,, ,, ...	31
26	,, ,, ...	32
27	,, ,, ...	33
28	,, ,, ...	34
29	,, ,, ...	35
30	,, ,, ...	36
31	,, ,, ...	37
32	,, ,, ...	38
33	,, ,, ...	39
34	,, ,, ...	40
35	,, ,, ...	41
36	Stewart's General Theorems	Prop. I.
37	,, ,, ,,	Prop. VI.
38	,, ,, ,,	Prop. III.
39	,, ,, ,,	Lemma II.

Such a system of wholesale piracy carries within it the seeds of its own condemnation, and full exposure must, sooner or later, inevitably follow. The practice, however, is by no means totally extinct, although of late years it has been publicly ignored by some of our most eminent mathematicians; nor can the noble example of citing authorities on all legitimate occasions, so fully carried out by Mr. Walton and others in their valuable collections of examples in different branches of Mathematics, fail to produce a powerful effect towards restoring literary and scientific *honesty* to its true position in the minds of those who are necessarily dependent upon others when engaged in compiling elementary treatises.

Art. XXXV. A problem on Friction,

with its investigation. By Mr. Colin Campbell.

Art. XXXVIII. An easy method of constructing an Azimuth. By Thomas Reik.

Art. XXXIX, LII, &c. Useful Propositions in Geometry. By M. A. Harrison.

*** These 17 Propositions with their corollaries belong to the same class as those relating to "Halley's Diagram" in *Burrow's Diary*, the *Liverpool Student*, and the *Horæ Geometricæ* in the *Lady's and Gentleman's Diary*. They are no doubt correctly attributed to the fertile pen of Mr. Lowry, by Professor Davies, in a list of the *propriæ personæ* belonging to several fictitious signatures with which he kindly favored me, a short time previously to his death, although another authority assigns them to the late Mr. T. H. Swale.

Art. XLVIII. Important connections of Simpson's and Emerson's determination of the height of the Tide. By John Landen, from the *London Magazine Improved*.

Art. LX. Animadversions on some remarks in the Preface to Mr. Howard's Treatise on Spherical Geometry. By John Lowry.

*** This is the paper alluded to in the remarks on Art. III, &c., and needs not be further noticed.

T. T. W.

Burnley, Lancashire, October 10, 1851.

(*To be continued.*)

ON THE "AMERICA" AND THE WAVE LINE.

The result of the recent sailing-matches, which proved the *America* faster than our best yachts, is one which ought to be made profitable to builders and yachting men, by turning their attention to the form of vessels, and by inducing them to investigate the principles upon which they are built. Each new yacht is nearly a repetition of old lines in new proportions, while any proposal to adopt new lines meets with little attention, or if they are tried, it is in some mutilated way; the result is not equal to expectation, and the system is condemned. This has been particularly the case with what is called the wave-line principle. It is stated in the papers that

the *Titania* is upon this plan; ‹
the views given of the *Amei*
appears to be much on the sam
ple. It will therefore be inter
investigate what the principle
more so as the numerous exp
made for the British Associat
brought to a successful issue
Scott Russell, are very im
known. It should also be m
that by a few simple rules of ca
cation, the form of each separ
can be accurately and minutel
mined, so as to make the whole
able to the principle. So far
1846 a small volume was publi
J. R. Smith, London, called "
on Naval Architecture by Capta
Fishbourne, R.N.;" to this w
indebted for my information,
give an extract, as well as refe
one interested in its subject, to
itself.

The following is "the genes
wave-line curve:" "Fig. 19 i
retic wave curve of a water-l
genesis of these curves is as
the length of the forebody, as c
with the length of the afterboc
3 to 2, therefore the whole l
divided into five equal parts, a
allotted to the forebody. A circl
diameter is equal to the half
determined upon, is described
circumference touching the cent
where the fore and after bodies j
circumference is divided into
equal parts, and the central line
fore and after bodies are each divi
eight equal parts; then, for the
the forebody, from the foremost
on the central line lay off the p
cular distance of the central li
the first or lowest division on the
ference of the circle, and fr
second division on the central
perpendicular distance of the
division on the circle, and so
eight divisions; then through thes
draw a line, and it will be the w
curve forward . . . I say a t
wave curve, because as it is gener
case, there is a long straight o
breadth inserted between the f
after bodies. Fig. 19 only repres
curves of the extremities." Th
for the afterbody is drawn in th
way from the points on the oth
circle, and remainder of central

Fig. 19.

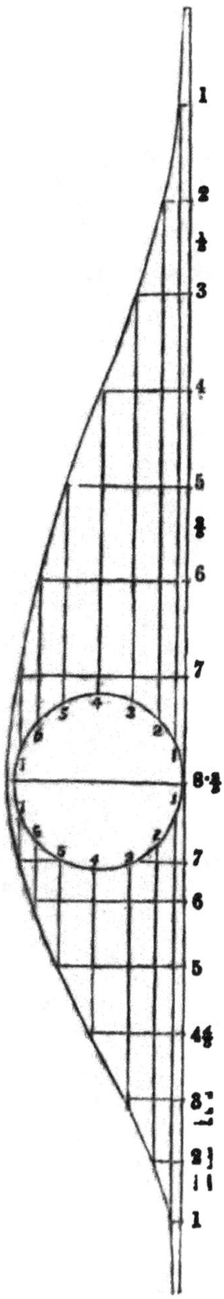

It would be instructive to have an account of the experiments and models by which the wave curve was ascertained; but the proceedings of the Association are not easily met with, and are understood to be very long. At different times other experiments have been published, particularly a detailed account of those made by Colonel Beaufoy; the object of these was to discover the form which offers least resistance to a fluid when moving in it. This is certainly the great primary question, and if it could be solved, modifications could be made to suit particular purposes; but all in accordance, as far as possible, with that form. There is a theory of Newton's, proving that of different cones, formed on the same base, according to certain directions, there is a proportion ascertained in which the larger cone is less resisted than the smaller cone within it, and he determined that a certain curve formed on the larger cone is the curve of least resistance. This certainly bears on the subject, but I believe the proposition is disputed by scientific men, and I mention it more for the purpose of showing that science allows of a form of least resistance, than for applying it to the form of the vessel. It has also been allowed that there is a particular curve which fluids take: "Whatever the pressure, the curves described by the particles are the same." It is therefore reasonable to expect to find in the wave-line curve the form of least resistance. Besides the difficulty of accurately observing what takes place to a body moving partly immersed in a fluid, and to the fluid itself, there are many difficult questions in the resistance of fluids, perplexing even to the scientific; we accordingly find that great difference of opinion exists as to where the greatest beam should be placed, and on this chiefly depends the question of resistance. Taking the annexed line as the total length, the marks show the places fixed upon by the different authorities. For cutters it is generally allowed to be one-third from the bow. Colonel Beaufoy makes it two-fifths from the bow, Chapman and the French one-twelfth before the middle, and the wave-line three fifths from the bow. The subject, then, is one open to much discussion. The resistance of air and water may not be compared, but I may just refer to the great range

Fig. 1.

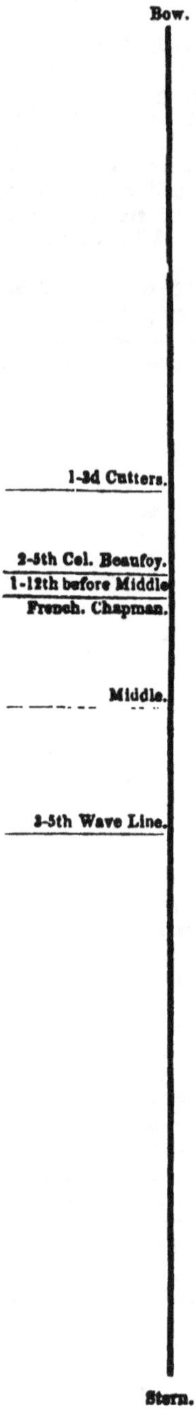

of the conical ball over that of the round bullet; may not it be referred to Newton's proposition? And we all know the different effect of a bluff or a sharp bow as to speed, though the proportion is not brought to an exact scale. Following Captain Fishbourne's lectures, let us look to the resistance to be overcome by a vessel passing through water. There is the pressure on the fore-part, the tenacity of the water, and its adhesion to the sides and after part. Some authorities rate this latter as very great; one French author states that a sphere in motion drags along with it about six-tenths of its own bulk of fluid with a velocity equal to its own. Colonel Beaufoy ascertained by experiment, that the pressure on the hinder part was reduced to one-fourth by making the length triple the breadth, and that by increasing the length of the body by an addition in the middle, of the same diameter, the resistance to its motion is diminished. This fact Captain Fishbourne alludes to in the genesis of the curve: "The general action of the water at the fore part is a gradual rise from the stem towards the centre of the bow, and a gradual subsidence from the centre to the after part of the bow. The water heaps up, and begins to turn aside a little before coming in actual contact with the stem. The velocity of the filaments of water is increased, and they acquire a divergent motion, by which they push aside the surrounding water on each side of the vessel; but as they are on all sides pressed by the fluid without them, their motions gradually approach parallelism, then bend inwards to fill up the trough left behind the body. The more divergency and rapidity that is given to the filaments, the longer they will be of filling-in behind the vessel. The rarification of the air behind a body moving rapidly I am unable to enter upon. There is another point to be noticed, that as the pressure of the surrounding water to fill up the space must depend on the depth of immersion of the body, when that is great, it will be filled up the quicker, as the adjoining pressure will be greater. It is stated that the water fills this space at a determinate velocity; the proportional lines of vessels, and especially of steamers where the power of propulsion is known, should then be formed with reference to this. From what has been ve appear to have before us the of the water against which, in t complicated form, a vessel be made to conform. The subght have been made more clear lings, but it will be thought that at minutiæ have already been o. It may be deduced that the ould be of such a form as will adily, and with least disturbance, e water, causing a gradual diy of the filaments; while the ght to be of such a length as to e for them to attain only such a of parallelism as to leave the the outer water still acting on press in the water behind, yet it in motion to clear the after the tendency of the water to it. In this the authorities apagree. In comparing the wave nes with the above, they seem ited than any others to meet the encies. If we make convex bows, er must be more disturbed: if oncave, there will be spaces left the water will either not have fill up, or a convex part coming y to the wave, stops her way. In bows we have a great divergency o the water, while a long after ags with it much of its own water. To take the wave line t the lowest estimate, it is natural ose that a vessel with curves to those which the water takes, oat more easily, and meet with stance, than one whose lines are variance with the curves of the It will naturally be said that it not carry the sails or afford stown regard to the sails, there is no why, if the hull is altered, we not alter the sails to suit; as to , it is said as if the wave-line e was simply one set of lines, s the principle is applicable to imension: but it is not the intenthese notes to enter particularly ese points.

one who has observed the boats sels in the Mediterranean, and east, must have been struck with culiar forms. There is much to in many of them, and there is no ut that they are moulded on cerht principles, though not sufficarried out in practice to be as t as they *ought to be*. They appear to attend too exclusively to the floating part, without giving a proportional holding in the water; and this would naturally be expected where the means of propulsion were generally by manual labour, and the change to sails was not provided for by a corresponding provision in the form of the vessel. This is a distinction of the parts of a vessel which I propose entering more fully upon by-and-bye. The great feature of the wave-line principle is, placing the beam far aft: frequent instances of this are to be found in boats celebrated for their fastness, and which have been built of the same form from ancient times. Figure 3 is an imperfect sketch of the lines of the exact model of a Chinese boat, well known to naval men as it comes racing out at a considerable distance from land, with pilots and provisions. The model was brought to this country upwards of thirty years since, for the purpose of introducing the form to this country, as capable of great velocity. They are made of all sizes up to 1,000 tons. This model is of one for seventeen oars, and carrying two of the large lug sails of the country. There are several peculiarities of these boats, which may be noticed. They are divided into many separate compartments—one or any number of which can be taken by a passenger. The rowlock pins are of some height, and several notches in them enable the rope grummets to be shifted, so that the oars may be made to work freely, according to the height of the vessel out of the water. As it is intended to go into shallow rivers as well as into deep water, to be either rowed or sailed, the rudder is made to slide up into a cavity in the after part for the purpose: the entire lines of the vessel are thus preserved, and while the whole floating part, to illustrate what was stated above, has a most buoyant appearance, and the lines all very full, particularly so at the stern, she appears very well adapted for rowing, while, by sliding down the large rudder, she is enabled to hold her way when sailing. The false keel is attached externally, not built into the vessel, and has a considerable gripe forward. For further exemplification, I will suppose that, exclusively for sailing purposes, the space to heel of rudder is filled up by a keel in the dotted line. The beam is placed

Fig. 3.

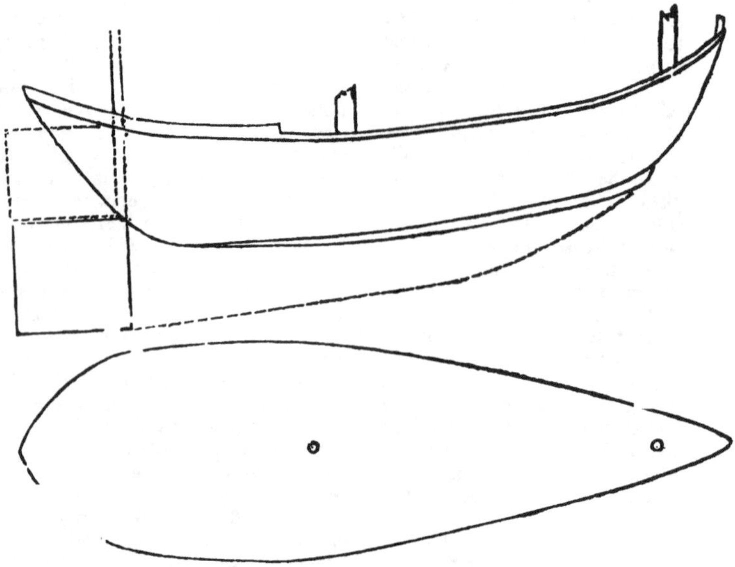

very far aft, and forward is very sharp, though the stem itself, as it rises above the water, is broad and flat. There is a considerable difference between the draught of water forward and aft, especially if we look at it when the rudder is lowered.

The Turkish caique is another example of placing the beam far aft, with a long sharp bow. These are said to be very fast, but are intended only for pulling. To fit this again for sailing, a keel would have to be added, following the shape of this floating part:

Fig. 4.

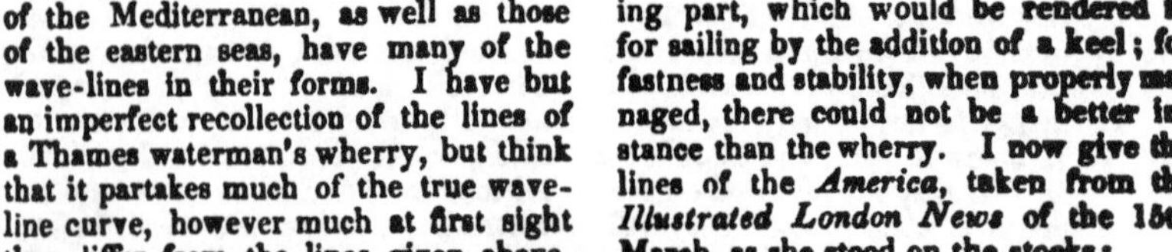

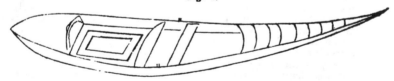

In fact, most of the lateen sail-boats of the Mediterranean, as well as those of the eastern seas, have many of the wave-lines in their forms. I have but an imperfect recollection of the lines of a Thames waterman's wherry, but think that it partakes much of the true wave-line curve, however much at first sight they differ from the lines given above, which are chiefly as examples of placing the beam far aft, and of an efficient floating part, which would be rendered fit for sailing by the addition of a keel; for fastness and stability, when properly managed, there could not be a better instance than the wherry. I now give the lines of the *America*, taken from the *Illustrated London News* of the 15th March, as she stood on the stocks:

Fig. 5.

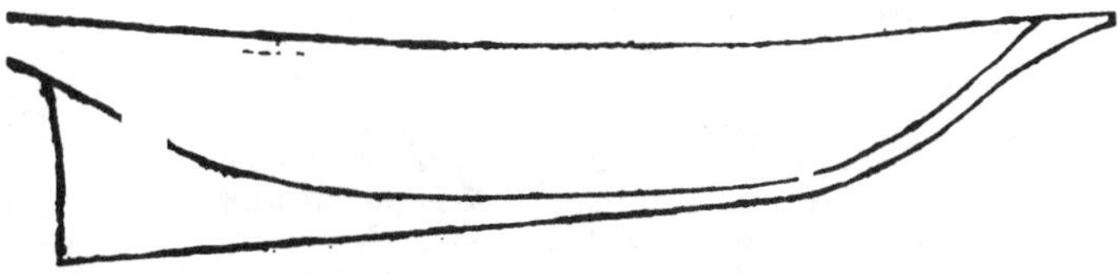

-llowing is an outline of her deck view, from the same paper of the -ust ;

Fig. 6.

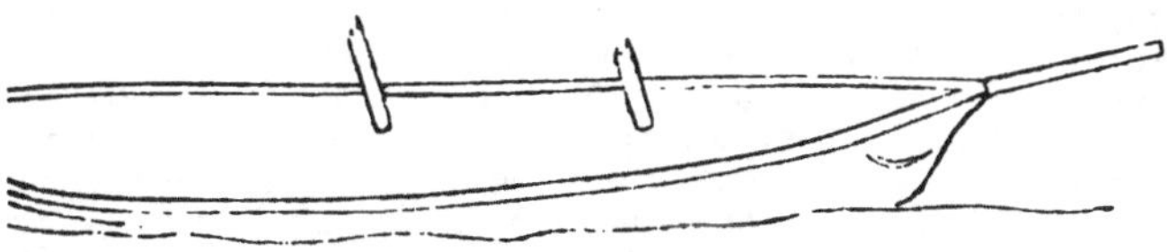

-mensions of the *America* are -

feet length over all.
feet length of keel.
feet 6 inches extreme width.

of water forward, 6 feet; aft, -t is stated: "In lieu of straight have curved and hollow lines." -nsions of the masts: foremast, inches; mainmast, 81 feet; gether with the wish to have -eezes in which to prove her -vince her capability for work. has yet to be added, I will not further illustrations on the identity of the *America* to the -ve. As some apology for an -le person writing so much on -t, I would remind the reader are chiefly notes from other -nd that there seems to be in of a mechanical turn a sort of feeling and appreciation of and fitness for a purpose, the musical ear, and the eye -ing, forms a rule for itself, waiting for the definitions of which slowly lays down rules in what order, these propor- -end upon each other. The has a strong sense of fitness, or of power in any object, part is proportional, and when is properly blended together; and a feeling of almost personal injury, when these are disproportioned. On this some claim is made to have one's say on the subject in hand, particularly when there exists so much difference of opinion among those who ought to decide on the best lines for vessels of all descriptions.

The parts or lines of a vessel may be divided—1st. Into the horizontal water-line, or that, as here chiefly referred to, of least resistance. 2nd. The floating lines, or those on which depend her buoyancy, her stability, and sea worthiness; this includes the lines lengthways, as well as the cross sections. 3rd. That portion which is immersed and gives both power to hold way, and to resist the wind acting on the sails in an opposite direction. It is evident that the divisions 2 and 3 are properly one, as on the blending of them together depends the capacity of the vessel, still by separating them we can more clearly distinguish that if a preponderance in any one is given over the others, there will be to the loss of one, or both the other qualities ensured by attention to the due proportion of each. In river steam vessels, where sails are little used, the floating lines are evidently of the greatest consequence, because being acted upon by the paddles low down, it is not necessary to give so much depth to the immersed part, the same in pulling boats; but in sailing vessels it is different. It appears to be too much the case in yacht building to disregard the natural floating lines, and to remedy this by the artificial one, of heavy ballast; this is unsafe, and certainly, though it may answer for racing purposes, it is a defect in the art of ship-building, as the most perfect form must be one which depends upon her natural, I may call them, floating qualities, brought to better bearings by the ballast or cargo which she is to carry. Greatest stability with least resistance, and I may add, least ballast, is the right question to be solved. On this I would again refer to the "Lectures." Yachts are, I think, built on too even a keel, or rather, with too little difference between the draught of water forward and aft. This is an instance of giving too much to the lower lines, and the consequence is, either, that to get the run aft properly carried out, the lines of the floating part are so much cut away as to injure the bearings of the after part; or, the run is brought to dead wood too low, while the upper part projects in a disproportioned manner: this is often to be noticed in wedge-shaped vessels, and there is a want of harmony in the lines in this manner when ballasted down. This must drag upon the water. There

Fig. 7.

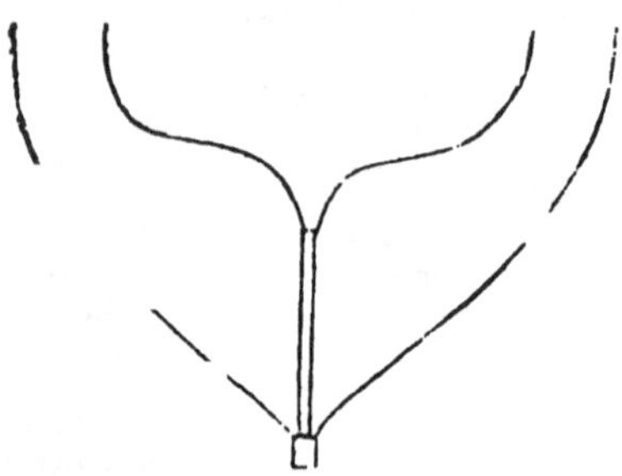

is also a very objectionable length given to the false stern; it proves, however, that there is a feeling of what is an eye-sweet proportion; but the objection is, that the real vessel is not made to partake more of this upper addition, and thus it be made serviceable to her. The stern-post should then be placed at a greater angle to the keel, and a greater buoyancy given to the stern lines. The wave line brings this part particularly well out; and the sketch of the *America* points to this being the case with her. The Chinese have this in the model very full. If the draught forward is considerably less than aft, it enables the builder to bring out the stern water line at the right place; and though contrary to our ideas of form, if the stern had even a slight turn up, it brings them out better. The *America* appears a good example of what Captain Fishbourne gives as a section of the form of greatest stability with fastness; in fact, she is little more than a skeleton vessel, having a floating part of stability, and a lower part, as before indicated, which for sailing purposes need be little more than a huge keel of dead wood; but to combine accommodation with keel, must be the object of our builders. Another reason for the difference of draught forward and aft is, that it brings the greatest depth, like the greatest beam, more gradually to meet the resistance of the water, while the raking stern-post brings the water in a similar way to the water-line, and best clears it from the following water—a matter of consequence, as before stated. I have not met with definite rules for applying the wave-line curve, except what has been already stated in regard to the curves formed on the whole length for the forms of the bow and stern; and that an additional body may be advantageously inserted in the middle, we also found that this insertion might be about one-third of the whole length.

(*To be Continued.*)

THE ELECTRIC LIGHT AT WORSLEY HALL.

The public are already acquainted with the leading features of Messrs. Staite and Petrie's inventions in electric illumination. On Friday evening, the 10th inst., Mr. Staite, by Command, had the honour of displaying before the Queen and the Royal and distinguished visitors at Worsley Hall, three illustrations of unparalleled splendour and power. A Manchester correspondent has favoured us with the following particulars, which we think will be interesting to our readers:

One light, of extraordinary brilliancy,

ed in the flower-garden on the end of the terrace, and facing ing-room windows. The light osed in a Fresnel dioptric lens, g somewhat elevated, must have n for many miles around. The f the surrounding scenery ap- s if bathed in sunshine, and the d terraces were as light as day. er lights, furnished with suitable of great size and power, were eously displayed at the two ends of the front terrace, and ted rays were directed on to the in the centre, which was kept the whole night. At intervals e reflectors was turned about in directions, rendering distant ob- tinctly visible,—some of them way. The Queen was highly and took much interest in the effects produced, particularly at tain, where a bow could be seen about, radiant with the prismatic Certainly never before did such nt light, or rather series of hine on so brilliant a company. nportant to observe that the s *self-sustaining*, and remained, any kind of interference, in action the whole night, until the nd visitors retired.

ompany at the Earl of Ellesmere's ed, in addition to the Queen and Albert, Field-marshal the Duke ngton, Earl and Countess Wilton, ay, Duke and Duchess of Norfolk, s and Marchioness of Westmin- arl and Countess Derby, Lord Sir John Potter, mayor of Man- the Bishop of Manchester, and fty of the gentry of the sur- g counties.

—◆—

'S "FIRST THREE BOOKS OF EUCLID."*

former editions of Euclid by tts, which have been noticed ongly recommended in this Ma- have taken their place amongst the standard educational works at the Universities, and in schools of all classes since their first appearance; however, a totally new description of mathematical students has been created, by the recent army regulations, and requirements of the medical profession. Both officers in the army and candidates for the diploma at Apothecaries' Hall, are required to pass an examination in a certain portion of "Euclid's Elements," and it is to meet the wants of this new class of readers that Mr. Potts has printed the First Three Books of Euclid, separately, at a reduced price; accompanied, however, with notes and copious examples and problems for practice. The solutions and hints, which were found so accept- able to the student in the larger editions are also preserved in this; together with a short notice of the Ancient Geometri- cal Analysis.

We can, therefore, recommend this edition as cordially as its predecessors to those for whom it is specially intended. It is printed at the Cambridge University Press, in the same neat and admirable form which characterised the former editions.

We are sorry to learn that a very mean and unprincipled attempt is about to be made, to rob Mr. Potts of the fruits of his labours, by what can be looked upon as nothing better than a pirated copy of this edition. We have reason to believe that Mr. P. risked a considerable sum in bringing out his former editions; and that his motives were of the most disinterested nature; viz., an ardent wish for the advancement of mathematical education—much more than any regard to private advantage or profit. When, after some time, Mr. P.'s book became firmly established, and likely to become a source of pecuniary profit, other members of his own Uni- versity were shabby and dishonest enough to step in with an edition of their own, which was little better than a mere piracy of his.

We sincerely hope, however, that these disgraceful attempts will not suc- ceed in their aim, and that Mr. Potts will at last be permitted to reap some small advantage from his zealous un- tiring efforts to promote the cause of sound education, both at the University and elsewhere.

* lid's Elements of Geometry, the First oks; chiefly from the Text of Dr. Ermin, anatory Notes; a Series of Questions on t; and a Selection of Geometrical Exer- n the Senate-house and College Examina- s, with Hints, &c. By Robert Potts, M.A.; llege. John W. Parker, West Strand.

SPECIFICATIONS OF ENGLISH PATENTS ENROLLED DURING THE WEEK ENDING OCTOBER 15, 1851.

WILLIAM BARKER, of Hulme, Manchester, millwright. *For improvements in machinery for chipping, rasping, and shaving dyewood and other materials, and in apparatus connected therewith.* Patent dated April 7, 1851.

The improvements claimed under this patent comprehend:

1. The application of knives for chipping and shaving dyewood to the same face-plate as the cutters used for rasping.

2. An improved arrangement of machinery for bringing the dyewood or other material in contact with the face-plate to which the knives or cutters for chipping, rasping, and shaving are attached.

3. Certain apparatus for grinding the knives or cutters used in machinery for chipping, rasping, and shaving dyewood. The chief peculiarity of this apparatus is, that the knives whilst under the action of the grindstone have a sideway movement alternately in opposite directions, by which means the sharpening is more rapid, and the wear of the grindstone equalized.

CHRISTOPHER CROSS, of Farnworth, cotton spinner and manufacturer. *For certain improvements in the manufacture of textile fabrics, and in the manufacture of wearing apparel from textile materials.* Patent dated April 8, 1851.

The present improvements have relation to the weaving of cloth of irregular forms, suitable for being made up into various articles of wearing apparel. Several methods of effecting this object have at various times been proposed, and one of these formed the subject of a patent granted to M. Jerome Andre Drieu, Aug. 1, 1849 (see vol. lii., p. 117). M. Drieu's method was to produce the fulness or irregularity in the cloth from an independent weft put in by hand, after the full width of cloth had been woven by power up to the point where the fulness was to commence; the hand-woven or irregular part was then brought parallel to the breast beam by a suitable take-up motion, and the weaving recommenced by power until another portion had to be let in. Mr. Cross now proposes to employ instead of the above-stated method a Jacquard or other suitable apparatus by which a determinate number of the warp threads, according to the extent of gore or inlet to be produced, will be prevented from opening a shed for the shool of weft, and thus the required irregularity be produced in the fabric. According to this mode of procedure, the shuttle can be thrown from box to box across the whole width of the loom, although a shed is opened and cloth will consequently be produced only in that part of the weft which is not under the action of the Jacquard. The take-up motion may be similar to that of M. Drieu, or strips of any material corresponding to the extent of fulness may be introduced between the folds of cloth on the take-up roller, or any other suitable mechanism capable of like service may be employed.

Claim.—The application of a Jacquard or other suitable apparatus for preventing certain of the warp threads from opening a shed when shapes for being made up into articles of wearing apparel are produced by the weaving of irregular quantities of work in the direction of the length of the warp thread.

JOHN GEORGE APPOLD, of Finsbury-square, gentleman. *For improvements in machinery for regulating and ascertaining the labour performed by manual or other power.* Patent dated April 9, 1851.

Mr. Appold's improvements have relation to that class of machines in which the friction of a band or strap, encircling a revolving shaft, is caused to indicate in connection with suitable apparatus the amount of power expended in producing rotation of that shaft. The object sought to be obtained is to equalize the amount of power required to be exerted under varying friction, and this is effected by the application of a compensating lever to the strap or friction break, by which the tightness of the band or break upon the revolving shaft is increased or diminished in a ratio corresponding to the increase or diminution of friction.

CHARLES MCDOWALL, of Hind-street, Bloomsbury, watch-maker. *For certain improvements in the construction of time-keepers.* Patent dated April 10, 1851.

These improvements have for their object the simplification of the construction of that part of time-keepers of various descriptions known as the "escapement," and this is effected by the employment in lieu of the ordinary "scape wheel" of an eccentric pin or one toothed wheel (driven by the ordinary train) which is made to act on a peculiar construction of pallet. The operation of the whole is that of a dead-beat escapement, and the peculiar feature of the improvements is that an impulse is given to the pendulum at the middle of its vibration by a direct instead of an oblique action.

HENRY JOHN BETJEMANN, of Upper Ashby-street, Northampton-square. *For* ments in connecting parts of bed- nd other frames, and in machinery d therein. Patent dated April 15,

mprovements here claimed compre-

e construction and application of a dovetail, or its equivalent, for the of fastening or securing together es of bedsteads or other structures, metal, wood, or any other mate- sides of which dovetail are, and act ned planes, they being so formed as t at an angle from the face of the o that when the male dovetail is , and driven down or into the reced- etail by a blow or pressure, it is traverse down the inclined planes ceding dovetail, whereby it is drawn home, until the shoulder of the rail h is or are formed the male dovetail tails, is closely brought up against of the post.

n arrangement of machinery for the dovetailed mortise-holes in the bedsteads and other structures, as the employment of rotatory cutters ical heads and cylindrical necks, in tion with a rest or moveable table eception and attachment of the bed- said table, while being advanced the cutters, being conducted by guides either upon the moving table onary bench, in a course which is at ight angles to the face of the post, ce, as soon as the cylindrical cutters gun to act, in a longitudinal course, sufficiently from the face of the form a mortise which shall bind the d tenons of the rail as they are down in their sockets.

e combination with the guide of stops, or their equivalents, whereby is limited in its course to the par- ange of cutting action for the time quired.

IAN SCHRODER, of Bristol, gentle- For improvements in manufacturing ing sugar. Patent dated April 15,

invention has relation to apparatus orating and granulating cane-juice ds containing saccharine matters of crystallization, and consists in ng with a pan or vessel, heated by other heated fluid contained in or through a pipe or series of pipes ges inserted in or through the pan employed, a series of discs or sur- able of moving or revolving, partly d in the liquid to be evaporated or ed. The operation is performed by bringing the heated particles of cane-juice or saccharine liquid into contact with air by the revolution of the discs, whereby evaporation to a great extent is carried on at a temperature below the ordinary boiling point of the liquid. The patentee prefers the pan or vessel to be of the sectional form of a segment of a cylinder, with the discs (which are of a circular form) mounted on a horizontal axis above the pan, and revolving partially immersed in the liquid, and with the heating pipes arranged in tiers across the pan; and occupying a position intermediate to the discs; he does not, however, confine himself to that particular form of vessel, nor to any particular shaped pipe or arrangement of pipes or passages for conveying heat to the liquid. The temperature is regulated by cocks, or other suitable means.

Claim.—The combination of moving or revolving discs or surfaces with pans or vessels, heated by pipes, or tubes, or passages containing steam or other heated fluid.

ROBERT MILLIGAN, of Harden-mills, Bingley, manufacturer. *For a new mode of ornamenting certain cloth fabrics.* Patent dated April 26, 1851.

This "new mode of ornamenting," consists in printing designs or patterns of any description in one or more shades or tints on or over flock-printed coloured cloths composed, or partly composed of alpaca, wool, or worsted, or wool and cotton mixed (such as are commonly called "Coburgs,") the paints or printing materials employed being white or coloured zinc paints mixed with oil in the usual manner.

Claims.—1. The use of zinc paint for printing upon fabrics such as described.

2. Printing with zinc paints or with ordinary oil paints, or with opaque pigments mixed with any suitable oil, varnish, or gum insoluble in water, or upon or over flock, printed coloured cloths, or coloured fabrics composed of the materials above mentioned.

3. Printing on those parts of such cloths or fabrics which are not covered by the flock with such paints or colours, as described.

WEEKLY LIST OF NEW ENGLISH PATENTS.

Robert James Maryon, of York-road, Surrey, gentleman, for improvements in obtaining and applying motive power, and in signalizing. October 10; six months.

Richard Archibald Brooman, of the firm of J. C. Robertson and Co., of 166, Fleet-street, in the City of London, patent agents, for certain improvements in the preparation and treatment of fibrous and membraneous materials, both in the raw and manufactured state, whereby they are rendered more durable, are contracted or expanded, are cleaned, and are more capable of resisting decomposition and of receiving and retaining colours. (Being a communication.) October 10; six months.

Hubert Sommelet, of Paris, in the republic of France, manufacturer, for certain improvements in the manufacture of scissors. October 10; six months.

Thomas Lightfoot, of Jarrow paper-mills, South Shields, Durham, paper manufacturer, for improvements in machinery applicable to the manufacture of paper. October 16; six months.

Thomas Henry Fromings, of the firm of Lomas, Fromings, and Co., of Sheffield, York, manufacturers, for improvements in forge hammers. October 16; six months.

Matthew Gibson, of Wellington-terrace, Newcastle-upon-Tyne, for improvements in machinery for pulverising and preparing land. October 16; six months.

William Onions, of Southwark, Surrey, engineer, for improvements in the manufacture of nuts and bolts, also of steps, bearings, axles, and bushes, also of mills and dies for engravers, also of bells, lathe, and other spindles, also of weft forks, shuttle tongues and lips for looms, also parts of agricultural implements, chains, roller guides, and throstle bars, by the application of materials not hitherto used for such purposes. Oct. 16; 6 months.

Thomas Perry, of Tower-street, Leicester, machinist, for improvements in the manufacture of looped fabrics. October 16; six months.

Richard Dover, of New-street, Spring-gardens, Westminster, merchant, for improvements in treating sewage or obtaining products therefrom, and combining such products with other matters. October 16; six months.

WEEKLY LIST OF DESIGNS FOR ARTICLES OF UTILITY REGISTERED.

Date of Registration.	No. in the Register.	Proprietors' Names.	Addresses.	Subjects of Design.
Oct. 9	2975	John Whitehead	Midland Junction Foundry, Leeds	Faller.
10	2976	John Tylor & Sons	Newgate-street	Moderator lamp.
,,	2977	John J. Peile	Whitehaven	Screw-jack.
,,	2978	Miller & Sons	Piccadilly	Parts of a signal-lamp for railways.
13	2979	John Chesterman	Sheffield	Double expanding and contracting spanner.
14	2980	Chadburn, Brothers	Sheffield	Barometer tube.
,,	2981	Edwin Rose	Manchester	Double-acting safety-valve.
15	2982	John Symonds	(Circus), Minories	Gold-washing machine.
,,	2983	Cartwright & Hirons	Birmingham	Cruet-stand.
16	2984	Henry Batchelor	Kennington	Candle shield.

CONTENTS OF THIS NUMBER.

LONDON: Edited, Printed, and Published by Joseph Clinton Robertson, of No. 166, Fleet-street, in the City of London— Sold by A. and W. Galignani, Rue Vivienne, Paris; Machin and Co., Dublin; W. C. Campbell and Co., Hamburg.

Mechanics' Magazine,

[MU]SEUM, REGISTER, JOURNAL, AND GAZETTE.

[No. 1]472.] SATURDAY, OCTOBER 25, 1851. [Price 3*d*., Stamped, 4*d*.

Edited by J. C. Robertson, 166, Fleet-street.

[...]UTANT'S PATENT WHEEL TYRE AND RAIL CEMENTING KILNS.

Fig. 1.

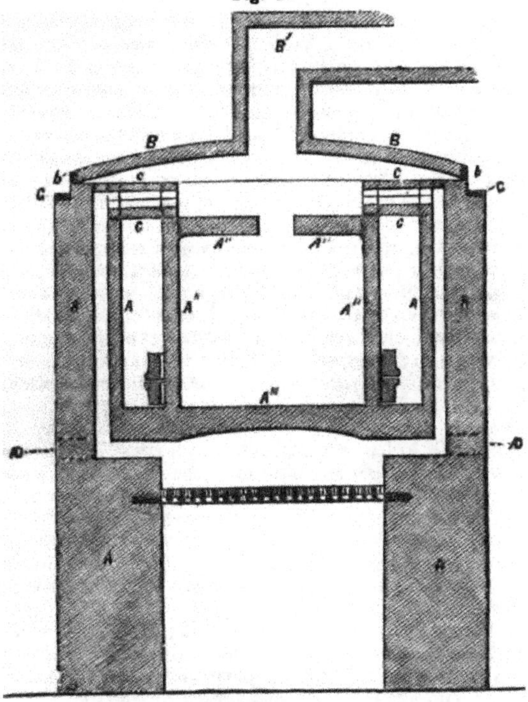

Fig. 2.

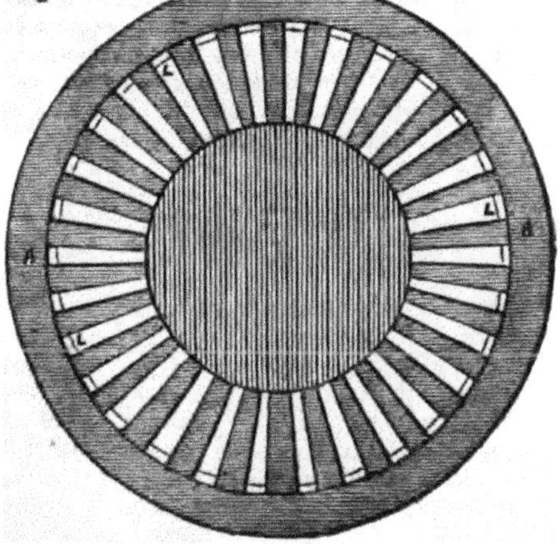

COUTANT'S PATENT WHEEL TYRE AND RAIL CEMENTING KILNS.

(Patentee, Antoine Victor Coutant, of Paris. Patent dated April 15, 1851. Specification Enr October 15, 1851.)

Specification.

My invention has for its object to remedy the serious loss and inconvenienc constantly arising from the rapid and unequal wear of the rails and switches o ways, the tyres of railway-carriage wheels, and of other iron surfaces exp much friction, and consists in an improved apparatus for partially hardenin surfaces, that is to say, hardening them on those sides only which are exp the friction. The usual process of hardening iron surfaces is by cementatio that is a process which has been hitherto applied to articles of small dim only, from the want of an apparatus of sufficient capacity and provided with ad appliances for the treatment of pieces of such magnitude as railway wheel-tyr In the process of cementation too, as followed even on the smallest scale, it is necessary to protect the parts not required to be hardened by cementati coatings of clay or earth. Now, by the said improved apparatus of my inv articles of any magnitude, or weight, or shape, may equally with the small hardened by cementation, and the cementation may be confined to the part required to be cemented, without the use of any coating of clay or earth. apparatus consists of four principal parts, or distinctive features—

First. A circular chamber, within which the articles to be hardened or cem are placed. This chamber is set over a hearth, the flame, smoke, and othe ducts of combustion from which circulate round the exterior of the chamber.

Second. A circular kiln from near to the grate of which a number of flues r outwards, by which the flame, smoke, &c., pass, in order to circulate rou chamber of cementation.

Third. A moveable cover or dome, which serves to concentrate the heat cementing chamber, and which is surmounted by a crane for lowering the a to be cemented, and lifting them out again after the process has been accomp

And, *Fourth.* A peculiar mechanical arrangement of the parts, wherel cementation of the articles is confined to any particular part thereof desired, out the use of clay, or earth, to protect the other parts, and whereby, als depth of the vertical or horizontal bed of charcoal in the cementing chamb be augmented or lessened at pleasure, and admitted into contact with the pa which is to be cemented.

Fig. 1 of the engravings is a sectional elevation of this apparatus adapt wheel tyres, and fig. 2, a horizontal section on the line AD, in fig. 1. Figs. 3 relate exclusively to the kiln for rails and bars, of which fig. 3 is a vertical through the middle, and fig. 4, a horizontal section on the line AD, of fig. 3. the exterior wall of the apparatus, which is here supposed to be of a circular but may be either circular or square, or rectangular, according to the shape objects to be cemented. It may also be of any size suitable for the articles in to be cemented. A′ A″, fig. 1, are two inner walls, which form, with the one, three spaces, or compartments, through the outermost of which the smoke, and other products of combustion circulate, in the second of which, the c tation is effected, and the third, or central one of which is left empty. A‴ vault over the hearth, on each side of which is the sole AA (circular, squa rectangular, as the case may be), on which are laid the rails, wheel-tyres, sw or other articles that have been introduced in the space reserved for the pro cementation. In the wall A there are four openings, or eye holes, for the p of observing the progress of the cementation, and seeing that the flame is circ equally all round. These openings may also be made use of for the introduc a smaller or greater quantity of air by lessening the openings, or making wider. In A′ is the mouth, through which the fuel is introduced on the B is the moveable dome or cover. B′, the chimney, which may be made i parts, one attached to the dome, and the other fixed at the outside of it, an ported by stays. The dome is made moveable, in order to permit of the int tion into the interior of the pieces intended for cementation. It is hooped at n ring *b*, and rests on friction-rollers, which move on rails, CCC, there s rail on each side, and a third in the centre, and supported at the outside rickwork by props or stays, bound to each other by means of cross-pieces. aid of these rollers and rails the dome is easily taken off, to throw open rmost orifice of the apparatus, through which the articles to be cemented ed and lifted out either by the assistance of a crane or some other suitable : *c* are hollow bricks, or moveable vent-holes. The dome being taken off, ratus there remains only to introduce in the space kept free for that pur- articles to be cemented; say for example, wagon-wheel tires. The number s superimposed one upon another must be according to the height of the ı. As for the introduction of the charcoal necessary for the cementation, e made through the centre of the apparatus if there is not a second inner ted beforehand. But if the contrary is the case, then it may be introduced

Fig. 3.

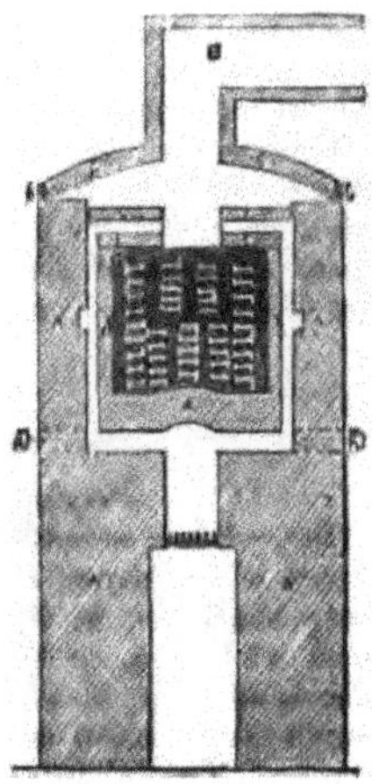

Fig. 4.

Fig. 5.

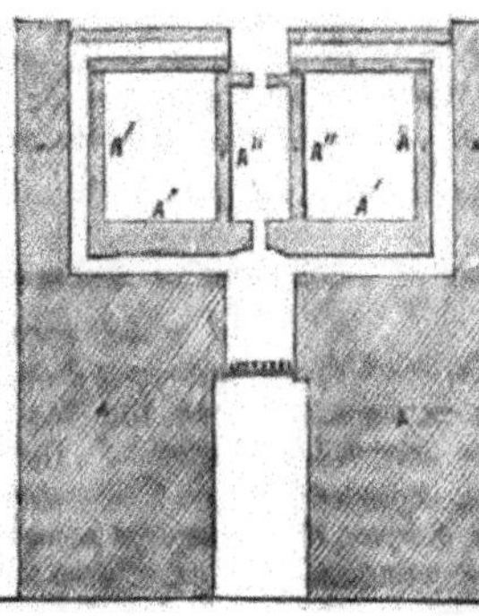

the top of the apparatus by means of a rammer acting between the wall wheel tyre, or other article to be cemented, a sufficient space being kept that purpose. Although the walls near to which the parts not to be l are laid, are sufficient, and although the superposition of the two pieces by not to be cemented, be a warranty against the cementation, it will be more to cover these parts with common clay, in order to be more certain of the n of any carbon. The apparatus admits, of course, of many more or less modifications without involving any material departure from the general or system on which the same is constructed. For instance, the kiln for switches may be made double, as exemplified in fig. 5. Or the inner wall, f being made up like the first one, I mean at the same time with the rest pparatus, may be constructed after the introduction of the articles to be l, and may also be composed of common clay or earth, instead of bricks, red with a bed of earth on either face, or on both faces. This second wall, y be more or less removed from the first inner wall, according to the or the thickness of the article to be cemented. Again; the openings or s made in the outer wall of the apparatus may be multiplied in num- l even their number may be increased to that of the flues, LL, which ram near the centre of the hearth. And the openings once made may up with moveable bricks, iron trap-doors, or any other means proper st or diminish the action inside. The equal distribution of the heat, and regulation of its intensity might also be made easy by augmenting in numl lower openings, by which fuel is supplied to the hearth. Moreover, in c render more certain in some cases the exclusion of the action of the carbon parts that are not to be cemented, a current of cold air might be made through the interior of the apparatus and introduced in the void and central the apparatus. It would act on the surface of the inward wall, to keep it coo same means could be applied, if need were, to any hollow and horizontal pa which might be erected in order to separate partially the wheel tyres or other from one another. Farther, the central or vacant part of the apparatus mi made use of for cementing articles of a smaller size by arranging that part, the caloric should come into it, either vertically or horizontally. Or it mi wholly shut up at the top, in order to prevent any communication between it space where the flame, &c., circulates.

At the circumference, and at several heights, openings might also be m receive trial pieces suitably enveloped with charcoal, or such articles as matri moulds, wedges, &c., that require to be but partially cemented. The inner or walls of the apparatus may be made of earth according to the nature operation, that is to say, according to the shape of the pieces to be cement also according to the parts desired to be cemented, and no other. Any re degree of cementation through my apparatus is effected in a more certai efficacious way than by any other mode known to me. Finally, it may be ob that in the kilns for rails, switches, and such like articles, the cementation made on one of the projecting parts alone, or on both, according to the cove earth or common clay round the parts that need not be cemented. The op are of course filled up with earth as fast as the matters are placed in the kiln

THE BOARD OF ORDNANCE IN OTHER DAYS.

Prefatory.

Sir,—Some recent proceedings of the Board of Ordnance, of a very extraordinary character, in respect to scientific men who hold, or have held, appointments under it, have been the subject of much conversation and severe condemnation. The following extract from the "Diary" of Reuben Burrow will prove that the meanest conduct was not more incompatible with the ducal coronet and an office of high power and trust at that period, than in recent times.

To the mathematicians of the English school, the name of REUBEN BURROW is as a household word; and as a geometer, there exists no question that he was only second to Dr. Matthew Stewart of all his contemporaries, Wildbore himself not excepted. Of his early life little is known beyond that he was of humble but reputable extraction, and a native of the neighbourhood of Leeds—Horsforth, I believe. The earliest scientific employment in which he is known to have been engaged, was that of Assistant at the Royal Observatory, under Dr. Maskelyne; and to him was entrusted the entire superintendence of the celebrated observations at Schehallien, of which survey a very curious i teresting diary is still in existence was the first to contest the ex right of the Stationers' Comp print almanacks; and the "D (known sometimes as "Carnan's I from Carnan being the trade p of them) are now very scarce tremely valuable. This, howe to a rupture with Dr. Hutton, th of the legitimate *Lady's Dia* general editor to the Stationers pany. Burrow's "Restoration Treatise of Apollonius on Incli is a most elegant and able wo furnishes a remarkable contrast jejune attempt of Bishop Horsle writings in the English periodi extremely numerous, and some dissertations by him appeared in volumes of the *Asiatic Res* His measurement of a degree meridian in India was drawn his papers by his friend Isaac (Professor of Mathematics in th Military College, Sandhurst),– sedly incomplete, however, fr loss of certain parts of the pape Burrow's death in India.

Reuben Burrow was a man c

ding, and more conversant
fic history than is usually the
nen of his class. He had a
miliar acquaintance with seve-
ges; but he only considered
dge valuable as furnishing a
scientific works locked up in
l languages. With the lite-
is own country he appears to
well acquainted, as is evinced
f the entries in his diaries.
e time, it is not pretended
a person of cultivated taste,
lly of that taste which shows
tention to terms of conven-
ment when stronger, harsher,
phatic ones existed in his own
Under provocation, his lan-
harsh; but the accompanying
will show to what unbearable
he was subjected.
ristic of his mode of studying
, and of his "killing two
ne stone," it may be men-
when in India, as Chief Sur-
he East India Company, at
of age, he studied the Sans-
hat the work he used was a
of the Beejgunneet; a trans-
ich, interlined in pencil, was
after his death to Professor
hat MS. is still preserved.
man's public services were
by the Board of Ordnance as
£100 a year, or not more
ird of that of a third-class
Ordnance-office at the same
le never received a sixpence
eys.

SENEX.

Diary.

of April, 1782, I left the Draw-
the Tower, in order to go to the
to Colonel Watson. Some
I left the Drawing-room I called
wnshend, and told him of my
He said he was sorry for it; not
much for yourself as on account
wing-room, because I'm told
are so much superior to the
of the world, and that you are
ommonly qualified for it, &c.,
ised me to call on the Duke of
nd tell him my reasons why I
ve the Tower, "not that you
e it when you please, but be-
piece of respect that ought to
Grace," says he. Accordingly
he Duke, and sent up the following note, viz. :—"The Mathematical Master of the Drawing-room in the Tower begs leave to speak with his Grace the Duke of Richmond about resigning his place." On this I was admitted, and he asked my reasons for wishing to quit it. I told him that I had the honour of being put into the Drawing-room by Lord Townshend, and, I believed, contrary to the desire of Sir Charles Frederick; and whether that or something else was the cause, I could not tell, but the fact was, that I was not a favourite of that gentleman,—that I since had incurred his displeasure by endeavouring to get one of his favourites dismissed who had been guilty of theft and perjury, and almost every kind of villany;—that Sir Charles F. was not a person to forget or forgive affronts of that kind, and that I supposed he would now have sufficient interest to be revenged by getting me turned out of my place, or at least he might give me more trouble about it than I should choose to put up with, and therefore I should rather choose to resign;—that the place was likewise disagreeable on some other accounts, particularly the dirty behaviour of the chief draughtsman; and that I had an increasing family, and 100*l*. a year was not sufficient for a man to save anything by, &c., &c. In answer to this he only made a number of promises—that if I would stay, neither Sir Charles Frederick nor anybody else should injure me, and that I might depend upon my place, and such like, &c. I told him it was impossible for him to do it, considering the number of enemies I had, and the pains they would take to injure me, and that I did not choose to give myself the trouble of counteracting every spiteful plot that might be raised against me;—that I was obliged to him for his favour, and should be glad if he would transfer it to a person in the Drawing-room who had been very ill-used by the chief draughtsman, as well as myself, viz., Mr. Gilder; but that the only favour he could do for me, was to accept my resignation. He repeated again his desire that I would stay, and talked of the length and dangers of a voyage to India, &c., but not a word of increasing my salary; and at last he gave me leave to resign. I then made several attempts to get my money that I had laid out in making the survey at Woolwich, as I had done for nearly half a year before, but to no purpose; and about a fortnight after, I wrote a long letter to the Duke, giving my sentiments on the state of the Drawing-room, and the means of improving it, which I proposed by superannuating the chief draughtsman and his deputy, and filling up their places with Mr. Gilder and Mr. Gould; I also recommended

Mr. Dalby as a proper person to fill up my place, and mentioned the 30th of April, 1782, as the day I should give up my office. Accordingly, I gave it up on that day; but as there were several of the mathematical instruments belonging to the Drawing-room at my lodgings in Throgmorton-street, I applied to Haines for some of the draughtsmen to fetch them, but he refused it.

Some time after, when I found I could not get the 7*l.* and odd money that I had disbursed at Woolwich, I went to Boddington, and Vidgen, and Parish about it, as I had done many times before; but as they only laughed at me, and behaved with impertinence (Parish, at least), I wrote a note to the Board, and carried it myself. After a little while—perhaps ten minutes—I was sent for, and the Duke behaved like a mean, dirty fellow, and told me he would advise me for the future not "*to abuse the officers*," and seemed angry, and looked as black as could be, as if somebody had been setting him very much against me. On this I looked at him with all the insolence and blackness I could assume, and told him that my behaviour was very proper for theirs, and that I did not choose to put up with impertinence from anybody. "It is best let alone," said he. I was going to reply, but he directly began to ask about the account; so I pointed out an abstract of the business, and said—"That's my writing; if you read that, you'll know the whole of the business at once." He then read it out, and looked over the bill, and found that in one place there were four dinners charged, when there were only two allowed. He seemed angry at this, and objected. I said—"There were two of my acquaintances helped me to survey, without wages; and as the Board had the benefit of their work, I thought the least I could do was to give them their dinner." At this he growled, but seemed to say nothing. He then asked me over and over whether that 7*l.* &c. was all that was *due* to me?—and as he seemed to lay a particular stress upon "*due*," I thought he wished to exclude me from any expectations upon the money I had before petitioned for as extra payment for making the surveys. I therefore answered him that my bill was nothing but money laid out of my own pocket. He repeated the question twice, and I as often answered him in the same words, or words to the same effect; only the last time I spoke rather resentfully, as I considered his repetition as an injury. He then told them to pay me the money; but they hesitated, and made pretence that it could not be done without an order; but he directly insisted it should be paid, and told them to call Lauzun to pay it. On this, the scoundrel Parish sneakingly
me, as I stood by the Duke. "*W*
call him, Mr. Burrow?" I answ
"Call him yourself; I shan't call
assure you." The sneaking rasc
said—"*I'm sure it was not my fau*
you was not paid; I did all I coul
told him, in a contemptuous tone,
was his fault. I then got the mon
wrote a receipt of half a page, settin
that what I had there received was o
own money, and that I had not rec
single farthing for my trouble, &c.

After this, I set off to Portsmo
believe it was on the 1st or 2nd o
1782: and after I had been there
while, at a great expense, and spent
all my money, the ships delayed to s
I went up to town, and wrote the fol
letter to the Duke of Richmond:

To His Grace the Duke of Richmo
Master-General of the Ordnan

May it please your Grace,

In 1777, I was ordered t
several of the gentlemen of the Dr
room and Woolwich cadets, and to
complete survey of the sea-coast fr
Naze in Essex to Hollesly Bay in
including the three large rivers, Sto
well, and Deben, up to Manningtre
wich, and Woodbridge, together w
islands, sands, and soundings in and
Handford Bay, Harwich Harbour, &
there had never been a plan of th
before but what was excessively err
I did the whole with great care and
ness, and plans were delivered to th
and Lord Townshend; but I never r
a farthing for my trouble.

I was likewise ordered, last year, t
a survey of Woolwich Warren; and i
it, I was obliged to lay out about 7
my own pocket. This 7*l.*, with gre
culty, and after a long and ineffectual
cation, by your Grace's interposition,
returned; but I never received a
farthing for doing the business itself.

Now if everybody else were to be
with their bare salaries, I do not see
should not be the same; but the *fact*
all other officers belonging to the Or
in the Tower have always an add
allowance for doing extra duty; a
claim to such allowance is certainly
(at least) as theirs. I had petition
Board for this allowance before your
came into the Ordnance, but have re
believe that my memorial was nev
before the Board. I have stayed so
expectation of the sailing of the Indi
that I am almost without money, and
fore shall be glad if your Grace will

I may expect this allowance or

our Grace's most
Humble Servant,
REUBEN BURROW.

r to this, I received the follow-

Office of Ordnance,
2nd August, 1782.

e Master - General and Board
idered your memorial requesting
e for your services in making a
e Essex and Sussex coasts in the
and also of Woolwich Warren
atter account you had with diffi-
paid your disbursements of 7*l.*),
heir commands to acquaint you
ll inquire into your pretensions;
s a long time since the business
ied, they wonder when you re-
l. you did not then make your
e above account, and that they
will produce the order by which
e survey.

I am, Sir,
ur most humble Servant,
JOHN BODDINGTON.

ben Burrow.

r to this I directly wrote the fol-
r, and gave it to John Harrison
e Duke of Richmond; that is,
t his house, but as I went into
and the Duke was out of town,
l giving it till the 27th August;
ed him to date it, and he dated
This neglect I had no notion
l been some time at the Mother-
e had informed me in a letter.

ght Honourable and Honour-
ble Board of Ordnance.

ds and Gentlemen,—In conse-
my former letter to his Grace the
ichmond, I have received a letter
Boddington, and find from the
it, that the Honourable Board
to consider the substance of my
s nothing more than "PRETEN-
nd that they "WONDER" I did
ny claim at the time I received
ements, and likewise require that
roduce the ORDER by which I
urveys.

the application of so peculiar a
retensions" is meant to be made
eys themselves: to the *manner*
ing them; or to my *claim* on
at is not for me to determine.
ence of the surveys is questioned,
ard are pleased to be of opinion
: to extort money from them by
nsions of *imaginary surveys*, I
efer to a *hundred* people who
know the contrary; or to *one* as good as a
hundred, Captain Congreve, of the Artillery,
who saw me deliver the general plan of the
first survey to Lord Townshend, and was
pleased to think that the Board ought to
make me a handsome recompense for my
trouble. If it be the *mode of performance*
that is intended to be censured, I think the
above epithet might have been spared; as
the method I took was not only the easiest,
but also the most exact and expeditious;
which I am ready to prove to any that are
of a contrary opinion, and have knowledge
enough to know when they are confuted.
As to the *claim* itself, whether it be considered as a matter of *right* or a matter
of *favour*, I hope the Board will be of
opinion that *mine* is equal to that of any
other person; especially if similar cases
require similar treatment; and if the chief
draughtsman, for instance, is allowed every
time he walks up to the Board, to make his
own charge for coach-hire exclusive of his
salary; or if the porter of the Drawing-
room (to go no further at present) is to be
paid eight or nine shillings a day besides his
usual wages every time he carries a letter
as far as Blackheath, I see no reason why
the mathematical master of the Drawing-
room should work day and night, Sundays
and all, for six months together on a business of real importance and public utility,
and yet be allowed nothing; when at the
same time that wretched compiler of other
men's productions, the mathematical master
at Woolwich, is paid with profusion for his
extra services; though he has more than
double the salary that I had, and his scholars never made half the improvement that
those of the Drawing-room did in the same
interval.*

* The Royal Military Academy at Woolwich, now more than a century old, was not
originally established for the education of
the gentlemen-cadets for officering the Royal
Artillery; but for that of a then subordinate body of persons—the officers—(who
rather belonged to the non-commissioned
than the commissioned class, in the modern
sense of the word)—of the corps of "Royal
Artificers." A school for the Artillery
officers had been established in the Tower
as far back as the reign of Charles the Second by Sir Jonas Moore, the then surveyor
of the Ordnance; and he wrote (*or caused
to be written under his name*) a course of
mathematics for the use of this school.
The school in the Woolwich Arsenal was,

Had the Honourable Board either read the memorial I delivered in on the day I received my disbursements, or the receipt I gave at the same time for the money, it might have prevented their "WONDER;" as both very plainly imply that I had already made application to the Board for some allowance, but had not received any. I had petitioned the Board for it as a matter of *favour* before that time; but as no notice was then taken of it, I concluded (perhaps too precipitately) that they did not choose to allow me any more than what I could *legally* claim as my bare salary; and it was on this supposition that when his Grace the Duke of Richmond repeatedly asked me at the Board whether the trifle of 7*l.*, &c., in my bill was "*all that was* DUE" to me; I as repeatedly answered that my bill contained nothing but the *money laid out of my own pocket* in *making the survey at Woolwich*. The whole Board must remember the circumstances. His Grace was displeased, and found fault with me for speaking contemptuously of two impertinent clerks, and called it "abusing the officers," and likewise seemed dissatisfied about "two dinners" that had been given to two men that had assisted in the survey without wages; which "dinners" I had charged to the Board in my bill. Now to solicit his Grace for favours at the moment I had the misfortune to incur his displeasure, or to think of claiming as my due what I could only expect as a gratuity, I could not but consider as equally imprudent; especially after such an instance of his Grace's economy as that of the two dinners; and therefore I gave the answer aforesaid, that I might resume my claim hereafter if I thought proper. This will account for my silence at that time, and my former letter contains the reasons for breaking it at present.

As to the "orders," that for the survey at Woolwich is already in the possession of the Board; it was sent with the bill and the vouchers, and I saw it there myself when the money was paid. The survey of the coasts of Essex and Suffolk was done without any written order by Lord Townshend's verbal commands, which were given me by his Lordship himself at the time he was down at Landguard Fort. The order for my going down there to attend the experiments I received from Mr. Boddington in writing, and though I have it not by me at present, I suppose a copy of it must be in the order-books of the office, which I presume will be sufficient. I have been thus particular, because Mr. Boddington's letter seems to throw a suspicion on my character which I am not conscious of deserving, and I hope the Honourable Board will on that account excuse the length of this letter.

I am the Board's most Faithful and
most Obedient Humble Servant,
REUBEN BURROW.

I sent the following letter to the Board the 9th of September:

To the Right Honourable and Honourable Board of Ordnance.

May it please, &c.,

I beg leave to inform the Honourable Board that there are some instruments belonging to the Drawing-room at one Mr. Breach's, a peruke-maker, at No. 18, Throgmorton-street, where I had a room when I was in town; they were left there for the convenience of the gentlemen of the Drawing-room when they went to survey; and when I gave up my office, I applied to Mr. Haines, the chief draughtsman, to give

---

however, by degrees converted to the use of the cadets; and the mathematical school in the Tower was merged into the "Drawing School" of the same fortress, as a subordinate part of the general scheme. The whole of the Tower-part of the establishment was then designated as "the Drawing-room in the Tower;" but whilst the Woolwich Academy was gradually absorbing the cadets, and eliminating the non-commissioned officers, the "Drawing-room" was the school in which the aristocratic portion of the cadets was still privileged to acquire their modicum of preparation for a commission. By degrees, however, the importance of the services of the subordinate class of men, the "Royal Artificers," became better understood; and they were constituted into a distinct regiment under the Board of Ordnance, their designation being now the "Sappers and Miners," and their officers bearing the name of the "Royal Engineers." These officers are chosen, as a *general* rule, from the cadets who have most distinguished themselves in the Royal Military Academy; and are classed *par excellence* as the scientific branch of the military service. What their scientific acquirements and requirements are, may be a matter of future remark.

The passage in the *Diary* relates to the period when there was strong hostility between the Tower and the Woolwich schools.

f the draughtsmen leave to fetch ray as was usual; this Mr. Haines proper to refuse; and as I had time nor inclination then to make a porter, I find the instruments are re still, and I am liable on that to pay rent for the place.

rmed the Drawing-room in general se instruments were deposited before town; and I cannot, from Mr. s well-known disposition, impute it ut a dirty motive that he refused to n be sent for. I therefore desire Board will be pleased to order them soon as possible; for though Mr. nay choose to have his *examinations* nem about stealing legs of beef, I do to be under any such predicament ose instruments.

I am the Honourable Board's most Faithful and most Humble Servant,

REUBEN BURROW.

ber 9, 1782.

es not appear that any answer was to these applications, as the subse- rts of the *Diary* and several letters the feelings of bitter disappointment ep sense of the wrongs done him.]

SENEX.

THE ROTARY BALLOON.

-It was not my intention to im- Sir George Cayley in any ques- novelties put forth in my pam- n Air Navigation,* to which Sir refers in your Number for last As Sir G. Cayley's papers were scientific treatises I had read on l aëromotion, and as I obtained valuable information therefrom, ly with respect to the necessary l power of balloons, and more- roposed nearly the same scale of ide, I considered an acknowledg- le to their author. Some of the ations" thus derived were rather than otherwise. For instance, ng that Sir G. Cayley allowed bs. for the weight of his propel- hought it desirable to avoid this rance, by using the surface of the for propulsion. It further ap- to me, that the tapering and form recommended might be by the screw-shape without in- ent length.

of Sir G. Cayley's objections to my scheme is founded on the assumption, that the resistance of the screw portions of the balloon would cause the eccentric rotation of the cylindrical part, thus occasioning great waste of power. But in this I think Sir George overlooks the fact that the two screwed ends exactly balance each other, and therefore counteract their opposite tendency to destroy the concentric movement of the cylinder. To estimate the effect of one screw only will give as imperfect a view as to consider the operation of a single wing of a bird, or a single paddle of a boat.

The other objection, that the circular form of the propellers is less advantageous than a flat surface, I anticipated by making them protrude so far beyond the cylinder. Whereas, the flattened tails of fishes seldom extend beyond the breadth of their bodies, the screw points of the rotary balloon protrude more than a third of its diameter beyond its sides, and are thus equivalent for propulsion to flat propellers of an area one-third less.

Though the screws I have proposed cannot strictly follow their own lead, being checked by the cylindrical parts of the balloon, I have found by experiment that the effect of their rotation is nearly equivalent. As experience, however is more satisfactory than testimony, I shall be happy to furnish any of your readers with the patterns (there are only two), which form the shape I have proposed, in order to try for themselves the efficiency of the propellers.

Should Sir G. Cayley's objections be substantial, the addition of the flat-screw fin he kindly suggests might be adopted to correct both the faults imagined, but it should not, I submit, supersede the screw form of the balloon.

The several parts of my scheme are so intimately connected that I am unwilling they should be judged separately. Thus, the plans for condensing the steam within the balloon, thereby adding to its buoyancy, and for using gas as fuel, which will entail no burden, are much facilitated by the rotary action. But I must not introduce topics not noticed by Sir George Cayley.

Thanking you for allowing space for the discussion of this important subject,

I have the honour to be, Sir,

Your very obedient Servant,

JOHN LUNTLEY.

New Broad-street Court, City, Oct. 18, 1841.

---

* Navigation, by means of the Rotary Balloon. Ralston and Stoneman. Paternoster-row.

ON THE "AMERICA" AND THE WAVE-LINE.

(Continued from p. 316.)

Taking the proportions of the *America*, they will exemplify the wave-line as well as any other. No. 1 is the wave-line, on the whole length of 92 feet. No. 2 is the wave-line for bow and stern, on a length of 60 feet, 32 inserted in middle.

No. 1.

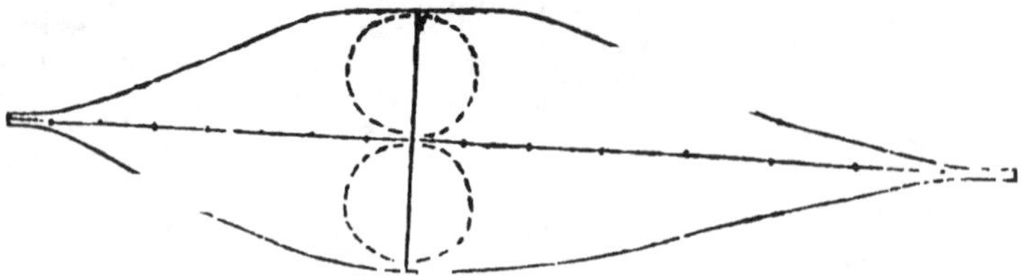

No. 2.

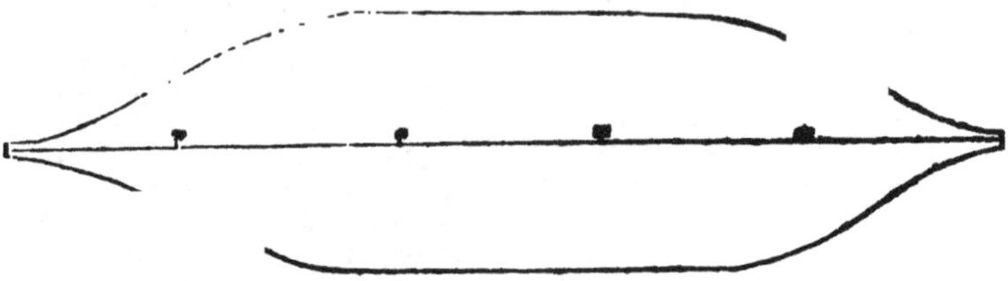

Such are examples of the horizontal water-lines; and by looking at the lower half sections, we have the floating lines of the whole depth. For racing purposes, this line would be taken higher up, supplying the place of this lower line with keel only. In the examples given, the false keel would be added, and, to follow out the principle, it should partake of the same curves. Supposing that the length and greatest beam, as well as the length of main body to be inserted, are determined on, also the length on which the wave curves of bow and stern are to be drawn, the horizontal water-line can be laid down as shown. To ascertain the floating line, different ways occur. In No. 1 and 2 the half section is simply taken, which seems to correspond pretty well with the depths forward and aft of the *America*, 6 feet and 11 feet: by taking from this half section the depth from centre line to water line, we get the depth for the cross section at each place, while the width is taken from the horizontal water line; thus, in No. 2, the depth at three-fifths of whole length from bow C, is 11 feet, the total width 22 feet; at one-fifth B, the depth to floating line is 7 feet; the total width is 13 feet; at four-fifths D, the depth is 10 feet, the width 20 feet; while E, at 9 feet from the stern-post, the depth is 4 feet, the width 7 feet; A, at 9 feet from bow, the depth is 2½ feet, width 5 feet. These cross sections are as follows, having reference, however, to what was stated, that a part of the midship side should be immersed. To obtain this, a proportional less depth of centre line is taken. To show these more distinctly a larger scale is used.

Fig. 9.

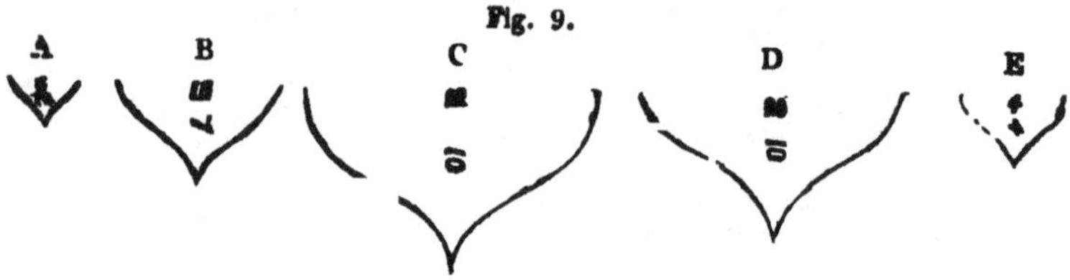

ier way of obtaining the floating the other side lines, is, to take surement from water-line, having rmined where that is to be, both and depth, and so lay off the nes fr. m keel to upper works. trate fully what it is wished to it, would require the whole lines el to be given. Any one accus- o this, who will apply the rule fter a few experiments, will find y application. What has been ers chiefly to the form of the arts. In yachts, the object is y to have as little upperwork as ; for vessels of other descrip- e purposes for which they are l must all be considered. Due ation should be given in all to aptain Fishbourne states about icular and flat sides, as well as to he upper lines blend properly lower. I would again urge the y of doing away in some degree g false sterns, by making them the real body of the vessel, and rry out below the flowing line very one feels is necessary above.

few more remarks on drawing s. For the midship section, the ine is taken a foot less than the lepth to keel, to allow for the side immersed. It will also be practice that it brings out the es in particular, best by increas- length of this centre line by rising e water line. In fact, it appears y to draw a sort of imaginary ping below the actual water line ides, and rising above it towards n and bow. I cannot give a defi- e for this line; but a few trials ble any one to judge of it, and to take depth for the cross sec- If the vessel were for racing would keep nearly equal distance at I call the lower floating line, entre line in the outline of the s. The wave line gives the rela- oportions of the parts to each ut there is still wanting a gene- for obtaining the best dimensions length, as to breadth and depth; ance, is 92 feet in length, 22 , and 11 deep, the best? The this does not affect the applica- the principle here advocated, in itself brings us nearer than er *system to the proportion of* the whole. It is scarcely necessary to add, that horizontally the part inserted in the middle should partake in a slight degree of the general curve.

These notes are remodelled from others made in 1847, when few seemed to receive the subject favourably; and I would now wish to have been able to condense and render them clearer, had time permitted. I trust attention will be drawn to the subject, and that we shall next season see that those who have the means and opportunity have successfully ventured to deviate from old-established lines.

I have just received your last monthly Number, which contains useful articles on sails, and on the *America*, confirming that both the *America* and *Titania* are on the wave principle. I have also got a beautiful print of the former, published in London, by Rowney and Co., Rathbone-place, to which I would direct attention, being drawn by an eminent marine artist. G.

October 9, 1851.

THE "OFFICIAL DESCRIPTIVE AND ILLUSTRATED CATALOGUE." — PARTS III., IV., AND V.

"After the beef, the mustard." During greater part of the period when the Exhibition was open, and when a complete "Official Descriptive and Illustrated Catalogue" might have been of infinite service, no such Catalogue was forthcoming, nor even for a considerable time any portion of it; but just as the Exhibition is closing, and catalogues of any sort are in no request, the thing that was wanted, or at least something which professes to be so, makes its appearance. The "now completed" work consists of five parts, for which the *moderate* sum of 3*l*. 15*s*. (!) only is charged. Two of these Parts we have already noticed, and with less commendation than it would have been our pleasure to bestow (see *ante* vol. liv., p. 394). The three additional Parts now before us are much of the same stamp; meagre, common-place, and unsatisfactory. As a whole, the work is quite unworthy of the Royal Commissioners, under whose sanction it appears; and of the numerous body of compilers and annotators (some thirty and upwards), by whom it has been produced. Of the latter, there are but four who can be said to have done their self imposed tasks with a passable degree of care and attention, —Mr. Ansted, Mr. Hunt, Mr. Glaisher, and Mr. Tomlinson; and all these gentlemen have shown by their writings elsewhere that we have but a very sparing taste of the talent which they could have brought to bear on the improvement of the Catalogue, had there been inducement enough for them, to devote themselves heartily to the work.

No doubt much of this, if not all, is to be ascribed to the vile contract system on which the Catalogue has been got up. The contractors had a large sum to pay for the privilege of publishing the Catalogue, and, in order to indemnify themselves, went of course the cheapest way to work. Poor pay for a world of toil became the order of the day. We are confirmed in this view of the case by a certain *typographical* economy apparent throughout the work, which is a thing which none but contract printers would have thought of, and but few persons unacquainted with the "black art" are likely to detect. Where titles of things should have been in Italics, and the descriptive matter in Roman (or *vice versâ*), both titles and descriptions are given in the same type, and in continuous lines, too, making it difficult for the eye to distinguish between one and the other. No catalogue-maker, looking to his own literary credit, would have permitted a work to be so marred; but a contract printer does so, because in the course of a long work, he may save a good many pounds by the difference in journeymens' wages, between setting up all in one type and setting up in two or several sorts of type.

Some part of the defects of the Catalogue must also be ascribed to the faultiness of the system of classification on which it is based, and which we have before shown to be most absurd and fantastical. (See *ante*, vol. liv., p. 394.) For this Dr. Lyon Playfair, and not any of those immediately concerned in the preparation of the Catalogue, is responsible, or, rather, those who conferred so important and difficult a duty—requiring great mental grasp, the nicest philosophical discri[illegible] tion, and large experience—on a gentl[illegible] fresh from Giessen, who has yet his [illegible] as a man of science and philosopher to [illegible]

THE GREAT EXHIBITION.—THE AWA[illegible]

The announcement of these A[illegible] has been received by the public [illegible] such a general burst of ridicule and [illegible] tempt that it may seem of little [illegible] encumber our pages with any noti[illegible] them. Viscount Canning, on whom [illegible] inglorious duty fell of presentin[illegible] Reports of the Councils and Jur[illegible] the Royal Commissioners, appea[illegible] have had a shrewd anticipation o[illegible] reception which awaited them, whe[illegible] deprecatingly observed in regard [illegible] distinctions made between Counci[illegible] Prize Medals, "The award of a Co[illegible] Medal does not necessarily stam[illegible] recipient as a better manufactur[illegible] producer than others who have rec[illegible] the prize medal"! His lordship [illegible] with equal truth have added, that n[illegible] does the award of a *prize medal* [illegible] sarily stamp the recipient as a [illegible] manufacturer or producer than [illegible] who received only an "*Honourable* [illegible]*tion*." In fact all the three mar[illegible] distinction seem to be very much [illegible] par. No broad and clear lines [illegible] markation are observable or tra[illegible] between them. We see "C[illegible] Medals" granted for things of no[illegible] the merit of others to which "[illegible] Medals" or "Honourable Men[illegible] only are awarded; and invention[illegible] discoveries of the very highest im[illegible] ance dismissed with a mere "Ho[illegible] able Mention," when the first pla[illegible] honour was manifestly their due. [illegible] assignment of an article to one [illegible] rather than to another has been [illegible] ably an affair of sheer caprice, or [illegible] —corrupt influence. A classifi[illegible] has been adopted implying three de[illegible] of merit, where no such degre[illegible] merit exist in reality; and all tha[illegible] been gained by it is the unworthy [illegible] of complimenting some and of depr[illegible] ing others. Even the *Times*, whic[illegible] been the zealous champion throu[illegible] of the Exhibition and of the Com[illegible] sioners (albeit it had in times [illegible] placed on record its convictio[illegible]

lity of all such displays) led to give up this affair ards, as altogether indefen- We have received," it says, us protests and complaints e jury awards, which, accord- usual custom, we should have to publish; but our corre- will feel that the demands ıld thus be made upon our d exceed all ordinary bounds, ıp questions *which it is now to settle satisfactorily. We and over again repeated that system of prizes has been s a mistake; and it has been bvious as it can be that the mmission, while committed to s, have sought in every possi- reduce its significance.* More- ng nearly six months the pub- ıd ample opportunities of im- l unbiassed observation. We rust that the agitation on this ll be allowed to subside, and pitors who have been disap- und believe themselves ag- rill not, by bringing their ominently forward, *provoke f attention to the decisions of which at present there seems tion to accord to them.*" For ırts, we shall decline to lend ours to give currency to the nents involved in these awards. ıot say who has been exalted based (attempted to be)—re- e decisions of the jurors, whe- aise or dispraise, as alike un- of regard.

ee with the *Times*, that "the em of prizes has been founded ke;" and this we said before had ventured on a word in aragement. If the reader to volume lii., page 55, of zine, he will find our rea- coming to this conclusion length. We contended for *s at all*," and predicted very ıat is now taking place should ; be persisted in: "For every voice there will be at least e angry remonstrants; and ı long day after the Exhibition ; ear will be filled with the the disappointed, and the of the winners and losers of among themselves."

However, there are certain national aspects in which this affair of the prizes presents itself, which may make it use- ful to preserve a record of them *limited to these aspects*. Leaving individuals out of account, it becomes a matter of some interest to know in what relative proportions the different nations are sup- posed to have contributed to the splen- dour and utility of the Exhibition. With this view the following abstract has been drawn up, showing the number of Me- dals and "Honorable Mentions" accorded to each country under each head of the Catalogue. The correctness of these general results must, of course, be affected more or less by the caprice, partiality, and injustice shown towards the individual Exhibitors; and this ab- stract can at best, therefore, be accepted only as an approximation, even in the cosmopolitan point of view, to the truth.

Abstract.

The number of awards of all classes—"Council Medals," "Prize Medals," and "Honorable Mentions"—is 5084; of this gross number 3045 distinctions have been given to foreign exhibitors, and 2039 have been received by our own countrymen.

JURY I. *Mining, Quarrying, Mineral Products, &c.—Council Medals:* France 2, United Kingdom 2, Prussia 1, Austria 1. *Prize Medals:* United Kingdom 26, France 10, Belgium 10, Prussia 9, Austria 7, United States 4, Russia 3, Canada 2, Tus- cany 2, Nova Scotia 1, South Australia 1, Hesse 1, Sweden 1, Spain 1, Nassau 1, Switzerland 1. *Honourable Mentions:* 99. Exhibitors of all Nations.

JURY II. *Chemical and Pharmaceutical Processes and Products.—Council Medals:* France 2, United Kingdom 1, Tuscany 1. *Prize Medals:* United Kingdom 28, France 20, Prussia 12, Austria 5, Sardinia 2, Bel- gium 2, Wertemberg 2, Russia 2, Nether- lands 2, Hesse 1, Tuscany 1, Bavaria 1, Mecklinburg Strelitz 1, Frankfort-on-Maine 1, United States 1. *Honourable Mentions:* 72.

JURY III. *Substances used for Food.—Council Medals:* France 4, United States 1, United Kingdom 1. *Prize Medals:* United Kingdom 35, France 18, United States 13, Canada 10, Spain 10, Russia 8, India 4, Portugal 3, Van Diemen's Land 3, Austria 2, Cape 2, Borneo 2, Tuscany 2, British Guiana 1, Sweden 1, Belgium 1, Greneda 1, South Australia 1, Jersey 1, Algeria 1, New South Wales 1, Trinidad 1, China 1, Turkey 1, Hesse 1, Prussia 1, Mauritius 1. *Honourable Mentions:* 127.

JURY IV. *Substances used in Manufactures.—Council Medals:* France 3, United Kingdom 2. *Prize Medals:* France 41, India 33, United Kingdom 29, United States 20, Austria 14, Spain 13, Prussia 11, Belgium 10, Russia 10, Algeria 9, Tuscany 8, British Guiana 8, Van Diemen's Land 8, Cape 6, Turkey 6, Portugal 3, Ceylon 2, New Zealand 2. Canada 2, New South Wales 2, Sardinia 2, Netherlands 2, Labuan 1, Trinidad 1, China 1, South Australia 1, Wurtemberg 1, Bavaria 2, Tunis 1, West Africa 1, St. Domingo 1. *Honourable Mentions:* 406.

JURY V. *Machines for Direct Use.—Council Medals:* United Kingdom 4, France 1, Belgium 1. *Prize Medals:* United Kingdom 39, France 6, Netherlands 1, Belgium 1, Canada 1, Prussia 1, Austria 1. [Class VA. *Carriages.*—United Kingdom 13, France 2, Belgium 2, United States 2.]

JURY VI. *Manufacturing Machines and Tools.—Council Medals:* United Kingdom 15, France 4, United States 1, Prussia 2. *Prize Medals:* France 24, United Kingdom 51, United States 7, Prussia 4, Tuscany 1, Switzerland 1, Austria 1, Belgium 2.

JURY VII. *Civil Engineering.—Council Medals:* United Kingdom 3. *Prize Medals:* United Kingdom 17, Netherlands 1, United States 1, Switzerland 2, France 2. *Honourable Mentions:* United Kingdom 6, Switzerland 2.

JURY VIII. *Naval Architecture and Military Engineering.—Council Medals:* United Kingdom 5, France 3, Austria 1. *Prize Medals:* United Kingdom 45, Belgium 8, France 16, Lubeck 1, Switzerland 1, Spain 1. *Honourable Mentions:* 26. Money awards, 4; amount, 160*l.*

JURY IX. *Agricultural Implements.—Council Medals:* United Kingdom 4, United States 1. *Prize Medals:* United Kingdom 29, Belgium 4, France 3, Netherlands 1, United States 1. *Honourable Mentions:* 1.

JURY X. *Philosophical Instruments.—Council Medals:* United Kingdom 16, United States 1, France 9, Switzerland 1, Tuscany 1, Holland 1, Bavaria 1, Prussia 1. *Prize Medals:* United Kingdom 40, France 16, Prussia 9, United States 7, Austria 3, Switzerland 3, Belgium 2, Bavaria 1, Saxony 1, Russia 1, Denmark 1, Zollverein 1, Hesse 1, India 1. *Honourable Mentions:* 54.

JURY XA. *Musical Instruments.—Council Medals:* United Kingdom 4, France 4, Munich 1. *Prize Medals:* United Kingdom 25, France 11, United States 6, Prussia 2, Wurtemberg 2, Belgium 2, Saxony 1, Tuscany 1, Spain 1, Nassau 1, Bava[...] *Honourable Mentions:* 56. Money a[...] 2; amount, 100*l.*

JURY XB. *Clocks, &c.—Council M[...]* United Kingdom 1, France 2, Switzer[...] *Prize Medals:* United Kingdom 10, [...] 9, Switzerland 9, Sardinia 1, Denm[...] *Honourable Mentions:* 17. Money [...] 1; amount 50*l.*

JURY XC. *Surgical Instruments.—[...]cil Medals:* None. *Prize Medals:* [...] Kingdom 19, France 5, Tuscany 1, S[...]land 1, America 1, Portugal 1.

JURY XI. *Manufactures of Co[...]* *Council Medals:* None. *Prize M[...]* United Kingdom 16, Switzerland 7, [...] 7, United States 2, Prussia 2, Sax[...] Belgium 1, Austria 1, Portugal 1, W[...]berg 1. *Honourable Mentions:* 19.

JURY XII. *Manufactures in W[...]* *Council Medals:* None. *Prize M[...]* United Kingdom 73, France 34, Pru[...] Saxony 14, Russia 6, Austria 5, Belg[...] Canada 2, United States 1, Netherl[...] *Honourable Mentions:* 26. Money [...] 1; amount, 10*l.*

JURY XIII. *Silks and Velvets.—[...]cil Medals:* None. *Prize Medals:* [...] 49, United Kingdom 31, Switzerla[...] Austria 8, Sardinia 5, Prussia 4, Zol[...] 3, Russia 2. *Honourable Mentions:* [...]

JURY XIV. *Flax and Hemp.—*[...] *Medals:* None. *Prize Medals:* [...] Kingdom 24, France 8, Belgium 7, [...] 5, Austria 1, Russia 1, Saxony 1. [...]*nourable Mentions:* 60.

Money awards of 10*l.* each are [...] favour of Ann Harvey and Jane [...] both of Belfast, and of "a little [...] years old," in Heepen Spinning [...] Prussia, for hand-spun flax yarn.

JURY XV. *Mixed Fabrics.—* [...] *Medal:* France 1. *Prize Medals:* [...] Kingdom 27, France 13, Belgium 4 [...]*nourable Mentions:* 40.

JURY XVI. *Leather, Fur, &c.—* [...] *Medals:* None. *Prize Medals:* [...] Kingdom 43, France 25, United St[...] Belgium 4, Prussia 4, Russia 3, H[...] Austria 1, Canada 1, Turkey 1, Swit[...] 1, Nova Scotia 1. *Honourable Me[...]* 74.

JURY XVII. *Paper and Stat[...] Printing and Bookbinding.—Counc[...]dal:* Austria 1. *Prize Medals:* [...] Kingdom 34, France 25, Prussia 6, [...] 4, Belgium 3, Wurtemberg 2, Sard[...] Denmark 1, Van Diemen's Land 1, [...] 1, India 1, Egypt 1, Bavaria 1, Sa[...] Canada 1, Hesse 1, Russia 1, Neth[...] 1. *Honourable Mentions:* 79. [...] awards, 2; amount, 20*l.*

JURY XVIII. *Woven, Spun, Felt[...]*

ries, when shown as Specimens of or Dyeing. — *Council Medals: Prize Medals:* United Kingdom ce 17, Switzerland 3, Austria 2,

.

XIX. *Tapestry, including Carpets rcloths, Lace and Embroidery, id Industrial Works.* — *Council* France 1, United Kingdom 1. *dals:* United Kingdom 55, France am 16, Switzerland 9, Saxony 4, , Spain 2, Sardinia 1, Sweden 1. *ile Mentions:* 82.

XX. *Articles of Clothing for Im- Personal, or Domestic Use.*— *Medals:* None. *Prize Medals:* Kingdom 46, France 26, Switzer- Austria 7, Saxony 7, Tuscany 2, , United States 1, Luxemburg 1, , Russia 1. *Honourable Mentions:*

XXI. *Cutlery and Edge Tools.*— *Medal:* United Kingdom 1. *Prize* United Kingdom 63, France 8, tates 3, Zollverein 5, Austria 4, Turkey 2, Sweden 1, Wurtemberg *urable Mentions:* 85.

XXII. *Iron and General Hard- ouncil Medals:* United Kingdom 4, Belgium 1, Bavaria 1, Prussia *Medals:* United Kingdom 109, 5, Zollverein 21, United States 8, , Belgium 5, Wurtemberg 3, Rus- etherlands 3, Spain 2, Saxony 1, 1. *Honourable Mentions:* 191.

XXIII. *Working in Precious Me- heir Imitation, Jewellery, and all f Vertu and Luxury not included r Classes.—Council Medals:* Uni- dom 6, France 6, Zollverein 3, *Prize Medals:* France 31, Uni- om 14, Zollverein 5, Switzerland urg 2, Spain 1, Netherlands 1, Sar- Belgium 1, Russia 1. *Honourable :* 51.

XXIV. *Glass.—Council Medals: Prize Medals;* United Kingdom ce 8, Austria 3, Switzerland 1, ds 1, Zollverein 1. *Honourable :* 35.

XXV. *Ceramic Manufacture— Porcelain, Earthenware, &c.* — *Medals:* United Kingdom 1, France *Medals:* United Kingdom 12, Zollverein 4, Austria 2, Portugal 1, India 1, Bavaria 1, Denmark *urable Mentions:* 27.

XXVI. *Decoration, Furniture, and y, including Paper-hangings, Pa- té, and Japanned Goods.—Council* France 4, Austria 1. *Prize Me-* ited Kingdom 23, France 22, Aus- Tuscany 3, Russia 3, Belgium 3, Bavaria 2, China 2, India 2, Prussia 2, Sardinia 1, Hamburg 1, Netherlands 1. *Honourable Mentions:* 68.

Jury XXVII. *Manufactures in Mineral Substances, used for Building or Decoration, as in Marble, Slate, Porphyries, Cements, Artificial Stones, &c.* — *Council Medals:* United Kingdom 2, Papal States 1, Russia 1. *Prize Medals:* United Kingdom 48, France 8, Tuscany 4, Malta 3, Russia 3, Rome 2, Austria 2, Belgium 2, India 1, Sweden 1, Prussia 1, Bavaria 1. *Honourable Mentions:* 97.

Jury XXVIII. *Manufactures from Animal and Vegetable Substances, not being Woven or Felted, or included in other Sections.—Council Medals:* United Kingdom 2, United States of America 1. *Prize Medals:* United Kingdom 26, France 12, United States 5, Austria 5, Switzerland 4, Spain 4, Prussia 4, Canada 3, Mauritius 2, Bahamas 2, Russia 1, Sardinia 1, Nassau 1, Portugal 1, China 1, Belgium 1, Sweden 1, Turkey 1. *Honourable Mentions:* 15.

Jury XXIX. *Miscellaneous Manufactures and Small Wares.—Council Medals:* France 2. *Prize Medals:* United Kingdom 45, France 32, Austria 11, Prussia 10, Wurtemberg 5, Spain 5, United States 3, Belgium 3, Turkey 3, Russia 3, Portugal 3, Sardinia 3, Hamburg 2, Bavaria 2, India 2, Brazils 1, Switzerland 1, Netherlands 1, Tuscany 1, Sweden 1, Tunis 1. *Honourable Mentions:* 74.

Jury XXX. *Sculpture, Models, and Plastic Art.* — *Council Medals:* United Kingdom 2, France 1, Prussia 1. *Prize Medals*: France 28, United Kingdom 27, Austria 6, Belgium 5, Prussia 5, Bavaria 3, Rome 3, Saxony 2, United States 1, Spain 1, Russia 1, Denmark 1. *Honourable Mentions:* 87.

ROSENBORG'S CASK-MAKING MACHINERY.

A novel method of constructing casks and barrels, and all vessels connected with cooperage, may be seen in operation at the Patent Cooperage Works in Wenlock-road, City-road. By the employment of the steam engine, the circular saw, and a recently-invented jointing and backing machine, a cask of the largest dimensions can be completely formed and made ready for use in the short space of five minutes from the raw material—viz., a piece of oak. The staves of the cask are first cut with straight sides, the circular saw being placed at a right angle with the oak plank. The stave is then placed horizontally, and bent into a curve by a powerful machine, and brought into contact with a circular saw on each side of it, placed at an angle. This process gives the proper shape to the stave, the sides being gradually tapered at the ends, and made to bulge in the middle. The jointing and backing-machine, the new invention, is also used for this purpose, and is more rapid in its execution than the angular saws; it in fact works with the most marvellous rapidity and precision. The staves and one end of the cask are then placed in a machine formed of iron rods, called a trussing machine; each rod acts upon a separate stave, and the whole of the staves being equally compressed into a circle, the hoops are placed around them, and the cask is complete. The neatness and finish of the work is equal to what a good cabinetmaker can produce, every part being true and accurate. The calculation is, that 15 workmen, with the use of this machinery, can make 150 casks a day; whereas the same number of persons, using only manual labour, could scarcely produce a seventh part of that number. The importance of the invention, and the application of steam power to it, may be imagined from the fact that the great brewing firms of the metropolis alone expend many thousand pounds annually in cooperage, that the expenditure of the navy is still greater, and that the demand of the vintages of the continent is so great that a great deal of wine is lost from the difficulty of furnishing vessels to hold it. The process of this invention will repay the time of a visit to the works.—*Times.*

PROGRESS OF SCREW PROPELLING ON THE ATLANTIC.

We will soon have five lines of screw propellers running between our country and Britain. At present we have three, namely the Philadelphia and Liverpool, the New York and Glasgow Lines, and Boston and Liverpool Line. From what we have heard about the vessels of this latter line, we anticipate very successful results: the *S. S. Lewis*, the pioneer of it, will soon make her first Atlantic voyage; she was built in Philadelphia, and is a splendid vessel. Her hull was built by Messrs. Birely and Sons, and is most substantially constructed. Her frame is almost entirely of white oak, and the planking and ceiling principally of the same material. The frame is bound together with diagonal iron braces, each 60 feet long, 5 inches wide, and 1 inch thick. These braces cross one another, and let into the timbers, being bolted through them and riveted on the outside, and at every intersection are bolted together. The hull was planned and superintended by the ingenious Captain Loper, and she is driven by one of his propellers—the wheel being 18 feet, 4 inches, with four fans. Her engines were designed, we believe, by Captain Loper, and for compactness, beauty, and power, are said to be superior to those of any other propeller steamship afloat. This line will be composed of four fine vessels. Next year we will have four screw steamers for freight and passengers, belonging to Messrs. Burns and Co., of the Cunard Line; these, with the four of the Philadelphia Line, and the Glasgow Line of two, together with the *Great Britain* and *Sarah Sands*, which, we understand, are to be put on the route between New York and Liverpool, will make twelve large screw-propelling steamships that will be running between Europe and our country next year. This looks like doing business in an improved way; for a year ago there was but one such vessel making Atlantic voyages; and from what we have heard from a number of sources, we may confidently assert that half as many more will be added to this list before the first of 1853.

Our fine packet ships will soon be looked upon like the old packets on the Erie Canal, for assuredly the propellers will very quickly take all the passenger trade out of their hands.—*Scientific American.*

SPECIFICATIONS OF ENGLISH PATENTS ENROLLED DURING THE WEEK, ENDING OCTOBER 23, 1851.

FREDERIC WILLIAM EAST, of the firm of Thomas East and Son, Bermondsey, leather dresser. *For improvements in dressing, embossing and ornamenting leather.* Patent dated April 15, 1851.

This invention consists in dressing, embossing and ornamenting the flesh side of leather. It has heretofore been usual, when leather has been dressed and embossed, in order to ornament it, to dye and emboss the grain side thereof, by which a close smooth and comparatively inferior effect is produced; but by dressing, dyeing and embossing the flesh side, a very improved effect is obtained. The invention applies to sheep and other skins tanned with shumac, or other tanning matter, but not to oil tanned skins; and for carrying it into effect, the patentee prefers to employ sheep skins tanned with shumac, but other skins, and differently tanned may be used. The skins are to be shaved just sufficiently to cut out flaws, and render them of uniform substance throughout, and after immersion in warm water of a temperature of about 120° Fahrenheit, they are to be brushed on the flesh side to remove

...d open the fibre, preparatory to ... Each skin is then to be folded with ... in side inwards, and the flesh side ...ls, and the edges are to be sown ...r, to prevent the dye getting to the ... of the bag so formed, or in place of ...each skin separately, two skins may ... together with their grain sides in ... and then have the edges sown to ...t the dye. They are then "scoured" ...wetened" as is well understood, and ... to be dyed, but the dyeing process ... found to require a longer time than ...ying on the grain side of skins, which ... the usual practice heretofore; and ... of using strong dye liquors, such ... for dyeing the grain side of skins, ...ferred to employ weaker dye liquors, ... repeat the process oftener, by which ...he dye penetrates the fibres and pro... more even colour. The skins are ...sed, opened out and dried. When ...y are to be "perched" on the flesh ...e perching knife, however, should ...so sharp as to cut the flesh off, the ...ing to loosen and open the fibres ...oduce a nap-like surface. Each ... again to be folded with the flesh ...twards, and is then to be pas...ough a solution of glutinous mat...ich the patentee prefers to compose, ... part by measure of size, or glair, ...d in three parts of water. Whilst ...t, the skins are to be strained on ... to dry; after which the edges are ...immed, and the surfaces bruised with ... render them soft, the flesh side being ...pt outwards, instead of the grain side. ...he process of embossing is to be per... clean water is applied evenly and ...ly on the grain sides of the skins, ...y are laid with their grain sides flat ...r for about two days, covering them ...ide the air, by which means the mois...ll pass through, so as to act on the ...us matter, and make it of service in ... a gloss on the embossed, or pressed ...f the surface, and also in producing ...r tone of colour to those parts. The ...ing is preferred to be performed by ... rollers or surfaces suitably engraved, ...etofore employed when embossing ... and heated to about 250° Fahrenheit. ...ition to dyeing, pigments and metals ...e used to give increased ornamental ...o the flesh side of tanned skins; such ...ts and metals being applied in the ...f powder, by sifting them over the ..., or parts thereof, by which means, ...he process of embossing has been ac...shed, the pigments or metals will be ...y the glutinous matter, and the pressure on those parts, where such pressure has been applied, and may be dusted off, or removed from the other parts of the surface. This mode of applying pigments and metals is also adapted for being used on the grain side of tanned skins when ornamenting the same by embossing.

Claim.—The mode of dressing, embossing and ornamenting leather above described.

Robert Newell, of New York, lock-manufacturer. ***For certain new and useful improvements in the construction of locks.*** Patent dated April 15, 1851.

The object of the present improvements is the constructing of locks, in such manner that the interior arrangements, or the combination of the internal moveable parts may be changed at pleasure according to the form given to, or change made in the key, without the necessity of arranging the moveable parts of the lock by hand, or removing the lock, or any part thereof from the door. In locks constructed on this plan the key may be altered at pleasure, and the act of locking, or throwing out the bolt of the lock produces the particular arrangement of the internal parts, which corresponds to that of the key for the time being, while the same is locked this form is retained until the lock is unlocked or the bolt withdrawn, upon which the internal moveable parts return to their original position with reference to each other; but these parts cannot be made to assume, or be brought back to their original position, except by a key of the precise form and dimensions as the key by which they were made to assume such arrangement in the act of locking. The key is changeable at pleasure, and the lock receives a special form in the act of locking according to the key employed, and retains that form until in the act of unlocking by the same key it resumes its original, or unlocked state. The lock is again changeable at pleasure, simply by altering the arrangement of the moveable bits of the key, and the key may be changed to any one of the forms within the number of permutations of which the parts are susceptible.

The following are the claims made by Mr. Newell.*

1. The constructing, by means of a first and secondary series of slides or tumblers, of a changeable lock, in which the particular form or arrangement of parts of the lock imparted by the key to the first and secondary series of slides or tumblers is retained by a cramp plate.

2. The constructing by means of a first

* The partner of Mr. *Hobbs*, by whose name the lock goes.

and secondary series of slides or tumblers, of a changeable lock, in which the peculiar form or arrangement of parts of the lock imparted by the key is retained by means of a tooth or teeth and notches on the secondary series of slides or tumblers.

3. The application to locks of a third or intermediate series of slides or tumblers.

4. The application of a dog with a pin overlapping the slides or tumblers, for the purpose of holding in the bolt when the lock is locked or unlocked.

5. The application of a dog operated on by the cap or detector tumbler for holding the bolt.

6. The application of a dog for the purpose of holding the internal slide or tumbler.

7. The application to locks of curtains or springs turning and working eccentrically to the motion of the key for preventing access to the internal parts of the lock.

8. The application to locks of a safety plug or yielding plate at the back of the chamber formed by such eccentric revolving curtain or ring.

9. The application to locks of a strong metallic wall or plate for the purpose of separating the safety and other parts of the lock from each other, and preventing access to such parts by means of the keyhole.

10. The application to locks of a cap or detector tumbler, for the purpose of closing the keyhole as the key is turned.

11. The constructing a key by a combination of bits or moveable pieces, with tongues fitted into a groove and held by a screw.

12. The constructing a key having a groove in its shank to receive the detector tumbler.

Charles Hardy, of Low Moor, engineer. *For certain improvements in the manufacture of scythes.* Patent dated April 15, 1851.

Mr. Hardy's improvements consist substantially in manufacturing scythes from a single bar of steel instead, as heretofore, of iron and steel welded together. The steel employed for this purpose may be either natural or cemented, or hammered or cast steel; and the bars having been marked off in lengths, each containing a sufficient weight of metal for the formation of a scythe, each length is cut off and drawn out under a hammer at one heat to about the full required length, of a blade shape, and a thickness equal throughout to that of the back of the scythe. The next operation is that of drawing out the handle, after which the proper curvature is given and the point formed; the extremity of the handle is then turned up, and the partially-formed scythe widened, but so as to retain the rib, and thinned off gradually to the edge. This is generally performed in about four successive manipulations, varying, however, with the length of the scythe and the width required to be given to it. The rib is next set off, and the required appearance produced, when it is planished cold to trim it, and give it a regular form. The edge is then cut out, either by the beam-cutting machine or by the hand shears, and the scythe is so far in a finished state, and ready for hardening. This operation is performed with coal, in a peculiarly constructed furnace in which the air is prevented from acting on the scythe during the process, or with coke or wood charcoal in an open fire. In all cases when brought to a red heat (more or less according to the desired degree of hardness) the scythe is immersed in a mixture of beef suet and mutton fat with equal parts of resin deprived of water. On removing it from the hardening bath, it is dried in powdered charcoal, and beaten in water in a slightly heated state; or still better, it is washed in boiling water to remove the grease, slightly heated, and beaten in cold water. In this state it is re-heated and annealed by covering every portion of it successively with red-hot sand, until it attains a violet, blue, or other colour indicating the required extent of temper. The process of sharpening is effected on a grindstone or by hammering, and the re-heating and annealing will require to be repeated more or less frequently, according as either of the above methods are adopted, and according also to the degree of perfection desired. The patentee prefers to cover the scythe, previous to its being used, with copal varnish, to preserve it from the effects of moisture. The edge of the scythe when worn away, may be renewed by gentle hammering and a subsequent application of the whetstone.

William Benson Stones, of Warwick-street, Golden-square. *For improvements in the use and treatment of peat and its products, and other carbonaceous matters; and also for apparatus applicable to such and other chemical purposes.* (A communication.) Patent dated April 15, 1851.

The improvements which form the subject of the present patent comprehend—

1. A method of charring or distilling peat or wood, or tanners' bark and coal together, whether for the purpose of making a peculiar sort of coke, or with the twofold object of producing gas and making such coke.

With a view to obtain a compact description of fuel, and to obviate the necessity of compressing peat previous to its being charred, the patentee has contrived the present system, ac-o which equal quantities of dry powat and pulverized coal (which should the property of caking well, and be s possible from sulphur) are mixed by means of an apparatus hereinmtioned, and exposed to heat in retorts. The expansion of the coal raction of the peat under the appliheat produce a considerable soliditct on the coke, but under circumwhere a more dense and compact quired, the mixture is subjected to al pressure in the retort, by means ding plate fitted to the interior and acted on by a screw passing the end of the retort, and worked . The gas produced by this dismay be applied to illuminating purnd where this is done, it is recomto pass the gas in contact with eant peat, for the purpose of combe distillation of the gaseous prothe peat, and causing the gas to arbon, by which its illuminating increased.

e application of the greases and lipose wax obtained from peat by m, and which, in the specification ner patent were called adipolain and , to certain useful purposes, such rrying of leather (for which purpose tee considers them to be superior to her materials commonly used) and aration and treatment of textile , to render them impervious to elastic. It is better to apply these n a warm state, as they will then adily penetrate the pores of the operated on, and will also carry m any portion of wax which they tain. By subjecting the grease to a second distillation, an oleous ll be obtained together with a resia resino-adipose wax which the calls peacerine), which produces a k polish when applied to leather, ected to rubbing with a smooth and ace. These greases and wax may be ted with gutta percha, the latter solved in bisulphuret of carbon. utta percha thus prepared may be rk, turf, or rags reduced or pulverthe material thus produced while ay be rolled into sheets or tubing, be found suitable for making varies.

manufacture of manures by comat, turf, humus or other vegetable with other substances of different ch as urine, stable dung, blood, marl, ashes, sand, burnt clay, or ters adapted for manuring purposes. These manures may be applied broad cast, or in any usual manner, or the seeds, grains, or roots, having been wetted with a liquid of an adhesive nature may be covered or sprinkled therewith, and then sown. Strong smelling matters, such as shale oil, or coal tar oil, may be also combined with them for the purpose of preserving the seeds or plants to which they are applied from the attacks of insects. For the purposes of this invention, the patentee recommends the collection into reservoirs of the urine from stables in towns, and in consequence of this suggestion he "claims the right" of collecting and carrying away urine from stables for the purpose of making manures by preparing and mixing them with vegetable organic matter, or of depositing the vegetable organic matters in the reservoirs to be saturated and afterwards carried away for the purpose of being made into or serving as manure.

4. An arrangement of apparatus for mixing, digesting and otherwise treating dry substances, fluids, or gases. This "churn" consists of a cylinder capable of revolving in a horizontal position and provided internally with tilting shelves or leaves projecting towards the centre. In certain cases where the application of heat is desirable, it is placed over a stove in such a position as to permit the heat therefrom to act upon it.

Claims.—1. The charring of peat or wood, or tanners' bark and coal together, whether for the twofold purpose of producing gas and making the above-mentioned decription of coke, or for the sole purpose of making the coke.

2. The greases and resino-adipose wax obtained from peat, and herein denominated adiposole, adipolein, and peacerine, for currying and polishing leather, and also for strengthening it and rendering it and other materials waterproof, and the various other uses herein mentioned and alluded to.

3. The use of peat, or turf, or humus, or other decomposed, or partly decomposed, vegetable matter to be applied by attaching it to the seed, or grain, or bulb, or root, &c., either alone or with other matter or ingredients, and reduced to a powdered or pulverised state for the purpose, and so prepared for the purposes of manuring generally; whether to be used alone or in combination with other fertilizing matter or substances, as described, and so as to form a manure resembling farm-yard manure in a concentrated and condensed state only, or the like manure with the addit on of other matter and inorganic substances combined, so as to give it the twofold character of an organic and a mineral and mechanical ma-

sure at the same time; and otherwise prepared or modified.

4. The system of apparatus described for mixing, digesting, bleaching, washing, purifying, and otherwise chemically matters, and fluids, and gases, and cation to other useful purposes.

WEEKLY LIST OF NEW ENGLISH PATENTS.

Richard Roberts, of Manchester, engineer, for improvements in machinery or apparatus for regulating and measuring the flow of fluids, also for pumping, forcing, agitating, and evaporating fluids, and for obtaining motive power from fluids. October 17; six months.

Ephraim Hallum, of Stockport, Chester, cotton spinner, for certain improvements in preparing and spinning cotton and other fibrous substances. October 22; six months.

John Ramsbottom, of New Mills, Derby, engraver, for certain improvements in machinery or apparatus for measuring and registering the flow of water and other fluids or vapours, which machinery or apparatus is also applicable to registering the speed of and distance run by vessels in motion, and for obtaining motive power and other similar purposes. October 22; six months.

Joseph Beattie, of Lawn-place, South Lambeth, Surrey, engineer, for improvements in the construction of railways, in locomotive engines, and other carriages to be used thereon, and in the machinery by which some of the improvements are effected. October 22; six months.

William Boggett, of St. Martin's-lane, gentleman, and George Holworthy Palmer, of Westbourne-villas, Paddington, civil engineer, for improvements in obtaining and applying heat and light. October 22; six months.

John Platt and Christian Schiele, both of Oldham, Lancaster, machinists, for certain ments in machinery or apparatus for the p and manufacture of fibrous materials, provements, or parts thereof, are also for the transmission of fluids and aërifo October 22; six months.

Donald Henderson, of Glasgow, ironm an improved apparatus for generating apparatus may be used for heating and lar useful purposes, and other apparatu ing and ventilating. October 23; six m

John Henry Pape, of Paris, France, f ments in ploughs. October 23; six mo

Jonathan Sparks, of Conduit-street, B Middlesex, surgical bandage-maker, fo ments in or substitutes for laced stockin dages for the legs. October 23; six m

Henry Adcock, of Northumberland-str Middlesex, civil engineer, for improvem manufacture of pipes, chimney-pots, vessels; also bricks, tiles, copings, col other articles used in building houses structures. October 23; six months.

Moses Poole, of the Patent Bill-offic gentleman, for improvements in axle railway carriages (Being a communica tober 23; six months.

Allen Searell, of Tanybwlch, Merio neer, for improvements in sawing-mach tober 23; six months.

WEEKLY LIST OF DESIGNS FOR ARTICLES OF UTILITY REGISTERED.

Date of Registration.	No. in the Register.	Proprietors' Names.	Addresses.	Subjects of De
Oct. 17	2985	William Walker.........	East Bridgeford	Winnowing and d chine.
,,	2986	Holliday & Clementson	Watling-street	Royal shawl mant
,,	2987	J. H. Beaumont..........	Oxford-street	Boot upper-leathe
18	2988	William L. Gilpin	Manchester-street..................	Screw capsule for b
20	2989	Chubb and Son	St. Paul's Church-yard............	Lock.
,,	2990	Charles Hart.............	Wantage	Skim plough.

CONTENTS OF THIS NUMBER.

LONDON: Edited, Printed, and Published by Joseph Clinton Robertson, of No. 166, F in the City of London— Sold by A. and W. Galignani, Rue Vivienne, Paris; Machi Dublin; W. C. Campbell and Co., Hamburg.

Mechanics' Magazine,
SEUM, REGISTER, JOURNAL, AND GAZETTE.

1473.] SATURDAY, NOVEMBER 1, 1851. [Price 3d., Stamped, 4d.
Edited by J. C. Robertson, 166, Fleet-street.

ROOMAN'S PATENT ROPE AND CORDAGE MAKING MACHINERY.

Fig. 1.

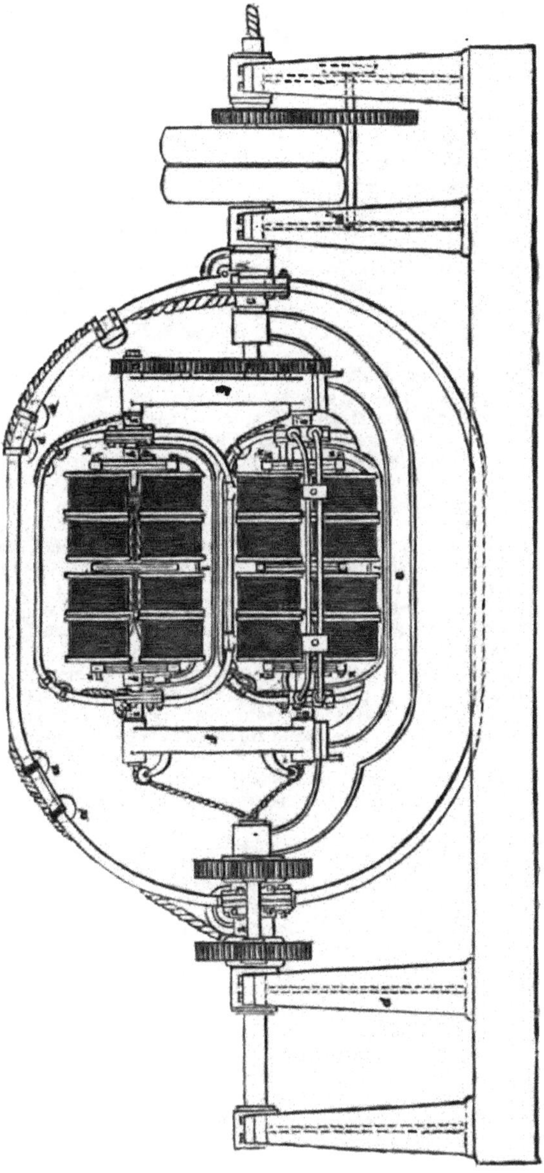

BROOMAN'S PATENT ROPE AND CORDAGE MAKING MACHINERY.

[Communication from Abroad. Patent dated April 2, 1851. Specification enrolled, October 2, 1851.]

Specification.

THE nature of this invention consists in a peculiar manner of laying and twisting the fibres in the manufacture of rope and cordage, whereby each and every fibre composing any rope or cord is made to bear its exact proportion of strain to resist fracture under tension. In cordage, as manufactured under the processes in common use this result is not attained, for it is well known that in all such cordage the exterior strands are required to sustain the principal part of the strain, as may be easily proved by applying (very gradually) sufficient force to break the rope when the outside strand will be seen to part first. Now, according to the said invention, the method of tubing, or forming the strand, is such that an exact parallelism is ensured in each thread composing it, and this is carried on from the core to the outward thread. Another valuable feature of the invention consists of a method of giving a double twist first of all to the strand itself; and, secondly, to the rope while laying the strand.

The means employed to obtain the above novel results consist of a train of mechanism so contrived as to put in revolution a number of metal rings, upon which the stock is delivered as fast as prepared, and which is from these conveyed to other rings as the work progresses to completion. Figs. 1, 2, 3, 4 and 5, of the annexed engravings, are representations of the different parts of the machinery employed for this purpose. *a* represents a large cylindrical ring, which for strength and other advantages is made of metal. This ring is attached from two of its sides, and directly in a line running horizontally through it to two short hollow shafts *b*, *c*; the said attachments being effected by flanges, to which the rings are screwed, or by welding, directly to the end of the shaft, as may be thought best.

These shafts rest upon standards *d'*, *d'*, *d'*, terminating in pillar blocks fitted to receive the journals. On the shaft *c*, there is a pulley to receive a belt for driving it round, and also a pinion which gears into a wheel, for driving the reeling machinery, to be more fully explained hereafter. Around the circumference of one-half of the ring *a*, there are attached several small sheaves, properly grooved, to convey off the cord received from the interior rings; which sheaves are shown at *d*. The shafts extend, also, a short distance within the ring *a*, and serve as bearings to sustain an interior frame, upon which rest the primary rings and their attachments. At *e* is seen the lower part of the interior frame, which consists of a semicircular piece of metal, which is of less radius than that forming the ring *a*, and terminates at each end in eyes fitted so as to play upon the shafts *b*, *c*. At the lower part this interior frame has much greater weight given to it than at the eyes; the object of this is to prevent it from being carried round with the shafts *b*, *c*, when they are put in rotation by the friction of the parts in contact; and this weight must also be sufficient to counterpoise the weight of the bobbins and primary rings which rest upon it.

The primary rings with their bobbins are next introduced in the machine. These are sustained upon triangular frames *f*, *g*, affixed to the belt *e*, and at right angles to it, which are also placed at such a distance apart as to give the necessary room to insert the primary rings, as seen in the engravings. These frames are set in such a position that their centres are the same as the centres of the shafts *b*, *c*. In the frame *g*, the three arms are so spaced as to divide it into three equal compartments; the use of which is to support the several trains of wheels which propel the primary rings, as will be hereafter more fully described.

The three primary rings are not suspended on the frames *f*, *g*, (the letters *h*, *h'*, *h''*, representing said rings); they are constructed much in the same manner as the large ring *a*, two short hollow pivots *i*, *i'* are affixed to them, as shown; by these they are suspended with the frames *f*, *g*, the pivots passing through proper holes in the same. In the frames *f*, *g*, the pivots come through at the termination of the three arms before named, and consequently are equidistant from each other. The pivots *i*, *i'* extend inside of the rings a short distance, and terminate at journals, as seen at *k*. On each of these points or journals, a semicircular bail or hoop *l*, *l*, *l*, is hung, very similar to the large bail *e*, before described, and, like the large bail *e*, they are all weighted on their lower parts, to prevent them from rotating with the friction of the revolving pivots *i* and *c*. On each of these bails two frames or standards are affixed precisely in the same manner as those at the letters *f*, *g*, on *e*; the use of these standards is to support the bobbins *m*, *m*, which hold the stock from which the cordage is formed.

The bobbins, are supported upon pins *x*, passing through holes in the standards, as shown. The distance each standard is apart from the other is sufficient to receive several bobbins on a pin. In the drawing each pin has two bobbins upon it. And there are in this case four sets on each frame or standard. It may be remarked here, that other standards, than such as are here described can be used to hold the bobbins; neither is it necessary to keep within this number of bobbins, for, on the contrary, the number may be varied at pleasure to suit different kinds of work. The primary rings have also on one of their sides a row of small pulleys or sheaves, similar to the large ring, to be used for conveying off the strand as it is formed from the bobbins.

Fig. 3.

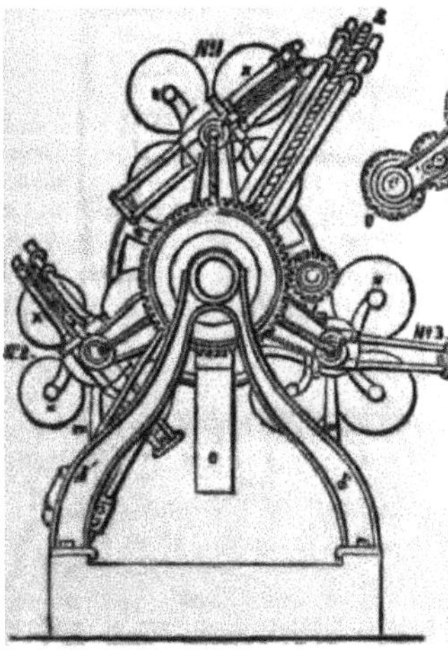

Fig. 5.

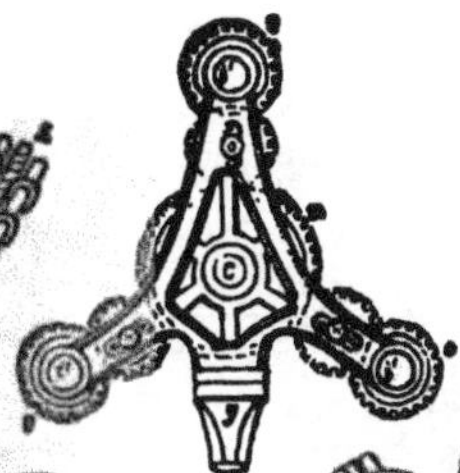

Fig. 4.

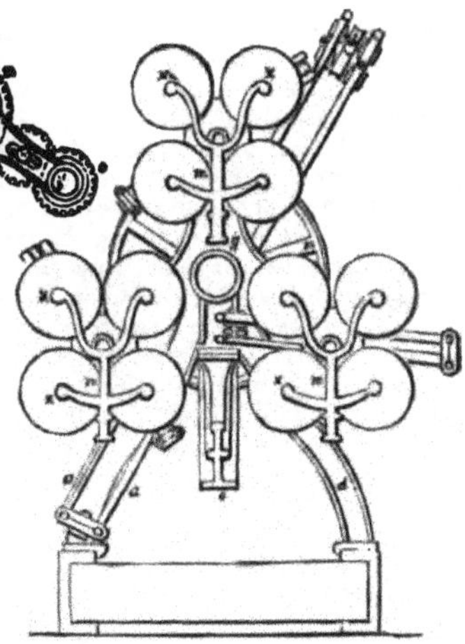

To describe now the gearing for giving motion to the primary rings. These all receive motion from one wheel placed at the centre of the large frame on g; this wheel is represented at n, and is keyed on the end of the shaft c, and receives motion from it; it also rotates in unison with the ring a. Each of the pivots i' passes through the hoop g sufficiently to allow of the pinion wheels o being keyed to them, and these are connected to the wheel m by intermediate wheelwork. The size of the pinion is such that it revolves with greater speed than n, which speed is to be varied to adapt itself to work different kinds of stock. In fig. 5 an end view is given, showing clearly the whole gearing in connection. The reason why a set of intermediate wheels is used, is in order that the primary rings may rotate in the same direction as the large ring a. The tubes for forming the strand are composed of metal, and have an interior opening sufficiently large to receive all the lines collected from the several bobbins. In fig. 3 this part of the arrangements is more clearly shown. The letters $p\,p$ show the tubes in their proper positions; they are held in place on a horizontal bar, which is sustained by a vertical post rising from l; the tubes are also in line with the centre of rotation of the ring h. In forming the strand (which consists of several layers), a series of threads from the bobbins No. 1 are passed through the tubes p'', p', and p; next, from the row of bobbins No. 2, a like number is passed through the tubes p' and p, these forming a layer over the first. Row No. 3, is next treated in the same way, passing the thread through p, outside of all. The threads thus collected form one of the strands of the rope, as yet untwisted. They are then all taken together, and put through the end hollow pivot k, and passed on through the bore of the same until they are brought out at the side opening at i'. Then from i' the strand, thus partly formed, is passed round so as to lay on the little sheaves placed on one-half the ring h, and thence through the opposite hollow pivot i to the rear-end of the machine. The tubes on the other two primary strands being filled in like manner, and the strands brought to the same point, finishes the three strands of which a rope is made. The materials of the three strands thus collected are all to be put through the bore of the hollow shaft *b* at the rear-end of the machine, thence they are passed over the sheaves *d*, placed on the large ring *a*, and thus carried to the opposite shaft *c*, through which they are also put, in the manner shown in the several figures. From the end of this shaft the collected strand is carried to the reeling-machine, which is of common construction.

So far as the machinery has been now described, it relates simply to the putting in of the stock preparatory to the process of twisting. The stock, when so put in, exhibits merely a mass of fine lines, brought altogether from the various bobbins to the place of final delivery at the end of the shaft *c*. The manufacture of perfect rope is now proceeded with by simply putting the shaft *c* in rotation, but not the *stock*, which is carried from place to place to effect the twisting, and remains all the time in a quiescent state. When all the parts are thus set in motion, the primary rings revolve rapidly around their respective sets of bobbins; the reeling-machine also commences drawing off the rope as it is formed, and the twisting process is going on throughout the several parts. The twist given by this machinery is peculiar, inasmuch as it consists in giving a double twist to each individual part composing the rope, and also the rope itself. This is effected by the peculiar action of the rings. By means of the primary rings, the threads from the several bobbins are twisted as they are collected through the tubes *p*, *p′*, *p″*. Two distinct twists are put in by each ring. The first twist at the place where the thread comes first upon it, as at *i*, and the second twist at the place where it leaves it as at *i′*, thus a full twist is put in at each end at every complete revolution. The effect of the first twist is to cause each layer of threads, as they are collected in the tubes, to be wound one upon another, and thus each thread is made to draw or bind upon the other with equal tension. The second twist is given at the outlet *i′*, and this is for the purpose of more perfectly hardening and combining the strand previous to laying it into rope. The laying of the strand into rope is now effected by bringing the large ring *a* into action. All the strands being brought together, are thrus into the hollow shaft *b*, and passed over the ring *a*, and out at its opposite side, through *c*, as before described. The principle of laying or twisting the strand into rope by the large ring, is the same as that produced by the primary rings giving two twists in the same manner; the rope is then formed, and nothing more remains to be done but reeling it.

Fig. 2.

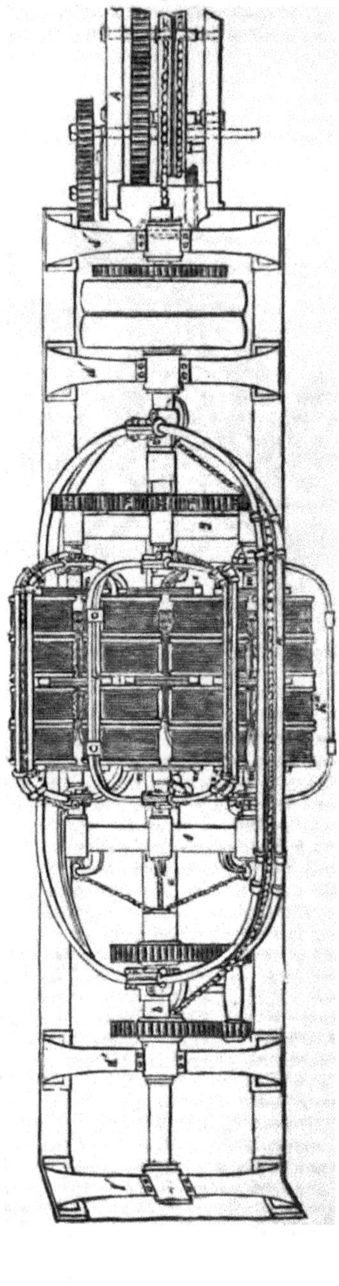

It may be proper to give various degrees of hardness to the rope according to the service it is to perform. This is effected by changing the speed of the drawing-off wheels on A. Accelerating the speed will diminish the hardness of the rope by putting less twist in a given quantity; and diminishing the speed will increase the hardness of the rope by contrary action, or putting more twist in less length.

To manufacture shrouding, or that kind of rope which has a central strand or core, by this machine, it is only necessary to prepare a reel of the proper-sized strand; this is then to be put upon a spindle placed vertically on the belt *e*, a short distance in front of the opening in the shaft *b*, as shown, and the strand being then inserted along with those received from the primary rings, the latter will be laid or formed upon it in the usual manner.

THE "AMERICA."

The *America* affords an example of the need for *singleness* in experiments, and in estimating the value of improvements. Some attribute her success to the wave-line form of her hull; others, to her being furnished with a sliding keel; a third opinion is, that she is indebted for her speed to the flatness of her sails. Probably all of these peculiarities contributed to her triumph, but in what degree respectively is quite unknown—quite incapable of ascertainment, seeing that they are conjointly exemplified in one and the same vessel.

In order to determine the value of each of these peculiarities, it would be desirable to institute experiments on each of them separately. For those on the form of the hull, vessels as small as boats would well afford preliminary indications of desirable forms, so that larger vessels would afterwards be to be constructed, of those forms which the little costly experiments on boats had shown to be advantageous.

In constructing these larger experimental vessels, provision might be made for a future insertion of sliding keels; so that a set of experiments to ascertain their value might be instituted without building vessels expressly for this purpose.

The same vessels would also suit for making experiments on sails—that is, after the experiments on the form of the hull, and those on sliding keels, had each been *separately* made.

Unfortunately, in all comparative experiments that have hitherto been undertaken on navigable vessels, many influencing circumstances have been disregarded—such, for instance, as their *relative bulk, their actual* weight, their *draught of water.*

In regard to sliding keels, absolutely nothing seems to be known of what their most advantageous form would be—whether uniformly quadrangular, like those of their inventor, Admiral Schenk, or whether like those of Commander Shuldham, and of the *America*, almost triangular. It remains also problematical whether the slide is most advantageous when in one single piece, or whether when divided into two or any other number of portions, as they were in many of the vessels that were provided with them towards the end of the last century.

It would seem that the rig of the *America* differed from that of other vessels in various particulars besides the *flatness* of her sails; for example, they were brought down much nearer to the deck than usual, so as to have precluded the use of bulwarks. The form and the dimensions of sails, their position in respect to the length of the vessel, are amongst particulars that are in much need of experimental investigation. So also do the best means of supporting and the easiest of manœuvring them, especially with a view to the getting rid of superfluous cordage. Even the best position for the masts is not yet known.

But supposing the best form of a vessel in regard to swiftness to be ascertained, other considerations have to be taken into account in deciding on the lines to be given to a sailing vessel, so as to suit her for the particular service for which she may be destined. For an advice boat, for a vessel for communicating signals, swiftness is a quality of first-rate importance. The *America* has shown that she is pre-eminently well-suited for such services, though not adequate to the requirements of many others, it is said. Thus, as a pleasure yacht, were it only her want of bulwar and the closeness of her sails to the deck, she would be less eligible than many other yachts; and it is also affirmed that she is deficient in point of internal accommodations, and of stowage-room for provisions. It would be well that her capabilities in these respects should be made as generally known as her speed has been.

The *America* herself would afford means of making experiments that could not fail to be extremely valuable in their results. Supposing the first series to be made with her in her present state—not in a single sailing-match, and under stipulations that it should not take place but when there should be a specified velocity of wind—but in a succession of trials in different seas, and in different states of wind and weather. How would she behave during winter in the Bay of Biscay? Could she work well off a lee-shore in a gale of wind? How are her decks for dryness in foul weather, &c., &c.

After having been sufficiently tried in her present state, her rig might next be altered, in order to ascertain the degree in which her success had depended on her sails. She might be furnished with the rig that in general opinion is the most appropriate for a vessel of her tonnage, the sails bellying as usual, and in every respect conformable to that customary mode. As in the first case, her officers and crew should be those accustomed to the use of her original rig, so in that for which it might be changed, the commander and the crew should be accustomed to that ordinary mode which might be adopted, and moreover persons thinking favourably of that particular mode.

The next set of experiments would be for the ascertainment of the advantages of sliding keels, and the comparative advantages of different forms and numbers of them, as also their effect in different parts of the length of a vessel, and at different depths. Here again it would be desirable to make these experiments under different rigs of the vessel the ordinary one, and that peculiar to the *America* at least.

Thirdly, the advantages resulting from the *America's form* would have to be ascertained. This would easily be done *by taking away her sliding keel, giving her a shallow lined* one in lieu, and then trying her with different rigs, as in the preceding sets of experiments.

On reference to the *Mechanics' Magazine* of May 27th and June 3rd, 1848, it will be seen that a series of experiments "with a view to the improvement of navigable vessels," as proposed by Sir Samuel Bentham in the year 1828, was shortly after authorised, and that they actually were commenced previously to his death, but were from that time lost sight of. The whole of that communication seems still well worthy of the attention of all who are interested in the improvement of naval construction.

In the Magazine for June 3rd, 1848, it will be seen, p. 536, that the sketch D A C, as also the section G, show a form of hull nearly identical with that of the wave-line of the *America*, and which apparently he conceived to be the most promising form, since it is the only complete one delineated. So evidently, he considered that *flatness of sails* is a quality essential to their efficient action; unfortunately that page of his manuscript is missing in which his observations on this subject are likely to have been written; but in a note to article ix., runs thus, "Could a vessel be made to expose to the wind no other surface than that of sails composed of materials *perfectly flat*, such as a metal *perfectly flat*, polished, and impervious to wind, in that case the angle to which such sail might be set might be the most acute," &c., for which see *Mechanics' Magazine*, June 3rd, p. 534.

M. S. B.

LUND'S TREATISE ON THE LAW OF LETTERS PATENT FOR INVENTIONS.*

The present work deals exclusively with what the author calls the "Substantive Law" of Patents, as distinguished from the "Law of Procedure," that is, in (perhaps) plainer terms, with the *principles* of the law which are vital and fundamental, which may be modified or expanded, but never entirely set aside, apart from the *forms* encumbering

* "A Treatise on the Substantive Law relating to Letters Patent for Inventions. By Henry Lund, Esq., of Trinity College, Cambridge, M.A., and of Lincoln's-inn, Barrister-at-Law. 12mo, pp. 248. Sweet.

ich are arbitrary, accidental, and to be at will. Mr. Lund has done wisely uct his work upon this distinction; h of the existing "law of proce- ay be safely considered as doomed abolition, on account of its acknow- exatiousness and absurdity; and it ave been useless to load his work it might, by the time of its publi- ave ceased to exist. We cannot, accord the same approbation to ose which Mr. Lund professes to iew in publishing his work; namely, are (the way) for the *consolidation* ubstantive law relating to patents ntions." We are not aware that h consolidation is wanted, or has led for. The decisions of the Courts g patent rights have been already , and the principles deducible from esented in forms sufficiently me- and compendious, by several pre- riters. Need of consolidation im- ent dispersion, multifariousness, fusion, which certainly is not the the law of patents at the present Although, therefore, we must look professed "purpose" of the work as t idle and superfluous, we are happy at there our slight censure stops; performance itself is deserving of aise for the style of its execution, have uses which, though not per- hin the contemplation of the author, e present time of the utmost im- . We shall be doing no injustice r works on the same subject in say- this will do more to popularise the atents than any one which has yet . Other writers on the Patent Laws itten for lawyers, and like lawyers; nd has written for the public at d produced a work which the public will understand and relish. It is, , not only a very readable, but most ning book. Mr. Lund has adopted py expedient of making the cases g patents decided by the Courts, tell e law is on every point—skilfully in each case its leading features, and eliminating all that is immaterial or irrelevant. The reader's curiosity is first excited by the statement of some interesting practical question which is in issue; and then, by seeing how principles are applied to its solution, he has these principles fixed in his mind in a way which no abstract enunciation of them, however distinct or clear, could possibly have accomplished. Among the good effects to be anticipated from this mode of teaching, we may confidently rank this—that it will make the English people discern a degree of reasonableness and propriety in the existing law, in respect to patent inventions, for which it has hitherto had but small credit in popular estimation. The vicious forms of the law have been confounded with its more essential elements, and much virtuous indignation wasted in denouncing as a whole, a system which, in all else but the outside forms, is excellent, and has served as a model for imitation to the rest of the civilized world.

Mr. Lund starts with this general statement of the basis of our present Law of Patents:

"The right of the Crown to grant to an inventor, by the issue of letters patent, the exclusive profit, for a limited time, of any improvement in the mechanical or useful arts, rests upon the common law, modified by a statute passed in the reign of James I., and these, again, are explained and illustrated by decisions in the Courts of Law." (P. 1.)

We cannot accept this as being at all a correct view of the case. The "common law" of England has from the earliest periods most emphatically repudiated monopolies of every kind. Coke, in his Commentary on Magna Charta, (2 Inst., 63,) states that *this* is necessarily to be inferred from it, that "*all* monopolies concerning trade and traffic are against the liberty and freedom (thereby) granted and *declared.*" And by the very statute of James, which Mr. Lund refers to (commonly called the Statute of Monopolies), it is again expressly declared, that "all monopolies, commissions, grants, licenses, charters, and letters *patent*, heretofore made or granted, or hereafter to be made or granted for the sole buying, selling, making, working, or using of any thing within this realm . . ARE ALTOGETHER CONTRARY TO THE LAW OF THIS REALM." The same statute does afterwards enact, that "the *declaration* before mentioned shall not extend to any Letters Patents and grants of privileges for the term of *fourteen years* or under, *hereafter* to be made of the sole working or making of *any new manner of manufactures* within this realm to the true and just inventors of such manufactures, which others at the time of making such Letters Patent and grants shall not use." But this, be it observed, was no *modification* of the common law, which admitted of no monopolies whatever, even for the shortest periods, but an excellent new piece of statute law, by which, in consideration of the advantage to be gained by the encouragement of "new manner of manufactures," the Crown was (for the first time) authorized to grant Letters Patent for the exclusive use of the same to the "true and just inventors" for the term of fourteen years and under. The import of the word "hereafter," used in the statute, deserves to be particularly noted. If the old common law had had the smallest corner in its toleration for monopolies of new inventions, the statute would not have pronounced all such monopolies, whether for long or short periods, as illegal, but nullified those only which exceeded the modern term of fourteen years.

Besides, it must not be forgotten when speaking of the rights of inventors that, according to a well-known decision of the highest legal tribunal in this country, (*see* the House of Lords case of Becket *v.* Donaldson, 1774,) inventors are in the same case with authors, and never had any rights at all at common law. Property in inventions, and property in literary productions, are equally the creatures of statute law, and have no existence out of or independent of it.

We are aware that the decision just referred to was not unanimous, and that it was opposed to the opinions of some of the ablest and wisest lawyers and statesmen of the time (including Blackstone, himself a host); but it would be useless to discuss now a point which is no longer open to discussion. The law on the subject has been long ago settled, and we but state what that law is.

We have thought it of importance in noticing Mr. Lund's book, to place the real history of our Patent Laws thus distinctly before our readers, because it is a book which, for the reasons we have given, is likely to get into extensive circulation, and because very false and extravagant notions are afloat among inventors on the subject of their rights, which have already done much harm, and are likely to do more; tending strongly to produce a re-action among thinking men—of the commercial class especially—hostile to the existence of patents altogether. Inventors and patentees cannot learn too soon that to take their stand on "innate rights," "mind's inalienable birthright," and other such high-flown nonsense, is to take the battle-field against not only the common law but the common sense of the country; they have, and can have, no other rights than such as it is for the public interest (that alone) to concede to them; and if ten, five, or even one year's protection (instead of fourteen) would serve the purpose of enlisting the greatest possible amount of inventive talent in the public service, the public would act foolishly if they conceded more.

Mr. Lund has not confined himself to the reports of English cases for authorities in support of the legal propositions he lays down, but judiciously had recourse to the reported decisions of the Scotch Courts, which are fully as remarkable as any to be met with in English law books, for clearness and largeness of view. We quote a striking example:

"Precisely the same view of the right to a patent for an applied principle, or for the spirit or substance of an invention, and not merely for the particular form or mode of ication, was very ably pressed upon by Lord Justice Clerk Hope, in an ought by Mr. Neilson, the patentee t-blast patent, in the Court of Ses- Scotland, against the Househill Coal Company for infringements of his Webs. Cases, 683—85): 'It is quite a patent cannot be taken out solely stract philosophical principle—for for any law of nature, or any pro- matter, apart from any mode of t to account in the practical opera- anufactures, or the business and arts ies of life. The mere discovery of inciple is not an invention, in the w sense of the term. Stating such le in a patent may be a promulga- e principle, but it is no application rinciple to any practical purpose: out that application of the principle ctical object and end, and without ation of it to human industry, or rpose of human enjoyment, a per- ot in the abstract appropriate a to himself. But a patent will be ugh the subject of the patent con- e discovery of a great, general, and prehensive principle in science or ature, if that principle is by the ion applied to any special purpose, eby to effectuate a practical result fit not previously attained.

main merit, the most important he invention, may consist in the n of the original idea, in the dis- the principle of science, or the law , stated in the patent, and little or may have been taken in working est manner and mode of the appli- the principle to the purpose set he patent. But still, if the prin- tated to be applicable to any spe- ose, so as to produce any result y unknown, in the way and for the escribed, the patent is good. It ger an abstract principle. It comes principle turned to account, to a object and to a special result. It then, not an abstract principle, eans a principle considered apart y special purpose or practical ope- ut the discovery and statement of a for a special purpose; that is, a invention, a mode of carrying a into effect. That such is the law, known principle is applied for the to produce a practical result for a rpose, has never been disputed. It very strange and unjust to refuse legal effect, when the inventor has ional merit of discovering the prin- well as its application to a practical The *instant that* the principle, although discovered for the first time, is stated, in application to, and as the agent of, producing a certain specified effect, it is no longer an abstract principle; it is then clothed with the language of practical application, and receives the impress of tangible direction to the actual business of life. Is it any objection then, in the next place, to such a patent that terms descriptive of the application to a certain specified result include every mode of applying the principle or agent, so as to produce the specified result, although one mode may not be described more than another; although one mode may be infinitely better than another; although much greater benefit would result from the application of the principle by one method than by another; although one method may be much less expensive than another? Is it, I next inquire, an objection to the patent, that, in its application of a new principle to a certain specific result, it includes every variety of mode of applying the principle according to the general statement of the object and benefit to be attained? You will observe, that the greater part of the defenders' case is truly directed to this objection. This is a question of law, and I must tell you distinctly, that this general claim, that is, for all modes of applying the principle to the purpose specified, according to or within a general statement of the object to be attained, and the use to be made of the agent to be so applied, is no objection whatever to the patent: that the application or use of the agent for the purpose specified, may be carried out in a great variety of ways, only shows the beauty, and simplicity, and comprehensiveness of the invention. But the scientific and general utility of the proposed application of the principle, if directed to a specified purpose, is not an objection to its becoming the subject of a patent. That the proposed application may be very generally adopted in a great variety of ways is the merit of the invention, not a legal objection to the patent.

"'The defenders say; you announce a principle, that hot air will produce heat in the furnace; you direct us to take the blast without interrupting, or rather without stopping it; to take the current in blast; to heat it after it leaves the blast, and throw it hot into the furnace. But you tell us no more; you do not tell us how you are to heat it. You say, you may heat it in any way in any sort of form of vessel; you say, I leave you to do it how you best can. But my application of the discovered principle is, that if you heat the air, and heat it after it leaves the blowing engine (for it is plain you cannot do it before), you attain the result I state; that is the purpose to which I apply the principle. The benefit will be greater or less. I only say, benefit you will get; I have disclosed the principle; I so apply it to a specified purpose by a mechanical contrivance, viz., by getting the heat when in blast, after it leaves the furnace; but the mode, and manner, and extent of heating I leave to you, and the degree of benefit, on that very account, I do not state. The defenders say, the patent on this account is bad in law. I must tell you, that taking the patent to be of this general character, it is good in law. I state to you the law to be, that you may obtain a patent for a mode of carrying a principle into effect; and if you suggest and discover, not only the principle, but suggest and invent how it may be applied to a practical result by mechanical contrivance and apparatus, and show that you are aware that no particular sort, or modification, or form of the apparatus is essential, in order to obtain benefit from the principle, you may take your patent for the mode of carrying it into effect, and are not under the necessity of describing and confining yourself to one form of apparatus. If that were necessary, you see, what would be the result? Why, that a patent could hardly ever be obtained for any mode of carrying a newly-discovered principle into practical results, though the most valuable of all discoveries. For the best form and shape or modification of apparatus cannot, in matters of such vast range, and requiring observation on such a great scale, be attained at once, and so the thing would become known, and so the right lost, long before all various kinds of apparatus could be tried. Hence you may generally claim the mode of carrying its principle into effect by mechanical contrivance, so that any sort of apparatus applied, in the ways stated, will, more or less, produce the benefit, and you are not tied down to any form.' This part of the learned judge's summing-up was not objected to, on the subsequent appeal to the House of Lords."

Mr. Lund might much enrich the future editions of his work, by extracting as freely from the decisions of the American judges as he has done from the Scotch. We have been often much struck with the ability shown by the American judges in their charges on patent cases; and have transferred not a few of them to our pages, as equally deserving attention on both sides of the Atlantic.

MR. HODGES'S PATENT POWER ACCUMULATORS.

Sir, — In No. 1375 (December 15, 1849,) of your Magazine, some diagrams with description, were given of some of my patent applications of India-rubber, and amongst others as an accumulator of power in the hands of a single man. You there state:

"According to our calculations, a thousand of such tubes as he describes (viz., 1 foot long and three-quarters of an inch in diameter,) would weigh no more than 250 lbs., and would pack in a space equal to six cubic feet, and if placed on board of a ship of war would, in many cases, be equal to an additional force of 1000 men."

The accompanying sketch illustrates the form of a power accumulator equal to 4 tons, 7 cwt., and 71 lbs. It contains within an area of one foot superficial, 151 simple accumulators, each a foot long, and equal to 65 lbs. when stretched to its *working* maximum; viz., 6 feet stretched to 7 feet, its power would be about double, or 130 lbs., and to 8 feet (which is close upon the breaking point), its power would again be doubled, or 260 lbs.

It will be perceived that these compound accumulators may be serviceable for many purposes, as they can be made of any required power, and to work through any distance, and the caoutchouc being vulcanised, has become permanently elastic, and unaffected by climatic temperature or moisture. Besides, they float in water, and their weight per ton of power would not exceed 10 lbs.

To move heavy bodies, as hauling off a stranded ship, the crew, with capstan and windlass, could extend accumulators of (say) 10 tons power each, until the object was attained, or the supply of anchors and cables exhausted.

In erecting and fitting machinery, or in bedding masonry, an accumulator of proper power may be attached to the hook of the usual lifting chain, and again made fast to the body to be so fitted or bedded; and supposing this body to be a stone of a ton weight, the mason, with his assistant, could depress it at will by the mere weight of their bodies, and effect quickly a perfect bedding.

The application to relieve the strain, or rather sudden jerk, on ships' cables, standing rigging, and tow-lines, can be by the insertion of an accumu- a spring in any part of the said igging, or tow-line, or by the tow-line being made fast to an lator fixed on the deck, or to rt of the vessel.

Fig. 2

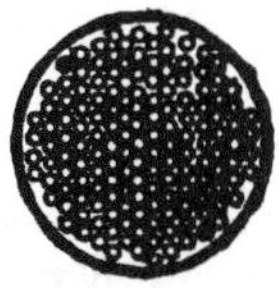

Fig. 4.

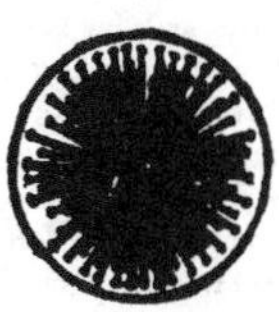

Fig. 3.

Fig1.

the foregoing explanations the on of the accumulators to rail- lers will be readily understood; ust be borne in mind that they are less weighty and bulky than the ordinary appliances, and may be made of any power, and to work through any distance.

If required, one or more ropes made fast at D, and pressed to, and again made fast at D[1], fig. 2, would serve as a guard or check to prevent the accumulator being stretched beyond its working power.

I am trespassing rather freely on your pages, but would make one remark, in conclusion, on the chief peculiarity of this new mechanical power, and in which it differs from the general appliances for moving bodies. It is, that it enables us to accomplish the *quick and sudden removal* of bodies, by accumulating the power at convenient times, and keeping it in a state of tension until attached to the body to be moved, when the attaching rope is *eased off*, and the body lifted or moved. This, indeed, is the principle of my patent projectors for throwing the harpoon, spears, arrows, ball, shot, &c., which you noticed in No. 1375, of your Magazine, before referred to.

Your obedient Servant,

R. E. Hodges.

44, Southampton-row, London, Oct. 25, 1851.

COLT'S REVOLVER AN ENGLISH INVENTION.

Sir,—A great deal has been said lately respecting the claims of Mr. Colt to the invention of the revolving pistol; it will, perhaps, throw a further light upon the subject when we state that in the year 1822, between the months of February and September, we made the barrels of 200 muskets and 200 pistols, upon precisely the same principle as those exhibited by Mr. Colt, for a gentleman named Collier, of Fountain-court, Strand, upon which occasion the lubricating fluid, now so universally used by engineers, viz., soap and water, was first introduced by us; one of these very barrels was, we are informed, exhibited in the English Firearm department of the Exhibition. We have one also in our possession, and can easily prove our assertion by our books, which we shall be happy to show to any gentleman upon application, so that the matter may be set at rest as to Mr. Colt being the original Inventor.

We are, Sir, your obedient Servants,

John Evans and Son.

Engine Lathe and Tool Manufactory,
104, Wardour-street, London, October, 1851.

THE "BERZELIUS" SWEDISH STEAMER AND HER BOILER TUBES.

Sir,—I have observed in a scientific journal of this month, a notice of the Swedish steamer *Berzelius*, in which the editor gravely observes that, "The boilers are tubular, and the tubes are fitted in the tube-plates in rather a peculiar way. In addition to the ends being upset into a countersink, the tubes are slightly enlarged just behind the tube-plate by a suitable tool, after they have been fixed in the ordinary way."

Now, Sir, with the exception of the "ordinary way," tubes are fitted in the tube-plates in no other than this "peculiar way" in America, nor have they been for many years past. I obtained a patent for the tool by means of which the operation is performed at once; that is to say, the end of the tube is expanded on either side of the tube-plate at one operation.

This tool is now in universal use in the United States, and with it, and under my own immediate inspection, all the tubes of the boilers in Collins's rather fast line of steamers were fixed. The number of these tools which I have sold is enormous, and yet to my astonishment I find that tubes are still being put in here in the old tinkering manner which no engineer in the United States would now tolerate for one moment.

I would observe, that every tube being a stay-bar, no others are used, whatever the pressure of the steam may be, and I have known of it being rather high out West.

I am, respectfully,
THOMAS PROSSER,
Of New York.

Birmingham, Oct. 22, 1851.

SPECIFICATIONS OF ENGLISH PATENTS ENROLLED DURING THE WEEK, ENDING OCTOBER 30, 1851.

THOMAS GREAVES BARLOW, of Bucklersbury, civil and consulting gas engineer, and SAMUEL GORE, of Park-road, Old Kent-road, engineer. *For improvements in the treatment of certain substances used in the production of gas for giving light and heat, and of some of the products of the said substances; as also in the apparatus employed in the manufacture of such gas, and in discharging and giving motion to gas.* Patent dated April 15, 1851.

This invention consists, firstly, in certain improved modes of treating certain substances for producing gas, and some of the products thereof.

Secondly, in improvements in the arrangement of retorts, valves, and apparatus adapted for the manufacture of gas.

Thirdly, in an improved apparatus for discharging and giving motion to gas, and adapted for withdrawing the gas from the retorts, or forcing it through the mains.

The patentees describe an arrangement of three retorts, &c., adapted for the purposes of their invention, in operating with which the retorts are, in the first instance, all charged with cannel or other rich bituminous coal, and the charges are worked off in the usual manner. The charge of the first retort is then withdrawn, and it is recharged with coal. The other retorts are not recharged, but the residue of the first distillation is left in them. Steam is now allowed to enter the third retort, where it will be decomposed by the incandescent coke, and converted into hydrogen, carbonic oxide, and carbonic acid gases, which, with any excess of steam, are conducted into the second retort. The excess of steam will, however, be condensed in its passage, and the gases will pass through the coke in the second retort, where the carbonic acid will become converted into carbonic oxide. The removal of the undecomposed steam in this manner is important, as it permits of the conversion of the carbonic acid into carbonic oxide in the second retort, which is not easily effected when an excess of steam is present. The gas passes from the second retort into the first retort, where it mingles with the rich gas from the fresh charge of coal, at the same time becoming itself carburetted by the bituminous products of the distillation of the coal. The whole of the gas then passes into the hydraulic main.

When the charge in the first retort is worked off, it is not withdrawn, but left in the retort, and the charge in the third retort is withdrawn, and it is recharged with coal. The steam is then allowed to enter the second retort, where it will be partially decomposed, and will then pass through the pipes and valves, where the excess of steam will be condensed, and the remaining gases will then pass over the coke remaining in the first retort as the residue of the first operation. The carbonic acid gas will thus be converted into carbonic oxide, and the mixture of carbonic oxide and hydrogen thus obtained will pass into the third retort, where it will unite with the hydro-carbons resulting from the decomposition of the fresh charge of coal, and the whole of the gas will pass into the hydraulic main.

; manner, when the charge in the tort is exhausted, it is not withut allowed to remain in the retort, second retort is then emptied, and d with coal. The steam is then o enter the first retort, where it is decomposed, and the excess of condensed in its passage through and valves, and the gases will then third retort, and, after traversing rt, will enter the second retort, and th the gases and other products from the new charge of coal. The the gas will then pass off into the main. When the charge in the etort is worked off, the first retort ed and recharged, as already deand the operations thus follow one n continual succession.

s mode of operating, the residue h charge in place of being immeithdrawn and replaced by a fresh s retained in the retort, and emn the production of gases of small ing power, for mixing with the ses given off from the fresh charge introduced into another retort. de of treatment is applicable to coal, especially the richer or more sorts, cannel coal, bituminous r schists, asphaltes, lignites, resinhs, and other similar substances res of the same. In place of emthree retorts, as above described, be employed, and in that case each emptied and recharged when the the other retort is worked off.

er mode of conducting the process in employing a single retort, each e length of which is charged alterwhen the charge in the other half sted, and the steam is always adt the end containing the exhausted and the gases allowed to pass off ydraulic main from the opposite he retort. The steam is thus ded by the exhausted charge, and the gases unite with those evolved fresh charge.

e of employing a long retort, and it alternately at each end, a shorter urnished with a partition or diamay be used, and each compartry then be charged alternately, when ge in the other compartment is d.

works conducted on a large scale enerally be found most advantagemanufacture the mixture of hydro- carbonic acid gases in separate devoted to that purpose. In this steam is first passed through a er retorts, charged with coke, or other solid carbonaceous material, from which the gas has been extracted, either in their natural state, or sprinkled with a solution of carbonate of soda or of potash, and the resulting mixture of gases and steam is then passed through a condenser, so as to remove the excess of steam, and the gas is then deprived of its carbonic acid by passing it through another heated retort or retorts containing coke or other suitable residue. When gas is required merely for heating purposes, the process may be stopped at this point; but when gas for giving light is required, the mixture of hydrogen and carbonic oxide thus obtained is then conveyed into another retort or retorts, in which fresh charges of coal, cannel, bituminous shale or schist, asphalte lignite, resinous earth, or other similar substances, or mixtures of the same, are undergoing distillation, or in which tar, melted resin, resin oils, oils, fatty liquids, or other liquid or fusible hydrocarbons are undergoing decomposition. The quantity of the mixture of hydrogen and carbonic oxide employed must depend upon the richness of the coal, or other carbonaceous material undergoing decomposition, and on the illuminating power of the gas required to be produced.

In lieu of passing the mixture of hydrogen carbonic oxide and carbonic acid gases, after the condensation of the steam, through a second retort, as above described, the carbonic acid gas may be separated before their introduction into the carbonizing retort, by passing them through lime, carbonate of soda, protoborate of soda, or other substances capable of absorbing carbonic acid. The lime may afterwards be deprived of its carbonic acid by calcining it, and may then be used over again. The carbonate of soda and protoborate of soda, after becoming saturated with carbonic acid, may be rendered again fit for use by simply boiling their solutions, which will expel the carbonic acid. The mixture of gases is always to be cooled so as to condense the excess of steam which is contained in it.

A method of treating tar, melted resin, resin oil, oils, fatty liquids, and other liquid hydrocarbons is next described. For this purpose the patentees employ the same arrangements of apparatus which have been already mentioned, with the addition of a pipe and cock communicating with each retort for the admission of the tar or other liquid. In operating with this apparatus the retorts are all charged with coke, or the solid residue of the distillation of cannel coal, bituminous shale or schist, or of asphaltes lignites, or resinous earths, or other solid carbonaceous substances, from which the gas has previously been extracted. Steam is then allowed to pass through one of the retorts, through the cooling-pipes or condenser, then through a second retort, and, lastly, into the third retort, into which tar or other liquid hydrocarbon is injected or suffered to drop, and for this purpose it is preferable to inject the tar or other liquid under pressure at regular intervals, so that it may fly over and fall upon the incandescent mass in the retort.

After a time, the coke or other residue in the third retort becomes saturated and incrusted with a carbonaceous deposit, and is rendered less fit for decomposing the tar or other liquid. The tar is then shut off from this retort, and admitted into one of the others, and the residue, with the carbonaceous deposit, is used in the production of the mixture of hydrogen and carbonic oxide gases; and the operations are thus conducted in a regular succession.

The operation of treating coal, cannel, or others of the solid matters before mentioned, may be combined with the treatment of the tar, or other liquid hydrocarbons, which operation is conducted in the manner already described with regard to coal and similar substances, until the charge in one of the retorts is exhausted, and tar or other liquid hydrocarbon is then injected or allowed to drop upon it. After a time the supply of tar is stopped, and another retort is then emptied and recharged, and the operation conducted as before, without the admission of tar. When this charge is worked off, tar or other liquid hydrocarbon is introduced upon it, and the operations thus follow one another in regular succession. The steam employed in operating on tar or other liquid hydrocarbons may be employed at its ordinary temperature, or it may be overheated by previously passing it through pipes heated in the furnace which contains the retorts, or in a separate furnace.

Another of the improvements consists in the production of gas for giving light and heat from tar, melted resin, resin oil, oils, fatty liquids, and other hydrocarbons, by subjecting them to destructive distillation, in heated retorts, partially filled with coke or other residuary solids, products of the distillation of coals, cannels, bituminous shales or schists, asphaltes lignites, resinous earths, or other similar substances, or mixtures of the said substances, upon, through, or over which a stream or an intermittent jet or jets of overheated steam is or are permitted to flow.

The gases produced in the various modes above described may be purified by lime, or otherwise, in the ordinary manner, as is well understood.

The patentees have found that some ties of the solid matters before men particularly Boghead cannel coal, aft ing been distilled, as before describe nishes an ash which contains a consi quantity of alumina, in a state in w is readily dissolved by acids, and w capable of being used in the manufac sulphate of alumina; and for this p the ash is introduced into a pan or with a sufficient quantity of dilute su acid, and it is heated by a fire, or oth so as to dissolve the alumina, and solution of sulphate of alumina. Th which they employ for this purpos vertical partition, perforated with which divides it into two compart The ash of the Boghead cannel coal is in one compartment, and heat is under the other compartment. T liquor is thus caused to circulate t the ash, extracting the alumina fr while the undissolved silica containe ash remains in the compartment w not exposed to the direct heat of t and the danger of burning the bot the pan, which would arise from it exposed to direct heat, and covered w ash, is thus avoided. The sulphate mina thus obtained is mixed with s of sulphate of potash or ammonia, alum crystallized in the ordinary r Sulphate of ammonia obtained in t nufacture of gas may be used for th pose.

The residue of the distillation of B coal may be employed in the manu of alum without treatment by steam, that case the patentees employ such as fuel for heating the pans or boile when it has thus been reduced to th of ash, they place it in the pans or and treat it with dilute sulphuric a above described.

By treating the ash in this manne obtain a residue of silica in a finely- and nearly pure state, suitable for bei with advantage in the manufacture o tery, porcelain, and glass, and whic be used as a dentifrice and as a po powder.

The residue of the distillation of B cannel coal and similar substances o ing alumina may be treated with the sulphuric acid, without previously re all the carbonaceous matter by comb or by treatment with steam. In th the residue, after the extraction of t mina, will be a porous carbonaceous rial, which may be advantageously us decolorant, in a similar manner to which animal charcoal is now employ

The second part of this invention

construction and arrangement of uster and regulator for relieving the on the retorts, and forcing the gas the mains.

s exhauster is a semi-rotary pump, ectangular piston fixed to a shaft the centre of a horizontal cylinder. t of one-sixth part of the cylinder off at the lower part of the cylinis again subdivided by a transverse in the centre, so as to form two es. A similar segment at the upper he cylinder forms two other valve-Each valve-box contains two valves, eating with opposite sides of the The valve-boxes at one end of der contain the inlet valves, and the other end contain the outlet The two valves in each box are a connecting rod, so that they balance one another, and thus e power required to work them.

xhauster is placed between the d the gas-holders, and serves to gas from the retorts and deliver it as-holders. To regulate its action, valve is placed in a "by-pass" ecting the inlet and outlet, so as f the return of a portion of the the production in the retorts dimi-A small gasometer is placed be- exhauster and the retorts, and by a beam and balance weight. production of gas diminishes in s, the pressure decreases slightly, gas returns through the balanced at if this valve should adhere to its gasometer falls, and presses, by a rod, on a long lever connected ve, which is thus lifted off its seat, tion restored. At the same time, a in the gasometer enters the water it works, and seals the passage for The stroke of the pump can be y a variable lever.

r arrangement of two semi-cylinausters, with balanced valves on a nstruction, is also described.

—1. The modes before described g the various solid and liquid subfore mentioned for the production giving light and heat.

arrangement and combination of rts with three-way valves or cocks, appurtenances, as described.

application of a jet pipe and a an elevated cistern with a cock or ved by clock-work or other regu-s, for the purpose of injecting tar iquid hydrocarbons into retorts.

mode or modes of treating Bog-nel coal, and similar substances the *production of gas for* giving light and heat by distillation in retorts, through which is passed a current of the gases resulting from the decomposition of water by carbonaceous matter, and the subsequent operations before mentioned, so as to produce sulphate of alumina, alum, silica, or decolorizing substances, as before described.

5. The combination, arrangement, and application of certain parts of machinery, as before described, to and for discharging and giving motion to aëriform fluids, and particularly the connecting and balancing the valves by a connecting-rod, with or without a weight and set screw, in all the different arrangements of such apparatus, when used with one or more shafts.

6. The placing of a regulator or adjuster to a gas-exhauster upon any part of the pipe (or dividing that pipe for that purpose in any part) between the retorts and the junction of the by-pass pipe with the inlet-pipe of the exhauster, or in any other position between the exhauster and the retorts, except on the by-pass pipe; or passing the gas through a regulator or adjuster, or causing a regulator or adjuster to act on the gas in any part of its passage from the retorts to the gasholder before it reaches the junction of the by-pass pipe with the inlet-pipe, as before described.

7. The application and use of a return pipe direct from the gasholder, to supply the exhauster when the retorts are not making sufficient gas for that purpose.

8. The partitioning and dividing a gasholder or gasometer into two or more compartments, to be used for a regulator or adjuster to a gas-exhauster; and also a peculiar arrangement of apparatus for that purpose.

Frederick Puckridge, of Kingsland-place, merchant. *For improvements in the preparation or manufacture of materials or fabrics suitable for ornamenting furniture and other articles.* Patent dated April 17, 1851.

These improvements consist in covering thin, transparent, or semi-transparent materials, such as prepared gut or skins of animals, weazens, bladders, gold-beaters' skin, or other thin membranous materials, either alone or in combination with light fabrics, such as silks, satins, and fine linens, with gold, silver, or metallic leaf, or with gold, silver, bronze, or other metallic powders. The material to be covered with the metallic leaf or powders is prepared by coating it with gold-size, japanners' or burnish gold-size, or other suitable adhesive composition, after which the leaf or powder is applied to the prepared surface, and will be found to adhere firmly.

The membranous material, when thus ornamented with metallic leaf or powder, is suitable for being applied to various purposes, and in this state may be spun or twisted into music strings; it may also (either before or after being so coated) be attached by means of any suitable adhesive composition to silks, satins, and other light woven fabrics, and in this state will be found peculiarly applicable for ornamenting curtains and articles of furniture. The membranous material (when unattached to woven fabrics) will present on both sides the appearance and lustre of the metal with which it is coated, with the further advantage of great strength and tenacity.

Claims.—1. The covering of the surface of gold-beaters' skin, gut, bladder, or other thin, translucent, or transparent membranous material or substance with gold, silver, or other metallic leaf, or gold, silver, or other metallic powder, so as to impart to the material or substance thus prepared the lustre and appearance of real metal.

2. The application of such prepared membranous substance or material, either alone or attached to or in combination with various thin fabrics, such as silk, satin, linen, or other light fabrics, for ornamenting furniture, articles of dress, or other articles for which it may be applicable.

Thomas Keeley, of Nottingham, manufacturer, and William Wilkinson, of the same place, frame-work knitter. *For improvements in machinery for manufacturing textile and woven fabrics, and other articles composed of fibrous or filamentous materials: also for improvements in the said fabrics and articles.* Patent dated April 17, 1851.

The principal particulars of this invention we shall give in an early Number.

William Andrews, of George-street, Westminster, mechanic. *For certain improvements in steam-engines and in boilers, in pumps, in safety-valves, and in wheels and axles.* Patent dated April 24, 1851.

1. The "improvements in steam engines" have relation to a rotary engine of the emissive class, consisting of two revolving parts, the steam acting on the external moving part, after having exerted its force on the inner revolving part, and the motion of the two being conveyed away by friction rollers in such manner as to produce a uniform continuous rotation of the engine-shaft.

2. The "improvements in boilers" consist in constructing tubular boilers with a series of vertical tubes rising from the furnace and communicating with a horizontal flue and a second series of horizontal tubes through which the products of combustion are conveyed away to the chimney.

3. The "improvements in pumps sist in constructing them with two b and dispensing with the foot-valve delivery-valve. The piston-rod of th bucket is made hollow to admit of t of the second and lower bucket through it.

4. The improved "safety valve" posed of a ring of metal acting as th and fitted to a valve-seat of an annul the parts of which are held toget stays.

5. The improved "railway wh constructed with a tyre formed by on a piece of metal in such manner grain shall lie in a direction transv that of the metal composing th which is done with a view to incre adhesion between the surfaces of and wheel.

6. The "improvements in axles" in welding on at the end, or bearing ring of iron, the grain of which is l direction corresponding to that of tion of rotation.

Claims.—1. The improved const of rotary engine.

2. The improvements in the cons of steam boilers.

3. The improvements in pumps.

4. The improved construction of valve.

5. The improvements in railway w

6. The improved railway-axle.

James Bagster Lyall, of T square, Brompton, gentleman. *improved construction of public c* Patent dated April 26, 1851.

The improvements sought to be under this patent comprehend,

1. The position and arrangement seats in omnibuses, cabs, and railw riages, whereby increased comfo accommodation are obtained for pa with the greatest economy of spac "omnibuses" constructed according invention, the seats for the inside pa are placed back to back, and the c is thus divided into two compar access to which is obtained from be a separate door to each. The roo are so arranged that the passengers f other, and the steps for ascending roof are placed at the back part vehicle between the doors.—The "in cab," is so constructed as to be enter the front, there being two doors, each side of the driver, for the conv of allowing him to open the doors moving from his seat.—The "railw riages" are divided longitudinally in compartments, each provided with a row of seats, the passengers in each c

ing each other.—A funeral car- also specified, in which the seats ourners are placed back to back, r-tight casing with a door behind, lso are the entrance doors), is con- under the double row of seats, for se of containing the coffin, to faci- introduction and removal of which or casing has a series of rollers to its floor.]

node of constructing such carriages ine strength and durability with This consists in the employment purpose of sheet-iron riveted to a T or angle-iron. The floor of the s composed of wood, and the hang- furniture may be of any ordinary n.

articular description of break for to be used on common roads and [Three or four arrangements of e specified. As applied to road the break-blocks are brought into ainst the fore or hind part of the ont wheels by the drag or tension n attached to the horses' collars, by their stoppage; the railway e actuated through the medium of r-rods when the carriages come ion with each other and any other

s Clinton Robertson, of the J. C. Robertson and Co., 166, et, patent agents. *For improve- musical instruments.* (A commu- Patent dated April 24, 1851.

. This invention consists in con- brass instruments on a new prin- d with new harmonic divisions, the well-known defects of keyed l instruments are remedied. Sup- h an instrument—the ophicleide, ple—to be in the key of C, and all and keys closed, the tone is then sonorous, and the notes CG, CE e produced in tune. If, however, to produce the note E following C, you must open four keys, and tube of the instrument is shortened tent of 2 feet 8 inches by the various the holes bored in it. The conse- f this is, that the tone is weakened, notes produced extend only to the ove. In continuing to ascend the e defect increases, and the notes weak and disagreeable. Again; ament in present use has but nine ereas it ought to have eleven—F nd G natural being wanting; and n these added, it would be an im- strument. It is now proposed to e ophicleide, and all such instru- upon *entirely different principles,* and to render them capable of producing any note in the scale according to the laws of harmony; their weakest notes will be equal to the best of those made on the old principle; for, instead of shortening the tubes, as at present done, the inventor lengthens them, so to speak, and the tones, instead of decreasing, increase in strength and volume. Supposing, for example, all the keys to be open, and that the performer commences with the upper C, then, by closing the key No. 1, he does not produce the note C natural, but descends to B, and so in like manner with the numbers 2, 3, 4, 5, 6, the notes thus acquiring, strength, volume, and equality, and being in perfect tune. Moreover, such instruments having no notes artificially produced, and fewer keys, the fingering of them is much more simple. For basses, counter-basses, and bombardos, four keys would be sufficient. The shorter and smaller instruments will require more keys, their compass being increased by six notes in the lower parts of the scale. The number of keys may be, of course, increased or diminished.

In regard to this branch of the invention, what is claimed is the producing of notes in a descending instead of ascending order, whatever the number of keys employed, and with that view, lengthening the tubes (in effect), instead of shortening them.

Secondly. The invention consists of the construction of wind instruments, partly in wood and partly in brass or other metal, the object being to modify their tones sufficiently to render them adapted for performances in a room. This the inventor has effected by doing away with all the brass parts of the instrument, from the pistons to the bell, and by replacing them with wooden tubes. The remainder of the tubes of these instruments, from the pistons to the mouthpiece, only are of brass or other metal. The part of the tube which is composed of wood is covered with calf-skin, or any other suitable kind of leather, in order to give it the necessary degree of solidity. Amongst the instruments to which this system is especially applicable, are the sax-horns and bugles in F and E flat soprano, C and B flat contralto, F and E flat altos, C and B flat tenor basses, C and B flat basses, all with four pistons or cylinders, and F and E flat double or counter basses. By giving proper proportions to the different tubes, and making them of the above combination of materials, sounds may be obtained analogous to those of the human voice. Either the ordinary mouth-pieces or reeds may be employed.

What is claimed under this head is, the combination of wood with brass or any other metal in the manufacture of that class of wind instruments, as above named.

Thirdly. The invention consists of an improved quartet horn. Of all wind instruments, the horn is one of the richest in tone and compass, and it is also susceptible of great variations by changing the keys. However, although it possesses a change of keys of a complete octave in extent, viz., from B flat in alto to B flat below, a composer writing for this instrument is restricted within these limits in the parts for the first, second, third, and fourth horns; he is therefore often compelled to take the deepest notes of a higher key to form the accompaniments of another key, which is not effected without difficulty, and always with indifferent success. To remedy this inconvenience, and open a wider field to composers, it is proposed to complete the class of these instruments, and form a quartet—not as it at present exists, with the four instruments all of the same dimensions, but by giving the soprano part tubes suitable to its diapason, and changing the proportion of the tubes for the contralto, tenor, and bass. The horns may be varied in respect of form at pleasure, and constructed either with or without cylinders and pistons.

What is regarded and claimed as the essential feature of this improved quartet horn, is the giving of the soprano part, tubes suitable to its diapason, and changing the proportion of the tubes for the contralto, tenor, and bass, as before described.

Fourthly. The said invention consists of a new valve, applicable to church and other organs, to the instruments called expressive organs, harmoniums, melodiums, and all others wherein valves are used, which affords a ready means of swelling and diminishing each note, so as to obtain a perfect forte or piano effect, and thus enable a performer to give that expression so desirable in music. The inventor has succeeded in obtaining the power of modulating each note at pleasure, by abolishing the flat valves which are used to shut the orifices of the pipes; these valves opening always to the same extent, and admitting always the same quantity of air, produce the monotony observable in these instruments. He has replaced this valve by a different one, of a conical form, the cone of this valve entering into the mouth of each tube; and the piano and forte effects being produced and regulated by the extent of such entrance, the valve opens according to the degree of pressure exercised on the keys, thus producing a crescendo effect, and the full power or forte is not obtained until the cone has quite disengaged itself from the mouth of the pipe. It results from this arrangement that, by gradually increasing the pressure on the keys, the val gradually disengages itself from the air passage, admits an increasing q of air (or *vice versâ*), and thus ea can be swelled or diminished at merely varying the pressure of the fi

The inventor reserves to himself t of placing these valves in any positi may be required to occupy in diffe struments, either in or out of the re as may appear best; also to employ kinds of levers, or means of commu between the touch or keys and the v

The claim in regard to this par invention consists of the conica affording a very small passage to when it is first opened by the of the fingers on the key with which municates, and opening gradually increased pressure, until it leaves th of the tube entirely free, whereby bo and pianos are produced at will.

Fifthly. The invention consist improved mode of constructing pi description of which will be given next.

The claim in respect of this br the invention, is the producing in single notes from two strings arra above described, the employment bridges, and the reducing and equali pressure on the sounding-boards, described.

Sixthly. The said invention con another improvement in pianos, applicable more especially to the upright pianos. The inventor pro employ two sets of strings in on both sets being sounded or struck same time by the movement of on keys.

In order to convey the movement keys to the back set of hammers, levers is provided in connection wi extending downwards to the botto frame of the instrument. At this front levers or rods are connect those at the back of the instrumen horizontal balanced levers, the fron which, together with the whole set of the same kind (there being o lever to every key), is mounted on which is connected with a suitable raising it up at any desired instant, disconnect the front levers from th and thus throw the back hammer action.

What is claimed under this sixth the invention, is the performing u actions or sets of strings by one set as above described.

Seventhly. The invention consis method of fixing the tuning ends

· strings of pianos and harps to the
f the instrument.
ends of the wires are attached to
carried by screws passing through
ormed in suitable pins fixed to the
. The ends of the screws are formed
for the purpose of being laid hold of
tuning key. Another method of
consists in compressing the wire by
of a screw between two bridges, over
t is stretched. By these arrange-
the continued striking of the wires
so easily slacken the screws as when
in any of the old methods of fixing
s.
laim in respect of this seventh branch
nvention is, the two several methods
d for tuning harps and pianos.
thly. The invention consists of a
of more easily tuning the reeds of
ns, seraphines, concertinas, and
y-reeded instruments. The spring
e of the reed is secured to the reed-
means of a second plate, between
nd the reed-plate the spring is held.
is passed through the second plate,
er end of which screw presses upon
gue immediately over a hollow or
rmed in the reed-plate. By causing
w to press the tongue more or less
recess, the pitch of the note pro-
raised or lowered accordingly.
laim here is for the method of tuning
s of accordions and other like instru-
bove described.

LIAM SMITH, of Snow-hill, gas-meter
and THOMAS PHILLIPS, of Brighton,
r. *For certain improvements in
ventilating, and cooking by gas.*
dated April 24, 1851.

s.—1. A peculiar construction of
apparatus to be heated by gas with
gement of hot and cold air cylinders,
ing or telescopic-burners; and the
ion of the same principle to the
of water.
he construction of swing and tele-
gas-burners to be used in conjunction
e patentee's heating and cooking
us.
peculiar construction of boilers for
water by gas for baths and other
s.
n arrangement of apparatus or fire-
e heated by gas from burners drilled
ath, for warming soldering irons and
ols.
he application of the heating stove
ratus to the ventilating of dwelling-
and other places.
peculiar arrangement of parts con-
g a stove or range for cooking by
gas, with swing or telescopic-burners drilled underneath.

7. The drilling or perforating of the tubes for constructing burners for the combustion of gas to be used for cooking and other purposes in the manner described, that is, so as to project the gas in lateral jets.

ROBERT HAWKINS NICHOLLS, of Pimlico, gentleman. *For improvements in machinery for giving motion to agricultural and other machinery.* Patent dated April 24, 1851.

These improvements, which are exemplified as applied to dibbling-machines and hand dibbles, consist; firstly, of an arrangement of mechanical parts by which by the continuous rotation of a winch handle in one direction, the machine is caused alternately to advance and remain for a certain interval in a quiescent state, during which time the seed is deposited; and, secondly, of an arrangement of levers, which when applied to hand-dibbles constitutes what is termed a "walking dibble;" which levers, when acted on by the alternate elevation and depression of the handles, cause the advance of the machine and the penetration of the dibbles into the ground to allow of the deposition of the seed.

Claim.—The mechanical arrangements and combinations of parts described and constituting improvements in machinery for giving motion to agricultural and other machinery.

WEEKLY LIST OF NEW ENGLISH PATENTS.

William Adolphus Biddell, of St. John's-square, Clerkenwell, Middlesex, founder, and Thomas Green, of No. 4, Trafalgar-square, for certain improvements in moulding, casting, ornamenting, and finishing articles and surfaces. October 29; six months.

Michael Scott, of John street, Adelphi, civil engineer, for improvements in punching, riveting, bending, and shearing metals, and in building and constructing ships and vessels. October 30; six months.

Frederick Crace Calvert, of Manchester, Lancaster, professor of chemistry and analytical chemist, for improvements in manufacturing iron, and in manufacturing and purifying coke. October 30; six months.

LIST OF SCOTCH PATENTS FROM 22ND OF SEPTEMBER TO THE 22ND OF OCTOBER, 1851.

Richard Archibald Brooman, of the firm of J. C. Robertson and Co., of 166, Fleet street, London, patent agents, for an improved method of making screws. (Communication.) Sept. 8; six months.

John MacDowall, of Walkinshaw Foundry, Johnstone, Renfrew, for improvements in cutting wood and other substances, and in the machinery or apparatus employed therein, and in the application of power to the same. September 22; six months.

Henrietta Brown, of Long-lane, Bermondsey, for improvements in the manufacture of metallic casks and vessels. (Communication.) September 24; four months.

Robert Newell, of the city of New York, U. S., lock manufacturer, for certain new and useful improvements in the construction of locks. September 24; six months.

John Baker Pitcher, of Syracuse, U. S., gentleman, for improvements in apparatus for regulating motive power engines. September 24; six months.

John Wormald, of Manchester, Lancaster, for improvements in machinery or apparatus for spinning and doubling cotton, wool, silk, flax, or other fibrous substances. September 29; six months.

Charles Watt, of Kennington, Surrey, chemist, for improvements in decomposing of saline and other substances, and in separating their component parts, or some of them, from each other; also, in the forming of certain compounds or combinations of substances; and also in the separating of metals from each other, and in freeing them from impurities. September 29; six months.

Thomas Kennedy, of Kilmarnock, Ayr, N. B., gun manufacturer, for improvements in measuring and registering the flow of water and othe September 29; six months.

Elijah Galloway, of Southampton-buildi dlesex, civil engineer, for improvements engines. September 30; six months.

William Johnson, of Millbank, Westmi improvements in ascertaining the weight October 1; six months.

William Barker, of Hulme, near Ma millwright, for improvements in mach chipping, rasping, and shaving dyewood, materials, and in apparatus connected t October 6; four months.

Henry Curson, of Kidderminster, civil for improvements in the manufacture o and rugs. October 10; six months.

Thomas Lightfoot, of Jarrow Paper M ham, paper manufacturer, for improve machinery applicable to the making of pa tober 10; six months.

George Robins Booth, of Portland-place worth-road, Surrey, for improvements in g and applying heat. October 15; six mont

William Onions, of Southwark, engineer provements in the manufacture of steel. 15; six months.

WEEKLY LIST OF DESIGNS FOR ARTICLES OF UTILITY REGISTERED.

Date of Registration.	No. in the Register.	Proprietors' Names.	Addresses.	Subjects of Desi
Oct. 23	2991	Frederick Lack	Strand	Anuphaton.
,,	2992	C. & J. Clark	Street, Somerset	Part of an elastic shoe or other cov the feet.
24	2993	James Coate and Co.	Brewer-street, St. James's	Diagonal semi-obliq trating hair-brush.
,,	2994	John Kerslake	Birmingham	Button boot.
27	2995	R. Timmings & Sons	Birmingham	Loose heater or Ital ing tongs.
,,	2996	Bathgate and Wilson	Liverpool	Cask-head and fasten
,,	2997	George Gotch	Islington	Window flower-pot
28	2998	George Wells	Bermondsey	Disc valve.
29	2999	W. Perks	Birmingham	Tap.
,,	3000	J. T. & H. Christy & Co.	Gracechurch-street	Ventilating button.
30	3001	Robert Harcourt	Birmingham	Blind furniture.

WEEKLY LIST OF PROVISIONAL REGISTRATIONS.

Oct. 10	308	William Dicks	Leicester	Elastic side-piece fo
23	309	Richard Barratt	Great Russell-street, Bedford-square	Secured handle brus
,,	310	Robert John Smith	Islington	Steering apparatus.
28	311	Gerrit Sorders	Holland	Double-acting safety
29	312	P. A. L. De Fontainemoreau	Finsbury	Wire cigar light.
30	313	John Roberts and W. Winter	Carlton-hill	Glove-fastener.

CONTENTS OF THIS NUMBER.

LONDON: Edited, Printed, and Published by Joseph Clinton Robertson, of No. 166, Fle in the City of London— Sold by A. and W. Galignani, Rue Vivienne, Paris; Machin *Dublin*; *W. C. Campbell and Co., Hamburg*.

Mechanics' Magazine,

SEUM, REGISTER, JOURNAL, AND GAZETTE.

1474.] SATURDAY, NOVEMBER 8, 1851. [Price 3d., Stamped, 4d.

Edited by J. C. Robertson, 166, Fleet-street.

ROBERTSON'S PATENT IMPROVEMENTS IN PIANOFORTES.

Fig. 1.

Fig. 3.

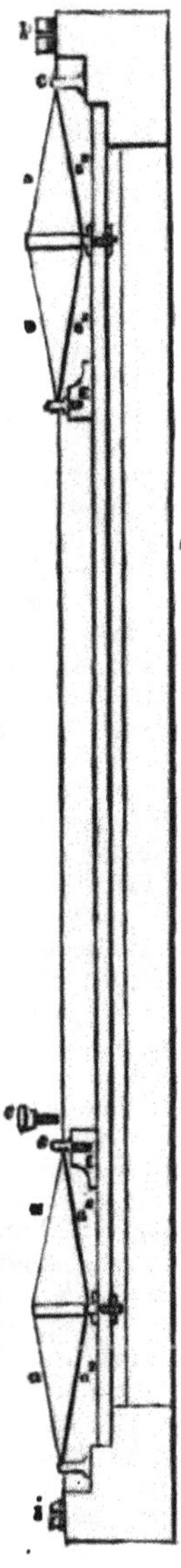

Fig. 4.

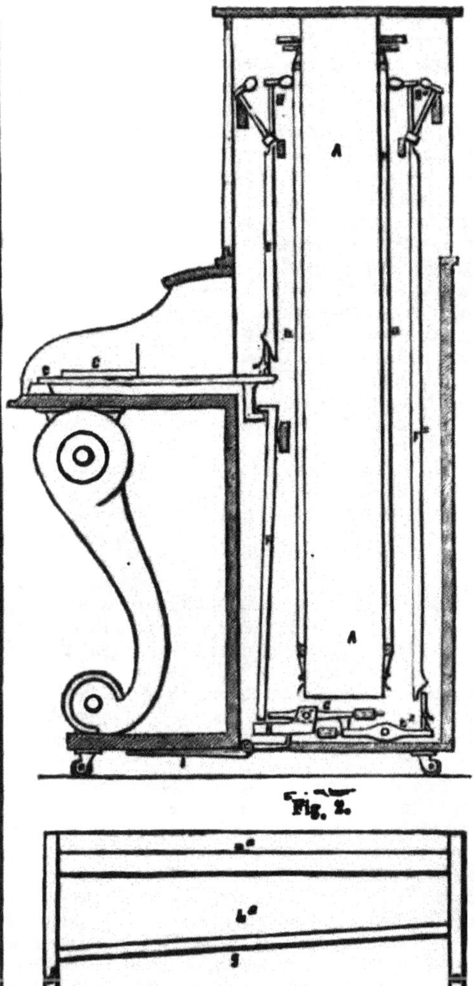

Fig. 2.

ROBERTSON'S PATENT IMPROVEMENTS IN PIANOFORTES.

[Communication from Abroad. See ante, pp. 357, 358.]

[In accordance with the promise contained in our last Number, we now proceed to lay before our readers a full description of these improvements, extracted from the Patentee's Specification.]

Specification.

The invention consists of an improved mode of constructing pianos, whereby the following advantages are obtained. *First,* The volume of sound is greater, the notes are fuller and stronger, they extend to a greater distance and fill a greater space. The quality of the tone is superior and more homogeneous, it comes more freely from the instrument, and the vibration is more prolonged. *Second,* The bass strings, which are very commonly deficient in volume and quality of tone, particularly in upright or vertical pianos, will now be of a sonorousness equal and proportioned to those at the middle and upper part of the instrument; and, *Third,* The longer the instrument is played upon the better it will become,—it will improve by time, like a violin, the strings being arranged so as not to press on the sounding-board; but, on the contrary, to keep it constantly in a state of equilibrium, and so also as to facilitate the vibration.

Fig. 1 is a vertical section of a piano, on this improved plan of construction, on the line of the 3^d *a* in scale, showing two wires or strings for that note; AA are the two strings, stretched parallel to each other between the two bridges EE, and passing through holes in the same; *a* is a continuation of one of the two strings, forming an angle between the bridges C and E, by resting on the top of the bridge D. The string, after passing through a hole in the bridge C, is fastened at the point *i* at one end, and at the other end it is fastened in a similar manner to the tuning screw, or stud *h*; a^2 is the other string, making an angle in an opposite direction to that of *a*, by being passed through a hole in the lower end of the bridge D, after which it is fastened in the same manner as the first cord at *i* and *h*. This arrangement admits of the bridges being much higher than usual; *b* is the lower table fixed on three sides to the frame of the piano, and to the cross beam *f*; *g*, an interval or separation between the two tables *b* and b^2. Fig. 2 is a part front view of the sounding-board of the piano reduced to a scale of one-half the size of the sectional view; a^3 framework of piano; *b* lower table; b^2 upper table; *g* separation between the two.

Fig. 3 is a longitudinal section of a piano with one table or sounding-board. The strings being arranged on the same system as in fig. 1, differing however as regards the construction of the bridges EE in this, that instead of the strings being passed through a hole in the bridge, they are passed through small studs *e e*, which are screwed into the bridges.

The claim in respect of this branch of the invention is the producing in pianos of single notes from two strings, arranged as above described, the employment of high bridges, and the reducing and equalizing the pressure on the sounding-boards, as before described.

The said invention consists also of another improvement in pianos, which is applicable more especially to the class of upright pianos. The inventor proposes to employ two sets of strings in one piano, both sets being sounded or struck at the same time by the movement of one set of keys. Fig. 4 is a vertical section of a piano thus constructed; AA is the frame upon which the wires or strings *a a* are stretched; B B^2 are the hammers by which they are struck; CC the keys, and E one of the levers by which the front hammers BB are actuated; E F^2 are levers, by which the movements of the keys are conveyed to the back set of hammers. At the lower end of the instrument the levers F F^2 are connected together by two horizontal balanced levers, G G^2. The front lever G is mounted upon a frame H, together with the whole set of levers of the same kind, there being one such lever to every key. I, is a pedal by which the frame H, together with the levers F and G, can be raised up at any desired instant, and the hammers on the back thrown out of action. What is claimed under this sixth head of the invention is the performing upon two actions or sets of strings by one set of keys, as above described.

MATHEMATICAL PERIODICALS.

(Continued from p. 316.)

—*The Mathematical Reposi-*
Miscellaneous Papers.—(*Con-*
)

. Volume II. Geometrical Pro-
By Mr. John Lowry.

is paper contains five theorems
five corollaries relating to the
of the circumscribed, in-
and described circles of any
all of which have since been
in the *Horæ Geometricæ* by
Davies and Weddle. Mr.
so gives a collection of thirty-
oblems in plane geometry, to
se theorems may be rendered
nt, but he has omitted their
as he "did not wish to deprive
geometer of the pleasure of
ing them himself."

'II. Demonstrations of Mr.
ethod of Constructing an Azi-
y Mr. Lowry, and the pro-
he Method.

III. Demonstrations of Emer-
orms of Fluents." By the
vans.

I. Reply to Mr. Lowry's Ani-
ns on Mr. Howard's Spherical
. By the Author.

is is the castigation alluded to
te to Art. III., and appears to
not wholly undeserved. By
to a letter bearing date
22, 1794," Mr. Lowry is
nvicted by Mr. Howard of a
confidence; who further com-
it when he "remonstrated
ehaviour so unprincipled [he]
eive an answer to [his] letter.
acknowledgment of hints re-
). 326, No. ix., of the *Scien-*
rticle, was all the reparation
ugh the offence [had] been
repeated till all the store
hausted." In reference to the
that Emerson's "Geome-
had furnished Mr. Howard
tending to establish the cor-
spherical analogues, Mr.
eminded of *Columbus and the*
in answer to the "wish for a
between some theorems in
the *Repository*, and book iv.,
reatise,' Mr. Howard refers
e fable of the Frog and the
ing him at the same time

that there are propositions as well as modes of demonstration in that book not to be found in Simpson's 'Geometry,' nor in any other work that [he knew] of." At the close of the article, a number of questions are put to Mr. Lowry respecting some of his demonstrations and conclusions, but we do not find that he ever replied to the charges brought against him by Mr. Howard.

Art. XII. On the Negative Sign in Algebra. By William Frend, Esq.

*** In this paper Mr. Frend contends for the *non-existence* of negative quantities in algebra and its applications to geometry, and after stating his peculiar views, expresses a hope that some of the readers of the *Repository*, "instead of troubling themselves with negative quantities and impossible roots, will assist in restoring algebra to its true principles."

Art. XVI. On the Description of Parabolic Trajectories. By Mr. William Wallace.

*** The application of the property, that "if three straight lines touch a parabola, a circle through their intersections will also pass through the focus," is here applied to various problems requiring the description of a parabola:—

1. When its focus and two tangents are given.
2. When it has to touch four right lines.
3. When three tangents and a point of contact are given.
4. When two tangents and their points of contact are given. And
5. When it has to touch three straight lines, and have its axes parallel to a given line.

From the frequency and earnestness with which Mr. Wallace repeated his claims to the discovery of the property above noticed, it may be inferred that he set a high value upon it, and the present paper was no doubt composed for the purpose of pointing out a few of its applications. The theorem, however, had been given by M. Lambert, in his *Insigniores Orbitæ Cometarum Proprietates*, nearly forty years before Mr. Wallace republished it in the *Repository*, (*Davies's Hutton*, vol. ii., p. 387).

Art. XIX. Mathematical Lucubrations. By Mr. William Wallace.

*** This valuable geometrical paper commences with the now well-known property that, "if there be a triangle inscribed in a circle, and from any point in the circumference there to be drawn, perpendicular to the sides, their points of intersection with the sides lie in a straight line. Several Loci, Local Theorems, and Porisms follow in order, all of which are well worthy of the student's attention.

Art. XX. On the Trigonometrical Tables of the Brahmins. By John Playfair, Esq., F.R.S., Editor.

*** This article is an abridgement of Playfair's elaborate paper in the *Edinburgh Transactions*, vol. iv., in which he shows that the foundation of the trigonometry of the Hindoo is based upon the following theorem: "If there be three arcs in Arithmetical Progression, the sine of the middle arc : the sum of the sines of the two extremes :: the sine of the difference of the arcs : the sine of twice that difference;—a property which was not known to the mathematicians of Europe till about 250 years ago. The professor hence concludes that, "even if we had no knowledge of the antiquity of the Surya Siddhanta, that the trigonometry contained in it is not borrowed from Greece or Arabia, as its fundamental rule was unknown to the geometers of both those countries, and is greatly preferable to that which they employed."

Art. XXIV. On Fluents, with an Investigation.

*** The *propria persona* of "A. B. L.," by A. B., was the late Miss Lousada, a Jewish lady, to whom Mr. Gompertz dedicated his first "Tract on Imaginary Quantities." She left a translation of Diophantus, in manuscript, with notes, which was at one time intended for publication. From some cause or other this project appears to have been abandoned, nor am I aware in whose possession the MS. now remains.

Art. XXVI. An *Old* Problem with a *New* Solution. By Mr. John Surtees.

*** This problem relates to the *minimum* duration of twilight, and was intended by Mr. Surtees as a *correction* of Emerson's solution, as it appears in his *miscellanies*. The late Dr. Gregory, in his "Astronomy," p. 93, points out the fallacy of this "*new* solution," by showing that, for all latitudes greater than 10°, the declination given by it is such as greatly to exceed the radius at the time of shortest twilight, or, that a *part* must considerably exceed the *whole*. The subject has since been very fully investigated by "Astronomicus" (*Sir James Ivory*), in vol. i. part ii., pp. 48—52, *Math. Repos.*, new series; by Mr. W. Wallace, Leybourn's *Diaries*, vol. iv., p. 314; by "β Cygni" (Mr. Skene), *Ibid.*, p. 315; by Delambre, in his "*Astronomie*," tome i.; also in the *Connaissance des Tems* for 1818; and, lastly, by the late Professor Davies, in the *Phil. Magazine* for September and October, 1833; also *Hutton*, vol. ii., p. 386.

Art. XXIX. Hints Relative to Friction. By Reuben Burrow.

*** This paper is extracted from the first volume of the *Asiatic Researches*, and has already been noticed in the account of the *Gentleman's Mathematical Companion*, where it was also reprinted.

Art. XXX. and XLVI. Four Propositions on the Sun's Distance from the Earth. By Mr. John Dawson, of Ledbergh.

*** These propositions were first published anonymously, in 1768, and fully proved "not only that the distance of the sun, *as attempted to be determined from the theory of gravity*, by a late author (*Dr. Matthew Stewart*), is, *upon his own principles, erroneous*, but also that it is more than probable this important question *can never be satisfactorily answered* by any calculus of the kind. Mr. Dawson not only succeeds in his object, but also points "out *how* an error of such magnitude originated, and *how* it should escape the notice of so exquisite a geometrician as Dr. Stewart." In the accuracy of these investigations he is supported by Mr. Landen, who, in his "Animadversions," expresses his opinion "that the distance of the sun will scarcely ever be determined by the theory of gravitation;" and a reference to Sir John Herschel's "Outlines of Astronomy," Art. 479, will show that the transits of Venus "afford the best and most exact means we possess of ascertaining the sun's distance."

Art. XXXV. and XL. A Method of Discovering the Number of Negative and

Impossible Roots in any Equation. By William Frend, Esq.

*** In the introductory remarks to this paper, Mr. Frend treats those "algebraists who deal in negative or impossible numbers" with much contempt; and although he considers, that "to discover these negative roots, or impossible roots, *is of no use whatever*, yet they who are curious after them may find their number in the easiest way," by the rules here laid down.

Art. XXXVI. Remarks on Mr. Landen's Correction of Emerson's Theory of the Tides. By Mr. B. Gompertz.

Art. XXXVIII. On Pressure. By Mr. John Lowry.

*** This paper contains nine problems relating to the magnitudes, directions, and resultants of forces, together with the pressure of spheres upon planes supporting them, and three spheres supporting a fourth. The latter problem includes Ques. 329, *Ladies' Diary*, as a particular case, which is also answered in Burrow's *Diary* for 1779; but as the method of treatment is entirely geometrical, it is consequently not well adapted to the present state of mechanical science.

Art. XXXIX. A Problem in addition to the Lemmas of Sir Isaac Newton, in his *Arithmetica Universalis*. By Mr. John Wallace.

Problem.

"Given one of the angles at the base ABC, the difference between the base and one of the sides, BC – AB, and one of the segments of the other side AO or OC, made by the line BO, drawn from the angular point B, to make given angles AOB, BOC, with AC at the point of intersection O, to construct the triangle."

Art. XLI. Geometrical Sections. By Mr. John Lowry.

*** These problems relate to the determination of certain lines, so that when added, subtracted, or added to one and subtracted from another, the lines so compounded may have given ratios, or that their rectangle, sum, or difference of their squares may be given. The three problems are thus divided into *eleven* cases, all of which are treated by Mr. Lowry in his usual elegant manner. The late J. H. Swale has also treated the *same series of inquiries* with great originality in the eleventh volume of his MS. remains.

Art. XLVIII. A New Method of finding Fluents. By Mr. James Cunliffe.

*** Mr. Cunliffe's paper contains nine examples of finding the fluents of fractions in which the numerators are *rational*, and the denominators are *irrational* functions of the variable and known quantities. By making suitable assumptions for the irrational parts of the denominators, he transforms the functions into others whose fluents are either known or readily determined, and this in the *then* state of the fluxional calculus appeared of sufficient importance to be entitled a *new* method.

T. T. W.

Burnley, Lancashire,
Nov. 1, 1851.

(*To be continued.*)

Errata.

Page 266, col. 1, line 7 from bottom, after "comments," add "are subsequently offered."

Page 307, col. 2, line 14, *for* "spheres," *read* "spherics."

Page 308, col. 2, lines 4 and 14 from bottom, *for* x *read* X.

Page 310, col. 1, lines 5 and 22, *for* "Reik," *read* "Keith;" also, *for* "T. H. Swale," *read* "J. H. Swale."

HARRISON'S ELECTRO-MAGNETIC ENGINE.

(Provisionally Registered under "The Protection of Inventions Act," 1851.)

This engine acts on the principle of the induced magnetic power of a compound coil or coils of insulated wire conveying a current of galvanic electricity, which acts upon and draws within a suitable aperture, or repels therefrom, a plate or a series of plates of soft iron, or a body of wire, or permanent steel magnets. Two highly important advantages are gained by this arrangement: 1. We succeed in altogether avoiding the greatest impediment which has hitherto existed in the economical use of electro-magnetism as a motive power, namely, the retarding influence of electro-magnets acting on each other after the battery current has been cut off: for as there is but one body of iron or steel in connection with each coil, no such counter-attraction can possibly occur. 2. Another great superiority of the principle is, that

... currents is very ... where permanent ... pass within the ... augments the ... a considerable ... of the mate- ... fitted.

... by our ...:—

... stroke ... and ... the ... an ... the

[illegible]

L... sai... per... Imag... lation ... with ... tended ... cause or oth... have been ab... in whose pos... remains.

Art. XXVI. A... *New* Solution. By ...

, This problem ... mum duration of twili... tended by Mr. Surtees a... Emerson's solution, as it a... *miscellanies.* The late Dr. ...

for ever since that patriotic and ... spirited nobleman, the late Lord ... tern, brought this clever mode of construction under the notice of the ... lic, I had taken great interest ... fortunes, and felt the strongest ... conviction that it was as sound in ... ciple, as cheap and convenient in pr... The facts of the case I ascertained simply these:—The centre links of the main chains broke, from a d... the metal; but so little did this affe... general stability of the structure persons who were on the bridge ... time did not observe the circums... and the traffic over it went on from ... to day without the slightest interru... Mr. Dredge has since done away the centre links altogether: they ... not, in fact, wanted, and the bri... better without them. As it is, it ... in my humble opinion, challenge ... parison with any suspension bridge ... dimensions in the United Kingdom.

I am, Sir, &c.,

AN ENGLISH TOU...

Glasgow, October 20, 1851.

MR. HOBBS'S PICK-LOCK EXPLOITS.

Sir,—Will you allow me a cor... a few words about Mr. Hobbs ... lock picking," as it is called. I ... essayed to have opened the ... ck years ago if it had been pos... ave done so with *one* instrument. ... understand was what the ... Bramah challenged the world to ... does away with all the difficulty, t... use of several instruments, as ... each representing a part of t... This is not what we call "pickin... opening as with a key. Beside... Hobbs was allowed to inspect the ... his leisure; and any one having ... memory could carry in his mind ... nearly the positions and depths ... slits, which, if know... secret of the affai... say, that there ... smiths in London ... ashamed to have ... (sundry parts) as ... who spent days, a... lities allowed them...

I am, Mr. Editor, yo...

'AMERICA" AND "TITANIA."

In page 193, of your Magazine,
educe an extract from the
n the trial-match between the
America and *Titania*, with the
hat both vessels are built upon
e-line principle, and that the
' the theory of construction is
hat of best carrying it out,
elonging to our transatlantic

g carefully read the account of
er in the *Times*, or so much of
iven in your Magazine, I see in
dence that this theory has been
rried out in the *America* than
nia, for so far as the cut of the
concerned, to which the writer
Times justly attributes much of
eriority of the *America*, her
is clearly irrespective of the
e theory.

simple accident of the much
ize of the *America*—her tonnage
more than double that of the
,—appears to me another reason
success of the former, for the
be vessel (*cæteris paribus*) the
he will sail upon a wind, owing
circumstance of the stability or
of carrying sail increasing with
d dimensions in more rapid ratio
e tonnage or weight to be moved.
rinciple is acknowledged by the
n the *Times*, who states, that in
lish yacht race the time allowed
rence of tonnage would be fifty-
nutes. Whether such allowance
he value of the difference I know
t supposing it to be so, the *Titania*
e the winner, seeing she was only
e minutes behind the *America*.
e rightly understood the meaning
writer in the *Times*, in his state-
garding tonnage, I think he has
sufficiently on this fact, which
to the conclusion that a yacht
Titania, if enlarged
ica, and having
decidedly take
manifest the
practice,
tecture.
much
of

sails cut after the fashion of those of the *America*.

The publication of the lines of these two vessels would, I am convinced, tend greatly to the advancement of naval architecture, and I should hail such an event as a boon to the country.

WM. CUDWORTH.

Darlington, 10, 20, 1851.

THE "AMERICA."—PRACTICAL EXPERIMENTS.

Sir,—Your correspondent "M. S. B." offers, in last week's Number, some suggestions which are of the kind we might expect from one, the practical nature of whose sound sense is only equalled by its versatility.

Before reading the communication alluded to, I had taken a walk by the Serpentine, and was much pleased to find several model boats, built and rigged like the *America*, scudding about upon the water under a stiff breeze. The following idea occurred to me:—Let us take two boats *made of gutta percha*, six feet long, cast in the same mould, and representing, as accurately as possible, the form of the *America's* hull. Call them A and B, and rig them precisely the same.

Let them be tried in easy winds, until one shows a decided superiority over the other. Suppose A proved to be faster than B. Now, softening B, and altering it, let it be matched with A, until it is so improved as to be decidedly superior to its former conqueror. Then begin again with A, until B, under its new form, is beaten; and thus proceed, until it is found impossible to make the worse of the two exceed the other in swiftness. Let the last form given to the victor be called C, and make a mould of its hull (in plaster or other substance).

Construct two new boats from this mould, D and E, and let the object of their trials be to ascertain the *best rig*, as that of the pair A and B was to discover the best hull.

Let a false deck be constructed for each vessel, precisely alike, and having the masts and bowsprit moveable fore
ft, by means of horizontal screws,
ving submitted D and E to a
s of trials and modifications,
rig which ultimately sails

not overrate the result.

the effect of secondary currents is very much reduced; and where permanent magnets are employed to pass within the coils, the induced current augments the primary current, and thus a considerable saving in the consumption of the materials of the battery is effected.

Other advantages possessed by our engine, we will briefly point out:—

1. We obtain great length of stroke in engines of reciprocating action, and in rotary engines an almost unlimited amount of power may be obtained; the soft iron plates, or permanent magnets, being arranged in the manner of an endless chain, with intervening non-magnetic bodies passing through the coils, over drums.

2. The body of iron or magnet acted upon exposing a large amount of surface, an instantaneous and powerful induction of magnetism occurs; and thus the highest speed is obtained.

3. By employing a compound conducting material, we transmit a strong current of electricity, and obtain increased magnetic effect.

4. The larger the engine, the greater is its economy, which is directly reverse to the fact in all other modes of applying electro-magnetism hitherto adopted.

After many years of experiments, we feel great satisfaction in having accomplished so much, and ascertained the true direction in which improvements are to be made.

We have no hesitation in asserting, that we shall be enabled to obtain motive power by our electro-magnetic engine at as cheap or cheaper a rate, and much more advantageously, than can be done by steam.

C. W. Harrison.
J. J. Harrison.

Larkfield-lodge, Richmond, Surrey, May 1, 1851.

THE DREDGE SUSPENSION BRIDGE OVER THE LEVEN.

Sir,—In the course of an autumn tour through Scotland I crossed the suspension bridge on Dredge's principle, over the Leven near Lochlomond, and was led to make some inquiries about it, having a lively recollection of a report which appeared not long ago in the newspapers, that this bridge had, shortly after its erection, given way. I was much pleased to find *that there was no* truth in the report; for ever since that patriotic and public-spirited nobleman, the late Lord Western, brought this clever mode of bridge construction under the notice of the public, I had taken great interest in its fortunes, and felt the strongest possible conviction that it was as sound in principle, as cheap and convenient in practice. The facts of the case I ascertained to be simply these:—The centre links of one of the main chains broke, from a flaw in the metal; but so little did this affect the general stability of the structure, that persons who were on the bridge at the time did not observe the circumstance, and the traffic over it went on from day to day without the slightest interruption. Mr. Dredge has since done away with the centre links altogether: they were not, in fact, wanted, and the bridge is better without them. As it is, it may, in my humble opinion, challenge a comparison with any suspension bridge of its dimensions in the United Kingdom.

I am, Sir, &c.,

An English Tourist.

Glasgow, October 20, 1851.

MR. HOBBS'S PICK-LOCK EXPLOITS.

Sir,—Will you allow me a corner for a few words about Mr. Hobbs and his "lock picking," as it is called. I would have essayed to have opened the Bramah lock years ago if it had been possible to have done so with *one* instrument, which I understand was what the Messrs. Bramah challenged the world to do. It does away with all the difficulty, to make use of several instruments, as Hobbs did, each representing a part of the key. This is not what we call "picking," but opening as with a key. Besides, Mr. Hobbs was allowed to inspect the key at his leisure; and any one having a good memory could carry in his mind's eye nearly the positions and depths of the slits, which, if known, constitute the whole secret of the affair. I mean, in short, to say, that there are thousands of locksmiths in London that would have been ashamed to have spent (cutting a key in sundry parts) as many hours as Mr. Hobbs spent days, and having the same facilities allowed them as he had.

I am, Mr. Editor, yours truly,

An Old File.

"AMERICA" AND "TITANIA."

·In page 193, of your Magazine, ·oduce an extract from the on the trial-match between the *America* and *Titania*, with the that both vessels are built upon e-line principle, and that the ' the theory of construction is hat of best carrying it out, elonging to our transatlantic

g carefully read the account of :r in the *Times*, or so much of iven in your Magazine, I see in dence that this theory has been rried out in the *America* than *nia*, for so far as the cut of the :oncerned, to which the writer *'imes* justly attributes much of :riority of the *America*, her is clearly irrespective of the e theory.

simple accident of the much ze of the *America*—her tonnage ore than double that of the —appears to me another reason success of the former, for the e vessel (*ceteris paribus*) the e will sail upon a wind, owing ircumstance of the stability or carrying sail increasing with dimensions in more rapid ratio tonnage or weight to be moved. nciple is acknowledged by the the *Times*, who states, that in sh yacht race the time allowed ence of tonnage would be fifty- ites. Whether such allowance e value of the difference I know supposing it to be so, the *Titania* : the winner, seeing she was only minutes behind the *America*. rightly understood the meaning riter in the *Times*, in his state- garding tonnage, I think he has t sufficiently on this fact, which d to the conclusion that a yacht lines of the *Titania*, if enlarged ale of the *America*, and having milar cut, would decidedly take of the latter, and manifest the ty of our country in the practice, theory, of naval architecture. ical and scientific world is much to the spirited proprietors of yachts for this truly interesting ch; I hope, however, they will se subject further, and try what *nia* can do with a set of new sails cut after the fashion of those of the *America*.

The publication of the lines of these two vessels would, I am convinced, tend greatly to the advancement of naval architecture, and I should hail such an event as a boon to the country.

WM. CUDWORTH.

Darlington, 10, 30, 1851.

THE "AMERICA."—PRACTICAL EXPERIMENTS.

Sir,—Your correspondent "M. S. B." offers, in last week's Number, some suggestions which are of the kind we might expect from one, the practical nature of whose sound sense is only equalled by its versatility.

Before reading the communication alluded to, I had taken a walk by the Serpentine, and was much pleased to find several model boats, built and rigged like the *America*, scudding about upon the water under a stiff breeze. The following idea occurred to me:—Let us take two boats *made of gutta percha*, six feet long, cast in the same mould, and representing, as accurately as possible, the form of the *America's* hull. Call them A and B, and rig them precisely the same.

Let them be tried in easy winds, until one shows a decided superiority over the other. Suppose A proved to be faster than B. Now, softening B, and altering it, let it be matched with A, until it is so improved as to be decidedly superior to its former conqueror. Then begin again with A, until B, under its new form, is beaten; and thus proceed, until it is found impossible to make the worse of the two exceed the other in swiftness. Let the last form given to the victor be called C, and make a mould of its hull (in plaster or other substance).

Construct two new boats from this mould, D and E, and let the object of their trials be to ascertain the *best rig*, as that of the pair A and B was to discover the best hull.

Let a false deck be constructed for each vessel, precisely alike, and having the masts and bowsprit moveable fore and aft, by means of horizontal screws, and having submitted D and E to a similar series of trials and modifications, let F be the rig which ultimately sails the fastest.

Now we must not overrate the result

so obtained, or suppress the fact that it is quite possible that the hull C and the rig F may not, after all, be the best combination, even though the hull and rig have respectively been proved the best, and it may be that some one of the modifications of A or B, as a hull rigged with some one of the forms of D or E, would be superior to F itself.

However, we may be pretty sure that were such a series of experiments entered upon, the boat F would be at least worth improving upon; and supposing that a prize of 50*l.* were offered for a faster model, every chance of fairly testing the excellence of F would be given.

Experiments like these have become possible only since we have had at command a material such as gutta percha,—cheap, waterproof, light, and easy moulded; and I am convinced that a few pounds expended in the mode now suggested would not be thrown away.

I am, Sir, yours, &c.,

JOHN MACGREGOR.

Temple, November 3, 1851.

BOARD OF HEALTH REPORT ON THE WATER SUPPLY OF TOWNS.

(Continued from p. 246.)

After consideration of the sources from which it would be most advantageous to obtain a supply of water to the metropolis, the question next in order seems to be the manner in which that water should be distributed to the inhabitants. The Board say in their Report, that "when making provision for a supply of water to a town population, we deem ourselves bound to consider th peculiar privations of the poorer classes;" accordingly they have obtained an immense mass of evidence on the subject, all of it showing the mischievous consequences of the present defective system. It is not the least of our obligations to the Board that they have not shrunk from entering fully into particulars of inquiry, disgusting as are many of its details.

There are parts of the town where no supply whatever is afforded by the water companies; in other parts the only delivery is from stand-pipes; in the greater part of the metropolis a day or two days' supply is received into butts or cisterns to be drawn from them for immediate use.

Jacob's Island, other parts of Bermondsey, and much of Rotherhithe, are instanced as being in the first named predicament. Mr. Bowie, in his report on Bermondsey, mentioned, in speaking of Jacob's Island, that "in this island may be seen at any time of the day women dipping water, with pails attached by ropes to the backs of houses, from a foul, fœtid ditch, its banks coated with a compound of mud and filth, and strewed with offal and carrion,—the water to be used for every purpose, culinary ones not excepted, although close to the place where it is drawn filth and refuse of various kinds are plentifully showered into it from the wooden privies;" and he goes on describing many other "abominations highly injurious." Dr. Milroy states, "that the people in this wretched locality do not willingly submit to this horrible grievance I can testify from my own inquiries," and mentions the instance of a woman who had expressed her willingness to pay an additional twenty or twenty-four shillings addition to her rent if water were laid on. Mr. Grainger thus reported of Rotherhithe, "it is also particularly characterised by the supply of water in it being most deplorable. Many of the poor depend for their supply on ditches which receive the Thames water."

As to the second mode, stand-pipes, the Board observe, "it is commonly considered that a sufficient and even liberal arrangement is made for the poorer classes if supplies of water be brought within their reach in public fountains, or in stand-pipes placed in their streets and courts, at which they are allowed to help themselves gratis. The notion, however, that such a supply can be gratis, is wholly erroneous. As compared with what may be accomplished by the supply of the water under natural or artificial pressure, they are highly expensive, and to the poorest classes, who can least afford waste of what is their only property,—namely, their labour; the arrangements in question are the most oppressively costly—so costly as to act as prohibitions of the full and free use of such a quantity of the natural element as is requisite for health and comfort."

"To illustrate this, we might take the case of a family living at one end of a court, having to fetch their water from a stand-pipe at the other end of the court. The more frequent case, however, would be that of families living in upper stories of the old and higher houses in towns, and having to carry up water into their own rooms. We will suppose the case of a family living on a third or fourth floor, and actually consuming fifty gallons per diem." "The weight of water alone to be carried per diem would be 500 lbs.;" "add one-fourth to this as the weight of the pitcher or pail, and the actual weight carried is 4,375 lbs." per week. "To fetch the water, and carry such a weight, would probably

the amount of two days' labour week." "Suppose the labour to rmed by the labourer's wife, and could otherwise gain only sixpence , the expense of fetching and car-ll be a shilling." "The conse-f this excess of labour is, that he water be brought to the very into the yard, there to be distri-tis, the labour, that is, the expense, restriction on the use of the com-the smallest quantity that can

ears that in the poorer districts of , especially in the eastern parts of pply is by means of stand-pipes. n during his house-to-house visita-d that, with some few exceptions, orer and middle-class houses in , Bethnal-green, and Shoreditch, are as I have mentioned,"—that is by es. He goes on to say, that "the serve the water either; first, in sterns are almost unknown (I only two attached to the dwellings of and middle-classes in Bethnal-one was a large covered butt, the open wooden cistern containing ns of cooked fish); or, second, in s, or earthenware jugs, pans, or s in small crockeryware bowls, etimes soup-plates. Water pre-such tubs or pans is nearly always -doors. Such vessels are never Even where there are small butts, is still preserved in-doors in small els; when there are tubs, the tubs ntly stowed away below the beds. stance a child fell into one of these ch projected from below the bed, lrowned."

ter kept in cisterns, even those the fully attended to, is universally so hat in the more wealthy families drinking is habitually sent for nearest pump. Dr. Milroy's ters extensively into the bad effects water at private dwellings in the istricts of the metropolis; he stated mongst the most serious evils the present system of intermit-ly of water to the dwellings of the s filthy and polluted state alike of ns and butts into which it is re-nd of the vessels in which it is kept. Whether the cistern or in or out of doors, it is usually l, and consequently exposed to all and smut that are continually fly-t even in the cleanest parts of a ." He mentions "the common mongst the poor of dipping their no matter of what sort, and whether clean or otherwise—right into the cistern or butt every time they require water." "It seems that when the cistern is outside of the house it is liable to the same horrible pollution from the common practice of par-ties who live in the upper rooms emptying everything out of their windows into the court behind." So disgusted are even the poorest with the "water in their butts or cisterns that they will use it only for the purpose of washing, and unless they can catch the water directly from the pipe when it is *on*, they are obliged either to beg it from some neighbour, or (as is frequently the case) get it from the public-house where they deal."

Dr. Milroy instances many other facts, demonstrating abominations to which water in cisterns is exposed,—most of them, it is true, resulting from the filthy habits of many amongst the poor, not on the mode of the water-supply; and observes, that "surely it but requires the fact to be made generally known to convince all of the urgent necessity of a Government surveillance of the dwell-ings of the poor. It is utterly hopeless to expect the correction of such monstrous evils in any other way."

Dr. Gavin was asked—"What is the effect of the retention of water in close atmo-spheres?" "In nearly every place whenever the water has been retained for a day, it has become offensive to the taste." The water in butts or cisterns, as generally situated, is "always exposed to the impurities floating in, and liable to be deposited from, the atmo-sphere, and always liable to be contaminated by the absorption of foul and malarious air."

The Report states, "that waters of the same source, and composed of the same che-mical constituents, may be largely varied in quality by exposure to different atmospheres; and it rests upon the evidence of medical observation, that impure water taken into the stomach occasionally produces more sudden and violent effects than impure air introduced into the system by the respiratory apparatus." "It may be stated as an aphorism, that they who drink water which has stood for a time, or been exposed in a town, drink *town air*, whilst they who drink water brought direct from an elevated rural district, without exposure, are drinking *country air*."

A great variety of other facts prove that the present mode of distributing water is prejudicial to health, and onerous to the humbler classes; besides which the Board adduce many examples of the costliness of the apparatus requisite in the present sys-tem. "Besides the great importance of the distributory apparatus in respect to the salubrity of the supplies, the very defects of the existing system of distribution in the metropolis, and the necessity for the construction and maintenance of tanks and cisterns to receive intermittent supplies, often occasion an expenditure to owners or occupiers equalling, and exceeding, the rates of charge for the water itself, high as they are. Thus, in new houses of the first-class, the expense of cisterns, together with their supports and connected apparatus for water-supply and distribution, the greater part of which, under a proper system, would have been dispensed with, generally exceeds 100*l.* per house. At the usual rate or interest for such property, the annual charge for that outlay would be 6*l.* 10s. But to this must be added the plumbers' annual bills for repairs of dilapidations, which will augment the charge to 9*l.* or 10*l.* per annum, apart from the occupation of space, the pollution of water, and the expense of filtration or of obtaining spring water—the charge for the water itself being between 5*l.* and 6*l.* per annum."

To do away with these several evils, the Board strenuously advise that the supply of water to the metropolis should be *constant* in its delivery into every house, and to the highest of its floors.

Water is already supplied on the constant system to several towns in Great Britain; Preston and Wolverhampton are particularly mentioned in the Report; also, on a small scale, in London, to Rose-court, Dockhead, and a small part of Jacob's Island, Bermondsey.

The great objection made to this mode of supply has been the waste of water that would attend it; the Board, therefore, made much inquiry on this head. Mr. Coulthart states, that at Ashton-under-Lyne, "daily experience is establishing in this part of England—and an annexed Table corroborates the affirmative of the proposition—that a constant and unlimited supply is by far the cheapest method of supplying a given number of dwelling-houses with water; and that the system is not practically attended with that wasteful consumption of the precious fluid which the term unlimited would seem to imply." "Experience has satisfied me" that "what the public voluntarily use" is "forty gallons per house, or seven gallons per head per day, which is amply sufficient on the average to meet all requirements of a domestic nature." Mr. Quick says, in regard to Rose-court, "the Company have also been gainers" by a constant supply, "as they save at least 200 or 300 gallons of water daily."

The Board have ascertained that, by the *present system*, the waste of water in the metropolis is enormous; the quantity delivered by the several water companies amounting to 50 *pails* full per house per diem. "140 *gallons* of water weigh 1,400 lbs., or more than half a ton, which it certainly exceeds the power of one servant-of-all-work to carry about a middle-class house from day to day. In the higher class of houses a consumption of more than 100 pails full of water from day to day, according to the returns, appeared to be impossible."

"Upon the consideration of the state of things brought to light up to this point by our investigation, the question will naturally arise how, with such extraordinary waste, and consequently loss of pumping power and expense, the Companies themselves do not, out of mere regard to their own interests, adopt the constant system of supply?"

"The answer is, that it would be wholly subversive of their present system of charges and their immediate commercial interests. If the constant system of supply were adopted it would be impracticable to have numerous variations of the heights or rates of delivery, or of charges for delivery at high or low pressure. Mr. Wicksteed, the engineer to the East London Water Company, estimates the expense of pumping by the Cornish engine, in London, at 1s. for raising 80,000 gallons 100 feet; or, in other words, 9d. per house for lifting a year's average supply 100 feet.* But his Company's charge for raising the water more than 14 feet high above the ground-floor of a lower-class house would be 14s., or eighteen times the acknowledged cost to the Company; and there would be no pretext for maintaining such rates of charge under the system of constant supply at high pressure."

It might be useful to inquire what the quantity of water may be which a gentleman's family on the Continent consumes, the whole being brought into the house by hand. To adduce one example, in the south of France, where the family consisted of five persons, servant included, the daily provision of water was usually two cans, holding each of them six or seven gallons. There was no washing of linen or floors.

The rare circumstance has happened to the writer of this of having twice inhabited houses in London where the supply of water from different companies was constant, and in a third house where there was a constant supply of spring water; in no instance did it appear that servants were wasteful of water because the supply was unlimited, or that, in ordinary circumstances, there was a greater expenditure of water than where

* "This cost, however, is exclusive of interest on capital."

y was intermittent; but at the same erience in those houses indicated y causes of waste would be to be gainst were the constant system l generally. To say nothing of ievous propensities of ill-educated er of malicious tendencies in their ccidents are liable to produce waste n the most careful families. In a room of one of the above-menouses, there was a tap within a d basin, closeable at pleasure with Supposing that plug to be withd the tap left by accident open, uld, of course, run away to waste basin; it would be the same from communicating with the houseut an accident that happened in e suggests one mode, at least, of g waste from heedlessness. On ion the tap was left open, the basin its plug; some little time afterservant, in the pantry below the room, saw water running through g; the dressing-room was found From a common sink the water ive run away to waste; but were nt supply adopted, with the simple on of placing sinks under water- nuisance of a wetted floor would tention to the turning off the water ugh for use were drawn. Doubter equally simple expedients will hemselves to mechanics for obviat- in this way. In another of those e tap had become corroded or worn er dribbled away continually over a uch waste would be usually disrebut, considering the imperfection and that water would be at high upon them, the loss of it from this would doubtless be considerable. ils are not brought forward as obto a constant supply—far from it, mentioned merely with a view to ttention to them, so that a remedy levised.

works for the distribution of water, d report that, "though the elements ience of hydraulics are fixed and yet the experiments on which the ons of the science to this particular are founded have been made on a ited scale, and on scanty observa- The practice at the commencement investigations was extremely loose irical. This is evident from the arying arrangements of works under conditions. The economic, as well ering failures, in part owing to a m of works, are displayed in the of inquiries under the Health of Commission, where, out of 50 sets of waterworks generally constructed by water companies under local acts, in only six instances could the arrangements be deemed in any comprehensive sense good, while in 13 they appear to be indifferent, and in 32 so deficient as to be pronounced *bad*." "Such, also, is the public inattention and ignorance in relation to this class of works, that those who find the means have no standard by which to judge of the extent of failures, which are therefore but little heeded, and affect but little the professional reputation of engineers employed on such works."

"It has been stated, that the quantity of water now delivered into the metropolis is nearly forty-five millions of gallons per diem." "We believe these returns to be, on the whole, correct. Believing this to be so, it follows from the various examinations of the quantity of water actually consumed, that nearly thirty millions of gallons are daily pumped into the metropolis to waste." "If this large body of water could only be considered as so much wasted, it would represent a certain amount of loss of money alone." "But the actual results are far worse; this water is not only waste, but a positive injury," "by saturating the whole subsoil with fluid refuse, tending to generate foul and highly dangerous gases; as also by rendering the basement floors, the walls, and yards unduly damp, producing all those ill-effects known to exist in connection with swampy, undrained districts."

Dr. Angus Smith, in his evidence on water-pipes, says—"The use of iron seems essential for the conveyance of large bodies of water; the use of other materials may be better suited for smaller bodies. It is the most wholesome that can be used, although it has several disadvantages connected with it, arising from its tendency to oxidise on exposure to water having air in it." "Pipes which are well supplied with oxygen by filling and emptying with water, are more apt to oxidise, according to experience." "When iron is covered with tin, the two metals unite and form an alloy some depth into the surface of the iron; this alloy gradually decreases in its amount of iron until, at the surface, there is a covering of pure tin;" "but, from the imperfection of the surface, and the method of covering, holes are left excessively minute, but sufficient, after a long time, to make themselves perceptible by an accumulation of oxide of iron." "It must be a small hole, indeed, which will not allow of the passage of an atom of oxygen." "Enamelled surfaces, I suppose, have been tried for large pipes. I proposed a mode of covering with pitch for the Manchester pipes, which seems to be useful, although there is room enough for invention still. It covers them with a black varnish, which is also a great preservative to the outside of the pipe if the covering is well made."

Dr. Angus Smith described this mode of coating water-pipes as follows:—"The pipe, is made clean, free from all rust and earth, which clings to it on coming from the mould; the cleaning is a very important thing, as the success very much depends upon it. The surface is then oiled with linseed oil, in order to preserve it till it is ready to be dipped. When the coating is to be made, the pipe is heated in an oven to about 300°, which should also be measured* in such a manner as to prevent soot from coming on it. It is then dipped into a pan of gas pitch, and kept in it for a short time, until it shall have taken up the pitch as intimately as is possible. The pitch should not be too hard, so hard as to be brittle; nor should it be too soft, so as to adhere to anything; when it becomes too hard it may be softened by adding more oil."

"When the pipe is taken out it is covered over with a fine black varnish, and looks exceedingly well." "I was at one time not inclined to think it of much value, but I saw lately one which had stood out every season, and was still in very good order."

The Report says, that "the use of lead piping and lead cisterns has long been objected to, and the remedy would be the disuse of that metal." Many instances are given of the corrosion of lead, and of the poisonous quality thereby imparted to water preserved in lead cisterns, or passing through lead pipes. "Iron piping is altogether better and cheaper than lead, and may now, it appears, at no great additional expense, be protected from oxidation by an earthenware glaze. But for the obstinate prejudice of professional men, and builders in favour of the most expensive materials, earthenware would have been manufactured and used for the distribution of water. It was so used by the Romans."

Might not a disregard of earthenware pipes be attributed rather to the imperfection of their manufacture than to obstinate prejudice? Of late they have been well made, and are coming into general use, though the mode of connecting their lengths together seems yet to call for improvement.

"In France we are informed that earthenware pipes have been laid down, and have been long in action for the conveyance of water at 160 feet pressure. We have had, for purposes of experiment, earthenware pipes made which have only broken at 1,500 feet of pressure. Earthen however, more frangible than iron, hydraulic shocks of the intermitt plies, which often shiver the strong iron pipes, will, without precaution earthenware. On the constant system ply these concussions are diminish by the expedient pointed out by vius, which would in the present thought a novelty, by the simple u air-vessel, the effects of those sh effectually obviated."

The expedient of Vitruvius w "over the venter long stand-pipes be placed, by means of which the of the air may escape."

The water and fire-extinguishing devised by Sir Samuel Bentham fo mouth, Plymouth, and Chatham Do were not provided with air-vessel than those attached to the force- but Mr. Mitchell, late engineer at ness, when he introduced analogou in that dockyard, added air-ve described by himself as follows, "' vessels are placed on different part lines of pipes, and are each sixtee the capacity of the forcing-pumps, either of the steam-engines may st minutes without being materially fel

The Board "have obtained speci earthenware pipes in use for water tion in Switzerland, having screw ware stoppers, and earthenware screw The practical advantages stated fr use of earthenware pipes for water d tion are derived from the higher n ducting power of the earthenware, th quent delivery of the water in a cool the comparative freedom from changes of temperature and sever freedom from oxidation, and, last cheapness."

Professor Clark appears to h covered a simple mechanical preven the poisonous effects of leaden pi cisterns. Some of the Bagshot w poisoned the Queen's hounds; "I ob he says, "a specimen of this water a few days came to the unexpecte that filtration would separate the lea a marine villa of Lord Aberdeen's, the servants suffered in health from water derived from pipes. Sand filt put up under my direction at th and subsequently at Haddo-hou making inquiry recently of his lo agent at Aberdeen, I learn that th have been in use ever since, and waters have been tested from time without any lead having been disco them."

It is stated that oxidation, both

* Query, Managed.

s less where there is a constant
r through pipes than when it is
t.

es have been proposed, and that
es should be made by means of
d by the blow-pipe; but they
yet been subjected to sufficient
rtain the expediency of their use.
rat is of opinion, that "the
ow Water-works are the most
d perfect in the kingdom, being
throughout by the self-acting
nted by James Macinlay, who
l the Shaw's Water-works at
Thus supposing a great fire to
n Glasgow (south side of Clyde),
five times the quantity of water
ed that is usually given off, the
ch extra quantity is drawn off
pipe, these sluices act from the
reservoir (five miles distant), and
uantity as is drawn off the main
e city is discharged from the
nto the distributing-tank, and
ge will continue for any length
quired, without the aid of any
manual labour whatever." "This
however, would require to be seen
ed on the spot to be duly appre-

dering an adequate supply of
mains self-acting seems a desi-
f first-rate importance in regard
e efficiency and to the economy
ve water-works; Mr. Macinlay
efore render essential service to
ers of such works were he to
awings and descriptions of his
ich appear to answer so well their
urpose.

ains, and as far as practicable
es, are recommended to be laid at
n three feet below the surface of
; this to protect them from frost,
streets from the jar of carriages
r them.

rs that it would be cheaper to lay
set of mains along each side of
n a single main along the middle
aving would arise from the lesser
f service-pipe required from the
from the middle of the street;
uld of course be materially in-
y local circumstances, such as the
street.

rd, and many of their witnesses,
he carrying mains (as well as
ng the back yards of houses, in-
e street in front. This innova-
unded on the custom of having
greatest quantity of water at the
ouses, consequently service-pipes
carried at considerable expense, and often inconvenience, across a front area, then through the house from front to back. Some of the witnesses seem well aware that inhabitants would be averse to the necessary invasion of their premises whenever repairs of the mains or sewers might be needed. Repairs, it is true, are of rare occurrence in regard to water mains, whatever may be the case as to sewers; but this innovation need not at present be adverted to.

The requisite diameter of service-mains seems to be very much less under a constant supply of water than when it is intermittent. This is accounted for by its being necessary in the latter system to give, in the course of an hour or two, the whole supply required for a day's, or two days' consumption, instead of spreading it over the whole of the time. It appears from Mr. Martin's evidence that, under the present system, for courts containing about thirteen houses, "it is the practice to lay down a 1½ or 2-inch pipe;" whereas, for such courts he finds, under the constant supply at Preston, "a three-quarter inch branch main more than ample." "I have not been in the habit of using anything smaller than half an inch, although I know such to be the case at other works." He would "not hesitate to supply 20 houses from the same pipe" of three-quarter inches. Mr. Quick said also that, at Preston, a 1-inch pipe served "the entire side of a long street of some 35 or 40 houses." "That in the middle of the day we went to the last house we examined, and asked the person whether the water was on, and we were told it was never off. That was a house at the highest end of the street, and the flow was very free from a half inch pipe." Mr. Stirrat said that the service pipes "in Paisley are a half and three-quarter inch; in Glasgow, half-inch in almost every case of domestic supply; and in Stirling, only quarter-inch lead pipe—the pressure in the latter place being 450 feet: these small pipes are found to be quite large enough." To what amount difference of pressure may in other instances have influenced the size of pipes required, does not appear.

The Report states collateral advantages that would result from the system of constant supply—the application of it to ventilation, as engine power for industrial purposes; for the extinguishment of fire; for washing streets and the fronts of houses.

As to ventilation, the Report runs thus: "In a consideration of the means of improved ventilation, the want has been much felt of a convenient power for the continued and regular application of small forces, as one-horse, half-horse, man, or boy-power. Ventilation dependent on the ascent of heated air, or on thermometrical conditions of the atmosphere, is irregular and uncertain in its action, and, at the time when it is most needed, utterly fails. For this reason we have found it necessary to recommend the application of mechanical power for ventilation, as indispensable for large buildings, or places containing considerable numbers, when regularity in the supply of pure air, or the discharge of vitiated air is to be ensured with certainty. A small power was required to work a pump for ventilating the Hospital for Consumption; hydraulic power was recommended, but in consequence of the expense or difficulty of obtaining a sufficient quantity of water delivered at high service, a small steam engine has been used, to which has been attached a new contrivance for self-regulation. Steam engines require to be examined at short intervals, and, for continuous work through the night, require two attendants. "Very small power would suffice for the regular ventilation of private houses; to such purposes hydraulic power would seem peculiarly applicable."

(*To be continued.*)

STIRLING'S PATENT IMPROVEMENTS IN THE MANUFACTURE OF METALLIC SHEETS, IN COATING METALS, IN METALLIC COMPOUNDS AND IN WELDING.

(Specification enrolled July 31, 1851. See *ante* p. 134)

The first of these improvements in the manufacture of metallic sheets is the use of polished rolls to such sheets as are either intended for being coated with other metals, or after such sheets have been so coated; and this improvement is more particularly applicable to iron plate, either coated or to be coated with tin, zinc, or other of the more fusible metals. After the plates or sheets of iron have been cleaned by pickling or otherwise in the usual way, they are to be passed between polished rollers, using sufficient pressure to smooth the surface without injuring the quality by producing brittleness; and as iron is of such different qualities as regards its ductility, both when hot and cold (according to the district from whence the ore is procured, and peculiarities of make,) no absolute rule respecting the amount of pressure can be given, but a little practice will enable a workman to judge, and care is to be taken that the rolls are clean. The plates so polished are then to be dipped in the usual manner into the metal or alloy intended for the coating. After the plates or sheets have been coated with any metal or alloy, they are, where a high degree of smoothness is desired, again passed between polished rolls, the degree of pressure being carefully regulated so as to avoid producing brittleness. It is not essential that the sheets of metal should be passed between the smooth rolls before coating, but it is preferred that such should be the case.

The first improvement in coating metals or alloys of metals, with other metals or their alloys, relates to coating iron with tin or its alloys after the iron has been coated with zinc. For this purpose the sheet, plate, or other form of iron, previously coated with zinc, either by dipping or by depositing from solutions of zinc, is taken, and after cleaning the surface by washing in acid or otherwise, so as to remove any oxide or foreign matter which would interfere with the perfect and equal adhesion of the more fusible metal or alloy with which it is to be coated, it is dipped into melted tin or any suitable alloy thereof in a perfectly fluid state, the surface of which is covered with any suitable material, such as fatty or oily matters, or the chloride of tin, so as to keep the surface of the metal free from oxidation; and such dipping is to be conducted in a like manner to the process of making tin plate or of coating iron with zinc. When a fine surface is required, the plates, or sheets of iron coated with zinc, may be passed between polished rolls (as already described) before and after, or either before or after they are coated with tin or other alloy thereof. It is preferred in all cases to use for the coating pure tin of the description known as grain tin.

Another part of the invention consists in covering (either wholly or in part) zinc and its alloys with tin, and such of its alloys as are sufficiently fusible. To effect this, the following is the process adopted:—A sheet or plate of zinc (by preference such as has been previously rolled, both on account of its ductility and smoothness) is taken, and after cleaning its surface by immersion in hydrochloric or other acid, or otherwise, it is dried, and then dipped or passed in any convenient manner through the melted tin, or fusible alloy of tin. It is found desirable to heat the zinc as nearly as may be to the temperature of the melted metal previous to dipping it; and to conduct the dipping or passing through as rapidly as is consistent with the thorough coating of the zinc, to prevent as much as possible the zinc becoming alloyed with the tin. It is recommended also that the tin or alloy of tin should not be heated to a higher temperature than is necessary for its proper fluidity. The metal thus coated, if in the form of sheet, plate, or cake, can then be rolled down to the required thickness; and should the coating of tin or alloy be found insufficient or imperfect, the dipping is to be repeated as

scribed, and the rolling also, if de-
ther for smoothing the surface or
educing the thickness.
er part of the invention consists in
ead or its alloys with tin or alloys
The process is to be conducted as
scribed for the coating of zinc, and
ice of the lead is to be perfectly
The lead may, like the zinc, be
more than once, either before or
ig reduced in thickness by rolling.
aulic press may be advantageously
in the process of coating lead or
with tin or its alloys; and as this
already practised and well under-
applied to the coating of lead-pipe
it is only necessary to remark that
orifice must be used of such length
as will allow an ingot, cake, or
be formed. On both sides of this
heet melted tin is to be poured into
receptacle, as is well understood
aking of pipe; but where only one
ortion of the cake, ingot, or sheet
inned, a partition or division should
to confine the melted tin, so that
nly be applied to that portion of
which is required to be tinned.
smooth surface is required, the
ther form of lead is to be passed,
heated state, through a collar of
ard and smooth material, such as
steel or iron, kept as cool as may
ere a strong coating of tin is re-
he lead so coated is to be passed
melted tin. Such coated lead, or
, may be reduced by rolling; and
lead so coated is to be reduced to
thinness, the further coating is
eously given after the coated metal
reduced to some extent by rolling.
ber of additional coatings may, in
manner, be given, according to the
for which the coated lead is re-
In coating lead or its alloys with
recommended that, for purposes
surface of lead is to be avoided,
should be used. When lead is
ith antimony, zinc, tin, or other
render the lead more hard than
ts ordinary state, the tin coating
be somewhat hardened by alloying
or other suitable hardening metal.
and its alloys may also be coated
or its alloys of greater fusibility
metal to be coated, as follows:—
, or other form to be coated, is to
as soon after casting as may be
on, gun metal or other suitable
or if this cannot be conveniently
surfaces are to be cleaned and
for the reception of the coating
her by previously tinning the sur-

face or by applying other suitable material to facilitate the union, as heretofore practised. At one end of the mould is to be attached chambers of more than sufficient capacity to contain the quantity of metal to be used for coating, which may with advantage form an integral part of the mould, or such chamber may surround the mould, and by one or more sluices or valves in such chamber or chambers, the melted metal is to be allowed to run on to the surface of the metal to be coated, when the metal is to be coated on one side only. When it is intended to coat the metal on both sides, the vertical position will be found convenient, and the coating metal is to be formed into a chamber or chambers attached to the mould, and to be introduced into the lower part of the mould by opening a sluice or valve, sufficient space being left on each side of the cake or other form to allow of the coating being of the required thickness; the sluice or valve should be of nearly the width or length of the cake or other form, and the melted metal should be allowed to flow into the bottom of the mould. (Mr. Stirling here observes, that he is aware that lead has been previously coated with tin by pouring tin upon the lead, and also by pressure, and that he does not therefore claim the coating of lead by such means.) The surface of the plate or cake ought to be smooth and true, and the mould, if horizontal, to be perfectly so, and if upright, quite perpendicular, so as to insure in either case an equal coating. The surface of the lead should also be clean, and it will be found advantageous to raise its temperature to a point somewhat approaching the melting point of tin or of the alloy employed for coating, as by this means the union of the two metals is facilitated. It is recommended also, that a somewhat larger quantity of the tin or alloy than is necessary for the coating of the lead or other metal, or alloy, should be employed, and that when the requisite thickness of coating has been given, the flow of the coating metal be stopped, as by this means the impurities on the surface of the tin will be prevented passing through the opening on to the surface of the cake; the chamber or chambers, should be kept at such temperature as to ensure the proper fluidity of the coating metal. Zinc and its alloys may in like manner be coated with tin and its alloys, by employing like apparatus to that just described for coating lead and its alloys, and it constitutes a part of the invention thus to coat zinc. The coating of zinc with tin, however, is not claimed, that having before been done by pouring on tin.

Another part of the invention consists in

coating zinc and its alloys with tin and its alloys by pressure. For this purpose Mr. Stirling takes a suitable piece of zinc or alloyed zinc (by preference previously rolled), and when it is desired to coat it on both sides with tin or alloyed tin, he takes a piece of rolled tin or alloyed tin, of sufficient dimensions to completely cover the zinc. He then subjects the metal so placed to pressure, to obtain perfect contact: and for this purpose, when making sheets, he employs rolls, and rolls out the two metals to the extent desired.

The last part of the invention relates to the employment of zinc when welding together plates or other forms of iron, which is principally applicable when piling iron. Thin sheet zinc, placed between the layers, has been found to answer well; but the use of calamine, in the form of powder or paste, is preferred. In the latter case the paste may be formed with water, to which a small quantity of borax may be added; the paste can then be applied with a brush or otherwise, to the surfaces of the plates or other forms of iron. Additional stiffness and toughness are produced by this process, and cold short iron is believed to be more particularly benefited thereby.

MR. HODGES'S PATENT POWER ACCUMULATORS.

The following description of the engravings which accompanied Mr. Hodges's communication in our last Number was accidentally omitted:

Description.

Fig. 1. The Accumulator in its normal state, 1 foot long.

Fig. 2. The Accumulator stretched to its *working* maximum; viz., 6 feet.

Fig. 3. End view.

Fig. 4. Section at A:

A, The 151 India-rubber tubes, extending from B to B, and looped at their ends.

BB, End boards.

CC, Cords which pass through the boards, and are attached to the loops at the ends of the tubes.

SPECIFICATIONS OF ENGLISH PATENTS ENROLLED DURING THE WEEK, ENDING NOVEMBER 5, 1851.

JONATHAN WRAGG, of Wednesbury, coach and axletree smith. *For certain improvements in railway and other carriages.* *Patent dated April 26, 1851.*

The first part of this invention relates to a method of constructing breaks for r carriages and trucks of all descriptions, are brought into action against the wheel by the buffer-rods being push The break blocks are suspended the framing of the carriage or tr such a position that when the buffer carriage is pressed back, it is caused directly on a lever connected to the blocks, and thereby force them into c with the rail or wheel, so as to eff stoppage in rapid succession of each riage in the train, and thereby dimini serious consequences resulting from sion of the train with another train, c any object on the line.

The second part of the invention co in constructing railway and other c axles, by casting them of metal, wh capable of being rendered malleable annealing. For this purpose the patente fers to employ that description of st the making of which a patent was gr on the 7th February, 1851, to W Onions, of Southwark (see *ante*, p. and to use a core for the purpose of m the axles hollow. The process of an is performed in the customary mann the time occupied for it will vary acc to the thickness or substance of the be annealed.

Claims.—1. The mode described c ing breaks to be brought into action the buffers of railway carriages, trucks, and other railway carriag pressed back.

2. The making of the axles of and other carriages by casting th metal, and annealing the same.

DANIEL DALTON, of Spon-lane, Bromwich. *For improvements ap to railways.* Patent dated April 26,

These improvements consist of constructions and combinations of ra longitudinal iron bearers or sleepers laid end to end on the ballast, so as in effect firm continuous iron bearin rails. The drawings exhibit various of rails, H, T, and U-shaped, co with wrought or cast iron longi sleepers, and secured by keys, or bo nuts, the sleepers being in every cas of forms adapted to the particular to be employed of securing the rails sleepers or bearers for bridge-rails, w wrought iron, are rolled, and when iron, are cast with a hollow central longitudinal projection, which fits hollow of the rail; and the rail secured to them by vertical bolts through the flanges on the edges of t or by horizontal bolts through the r the rib of the longitudinal bearer he line is preserved by tie-rods, in manner, and the rails and bearers aid so as to break joint with each

—The several constructions and ons of longitudinal bearers and ibed.

MIN HYAM, of Manchester, tailor er. *For certain improvements in d of fastening down trousers or icles of wearing apparel.* Patent il 26, 1851.

proved method which forms the this patent is equally applicable s, gaiters, overalls, &c., and cone employment of an elastic band o the lower back part of the artipassing around the front of the boot, so as to keep the trousers he heel, and prevent their rising.

—The fastening down of trousers similar articles of wearing apparel, ployment of a strap of India-rubher suitable elastic material at- the lower extremity thereof in ter as to draw the back part of close to the heel of the boot, the acting against the front part of the e boot or shoe, instead of under customary.

OOPE HADDAN, of Bloomsbury- vil engineer. *For improvements manent way of railways, in railother carriages, and in the manupapier maché to be used in makges and other articles.* Patent il, 26, 1851.

"improvements in the permanent ways," consists in constructing it ms of rails or bars of iron to be each other, one form constituting nd part of the sleeper, and the n the remaining portion of the The two portions are connected t an obtuse angle, at the apex of the rail, and the tie-rods are so s not only to preserve the gauge e of rails, but also the angle at two portions which constitute the earer are in the first instance set. cting permanent way according to tion, care must be taken to make on of the two "forms or bars of ak joint with each other respec-

improvements in "railway caronsist in constructing their sides or other parts by combining or pieces of any suitable composiher material squeezed, pressed, or so as to constitute the framing ing, or the framing covering and Mr. Haddan prefers to employ for this purpose papier-maché pulp, which he moulds to the required shape, forming also at the same time any mouldings, or ornaments which may be desired. The moulded framing, panels, &c., formed in this manner are, when dry, to be saturated with linseed oil, and baked, in order to render them waterproof, as is well understood. Or, instead of producing the framing, &c., by the method just described, sheets of papier maché may be employed, and the mouldings, cornices, or ornaments produced by cementing on successive slips of paper, until the required form is obtained.

3. The improvements in "other carriages," are improvements upon a design provisionally registered by Mr. Haddan, February 1, 1851, for "a handle apparatus for omnibus roof," and consist in connecting the upright and horizontal portions of the hand rail, by a curved or elbow piece, so that the hand of a person mounting the roof of an omnibus may be slided freely along the said rail, and in jointing the rail to the roof by a hinge, or other contrivance, so that when not required for use, or when the carriage is passing under a gateway, the rail may be folded down flat upon the roof.

4. The "improvements in" the manufacture of papier maché, consist in forming sheets thereof by cementing together successive layers of paper, by means of Jeffrey's marine glue, or other suitable composition, each sheet as laid on is subjected to pressure under a steam-heated surface, to express the superabundant cement, and evaporate the volatile portions of the same, when the pressure is removed, another sheet of paper is cemented on, and this process is continued until sufficient thickness has been obtained. When the paper is of so absorbent a nature as to neutralize the effect of the cement, the sheets should be saturated with coal tar, naphtha, or other suitable material, previous to their being cemented together.

JAMES NASMYTH, of Patricroft, engineer, and HERBERT MINTON, of Stoke-upon-Trent, China manufacturer. *For improvements in machinery or apparatus to be employed in the manufacture of tiles, bricks, and other articles from disintegrated or pulverized clay.* Patent dated April 26, 1851.

The machinery which forms the subject of this patent is so contrived as to effect the supply and compression of the clay-dust between the dies and the removal of the finished brick by the continuous rotation of the main shaft. The upper die, by which the compression is effected, is actuated by an eccentric, and thus a pressure is obtained which is well calculated to cause the complete expulsion of air from amongst the particles of clay; and two cams on the same shaft, with the aid of lever connections, cause the movements of the lower die and moulding plate necessary for raising the finished brick, and bringing the moulding plate along the surface of the table in a fit position to receive a new supply of pulverized material. The continuous rotation of the shaft then causes the descent of the upper die upon the lower one, which will by this time have arrived at its position of rest, and a repetition of the several movements just described again takes place. The whole apparatus may, if thought desirable, and in order to effect the more perfect exclusion of atmospheric air from the brick or article in process of formation, be placed in a chamber from which the air has been exhausted. The hopper would, in that case, require to be elongated to such an extent as to allow its top to project above the air-tight chamber.

Claim.—The employment or use of mechanical power, in the condition of continuous rotary motion, to produce a reciprocating action for the purpose of consolidating or compressing disintegrated or pulverized clay into the form of bricks, tiles, and other like articles, and to perform the other several functions of the machine, successively as described, and also (if preferred) the placing of the said press, when employed for the purposes named, in a chamber from which the air has been extracted.

Benjamin William Goode, Richard Boland, and James Newman, of Birmingham. *For improvements in chains, chain pins, swivels, brooches, and other fastenings for wearing apparel.* Patent dated April 29, 1851.

1. These improvements consist in employing a chain in combination with a cap or shield of a pin. The chain is attached to the pin by means of an eye of sufficient size to admit the chain to pass freely through it, and an ornamental drop is fixed at one end of the chain, which drop has a hook or other fastening to connect it to a part of the chain intermediate of the two ends, so that when so connected the cap shall be held secure on the point of the pin, and the pin will be retained in the dress till the hook is released. The novelty of this part of the invention consists in combining a cap or shield by means of a chain to a pin, as described.

2. The invention consists in raising pieces of sheet metal into such forms that when soldered or brazed together they shall form connected links of a chain; and it is the forming chains in such manner which constitutes this portion of the invention.

3. With relation to "swivels," several improved or novel modes of const are described. The first impro consists in forming the swivel with a its ring attached to a sliding bolt mc a spring, which has a tendency to k ring of the swivel in an unbroken st pressing down the sliding bolt, the s opened, and again closes when the p is removed. Another improvem swivels consists in using a sliding pin to lock that part of the ring opens; in this case also a spring ployed to keep the parts together. improvement consists in causing par ring to move laterally on an axis, portions being held in their closed p by a screw or spring; and a fourth in ment consists in forming one portion ring of a swivel to slide within th part in the manner of a bolt, the portion being kept in its closed s means of a coiled spring.

4. The improvements have relat brooches, and consists in protecting the of the pins by shields which are acted springs, several modifications of t rangement being shown. Under this of the invention, the patentees also an improved fastening which con two ornamental buttons or surfac provided with pins like brooches, a nected together by a strip of elastic rial. The use to which this faste applied is to connect together two p a dress or other article of wearing ap

5. An improved fastening for br the ends of chains, &c., is described fastening consists of a hook-and-ey slide, by which the parts are prevente becoming disconnected.

6. The improvements have relat covered or other buttons in which the consists of a cross-bar of metal, th of the button being in the form of The bar is formed with forked or ope and the connection between the sha shell is effected by compressing th ends of the bar, so as to embrace th of the ring or shell which has been between the forked ends.

Henry Lund, Esq., of the Temp gineer. *For improvements in prop* Patent dated April 30, 1851.

This invention consists of an im under-water propeller, a full descrip which, with engravings, will be given early Number.

Philip Webley, of Birmingham, facturer. *For improvements in the facture of boots and shoes, and in ren the said manufacture waterproof; the machinery and materials to be therein.* Patent dated April 30, 185

aprovements comprehended under
it are—

method of fastening on the heels of
shoes by means of a double-act-
, by which the holes are pierced,
nails necessary for securing the
ather inserted.

ew manufacture of rasped leather
a-rubber or gutta percha, which
atee calls "compound leather."
igs and waste pieces of leather are
for this purpose, which, having
hed, are formed into blocks with
adhesive matter, rasped, again
to remove the glue, dried, and
a suitable apparatus with gutta
r India-rubber, and then pressed
ts, which are afterwards to be em-
a the same manner as ordinary

method of waterproofing boots and
means of thin sheets of gutta per-
h are warmed on a last to bring to
shape of the boot or shoe to which
to be applied.

.—1. The application and use of a
ting press for the purposes named.
application and use of perforated
slides containing piercers and man-
ier separately or combined.

application and use of a press, in
ion with perforated pieces or slides
g piercers and mandrils, or either.

system or mode of treating the
parings of leather in the manner
, and the mode of reducing them
r raspings.

use of scraps or parings of leather,
with or united by gutta percha or
ic.

system or mode of waterproofing
shoes.

ARMAND LECOMTE DE FON-
REAU, of Finsbury and Paris. *For*
ents in the manufacture of fuel.
unication.) Patent dated May 3,

ivention consists in manufacturing
rom the small branches of trees,
ants, all kinds of refuse of wood,
n, shavings, saw-dust, &c., and in
g the same with other inflammable
s.

anches are, in order to their con-
charcoal, subjected to several dif-
rations; the first of which consists
y carbonising them to facilitate their
it pulverisation; the next opera-
pulverize the carbonized branches,
ffected by means of rollers revolv-
d a centre in a circular trough;
then impregnated with coal-tar, or
ilar *bituminous or resinous* sub-
ed *well mixed into* a paste; the
mixture is then compressed into blocks, which are finally subjected to carbonization in close pots in a suitable furnace, into which the pots containing the moulded charcoal are introduced on carriages running on rails laid down in its interior. The pots are placed on the carriages, two of which are introduced into each furnace, and the process of carbonization is effected by the heat generated by the ignition of the gases arising from the moulded charcoal. When the charcoal ceases to give off any gas, the process is considered completed.

Coke-dust may be also treated in a similar manner, and a good fuel thereby produced, the quality of which may be improved by the addition of coal-dust in quantities varying according to the nature of the product desired. A charcoal having a metallic sound may be produced by treating coke-dust of good quality as above described, and then immersing it in coal-tar, and subjecting it to carbonization in closed vessels, repeating the immersion and carbonization until it ceases to increase in weight.

In order to obviate the difficulty which occasionally exists in igniting charcoal of a close texture, the patentee sprinkles the blocks, previous to the final carbonization, with a solution of some salt capable of melting, crackling, or decomposing under the influence of heat; or he mixes the salt with the charcoal in a state of powder, and thereby produces in the carbonized blocks an artificial porosity.

Claims.—1. The carbonizing the small branches of trees and other animal plants, and the refuse of all ligneous substances, by means of the apparatus described.

2. The pulverizing of carbonized small branches of trees and animal plants, and the refuse of all ligneous substances, as described.

3. The mixing of carbonized and pulverized small branches of trees and animal plants, and the refuse of all ligneous substances by means of the apparatus described.

4. The moulding of carbonized, pulverized and mixed small branches of trees, animal plants and refuse, and all ligneous substances, by means of apparatus described.

5. The mode of carbonizing in close vessels the carbonized, putrefied (*sic* in orig.) mixed and moulded small branches of trees, animal plants and refuse of ligneous substances by means of the apparatus described.

6. The mode of rendering charcoal more easily inflammable by means of an artificial porosity. (By a clerical error the word "animal" is throughout the claims substituted for "annual.")

WEEKLY LIST OF NEW ENGLISH PATENTS.

Thomas Greenwood, machinist, and James Warburton, worsted spinner, both of Leeds, York, for certain improvements in machinery for drawing and combing wool, silk, flax, hemp, and tow. November 3; six months.

George Fergusson Wilson, manager of Price's Patent Candle Company, Vauxhall; David Wilson, of Wandsworth, Esq.; James Childs, of Putney, Esq.; and John Jackson, of Vauxhall aforesaid, gentleman, for improvements in presses and matting, and in the process of and apparatus for treating fatty and oily matters, and in the manufacture of candles and night-lights. November 3; six months.

Francois Marie Lanoa, of Paris, for improvements in apparatus for holding and drawing off aërated liquors, and in machinery for filling vessels with aërated liquors. November 3; six months.

Henry Vigurs, of Camden-town, Middlesex, engineer, for improvements in buffers, grease-boxes, axle-boxes, and springs, and in appendages to railway engines and carriages. November 4; six months.

Jules Francois Dorey, of Havre, in the Republic of France, gentleman, for improvements in illuminating the dials of clocks and other instruments in which dials are employed. November 4; six months.

Theodore Kosmann, of Cranbourne-street, Middlesex, for improvements in brooches and other dress fastenings. November 4; six months.

Henry Hussey Vivian, of Llangellen, Glamorgan, Esq., for improvements in obtaining nickel and cobalt. November 4; six months.

Joseph Robinson, of the Ebbw Vale Iron Company, and Charles May, civil engineer, of 6, George-street, Westminster, and William Thomas Doyere, civil engineer, of Euston-square station, for improvements in the permanent way of railways. November 6; six months.

George Dismore, of Clerkenwell-green, Middlesex, jeweller, for improvements in locks. November 6; six months.

Robert Beswick, of Tunstall, Stafford, builder for certain improvements in the making or manufacturing brick and tiles, or quarries, and in constructing ovens or kilns for burning or firing bricks, tiles, and quarries, and other articles of pottery and earthenware. November 6; six months.

Alexander Douil, of Greenwich, Kent, civil engineer, for certain improvements in railway construction. November 6; six months.

Michael Leopold Parnell, of 32, Little Queen-street, Holborn, Middlesex, ironmonger, for certain improvements in locks. November 6; six months.

William Thomas, of Exeter, Devon, engineer, for certain improvements in the construction of apparatus and machinery for economising fuel, and in the generation of steam, and in machinery for propelling on land and water. November 6; six months.

WEEKLY LIST OF DESIGNS FOR ARTICLES OF UTILITY REGISTERED.

Date of Registration.	No. in the Register.	Proprietors' Names.	Addresses.	Subjects of Design.
Oct. 31	3002	Deane, Dray, and Co.	London Bridge	Enamelled gas cooking apparatus.
,,	3003	W. Hamill, J. Kelly, and N. D. Maillard	Dublin	Portable flax-breaking and scrutching mill.
Nov. 3	3004	W. Forbes	Ellon, Aberdeenshire	Drain pavement.
4	3005	Edward Phipson	Birmingham	Metallic bed-sacking.
,,	3006	W. Reichenbach	Borough-road	Reflector gas-lamp.
,,	3007	W. King	Littlebury, Saffron Walden	Bee-hive.
6	3008	F. S. Bremmer	Camden-town	Oblique pen-holder.
,,	3009	Henry Woolf	Houndsditch	Easy cap.

WEEKLY LIST OF PROVISIONAL REGISTRATIONS.

Oct. 31	314	William Beales	Arlington-street, Camden-town.	Portable colour-box.
Nov. 3	315	M. A. Holden	Birmingham	Double signal-lamp.
6	316	Lambert and Co.	Portman-street	Vertical pianoforte-brace.

CONTENTS OF THIS NUMBER.

LONDON: Edited, Printed, and Published by Joseph Clinton Robertson, of No. 166, Fleet-street, *in the City of London*— Sold by A. and W. Galignani, Rue Vivienne, Paris; Machin and Co., *Dublin*; W. C. Campbell and Co., Hamburg.

Mechanics' Magazine,

SEUM, REGISTER, JOURNAL, AND GAZETTE.

1475.] SATURDAY, NOVEMBER 15, 1851. [Price 3*d*., Stamped, 4*d*.

Edited by J. C. Robertson, 166, Fleet-street.

OSS'S PATENT IMPROVEMENTS IN WOOL-COMBING MACHINERY.

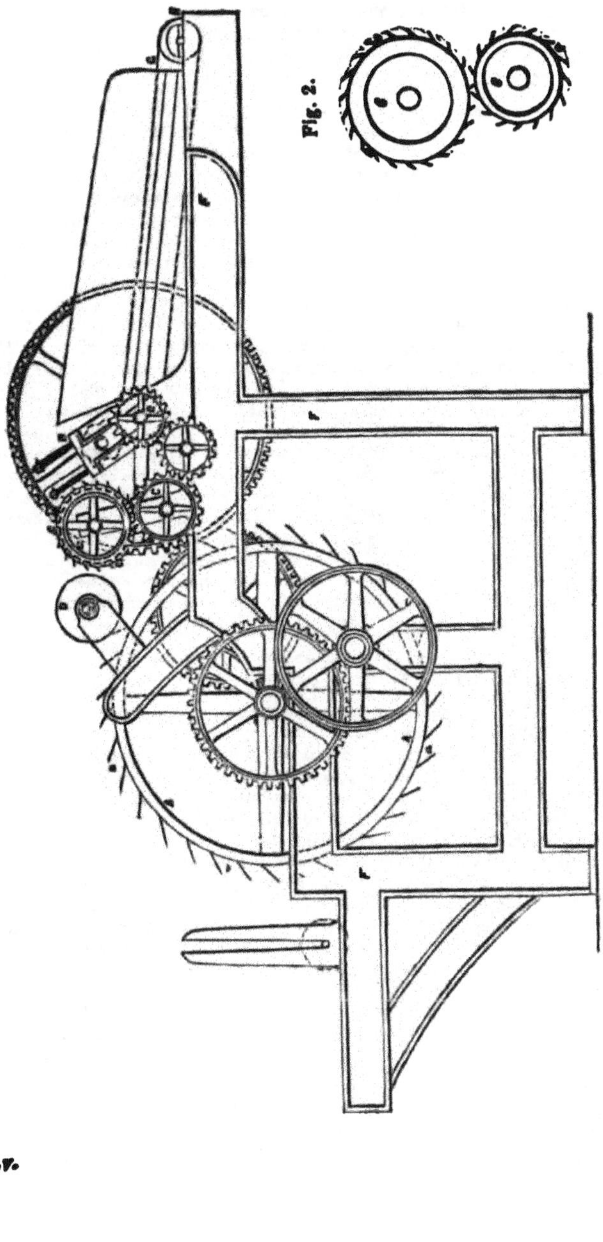

Fig. 2.

Fig. 1.

ROSS'S PATENT IMPROVEMENTS IN WOOL-COMBING MACHINERY.

(See *ante* pages 256 & 280.)

The importance of the improvements specified under this patent to the wool-combing business, calls upon us for a fuller description of them than we were able to give at the time of the enrolment of Mr. Ross's specification.

I. The first improvements described have relation to the machine for forming the wool into sheets of a nearly uniform thickness, technically known as the "sheeter," and consist chiefly in combining with the ordinary sheeting-drum or cylinder rollers, designated, from their resemblance to porcupine quills, "porcupine rollers;" these rollers having their teeth or quills set in rows, and the rows of one roller gearing or taking into the spaces between the rows of the other.

Fig. 1 is a side-elevation of a sheeting-machine thus constructed:—FF is the general framework upon which the several working parts of the machine are mounted. A is the main or sheeting-drum or cylinder which is studded with rows of combs or "porcupine" teeth *a, a, a*, the length and fineness of which are varied according to the length of the staple of the wool or other material to be operated upon. Instead of the rows consisting each of a single set of teeth, two, three, or more sets may be combined together. The number of wires which may be placed on one line should vary with the quality of the wool or other material. In long staple machines, the number may vary from four to ten or more, and in short staple machines from five to twenty or more per inch. BB are two fluted feed-rollers. CC two porcupine combing-rollers, by which the wool is partly combed while passing from the feed-rollers to the surface of the sheeting-drum; an end elevation of the porcupine combing-rollers on an enlarged scale is given at fig. 2. The teeth *cc* are set in rows, and the rows of one roller take or gear into the spaces between the rows of the other. D is a grooved guide-roller for preventing the wool or other material escaping the combing action. The wool or other material is laid by the attendant evenly upon the upper surface of an endless web G, which works over the under feed-roller, and a plain roller H which is mounted in bearings on the front of the machine. The feed-rollers gradually supply the wool thus spread upon the endless web to the two porcupine-combing rollers, where it is partly combed and separated, and being so prepared it is laid hold of by the teeth of the sheeting-drum, by which it is still further drawn out on account of the greater velocity with which the surface of the sheeting-drum travels. When a sufficient quantity of the wool or other material has been thus collected on the surface of the drum, it is removed by the attendant passing a hooked rod across the surface of the drum, and raising up one end of the sheet, when the whole may be easily stripped off and removed, being then in a fit state for being supplied to the comb-filling machine next to be described.

A modification of this sheeting-machine is represented in figs. 3 and 4; which differs from it in this, that it is fed from *both* ends:—

In this modification a double set of feeding-rollers is employed, so that the machine may be fed from both ends. These rollers are grooved and gear into porcupine combing-rollers similar to those before described, which are followed by brush cylinders or grooved guide-rollers. A is the sheeting-drum as before. BB the fluted feed-rollers. CC the porcupine combing-rollers, which gear into the fluted ones. DD are the grooved guide-rollers. FF are brush cylinders, which may in the case of long work be dispensed with. GG are the endless webs upon which the wool is laid. The framing and gearing by which the several parts are put in motion are omitted in the drawings for the purpose of clearly exhibiting the more important working-parts of the machine. The arrangement of sheeting-machines just described, in so far as regards the employment of a fluted feed-roller in conjunction with a porcupine combing-roller, and grooved guide-roller is more especially applicable to sheeting fine short wool, but may also be applied with advantage to wool or other material of a longer staple. In the case of fine short wool, the sheet may be drawn off by means of rollers, in the manner represented in fig. 4. HH are the drawing or straightening-rollers, and I the receiving-rollers. During the operation of drawing the wool and winding it on the receiving-roller, the sheeting-cylinder must have a motion imparted to it in the reverse direction.

II. The next head of Mr. Ross's specification embraces several improvements in comb-filling machines, which have for their common object the partial combing of the wool while it is in the course of being filled into the combs. We select, for exemplification, what the patentee regards as the best of these arrangements:

Fig. 5 is a side elevation of a comb-filling machine as thus improved. A A is a skeleton drum, which is composed of two rings *a a*, affixed to the arms *b b*, which last are mounted upon the mainshaft of the machine, which has its bearings upon the general frame F F. B^1 B^2 are the porcupine combing-rollers, and C^1 C^2 brushes by which the porcupine combing-rollers are cleansed from the wool that collects upon them, and by which the wool is again delivered to the combs *e e*. D D are the feed-rollers, and E an endless web which runs over the lower feed-roller and the plain roller G, which is situated at the

Fig. 3.

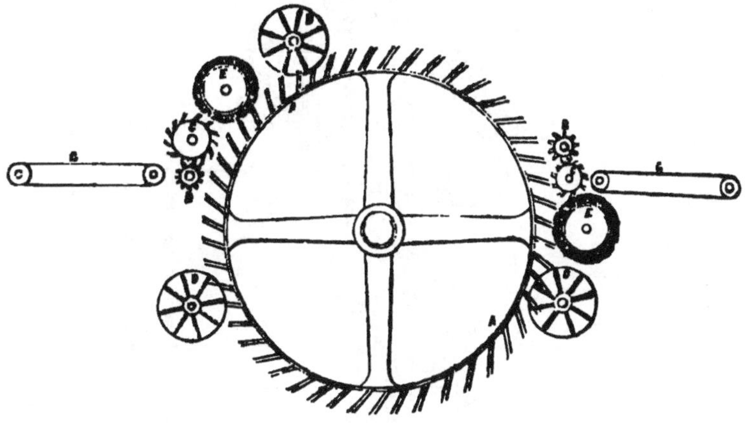

Fig. 4.

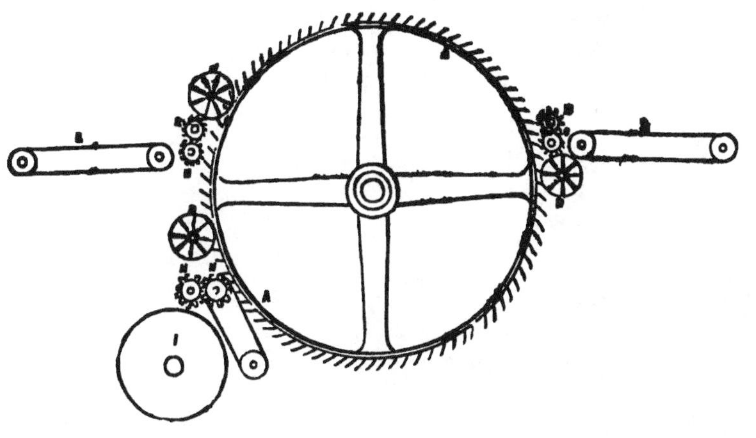

front of the machine. H H are the driving pulleys, by which the power is applied to the machine, and I, I, I, the wheel gearing by which motion is communicated to the different parts. The wool which has undergone the process of sheeting in the machine first described is spread upon the endless web E, and in passing between the feed-rollers, and between or under or over the porcupine combing-rollers, is taken hold of by the combs *e e* as they revolve, and, being drawn under the first porcupine-roller B^1 and the brush C^1, the continued revolution of the drum and combs causes the wool to be brought into contact with the other porcupine combing-roller B^2 and brush C^2. As the combs get filled the wool is thus

continuously being brought under the action of the porcupine combing-rollers and brushe
and each new portion of the wool taken up is instantly combed out. For some purpos
the combing will be found carried so far by this operation that the wool will require n

Fig. 5.

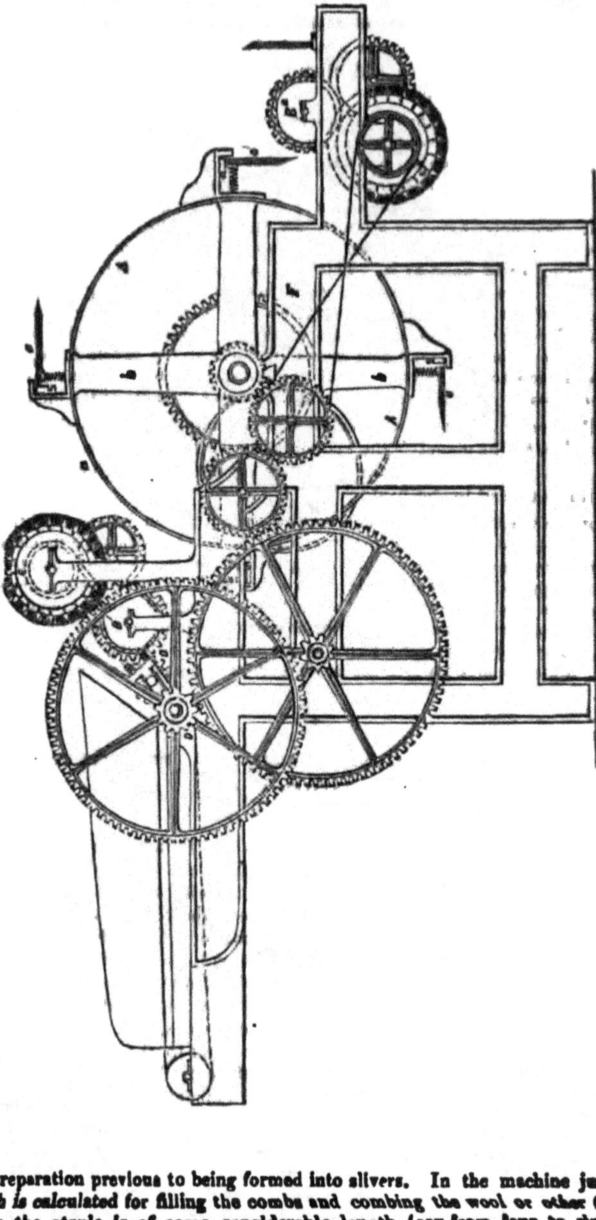

farther preparation previous to being formed into slivers. In the machine just describ
and which is calculated for filling the combs and combing the wool or other fibrous ma
rial, when the staple is of some considerable length (say from four to sixteen inch

two porcupine combing-rollers, with their brushes, employed; but I do not con- elf to that number, as in some cases a single porcupine combing-roller and brush ound sufficient. For the purpose of facilitating the process of combing and filling s, three or more rollers and brush-cylinders may be used with advantage; such as e staple is short, or where the fibrous material operated upon is very close, and l with difficulty.

Mr. Ross next describes some improvements in the combing-machine of his n, patented in 1841, and now extensively used. The following general de- n will indicate with sufficient distinctness to those familiar with the machine ire of these improvements:

I give to the saddle-combs in the said machine a compound to-and-fro and up-and- ovement, whereby they recede from and advance towards the comb-gates, and simul- y therewith alternately rise and fall, so that each time the comb-gates pass the ombs they do so in a different plane, and thus the position of the combs in relation other, as well as to the hold they take of the wool or other material, is constantly anged. Secondly, I employ a fan to lash the wool in the comb-gate or flying comb at the saddle-comb, which renders it impossible for the wool to pass by the saddle- thout being acted upon by it. Thirdly, I attach the springs by which the gates ated to the lower arms of the comb-gates, instead of their being placed parallel to ght shaft of the machine, as formerly, whereby a considerable gain in space and ness is effected; and, Fourthly, I use breaks to prevent the sudden jerk which is hen the wool in the comb-gate leaves its hold of the saddle-comb, or inclined nd also to counteract the sudden recoil of the springs, by which the comb-gates are in when these springs are released from the grip or pressure of the inclined plane.

Mr. Ross concludes with a description of an improved method of heating the which has for its object "the economizing of fuel, the better heating of the and the prevention of mistake in removing the combs before they have been ient time exposed to the heat:"

ody of the heating box or stove is divided by a partition into two portions, which nicate together at the back or further end of the stove, so that the flame and heated , after having circulated under and along the sides of the two lower comb-chambers, nto the upper portion of the stove, where they have to traverse along the sides and top of the two upper chambers, ultimately escaping into the chimney through a The length of the heating-box, or the chambers, should be about double the length ombs. The cold combs are inserted at one end, and, on being put into their places, more heated combs towards the other end of the chambers, from which they are l.

S ON THE THEORY OF ALGEBRAIC EQUATIONS. BY JAMES COCKLE, ESQ., M.A., BARRISTER-AT-LAW.

(Continued from page 178.)

ird and Concluding Series.

VII. CUBICS.

n two of the roots of a given re equal, and *vice versâ*,

$$\sqrt[3]{n\rho}=\sqrt[3]{\frac{n^2}{\rho}},\text{ or }\rho=\sqrt{n};$$

bstituting for n, a, and a^2, we see, ns of (5.), that the value of each equal roots is

$$-\frac{1}{3}\left\{a-\sqrt{a^2-3b}\right\},$$

er root being

$$-\frac{1}{3}\left\{a+2\sqrt{a^2-3b}\right\};$$

and when, in addition to the preceding condition, $a^2=3b$, all the roots are equal.

The reader will find a discussion of the nature of the roots in my remarks "On a Solution of a Cubic Equation" at pp. 95—6 of vol. iv. of the *Cambridge and Dublin Mathematical Journal.*

At page 186 of the sixth volume of the same work, Mr. Cayley has applied my method to the case in which x^2 has a coefficient. His result, like everything which emanates from his mind, is well worthy of contemplation

Let us now consider the relation which the foregoing method of solving a cubic bears to the general theory of Lagrange. For this purpose, if x_1, x_2, and x_3 denote the various values of x, we have, by means of the relations given at line 6 of the fifth Note of this series (vol. liii., p. 329),

$$x_1 = -\tfrac{1}{3}\left\{a + \sqrt[3]{n\rho} + \sqrt[3]{\frac{n^2}{\rho}}\right\},$$

$$x_2 = -\tfrac{1}{3}\left\{a + a^2\sqrt[3]{n\rho} + a\sqrt[3]{\frac{n^2}{\rho}}\right\},$$

$$x_3 = -\tfrac{1}{3}\left\{a + a\sqrt[3]{n\rho} + a^2\sqrt[3]{\frac{n^2}{\rho}}\right\}.$$

Multiply the second and last of these equations by a and a^2 respectively, and add; then, since

$$a^3 = 1, \text{ and } 1 + a + a^2 = 0,$$

we have

$$x_1 + ax_2 + a^2x_3 = -\sqrt[3]{n\rho} = -\sqrt[3]{n(a+3z)} \ldots. (6.),$$

and consequently

$$z = -\tfrac{1}{3}\left\{a + \frac{1}{n}(x_1 + ax_2 + a^2x_3)^3\right\},$$

and the root of the auxiliary equation is a rational function of

$$(x_1 + ax_2 + a^2x_3)^3,$$

which (LAGRANGE, *Notes** *sur la Théorie des Equations Algébriques*, Note XIII.) has only two values, and the cube root of which I shall call y_1.

Let

$$x_1 + a^2x_2 + ax_3 = y_2,$$

then

$$y_1y_2 = a^2 - 3b = n \ldots\ldots.. (7.),$$

and, by (6.),

$$n\rho = -y_1^3,$$

consequently

$$\sqrt[3]{\frac{n^2}{\rho}} = -\sqrt[3]{\frac{n^3}{y_1^3}} = -y_2, \text{ by } (7.);$$

hence

$$x = \tfrac{1}{3}\left\{-a + (1)^{\frac{1}{3}}y_1 + (1)^{\frac{2}{3}}y_2\right\}.$$

On solving my equation (b), at p. 95 of vol. iv. of the *C. and D. Math. Jour.*, with respect to z, it will be seen that the resulting expression for z has only two values however we interchange the x's. The possibility of our solution is thus seen, *à priori*.

In Art. 21, p. 14, of DR. HYMERS's Theory of Equations the relations between the coefficients and the roots of equations are alluded to as "not leading to the determination of the roots." This statement requires to be qualified by the word "immediate" used by DR. HYMERS in the preceding article; for, at pp. 191, 2 of that work, cubics and biquadratics are solved by means of those relations, and at p. 166 also DR. HYMERS has availed himself of them. The inaccuracy, however, is but trifling. I speak of the first edition of that excellent work—(Cambridge, Deighton, 1837.)

I have lately found a letter addressed to me some three years ago by the distinguished mathematician DR. RUTHERFORD. As it contains nothing of a private nature I enclose it to the Editor of the *Mechanics' Magazine* for publication, merely striking out the solution and two examples, which it is unnecessary to reprint, inasmuch as DR. RUTHERFORD published them at pp. 180—2 of the *Mathematician* for November, 1848; and, subsequently, in the Note C to his "Complete Solution of Numerical Equations,"* an interesting and valuable work.

[*Letter from Dr. Rutherford, to Mr. Cockle.*]

"Royal Military Academy,
Woolwich, September 2, 1848.

"Dear Sir,—I have recently received from you some pamphlets on equations extracted from the *Mechanics' Magazine*, and for which I have to request your acceptance of my best thanks. It is not often that I look at the subject of equations now, especially as the beautiful method of Horner enables me to find the roots of numerical equations of all degrees, whenever I want them.

"Some time ago I discovered the following method of resolving a cubic equation complete in all its terms. Cardan's method applies only to the form $x^3 + ax + b = 0$, but the method I now submit to your notice dispenses with the process of transformation necessary to bring the general form $x^3 + ax + bx + c = 0$ to Cardan's form.

[Here follow the solution and two examples already published, and consequently here struck out, as above stated.]

---

* These "*Notes*" are appended to his great work the *Traité de la Résolution des Equations Numériques*. The 3rd edition bears on its title-page, Paris, Bachelier. 1826.

* London, Bell, 1849. Price 2s. 6d.

at is your opinion of this mode, pplies to the general form as well as ther form?

"I am, dear Sir, &c.

Cockle, Esq., &c., &c."

get the precise nature of my anat the elegance and beauty of the would obviously admit of but one the inquiry which Dr. Rutherld me the high honour to address If I recollect right, my first imn was that, scientifically speaking, sification would be somewhere the methods of Bezout. By the the method of Lagrange we of course, have no difficulty in it to that common source from ll solutions diverge. Its relation range's process will be best ed by comparing it with my own. ay be done by an easy transfor; for, adopting the notation of of vol. liii., of this Magazine, the n (2) of p. 124, *ib.*, becomes, on ting for y,

$$\frac{\ldots(s-z)+B\}^3-D^3(s-z)^3}{A^3-D^3}=0,$$

division of the numerator and nator by D^3 would still further the ultimate identity of Dr. Ruord's solution and my own. Since cussion of Lagrange, this is no han we might have anticipated, *ri*, and it in no degree detracts he originality or value of Dr. rford's solution.

Rutherford's y and z are the lues of my z with the signs d. The functions

$$a^2-3b,\ ab-9c,\ \text{and}\ b^2-3ac$$

most conspicuous part in the of cubics. [*See* Equations (7), (8), Dr. Rutherford's "Complete m," &c., and elsewhere.]

James Cockle.

p-Court, Temple, September 23, 1851.

AROMETER WITH AN ENLARGED SCALE.

—In this barometer a light fluid is ced upon the top of the mercurial of the common barometer, the the instrument being enlarged at nt of junction of the two fluids, ch device an instrument of equal, *rior, extent of scale* may be obtained without the expense and difficulty attendant upon the construction and erection of a barometer, of which the whole tube is occupied by a light fluid.

R. S. N.

November 5, 1851.

SIMPLE APPARATUS FOR RAISING WATER.

Sir,—The following plan for raising water was suggested to my mind a short time ago; it appears to be practical, and certainly merits attention from its simplicity.

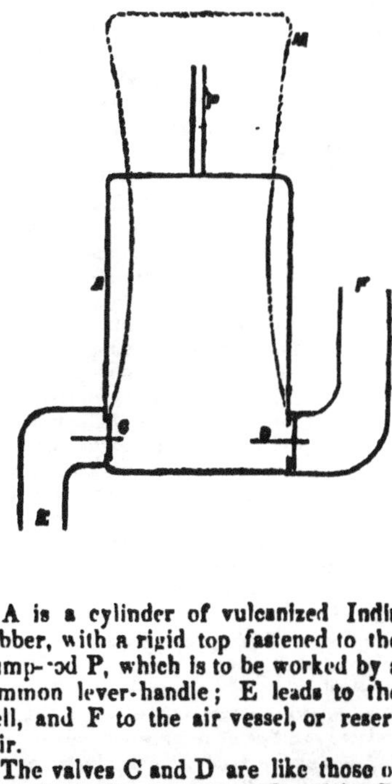

A is a cylinder of vulcanized India rubber, with a rigid top fastened to the pump-rod P, which is to be worked by a common lever-handle; E leads to the well, and F to the air vessel, or reservoir.

The valves C and D are like those of the ordinary forcing pump. On raising P, the partial vacuum produced inside of A will cause the cylinder partly to collapse, as at M, and this is the only reason why it should be made of *thick* vulcanised India rubber. The resiliency of the cylinder will greatly assist the down stroke of the rod P. And there being no piston or plunger, the only wear and tear will be in the valves.

This plan, it may be said, resembles that of Shalders' Pump, but the difference between a cylinder and a diaphragm expanded is too manifest to be insisted upon.

Yours, &c.,
JOHN MACGREGOR.

Temple, November 7, 1851.

BOARD OF HEALTH REPORT ON THE WATER SUPPLY OF TOWNS.

(Continued from p. 374.)

Much as water power has been employed in giving motion to machinery, it is believed that the raising artificially a large body of water at one spot, to be distributed from thence in portions as an occasional hydraulic power, was first proposed by Sir Samuel Bentham in the year 1812, in his official indicator of the Desiderata in a Naval Arsenal. Article 15 of that communication stated, amongst other items to be provided by waterworks, the "giving or transmitting motion in some cases to machinery along the range of the pipes." Although his plan for a naval arsenal at Sheerness was not adopted, it was known, to the private engineer subsequently employed there, yet neither water nor fire-extinguishing works were provided by him—but afterwards the official engineer (Mr. Mitchell)* proposed and carried into execution there both water and fire-extinguishing works, and provided plugs upon the mains ready for the application to them of hydraulic apparatus, such as would be suitable for the performance of such laborious operations as are of a temporary nature. He was not fortunate enough to convince a superior officer of the eligibility of this mode of employing hydraulic power; yet, as his provision of pipes and plugs remains in convenient parts of the yard, they would now afford opportunity, at little cost, of ascertaining the practicability and economy of this mode of employing the force of a powerful steam engine for temporary services at different and distant places.

"Mr. Philip Holland has illustrated the convenient application to industrial purposes of the hydraulic power derivable from the constant supply."

"At present many trades employ very small steam engines for purposes that may, almost as cheaply, be accomplished by hand; for instance, coffee-grinding. There are many purposes for which steam might be substituted for manual power with advantage, were it not for the cost of skilled labour required to attend to it, and the expense and trouble of keeping up the steam when the power is not wanted. If some hydraulic engines (such as the tourbine) were employed and worked by water from the pipes—which could be set to work and stopped in an instant, which consumes no power except when at work, which requires no skilful mechanic to work it, and is quite free from risk from fire or explosion; there is no doubt but that numerous applications of such power would be introduced which are as yet scarcely thought of. It would be easy to work cranes and hoists for raising or lowering goods or persons in warehouses, where the occasions for their use are not sufficiently numerous to have a steam engine economical. Such an instrument would work presses in the smaller printing-offices, where it is not worth while having a steam engine. For many purposes a simple hydraulic press, with a large cylinder acted upon by the direct pressure from the pipes, would be sufficient for packing. In others, Bramah presses might be worked by the hydraulic engine. Turners might work their lathes, and smiths their bellows, by water-power; chaff might be cut and oats and beans crushed by the same means; in fact, it is impossible to mention all the various uses to which it might be applied if water were supplied constantly and at high pressure."

Mr. Holland was aware that there would be power lost by lifting the water thus, "by the amount of friction lost. But large engines are employed for pumping, and small ones are got rid of." "At a very small cost any person wishing for the services of a one or two horse engine for an hour or two a day, might have it without trouble, risk, or uncertainty;" and it would be much used, "if the charge for the power were not very much more than the cost of raising the additional water required, and the expenses attending its distribution." The Board express their opinion, that "though existing steam engines, excepting those of the smallest class, would not probably be superseded by this water power;" "yet, having regard only to the public health, they submit that it is desirable that such applications of power may be promoted, as will in any degree tend to check the multiplication, even of small steam-engine furnaces and the aggravation of the smoke nuisance."

The question—In what manner are private employers of hydraulic power to be rated for its use? does not seem to have been adverted to. When Sir Samuel proposed a similar application of it at Sheerness,

* Mr. Mitchell from the year 1803, had been employed in Sir Samuel's establishment of mill-wrights in Portsmouth Dockyard, and was selected by him in 1809, to superintend the introduction of *water and fire-extinguishing* works in Plymouth *Dockyard.*

the consideration, that a Royal is, as it were, an individual manu- concern, carried on for the benefit d the same firm, so that increase iture in one branch, to save a still m in another, is a source of profit rm upon the whole—not so the as concerns of the metropolis, for re are as many persons implicated e households; it hardly seems just, e householders should, by an equal upon houses contribute towards manufacturers with a cheap mobile. It is true, that Mr. Hol- mplates a charge for the power so ; but water under pressure being every house, by what inquisition occasional use of it as a power be —The Report recommends a man ount of it, to be employed for ven- ivate houses. Amongst Sir Samuel s old papers is the project of a c steam engine" for washing, knife, and shoe, and saucepan- &c., &c.; doubtless, many ex- ould soon be afforded of the appli- hydraulic power to a variety of operations, mostly in the houses lthy and of persons in easy circum- hilst the poor would rarely possess enabling them to profit by such in- and it were hard to make the lower y extra for their water, in order to late others with hydraulic power. cies of this nature, if not foreseen led for at the outset of plans, are bring into discredit the most be- ies.

meters would ascertain the quan- ter actually consumed; possibly and cheap ones may be devised ed arises for them, such as would o costly for application to every

ard lay considerable stress on the e of combining with the water- the metropolis, arrangements for aishment of fire, and indicate many ere such works have been success- duced, but make no mention of he Royal Dockyards, where they established half a century ago, y have proved on all occasions of of fire efficacious in subduing it and which have been copied in all the same nature that have been tly introduced. The Board's evi- s to collect the fullest information oint under their discussion renders rance of Sir Samuel Bentham's in- fire-extinguishing works remark- hose works in the several dock- *of public notoriety*, have formed the subject of many of his official commu- nications commencing in the year 1797, and descriptions of his apparatus were published in the year 1827, in his "Naval Papers, No. 8," and subsequently, both in the *Mechanics' Magazine* and in *The Builder*. The Board had been, till last July, equally ignorant of his proposal in the year 1830, (submitted then to Sir Robert Peel; and also in the year 1844, to Lord Lincoln), suggest- ing the adaptation of the waterworks of the metropolis to the extinguishment of fire, in the same manner as the waterworks at Portsmouth were arranged by him for the double purpose of supplying water, and of extinguishing conflagration. That proposal has appeared at length in Number 1464 of the *Mechanics' Magazine*; its readers will perceive by the following quotations from the Report, that the Board and its witnesses employ the same arguments that Sir Samuel used in favour of similar works for the metropolis—namely, the importance of *immediate* application of water at the first outbreak of fire; the delay occasioned by waiting for turncocks and engines; the expediency of keeping hose and apparatus within short distances, and in such custody as to be immediately applicable; the giving to the police means of putting that apparatus to immediate use; the making collateral use of that apparatus with a view to keeping it in good working order, &c., &c.

The Report furnishes much information which shows the accuracy of Sir Samuel's views, and the efficacy of fire-extinguishing works like those at Portsmouth. Mr. W. Baddeley, "an engineer who acts as an inspector to the Society for the Protection of Life from Fire," stated that, "the num- ber of fires last year was 838, and if there had been means of applying water imme- diately, two-thirds of them would have been stopped immediately."

"Would you say that the means of ap- plying water in adequate quantity within five minutes of the commencement of a fire would prevent the progress of two-thirds of them?—Yes; about that time."

"In some places has not delay occurred from the turncocks being at wide intervals? —Yes; on the south side of the Thames par- ticularly. It is generally objected by the Companies that none but their own servants shall have command of the fire-plugs, in order to prevent confusion, as, though it is necessary to have one or two plugs open, it is necessary to close two or three others to get the supply."

"Under a high-pressure system, of course there would be a much stronger jet given than can be procured by an engine?—Yes, certainly?"

"Then from your experience you have no doubt that the *prompt* application of water would be most beneficial?—I have no doubt whatever."

The Board after investigating the time that usually elapses before an engine can arrive after an alarm of fire is given, say that "the men must be assembled and conveyed to the spot, the turncock sought, and the plug opened, all operations then performed in the best possible manner, occupying time," which time, "under the most favourable circumstances" will be 28 minutes, when the distance of a fire is a mile from the station, or for half a mile distance, not less than 20 minutes," "during which time the fire is rapidly extending."

The fire-extinguishing works at Hamburgh appear to be precisely similar to those in Portsmouth yard. Mr. Lindley (the engineer who laid down these works at Hamburgh), was asked, "How soon can a jet be applied?—In two minutes."

"Have there been fires in buildings in Hamburgh in the portion of the town rebuilt?—Yes; repeatedly. They have all, however, been put out at once."

There appears to be an excellent arrangement in Liverpool for the speedy conveyance of hose, devised by Mr. Newlands. "It is this—at each fire-station reels of hose, each mounted on a light hand-cart or barrow, are kept; the bottom of the barrow forms convenient lockers for containing the stand-pipes, nozles, and tools, and the whole apparatus is so light, that a man can *run* with it. On notice of a fire being received at the station a couple of men set out immediately with one of these reels On arriving at the fire the stand-pipes are put down, and the hose run off the reels in an instant, and it may be a dozen jets are playing, and the flames nearly extinguished before the engines have left the yard." "Water jets, under high pressure, are available over nearly the whole of the town. Serious fires are now seldom heard of, for before the flames can gain head the jets can generally be played upon them; and this is the only time that there is any chance of subduing them."

In the third Appendix to the Report, there is the transverse section of a street as rebuilt at Hamburgh, after the great fire of 1842, showing the distribution of water for use, and for the extinguishment of fire. Drawings, where men are introduced applying apparatus to use, realize to the *general* reader what otherwise might be difficult of comprehension, so that the drawing in question is likely to forward materially the introduction of fire-extinguishing works. The section of the street at *Hamburgh appers* to be in all respects accurate; it shows one man in the street throwing water over the roof of a house, by means of a hose affixed to a fire-plug, whilst another man in the attic floor of the same house is throwing water on its roof from the interior. In the Wood-mills at Portsmouth, provision is made in each floor for supplying a hose with water under a head by simply turning a cock; at Hamburgh, the hose for use in the interior has to be led through the ground-floor of the house and up the staircase,—an arrangement which seems unnecessary, since water, under high pressure, is served by, apparently, a large pipe to every floor of the house. At Portsmouth the fire-plugs on the mains are fixed at different distances, varying from about 50 to 200 feet, according to the description of buildings or works sought to be protected from fire; at Hamburgh such plugs are placed throughout the town at the uniform distance of 40 yards (120 feet). These seem the only variations of the Hamburgh works, from those that have so long existed at Portsmouth.

After the destruction by fire of the terminus of the South-Western Railway, Mr. Quick recommended works closely resembling those at Portsmouth. He states—"I recommended them to have a 9-inch main with 3-inch outlets leading to six stand-pipes, with joining screws for hose-pipes to be attached; and that they should carry a 3-inch pipe of the same description up into each floor, so that a hose might be attached in any room where the fire commenced."

"In how many minutes may the hose be attached?—There is only the time of attaching the hose, which need be nothing like a minute. I have, indeed, recommended that a short length of hose with a short nozzle, or branch, should be kept attached to the cock, so that the cock has only to be turne , which is done in an instant." Thus is recommended precisely the same arrangement that was from the first introduced in the Portsmouth Wood-mills.

The Board say, that "It appears that fire engines require twenty-six men to work each engine of two 7-inch barrels, to produce a jet of about 50 feet high. The arrangement carried out at your recommendation with six jets, is equivalent to keeping six such engines, and the power of 156 men in readiness to act at all times, night and day, at about a minutes' notice, for the extinction of fires?—It will give a power more than equal to that number of men."

The Board appear to condemn tanks on the roofs of buildings, though they say that "In many instances, a constant supply of water has been sought to be secured by the provision of large tanks. In some cases,

he apparatus connected with these
ppened to be ready, the efficiency
roposed system of a constant sup-
proved by the extinction of fires,
'a few minutes' delay had occurred,
ave inevitably destroyed the pre-
' yet, they add, "but heavy disas-
shown that such expensive prepa-
yield only a very imperfect secu-

opinion, adverse to tanks, seems
d solely on the evidence of one sin-
on, Mr. Samuel Holme, a builder
pool. Of their failure in case of
gives the following instance:
ie years ago, the owner of a cotton-
ich had been repeatedly burnt, took
is head to erect a large tank in the
lis idea was, that when a fire occur-
y should have water at hand; and
e fire ascended, it would burn the
tank, and the whole of the contents
scharged on the fire like a cataract,
. at once extinguish it. Well, the
in took fire; the smoke was so suf-
that nobody could get at the inter-
, and the whole building was again
d. But, what became of the tank?
not burn, because it was filled with
consequently, it boiled most admi-
No hole was singed in its side or
It looked very picturesque, but it
rly useless." A solitary instance of
—but is it an example of sufficient
nce, or described in sufficient detail
ps influencing circumstances, to jus-
rejection altogether of elevated

ny cases, as at Portsmouth, an ele-
ank, connected with water-mains,
ford an *instantaneous* supply of water
he building itself. Experience has
ed that, in the greater number of
has been practicable to enter a
in flames, and to apply a hose in
nner as to extinguish them, even
nflagration had become extensive;
posing it to be impossible to enter
ure, the flow of water from pipes
t might be rendered self-acting in
fire by forming of *fusible metal*
rt of the apparatus. Had part of
om of the tank Mr. Holme so ridi-
n formed of an alloy fusible at 212°,
ould really have been discharged in
upon the fire as soon as it had
e heat to that of boiling water.
orming within a building either part
ipes themselves of fusible metal, or
it in suitable places, water would
duced as soon as the heat should
12°; farther, by connecting a pipe
spreader with the pipe, just below
the fusible part of it, water would, self-actingly, be thrown over the whole interior of a chamber.

The preservation of life in case of fire has not failed to engage the attention of the Board; but the Report states that, "Arrangements for fire escapes or other efficient machinery are, if complete, excessively expensive, or, if inexpensive, generally inefficient. To make these arrangements thoroughly efficient as respects the time of application, one complete set of fire apparatus should be provided for every large street, or for each small sub-district." Mr. Braidwood said that "one, with a man to attend it, should be within a quarter of a mile of each house, as assistance, to be of *any use, must generally be rendered within five minutes after the alarm is given.*" At present, scaling-ladders are kept at all the engine-stations, and canvas sheets also at some of them: several lives have been saved by them; but the distance of the stations from each other renders them applicable only in a limited number of instances." Thus, so far, the recommendations of Sir Samuel Bentham have been adopted; but other apparatus, indicated by him as being in use in foreign countries, does not seem to be yet introduced in London. His attention had been particularly drawn, when last he was at St. Petersburgh, to a fire-brigade there, the men of which had charge, each of them, when called to a fire, of either a light ladder, or a hook, or some other of the various contrivances in use in that city for saving life or property.

The Board recommend cleansing the streets of the metropolis by means of jets of water. They observe, that,—"those whose duties have led them to visit courts and alleys in the more depressed districts, after they have been swept by the scavengers' broom, are aware that, though this may remove the larger collections of filth from the pavements, it frequently spreads the ordure over the surface. Such surfaces, with the walls and basements of buildings, can only be thoroughly cleansed by washing."

The trial of a *jet d'eau*, with a hose affixed to the water-mains, was recommended to the Metropolitan Commissioners of Sewers, and a number of careful trials were made by Mr. Levick, "It appeared that, taking the extra quantity of water required at the actual expense of pumping, the paved surfaces might be washed clean at one-half the price of the scavenger manual labour in sweeping."

The effect of this mode of cleansing in close courts and streets was found to be peculiarly grateful in hot weather. "The same expedient was resorted to for cooling the yards and outer courts of hospitals, and the shower thrown on the windows of the wards afforded great relief."

The Board affirm, that "the horse-dung which falls in the streets of the whole of the metropolis cannot be less than 200,000 tons a year. Much of this, under ordinary circumstances, dries, and is pulverized, and, with the common soil, is carried into houses as dust." "Dr. Arnott has long urged the consideration of the injurious effects of the excess of dust upon the public health in the metropolis."

"Judging by Mr. Levick's trial-works, the carriage-way, and also the foot-way of the Strand, might be completely washed on six days in the week, in one hour every morning, before the commencement of the traffic, at a charge of little more than fourpence per house per week: according to Mr. Lee's trials with a more powerful jet, it might be done for less than one-half that sum." "In Philadelphia, the water-jet is used for the purpose of cleansing the fronts of houses, as well as foot-ways and paved carriage-roads; and it is stated that its main streets, public buildings, and houses have all the cleanliness and brightness of a Dutch town, where the fronts of houses are washed by water thrown from scoops by hand."

"It is estimated that the proposed systematic street-cleansing by the jet would require an additional supply of water of about ten millions of gallons per diem."

"The constant use and practice in applying the apparatus" for the extinguishment of fire "to street-cleansing, would ensure its being at all times in a state of efficiency on the occurrence of a fire."

Constant use of such apparatus for ensuring its efficiency was also provided for by Sir Samuel Bentham so early as January, 1804, when, in his "Regulations for the Repairs of Buildings in the several Naval Establishments," article 31 specified that—"As it is very important that the readiness for use of the apparatus prepared for extinguishing fire should be frequently ascertained, this apparatus should be used at least once a month for washing the outsides of windows to buildings in general, according as they stand most in need of it;" and in his proposal to Sir Robert Peel* for the extinguishment of fire in the metropolis, he specified that the hose should be employed for watering streets.

It were vain to attempt concealment, that endeavours have been made above to obtain for Sir Samuel Bentham his just meed of credit for the invention, and successful adaptation of waterworks to the extinguishment of fire. The idea of cleansing streets by means of the jet is due to others; but, with this sole exception, his works in Portsmouth, Plymouth, and Chatham Dockyards, were examples of their utility, open to the public long before fire-extinguishing works were thought of by private engineers. Those same works have been closely imitated in all the many cities and towns where fire-extinguishing works have since been combined with those for supplying water; nor does the Report of the General Board of Health appear to offer any improvement whatever upon what he practised half a century ago. Indeed, it is remarkable, considering the recent great diffusion of science, and its application to useful purposes, that such rare examples should be found of improvement upon his contrivance. However, although his name be never mentioned as the originator of fire-extinguishing works, no other person appears to have claimed the merit of the first conception of them, or of having been the first to carry them into execution.

The difference of level between the higher and the lower parts of the metropolis does not seem to have been considered in the Report as influencing the distribution of the constant supply, though it be alluded to in the evidence of Mr. Mylne. This gentleman appears to advocate that system; for he said, "I think that where a domestic supply is required for a moral and well-conducted population, the constant system of supply is, under certain provisions, the most efficient, and in annual cost, I conceive, the cheapest. In expressing this opinion on the constant supply, I do not support the abandonment of the cisterns and tanks in respectable dwellings; one or more in every large house is essential for closet purposes, for steam apparatus, and to meet cases of unavoidable interruption of supply during repairs. A large tank in every cluster of houses occupied by the poor is equally necessary, from which each house should be supplied;" and he deprecates "what I have observed to be practised under the constant supply system, the rise drawing the supplies to wasteful districts and those of very low level, for the purpose of securing a more full discharge to other portions of the town at higher elevations."

The tanks for supplying clusters of houses in Jacob's Island have been found to answer well, and by filling them at regular intervals they might be made to serve as meters of the quantity of water consumed in a district. Considering that whilst south of the Thames the ground is below high water mark, the level is about 80 feet above it at

* *Mechanics' Magazine*, No. 1466, August 30th, 1851.

of Oxford-street, it does seem a general arrangement should be preventing the waste of engine raising water unnecessarily high. ulations that appear in the Report, h engine is supposed to raise water feet; but to deliver it under press tops of houses in many parts of the double of that height would d, consequently the cost would be of so raising it—a source of ex- large to be neglected in such an city as the metropolis. Even in th-yard, the site of which is level, ll worth providing for the raising water no higher than to the eleva- hich it was to be used.

M. S. B.

TIONS OF ENGLISH PATENTS EN- D DURING THE WEEK, ENDING BER 10, 1851.

ES COWPER, of Southampton- , Chancery-lane. *For improve- coverings for buildings.* (A com- n.) Patent dated May 3, 1851.

vention consists in constructing for buildings of plates of iron, to the shapes of tiles, and fur- th suitable contrivances for fixing and afterwards coated with a vitri- e or enamel.

es are shown in the drawings as ewhat the shape of a spade, with ndle, the end of which is turned as partially to lap round one of the which the tiles are to be sup- hey are also pierced with holes for ion of nails or bolts, for further them to the roof of the building, tamped with a ridge all round the the purpose of carrying off water nting its falling down between. form of tile is also shown, the sides are turned up at right angles to the l in combination with which a ridge- it be employed to cover the point o tiles meet. It is necessary that for the insertion of nails or bolts, contrivances for holding down the ld be formed therein or attached revious to applying the coating of enamel. For the purpose of so he tiles, two kinds of glaze are —one to constitute the body and the surface enamel. The body prepared by melting together, in l crucible, 100 parts of silica or 60 parts of borax; these ingre- then reduced to an impalpable and *mixed with 2½ parts* of clay ent water to bring the whole to a paste of the consistence of rather thick cream. The iron tiles, having had their surfaces well cleaned from dirt or oxide by immersion in a pickling liquor of muriatic acid and water, are plunged into this paste, and well shaken when removed, in order to spread it evenly over their surface: they are then dusted over with the second or finishing glaze, baked in a muffle or crucible, and cooled in an annealing chamber, in order to prevent the cracking of the enamel. The finishing glaze is composed of one part of silica, one-tenth of a part of borax, two-tenths of a part of minium or red lead, and one-eighth of a part of calcined bones, melted together and ground in water, then dried, and reduced to a very fine powder for use. Or the following enamel, composed of one-fourth of a part of silica, two parts and a half of white glass, one-sixth of a part of borax, and one-tenth of a part of scales of oxide of iron, melted together, ground in water, dried, and reduced to a fine powder, may be employed. In order to facilitate the fusion of the latter compound, from one-twentieth to one-fifth part of oxide of lead may be added, according to the fusibility of the glass employed. The melting point of the two enamels should be about the same, in order to ensure a uniform surface when baked, and prevent frothing of the finishing glaze. Tiles manufactured in this manner may be employed for other purposes than the roofing of buildings, and they may be ornamented by the employment of coloured glaze or enamels, or the enamels may be painted with colours, or with gold, or other metal, as is practised in the ornamenting of glass or porcelain. Cast iron may also be be in the manufacture of tiles according ing to this invention; but its use is not recommended, as tiles made from it are not so light or durable as those made from the sheet metal.

Claim.—The constructing of coverings for buildings of enamelled tiles, consisting of iron plates of the form of tiles, furnished with hooks or holes, or other means of attachment, and coated with enamel, as described.

PIERRE ARMAND LECOMTE DE FONTAINEMOREAU, of Finsbury and Paris. *For improvements in electric telegraphs.* (A communication.) Patent dated May 3, 1851.

The invention here sought to be secured consists in the application to electric telegraphic apparatus of a key-board similar to that of a pianoforte, in conjunction with a toothed cylinder combined with a ratchet-wheel and levers, put in motion by keys or hammers, by means of which it is merely requisite to place the finger upon a series of keys, on which signs, letters, or numbers are written, to effect the transmission of intelligence.

The arrangements of mechanism for carrying the invention into effect are as follows:—Underneath the key-board is set a cylinder or axis, from which project a series of radial rods equal in number to the keys, and set in a helical line around the cylinder, for the purpose of enabling each of the rods during the rotation of the shaft to be stopped by a catch attached to the particular key lowered. The lowering of any one of the keys is caused to take effect on a horizontal bar, also placed underneath the key-board (which is so arranged as to rise to its former position when the key is released from pressure), which bar, in its descent, liberates a ratchet, which gears into a ratchet-wheel on the rotating shaft, and thus allows the shaft (which is set in motion by clockwork) to revolve until a second rod corresponding with the key which has been lowered meets the stop on that key. On the lowering of another key a similar effect is produced, and the shaft is turned through an angle proportioned to the length of the arc of the helix between the two keys which successively stopped its motion; so that, if the cylinder is provided with an electric interrupter which opens and closes the circuit every time one of the teeth of the ratchet-wheel passes through, the effect produced will be identical with that produced by the rotation of a dial provided with as many signals as there are keys in this apparatus, but with increased advantage. The rotation of the cylinder being uniform, and regulated to the greatest speed that the efficient working of the receiving apparatus will permit, a communication once established between the receiver and transmitter, continues to subsist independently of any irregularity in touching the keys, provided time be given for the hand of the dial to run over its divisions.

The clockwork for setting the cylinder in motion must be wound up from time to time, but its use may be altogether dispensed with, and a spring substituted for it, on which the bar, actuated by the keys, may be caused to take effect so as to produce on the ratchet-wheel a propelling power which should slightly exceed the average force required to be exerted.

Claims.—1. The application to telegraphic apparatus of a key-board, combined with a series of stops disposed in the form of a helix around a cylindrical axis, so as to permit that axis to turn of a proportionate amount to the position of the key which is acted upon as described.

2. The application to telegraphic apparatus of a parallel bar underneath the key-board, for enabling indiscriminately each of the keys to put in motion the axis of the telegraph.

Edwin Ross, of Manchester, civil engineer. *For certain improvements in boilers for generating steam.* Patent dated May 3, 1851.

These improvements consist in combining with tubular boilers, whether cylindrical or otherwise, a chamber or flue between the tubular portion of the boiler and the fire-grates (of which there are two, side by side), for the purpose of effecting a complete combustion of smoke, and the gases which might otherwise pass off without being consumed. The fires are lighted and kept up in both grates at the same time; but they are fed with coal alternately, in order that the gases, &c., arising from the fresh charge in the one fire-place may be consumed by the heat from the incandescent fuel in the other at the time of introducing the feed. A throttle-valve is placed in the chimney flue, which is actuated by a rod having a handle at the front of the boiler, to enable the fireman to regulate the draught in the chimney without the necessity of quitting his station.

Claim.—Combining fire or smoke flues with a multitubular portion of boiler, having a chamber between them, whether made together in one boiler or any of the parts made separately, and subsequently united to the other parts, so as to act together in the manner described.

John James Greenough, Esq., of Washington, United States. *For improvements in obtaining and applying motive power.* (A communication.) Patent dated May 3, 1851.

This invention* consists of certain methods of obtaining motive power from the mutual action of magnets and helices or conductors conveying currents of electricity. The power thus obtained is denominated by the inventor "electro-dynamic axial power," to distinguish it from the electro-magnetic power produced by the mutual action of magnets. The elementary principle on which the invention is based is that of the axial force by which a helix conveying an electric current draws a magnet within it in the line of its axis, the magnet at the same time reciprocating the action of the coil in an opposite direction. This new power is intended to supersede the employment of steam and other prime movers in the actuation of engines of the reciprocating, vibrating, and rotary classes, and is also applicable when embodied in the simplest forms of machinery to the purposes of pile-driving, crushing, and breaking substances,

* Which is apparently that of Professor Page.

ing, pressing, punching and ham- etals, and pumping and forcing air, to which operations the mag- ıxial bars may be applied directly, ıut the intervention of other me- for the axial bars may be lifted series of coils, and then allowed their own weight, or, when lifted, r of the coils may be applied to the descent of the bars, and con- increase their effect in striking a d when employed in pumping, the or magnet may be caused to act as and the coils or helices as the r barrel of the pump. The draw- s various modes of carrying the nciples into effect, as also sundry improvements in the construction ils or helices, and of galvanic bat- nature of which will be readily rom the following claims and the ns appended thereto.

—1. The employment of a succes- oils or helices in line, through axial bar or bars are drawn or hether the helices be arranged in line to operate on a straight bar, rved line to operate on a similarly r, whereby any desired length of y be produced, or a continuous aintained, and an equable action either on the straight or curved so as to produce motive power.

employment generally of square, ectangular wires or any known the construction of electro-mag- ces, when such wires are wound ackwards and forwards over the gth of the coils (The patentee hat copper ribbands have been spirally so as to form flat spirals; e does not therefore claim, but he laim to the employment of such n wound helically and continu- k and forth, and over and over hole helix is finished.) Also the nt specially of square, flat, or r ribbands or wires of any pro- r dimensions in the construction or spirals when applied to operate ars for the purpose of obtaining wer, the use of such wires in this connection or combination being rith peculiar advantages.

ıcullar mode of operating with a of helices, by which a uniform ısured upon the axial bar or bars t the whole length of the stroke, its continuous rotary motion, and he *injurious action of secondary prevented, and the metal of the s a great degree preserved from d by the spark* resulting from the breaking of the connection; the said mode consisting chiefly in the employment of a number of sections or short helices in a line, and in using several of these at a time to operate upon the axial bar, (which several sections form what the inventor terms the "helix of actual operation,") and as the axial bar advances, certain of the sec- tions are cut off from one end of the circuit and others added to the helix of actual operation. In other words it may be stated thus: the helix of actual operation is a cer- tain portion of a long helix made up of a number of short helices or sections, which helix of actual operation preserves a con- stant relation to the extremity of the axial bar throughout its motion, and thus secures uniformity of action and the other advan- tages named.

4. A mode of effecting the peculiar action of the helix of actual operation, the said mode consisting in the manner of connecting the short helices or coils with each other and with the portion of the cut-off, and bringing these last in communication with the poles of the galvanic battery. That is to say, all the elementary helices are connected toge- ther, and also with the appropriate segments of the cut off, while the current is conveyed to the helices by means of communicating slides, which are connected with the poles of the battery in the manner described.

5. The employment of additional or out- side helices in the rotary axial engine, the said helices or coils being intended to com- pensate for the loss of space caused by the friction rollers employed in that engine.

6. The employment of a reciprocating cut off in combination with the recipro- cating axial movement—irrespective of any particular method which may be adopted for effecting the said movement. (It is ob- vious that there are many mechanical con- trivances belonging to steam and other engines which may be applied for moving the parts that direct the current of electri- city: these devices, however, are common and well known, and are not therefore claimed; the inventor wishing only to se- cure the advantages resulting from their employment with the axial engine in this special relation.)

7. The employment of hollow iron bars in the axial engine, the advantages of which are, that they are lighter, and give more power in proportion to their weight; and further, that they will admit of the employ- ment of guides within them if necessary.

8. The employment in the electro-dyna- mic axial engine of a vibrating or recipro- cating iron bar or bars, straight or curved, in combination with two sets of helices for each bar, operating alternately upon each

and thereof for the purpose of keeping up the magnetism.

9. A peculiar method of constructing helices for the mechanical application of electro-magnetic power, by enlarging the size or conducting power of the wire from within in an outward direction; the enlargement may be either gradual or irregular, but the claim extends to the employment of a larger wire enclosing a smaller one, or, in other words, an increase of the size or conducting power of the wire from within outwards, in any proportion or ratio found desirable for the purpose to which the helices are to be applied.

10. The facing of the points of the cut off, which are exposed to be burned by the spark resulting from the break of connection with strips of metal arranged in such manner as to be capable of being turned round or slid along, and thus renewing and exposing a fresh surface to the spark.

11. A method of insulating the square or rectangular wire employed in the construction of the helices by means of strips of cloth, or other suitable non-conducting material, applied longitudinally, and attached two or more contiguous sides of such wire, the insulating material being attached to the wire by some adhesive substance, thereby facilitating the operation of covering the wire, saving material, and especially making the helices more compact.

12. The employment of a helix or helices attached permanently to the axial bar or bars, and moving with it or them for the purpose of increasing the electro-dynamic effect, and keeping up the magnetism of the bar or bars.

13. The construction of a rotary axial engine, irrespective of the form of the cross section of the bars or their number, or the arrangement and number of the helices or coils, or the modes of gearing or of working the cut off.

14. The construction of a pendulous, or vibrating axial engine, irrespective of the form or character of the bars or helices employed therein.

15. A mode of securing the platina plates of galvanic batteries to the zinc plates thereof, by which soldering is dispensed with, and a considerable saving of the platina effected, combined with great facility in uniting the two metals. (A groove is cut in the edge of the zinc plate, into which the edge of the platina plate is inserted, and retained therein by compressing the zinc upon the platina.)

16. A method of compensating for the obliquity of the wire upon the ends of the coils or helices by means of what the inventor terms "risers." In consequence of the obliquity of the wire to the axis of the helix, a considerable vacant space is left on the ends of the helix at each layer of wire, where the wire has to rise upon itself to commence a new layer; this space is filled up with a piece of metal, termed a "riser," which is made to fit the space, and which is brought into metallic contact with the wire of the helix, which is readily effected when the above-mentioned method of insulating the wires employed in the construction of the helix is adopted; the insulating material being upon two sides only of the wire, leaves that side of the wire next the riser in metallic contact with it, and the riser thus comes into the circuit, fills up the vacant space, and adds to the conducting power of the wire. When the helices are constructed of flat spirals these risers are not necessary, but when a continuous wire is used, returning upon itself, their employment is of considerable importance.

17. A method of preparing carbon plates [by introducing plates of sheet iron into gas retorts, and allowing them to remain there until covered (except at the edge, which is to be protected by a coating of clay) with a crystalline deposit of carbon], and the employment of carbon plates so prepared in galvanic batteries.

18. The preparation of the outside cells of galvanic batteries by varnishing or saturating with oil, wax, resinous, bituminous, or other suitable insoluble matters, unglazed or porous earthenware.

19. A method of making the porous cell attached to or in the same piece with the outside cell.

20. A peculiar form of zinc plate, to be employed in conjunction with a Grove's battery.

21. The employment of the axial repulsive force (exerted by the magnet or axial bar or bars) either alone or in conjunction with the axial attractive force (exerted by a helix or coil), for obtaining motive power.

22. The employment of the helices and axial bars, for pumping water or air, upon the principles herein before mentioned.

William Cooke, of Gt. George street, Westminster, civil engineer. *For improvements in the manufacture of soda and the carbonate thereof.* (A communication.) Patent dated May 3, 1851.

The process adopted by the inventor of the present improvements is as follows:—In order to manufacture caustic soda he takes a solution of common salt and places it between two metals, iron and copper, connected together; decomposition ensues, at the expense of the iron, the chlorine of the salt, combining with the iron and forming chloride of iron in solution, while the of the salt is set free, and combin-h the oxygen of the water (hydrogen liberated) forms caustic soda, which towards the copper, and in order to there separated from the chloride of porous diaphragm is placed between tals, the salt being on the side where n is, and clean water to receive the ing placed on the side where the cop-The decomposition will not, how-ke place except the solution is main-at a temperature of from 70° to 150° uring the operation. It is necessary order to the obtaining of a perfect that the access of air should be pre-, that is, the vessel must be covered tight, to prevent the chloride of iron ing converted into oxide of iron, it is very apt to do; and in which se soda would re-combine with the se set free, and form salt again; and hen the air has free access to the solu-the salt is continually being decom-and re-composed, and the iron is without any beneficial result. To the process on a large scale, so as e a ton of soda at a single operation, or cistern must be constructed of stone, or other material capable of ng the action of caustic soda; the sions of which should be about 11 feet y 6 feet wide, and 3 feet deep. The is to be divided by porous dia-ms into three compartments, the cen-se being 1 foot wide, and the two end ach 2½ feet wide by 11 feet long. To te the compartments from each other diaphragms, composed of tiles or of biscuit-ware about one-eighth of h thick, set in frames of gutta percha window-sashes, are to be employed; should be secured to the interior of nk in such a manner as to render each rtment water-tight, and prevent its micating with that adjoining, other-han through the medium of the po-iaphragms. In each of the end com-ents must be placed a number of of pig-iron (Scotch being perferred), upon each other, but leaving space nt for the compartments to contain llons of salt water. The pigs of iron re in complete metallic contact, and ay be effected by placing a bright spot ch pig in connection with another spot, however small, on the pig next ing it. In the central compartment tank are to be placed two plates of r (which need not be thicker than r foil, as they are not subjected to ting action, and their surfaces only quired), each plate being brought into se contact with the end compartment next it, by connecting a strip of copper attached to the plate to a bright spot on one of the pigs of iron by means of a screw or otherwise. Now, to make a ton of soda-ash, 2,489 lbs. of salt are to be dissolved in 747 gallons (equal to 7467 lbs.) of water, which will make a saturated solution (equal to about 113 cubic feet) of salt water, half of which is to be poured upon the iron in one compartment of the tank, and the remaining half upon that contained in the other compartment. The middle compartment, which contains the copper plates, is to be filled with clean water to the level of the salt water in the end compartments. A cover is then placed on the tank, and luted on, to prevent the access of atmospheric air; and in order to admit of the escape of the liberated hydrogen, a bent tube is to be introduced in the cover, the end of which tube dips into a vessel containing water. The temperature being now kept at about 70° Fahr., the decomposition of the salt water will be accomplished in seven days, the salt water will then be found changed into a green liquid (a solution of chloride of iron) and in the middle compartment will be found caustic soda in solution, which contains also a small proportion of salt. The strength of the solution will of course depend on the quantity of water in the middle compartment, but if all the salt is decomposed, the solution will contain 1,327 lbs. of dry caustic soda. The caustic solution, if intended for soap-making, may be used at once for that purpose without any preparation; but if it is desired to manufacture carbonate of soda, it must be evaporated to dryness in an iron pot (the undecomposed salt being removed before the evaporation is quite completed) and when dry kept hot and stirred for an hour or two, during which time it will rapidly absorb carbonic acid from the atmosphere, increase very considerably in bulk, and will finally be converted into pure carbonate of soda (soda-ash of 59 per cent. alkali). 2,489 lbs. of salt will consume 1,161 lbs. of iron, producing about 1,327 lbs. of dry caustic soda, equivalent to about one ton of pure carbonate of soda, which is soda-ash containing 59 per cent alkali.

The patentee does not confine himself to the precise arrangement above described, as the iron and copper may be placed alternately; but in this case the last copper must be connected to the first iron plate, as in the ordinary galvanic battery.

The claim is for the obtention of caustic soda and carbonate of soda (soda-ash), by the means, or process, or processes, described.

James Pyke, of Westbourne-grove,

Bayswater. *For improvements in the manufacture of leather; also in the making of boots and shoes.* Patent dated May 3, 1851.

1. The "improvements in the manufacture of leather" consists in the application of heat when extracting tannin from bark and other substances used for making tanning liquors. The employment of free steam for this purpose is not new, but the novelty of the present improvements consists in the employment either of hot water circulating through pipes placed in the vats, and arranged on the well-known system of Mr. Perkins, or of steam, which is also caused to pass through a suitable arrangement of pipes, so as to raise the temperature of the liquid to about 150° to 210° Fahrenheit, in both cases suitable means being adopted for regulating the temperature at pleasure. The pipes are preferred to be arranged at the bottom of the vessel, underneath a false bottom, on which are placed the materials from which the tannin is to be extracted; and, as it is desirable to prevent the access of air during the operation, the cover of the vat (which is composed of galvanized iron) is turned down at the edges, and grooves are formed all round the top of the vat, which are filled with water, and in which the edges of the cover are inserted,—a water-joint being thus formed by which the atmosphere is entirely shut off from communication with the liquor. When cold, the liquor is drawn off by a tap from below the false bottom, and filled into pits in which the hides to be tanned are suspended from cross bars, which are fitted with a fillet at top, so that, when laid side by side across the pits, the fillets of the bars shall touch each other, and thus cover in altogether the top of the pits. Care must be taken that sufficient liquor is placed in the pits to immerse the hides or skins entirely; and it is not necessary to raise the hides from the pits, or move them at all, until the operation of tanning is completed; but the liquor must be changed as often as the quality of the hides may require, and until they are thoroughly tanned. They are then removed, and subjected, while in a damp state, to the operation of rolling; the method of performing which constitutes another of the improvements. Hitherto, leather has been rolled between two unyielding surfaces; but the patentee now proposes to employ one roller of metal or other unyielding substance, or a table composed of the same, in conjunction with a second roller composed of a cylinder of vulcanized India-rubber, covered with metallic rings placed close together side by side, the advantage of which is, that it yields to any irregularities in the thickness of the leather, and thus subjects the hide to a uniform degree of pressure in every part. The same result may be obtained by the employment of a table of vulcanised India-rubber, having its surface covered with bars of metal laid side by side, the leather being placed thereon, and subjected to the pressure of a solid metallic roller.

2. The "improvements in making boots and shoes" have relation to a method of securing the soles to the uppers and the heels to the soles of boots and shoes when metal brads or rivets are employed. The patentee makes use for this purpose of what he terms a "mould box," in which is placed a filling piece to support the last, the sole part being uppermost. Before placing the last in the mould box, the upper, previously damped, is stretched over it, and its edges are pressed down all round over the inner sole, which had been previously laid on the last. This pressing down of the edges of the upper is effected by the contraction of a block of India-rubber attached to the top of the mould-box, and through an aperture in which the last, and upper upon it, are introduced. The sole is now pressed down upon the last, so as to compress the edges of the upper between it and the inner sole. The holes for the rivets or brads are made by means of a "broaching frame," to which are attached suitable points for that purpose; and the brads are then put into these holes and subjected to pressure, to cause them to hold the parts securely together. The edges of the last are covered with metal, which is corrugated so as to form a series of ridges, and the points of the brads, when forced through the leather, come against the inclines, and are thereby directed upwards, by which means they will be clinched so as effectually to secure the parts together. The punches are then withdrawn from the mould-box, and the last, with the so far finished shoe on it, removed, after which the heel is attached by a similar method of proceeding. The latter method is also adopted for the purpose of securing the heels of boots.

Claims.—1. The improvements described in the manufacture of leather.

2. The improvements described in manufacturing boots and shoes.

William Smith, of Holloway, engineer. *For improvements in locomotive and other engines, and in carriages used on railways.* Patent dated May 3, 1851.

The improvements embraced under this patent are as follows:

1. The employment in locomotives and other engines of curvilineal valves, working on planes at right angles to the axis of the piston rod, instead of parallel with it, as in

of the ordinary four-way cock, or a slide. The cylinder of the engine louble, forming an outer and inner ', and having the apertures for the nade to extend all round the entire erence. The valves are concentric cylinder, and work around it, being l by rods passing through stuffing- the cylinder cover. A much greater thus obtained for the emission of and, in order that its exit may be ntaneous as possible, the space be- he cylinders is allowed to communi- ly with the atmosphere. The valve connection with the boiler is also on the same principle. The slide- nsists of a ring placed concentric to cylinder at each end of the box, he entire periphery of which are a series of apertures: these ring re connected together, and worked l passing through a stuffing-box in er plate of the valve box. By this nent of distributing the steam uni- over the cylindrical valve, the coun- sure of steam is removed, and the tion remaining is that of the rubbing of the metal, which is but trifling.

e application of elliptical eccentrics orking of pumps and "other hydrau- anical inventions," by which means imum effect of the power exerted is d during the lift, and the minimum he down stroke.

improved method of signalizing on by the simultaneous employment le signals produced by compressed visible signals or coloured lamps. riages of a train, or as many of them be necessary, are provided with contain the compressed air, which ied to them at the stations from a r, which is itself supplied by means mp, which may be worked by the ive while standing at the station the departure of the train. The employed are of the ordinary con- , and the lamps are shown simul- y with the sounding thereof by con- the axis on which the lamps are turn with the handle for turning on pressed air into the whistle.

s.—1. The double cylinders, which a free space around the entire peri- f the inner one, and the top and plates of the outer cylinder, which ree issue for steam above and below ns.

curvilineal valves extending around e circumference of the cylinder, and ith either a double cylinder or in- utside a single cylinder.

e *curvilineal slide* valves for the admission of steam to the cylinder, by which the pressure of steam is neutralized.

4. An improved rectilineal slide valve.

5. The elliptical eccentrics.

6. The combination of condensed air-reservoirs, whistles, and signals, as applied to railway carriages.

William Edward Newton, of Chancery-lane, civil engineer. *For improvements in the manufacture of woven and felted fabrics.* (A communication.) Patent dated May 3, 1851.

Claims.—1. A mode of producing a peculiar description of fabric by covering or coating felt or semi-felt (that is, felt which has passed through the process of manufacturing termed "hardening") or other woven fabrics with caoutchouc or similar gums, or compounds thereof, a bat or fleece of cotton, flax, silk, or other similar fibres being interposed between the felt and its coating of India rubber.

2. The covering or coating with caoutchouc or similar gums, or compounds thereof, the napped or fleecy surface (being of cotton, flax, silk, or other similar fibres) of the fabrics known as Canton flannel, or other woven fabric having a fleecy surface in manner described.

(Under the term "woven" the patentee intends to include that class of fabrics which may be produced on the circular loom, and which are known as "hosiery piece goods," and as "knitted piece goods.")

Gaetan Kossovitch, of Myddleton-square, gentleman. *For improvements in rotary steam engines.* (A communication from Alexis Khomiakoff.) Patent dated May 3, 1851.

This invention is described as consisting in maintaining the constancy or regularity of the inflow and outflow of steam, and also the uniformity of the vacuum produced, by arranging the mechanism of the engine in such manner that the introduction and escape of the steam shall take place through certain moveable parts or valves, which are caused to rotate with the piston and its axle, so as to present the necessary ports or apertures for those purposes.

WEEKLY LIST OF NEW ENGLISH PATENTS.

William Sinclair, of Manchester, Lancaster, engineer, for certain improvements in locks. November 13; six months.

Julian Bernard, of Green-street, Grosvenor-square, Middlesex, gentleman, for improvements in the manufacture of leather or dressed skins, and of materials to be used in lieu thereof, and in the machinery or apparatus to be employed in such manufacture. November 13, six months.

William Smith, of Derby, William Dickinson, also of Derby; and Thomas Peake also of Derby, for certain improvements in the manufacture of chenille and other piled fabrics. November 13; six months.

George Sheppard, of Stuckton Iron Works, Fording Bridge, Hants, engineer, for improvements in the construction of apparatus for grinding grain and other substances. November 13; six months.

Hugh Bowlsby Willson, of the York Hotel, Blackfriars, in the City of London, Esq., for improvements in the construction of rails for railways. November 13; six months.

LIST OF SCOTCH PATENTS FROM 22ND OF SEPTEMBER TO THE 22ND OF OCTOBER, 1851.

Henry John Betjemann, of Upper Ashby-street, Northampton-square, Middlesex, for improvements in connecting parts of bedsteads, and other frames, and in machinery employed therein. October 16; six months.

Daniel Dalton, of Spon-lane, Westbromwich, Stafford, iron founder, for improvements applicable to railways. October 16; six months.

William Jean Jules Varillat, of Rouen, France, manufacturer, for improvements in the extraction and preparation of colouring, tanning, and saccharine matters, from vegetable substances, and the apparatus to be used therein. Oct. 20; six months.

William Onions, of Southwark, Surrey, engineer, for improvements in the manufacture of nuts and bolts, also of steps, bearings, axles, and bushes, also of mills and dies for engravers, also of bells, and other spindles, also of weft forks, shuttle tongues and lips for looms, also parts of agricultural implements, chains, roller guides, and throstle bars, by the application of materials not hitherto used for such purposes. October 20; six months.

Thomas Sanders Bale, of Cauldon-place, Stafford, china manufacturer, for certain improvements in the method of treating, ornamenting, and preserving buildings and edifices, which said improvements are also applicable to other similar purposes. October 20; six months.

Robert Griffiths, of Havre, engineer, for improvements in steam engines and in propelling vessels. October 21; six months.

Frederick William Mowbray, of Leicester, gentleman, for improvements in machinery for weaving. October 21; six months.

William Fawcett, of Kidderminster, for certain improvements in the manufacture of carpets. October 21: six months.

George Fergusson Wilson, of Vauxhall, David Wilson, of Wandsworth, and James Child, of Putney, for improvements in presses and matting, and in the process of, and apparatus for, treating fatty and oily matters, and in the manufacture of candles and night lights. October 21; six months.

Donald Henderson, of Glasgow, ironmonger, for an improved apparatus for generating gas, which apparatus may be used for heating, and other similar purposes, and other apparatus for heating and ventilating. October 22; four months.

ERRATUM.—*Ante* p. 378.—In the heading of Mr. Lund's Patent, *dele* the word "Engineer."

WEEKLY LIST OF DESIGNS FOR ARTICLES OF UTILITY REGISTERED.

Date of Registration.	No. in the Register.	Proprietors' Names.	Addresses.	Subjects of Design.
Nov. 6	3010	Captain H. Toynbee & J. D. Potter	Poultry	Revolving parallel-ruler.
,,	3011	James Wilkes	Wolverhampton	Circular padlock.
7	3012	C. S. Vessey	Birmingham	Detection tap.
,,	3013	John Scarthff	Lincoln	Telegraphic bell-board.
8	3014	Augustus Smith	Whitechapel	Hand-protecting stove-brush.
11	3015	W. & J. Lea	Wolverhampton	Lock.
,,	3016	Francis Taylor	Westbourne Park-villas	Embossing press.
12	3017	J. Elce & Co.	Manchester	Apparatus for applying grease to gearing.

WEEKLY LIST OF PROVISIONAL REGISTRATIONS.

Date of Registration.	No. in the Register.	Proprietors' Names.	Addresses.	Subjects of Design.
Nov. 6	317	John Crosby	Fakenham	Safety sea-bathing machine.
8	318	Thomas Capps	Leadenhall-street	Brunswick parasol.
10	319	J. J. Cortins	Gt. Pultney-street	Extending boot tree.
,,	320	J. S. Cockings	Birmingham	Safety lever-bolt.
11	321	T. R. Grimes & Co.	New Bond street	Ring - head lamp cotton-holder.
,,	322	Joseph Kertchly	Ansty, Leicester	Chemical powder case.
12	323	J. W. Stephens	St. James's street, Dublin	Fire-protector for iron safes.

CONTENTS OF THIS NUMBER.

LONDON: Edited, Printed, and Published by Joseph Clinton Robertson, of No. 166, Fleet-street, *in the City of London*— Sold by A. and W. Galignani, Rue Vivienne, Paris; Machin and Co., *Dublin*; *W. C. Campbell and Co., Hamburg.*

Mechanics' Magazine,

MUSEUM, REGISTER, JOURNAL, AND GAZETTE.

No. 1476.] SATURDAY NOVEMBER 22 185 [Price 3*d*., Stamped, 4*d*.

Edited by J. C. Robertson, 166, Fleet-street.

CAPT. CARPENTER'S PATENT IMPROVEMENTS IN SCREW PROPELLING.

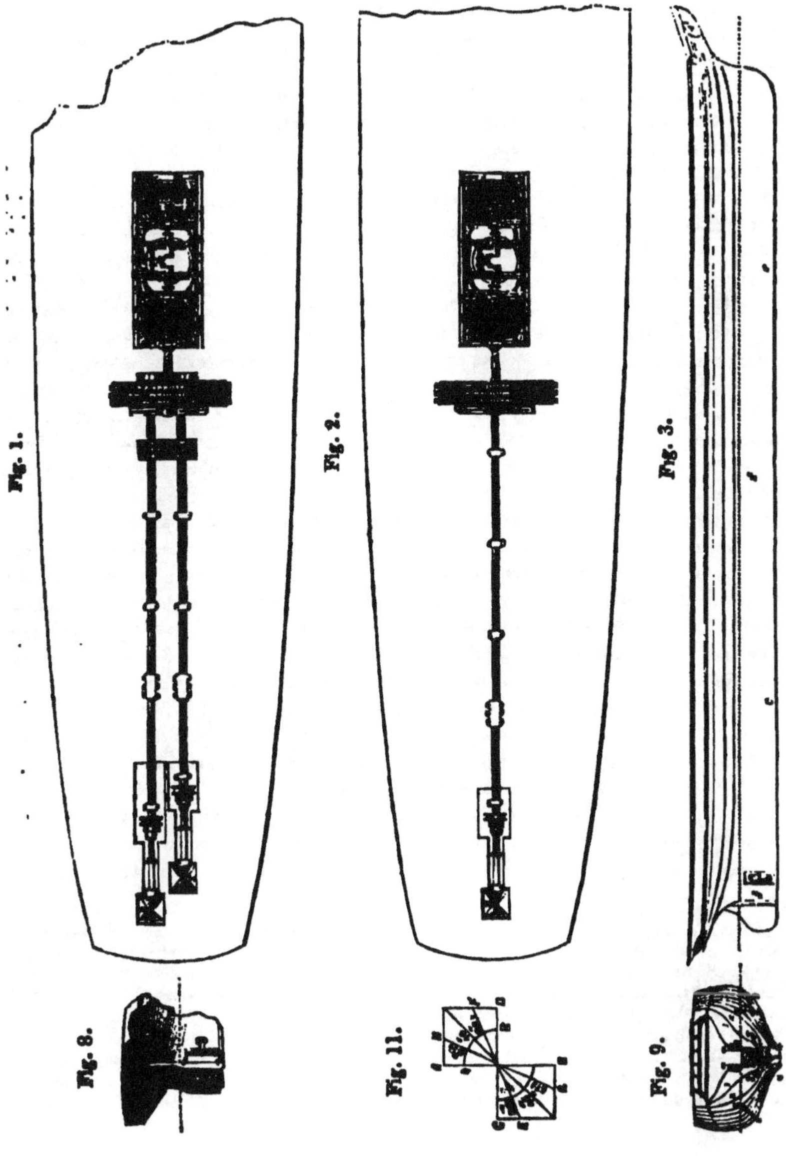

CAPTAIN CARPENTER'S PATENT IMPROVEMENTS IN SCREW-PROPELLING.

(Patent dated May 13, 1850. Specification enrolled November 13, 1851.)

Specification.

Firstly. My invention, in so far as it regards improvements in the construction of ships and vessels, has exclusive relation to ships and vessels which are intended to be propelled by means of screws or other submerged instruments, worked by steam power, either alone or in combination with wind or other power, and have for their object the adaptation of such ships and vessels, to be worked by two propellers, instead of by one only, as usual. Figs. 1, 3, 4, 8, and 9 of the drawings annexed, are different views of a vessel fitted with two screws (SS), and figs. 2, 5, 6, and 10 analogous views of a vessel constructed or designed for propulsion by one screw only (S). By comparing the one set of these figures with the other, the differences in point of construction between them will be readily recognised. The ordinary rudder, stern-post, dead wood, and after parts of the keel are removed, and the space in the hull hitherto appropriated to them is inclosed within a strong framework, open to the water both fore and aft and beneath, the two sides of which terminate at bottom in two additional keels or bilge pieces parallel to the main keel. It is intended that the two propellers aforesaid should work in the water space inclosed within this framework in the same way as the single screw now commonly works in a space cut out for it in the dead wood. In figs. 3 and 4, *ff* represent the above framework, and *f'f'* the water space within the same; *c* represents the starboard additional keel; *d* the port additional keel; and *e* the main or midship keel. The flooring immediately over the framework and water space has an arched form given to it for the sake of giving strength to the after frame of the vessel; and in order that the water may flow the more freely through the inclosed space beneath towards the propellers. The water space *f'f'* is constructed with a gradual inclination downwards from the water-line, between the two stern-posts, as far as the point where the main keel begins to change its original line of direction. The additional keels terminate each in a separate stern-post (*bb*, fig. 9), and each stern-post has a separate rudder attached to it. A steering wheel or apparatus should be used, capable of steering either with two rudders moving together in parallel lines, or separately. The distances between the additional side keels and the midship keel will of course depend on the diameter of the screws, and that again will depend on the quantity of propelling surface and steam power proposed to be employed. The exterior sectional lines (1, 2, 3, 4, 5, 6, 7, 8, fig. 9,) are carried from the midship section aft in the usual way followed in ship-building, allowing for the distance each keel and stern-post will be removed from the position occupied by these parts in ships on the old plan of construction; and the interior lines are curved from the flooring to each keel, so as to allow the water to pass away with as little obstruction as may be. The variation in these lines will be better understood by referring to fig. 10, where the numbers 1, 2, 3, 4, 5, 6, 7, and 8 represent the sectional lines of a vessel having one keel only, and adapted to propulsion by one screw only. From the midship section to the stem, or from the point where my alterations in construction commence to the stem, the construction would be the same as usual.

Secondly. My invention, in so far as it regards the machinery or apparatus for propelling vessels, consists in substituting for the one screw or submerged propeller now ordinarily employed, two propellers placed in the water space formed as aforesaid, the one a little in advance of the other, and which may be worked, either both at the same time or one only at a time. Each screw may be made of a diameter equal to that of a full-sized screw of the sort commonly used when placed in a space cut out of the dead wood, that is to say, of as large a size as the distance between the water-line and the bottom of the keel will admit of; so that when one only of the two screws is worked, the power obtained will be equal to that now obtained from single screw propellers, and when both are worked, an increase r more or less considerable will be realised, and that even though no more steam be expended. Why more speed should be obtained without the use of more steam,

Fig. 4.

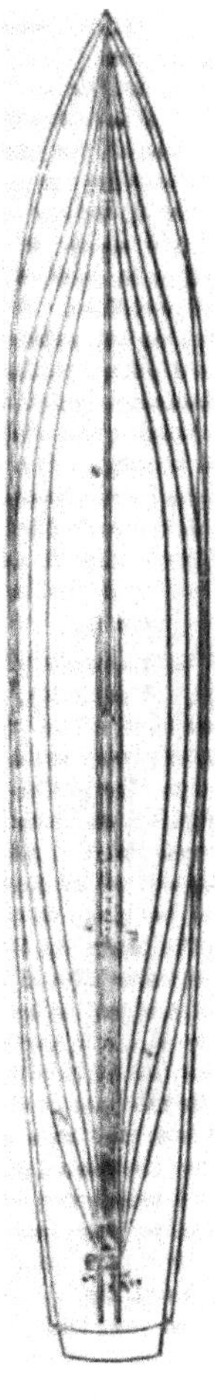

Fig. 5.

Fig. 6.

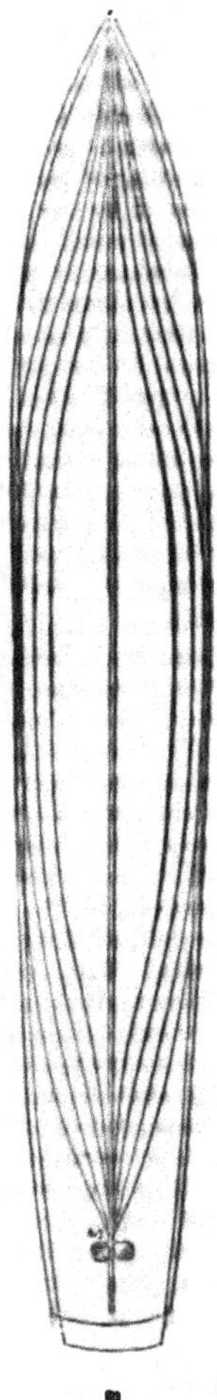

Fig. 10.

Fig. 7.

ecount for as follows :—A single screw placed in the deadwood, for the purpose of ... the vessel onwards, loses power in four different ways—First. By slip, which is owin the blades of a single propeller not having sufficient surface to overcome the heavy weight of a vessel without the water first yielding to the pressure of the screw, which water should act in the same way as a nut does to a screw. Secondly. From the single screw revolving in water which has previously been disturbed, and has not had time to settle into its ordinary density, the propeller is always working to disadvantage. Thirdly. A single screw has a tendency to turn the vessel away from a straight line in its course, and consequently the rudder is absorbing power which lessens the speed. Fourthly. When a vessel fitted with a single screw is pitching, the propeller is revolving at times without producing effect, from the blades being placed all on one shaft, and that shaft placed in the midship line of the vessel, while the propeller is at the extremity of the vessel. I have found, by my experiments, that these objections are very much overcome by the use of the two propellers, and that a considerable saving of fuel is effected by the second or auxiliary propeller picking up the power that is wasted by the single screw. I have also found, by my experiments, that two screw propellers, of much less size and weight than a single screw, will produce more speed with an equal amount of power; also, that two rudders in connection with the two screws will turn a vessel about in a much shorter space than heretofore. I have also found, by my experiments, that by disconnecting one screw and filling up the orifice, that the opposite screw may be used for propelling, as it formerly was when placed in the midship line of the vessel. Again; the vibration complained of in screw vessels is much diminished when two screws are employed; and, by having two screws at command, the risk of stoppage of the vessel is also greatly lessened; for though one might be damaged, the improbability is great of both being damaged at the same time. The screws being in advance the one of the other, revolve quite free of each other; but I wish it to be understood that I do not regard it as indispensable that they should be placed at the exact distance apart as represented in the engravings, for that they may, without material prejudice, be placed more or less forward or more or less aft.

Thirdly. My invention consists of an improved form of screw propeller, which may be used either where two propellers are employed, as aforesaid, or one only, and either in a space cut out of the deadwood or in a space formed as described under the first head of this specification. Fig. 7 is a side view, showing the manner of constructing this propeller. The vanes or blades are here represented as perfectly flat, and set vertical on the shaft, ready to be twisted into the required form; and the diagram, fig. 11, exhibits the geometrical construction necessary to be observed in the production of that twist. A B represents the propeller shaft; C C the propeller blades, which may be constructed of metal, or wood, or any other suitable material. The strength and superficies to be given to them will depend upon the size and weight of the vessel to be propelled. At the ends, on the line *mn*, these blades are made perfectly flat; but as it is necessary that the water should pass away freely towards the axis of the shaft in its rotation, they are on that account to be partially curved from the line *mn* towards the shaft or line E F; but the curve is not a curve which, if continued, would form a complete screw, but one which admits of the extremity of the blade or its flat surface being set at the angle of 22½°, and that part near the boss at the angle of 67½°, or nearly so. Suppose, for example, A B, fig. 11, to represent the axis of the propeller shaft, and C D blade set vertical on that shaft, and, in the first instance, perfectly flat surfaces; then, if the blades are turned round on their own axes towards the line E F, they will be set at the angle of 22½°, represented by the arc Q R. Again; suppose the upper parts of the blades to be fixed, and the lower to be gradually twisted on their respective axes towards the line G H, near to the shaft (equivalent to the line *mn*, fig. 7), they will on that line be bent to an angle of 67½°, the general result being, that the blades are part flat and part curved.

I do not confine myself to two blades, as three or four blades, formed in the same way, may be used on the same shaft. Neither do I confine myself to the precise angles above specified, or to the proportions represented in figs. 7 and 11, deeming this only to be of essential importance that the blades shall, as before said, be part flat and part curved.

Fourthly. My invention consists of certain improved arrangements whereby the improved propeller before described, or any other, may be wholly lifted out of the water, or the vanes or blades may be shifted or "feathered" at pleasure, without lifting the propeller shaft along with them. To lift the propeller wholly out of the water, a chock is placed between the two posts which form the frame for lifting the propeller in the usual way, having grooves in them to guide the propeller shaft. On this chock there is fixed a rod or apparatus for lifting it up or lowering it down to the blades. On each side of the chock there is a strong *metal plate*, which overlaps the edges of the posts, and serves both to guide the blades and to sustain their pressure. Underneath the chock there is attached a strong metal plate to admit of the blade abutting against it. Then, by means of a lever or levers grasping the blade firmly, the propeller may be lifted up by the chock, without having recourse to any such cumbrous machinery as that now commonly employed for the purpose. To enable the blades to be shifted at pleasure, it will be necessary that they should be made to put on and take off (which may be done by a key being let into the shaft, and continued so that it shall pass through the lower part of the blade, and in various other ways); so that, with the assistance of my improved mode of lifting the propeller, and auxiliary gripping lever or levers before described, the propeller blades can be lifted on deck separately, instead of lifting the shaft along with them. Flat blades may afterwards be attached to the boss, to lie in a straight line with the shaft, which will fill up the aperture, and in that way be of advantage to a vessel while sailing.

In addition to the advantages before stated as resulting from the employment of the two propellers, and the arrangements connected with them, the following are deserving of mention:—The vessel will be on a better construction for strength in the after part for steam propulsion. There will be increased breadth for accommodation aft; also for carrying guns in the stern and on the quarters. There will be more bearings for carrying heavy weights aft, and the engine or engines may consequently be brought farther aft. The vessel will be able to turn round quicker, and steer better going ahead or backing astern. She will hold a better wind when sailing, and descend into the water astern easier, scud safer, and roll easier. She will start quicker and stop sooner than with a single propeller. The propellers will have a more solid mass of water to revolve in, therefore they will have more resistance and less slip; consequently, they will give more towing power, less loss of power from the engine, and greater speed. As the motive power may be divided half on one side and half on the other from the engines, the weight of the machinery will be better disposed. The propellers and gearing will be of less weight, therefore they will be more easily handled, and there will not be such a dead weight lodged over the keel, or in one point in the stern, near the rudder. The propellers will be under water, therefore, out of the reach of gun-shot. The framework on one side will be a protection for the other against shot, or in gales of wind. In action, in the event of one rudder being shot away, or one propeller being disabled from the wreck getting foul of it, the other propeller and rudder could be used; while with a vessel fitted with a single rudder and a single propeller only, she would be disabled.'

The two-screw plan is farther peculiarly adapted to line-of-battle ships as an auxiliary power; also to boats or vessels having a shallow draught of water, and requiring an increased area in the screw propeller.

And having now described my said invention in the several parts thereof, and pointed out the advantages thereof, I declare that there are parts before described which are old and well known, and therefore not meant to be set forth by me as of my invention, but that the improvements which I limit myself to and claim are as follows:—

First. I claim the constructing of ships or vessels with a water space formed out of the hull, for screws and other submerged propellers to work in, by the removal of the ordinary rudder, stern-post, deadwood, and after part of the keel, and enclosing with a framework the space heretofore occupied by these parts, as before described.

Second. I claim the employment, for the propelling of ships and vessels, of two propellers placed in the water space so formed as aforesaid, the one a little in advance of the other, and capable of being worked either both at a time or one at a time only.

Third. I claim the improved form of propeller represented in figs. 7 and 11, and before described, having blades parts of which are flat and parts curved. And,

Fourth. I claim the improved arrangements before described for raising propeller blades out of the water, or, in other words, "feathering" them, without raising or shifting the propeller shaft along with them.

The following experiments were made on the Clyde Canal, during the month of September, last, with a view of testing the advantages of Captain Carpenter's two propellers an two rudders against the single screw propeller. Dimensions of model—length over all, 4 feet 9 inches; beam, 9 inches; draught of water, 3 inches.

TABLE OF EXPERIMENTS.

Propeller.	Diameter.	Extreme breadth of each vane.	Angle.	Power.	Distance run with the whole power of spring.	Force of wind.	Speed obtained in stated distances. Feet. 50	Feet. 70	Feet. 90	Feet. 110	Feet. 124	Remarks.
					Feet.							
Single ..	$2\frac{3}{4}$	$1\frac{5}{16}$	$22\frac{1}{2}°$	Full.	90	Calm.	17″	24″	33″	Minus.	Minus.	This experiment proves that the model goes 34 feet further with the two propellers than with a single one, with an equal amount of power—also that she has greater speed.
Duplex..	$2\frac{3}{4}$	$1\frac{5}{16}$	do.	do.	124	do.	16″	22″	31″	39″	45″	
					Feet.							
Single ..	$2\frac{3}{4}$	$1\frac{5}{16}$	$22\frac{1}{2}°$	Full.	90	Calm.	17″	24″	33″	Minus.	Minus.	This experiment proves that the model goes 20 feet further with the two propellers of less diameter and area than the single screw; also that she goes faster with an equal amount of power.
Duplex..	$2\frac{7}{16}$	$1\frac{1}{8}$	do.	do.	110	do.	$15\frac{1}{4}$″	$21\frac{1}{2}$″	29″	37″	Minus.	
					Feet.							
Single ..	$2\frac{3}{16}$	$\frac{13}{16}$	$22\frac{1}{2}°$	Full.	70	Calm.	16′	23″	Minus.	Minus.	Minus.	This experiment proves that the model goes only 70 feet with a single propeller, small size, and that she goes 90 feet with the two propellers of the same size, and has more speed.
Duplex..	$2\frac{3}{16}$	$\frac{15}{16}$	do.	do.	90	do.	15′	21″	30″	Minus.	Minus.	

Experiments with the Rudders.

With the helm put hard over to starboard, with a single rudder, the model turned completely round, in a circle measuring in circumference about 60 feet, in 30 seconds. With the helm put hard over to starboard, with the two rudders, the model turned completely round in a circle measuring in circumference about 24 feet, in 16 seconds. In backing astern the two rudders act alike both ways, whereas with the single screw the rudder acts only one way, which experiments prove the advantage of using the two rudders.

Judging from these experiments, therefore, it is not difficult to foresee that if the distances represented above by feet, were taken as miles, that a vessel with a given amount of power and fuel on board, and fitted with a single screw propeller, would go about 90 miles and stop, and the same vessel having the same amount of power and fuel on board, and fitted with the two propellers, would go about 124 miles at a greater speed and stop. It should also be borne in mind that each of the two propellers in the above experiments is working with only half its proper power of the engine, because the power is divided between two larger propellers which admit, each of them, of an equal amount of power being applied to them as the single screw has, which power would give a vast increase of speed.

HORNER'S METHOD OF SOLVING NUMERICAL EQUATIONS.

Sir,—I have lately met with a small treatise on the "Elements of Algebra," by "Robert Wallace, A.M., late Andersonian Professor of Mathematics, Glasgow." The work is dedicated to "Doctor George Birkbeck, F.G.S., President of the London Mechanics' Institution," and was published by "Basil Stewart, 139, Cheapside, London, 1828;" but would scarcely merit particular notice did it not profess to include "*A Simplification of the Rule for the Solution of Equations of all Dimensions.*" The method and its illustrations, which occupy pages 53—60, need not be transcribed, since they are in every respect

al with Horner's unmodified pro-
; but Art. 183 contains a statement
I could wish to see verified by
of your correspondents. It is as
:—

Art. 183.

he solution of equations of all
s by the method from which this
as derived, is generally ascribed
r. Holdred, of London, who pub-
a tract on the subject in 1820. A
method, by Mr. Horner, of Bath,
ed in the *Philosophical Transac-*
or 1819.* It is not given, how-
o one individual to accomplish
rk of ages. For, while we do
pute the originality of either of
uthors, we claim the priority of
ention for a Scotsman of the name
bert, schoolmaster at Auchinleck,
r as regards the solution of equa-
f the *third* degree. While Mr.
astle, in his elementary treatise,
l, so late as 1818, that the solu-
the *irreducible case* of cubic equa-
except by means of a table of
r by infinite series, had hitherto
the united efforts of the most
ted mathematicians of Europe;
e for solving cubic equations of
ds, whether *reducible* or *irredu-*
had been given by Halbert, so far
1789, in his 'Treatise on Arith-
published at Paisley in that year.
ventor, after giving his rule, and
variety of examples, says—'So
reckon this method a valuable
ry, when compared with the jar-
other authors about *transmuta-*
limitations, and *approximations*,
hich brings us never the nearer
pose.'"

e never seen Mr. Halbert's trea-
r is anything more than his name
ed by Professor De Morgan in
of authors in his "Arithmetical
' or elsewhere; hence, if any of
eaders can furnish the "rule,"
a "example," *as they stand in*
rk referred to, a connecting link
supplied of some value in the
of the problem of evolution.

I am, Sir, yours, &c.,

T. T. Wilkinson.

r, Lancashire, Nov. 1, 1851.

ioning Mr. Horner's paper *last*, appears
h like *putting the cart before the horse*.

ON CONICAL SHOT.

Sir,—Since you did me the favour to publish my first paper on projectiles, in the year 1843, the subject has received much attention and undergone continual developement. That paper was immediately read and carefully considered by the most eminent mathematicians in Europe; and from our first geometer it elicited, after a lapse of four years, some views of the law of atmospheric resistance which were far in advance of all preceding theories. Soon after the publication of his theorem, and of my subsequent papers, the revolution in France impelled the nations of the continent to attend to the problem in a very practical manner, and my theories were brought to the test on the field of battle.

Since Dr. Hutton made experiments at Woolwich, with cannon, as he observes, "very nicely bored, and cast on purpose," and discovered that the round shot thrown by these guns were "greatly deflected from the direction in which they were projected; and that, as much as 300 or 400 yards in the range of a mile, or almost one-fourth of the range," it has been made clear that, though a sphere may be the figure due to a projectile in a non-resisting medium, yet it is by no means the figure due to a projectile in a resisting medium. Indeed, this fact had been admitted long before, without being successfully applied to gunnery.

In the year 1843, after a train of experiments extending through the ten preceding years, I suggested that the rear of a projectile should contain a parabolic chamber, because all rays parallel to the axis of a parabola, after impinging on the curve, are discharged into the focus. For many years this principle has been applied to the reflectors used in lighthouses, and also to lamps for carriages, where the rays of light are required to go off parallel to each other. It is also applied to the patent chamber in guns, to drive the force of the powder straight up the barrel, instead of permitting it to expand in every direction. Shortly after the publication of this paper, the French adopted the suggestions contained in it, and added a little ingenious fancy of their own, to expand the lead by means of an iron capsule. The Germans followed, and also adopted the parabolic chamber. In the meanwhile the conical or sugar-loaf shot, which I had been using since the year 1836, was brought forward, independently of mine, by a London gun-maker; and I am informed that rifles made by the late Mr. Staudenmayer, and Mr. Egg, who lived at the commencement of the present century, were frequently fitted by them with what is now denominated a sugar-loaf or conical bullet.

At page 556 of No. 1296 of the Magazine, is represented the figure of an elongated projectile, which has been proved by experiment to be true to the direction of the rifle gun at a range of 1,200 yards. It is also found that, when this figure is copied in miniature, the diminished solid is equally true at comparatively diminished ranges Yet the great weight of the solid required an increased charge of powder, and this additional amount of powder and lead being more than common guns are constructed to sustain, the recoil is found to be so severe that great uncertainty ensues.

Fig. 1.

I also enclose a drawing of a German bullet, in section, of the actual size. This very heavy projectile is similar, as I am informed, to many that were used in the Holstein war. The deep circular groove that surrounds it is intended to receive what engineers call a packing, such as is put round the piston of a steam engine, to fit closely, and prevent the escape of the steam; and in this example the packing may be of hemp dipped in oil, or worsted, or any other material suited to prevent the escape of the gases set free by the combustion of the powder, and also to relieve the friction against the barrel. The centre of gravity of this projectile is exterior to *the parabolic* chamber, and, in fact, considerably in front of it, which I think an unfortunate arrangement; yet, if the chamber had been carried more forward, the lead might have been burst by the explosion, at the part weakened by the deep grove.

I could have forwarded other examples of a smaller calibre, but of similar design, yet this may be sufficient. The moulds were brought from Germany, and were of a complex and expensive construction.

Fig. 2.

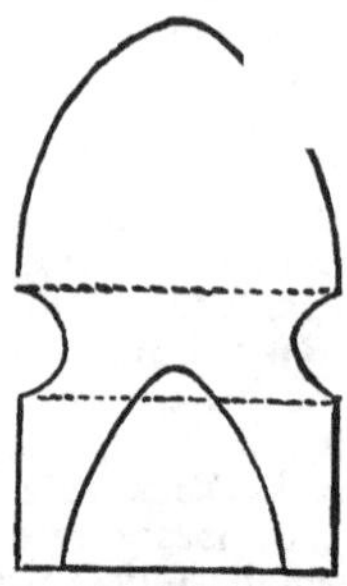

The French bullet of the army is figured below; and a Swedish officer has favoured me with the dimensions and weight:—The length of cylinder =18 millimetres; axis of spheroidal head, =12 millimetres; diameter of chamber at base, =$16\frac{1}{10}$ millimetres; weight, = 50 grammes; ditto, with iron capsule, =51 grammes; diameter of gun, =17 millimetres—the millimetre being ·03937 of an inch.

This is fired from a four-grooved rifle with excellent effect at long ranges, as the Roman unwashed assassins and flea-bitten highwaymen discovered, when the light-infantry of France opened their fire, and commenced the siege of Rome.

The little iron capsule which covers the entrance of the parabolic chamber is driven by the explosion down the chamber, and is intended to expand the bullet, and make it fit the barrel air-tight.

Fig. 3.

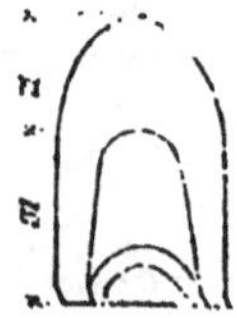

it sometimes fails, and is blown h the bullet near the base of the :r, passing out through the side spheroid. This, of course, sends .l wide of the mark; and these ; may, in time, convince the . that they will shoot better with- capsule than with one. A short ' wood might answer at least as the capsule, and be driven into .mber by the force of the powder. ve they ever thought of filling the :r with gun-cotton, and closing ice with thin paper, saturated with on of saltpetre? or of filling the :r with rocket mixture?

'y one that is acquainted with the of a ship's lines, is aware that iust be no resisting angle opposed water, and that the curvature e what Newton terms a "*curva- ntinua.*" A military projectile iing else than a corresponding naking its way through a different hough under altered conditions as ratio of velocity and resistance. ien this projectile consists of a r surmounted by a cone, as in the below, it is found to fly very

Fig. 4.

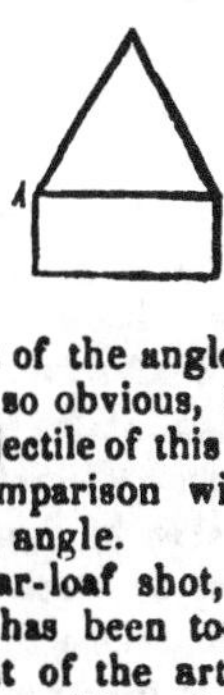

defect of the angle at the base of ne is so obvious, that it is sur- a projectile of this kind should be in comparison with others that o such angle.

ie sugar-loaf shot, the curvature solid has been too often left to dgment of the artisan who con- 1 the mould, and whose eye was he imitation of no known curve eys an algebraic law. Perceiving i my paper of the year 1846, I . the adoption of an elliptic curve, me that is well known, and supe- any tentative or empirical curve, being tangential to the sides of rrel. Yet, in the ellipsoid, the the curvature at the vertex is too out its way freely through the air. The satisfactory effect of conical shot distinctly suggests that the ellipsoid may be improved by extending the solid between tangents meeting above the vertex; and it is possible that these tangents should intersect at a very small angle, as in the figure, page 38, No. 1040 of the Magazine.

The very extensive ranges commanded by these projectiles demand a proportionate elevation of the rifle-sight near the lock of the gun. The Prussians now set up their targets at 800 yards; and the English rifle is good at 1,200 yards. The top of the rear sight, instead of being about an inch above the barrel, as before, requires an elevation at such ranges of from three to five inches in height. This brings the stock of the gun so inconveniently low at the moment of taking aim, that some alteration will in future be probably required in the stock, such as either to cut the wood absolutely straight, or to reverse the present angle.

Though the projectiles adopted by the English, the French, and the Germans, are at present very different from each other, yet it is plain that all are tending towards one general type which shall combine the advantages that each nation has obtained, and reject the defects that each is willing to acknowledge. What may be the ultimate figure of this common type, it would be injudicious and premature to assume in our present state of imperfect knowledge; yet, so far as we can discern, it will probably differ in no material respect from the figure here given, unless the vertex of the solid should admit a curve of contrary flexure, which is not impossible.

The next object is to apply such elongated projectiles to cannon, instead of the present round shot that sometimes fly a quarter of a mile wide of the mark. If the weight of metal is increased by employing a solid elongated shot, the charge of powder must be increased also, and therefore the guns must be made stronger at the reinforced parts than they are now; although Mr. Monk has effected great improvements in this respect within the last few years. To enable such cannon as are now in use to be fired with conical shot, the weight of metal in the projectile must be as nearly equal as possible to the weight of the common round ball, or to such a weight of metal as the gun is able to bear.

Fig. 5.

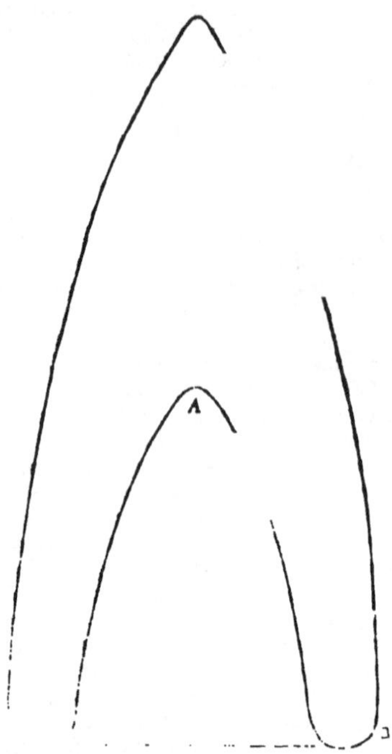

This object may be attained by expanding the chamber of the projectile until the metal that remains is of the weight required.

In a previous communication, I pointed out that the expression for the length of an elongated projectile is a function of the velocity. This circumstance suggests that high velocities require a projectile of greater length than those suited to the rifle gun, where the velocity is given by a small charge of powder.

The problem of the figure due to a projectile in a resisting medium may now be considered as practically solved. It has ever baffled the utmost resources of mathematical analysis, and it is only by a long train of experiments guided by theory, that we have attained thus far to a practical solution. The principle of Bacon, that theory must depend upon experiment, and not wander without so useful a guide, is verified once more. The artillery-officers of Europe will now institute experiments *with their cannon*, and determine all the details of the problem. It is certain that the nation whose officers are first able to do this will possess the means of victory over all armies that may stand before them in war.

Much obloquy is often cast by unreflecting persons upon those who propose or effect improvements in the art of war. Yet it is by superiority in arms alone that civilized men can protect their arts, their laws, and their homes against savages who would destroy all, and leave a desert behind them. Not by superiority in numbers, but by superiority in arms does the colonist at the Cape defend his wife and daughters against the wolfish and blood-thirsty Caffres. When the rifle fails, he will be swept into the sea.

The observations that I have now the honour to lay before your readers will be carefully considered, as before, by the most eminent mathematicians in Europe and America; and the suggestions that I have made will, as before, be tested by the most distinguished officers in Europe, and tested too upon the field of battle. In due season I intend to report upon their conclusions; and till then I remain, Dear Sir, yours faithfully,

C. A. HOLDSTOCK.

25, Upper North-place, Guildford-street.

EXPERIMENTS IN NAVAL CONSTRUCTION.

It is to be hoped that the victory obtained by the *America* will lead to a series of analytical experiments that may determine many doubtful, but important points in naval construction. It will be seen in Nos. 1294 and 1295 of the *Mechanics' Magazine*, that Sir Samuel Bentham in the year 1828, proposed that a course of such experiments should be undertaken by the naval authorities in a manner by which little expense would be incurred; since that time, Mr. Scott Russell, has in the way proposed, that is, in the first instance by small models, made many valuable experiments, which have led to improved constructions of sea-going vessels as to speed.

Sir Samuel Bentham, from early in the year 1828, to the month of February 1831, was strenuous in his endeavours to induce the naval authorities to institute a set of analytical experiments; but although

members of both the Admi-
Navy Boards were convinced of
leness of making them, it ap-
they had no competent indivi-
ir command who had sufficient
superintend such an extra-
ness. Sir Samuel, therefore, in
the Secretary of the Admi-
ber 11, 1830, said "I am in-
offer to devote my time and
the pursuit of the experiments
in *any manner* which may ap-
advantageous after consultation
rveyor of the Navy as to the
to various other particulars. I
re strongly induced to offer my
services towards the furtherance
nt proposal from the considera-
e expense of the models and the
for ascertaining the resistance,
y, and the weatherliness of ves-
t exceed a few hundred pounds."
this period, Mr. Henry Mauds-
t of the firm of Maudslay and
d various models to be made at
ense, such as were indicated by
, so that he was enabled on the
ry 1831, to acquaint the Comp-
he Navy, that so small a sum as
would now suffice for providing
e apparatus, and an expenditure
unt was sanctioned by the Navy
he 22nd of the same month.
ng to the petty impediments that
ed themselves to the instituti n
periments, it may be inferr that
es are more likely to be executed
irited private individual, than by
authorities; it is in this concep-
the enclosed copies of various of
's communications to the Admi-
ffered for insertion in the *Me-*
agazine, as they may afford use-
tions for the framing a suitable
xperiments, and tend to the res-
national superiority in naval con-
rom that state of humiliation in
as been thrown by the superiority
ries in point of speed.

M. S. B.

Lower Connaught-place,
Jan. 17, 1831.

Sir,—I would beg you to lay before the Lords Commissioners of the Admiralty the enclosed papers relative to the means of ascertaining the fittest form and equipment of vessels for the naval service, some of which papers, it will be seen, are dated so long as two or three years ago, and which have been perused and considered by several in the highest situations in the naval department who, while they expressed their conviction of the importance of the subject treated of, seemed to entertain no doubt of the practicability of the means proposed for the attainment of the object in view. I had also during the last Administration consulted with the Comptroller, and with the Surveyor of the Navy relative to the means of pursuing the proposed course of experiments, and as I have offered to occupy myself gratuitously in the conducting these experiments, the Navy Board seem ready to afford me the means of commencing them, it therefore seems expedient that I should submit these papers to the consideration of the present Board of Admiralty.

I have the honour to be, Sir,
Your very obedient Servant,
SAMUEL BENTHAM.

Captain, the Hon. George Elliot.

Papers enclosed with the above letter to Captain Elliot.

No. I.—In a course of experiments with a view to the improvement of vessels for navigation, the following particulars suggest themselves as requiring to be ascertained. They are drawn up on the supposition of its being admitted that these experiments should not be confined to the vessel itself, but should extend to the whole of the apparatus for locomotion, whether by steam, by manual labour, or by the natural force of wind.

The purposes for which a vessel requires to be navigated, whether with a view to war or to commerce, are evidently to afford a given quantity of empty space for the purpose of conveying various articles of a given weight and bulk from place to place with the greatest certainty and control as to place, direction, and velocity.

Consistently with these objects, the points to be ascertained are; 1st. In regard to the vessel itself whatever be its bulk, what is the form which will best facilitate its locomotion and guidance in all places where it may have occasion to be navigated? 2nd. In regard to the apparatus, what is that which in regard to each of the above-mentioned forces, will, with the least bulk and weight, enable that force to be applied with the greatest advantage in effecting the several above-mentioned purposes of the vessel?

In regard to the form of the vessel in general, whatever be the locomotive force, looking to the result of previous experiment and observation, all that appears to have been as yet ascertained is, that the form of the after part of the vessel influences the resistance to progressive motion as decidedly as that of the forepart, that the broadest part should not be in the middle, but nearer the foremost end, and that of two different forms, that which meets with the *least* resistance when moved with *one* degree of velocity, may meet with the *greater* resistance when moved with *another* degree of velocity.

The vessel, it is evident, when moving, must push the water it meets with out of its way; this, it will also be seen, may be done by giving the vessel such a form as shall cause the greater part of the water to pass under it, or by giving it such another form as shall cause the water to pass chiefly on each side; or by other variations in the form which may cause the water to flow partly under, partly on each side of the vessel.

Adopting in the first instance the parallelogram, as being the simplest form for the transverse vertical section of the vessel at its broadest part, the experiments would be directed to show how each end should be shaped so as to divide the water to the best advantage, and make it flow laterally, or under the vessel; and how much of its length the same breadth and depth may be continued without encreasing the resistance to progressive motion. All these questions will be easily solved by the proposed course of analytical experiments detailed in a separate paper. SAMUEL BENTHAM, 1828.

No. 2. This Paper has already been published in the *Mechanics' Magazine* for May 27, and June 3, 1848.

July, 1828.

No. 3. Enclosed to Captain Elliot, 17th January, 1831.

On the means of ascertaining the fittest form and equipment of vessels for navigation whether for commerce or for war, and whether the locomotive force be wind, steam, or manual labour. Also, the apparatus requisite for noting and registering the particulars which influence locomotion.

The advantages already obtained by the introduction of the expansive force of steam as a locomotive force for navigation, cannot but afford a strong indication of the very great advantages generally to be derived by applying to naval architecture that close investigation of cause and effect, mechanical and chemical, which has of late years enabled other arts to make such rapid progr of which this art, however essential prosperity of the country, will be f stand much in need. Naval Archi appears indeed of late years to ha looked to by many officers of his M Navy as a study on which their prof credit, their fortunes, and on many o their lives depend. There seems, th reason to expect when the attentio rected to the great variety in the ves ployed for the same purposes in t other countries, and to the yet gr versity in the opinions entertained c comparative merits of each variety, the causes upon which their more fitness depends, that the fixing up criterion by which all these differ opinion may be satisfactorily adju be looked upon as a most desirabl and it may then be hoped that the tion of a very simple means of attai object will be favourably received.

In regard to the attempts made the questions continually started subject, a few words may be suff show the unfitness of the mode of pr hitherto pursued for their solution or experiments, the insufficiency c has indeed now been so often exp as to render evident the need of so trustworthy mode of proceeding.

In the construction of vessels various classes and descriptions their shapes have indeed been co more or less varied, with a view, n among other objects to that of in their sailing. With the same v ships already in use have been a form, some of them extended in others in breadth or in depth; so of the parts exposed to the resistan water has been very materially varie more or less immersion either of ferent ends of the vessel, or of th so also the form and the bulk of t exposed to the action of the wind, h changed by cutting down one or mo But how far by any of these cha resistance to progressive motion l diminished, or the resistance to lat tion or lee-way, so essential to good has been increased, has in no ca satisfactorily shown; and therefore n are still built no less varying in the under water, and in their height abov than before.

In regard, however, to vessels to gated by steam or by manual labou which are under command as to d as well as to amount, experiments i this country, as well as in others, h more satisfactory in their result.

been found that several steam vessels first built of the usual form, and afterwards lengthened and sharpened, have attained a very great increase of velocity, the locomotive force remaining the same, and the altered form has also, on many occasions, been found as well suited to navigation in a troubled sea as in smooth water.

So far, therefore, the form of vessels to be navigated by steam may be considered as materially improved; but when the same vessel is to be adapted to navigation by the uncontrolable force of wind, a variety of questions will present themselves, such as the attempts at improvement hitherto pursued are in nowise adapted to solve.

(*To be Continued.*)

"THE CRYSTAL PALACE."*

Many as are the thousands who have personally surveyed the wonders of the Crystal Palace; there is still a far greater number who must be indebted for all their knowledge of them to hearsay or printed description. The proper way to have satisfied the curiosity of the millions who were unable to visit the Exhibition would have been for the Commissioners to publish a Descriptive Catalogue—good and cheap—which might have found its way to every fireside, and left nothing for volunteer note-takers and annotators to do. Certainly a few thousands of the immense balance remaining at their disposal could not have been expended to a better purpose. Instead of that, the only thing of the sort published by (their) authority is an Illustrated Catalogue, so enormously dear that few can afford to buy it, and so wretchedly ill-done (as a whole) that it is not worth buying (see *ante* p. 331.) The contractors, to whom the Commissioners, for the sake of a few pounds *certain*, albeit in amount inconsiderable, handed over the execution of this important portion of their duties, finding that their costly and worthless Catalogue did not make way with the public, have sought to indemnify themselves by the publication of what they style a "Popular Record of the Great Exhibition," *alias* "Hunt's Handbook." The move has been judicious. Mr. Hunt's book will sell when the big "Illustrated" blunder will only serve to record on trunk-linings the impolicy of making a public trust, a matter of private and mercenary jobbing. Mr. Hunt never writes anything but what is well worthy of being read, and our Catalogue contractors have done wisely in calling in his powerful pen to their aid. His "Hand-book" is an exceedingly clever, entertaining, and instructive work. Its chief fault is that it is in the ultra-laudatory vein—all rose-colour and varnish; a thing done well, but done so palpably *to order* as to inspire but small respect and less admiration.

All these circumstances have prepared us to give a hearty welcome to the book of the "Crystal Palace" now before us. Judging from so much of it as has been published, it will be by far the best, as it is assuredly the cheapest, account of the contents of the Palace yet given to the public, whether "official" or unofficial. Not better in point of mere literary style than Mr. Hunt's —which it would be difficult for any book to be—but better in point of comprehensiveness—of circumstantial detail—of profuseness of pictorial illustration (immeasurably so)—of historical and anecdotal reminiscences—of intelligent, independent, and impartial criticism. Some of the engravings seem to us as if they were old acquaintance; but they are none the worse for that. All are in the very first style of xylographic art, and rivalled only by the admirable embellishments of the *Illustrated News*. Whoever has not visited the Exhibition, and would know as much about it as those who have—nay, infinitely more—should add this account of it to his literary store. It will make him quite as wise as the wisest of his neighbours in regard to it, and at twenty times less expense, at least.

* "The Crystal Palace and its Contents. An Illustrated Cyclopædia of the Great Exhibition of 1851." Nos. I., II., III., IV. Clarke, London.

THE SUBMARINE ELECTRIC-TELEGRAPH BETWEEN DOVER AND CALAIS.

The deficiency in the submerged cable has now been made good, and the electric communication between London and Paris fully established. On Thursday, the 13th of November, this event, which should be ever memorable in the annals of the two countries, occurred. The communication was established in so effective a manner that the contents of a paper which marked the prices of the funds on the London Exchange were instantaneously transmitted from the English to the French coast. The success of the undertaking was also evinced in a manner which appealed still more to the imagination from the dramatic circumstances with which it was attended. The Duke of Wellington, on this memorable Thursday, was at Dover for the purpose of accomplishing some ceremony connected with the Harbour Sessions in his character of Lord Warden of the Cinque Ports. It was arranged that on his departure for London by the two o'clock train the illustrious old warrior should be saluted by a gun *fired at Calais*. A 32-pounder, loaded with 10 lbs. of powder was prepared, and brought into communication with the wire. When the church bell struck two at Calais, and when the train at Dover was just gliding from the railway shed, the spark was to be applied in France and the gun to be discharged in England. Punctually as the clock struck two a loud roar was heard along the Dover cliffs, and the ground quivered to the report. The feat had been accomplished. Nor was the future success of the experiment left in doubt. To celebrate the victory of science Calais continued to salute Dover, and Dover Calais—and the guns discharged at one port were fired from the other.

"Fate and Professor Wheatstone," as the *Times* happily observes, "have for some time past been at odds as to the future fortunes of the human race, but now the Professor has got the best of it. 'A man can't be in two places at once,' says the old proverb, but by the aid of electricity, any one of us may be in all the capitals of Europe at the same moment of time. We do not of course refer to the gross physical presentation of a man—to that portion of him which weighs fourteen stone and has the gout—but to his immortal essence, to his intellect, to his will. What matters it in what nook the *eidolon* may be thrust out of the way while the immortal spirit dabbles in shares on the Parisian Bourse, countermands the march of an armed host in Hesse, or gives its opinion of a painted window in the studio of a Munich artist? All these things a man can accomplish without moving from the Electric-Telegraph Office in Charing-cross. The sea is no longer an obstacle to the instantaneous transmission of information. The deep waters have been reduced by scientific skill to be the highway of human thought—not, as heretofore, painfully conve[yed in] lumbering packages, and freighted [in] large argosies which required half a y[ear to] put a girdle round the globe. S[o] thought on its first conception in a [...] human mind can thought now be [...] from one capital of Europe to anoth[er,] beget at the distance of a thousand mi[les the] same sequence of ideas which plays [at the] very moment round the fretwork [of the] parent brain. We know not why, in [...]ness, the transmission of the electri[c spark] 'full fathom five,' or five time[s five] fathoms, beneath the troubled surface [of the] ocean, should affect the mind with [more] awe than its instantaneous passage [through] the solid earth. Still this conquest [achieved] by science over the waves must ever [be] recorded as amid the greatest of [the] achievements since record has existed [of the] mighty feats accomplished by man. [It is] wonderful to reflect that while the [great] ships 'reel to and fro, and stagge[r like] drunken men,' far, far beneath their [keels,] amid the wrecks of former days, the [current] of human thought is evenly flowing o[n with]out disturbance or interruption. I[f a] ship begin to weigh her anchor in a[n Eng]lish port, and before three turns of th[e capstan] have been taken on the windlass, t[he an]nouncement of her departure may [be] flashed beneath the waters to the c[oast of] France. The next moment another [...] will convey the intelligence to the ci[ties on] the banks of the Seine, the Rhine, t[he dis]tant Danube, and far within the inhos[pitable] dominions of the Russian czar."

SPECIFICATIONS OF ENGLISH PATEN[TS EN]ROLLED DURING THE WEEK, E[NDING] NOVEMBER 15, 1851.

WILLIAM HENRY BROWN, of [...] End Steel - works, Sheffield, steel [...] *For certain improvements in the ma[nufac]ture of helves.* Patent dated May 6, [...]

The present improvements apply [to] helves in use in the manufacture o[f steel] for hammering the metal while in a [heated] state, and which have been hitherto [for the] most part made of wood (generally [...]) bound with iron hoops. These w[ooden] helves are subject to the disadvant[age of] becoming so much heated, owing [to the] high heat of the steel on which they op[erate,] as to involve the necessity of freq[uently] stopping their action, to allow time f[or the] wood to cool, and thus a loss is incur[red by] the manufacturer; they are also lia[ble to] char from the same cause. It has [been] proposed, as a means of obviating [these] disadvantages, to employ helves of [...] metal; but their use was discontinue[d] ice of sufficient elasticity to enable withstand the effects of long-conti- ommering. The patentee has, how- covered that hollow metal helves, composed of wood and metal com- ssesses the requisite degree of elas- l are at the same time very durable omical. These improved helves ructed of boiler plate, riveted to- the ordinary form, and stayed th a vertical partition extending at the whole or the greater part of th; and in order to preserve their being crushed by the pressure of ner-head and tail hoop, blocks or f wood are driven in when those of the helves are fitted on. The or about which the helves rock or situated rather nearer to one end other, and the helves are made der at that part of their length. may be formed in the same piece helve, or fitted on afterwards. itee also constructs helves of metal th blocks of wood interposed, or of wood faced with metal; and he makes them of hollow metal, in er case he prefers that they should lliptical or slightly ovate section.

.—1. The manufacture of hollow boiler plate.

manufacture of helves of boiler ombination with wood.

manufacture of hollow metal

Delemer, of Radcliffe, civil en- l machinist. *For certain improve- the application of colouring matter cottons, silks, woollens, and other nd to linen, cotton, silk, woollen. weft, and also in machinery or for these purposes.* Patent dated 851.

improvements consist in printing n woven fabrics or yarns, in such hat the greatest depth of colour it the centre of the pattern, the ing gradually shaded down from ght towards the edges; and this is y printing on the cloth or yarns a damp state, and then winding rollers in such an exact manner repeat of the pattern shall come ver and cover the corresponding f the pattern on the fabric last taken up, and subsequently apply- are to the same while in a damp which the colours are caused to t from the centre of the pattern, produce the effect above described. fication also describes a method of inted fabrics, by passing the same over a series of rollers above a steam-heated surface.

Claims.—1. The construction of a machine for printing on damp cloth or yarns, and taking up the said cloth or yarns, when printed, on a roller or cylinder in such an exact manner that every repeat of the pattern, or part of the same, shall come exactly over or upon its corresponding part or portion of the pattern last wound on or taken up by the cylinder or roller, so as to prevent the colours marking the whites.

2. An arrangement of apparatus for giving pressure to the goods so printed, while in a damp state, to produce the shading effect described.

3. The blocks employed in this process, when made or manufactured with raised portions corresponding to the design or pattern.

Thomas Robert Mellish, of Regent-street, glass manufacturer. *For certain improvements in instruments and apparatus for the admission and exclusion of light and air into and from buildings and carriages, and in the manufacture of reflectors of light; parts of which improvements are also applicable to the decoration of articles of furniture.* Patent dated May 7, 1851.

Claims.—1. The construction of blinds or screens, ventilators, and other articles used for the purpose of or in connection with the admission and exclusion of light and air into and from buildings and carriages, of glass, in the form of tubes or parts of tubes, or rods or strips, whether the same are plain or coloured, or a combination of both, and whether silvered or not silvered on their interior surfaces, as exemplified and described; also the application of such glass tubes or parts of tubes, or rods or strips, to the decoration of articles of furniture generally.

2. The construction of ventilating panes with strips of glass, notched or indented on their edges.

3. The construction of reflectors composed of glass formed double by blowing, and silvered on the inside. Or by uniting two concave or other glasses together, the interior surfaces having been previously silvered.

4. The construction of reflectors composed of glass tubes bent into the required forms and silvered on the inside.

5. The construction of reflectors with hollow segments of glass, silvered on the inside, and joined together, to form a circle or portion of a circle.

William Edward Newton, of Chancery-lane, civil engineer. *For improvements in apparatus for the generation and conden-*

sation of steam for various useful purposes; also improvements in certain parts of engines to be worked by steam, air, or gas. (A communication.) Patent dated May 8, 1851.

Claims.—1. A method of regulating the operation of blowing off the saturated water from steam boilers, in which salt or other impure water is employed, by the action of the same mechanism which controls or regulates the admission of the feed-water, so that the blow-off water will always bear a a relative proportion to the quantity of water fed into the boiler, which proportion must vary under varying circumstances, so as to cease entirely when the feed-water is cut off or ceases to enter the boiler. Also, the connection of the blow-off valve with the check-valve in such manner that the blow-off valve will be operated by the stem of the check-valve.

2. The employment in packing the stuffing boxes of steam and other engines of metallic packing arranged in a metallic matrix of a cup-shape, for the purpose of allowing the piston-rod or shaft when passing through or working in the stuffing-box, to vibrate laterally sufficiently to yield to the unavoidable irregularities of the engine to which it may be applied. Also, the combination of this conical-shaped matrix or stuffing-box with metallic rings contained therein, and through which the piston-rod or shaft passes or works, so that as the inner surface of the packing rings wears away, the rings may, by the action of springs or screws made to press them in the direction of the axis of the rod or shaft, be advanced along the conical or cup-shaped surface of the matrix, so as to maintain a perfect contact both with the piston-rod or shaft, and with the internal conical surface of the matrix. Also, the use of the said vibrating cup in combination with hempen packing.

3. The double connection of a condenser with an auxiliary boiler as well as the main boiler or boilers of a ship, in combination with the double connection of the engine with the auxilary boiler, and the main boiler or boilers. Also, the employment, in combination with the arrangements described, of a filter in connection with the condenser for the purpose of purifying the water of condensation.

William Longmaid, of Beaumont-square, gentleman. *For improvements in treating ores and minerals, and in obtaining various products therefrom; certain parts of which improvements are applicable to the manufacture of alkali.* Patent dated May 10, 1851.

Mr. Longmaid's present improvements are principally based on certain previously-patented processes of his for obtaining copper and silver, and for the manufacture of alkali, by the calcination and decomposition with common salt of iron pyrites and other ores or minerals containing sulphur: they have also relation to a method of employing anthracite coal in the production of iron from the oxide thereof.

The *first* improvement specified consists in the application of coke and anthracite coal in the decomposition of ores or minerals, by calcining them in mixture with common salt. When using coke for this purpose, the patentee prefers to employ a furnace closed by a door, and to supply the ashpit thereof with water, which, by becoming converted to vapour, rises and facilitates the combustion of the fuel. When employing anthracite, it is mixed with about one-sixth part of coking bituminous coal, and, being placed in front of the furnace door on a plate for that purpose, is partially converted into coke before being supplied to the furnace. In this case, also, the ashpit of the furnace is supplied with water. The condensation of the volatile products resulting from these operations will be found to be much facilitated, and the working of the processes generally improved by introducing steam into the flue leading from the furnace to the condenser.

The *second* part of the invention consists of an improved method of effecting the precipitation from their solutions of the salts of silver and copper, which are resulting products of all the operations of decomposing ores and minerals containing them by calcination with common salt. The process of calcination is conducted according to the directions given in previous specifications, and a solution of alkaline salts employed to dissolve the silver and copper, and other products from the calcined mass. In order to separate the silver, the patentee causes the solutions to pass through vessels containing metallic copper, by which the silver is precipitated, an equivalent of metallic copper being dissolved at the same time. When the whole of the silver is precipitated, the copper is obtained by metallic iron, in the usual manner. The alkaline solution is subsequently employed in the manufacture of sulphate of soda, or to dissolve fresh quantities of silver and copper.

The *third* improvement has relation to the manufacture of sulphate of copper from the sulphide thereof, and consists in calcining the regulus or other sulphide obtained by such processes as are above alluded to, at a low temperature with access of atmospheric air by which sulphate of copper

le salt of silver will be produced. ducts are then dissolved in water, ls of sulphate of copper obtained ipitating the silver by means of The precipitate is treated in the mer to obtain the silver in a metal-

urth head of the invention con- separating silver and copper from tions by means of the sulphide of alkali waste, and compounds of nd metallic salts, such as "black" a ash," both of which are com- sulting from the patentee's pro- the manufacture of alkali. These may be employed either in a pul- ate or in solution. The residual containing the precipitated silver er are fluxed and converted to regu- bsequent operations.

fth part of the invention relates to nent of ores containing a large pro- f silver with little copper or sul- olutions are obtained therefrom (by ary processes) which when diluted 30° Twaddle, and boiled, yield a of chloride of silver, which may be with lead ores in the manner ordi- apted for such purpose to obtain in a metallic state.

ixth branch of the invention con- separating silver and copper from us obtained from any of the above- d processes by gradually calcining with from five to ten per cent. of salt. The product is then treated er to dissolve out the metallic salts, ll precipitate, and may be collected ted in the usual way.

st improvement consists in obtain- from the oxide thereof by mixing in a granulated state with carbona- atters in sufficient quantity to de- it and with clay enough to provide being made into balls, which, when in a reverberatory furnace, yield iron quality. The carbonaceous matter d for this purpose is anthracite charçoal as free as possible from sul-

DING HALLEN, of Burslem, Stafford, turer. *For improvements in gas-*. Patent dated May 10, 1851.

invention consists of a method of ng together "pot or fire-clay," or lay composed of any mixture of materials with metal in the con- n of gas-burners, the external por- the burner being of metal, and that of the burner in which the holes are being made of clay, which is much alculated to resist the action of the flame or the corrosive action of the products of combustion, which speedily destroys gas-burners made entirely of metal. The improved compound gas-burners are also more accurately manufactured in the first instance, and insure a uniform and permanent flame, which will retain its size and shape for any lengthened period.

The drawings represent a fish-tail burner and an Argand burner constructed according to the invention, the external portions thereof being composed of metal, and the inner portions of a button, and a ring of clay respectively. The button for the fish-tail burner, and the ring of the Argand burner are each formed in suitable moulds while the clay is in a plastic state, after which they are dried and burnt till of sufficient hardness, when they are fixed into their place in the body of the burner either by cement alone, or by cement and by burnishing down the edge of the metal upon the upper surface of the clay. It is obvious that a great many varieties, if not all kinds of gas-burners, may be formed in the manner just explained for forming fish-tail and Argand gas-burners; that is to say, by the insertion of perforated buttons, rings, or pieces of pot or fire-clay into that part of the burner through which the gas issues.

Claim.—The constructing of gas-burners of a combination of metal and "pot or fire clay," or other clay composed of any mixture of potters' materials, as above described.

EMILIAN DE DUNIN, of Queen Charlotte-row, New-road, gentleman. *For improvements in the apparatus for measuring persons, and for facilitating the fitting of garments.* Patent dated May 10, 1851.

The "apparatus for measuring" is composed of two parts, which open in front like trousers and waistcoat, and are provided with buckles to fasten them together. They are formed of a series of graduated straps, connected together by strips of vulcanised India-rubber, having slides at the intersections of the straps to adjust to the more prominent portions of an individual in course of measurement. The person to be measured puts on the apparatus and buckles it tight: the degree of expansion, as indicated by the graduated straps, is then noted down, as is also that of the India-rubber strips, which is ascertained by a tape or other measure, and the measuring apparatus is then transferred to a mechanical figure, buckled tight, and the figure expanded in the required directions until the indications given by the graduated straps and India-rubber strips correspond to those noted down previously. The measuring cover is then removed, and the figure employed for the purpose of trying on garments.

THOMAS HAIMES and JOHN WEBSTER HANCOCK, of Melbourne, manufacturers, ALBERT THORNTON, of the same place, and JAMES THORNTON, of Leicester, mechanics. *For improvements in the manufacture of knit and looped fabrics, and in the raising pile thereon.* Patent dated May 10, 1851.

The patentees describe and claim—

1. Certain improvements in knitting machinery, whereby the thread or yarn is introduced under the beards of the needles, which are in some cases made capable of closing without the use of pressers, and whereby the loops are formed, and the work knocked over by certain peculiar apparatus for that purpose. (Under this branch of the invention, the patentees also specify improved methods of manufacturing plaited fabrics and terry and cut-pile fabrics, and also fabrics having two true surfaces.)

2. An improved means of taking away the work, as produced, in that class of machinery wherein the work is passed through a revolving tube or hollow axis at the centre of the machine.

3. Certain improved machinery for the production of warp fabrics, whereby two needle bars in the same machine are so worked, in combination with other suitable apparatus, that two separate fabrics may be made, or two fabrics combined together so that, when separated by cutting asunder, they may form two piled fabrics, or whereby two fabrics may be produced on the two sets of needles, and combined together by intermediate threads, so as to form one fabric.

4. Certain improvements in machinery for producing warp fabrics by the employment of two sets of needles intersecting each other, and the one set taking its supply of thread from the other set or range of needles.

L. CHARLES MOREY, of New York, gentleman. *For improvements in machinery for preparing, dressing, cutting, and shaping stone, and other materials made use of for building purposes and architectural decorations.* Patent dated May 10, 1851.

1. Mr. Morey describes a "preparing" machine for reducing rough blocks of stone, as brought from the quarry, the cutters of which are actuated by a series of cranks on the driving shaft, to which rotation is communicated from any suitable prime mover. The cutters are serrated on their edges, and the throw of the cranks by which they are worked is so regulated that the cutters shall come into action successively. The stone to be cut is supported on a carriage running on rails, which has a gradual progressive motion imparted to it by gearing from the main shaft.

2. The stone after being prepared by the machine just described, is submitted to the action of a finishing machine having serrated cutters or discs, which, by rolling over the stone, produces on it a smooth surface. These discs or "burrs" are mounted on radial arms projecting from a vertical shaft driven by a strap from a steam-engine, and as the stone under operation is moved gradually forward, the cutters in their rotation pass over its surface and describe a series of parallel arcs.

3. An arrangement is described for producing smooth surfaces on grindstones and other circular discs of stone. The cutters consist of serrated revolving discs, which are set at a right, or any other required angle to each other, and act on the stone, which is secured to a face-plate on a revolving shaft, the one cutter reducing the side of the stone while the other acts on its edge.

4. Mr. Morey describes an arrangement of cutting tools for acting by impact, in which the burrs or cutting discs are mounted in rows around the periphery of a drum, which is caused to revolve, and thus the cutters act on the stone which is supported on a travelling carriage placed underneath the drum.

5. For cutting fluted or grooved columns, concave cutters are employed, which are arranged around the periphery of a drum, and act by impact. The grooves or flutes are produced in succession, and the column must have a suitable amount of movement around its axis given to it as each flute is formed.

Claims.—1. The arrangement of machinery for effecting the preparatory dressing of stone, as described.

2. Mounting the burrs or cutting tools, so that they may work in a plane at right angles to the axis round which they rotate.

3. Mounting the burrs or cutting tools at a right angle, or any other angle to each other, so as to act on two sides of the stone simultaneously.

4. The arrangement of the burrs or cutting tools for acting by impact, as described.

5. The making of the burrs or cutting tools by casting them in metal moulds, and thereby giving them a partially hardened or chilled cutting surface.

EDWARD WILKINS, of 60, Queen's-row, Walworth, gentleman. *For improvements in labels or tickets.* Patent dated May 13, 1851.—*No claims.*

Smith, of Littleborough, mechanic, Smith, of the Sun Iron-works, l, power-loom maker, and Matith, of Over Darwen. *For im-ts in fabrics, in weaving, and in y and apparatus for winding, cutting, and printing.* Patent y 14, 1851.

—1 & 2. The manufacture of fabrics ollen cords and woollen velveteens g the weft for the back of cotton and an improved mode of weaving rics.

An arrangement for imparting a relocity to the spindles and fallers or cop winding machines, and for ach spindle of pirn or cop winding self-regulating.

tain improved machinery applicable in which two or more shuttles are each side of the loom, for bring ired picker into action, and for rom both sides of the loom at the e.

arrangement of apparatus for the shuttle without the aid of the or picker-rod, when applied to which two or more shuttles are l.

tain improved machinery for ap-e break and stopping the loom in ce of the weft.

improved arrangement of machi-taking up the cloth when applied in which the reed is supported in a ame, and having two or more shut-

0. The application of the weft-the machinery for delivering the the warp beam, and to the taking-p of looms.

improved arrangement of appara-olding the warp beam when the es home the weft.

e application of intermittent wheels motion to revolving shuttle boxes

e employment of a chain for giving intermittent wheels.

improved tappet for giving mo-e healds of looms, and machinery ing the same.

n improved method of reversing n of looms.

arrangement of apparatus for pro-e reed and keeping the cloth dis-the loom.

ertain improved machinery for and slackening the warps.

arrangement of apparatus for the motion of the healds.

arrangement of apparatus for keeping double fabrics at a suitable distance apart, and for cutting them asunder.

20. Certain improved machinery for giving motion to printing blocks and the fabric or material under operation.

21. An improved construction of printing table.

William Hemsley, of Melbourne, lace-manufacturer. *For improvements in the manufacture of looped fabrics.* Patent dated May 15, 1851.

Mr. Hemsley's improvements have relation to that description of warp machinery (patented by T. and J. Whitely, in 1840) in which the warp threads are divided into two rows, and traverse diagonally, and consist in—

A mode of combining the parts of warp machinery, wherein the two rows of threads into which the warp is divided are caused to traverse diagonally in opposite directions from selvage to selvage, in such manner that each thread as it arrives at either end of its course, after working from selvage to selvage in one row of threads, and after working with the outermost needle, may be transferred into the other row of threads.

Robert Oxland and John Oxland, both of Plymouth, chemists. *For improvements in the manufacture and refining of sugar.* Patent dated May 15, 1851.

This invention consists in the employment of phosphoric acid, in a combined state, for defecating and decolorizing saccharine liquids or solutions of sugar.

The patentees do not confine themselves to any particular phosphates or superphosphates, nor to any stated proportions thereof, but they claim the employment of phosphoric acid in a combined state as described.

Specification Due, but not Enrolled.

Hugh Barclay, of Regent-street, Middlesex. *For improvements in the means of extracting or separating fatty and oily matters; in refining and bleaching fatty matters and oils, animal and vegetable wax, and resins; and in the manufacture of candles and soap.* Patent dated May 19, 1851.

WEEKLY LIST OF NEW ENGLISH PATENTS.

William Charles Scott, of Camberwell, gentlemen, for certain improvements in the construction of omnibuses and other public and private carriages. November 15; six months.

James Lott, of Whitchurch, Southampton, saddler, for improvements in harness and fastenings. November 15; six months.

Charles Ewing, of Bodorgan, Anglesea, steward and gardener, for an improved method or methods of construction applicable to architectural and horticultural purposes. November 15; six months.

Claude François Tachet, of Paris, mathematical instrument maker, for improvements in preparing wood to prevent its warping or shrinking. November 15; six months.

Pierre Erard, of Gt. Marlborough-street, Middlesex, piano-forte maker, for improvements in piano-fortes. November 15; six months.

Antoine Dominique Lisco, of Slough, for improvements in the manufacture of chains, and in combining iron with other metal applicable to such and other manufactures. November 15; six months.

William Hamer, of Manchester, for certain improvements in weaving textile fabrics. November 15; six months.

Henry Bessemer, of Baxter House, St. Pancras, for improvements in producing ornamental surfaces on woven fabrics and leather, and rendering the same applicable to bookbinding and other uses. November 19; six months.

Frederick Joseph Bramwell, of Millwall, engineer, for improvements in working the valves of steam engines for marine and other purposes, and in paddle-wheels. November 20; six months.

Thomas Statham, of Sidney-street, City-road, pianoforte-maker, for certain improvements in pianofortes. November 20; six months.

Joseph Sharp Bailey, of No. 3, Victoria-terrace, Keighley, York, machine wool-comber, and Isaac Bailey, of Victoria-street, Bradford, York, bookkeeper, for certain improvements in preparing, combing, and spinning wool, alpaca, mohair, and other fibrous materials. November 20; six months.

WEEKLY LIST OF DESIGNS FOR ARTICLES OF UTILITY REGISTERED.

Date of Registration.	No. in the Register.	Proprietors' Names.	Addresses.	Subjects of Design.
Nov. 14	3018	T. and T. C. Robson	Liquorpond-street	Double-bevelled iron for solid hoop wheel-tyre.
15	3019	Arthur Craven	Stamford-hill	Feeding apparatus for steam boilers.
17	3020	Samuel Williams	Commercial-road, Lambeth	Derrick lift.

WEEKLY LIST OF PROVISIONAL REGISTRATIONS.

Nov. 14	324	T. E. Jones	Birmingham	Improvement in the solar shade.
,,	325	Alfred A. Hely, C.E.	Westminster	Anti-pickpocket, or pocket-protector.
,,	326	Walter de Winton	Lambsconduit-place	Brougham cab.
15	327	M. Cavanagh	Notting-hill	Adjusting lock-spindle.
,,	328	Thomas Harrison	Liverpool	Prince of Wales' pianoforte.

CONTENTS OF THIS NUMBER.

LONDON: Edited, Printed, and Published by Joseph Clinton Robertson, of No. 166, Fleet-street, in the City of London— Sold by A. and W. Galignani, Rue Vivienne, Paris; Machin and Co., Dublin; W. C. Campbell and Co., Hamburg.

Mechanics' Magazine,

USEUM, REGISTER, JOURNAL, AND GAZETTE.

1477.] SATURDAY, NOVEMBER 29, 1851. [Price 3*d.*, Stamped, 4*d.*

Edited by J. C. Robertson, 166, Fleet-street.

MR. W. D. SHARP'S IMPROVED SLIDE VALVE.

Fig. 1.

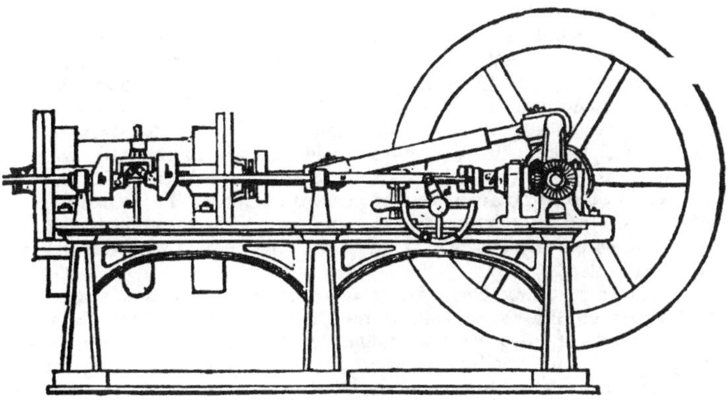

Fig. 2.

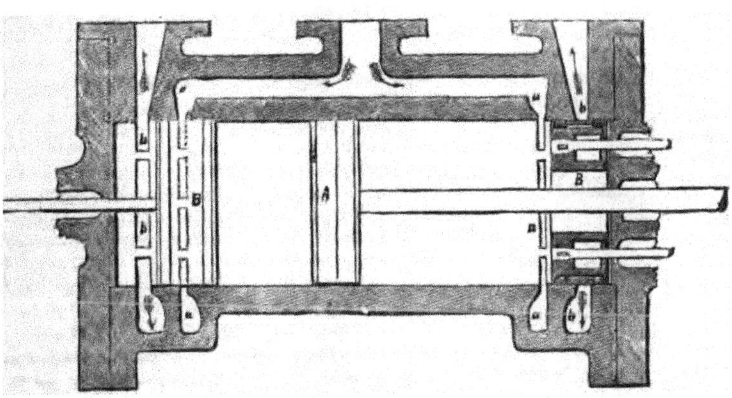

DESCRIPTION OF AN IMPROVED SLIDE VALVE. BY MR. W. D. SHARP.

The object and advantages of this improvement on valves will be better understood by referring to the speed and pressure at which steam engines are now, and have formerly been worked. The slide valve now generally used in locomotives and other engines is quite well adapted to the work required of it, when applied to engines working with low-pressure steam, and when the velocity of the piston seldom exceeds 200 ft. per minute; which may be taken as the average speed at which engines were worked until the introduction of the locomotive, in which the velocity of the piston is generally not less than 1,000 ft. per minute, when going at an average speed, or say five times the former velocity.

The pressure at which steam is now generally worked is also greatly increased; in locomotives it is seldom less than 100 lbs. per square inch,—being about double the pressure in use in locomotives not over ten years ago. Now this great increase of speed, and pressure, has rendered the slide valve as at present employed a very unfit agent for the emission of the steam from the cylinder.

This will be understood, that if the speed be five times greater than formerly, then five times the quantity of steam must pass through the same area of opening in the same time, or the velocity of the steam in making its exit from the cylinder will be five times greater than formerly. But this is not all. I have said the steam now worked in locomotives is seldom under 100 lbs. per inch; now steam of this pressure, on access being given it to the atmosphere (as by the opening of the valve) will expand to eight times its former volume, and consequently, to get the same free outlet as is obtained in low-pressure engines with the slide, we would require to have $8 \times 5 = 40$ times the area of ports or openings. Those openings have certainly been enlarged in relation to the cylinder to what they were made some years ago, but to obtain anything like the requisite enlargement with the slide valve is quite impracticable.

The disadvantage resulting from this is shown in the locomotive, where, if the pressure of steam on the acting side of the piston be say 100 lbs., a resistant pressure of upwards of 40 lbs. will be found on the other; thus nearly half the power being expended in expelling the steam from the cylinder. It may, however, be said that this is in consequence of the blast-pipe,—and no doubt to a certain extent it is; but then the area of the valve-openings rarely exceeds, and frequently falls under the area of the blast-pipe—that is, taking the valves when unclosing the greatest area; but then we must take into account, that with the slide-valve the piston makes one-half of the stroke with a mean area not exceeding one-half of the maximum opening. I have no doubt that with a more free and capacious outlet to the steam from the cylinder, instead of its being throtled in the small and tortuous opening the slide valve affords it, that a considerably larger area of blast pipe might be used, and with the same beneficial results as regards the draught. I believe, also, that much may yet be done in taking advantage of the motion of the locomotive through the atmosphere, which if done in a judicious manner would be the means of assisting the blast-pipe in creating the draught, and thus tend still further to its enlargement; but as the valves are at present constructed little or no advantage can be gained by doing so.

Another great evil attends the slide valve; viz., the pressure of steam on the back of the slide,—and this disadvantage has also been greatly augmented in consequence of the increase of pressure at which engines are now worked. Thus the slide of an 18-inch cylinder of a locomotive, with the steam at 100 lbs., will have to move under a pressure of 14,000 lbs., which, at a moderate computation, in a locomotive going at an average speed, will absorb 35-horses power in working the two slides: but this is not all, nor perhaps the greatest disadvantage attending it, for from the great force requisite to work the valve, the joints and links connecting it with the eccentric cannot work in that smooth and steady manner so requisite; and again, the apparatus is so much more liable to wear in consequence, that it is next to impossible to keep up that accuracy of movement in the valve which is above all so desirable.

The object and intention of the valve now to be described is to remedy those ets that have been pointed out, and under which the slide valve labours. Fig. 1 is levation of an engine with this valve, and other improvements; and fig. 2 is an rged section of the cylinder, showing the valves, &c. In fig. 2 the piston is reented at A, BB the valve pistons,—which it will be observed are placed within the nder at each end, the cylinder being made somewhat longer than usual for this ose, they are made the same diameter, and furnished with packing rings, similar e steam piston; one of them is shown in section, from which it will be observed the steam exhausts through the central opening, and that the quantity of steam fective each stroke, will be only equal to the depth of the valve piston, and the of the port or opening made therein. For the purpose of making the valve ns as light as possible, I would prefer forming them of thin plate iron, as there other strain on them, but that of retaining the packing rings in their place. letters *a a a*, denote the passages for the inlet, and *b b b* for the outlet of steam, while the arrows indicate the direction of the steam in these passages. se openings extend quite round the circumference of the cylinder, and are coned at suitable points by ribs for the packing rings to bear against, as shown he figure.

Fig. 4.

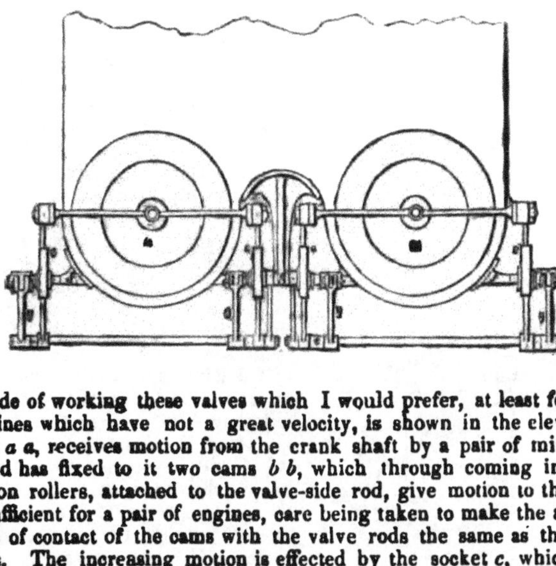

he mode of working these valves which I would prefer, at least for marine and r engines which have not a great velocity, is shown in the elevation, fig. 1. shaft *a a*, receives motion from the crank shaft by a pair of mitre wheels, as vn, and has fixed to it two cams *b b*, which through coming in contact with -friction rollers, attached to the valve-side rod, give motion to the valves; one t is sufficient for a pair of engines, care being taken to make the angle between points of contact of the cams with the valve rods the same as that formed by cranks. The increasing motion is effected by the socket *c*, which slides on a fixed to the shaft, and has a spiral slot formed in it, into which takes a key the mitre wheel, so that by moving the socket along the shaft, the position of cams is changed in relation to the crank; and the motion of the engine thereby rsed,—motion is given to the socket *c* by the lever *d*, as shown. The expansive ion is effected by the cross rod *e*, and double joint shown, which is attached to valve side rods; by raising the rod, the valve pistons are brought nearer together, the cut-off of the steam thereby accomplished. This cross rod may either have rect motion given it from a variable eccentric or cam, or it may be furnished a spring or weight, of sufficient energy to keep the valve rod up to the back of cam, by which a simple though limited expansive movement may be given.

he eccentric is quite as applicable to work this valve as it is to work the ordi- slide valve. Figs. 3 and 4 show an arrangement adapted to the locomotive this motion, and having an expansive movement. Fig. 3 is a sectional sketch taken through the centre of boiler or smoke-box, and fig. 4 is an end view of bottom part of smoke-box, showing the cylinders, &c. *a a*, the cylinders; *b* steam pipe; and *c c*, exhaust pipes; *d*, the valve side rod, jointed, as shown, to the link *e*, which moves on a fixed centre at *f*; the link *e*, is attached by the rods *g g g*, to the lever *h*, which through the lever *i* and rod *j* is attached to the hand gear in the usual manner. The sketch is shown as working with full steam, the dotted lines show the position the levers will assume when working expansively. Two sets of eccentrics may be used (in the sketch I have shown only one); the reverse motion to be made in a similar manner to that shown in fig. 1; viz., the eccentrics to be fixed to a socket having a spiral slot formed in it, and taking into a key fixed to the crank shaft, suitable means being provided for traversing the eccentrics on the shaft. This mode I would prefer, not only because it dispenses with the duplicate eccentrics, rods, &c., but also because it gives the advantage of adapting the lead of the valves to suit the various degrees of expansion.

Fig. 3.

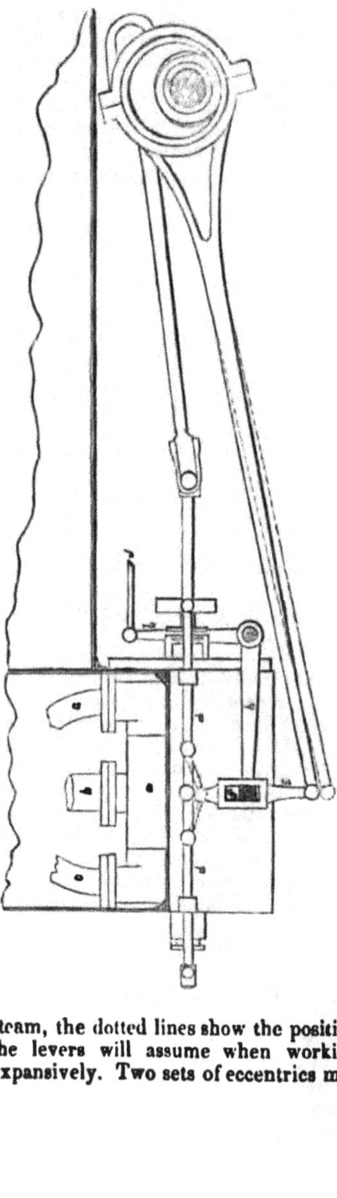

One disadvantage attending the eccentric motion, when applied to the slide valve, will be considerably diminished in its giving motion to this valve,—what I allude to is the slow cut off of the steam which the eccentric gives through the slide; now, as the area for the inlet of the steam does not require much enlargement, and as the length of opening will be at least four times that usually given by the slide valve ports, consequently it will have the same effect in giving a more instantaneous cut off of the steam, as if the eccentric had at this point been increased by four times its usual velocity.

The advantages of this valve may now be understood as consisting in affording a greatly enlarged area for the outlet of the steam as compared with the slide valve, and in moving without any other friction than that due to the elasticity of the packing rings of the valve pistons; and that where the eccentric may be preferred for communicating motion to it, its action on the valve will be improved (as compared with the same motion applied to the slide valve), in its more instantaneous acting on the steam. In engines with vertical cylinders the valves will require to be balanced in a similar manner to the D-valve in the same description of engine.

W. D. Sharp.

—The description of his "Fire work," or Steam Engine, which ırquis of Worcester has given in his ury of Inventions," is by no means r unambiguous; and consequently ɔer of different explanations of it een attempted since the days of great improvement, rendering engines such important things. one of those which have come my notice is satisfactory to my and from the style in which they itten, even their authors seem to een equally in a state of incerti- out it. I have, therefore, endea- to get over the difficulty myself; lieving that I have succeeded, now u my conclusions on the subject.

part of the description which con- he puzzle is as follows:—"So found a way to make my vessels t they are strengthened by the ithin them, and the one to fill he other, I have seen the water te a constant fountain stream 40 gh: one vessel of water rarified driveth up forty of cold water; nan that tends the work has but to o cocks, *that one vessel of water consumed another begins to force fill with cold water, and so suc- y.*" The words in italics are the e of the hobble;" to explain them iter alters "force and refill" to ce and empty;" another gets over atter by adding a word or two, force and refill the cistern," re- to the reservoir into which the is forced by the engine's work. lloway, in his "History of the Engine," gives a drawing of the as he supposed it to be, and re- s it with only one vessel acted on fire. I do not think the mar- words will bear either of these etations, and am of opinion that puzzle, as in many others, the lies much nearer to the surface generally supposed; I think what ırquis meant, was that he had d two vessels or boilers, so that them had a communication with sel which contained the water to be d, but yet could act upon it sepa- and could be filled independently other, the one while the other work,—an arrangement which is sily *conceivable;* and that the apparatus being thus arranged, he used to work one boiler until the water in it was "consumed," and then having got the other ready, it was set to work, and began "to force," the fire being lighted and the communication between this boiler and the water to be raised being opened; in the mean time, while this one was working, the man would turn the taps and "refill with cold water" the boiler first emptied, "and so succes- sively." This interpretation does not go far from the very words of the marquis; but his style of writing is so involved and ambiguous, that I think the amount of latitude here claimed may be safely taken, and I do not consider it is going too far to read thus: "having. . . . made my vessels . . . . so as to fill the one after the other . . . . the man that tends the work has only to turn two cocks, so that when the water in the one boiler is evaporated, the other shall begin to work; and then to refill with cold water (*i. e.*, refill the empty one by turning one of the taps), and so successively." It seems to be tolerably clear that there were two boilers (not cisterns, as some think), and that the vessel of water "consumed" was one of those "rarified by fire,"—for it is evidently one of those which "begin to force," *i. e.*, a boiler; and if it was not a boiler, why does the marquis state that he had found out a way to fill his vessels alternately in a manner indi- cating that that had a great deal to do with the mode of operation? Surely he would not be thus particular about so simple a thing as filling two cisterns one after the other; though if, as I suppose, the boilers were filled alternately, then it would be an important part of the whole operation and be necessary to be described. This, also, is to be observed, that he made his *vessels* to be strength- ened by the force within them,—which can hardly refer to the cisterns, though it might to the boilers, and implies the use of more than one.

This simple explanation of the mar- quis's words appears to me to be reason- able and fair; and as anything tending towards a fair apportionment of the honours awarded to the discoverers of the steam engine is interesting, I have sent it to you.

I am, Sir, your obedient servant,

INVESTIGATOR.

EXPERIMENTS IN NAVAL CONSTRUCTION.

Mr. Macgregor's recommendation of gutta percha (see *ante*, page 367) as a material for the hulls of experimental vessels, is likely to facilitate much the undertaking of experiments by private persons; it is so easily manageable, and loses so little of its intrinsic value by use, that the expense of experiments would with it, little exceed the cost of workmanship; models in wood, on the contrary, require much skilled labour in their formation, and, when done with, are only fit for firewood. It is the expense of experiments that is the usual bar to their commencement, and, when begun, to their continuance to a satisfactory termination.

Another of this gentleman's proposals would be found of great practical value in ascertaining the best rig of vessels; namely, that of "having the masts and bowsprits moveable fore and aft by means of horizontal screws." There might, in addition, be screws enabling those supports to be made to rake or incline to an angle more or less acute, as on this depends the perpendicularity of a sail whenever the vessel deviates from an even keel, and which may possibly be one of the reasons why vessels have so often been found to be the best sailors when their draught has been less by the head than by the stern, or *vice versâ*.

In the comparative trials proposed, motion is to be given to the vessels by easy winds. Mr. Macgregor being evidently anxious to ascertain the real comparative merits of models of vessels, will not take amiss suggestions relative to them; therefore, without hesitation, it may be asked whether the modes of trial recommended by Sir Samuel Bentham might not be preferable to wind? namely, the making the experiments in a stream of water, the current of which should alone be the motive power. That power, it may be said, is a constant one,—not variable, as is the wind, either in amount or direction. In addition to Sir Samuel's modes of ascertaining comparative excellence of different models, Mr. Macgregor's plan of trying vessel against vessel would be very satisfactory.

The superiority of one rig of vessel over another is a much more complicated question to solve than that of the form of hull. Though the hull of a seagoing vessel be exposed to a variety of influencing forces from waves and wind, her rig is subject to a still greater number: the position of her sails, besides being changeable by the motion of the hull, has also to bear the variable action of the wind upon the sails themselves; so that what would be favourable in the rig under certain circumstances, would be detrimental in others; thus absolute perfection cannot be hoped for in the equipment of any vessel, but that which, on the average, would be the best, is all that can be reasonably anticipated.

The complimentary phrase with which the above-mentioned communication commences, has neither been overlooked nor unappreciated; but "M. S. B." figures in borrowed plumes, having access to Sir Samuel Bentham's papers;—they are on a vast variety of subjects, and, being grounded on general principles, are applicable to very many of the subjects under discussion at the present day.

M. S. B.

LIFE-BOATS.

In the proceedings of the Committee for determining to which of the competing life-boats the Duke of Northumberland's Prize should be awarded, it is stated that all difficulties in coming to a decision were removed as soon as the qualities desirable in such a boat were distinctly stated, and that their respective degrees of importance were decided on.

It would, doubtless, tend materially to the perfecting of *all* vessels for navigation, were the *desiderata* in those for different services defined, and were a scale framed marking the respective degrees of importance of these several *desiderata*.

Such a scale in regard to all vessels would be in two grand divisions; namely, *form* of structure, and *strength* of structure.

Form of structure would include all of the different *desiderata* as to form and equipment for rendering a vessel a good sea boat, speedy, weatherly, capable of contending with adverse winds and storms, and of working off a lee-shore, capacious, dry, &c., &c.

Strength of structure would include the *desiderata* requisite for opposing resistance to the various strains and injuries to which a sea-going vessel is exposed; such as those from winds, from waves, from sinking into mud, or beating upon sand-banks, foul or rocky ground, from the strains occasioned by the rig of the vessel, and from those con-

the weight of guns, or the firing , &c., &c.
nation of such scales in any ;e of perfection would require the ship-builder, the experience ıan, and that of the naval officer en actual service in time of war; re many eminent men in all of hes, and without looking to abection in such a scale as is sug-:n one imperfectly framed, a ısemblage of the *desiderata* in essels, would lead to immediate nts in the art and science of tecture. M. S. B.

ITS IN NAVAL CONSTRUCTION.
(Concluded from p. 413.)

l to the latest of these attempts, teers in the pursuit of naval im- having either by mathematical cal in consequence of some practical s of their own, become confident troducing certain curvatures in of the hull, superiority in sailing ıssured, those in authority have ed very liberally to authorize the me ships of war conformably to roposed, and to cause these ves- built, to sail together in com- others: but in these cases, as in so long as the various effects : result from the different forms are not exhibited *separately* and rom each other, and from those produced by the differences in the pplying the locomotive force, no to the cause of any differences their sailing can have been satis- olved, and the proposer of each cely to remain as confident as ne superiority of his plan, attri- failures to the effect of some cir- over which he had no control.
ıe assistance to be derived from ms purely mathematical, the pro- g to determine the form to be body of a given weight to enable ropelled through the water by a with the greatest velocity, the cian is in want of data on which his calculations. The weight of and of its contents being given, o doubt readily ascertain the sary to give the requisite degree y; but he cannot ascertain how t bulk is to be made up in length, much in depth, or in breadth, so ative proportions of these dimen- h it is known exercise so much influence, not only on the velocity, but on the efficiency of a vessel, are still left to the discretion and guess of practical builders. If, again, the mathematician, trusting to received opinions as to the necessary data, should satisfy himself as to the fittest curvature for dividing the water, it does not appear that the course which is actually taken by water when forced out of the way by a body moving through it has ever been ascertained, and he must therefore again remain at a loss to determine the *direction* in which the curvature he has fixed upon is to be given to the vessel.*

In the instance of the few vessels I constructed whilst in office in this country, some modifications in form, many others in structure, were exhibited, some of which have been gradually adopted,† and others I have reason to believe will, when attended to, be found very suitable to the purposes they were designed for; but still I felt how imperfect were my ideas on the subject, and was fully aware of the expediency of devising a more certain, and less expensive mode of proceeding than the building a new ship of war to try the more or less resistance in

* The following paragraph is taken from the copy of Sir Samuel Bentham's paper as it was originally written, but in which it is crossed over as not to be inserted in the Paper presented to the Admiralty, lest, influencing as the example was, it might be thought too trivial to be laid before their Lordships. M. S. B.

"As to myself, having at a very early age been under the tuition of a distinguished mathematician (Cowley), I was impressed with a great deference for mathematics, and was then (nearly 60 years ago) eagerly pursuing the study of Naval Architecture. I looked upon the *curve of least resistance* indicated by Emerson in his "Treatise on Mechanics," as that which must undoubtedly be the best suited to the construction of ships; and having shaped a model 3 or 4 feet long, according to that curve, and rigged it in similar manner to others of different forms, I sailed them together, and did not fail in any case where the mathematical form had the advantage, to attribute it to the superior effect of that curvature; but while I was pursuing these experiments, being in the habit of observing the difference in sailing amongst the vessels of various shapes which passed Woolwich every tide, I happened to notice a sand-barge, which occasionally beat the Gravesend boats, generally considered as the best sailers, and having satisfied myself that of all the various vessels which came under my observation, there were none that differed so much from each other as these two which sailed nearly equally well, —the one being long, narrow, and shallow, the other short, broad, and deep; I lost all deference for the comparatively trifling effect to be expected from the curve of least resistance, and neglected my experiments on this subject, not being able then to devise any means by which I could hope to reconcile the various opinions I heard most decidedly pronounced on this most complicated question; and I have no doubt, but that others whose duty or inclination have induced them to attend to it had been under the same perplexity.

† The quickest steam vessels that I know of are those which have been brought almost as sharp, and to the form of the *Arrow* and the *Dart*.

the water, or in air, occasioned by every new modification that may be suggested of the form, or equipment of a navigable vessel.

With this view of the subject, after having in my "Naval Essays," noticed the desiderata in a vessel of war, and proceeded to treat of the efficiency of a vessel of war in regard to its means of locomotion, and to its form, in as far as it may increase or diminish resistance to locomotion, I was about to discuss the comparative fitness of different modes of structure, supposing the forms to resemble those in more general use, according to which the commodiousness of the interior arrangements, and exterior beauty, however important and desirable seem to be the objects too exclusively aimed at. Upon further consideration, however, it appeared to me highly expedient with a view to economy, as well as to efficiency, to postpone the consideration of the structure, combination, and fastening of the component parts of a vessel till some distinct idea should be obtained of the general form best suited to favour locomotion, and to resist all adverse action of the sea. It was on that occasion that I drew up the plan of a course of analytical experiments for determining these most important questions, and that I also indicated the mode of proceeding in making such experiments, and proposed an apparatus, showing in a satisfactory manner the effect of every force influencing locomotion and resistance, whether constant or accidental, as described in the Paper referred to in my first Essay (page 115), which I herewith inclose.

In order that the apparatus described in this paper may completely answer the purpose of measuring the effect of all the influencing forces, whether constant or accidental, and to reduce the result to the expression of weight in pound, there should be provided the undermentioned *meters*, or philosophical instruments; they should be so contrived as to ascertain and register the following particulars, in all the variety in which they take place at every instant of time, during each hour and day, to the extent of each cruize or voyage;—

1. The degree of resistance which any vessel meets with, not only while propelled steadily through still water, but also while advancing in an agitated sea, when the different parts of the vessel are alternately in and out of water, and struck by the occasional impulse of detached waves.

2. The degree in which a vessel is more or less adapted to afford resistance to leeway.

3. The force with which the wind acts against every square foot of the sails, and of all other parts of the vessel exposed to its *action*.

4. The direction in which the ves placed in regard to that of the wind.

5. The angle to which the sails are regard to the middle line of the vessel.

6. The angle at which the rudder i in regard to the middle line of the ves

7. The angle at which the vessel pi or oscillates in the longitudinal directi

8. The angle to which the vessel or is made to incline, or oscillate versely.

9. The depth to which the vessel i mersed in the water at each end.

10. The actual distance which the has passed through the water at any of its course.

11. The rate at which the vessel through the water progressively, as leeward.

The two first of the above instrume requisite for ascertaining the most tageous form of hull; the remainin are requisite for determining the of the locomotive apparatus, supposi hulls of all the vessels experimented u be perfectly similar.

Considering the perfection to whic losophical instruments have been br in this country, it is remarkable that have yet been proposed for indicating, less for registering, the important pa lars which the above-mentioned instru are contrived to register as well as to tain. Perhaps it may in no small deg attributed to the want of such instru that a variety of doubts relative to the rig, and other particulars of vessels been suffered so long to be entertai while there are no means of accurate surement, opinions are likely to v in regard to the most ascertainable culars relative to the locomotion of sel as they would in regard to heat o heaviness or lightness, if there we thermometers to ascertain the degrees one, or no scales and standard weig measure the other. It must not, the be considered, that by providing the i ments above proposed the trouble an would be no otherwise compensated fo by the advantages obtainable by the posed experiments; since instruments same kind are requisite to exhibit the re cumstances on which difference in the of the same ships at different times depe instead of trusting, as at present, more or less acute perception, or guess of the different navigators. Al *terms* made use of to indicate the in which the several influencing ci stances are stated to have existed, wo longer be such vague expressions as "l or "moderate airs," or "strong br "light, or heavy gale," "squally or steady hurricane, or storm;" but the different force of the wind would be as accurately and uniformly expressed by the *degree* marked on the wind gauge, as the difference of heat is expressed by the degree marked on the thermometer.

Particulars thus recorded in regard to any ship would afford important data to be studied at home. In the case of extraordinary expedition of a ship, for instance, it would be seen whether it had depended entirely on extraordinary favourable winds or other uncontrollable circumstances,—whether on uncommon skill in the commander in taking advantage of favourable circumstances, and in avoiding the mischievous effects of adverse ones—or, as relating more immediately to the subject of this Paper, whether it had been the consequence of the particular form, the stowage, the rig, or other locomotive apparatus peculiar to the vessel itself.

SLIDING AND REVOLVING KEELS.

Mr. Editor,—As it was observed in the 1470th Number of the *Mechanics' Magazine*, that my son, Lieut Shuldham's description of a revolving keel did not seem sufficient to enable any observation to be made upon that invention, I feel myself called upon to explain it, as the original inventor of *revolving* keels, not *sliding* ones, being perfectly aware that the invention of the latter is due to the late Admiral Shancks.

My son labours under a mistake in supposing that the yacht *America* was fitted with either sliding or revolving keels; something was said about them in the newspapers before her arrival in this country, but when seen here, she had nothing of the sort.

It is a misnomer to call my moveable keel a *sliding* one, as it revolves on a bolt fixed to the fore-end of the vessel. I invented it when I was prisoner of war at Verdun—I believe in the year 1809,—and the model which I then made may be seen at Mr. Ransome's Museum at Ipswich. As there are many objections against moveable keels in general, I should have allowed the plan to remain quiet, had not Commander Ellis, of the Coast Guard, Great Yarmouth, exhibited a model of a vessel fitted with one, at the Crystal Palace, which he denominated a *new plan*. Many persons must have noticed it. It only differed from mine by the lower part of the rudder partaking of the movement of the keel—an arrangement which I had thought of, and rejected as making the plan more complicated, requiring an unusually thick stern-post, and rendering the rudder liable to gather weeds.

I inclose a sketch of the revolving keel, and also of the boat which was fitted with it for the space of about eight years, and, if it does not occupy too much room in your valuable Journal, I cannot do better than copy a letter which I received from the late Captain Sir Fleming Senhouse, R.N., containing some remarks upon the keel in question.

The keel was composed entirely of lead, and constituted the whole of the boat's ballast, with a convenient and simple purchase for raising and lowering it; namely, a keel-rope passed round a sheeve fixed to the upper part of the keel, which made a double power; to this keel-rope, a tackle was affixed consisting of a very small fourfold and a treble block, which increased the power to twelve. Thus, the keel weighing 5 cwt., was easily hoisted up, until its upper edge touched the deck beams, when its lower edge was even with the boat's keel, so of course the centre of gravity was vertically moveable at pleasure, and as great part of the weight of the keel rested on the bolt, it may be conjectured with what ease it was hove up without the necessity of moving from the helm. Now, without any wish to bolster up the plan, I shall state with truthfulness my opinion of it, together with the result of my experience; but, in the first place, I must deny all imputations about my being the inventor of *sliding keels in general;* I know something about the late Admiral Shanck's plans, and should like to know more,—such as whether he used other keels than wooden ones; and what was the specific gravity of his keels. For such a heavy one as mine, it was highly necessary that it should work steadily, without the least play—which mine did; and which cannot be effected in a *sliding keel* which requires a purchase at each end; and if in raising it more power be given at one end than on the other, it gets jammed—especially by the help of mud and small pebbles; whereas mine, in the thousands of times it was hove up and lowered, never was jammed. The inside of the keel-case was furnished with ribs of copper, both for adding strength to the case, and for keeping the keel clear of the woodwork, and thus also decreasing the friction. Now for its good qualities as compared to a *sliding keel.* In the first place, it acts as a shoal warner; in the second, it performs the quality of a buffer in preventing the vessel from being injured when running aground under full sail, under which circumstance the vessel's headway lifts the keel up. Compare this with a *sliding keel*, which would have been carried away or the vessel torn by it. I may here mention that whatever be the mechanical power for hoisting the keel, it must be contrived so as to move upwards with the keel when it is lifted by the vessel running aground. Thirdly, on running aground the vessel vibrates on a central point, so that there is no difficulty in placing her head off shore, when, after raising the keel a little, off she goes. I have done this hundreds of times, and never but once stuck fast even on an ebb-tide, and that was when there happened to be a shoal outside too shallow for the boat to float over it with her keel up; in short, nothing could exceed the great use and handiness of such a keel; but this is the best side of the picture—the other side should be as fairly represented.

Fig. 1.

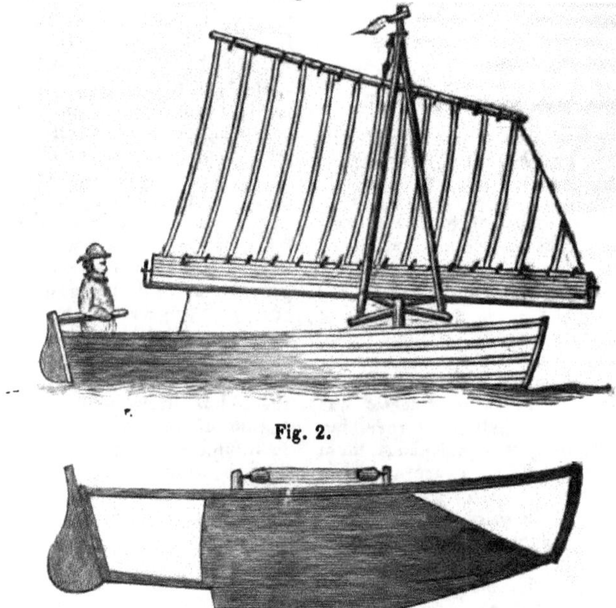

Fig. 2.

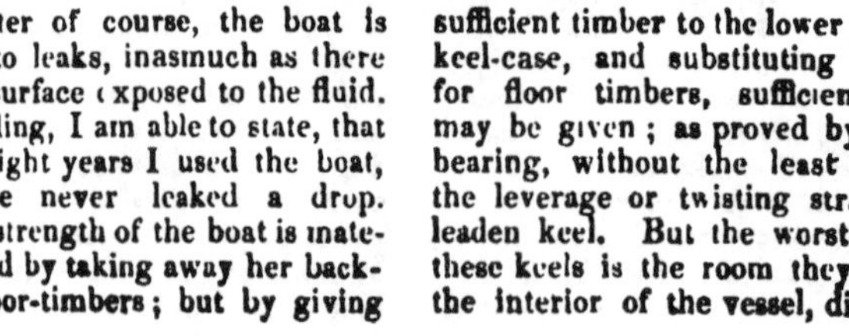

As a matter of course, the boat is more liable to leaks, inasmuch as there is a greater surface exposed to the fluid. Notwithstanding, I am able to state, that during the eight years I used the boat, the keel-case never leaked a drop. Again; the strength of the boat is materially lessened by taking away her backbone and floor-timbers; but by giving sufficient timber to the lower part of the keel-case, and substituting iron knees for floor timbers, sufficient strength may be given; as proved by my boat's bearing, without the least injury, all the leverage or twisting strain of her leaden keel. But the worst feature in these keels is the room they take up in the interior of the vessel, dividing her re into two longitudinal halves. ɔjection would tell against its ı in yachts or pleasure-boats; it refore be asked, in what vessels ı it might be usefully applied? I say to those boats belonging to ɩth very shallow bar-harbours,— Southwold, and many others on ɩt. Were I obliged to reside near hose, and had a boat, I should γ have her fitted with a *revolv-ɭ*, whether for fishing or sailing. sels in general, the choice must the owners, whether for the sake he good qualities which I have ated, they will put up with its niences.

e been told that revolving keels ımon in the United States, espe- ı New York, under the name of *ɔoards*. If it can be proved that re in use before the year 1809, I of course, relinquish my sup- riority of the invention. As mine ɪibited in the Adelaide Gallery in ɪe probability is that the idea was ɪp about that time. I may take portunity of stating, that during ɔle of my life, all the sails of my · vessels were made to stand as ɔoards, and were level to booms; ɩt I had a yacht rigged precisely r. Steven's yacht, the *Maria*, of ɤ York Yacht Club, viz., a sloop ɪly two working sails—mainsail esail. The foresail or jib—for it ː called either—was laced to a ʼhich worked on a horse, so that, ing two guys aft within reach of ɪm, I could steer her and work self. I discontinued the rig on . *only* of my having found it very ɪnient and troublesome in a heavy This may not be so in Mr. Ste- ʼell-manned yach ; but mine was anded.

ɪe accompanying sketch, fig. 1 he revolving-keel boat under sail, ;. 2 a longitudinal plan of the ng keel: the keel, as its case dis- a quadrangular in shape, the rea- ʼhich is, that it can be lowered ʼithout injuring the boat by its é or twisting strain.

I am, Sir, yours, &c.,

MOLYNEUX SHULDHAM,
Commander R.N.

ɪmitage, Melton, Woodbridge,
ɪffolk, November 11, 1851.

Captain H. F. Senhouse's Letter to Lieut. Shuldham.

My Dear Sir,—I remember well the surprise with which I witnessed the astonishing performances of your little skiff, about two years since, rigged with your revolving sheer mast and one square sail.

I had observed her manœuvres repeatedly, and remarked her weatherly quality, her quickness and certainty of staying, either in a seaway or in smooth water—the facility in wearing, and the small compass in which the evolution was performed, with the safety in jibing the sail in a fresh breeze, when compared with fore-and aft or sprit sails. But I more particularly observed these qualities, and the press of sail the boat appeared to carry with ease, on accidentally seeing her beating up the Salcombe Estuary, in a strong easterly gale, with a sailing boat of four or five tons, of the Cawsand rig, drawing about four feet water, when your little boat appeared to fore reach and weather equally as well as her large competitor. I am not prepared to say that the whole of these advantages were derived from the mode of rigging so ingeniously invented by you, as I think much must be attributed to the sliding leaden keel with which the boat was fitted, which gave the weatherly quality in a great measure, by increasing the draught of water at will, and also the capability of carrying sail, by lowering the centre of gravity at pleasure. But I can safely say, I do not think the boat would have possessed all the good qualities of sailing above mentioned, either on a wind or from a wind, with any other known rig, without much additional trouble in changing and trimming the canvass. And this brings me to the manual part of the business, in which this mode of rigging seems peculiarly to excel. I allude to the simplicity and ease with which the greatest quantity of canvass is set that can be carried consistently with lying as near the wind as the sail can possibly do good; and the facility with which all possible canvass can be immediately spread, at right angles with the hull, so as to do the most possible good in going from the wind. To this is to be added, the ease and expedition with which the sail may be reduced or increased by the reefs—more simply and expeditiously than any mode known to myself. All these things cannot be obtained without some inconveniences; and it remains with experience to prove whether what we shall call the increase of "top hamper," that is, of machinery, may not be in the way of other requisites, as well as the whole equipment resting entirely on the support of a single pivot. The latter objection may be obviated; and I will conclude by saying, that although the little boat in question did a great deal more than anything of the kind I had ever seen, or that could be done in every point by any mode of rigging known to myself; yet I am not prepared to say how equally well the same might answer applied for purposes of general utility to larger models. This must be determined by experience.

I am, Sir, yours, &c.,
H. F. SENHOUSE,
Captain R.N.

Salcombe, Kingsbridge,
Oct. 6, 1824.

THE LATE PROFESSOR DAVIES.

We quote the following interesting particulars respecting our late lamented friend and contributor, Professor Davies, from a memoir of him in the *Expositor* — commenced by Dr. John Cockle, and completed by his brother, James Cockle, Esq., M.A., both of whom were "happy in a long and intimate acquaintance with the deceased." The memoir is accompanied with a good likeness of the Professor by Mr. Robert Landells.

If, as our philosophic bard has observed, the good *men* do is oft interred with their bones, we venture to hope that Thomas Stephens Davies forms one of the exceptions of the poet's rule; for he has given an initial impulse to a chain of scientific observations which must undulate on to the vast ocean wave of future scientific discovery. At once a geometer of surpassing excellence, and a most distinguished disciple of the great father of the experimental method, his writings exhibit a method so inductive, that they must rank as models, and imitative ones, for all who seek distinction in this intricate department of science, a department in which laurels are not easily won, and, least of all, by those who want the aid of method.

Happy in a long and intimate acquaintance with the subject of this memoir, we have, perhaps, been able to appreciate, in some degree, the influence which a systematic and logical analysis of our ideas exercises in rendering a man both able to write succinctly, and impart successfully, the result of his labours to others; and this even when they are directed to a most recondite subject matter.

The fame of Professor Davies was not
limited to his country nor, w
to add, to his age. Not his
merit of either simply collati
old materials—his claims wer
vated in their nature. Th
the greatest originality and t
in those departments of the
which he has so elegantly s
treated. How flattering to
and how honourable must it
to the man, to think that o
visits paid, when in England,
ous contempory, Chasles, t
geometry, was to himself,—
appreciation of British geniu
tation to the professor of
work, "L'Aperçu Historique
en Géométrie!!" How delig
pathy of kindred genius!
to the true dignity of science
exhibited without the alloy
which lesser minds might feel
by the baneful influence of
bright the example thus se
rarely-gifted intellects! How
butive to the true republic of

Inexorable death has cut sl
career of Professor Davies a
his mind was in its very pleni
at a period, too, when he w
triously occupied with paper
culable value to geometric sc

. . .

It is in his character of
the eye of admiration has b
turned to Professor Davies, a
is best known to the world.
and profundity of his views
geometrical attainments, wi
the theme of many a future p
best discharge my present
enumerating his voluminou
searches (for that has been d
but by pointing out those wl
unfinished, some in the mids
course of publication.

Two series of "Geomet
tions concerning the Phenom
trial Magnetism," by Profess
appeared in the *Philosophic*
It is not unlikely that materi
ing, if not for completing,
tions, will be found among l

A number of papers on "
of the Line and Plane" ha
the *Mechanics' Magazine.*
papers is nearly complete.
sirable it so much of them
appeared were published sep

* The remainder of this ex
part continued by Mr. James Co

His "Geometrical Notes" were also in the course of publication in that work. The last of them appeared in the *Mechanic's Magazine* for December 7, 1850. The last of his "Problems on Railway Cuttings" appears in No. 1202 of that work. This set of papers also seems to be incomplete.

In the *Lady's and Gentleman's Diary for* 1851 appears one of a series of papers on "Radical Axes and Poles of Similitude." These papers were to have been continued.

Since the death of Professor Davies I have communicated to the *Philosophical Magazine* autograph papers of his entrusted to me by Mrs. Davies. One of these papers appeared in the supplementary Number of that work, published in June last. Another will probably appear next month.

He had it in contemplation to add notes to Mr. Potts's proposed work on Porisms.

As a geometer, Professor Davies utterly rejected the idea of a geometry founded upon definitions alone, without axioms. He had pondered carefully upon logic as applied to that science, and undoubtedly there is much that is peculiar and special in such application. He was disposed to affix this peculiarity in the logic of geometry to the *copula*, and to term it the *copula of equality*, in contradistinction to the ordinary *copula of identity*.

If Professor Davies had no other claim to the homage of the algebraist than the untiring zeal and energy with which he enforced the scientific claims of Horner, that alone would constitute a valid title to admiration. He contributed in a great degree to fix attention upon Horner's processes, of which his keen and piercing intellect at once detected the value; and he was probably the first mathematician of note who saw them in their true light, and perceived the place they were one day destined to occupy. I need scarcely say that these processes constitute an integral part of algebra, and are of immense *practical* utility in even the domain of arithmetic.

But still more must the name of Davies claim our respect, when we reflect that in his generous devotion to the rights of Horner, he forebore from prosecuting his own independent researches on Equations—forbore lest perchance his claims might conflict with those of his friend. What his success in such investigations might have been, it is of course now impossible to say; but, if we may judge from the useful results which have accrued to the *numerical* theory, from a suggestion which he appended to the first part of my "Analysis of the Theory of Equations," we cannot help inferring that brilliant results would have followed his cultivation of this *department* of science. In the character of Thomas Stephens Davies this ardour in the service of his friend was one of those fine elements without which even talent and accomplishment alike has failed to inspire a warmer feeling than admiring envy.

But it must not be supposed that his analytical claims are limited to what has just been stated. In illustration, let me refer to such papers as his discussion of Dr. Matthew Stewart's "General Theorems" in *The Edinburgh Transactions*. To the clearness of his view of the *relation* between algebra and geometry, his researches on the algebraical analysis of prisms bear ample testimony.

Those who are at all familiar with the writings of the subject of this memoir, will have observed the copiousness of historical knowledge which they display, and, in the quantity of historical facts which they bring before the reader, an example well worthy of imitation by all writers.

I feel it right to add here, that to the ever active mind and pen of Professor Davies must be attributed numerous able papers to which his name is not appended.

There is one point upon which I have refrained from touching, and that is upon the public and official character of Professor Davies. A long career of practical usefulness and honourable exertion, continued almost up to the time of his death, and in one of the public institutions of the country, may perhaps be deemed sufficient to constitute a claim to public gratitude; I mean a claim upon those who for this purpose represent the public, and to whom society confides the trust of awarding the tokens of its approbation. Such tokens cannot of course reach him who may be thought to have justly earned them; he is beyond reward or neglect, having closed a not inglorious life amid the most exemplary patience under bodily suffering and mental anxiety. Many, many would be gratified (and who could be otherwise?) should the Ordnance or the Treasury bestow some suitable mark of public approbation on the widow and only son of their able and faithful servant.

COTTON SAILS AND HORIZONTAL SEAMS.

(From McMakin's "Model Courier," U.S.)

We hear it stated, that at the recent Royal Yacht race at Cowes, the English yachts, to increase their speed with the *America*, had recourse to wetting their sails. Should Captain De Blaquiere, the present owner of the *America*, adopt the hemp duck, as used by the British Yacht Squadron, and have her sails cut on the old-fashioned balloon principle, there is fear that the laurels she so gallantly won might soon wither with the *Titania*, in a suit of cotton sails made properly. The English method of cutting fore and aft sails differs materially from ours. For instance, they give the foot of their sails a greater circular sweep, which hangs below the foot-rope. The leeches are exceedingly hollow, caused by the stretching of the bolt ropes, thereby sustaining an extra extent of spar. The *America's* sails, like all cut here, are straight in leech and foot.

The cotton canvas has now almost entirely superseded all other duck. It was invented by Mr. James Maull, of this city, and first manufactured for him by Mr. John Simpson, then residing at Wilmington, Delaware, during the late war with England, at which time Russian, or any foreign canvas, it is well known to those in the trade, was selling at forty-five to fifty dollars per bolt.

The canvas was at first made by the hand-loom, which rendered it exceedingly soft and pliable; this was obviated by Mr. John C. Colt, of New York, who some thirty years, since commenced its manufacture with the power-loom. Mr. Colt, and Messrs. Craig and Sergeant, were well aware of the difficulty Mr. Maull experienced in securing its introduction, and it was several years before it was at all noticed by other sail-makers, with the exception of Lambert Tree, who subsequently brought it into notice among our smaller vessels. Among the first who used the cotton canvas, was Captain Parker, for the sloop *Trial*, of Trenton, and Captain Stokes, now of the sloop *Planter*, of Wilmington.—After a few years' wear, Captains Stokes and Parker both became dissatisfied, particularly Captain Stokes, who stated that the disadvantages were that the cotton canvas was liable to continual ripping and expense of re-sewing; and notwithstanding its advantages in other respects, would renounce its use, if there was no method of obviating this defect—which was eventually a general objection. After some reflection, Mr. Maull suggested to Messrs. Craig and Sergeant — the then agents of Mr. Colt—the adoption of cotton twine as a ready means to remedy the objection, impressing on them the ill effects of hempen twine. They induced Mr. Colt, on these representations, to make the cotton twine for the first time. It was made, and used with the most complete success, not only for cotton canvas, but for Russian duck—its efficiency consisting in its superior durability. It was then considered as an innovation, and condemned by many as visionary. Its present and general adoption in the United States is the best commentary on the success of Mr. Maull's efforts.

Mr. Maull early imbibed the impression that a vessel sailing against the wind would sail faster if her sails were constructed upon the principle of his Patent Horizontal system, wherein the least resistance to the action of the wind is practically obtained—the seams being horizontal, or in the line of direction of the wind.

The celebrated yacht *Maria*, owned by John C. Stevens, Esq., of New York, has been provided with these sails, and, although nearly four years in use, they are admitted to be the best-fitting sails in New York. Her contest with the world-renowned *America*, the victress of Johnny Bull, has settled her superiority even over that famous yacht—a fact admitted by Mr. Schuyler and other members of the Yacht Squadron. Mr. Stevens has stated that he was under the impression, ten years before Mr. Maull obtained his patent, that the principle was the best method of cutting sails, and he was the first to introduce them in New York on the *Maria*. His other schooner, the *Uncle John*, of 150 tons, has been provided with the patent sails, which have been in constant use four years; and from a statement of Captain Baldwin, who commands her, we have learned that they have not been repaired, with the exception of roping, and that he expects they will last two or three years longer.

NEW PHOTOGRAPHIC PROCESS.

The following photographic process has been communicated to us, by Mr. C. J. Muller, from Patna, in the East Indies. We have submitted it to an experienced photographer; and he informs us that it offers many advantages over the Talbotype or the Catalissotype of Dr. Woods—which it somewhat resembles; that it is easy in all its manipulatory details, and certain in its results. We give Mr. Muller's own words:

"A solution of hydriodate of iron is made in the proportion of eight or ten grains of iodide of iron to one ounce of water. This solution I prepare in the ordinary way with iodine, iron turnings, and water. The ordinary paper employed in photography is dressed on one side with a solution of nitrate of lead (15 grains of the salt to an ounce of water). When dry, this paper is iodized either by immersing it completely in the solution of the hydriodate of iron, or by floating the leaded surface on the solution. It is removed after the lapse of a minute or two, and lightly dried with blotting-paper. This paper now contains iodide of lead and protonitrate of iron. While still moist, it is rendered sensitive by a solution of nitrate of silver (100 grains to the ounce), and

n the camera. After an exposure of
ation generally required for Talbot's
t may be removed to a dark room.
mage is not already out, it will be
peedily to appear in great strength,
a beautiful sharpness, *without any
application*. The yellow tinge of
ts may be removed by a little hypo-
of soda, though simple washing in
eems to be sufficient to fix the pic-
The nitrate of lead may be omitted;
n paper only, treated with the solu-
the hydriodate of iron, and acetic
y be used with the nitrate of silver,
enders it more sensitive. The lead,
, imparts a peculiar calorific effect.
l tinge brought about by the lead
changed to a black one by the use
ate solution of sulphate of iron—by
ndeed, the latent image may be very
developed. The papers, however,
keep after being iodized."

Fuller suggests that as iodide of lead
letely soluble in nitrate of silver, it
urnish a valuable photographic fluid,
ould be applied at any moment when
.

mall degree of interest attaches to
ocess — originating in experiments
on in central India. It appears
y applicable to the albuminized glass
odion processes.—*Athenæum*.

GERMAN PATENTS.

ntage will be taken of the meeting
ext Zollverein Congress, in Berlin,
ss the question of a general law of
or the whole Union. This very de-
bject has never yet been attained:
tor must apply for a patent in every
State, and the principles on which
lege is granted are so different in
parts of Germany, that often a
was granted for an invention in one
d refused in another.—*Berlin Cor-
nt of Times*.

RICAN GREAT PATENT CASE.—
MC CORMICK'S REAPER.

Circuit Court, Albany, N.Y., Octo-
Judge Nelson presiding.—This case
l the court for six days; the parties
Cormick *v.* Seymour and Morgan,
kport, N.Y. The action was one
r the infringement of the patent of
ginia Reaping Machine, the same
s cut such a glorious figure in Eng-
d for which the Council Medal of
at Exhibition was awarded.

The jury rendered a verdicts in favour of the plaintiff, with damages against the defendant of the great sum of 17,606 dollars!—*Scientific American*.

Telegraph Case of Morse v. *Bain*.

U. S. Circuit Court, Philadelphia.—Judge Kane gave his decision on last Monday morning (Nov. 3rd) in the case of Morse *v.* Bain, sustaining the claims of Morse throughout. The case is to be taken to the supreme Court.—*Ibid*.

MATHEMATICAL INQUIRY.

Sir,—May I beg that some of your mathematical correspondents will be so kind as to point out a simple method (if such be practicable) of determining the radius of a circle, *having giving the length of the chord of a segment, and the area which it cuts off?*

It is a useful practical problem to place a given area in the segment of a circle on a given chord; and as its solution is not to be found in any work on mensuration or geometry which I have seen, perhaps a simple solution will be interesting, if not practically valuable to others, besides,

Yours,
A CONSTANT READER.

November 15, 1851.

MESSRS. R. AND J. OXLAND'S PATENT IMPROVEMENTS IN THE MANUFACTURE AND REFINING OF SUGAR.

(See ante page 319.)

In the specification of a former patent, dated April 26, 1849 (*see* vol. li., p. 429), the patentees described a process for manufacturing and refining sugar by the employment of acetate of alumina, in which process lime was used for removing the alumina employed. It has been found, howe be that even with the most careful working, some portion of the alumina is liable to be left in the solution, and carried through the several processes to which the saccharine liquid is submitted into the molasses. When acetate of alumina and lime have been used, the remaining alumina has been separated by means of superphosphate of alumina or superphosphate of lime, by adding a small quantity of either of these ingredients to the syrup, and then boiling for two or three minutes, taking care to neutralize any excess of acid by the addition of aluminate of lime, or saccharate of lime, lime water, or milk of lime; and the operation being completed after the extraction of all the alumina, as directed in the above-mentioned specification. The patentees have now discovered that, in place of adopting this method of procedure, phosphates and superphosphates may be employed directly, and that they are capable of producing analogous effects, but with this advantage, that the whole of the agent employed is capable of separation from the saccharine matter.

In the treatment of saccharine liquids or solutions of raw sugar with phosphates, the sugar—say, for example, ordinary Mauritius sugar—is dissolved by blowing up with steam, in the usual manner, but without the use of blood, a soluble phosphate being added to the water employed. (If the ordinary crystallized phosphate of soda be used, the requisite proportion thereof is about 1¼ lbs. to the ton of sugar.) The solution is then brought to the boiling point, the excess of acid carefully neutralized by aluminate, or saccharate, or milk of lime, or lime water, and the syrup (which will be of a density of about 25° to 30° Baumé) passed through filter bags in the ordinary manner, when the defecation will be found to be completed. The residual matter left in the bags is subjected to washing with water, to remove all traces of saccharine matter, and the weak solution thus obtained may be used in blowing up fresh quantities of raw sugar. By this operation the syrup is partially decolourized—to an extent which may in some cases be sufficient previous to boiling it in the vacuum pan or otherwise, for crystallization; but its colour may be still further removed by the addition of 5 to 8 per cent. of hydrate of alumina, dried at 212° Fahr., and diffused through the water employed in the operation of blowing up; and thus the use of animal charcoal may be entirely dispensed with. The residual alumina left in the filter bags after the separation of all the saccharine matter may be dried, and the organic matter removed therefrom by ignition, and, after a further washing to remove any soluble saline substances, it may be employed in the manufacture of hydrate or of superphosphate of alumina; or, after washing, and previous to ignition, it may, by the addition of a further quantity of hydrate of alumina, be rendered fit for use over again.

When superphosphates are employed the following is the method adopted; the superphosphate, say, for example, of alumina, is mixed with the blowing-up water in the proportion of 6 lbs. of alumina dissolved in phosphoric acid to a ton of sugar. While the syrup, of a density of about 25° to 30° Baumé, is being brought to boiling, a sufficient quantity of either aluminate or saccharate of lime, or milk of lime, or lime-water must be added to neutralize the acidity; and the syrup should be passed through bag filters, and the clear liquor conducted to the receiver, from which the vacuum pans or other boilers are supplied; the subsequent operations are the same as in the old plan of working. The matters left in the filter bags are to be treated as before described.

The superphosphate of alumina is obtained by dissolving alumina in phosphoric acid, which the patentees manufacture in the following manner: They burn bones white, grind them to powder, and digest in sufficient muriatic acid to effect the solution of the carbonate of lime only. The residue after washing to remove the soluble matters, is dried, and to a given quantity of it, mixed in water to the consistence of thin paste, is added sufficient sulphuric acid to combine with the greater portion of the lime present, leaving free, however, about two to three per cent. of lime; the mixture is then well-stirred, and kept at a temperature a little above 90° Fahr. for 24 hours, when the mass is lixiviated in water until all the soluble matter is separated from the sulphate of lime which will be precipitated. The strong liquors thus obtained may be used for combining with alumina and the weak solution for lixiviating fresh quantities of phosphoric acid in the course of production. When alumina is digested in phosphoric acid, a phosphate thereof, insoluble in water, is produced; this is treated with just sufficient phosphoric acid to dissolve it, and the superphosphate obtained in this manner is filtered to prepare it for use as above described.

The aluminate of lime is prepared by dissolving alumina in caustic soda or potash, and adding milk of lime or lime-water to the solution; aluminate of lime is precipitated, which having been separated from the solution, is diffused in water and used in that state; and the patentees state that they prefer the employment of that agent to the use of the saccharate of lime, or milk of lime, or lime-water.

When manufacturing sugar from the cane, the defecation of the cane juice is effected with aluminate of lime in the usual manner; the excess of lime being neutralized with superphosphate of alumina or superphosphate of lime, and after filtering and concentrating the filtered liquid to from 25° to 30° Baumé, it is treated with phosphate of soda in the same manner as above directed for treating raw sugar. The same process is also observed when operating on beet-root juice, only that a larger quantity of aluminate of lime, or milk of lime, is employed in the first defecation.

COLT'S REVOLVERS.

At a meeting of the Institution of Civil Engineers, held on Tuesday last, the paper was read "On the Application of Machinery to the Manufacture of Rotating Chambered-Breeched Fire-Arms, and the Peculiarities of those Arms," by Colonel Samuel Colt, U.S., America.

The communication commenced with an historical account of such rotating chamber fire-arms as had been discovered by the author in his researches after specimens of the early efforts of armourers for the construction of repeating weapons, the necessity for which appears to have been long ago admitted.

The author, entirely unaware of any previous attempts to produce such weapons, made a series of experiments on skeleton fire-arms, which were very successful; but subsequently he fell into many of the errors of his predecessors, for, by covering the breach and the mouths of the chambers, simultaneous explosion of several charges constantly occurred. This induced the restoration of the arms nearly to their original skeleton form, and the result was the production of the present perfect arm, which has been so universally adopted in America that the author's large manufactory has proved quite insufficient to supply the demand.

The means for manufacturing these arms on so large a scale was the main point of the paper; for, unlike the system adopted in England and on the Continent, of making fire-arms almost entirely by manual labour, the several parts comprising these weapons are forged, planed, shaped, slotted, drilled, tapped, bored, rifled, and even engraved by machinery to such an extent that 10 per cent. only of the value of the arm was for hand labour in finishing and ornamenting; 90 per cent. being executed by automaton machines guided by women and children, whose labour was represented by 10 per cent., leaving 80 per cent. for the machinery.

The action of these machines was described, and it appeared that, though, like a cotton or flax mill, the manufactory at first sight appeared intricate, yet that each part travelled independently through its course until at length the finishing workman had only to put the several parts together, almost indiscriminately, and the uniformity was so precise that little or no fitting was required beyond removing the "burr," or rough edge left by the machines. This was a point of great importance, especially in a country of such extent as America, where the necessity for sending arms from one district to another for repair might be attended with serious consequences.

The arms now manufactured by the author, and of which numerous specimens were exhibited, were of the simplest construction; the lock consisted of only five working parts, contained in a lock-frame cut out from the solid metal, into which the breech arbor was firmly inserted, and by it rigidly attached to the barrel in such a manner as to regulate, with the greatest precision, the contact between the end of the barrel and the mouths of the cylinders, so as to prevent any serious escape of lateral fire.

The rotating of the cylinder was accomplished by a self-acting lever, to which motion was given by the act of drawing back the hammer; at half-cock the cylinder was free to rotate in one direction, for the purpose of loading and putting the caps on the nipples, the former operation being rapidly accomplished by the conversion of the ramrod into a jointed lever, attached to the barrel, by which means the bullets were rammed home so securely that no patch or wadding was required. The grooves in the barrel were of a peculiar spiral, commencing almost straight, near the breech end, and terminating at the muzzle in a curve of small radius. The bullets were either of cylindrical or conical shape, and from some diagrams of several practice targets sent from Woolwich, by Colonel Chalmers, R.A., for exhibition at the Meeting, it appeared that even by men unaccustomed to the use of this particular arm great precision of firing could be attained, as with a small revolving belt pistol, at a distance of fifty yards, out of 48 shots 25 bullets took effect within a space of one foot square, and of them thirteen hit the bull's-eye, which was only six inches in diameter, the whole number of shots striking the target.

SPECIFICATIONS OF ENGLISH PATENTS ENROLLED DURING THE WEEK, ENDING NOVEMBER 26, 1851.

PERCEVAL MOSES PARSONS, of Robert-street, Adelphi, civil engineer. *For improvements in cranes capable of being used on railways and in parts of railways.* Patent dated May 19, 1851.

The "improvements in cranes" consist in constructing them in such manner that the tensile strain exerted on the back-stays of the jib when the jib is deflected during the raising of a load may be transmitted through levers to act on a counter-balance weight or cylinder running on the girders forming the frame of the crane, which weight is caused to recede from the foot of the jib up an incline of an increasing curve, so as to produce a leverage in a contrary direction to that of the load, and increasing with the amount of deflection of the jib, by which the equilibrium of the crane will be preserved, whatever may be the weight lifted. When intended for use on railways these cranes are mounted on wheels to admit of their being readily transported to any place where their services are required; but for ordinary purpose they may be made stationary.

What is claimed under this branch of the invention is the constructing of cranes with self-acting or self-adjusting counterbalance-weights as described.

The improvements in parts of railways have relation to switches and crossings.

With regard to switches, the novelties are stated to consist,—1. In fixing the main rails with the slabs or bars for the support of the tongue rails on an iron plate extending throughout the whole length of the switch. 2. In fixing the supports for the tongue rails on or against the main rail itself, when the form of the main rail admits of this arrangement. 3. In constructing the main rail itself in such manner as to form a suitable platform or support for the tongue rail. 4. In fixing the main rails on separate cast or wrought iron plates to act as sleepers, which sleepers are also provided with slabs or bars for supporting the tongue rails. 5. In fixing the chairs for the support of the tongue rails to the main rails, when the latter are of suitable sections. And, 6. In certain methods of constructing tongue rails.

With reference to crossings, the improvements consist—1. In constructing crossings for railways, by fixing the point and wing rails on a wrought-iron plate extending the whole length of the crossing, and having junction-chairs attached to it at the ends. 2. In fixing the wing and point rails on separate cast or wrought iron sleepers; 3. In fixing chairs or saddles on or against the wing rails to support the point rails, or by forming the wing rails themselves so as to make a support for the point rails, or *vice versâ*; 4. In forming the wing and point rails and sole-plate in one solid mass of wrought iron; 5. In fixing the point rails, by wedging them in between the wing rails, which are fixed on cast or wrought iron plates or sleepers, extending the whole length of the crossing, in one piece or in separate pieces, and whether junction chairs are employed or not; 6. In cutting two point bars out of one parallel bar, one end of each of the point bars being of a trapezoidal, and the other end of a triangular shape; and 7. In certain peculiar forms of wing and check rails.

BENJAMIN BAILEY, of Leicester. *For improvements in the manufacture of looped fabrics.* Patent dated May 23, 1851.

Claim.—The combining parts of knitting machinery in such manner that the necessary motions may be communicated to the needle bar to admit of the sinker bar and presser bar remaining fixed and stationary, and so that the jack-sinkers may simply be required to fall and rise; and also the interposing of levers between the slur cock and the jacks.

GEORGE TATE, of Bawtry, York, gentleman. *For improvements in the construction of dwelling-houses and other buildings, including floating vessels; and for the adaptation and manufacture of materials for such uses.* Patent dated May 22, 1851.

Claims.—1. The construction of dwelling-houses with staves or pieces of timber, or other suitable material, fitted together and bound by hoops or similar binders, whether the shells of such houses be constructed singly or jointly, one within another, leaving spaces between. The construction of floors, roofs, and partitions, by wedging up pieces of wood or other material in concentric rows, surrounded by an external hoop, binder, or fastener, and having the interstices filled with glue or other viscous and siccative matter. The construction of doors and window-frames with a collapsible fastening, and of doors and window-shutters with angle and strut supporters.

2. The construction of water-closets in such manner that the soil shall be withdrawn and choking prevented by apparatus adapted thereto and working therein. The construction of cesspools with staves or pieces of timber, combined and adapted to receive apparatus for mixing and removing the soil; and the construction of sewers, drains, and pipes for the conveyance of fluids by means of staves and pieces of timber, or other suitable material, combined and held together by hoops.

3. The preserving of timber by means of carbonate of lime, carbonate of baryta, hydrated oxides of earths or metals, arseniates or arseinates, precipitated in the pores of the wood.

4. The construction of suspension bridges with chains made rigid by means of certain arrangements described; a peculiar construction of arch girders for bridges; and the construction of bridges with staves or pieces of timber, or other suitable material, combined and bound together with hoops.

5. The adaptation of hollow divided co-

for temporary purposes, and the con-ion of buildings.

The manufacture of bricks and tiles ated transversely as well as longitudi- the making of bricks, tiles, and pipes y by centrifugal action; the making cks, tiles, and slabs of brick-earth, ed to the required form, and faced lay of a finer description, or with ain, painted or otherwise ornamented, zed, and the adaptation thereof to the ental decoration of buildings and edi- the manufacture of pieces of metal of le forms, glazed, coloured, and ena- l, and the adaptation thereof to the ating of buildings; and the adapta- f channels, flues, orifices, pipes, or agms to buildings.

The construction of floating vessels staves or pieces of timber or other le material, fitted and bound together; h staves or pieces of corrugated metal rly bound together; with a combina- f revolving buoyant supporters, or combination of air and water-tight rtments and buoyant revolving band ters and rollers. Also the construc- f floating vessels capable of being ted or shortened at pleasure; and of ats with a combination of revolving rtments and buoyant band supporters ers.

F SCOTCH PATENTS FROM 22ND OF OBER TO THE 22ND OF NOVEMBER, 1.

n Deeley, and Richard Mountfort Deeley, of n Bank, Stafford, flint and bottle glass ma- rers, for improvements in the construction of s for the manufacture of glass. October 31; ths.

d Vincent Newton, of Chancery-lane, Mid- mechanical draughtsman, for certain im- ents in the construction of railways. (Com- tion.) November 4; six months.

am Smith, of Upper Grove Cottages, Hol- Middlesex, engineer, for improvements in ive and other engines, and in carriages used ays. November 4; six months.

rt Hyde Greg, of Manchester, Lancaster, cturer and merchant, and David Bowlas, of , Lancaster, manufacturer, for certain im- ents in machinery, or apparatus for manu- g weavers' healds or harness. November 4; ths.

el Scott, of John-street, Adelphi, civil en- for improvements in punching, riveting, , and shearing metals, and in building November 5; six months.

min Hallewell, of Leeds, York, wine mer- or improvements in drying malt. Novem- ix months.

w Gibson, of Wellington-terrace, Newcas- Tyne, for improvements in machinery for pulverising and preparing land. November 7; six months.

William Longmaid, of Beaumont-square, gentleman, for improvements in treating ores and minerals, and in obtaining various products therefrom, certain parts of which improvements are applicable to the manufacture of alkali. November 7; six months.

Antoine Dominique Sisco, of Slough, for improvements in the manufacture of chains, and in combining iron with other metals applicable to such, and other manufacture. November 11; six months.

Henry Lund, of the Temple, Esq., for improvements in propelling. November 12; six months.

Frederick Joseph Bramwell, of Millwall, Middlesex, engineer, for improvements in working the valves of steam engines for marine and other purposes, and in paddle-wheels. November 12; six months.

William Boggett, of St. Martin's-lane, gentleman, and George Holworthy Palmer, of Westbourne Villas, Paddington, civil engineer, for improvements in obtaining and applying heat and light. November 14; six months.

Henry Richardson, of Aber Hirnant Bala, North Wales, Esq., for certain improvements in life-boats. November 14; four months.

James Bagster Lyall, of Thurloe-square, Brompton, gentleman, for an improved construction of public carriage. November 14; four months.

James Pyke, of Westbourne-grove, Bayswater, for improvements in the manufacture of leather, also in making boots and shoes. November 17; six months.

Hugh Bowlsby Willson, of the York Hotel, Blackfriars, London, for improvements in the construction of rails for railways. Nov. 19; six months.

George Tate, of Bawtry, York, gentleman, for improvements in the construction of dwelling-houses, and other buildings, including carriages and floating vessels, and in the propulsion of said vessels, and in the adaptation and manufacture of materials for such uses. November 21; six months.

LIST OF IRISH PATENTS FROM 21ST OF SEPTEMBER TO THE 19TH OF NOVEMBER, 1851.

Samuel Holt, of Stockport, Chester, manager, for certain improvements in the manufacture of textile fabrics. September 24.

Henry Wimshurst, of Broad-street, Radcliff-cross, Middlesex, ship builder, for improvements in steam engines, in propelling, and in the construction of ships and vessels. September 30.

Charles Hardy, of Low Moor, York, Esq., for certain improvements in the manufacture of scythes. October 6.

Peter Robert Drummond, of Perth, for improvements in churns. October 20.

John Oxland, and Robert Oxland, of Plymouth, chemists, for improvements in the manufacture and refining of sugar. November 8.

James Webster, of Leicester, engineer, for improvements in the construction and means of applying carriage and certain other springs. Nov. 3.

Alexis Delemer, of Radcliffe, Lancaster, engineer, for certain improvements in the application of colouring matter to linen, cotton, silk, woollen, and other fabrics, and to linen, cotton, silk, and other wefts, and also in machinery or apparatus for those purposes. November 5.

Perceval Moses Parsons, of Duke-street, Adelphi, civil engineer, for improvements in parts of railways, and in cranes. November 18.

WEEKLY LIST OF NEW ENGLISH PATENTS.

Samuel Colt, of Bond street, Middlesex, for certain improvements in fire-arms. November 22; six months.

Thomas Marsden, of Salford, for improvements in machinery for heckling and combing flax and other fibrous materials. November 22; six months.

Enoch Statham, of Siddal's-road, Derby, for improvements in the manufacture of lace and other fabrics. November 22; six months.

Frederick Weiss, of the Strand, Middlesex, surgical-instrument maker, for improvements in certain surgical instruments; also in scissors and other like cutting instruments. (A communication.) November 22; six months.

Frederick Benjamin Geithner, of Camden-street, Birmingham, for improvements in the manufacture of castors and legs of furniture. November 22; six months.

Jean Baptiste Chaluren, of Rouen, merchant, for improvements in preparing and weaving cotton. November 22; six months.

William Armand Moreau Gilbée, of 4, South-street, Finsbury-square, London, gentleman, for certain improvements in the process of and apparatus for treating fatty and oleaginous matters, and in the manufacture of candles and other useful articles therefrom. (A communication.) November 22; six months.

George Mills, of Southampton, Hants, engineer, for improvements in steam-engine boilers and in steam-propelling machinery. November 22; six months.

Alexander Southwood Stocker, of Wandsworth, Surrey, gentleman, for certain improvements in the stoppering or stopping of bottles, jars, pots, or other such like receptacles. November 25; six months.

Henry Ellwood, of the firm of J. Ellwood and Son, of Gt. Charlotte-street, Blackfriars, hat manufacturers, for improvements in the manufacture of hats. November 27; six months.

Richard Whytock, of Edinburgh, for improvements in applying colours to yarns or threads, and in weaving or producing fabrics when coloured or party coloured yarns or threads are employed. November 27; six months.

John Lee Stevens, of Kennington, Surrey, gentleman, for certain improvements in propelling vessels on water. November 27; six months.

WEEKLY LIST OF DESIGNS FOR ARTICLES OF UTILITY REGISTERED.

Date of Registration.	No. in the Register.	Proprietors' Names.	Addresses.	Subjects of Design.
Nov. 21	3021	William Ashton	Louth	Universal sponging bath.
,,	3022	J. and A. Ridsdale	Minories	Fastening for ships' scuttle-light and ports.
,,	3023	S. M. Peary	Willingham	Wheel supporter.
22	3024	I. Biggs and Son	Leicester	Shirts made of looped fabrics.
24	3025	William Barwell	Birmingham	Metallic reel.
25	3026	Robert M'Connell	Glasgow	Water-closet.
,,	3027	J. W. and T. Allen	Strand	Despatch-box.
26	3028	Henry Watson	Newcastle-on-Tyne	Parts of a safety-lamp.
,,	3029	William Drary	Swan-lane, City	Bullock-tie.

WEEKLY LIST OF PROVISIONAL REGISTRATIONS.

Nov. 21	329	J. S. Long	Westminster	Perambulator.
25	330	Isidore Burnsteen	Essex-street, Strand	Travelling coat.
26	331	F. G. Yeates	East-road, City-road	Winder for boxes for string, twine, &c.

CONTENTS OF THIS NUMBER.

LONDON: Edited, Printed, and Published by Joseph Clinton Robertson, of No. 166, Fleet-street, in the City of London— Sold by A. and W. Galignani, Rue Vivienne, Paris; Machin and Co., *Dublin*; *W. C.* Campbell and Co., Hamburg.

Mechanics' Magazine,

MUSEUM, REGISTER, JOURNAL, AND GAZETTE

No. 1478.] SATURDAY, DECEMBER 6, 1851. [Price 3*d*., Stamped, 4*d*.

Edited by J. C. Robertson, 166, Fleet-street.

SHIPTON'S PENDULOUS ENGINE.

Fig. 5.

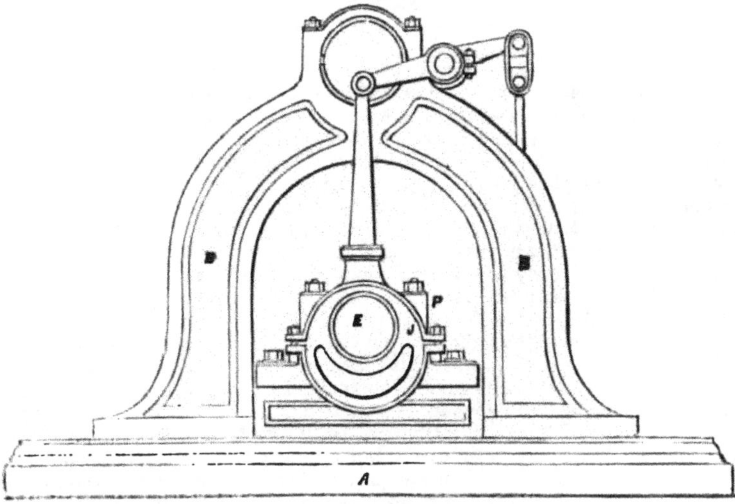

Fig. 6.

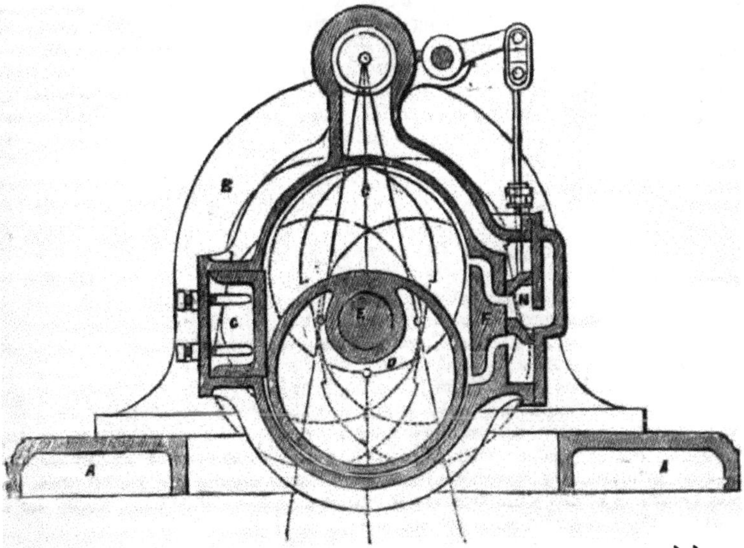

ON THE DIRECT CONVERSION OF RECTILINEAR INTO CIRCULAR MOTION IN THE STEAM ENGINE, WITH DESCRIPTION OF A PENDULOUS INVENTION OF NEW CONSTRUCTION. BY MR. JAMES A. SHIPTON, OF MANCHESTER.

(From Transactions of Institution of Mechanical Engineers, October 22, 1851.)

The years that have elapsed since the steam engine was first generally introduced as a prime mover, and the few alterations that have taken place, notwithstanding the many threatened invasions of a variety of ingenious inventions, must lead to the reflection that only a master mind could have combated with the difficulties that beset such an undertaking, and have sent it forth to the world in so perfect a form; the unskilled hand of the workman being the only drawback from its being then what it is now.

The costly work of the various parts subjected to immense strain, and also the searching and penetrating action of steam, depended then entirely upon the manual dexterity employed, and therefore the machinery of the steam engine was required to be of the most simple form to place it within the reach even of the most opulent; but as its use and value presented itself to the country at large, so the development of machinery took place to meet the demands of its manufacture; and, though it must be acknowledged that the same mind that brought the steam engine into commercial operation, contemplated also the reduction of its cost, by the use of self-acting machinery in the production of its parts, yet the task of bringing such into operation has been nearly as arduous as the former one. What the mechanic would then have looked upon as an impossibility, is now perfectly simple, and the cost in comparison trivial.

That the principle of the ordinary steam engine as regards the reciprocating action of the steam cannot be improved, is the author's opinion; but also that its mechanical construction may be materially altered, and improvements effected in this respect, owing to the many advantages possessed at the present time of having tools to meet every requirement, the rapid progress made in this branch of mechanical science having placed the steam engine in its present commercial position.

The subject of the present Paper is the, "Direct Conversion of Rectilinear into Circular Motion," and it also brings under notice a steam engine, not deviating in principle from the ordinary reciprocating engine, but simply in its construction; as the inventors feel convinced the nearer they approximate to the original the less liable they will be to err. The diagram, fig. 1, represents a piston and crank engine of ordinary construction, and although the whole area of the piston be exposed to the pressure of the steam throughout the stroke (supposing the valve be kept open), there are certain points when this pressure is useless; namely, when the crank is on the centre: thus, the circular motion of the crank restricts the piston from exerting its full force with regard to that circular motion, and thereby the velocity of the piston is constantly varying throughout the stroke, as also the power exerted and the steam consumed in like proportion. Thus the actual power exerted is the average velocity of the piston multiplied by the pressure.

Now, an eccentric being a mechanical equivalent for a crank, if the area of the piston of fig. 2 be equal to the area of the piston of fig. 1, and the throw of the eccentric B, equal to the stroke of the crank A, they are of like power. Then, by altering the mechanical arrangement, as in fig. 3, and placing a piston at top and bottom of the eccentric, or, in other words, placing the eccentric in a large piston, the area of piston and throw of eccentric being equal to B, an engine of like power is obtained. Therefore, A, B, and C, are equivalents of each other, differing only in mechanical construction; and the power obtained from each would be the same, not taking friction into consideration.

Dispense with these pistons, and admit steam alternately, top and bottom of the circle D, fig. 4, and this eccentric piston would be propelled, up and down, in a rectilinear direction, and this motion would be converted into a circular motion, during the propulsion of the piston. Here is obtained the amalgamation of the two motions of the ordinary engine in one body; the same body containing the properties of the reciprocating piston, and also of the crank.

The practical application of this principle is effected in the pendulous engine, shown in figs. 5, 6, 7, which represent a 20-horse power steam engine.

A is the base-plate, that carries the entire engine. The side framing BB, is fitted to it, and bolted firmly down, and upon this the cylinder C is suspended, and swings with a pendulous motion. The piston D is turned perfectly true, having the shaft E keyed eccentric in it, this shaft works in pedestals, PP, which are fixed on the bed-plate.

Fig. 1.

Fig. 2.

Fig. 3.

Fig. 4.

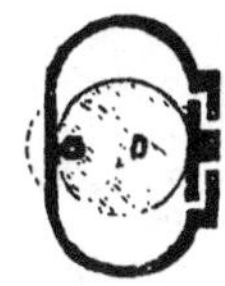

Fig. 7.

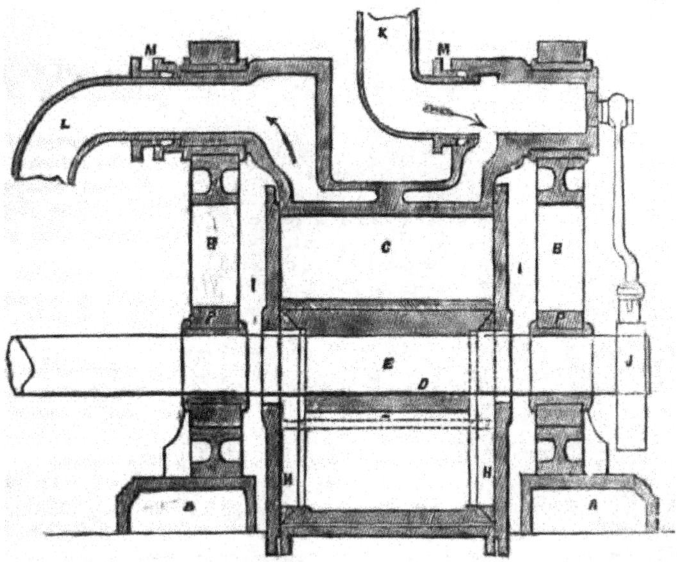

The piston works between two parallel surfaces, F and G; the surface at F being a plate dove-tailed into the cylinder, and the plate G is fitted into the recess prepared for it, and so arranged that, by means of adjusting screws, it follows up any wear that may take place on the periphery of the piston, and maintains a steam-tight joint. The piston is packed at the ends with the rings HH, these being fitted into a conical seating, and as the wear takes place, they are sprung open by means of a small wedge and bolt inserted where the ring is cut open, to allow it to expand. These rings work against the side plates I I, which are bolted to the cylinder, and have metallic joints. The peculiar motion of the ends of the piston against these surfaces causes a most beautiful wear, as the rings keep receding in their seatings, and never come over the same parts twice together.

Steam is admitted precisely the same way as in an ordinary engine, at top and bottom of the piston; but the valve N, is on the equilibrium principle, and exhausts through the back, being packed by a conical ring, in a similar manner to the ends of the piston; the valve is worked by the eccentric J, by means of levers and weigh shaft. The steam and exhaust pipes are shown at K and L, and are packed with the glands MM, to admit of the pendulous motion of the cylinder. The distances between the centres are calculated so as to allow the pendulous motion of the cylinder to coincide with the rate of revolution of the engine, and consequently, only a small portion of the weight of the moving body has to be overcome, with respect to the vibration.

The advantages of this engine are,—economy in first cost,—economy in space,—economy in foundations, being self-contained,—simplicity and economy in repairs, as the wearing parts are insertions, and may be renewed at a short notice,—direct application of the steam to produce the rotary motion of the shaft, without the intervention of joints or connecting rods,—and not being liable to derangement, as the moving parts are so few in number. As it contains less frictional surface than the ordinary engine, economy in consumption of fuel may be expected. High speed may be obtained, and thus gearing, wheelwork, &c., may be dispensed with; and from its compactness, this engine is most suitable for working the heavy class of machinery,—such as rolling-mills, &c., or screw-propellers in steamboats. A 20-horse engine is being constructed on this plan, and will be at work in a few days, and the author will be ready, at a future period, to lay before the Institution its performances, tested by a dynamometer.

Discussion on the preceding Paper.

MR. ELWELL inquired, whether a similar engine had not been at work for a considerable time at Wolverhampton?—and, what was the probable cost of such engines?

Mr. SHIPTON replied, that the engine at Wolverhampton was constructed on a modification of the plan, in which the piston oscillated instead of the cylinder. The cost would be 9*l.* per horse-power, exclusive of the boilers. The only comparative trial that had yet been made as to the consumption of fuel was with an engine at Manchester, which showed a saving of 10½ per cent., as compared with another direct-acting engine which worked from the same boiler.

Mr. SIEMENS inquired how the packing was made steam-tight? And Mr. Shipton explained that it was by an expanding ring, of triangular section, giving an equal pressure on both surfaces: the same plan was adopted for packing both the piston and the valve. He exhibited one of the packing rings.

Mr. CLIFT observed, that Mr. Shipton had argued that a considerable saving would be effected by avoiding the crank motion and the reciprocation of the piston; but he had against that the whole weight of the cylinder in motion, which was a large weight to be stopped and reversed at every revolution; and therefore he could not see what advantage the invention possessed over the ordinary engines.

Mr. SHIPTON replied, that the pendulous motion of the cylinder prevented the loss of power in reciprocation. The cylinder vibrated as a pendulum; and they found that one man could work the cylinder of a 20-horse engine in vibration, at full speed.

Mr. CLIFT inquired, whether the cylinder was made of correct length to vibrate, according to the law of a pendulum, at the actual working speed of the engine? as, if not correctly adjusted, it might require a large amount of power to force it into the required rate of vibration.

Mr. SHIPTON said, the remark would apply to all oscillating engines; but in this engine they had calculated the length of the centres of oscillation, so as to agree with the intended rate of working of the engine.

Mr. SIEMENS observed, that weight was certainly of secondary consideration, if the centre of oscillation could be made to agree with the corresponding length of pendulum; and such an arrangement would make the power more uniform throughout the stroke. But it would not be correct to consider the weight of the piston as a loss, in a reciprocating engine, as the momentum was gradually absorbed by means of the crank, and given out again in starting the return stroke. In some cases a heavy weight of piston and connecting rod was actually an advantage; an expansive engine, cutting off at half stroke, worked more steadily with a heavy connecting rod than with a light one, as it absorbed surplus power at one part of the stroke, and gave it out again when the moving power was deficient, tending to equalise the power. With respect to Mr. Shipton's engine, he considered the question to be one of comparative friction, compactness, and simplicity.

The CHAIRMAN (Mr. Slate) understood Mr. Shipton to bring forward his engine rather as one presenting advantageous points of construction, than as one which led to saving in fuel.

MATHEMATICAL PERIODICALS.

(Continued from p. 365.)

No. XXVIII.—*The Mathematical Repository.*—*Miscellaneous Papers*—(*Continued.*)

Art. I., Vol. III., Old Series.—An investigation of theorems relating to the sums of the powers of perpendiculars to all the sides of a regular polygon, from any point in the circumference of its inscribed circle. By Mr. James Cunliffe.

*** In this paper Mr. Cunliffe first finds as a lemma the sums of the powers of the roots of the equation

$$v^n - Pv^{n-1} + Qv^{n-2} - Rv^{n-3} + Sv^{n-4} - \&c. = 0;$$

n being a whole number; and by connecting these with the expression for the versed sine of nA, he thence deduces:

1. Sum of squares of perpendiculars $= \dfrac{3\,nr^2}{2}$;
2. ,, cubes ,, $= \dfrac{5\,nr^3}{2}$;
3. ,, 4th powers ,, $= \dfrac{35\,nr^4}{8}$;
4. ,, 5th powers ,, $= \dfrac{63\,nr^5}{8}$;

where n = number of sides of the polygon, and r = radius of the inscribed circle. These expressions obviously agree with the 5th, 22nd, 28th, and 39th of Dr. Stewart's "General Theorems," several of which had been demonstrated by Messrs. Lowry and Swale by other methods, in vol. i. of the *Repository*. The two last theorems are also elegantly investigated by Mr. Babbage, in the first Number of the *Journal of Science* (1816), and by Professor Davies, in the *Edin. Phil. Trans.*, vol. xv. p. 600.

Art. II. and XIII.—Improved Solutions to some curious Mathematical Problems. By Mr. James Cunliffe.

*** These two articles contain improved and generalized solutions to *nine* diophantine problems of considerable interest and utility. The first problem is, "to find three square numbers such that the difference of every two of them may be a square number;" to which *three* different solutions are given, the first of which is identical in principle with that given by Mr. Samuel Bills to the same problem in pp. 200-1, vol. iii. of the *Mathematician*. The second problem considers the case when "the *sum* of every two is a square number;" the *fourth* involves Ques. 823, in the *Gentleman's Diary* for 1802; the *sixth* corresponds to the 22nd question of Bonnycastle's "collection of diophantine problems;" and the *seventh* is the same as Ques. 55 of the *Liverpool Student*. Mr. Cunliffe's investigations, however, are much more elegant, comprehensive, and general than most of those to be found in the places cited.

Art. III.—On Continued Surds. By Mr. Benjamin Gompertz.

*** Mr. Gompertz here shows that "if $x^2-x=a$, then

$$x=\sqrt{a+\sqrt{a+\sqrt{a+}}}$$

ad infinitum;" and also that

"If $x^4=a^2bx$, then

$$x=\sqrt{a+\sqrt{b+\sqrt{a+\sqrt{b+}}}}$$

ad infinitum;" which demonstrate and confirm John Bernouilli's solutions, contrary to what Emerson has stated "in page 433 of his *Miscellanies*," inasmuch as the expressions are proved to be neither *diverging* nor of *infinite* value.

Art. VII.—Investigation of Dr. Waring's Theorem for Finding the Sum of *any* Power of the Roots of an Equation. By "A. B."

Art. VIII.—An Attempt towards a Restitution of the Lost Treatise by Eratosthenes, entitled "*Locis ad Mediatates*." By Mr. M (ichael) F (ryer).

*** This article was reprinted in the *Gentleman's Mathematical Companion*, and has been assigned to Mr. Fryer by the late Professor Davies, who was well acquainted with its author. It contains two theorems on the conic sections, without their investigations, but which their discoverer says "may be easily demonstrated by means of the modern analysis." This has been done by Mr. Davies, in his interleaved copy of Pappus's *Mathematical Collections*, whose demonstrations will shortly be given in a separate paper.

Art. IX.—On Vanishing Fractions. By "W. C."

*** This paper was written in order to controvert the opinions of those who maintained "that when a fraction appears, by assigning a particular value to the variable, under the form

$$\frac{0}{0},$$

the fraction in that instance ceases to exist;" and, in the course of his reasonings, the writer gives neat proofs that

$$0^{\circ}=\frac{0}{0}=1;$$

"when 0 does actually represent no number at all."

Art. XII.—Demonstration that the Area of a Parabolic Segment, contained under the Abscissa, the Ordinate, and *the Curve, is equal to* two-thirds of the Parallelogram contained under the Abscissa and Ordinate. By William Frend, Esq., of Jesus College.

Art. XVI.—Solution to Colonel Silas Titus's Arithmetical Problem. By Jas. Ivory, Esq.

*** This paper contains a solution of the following equations:

$$a^2+bc=16;$$
$$b^2+ac=17;$$
$$c^2+ab=18;$$

which appears to have been first proposed by "the famous algebraist, Dr. John Pell," to "Colonel Silas Titus, a gentleman of the bedchamber to Charles the Second." Dr. Wallis solves the question in pp. 225—256 of his "Algebra" (1685), and his solution has since been reprinted, with additions, by Baron Maseres, in his "Tracts on the Resolution of Equations" (1800) where also may be seen another solution by Mr. Frend. Mr. Ivory's solution is also included in the concluding volume of the *Scriptores Logarithmici*, by Maseres; and since his time the problem has passed through the hands of Whitley, Lettle, and Ryley, in the *Liverpool Apollonius*, No. II., pp. 120—1; Professor Davies, in his "Solutions to Hutton's Course," pp. 272—3; and Mr. Septimus Tebay, in the *Preston Chronicle*, Ques. 22, 1845. The last-named solution does not differ in principle from that by Mr. Frend, and hence I need not enter into any discussion on the merits of the different solutions of this "famous problem," since that has already been done in a clear and masterly manner by Mr. James Cockle, in No. XI. of his valuable "*Horæ Algebraicæ*," published in this Magazine.

Art. XXII.—On the Quadrature of the Circle. By Mr. Benjamin Gompertz.

Miscellaneous Papers.—New Series.

Art. I., Vol. I., Part II.—Demonstrations of some Propositions relating to such portions of the Surface and Solidity of a Sphere as may be exactly Squared and Cubed. By Mr. James Ivory.

*** This paper contains a co-ordinate investigation of several curious properties relating to the problem proposed by Viriani, as a challenge to the mathematicians of Europe, in the *Acta Eruditorum* for 1692. In the 12th volume of the Edin. Phil. Transactions, pp. 305—

essor Davies has considered the
oject by means of *spherical* co-
s; and a synopsis of his methods
lts is given in the Appendix to
tleman's Diary for 1836, and
Ques. 1564 and 1583 in the
Diary for 1834 and 1835 respec-

[.—Demonstration of a Theorem
's) respecting Prime Numbers.
James Ivory.

'heorem.—" Let *p* be any prime
and N any number *not* divisible
hen $N^{p-1}-1$ will be divisible

ory justly considers this theo-
'one of the most important in
ry of numbers," and believes
nonstration to be new and more
han that by Euler." *See* also
" Theory of Numbers," page
dy's Diary, Ques. 1550, and
ther works.

I.—Fagnani's Theorem respect-
tic Arcs rendered more General.
James Ivory.

n this paper " the curious
s by Count Fagnani, which
he difference of two arcs reck-
m the extremities of the quad-
n Ellipse," is extended to a
eral Theorem, which compre-
e former as a particular case,
signs the difference of two
arcs reckoned from any two
ints in the periphery of an
Mr. Ivory also deduces "a
property for all the Conic
' Another investigation of
Theorem may be seen in
Hutton, vol. ii., p. 514, and
autiful collateral properties due
sors Graves, MacCullagh, and
les are given in " Salmon's
tions," second edition, p. 296.

V. A Geometrical Porism,
examples of its Application to
on of Problems. By Scoticus.
is fictitious signature is one of
pted by Dr. Wallace, as ap-
a manuscript note in the late
Davies's copy of the *Mathe-
epository*, on the authority of
Leybourn, the Editor. The
tains the Analysis and Con-
of the following Porism :—
. and B be two given points,
a circle given in position,
iven a *point* P in the straight
line joining the points A, B, such, that if from P any straight line whatever be drawn, meeting the circle in C and D, the points A, B, C, D, shall be in the circumference of a circle." The resulting conditions are very elegantly applied in the construction of the problems: " To describe a circle that shall pass through two given points A, B, and touch another circle C, E, D, given by position;" and "from two given points A, B, to draw two straight lines AD, BD, to meet in a straight line MN, given by position, so that AD + BD may be equal to a given line." The first of these problems occurs in " Simson's Geometry," Prob. 45. The second forms Problem 15 of the same work, and also Problem 49 of his "Select Exercises." The porismatic point itself is obviously the *radical centre* of the given circle, and of the two points A and B considered as circles of *point* radius.

Art. V. Geometrical Propositions. By Mr. James Ivory.

*** The proposition here demonstrated is the first of the Porisms (LX.) published without Solutions by Dr. Stewart, at the end of his " General Theorems." At the close of the Analysis and Composition, Mr. Ivory gives a Lemma as "Proposition II.," which leads to a generalisation of the curious property of the point investigated in Prop. I., from Dr. Stewart. The more general Theorem is numbered as " Proposition III.," and includes (I.) as that particular case when DC and CE form one straight line:—the whole paper is well worthy of the student's close attention, inasmuch as it furnishes him with one of the best specimens of the Geometrical Analysis of Porisms. It is to be regretted that Mr. Ivory did not continue his investigations so far as to include the whole of the Propositions left undemonstrated by Dr. Stewart.

Art. VI. Solution of a Dynamical Question, viz., Prop. 3, p. 131, of " Simson's Miscellaneous Tracts " By Mr. James Ivory.

Art. VII. Of the Equilibrium of a very Long and Slender Cylinder floating in a Fluid. By Jac Rube.

*** This article relates to the " Question about the Exciseman's Staff," which may be seen in the account of the Gentleman's Mathematical Companion in the *Mechanics' Magazine*, vol. xlviii., pp. 401-2. This inquiry has lately been gone into at considerable length by Mr. Septimus Tebay, of Newcastle-upon Tyne,—one of our best authorities on Dynamics; and many important results have been elicited, which it is to be hoped he will soon make public.

Art. VIII. Of the length of an Arc of a Circle in Terms of the Tangent. By Mr. Benjamin Gompertz.

Art. IX. Geometrical Porisms. By Mr. Mark Noble.

⁂ This paper contains two Lemmatical Theorems, and the application of Mr. Noble's "New Method of conducting the Geometrical Analysis" to four Porisms of considerable interest. The *method* consists in obtaining from the Analysis, *by one statement of the hypothesis*, those conditions which Dr. Simson elicits by means of *two* or *more*, and when the word "given" is used to denote anything which the proposition affirms to be given, *and which is to be found*, it is printed in *italics*; a procedure which has since merited the approval of the late Professor Davies.

These improvements were "discovered towards the close of the year 1804," and appear to have been published almost simultaneously in the *Repository* and in the *Gentleman's Mathematical Companion* for 1806. Proposition I. generalizes "Porisms XXII., book iii." "Leslie's Geometrical Analysis," by extending it to the case of any three *parallel* lines, instead of the three perpendiculars as considered by Leslie and Playfair. Proposition III. has been noticed at some length by Mr. Davies in his "Historical Notices respecting an Ancient Problem,"(*Mathematician*, vol. iii., p. 76, and "Supplement,"p. 42); and Proposition IV. is a reconsideration of the Porism given by Scoticus in Art. IV. of this volume of the *Repository*.

Art. X Diophantine Problem. By Mr. James Cunliffe.

⁂ Mr. Cunliffe here gives very general solutions to three Diophantine Problems, the *first* of which finds "three square numbers, such that the sum of every two of them may be a square number;" the *second* determines "values for the sides of a triangle in whole numbers, such that the lengths of the three lines from the angles to the middle of the opposite sides may be expressed by rational whole numbers;" and the *third* finds "two isosceles triangles, such that their perimeters and areas may be equal."

Art. XI. Problems relating to the Twilight of Shortest Duration. By Astronomicus (Sir James Ivory).

Problem I. To find the day when the twilight is shortest in a given latitude.

Problem II. To find the latitude of that place where the twilight is a minimum on a given day.

Art. XII. Certain Fluents expressible by an Elliptical Arc. By Mr. James Cunliffe.

Art. I.—Part III. Solutions of some Problems relative to Spherical Triangles:—together with a complete Analysis of these Triangles. By M. J. Lagrange.

⁂ This valuable memoir is translated from the second volume of the *Journal de l'Ecole Polytechnique*, and contains a series of properties relating to the inscribed and circumscribed circles of any spherical triangle. The theorems of Gerard, De Gua, Legendre, &c., are also successively deduced in a much more elegant and consecutive manner than they were originally obtained by their respective authors.

Art. II. An Essay on Numerical Analysis; and on the Transformation of Fractions. From the *Journal de l'Ecole Polytechnique*. By Lagrange.

Art. III. The Inverse Method of Central Forces. From the fourth and fifth volumes of the "Memoirs of the Manchester Philosophical Society." By M. John Dawson. T. T. W.

Burnley, Lancashire, Nov. 29, 1851.

(To be Continued.)

NOTES ON THE THEORY OF ALGEBRAIC EQUATIONS. BY JAMES COCKLE, ESQ., M.A., BARRISTER-AT-LAW.

(Continued from page 387.)

Third and Concluding Series.

VIII.—BIQUADRATICS.

Let ϕ and ψ be symbols of combination only, and not of permutation. Then from the expression

$$\psi\left\{\phi(x_1, x_2), \phi(x_3, x_4)\right\}$$

which, so long as the functions are rational, has three values only, we may deduce a great variety of solutions of a biquadratic. Thus, one might be founded on the expression

$$(x_1-x_2)^2+(x_3-x_4)^2;$$

another might be obtained by an interchange of signs in the last formula, and so of other forms. We are not restricted to integral functions; but it is needless to give examples of the fractional ones.

Since

$$\sqrt{a} \text{ and } \sqrt[4]{a^2}$$

are arithmetically equal, we may convert the ordinary expression for the roots of a biquadratic into one involving biquadratic surds. But the surds are *reducible*. I restrict the symbol $\surd$ to the expression of arithmetic or quasi-arithmetic roots. The expressions

$$(a)^{\frac{1}{2}} \text{ and } (a^2)^{\frac{1}{4}}$$

are not symbolically equivalent.

I shall now give values of some of the symmetric functions of the roots of

$$x^4+Cx+D=0\ldots(1.).$$

It must be understood that all omitted intermediate functions are either equal to zero or will be found elsewhere. I use f to denote Mr. JERRARD's symbol of symmetry.

$$f0=4;\ f3=-3C;\ f4=-4D;$$
$$f2^2=-f4=4D;$$
$$f1\cdot2=3C;\ f1\cdot5=-f6;$$
$$f2\cdot5=-f7;\ f5^2=-f10.$$

But, by NEWTON's theorem on the sums of the powers of the roots (see pp. 48, 63 of JERRARD's *Mathematical Researches*) we know that

$$f6=3C^2;\ f7=7CD;$$
$$f10=-Cf7-Df6=-10C^2D.$$

In considering the problem discussed in the latter part of my paper, at p. 36 of vol. xlv. of the *Mechanics' Magazine*, we may pass at once from (1.) to

$$y^4+B'y^2+C'y+D'=0\ldots(2.)$$

through the relation

$$y=Qx+Rx^2+x^5;$$

and we shall have

$$B'=3CQR-3C^2Q+$$
$$2DR^2-7CDR+5C^2D$$
$$=(R-C)(2DR+3CQ-5C).$$

It may be as well to mention here that, universally, when

$$fabc\ldots=0, \text{ then also } f0^m abc\ldots=0.$$

Let $f0pq\ldots=f'pq\ldots$ then if, in my formulæ at p. 249 of vol. iii. of the *Mathematician* (which are deducible from those at page 13 of JERRARD's *Researches*), we make

$$n=4,\ m=3,\ \mu=2,$$

it is seen that, in the case before us, we have

$$f'pq=2fpq,$$

and hence the following values of symmetric functions:—

$$f'1\cdot2=6C;\ f'1\cdot5=-6C^2;$$
$$f'2^2=8D;\ f'2\cdot5=-14CD;$$
$$f'5^2=20C^2D.$$

The values of the latter symmetrical functions do not occur in the problem under discussion, but they may possibly be useful upon other occasions. Those which follow are of actual occurrence in the question:—

$$f1^2\cdot5=2f7;\ f1\cdot2\cdot5=2f8$$
$$=8D^2\ (\text{since } f8=-Df4);$$
$$f1\cdot5^2=2f11=-22CD^2,$$
$$\text{since } f11=-Cf8-Df7;$$
$$f2^3=2f6=6C^2;$$
$$f2^2\cdot5=2f9=-6C^3;$$
$$f2\cdot5^2=2f12=6C^4-8D^3;$$
$$f5^3=2f15=30CD^3-6C^5.$$

We are now in a condition to develope

$$C'=f\,(1Q+2C+5U)^3=Q^3f1^3$$
$$+3Q^2f1^2\cdot E+3Qf1\cdot E^2+fE^3,$$

in which expression

$$E=2C+5U,\ U=1,$$

and the development is conducted according to Mr. JERRARD's processes. For the values of certain functions, not here given, the reader is referred to pp. 32—3 of the *Researches* of that analyst. We now have

$$f1^3=2f3=-6C;$$
$$3f1^2\cdot E=3\,(-8+14)\,CD=18CD;$$
$$3f1\cdot E^2=3\,(16-22)\,CD^2=-18CD^2;$$

and, lastly,

$$fE^3=(6-18+18-6)\,C^5$$
$$+(30-24)\,CD^3;$$

so that we have

$$C'=-6C\,(Q-D)^3,$$

which, as we have already seen, is a nugatory result. It is not desirable to discuss here the solution depending upon the other factor of B'. JAMES COCKLE.

2, Pump-court, Temple.

Postscript.—We may, at pleasure,

change the *plus* on the left-hand side of the equations (1),.., (4) of my last paper but one (*vide sup.*, p. 172) into *minus.* Obstacles, similar to those alluded to at p. 172, interpose whichever form we employ, or whatever equation we select. Thus, if, from (4) transformed, we obtain

$$x^5 = x^2 + ax,$$

we cannot, by linear substitution, convert the right-hand side into a perfect square.

Let me add, that the use of the term *quintic*, to denote five-degreed functions, has the high authority of Mr. SYLVESTER (see *C. and D. Math. Jour.* for May last). In the nomenclature of equations, I shall adopt the Latin in preference to the English derivation.

Second Postscript. Temple, 19*th November.*—The reader will be pleased to make the following corrections in my last Note:—*Supra*, p. 385, col. 2, line 5 from the bottom, supply the omitted full stop, and the same at line 19, col. 2 of p. 386; p. 386, col. 2, note, line 2, for *Résolutions*, read *Résolution*; p. 387, col. 1, line 28, for *denomenator*, read *denominator*; at line 18 of Dr. Rutherford's letter, for $a\,x$, read $a\,x^2$. It would seem that for "and" at line 4 of that letter "&c." should be substituted.

J. C.

SAN TORINO EARTH.

Extract from a communication from the Consul of the United States at Trieste, published in the Report of the American Patent-office for 1850.

There is a cement now coming into general use here, which, if not already known in the United States, is worth being tried. It is called "San Torino earth," and has been found to be the best *under-water* cement ever used in these parts. It is mostly used in the building of piers, moles, docks, &c. The island, where it is found, and from whence it takes its name, is situated in the Grecian Archipelago, latitude 36° 23′, and longitude 25° 26′. The name given to it by the Turks, appears to be "Kameni," or, "the Burnt Isle," and it is generally considered to be the remains of a volcano not yet entirely exhausted. The earth is exceedingly dry and appears to consist of silicate of iron and alumina, with a large proportion of a light, porous and fibrous substance, which floats upon the water, and is supposed to be "pumice." It has been very extensively used in Syria and in Algiers, in the building of fortifications, &c., where it has been found to answer admirably, and also here in Trieste, and at Venice and Fiume, with equal success. These works have been chiefly under water in the sea, in which the cement sets very hard in a comparatively short time. Trials, however, have *been made above water*, and exposed to the action of the air, which are said also to have answered very well. For use, the following composition has been prescribed for works *under water*; viz., 7 parts San Torino earth —2 parts lime, and 7 to 9 parts stone rubbish; and for works above water, 6 parts San Torino—2 parts lime, and 6 to 7 parts stone rubbish. The rubbish stone should consist of pieces not too small, and as rough and irregular as possible, thus binding better with the cement. Where economy is greatly desired, a portion of sand may be used instead of wholly San Torino earth, in proportion of 4 parts San Torino, 2 or 3 parts sand, 3 parts lime, and 6 parts stone rubbish. This is said to be just as good as the first-mentioned composition, though requiring longer time to harden. For use, the San Torino earth, sand, and lime, must be well mixed, and made, with the necessary quantities of water, into a very consistent mortar, then heaped together, and placed under roofing for two or three days. In the mean time the foundation should be made with loose stones thrown into the sea at the spot required, and the caisson or form sunk on this. Into this caisson are to be thrown alternate layers of the mortar and stones. To every 2 or 3 feet of mortar the same quantity of stone rubbish, and so on to the water's edge.

The price of this earth at the quarry is said to be from 8 to 10 carantani, per stago beneto, equal to about from 6 to 8 cents. per cwt. If desired by Government, I will order a barrel for trial from here.

CHESTERMAN'S REGISTERED DOUBLE EXPANDING AND CONTRACTING SPANNER.

[James Chesterman, of Sheffield, Tool Manufacturer, Proprietor.]

Description.

Fig. 1 of the annexed engravings is a side view of this improved spanner in its half-expanded state; and fig. 2, a section of the same on the line *a b*. AA¹ are the two cheek pieces which terminate in the jaws BB¹. To the centre part of the cheek A, there is forged a lug or piece C, indicated by the dotted line, which takes into a corresponding recess cut out of the cheek A¹. Through the cheek A¹ and the upper portion of the lug C, there is drilled a hole, in order that a pin *c* may be inserted; which pin forms a fulcrum for the two cheeks to turn upon. DD¹ are two adjusting screws, the one, D, being tapped through the cheek A, and the point of it bearing upon the inner part of the cheek A¹,

Fig. 1.

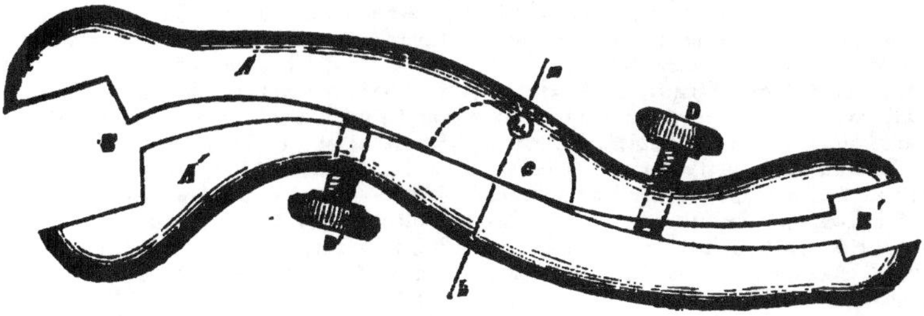

Fig. 3. Fig. 2.

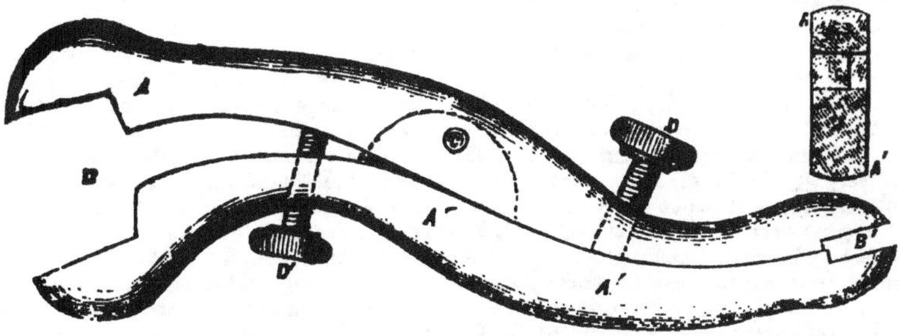

while the other screw D^1 is tapped through the cheek A^1, and acts in a similar manner upon the cheek A. In order to adapt the jaws of this spanner to the size of any nut required to be turned, it is only necessary to turn the screws DD^1, taking care to move them in contrary directions: that is, if the larger jaw B is required to be expanded, then the screw D^1 must be lightened, while the screw, D, is loosened; and *vice versâ*, if the contrary effect is desired. Fig. 3 represents this spanner with the larger jaw B expanded to its utmost limit, while the smaller one B^1, is contracted to a corresponding degree. By these arrangements each end of one spanner is enabled to be adjusted to suit nuts of different sizes.

PIN MANUFACTURE IN THE UNITED STATES.

During the war of 1812, in consequence of the suspension of importations, pins became very scarce. The prices asked for the few in the market, were many times the original cost—in some instances as high as a dollar a paper, by the pack. About this time an effort was made to introduce the manufacture in New York. Some pin-makers came from England, bringing the necessary implements, and commenced the business at the old States Prison at Greenwich (New York), employing the labour of the convicts. I think the establishment belonged to, or was managed by a man named Haynes. How much was done, I am not informed; but the low prices which prevailed very soon after the termination of the war, were fatal to the enterprise, and it was abandoned. In the year 1820, Richard Turman obtained the tools which had been used by Haynes. He made a contract for pauper labour, and undertook the manufacture in the alms-house at Bellevue. Mr. Turman carried on the business a year or two, when he died; having lost by the undertaking a considerable share of his property. Probably the trouble and perplexity of the business, together with the confinement consequent on attending to it, hastened his end. No further use was ever made of the tools. I recollect hearing Mr. Turman say at this time, that he had seen a *machine* for making pins, that it had made pins, but was too delicate, or intricate to be used with advantage. I suppose this machine was one which was invented and patented by Moses L. Morse, of Boston, during the war. I think Morse' machine had been worked to some small extent at that time; but it had passed into other hands, and was never used afterwards. His specification showed him to have been a man of good mechanical talents.

Lemuel William Wright, of Massachusetts, patented a machine for making "solid-headed pins," both in the United States and in England at an early period. I believe his specification and drawings are published in the London "Repository of Arts." He never attempted to put it to use in the United States, but in London he formed a company with a large capital, for the purpose of operating with it. The company built a large stone factory in Lambeth, and constructed some sixty machines at great expense. It is understood that the machines failed in pointing the pins, and for that reason never could be put into successful operation. To obviate this difficulty, Wright invented a machine for heading the shanks, pointed and cut in the ordinary way by hand. The company did not succeed, and broke up with the loss of a great part of the investment. D. F. Taylor, who had been ruined by this failure, afterwards came in possession of the machinery, and, by connecting himself with a capitalist, under the firm of D. F. Taylor and Co., was enabled to start a manufactory of "solid-headed pins" at Stroud, in Gloucestershire. This was in 1832 or 1833. Some pins of their make even sold as early as the year 1833; which were the first "solid-headed" pins ever sold in any market. They obtained a patent for the "solid-headed" pin by Act of Parliament. They used (principally or solely) the machine for heading only. Some account of Wright's machine is given in Mr. Babbage's work on the "Economy of Manufactures."

In 1832, a patent for a pin-machine was obtained for the United States, by John J. Howe, and in 1833 and 1834, patents for the same invention were obtained for England and France. This machine was designed to make pins similar to the English diamond pins, the heads being formed of a coil of small wire fastened upon the shank by pressure between dies. No arrangement was made to use this invention in Europe; but in December, 1835, the How Manufacturing Company was formed in New York, for the purpose of putting it in operation. This company removed to Birmingham (Derby); Connecticut, where its manufacturing operations are now carried on. In the spring of 1838, a second patent for the United States was obtained by John J. Howe, for a machine for making "solid-headed" pins in 1840; and this is the machine which is now in use by the Howe Manufacturing Company.

Samuel Slocum, of Rhode Island, obtained a patent in England for a machine to make "solid-headed" pins in 1835. His invention was not put to use in England; but he established the manufacture of pins, by means of it, in Poughkeepsie, in 1838, under the firm of Slocum, Gillson, and Co. His machine has not been patented in the United States, but has been, as it still is, run in secret. At this period, and till the Tariff of 1842, came into operation, pins (under the "Compromise Act") were free of duty; while brass wire, of which they were made, was subject to a duty of twenty to twenty-five per cent. Under this discouragement, the business made but slow progress. But under the encouragement given by the Tariff of 1842, the two companies above named went on increasing their production, and doing a profitable business till 1846. In the meantime, it having been found that pins *could* be successfully manufactured by machinery—and exaggerated ideas both as to the extent of the business and the profits to be made in it, having obtained extensive prevalence, — many persons in different parts of the country became engaged in scheming on machinery for making pins, and much capital was expended, and finally sunk in these projects. These attempts were attended with various degrees of success; in a few instances a good article was produced, but in most cases the article produced was more or less inferior in quality. The consequence was, that at this time, within but a few years after the manufacture had been commenced, and before it was fairly established (at least on its present basis), the market was overstocked with goods, importations were nearly or quite arrested, and the business was ruined by domestic competition. This depression continued about two years, from 1846 to 1848, and during this period, nearly every party engaged in the manufacture, or attempting to engage in it, excepting the two companies before named, suspended operations. Slocum, Gillson, and Co. sold out their establishment to the "American Pin Company," of Waterbury, Connecticut, and the machinery was removed from Poughkeepsie to Waterbury, where it is now used by the last-named company.

The "American Pin Company," and the "Howe Manufacturing Company," now manufacture nearly all the pins consumed in the United States. There is a party at Poughkeepsie doing a limited business, and a small amount imported. Since the depression of 1846 to 1848, the business at

mpanies named has been reasonable, having been rendered so reducing the cost of production pense of selling than by the small price which has been realized. panies manufacture the wire for ir pins. During the last year, ompanies have used principally rior copper for making their wire at consumption of copper amount at 250 tons per annum. The pre- ly production of pins by the two may be stated at about 8 tons.

nection with the improvement the manufacture of pins, by the m of self-acting machinery super- process which formerly required en different manual operations, improvements have been made in d of sheeting the pins, or sticking paper. This, as previously per- inserting a few pins at a time by a tedious process, at which some dozen papers were as many as a could do in a day. By the im- chinery now in use, one hand will 75 to 125 dozen a day, and do etter than it was usually done in y. There are three patents now r improvements in the machines this operation, viz., one granted Slocum, one to De Grass Fowler, John J. Howe. These patents intly by the "Howe Manufactur- any" and the "American Pin "

esent price of American solid- s is believed not to exceed two- he lowest price at which imported ual weight were ever afforded be- manufacture was introduced, and , they are undoubtedly better than of which they have taken the

erican improvements in both the g and pin-sticking machinery have everal years in operation in Eng- probably in other parts of Europe. *of American Commissioner of or* 1850.

AILWAY BRIDGE AND VIADUCT OVER THE WYE.

ak that occurs at Chepstow in the mmunication on the South Wales not likely to be soon filled up by letion of the great bridge now in erection over the Wye; but, though n be fixed with certainty as to the its completion, still it is rapidly g. The structure is now so far that an excellent idea of the bridge may be formed. Already numerous visitors have been to see it, and it will acquire a fame equal to that of the Britannia or Menai Bridge. The whole will be made of wrought iron, and will combine the principles of the suspension with those of the tubular bridges. Including the viaduct, the bridge is 623 feet in length; the span or suspended part being 290 feet. There are two separate roadways, each being perfectly independent of the other, and their height is 70 feet over the river Wye at high-water mark, so that vessels can pass under. The roadways of the bridge are formed of iron, put together in plates, and in form they are similar to the tubes forming the Conway and Britannia tubular bridges, but, instead of being roofed in with cellular divisions of iron, there is for each roadway, and suspended above it, and at some distance, a strong cylinder of iron. It is suspended on piers, and from the extremities of this cylinder a looped chain runs under pins placed on each side of the roadway, in order to brace and support it. Likewise strong iron braces pass from the cylinder to each side of the tube, and from the top of each of these side supports to the bottom of the other, chains are placed for additional strength. On the Chepstow side the roadways rest on six upright iron cylinders, which have been filled with concrete, and driven firmly on a foundation of rock. The roadways on this side are continued in the form of a viaduct for about 300 feet more, resting upon these upright cylinders filled with concrete, and firmly imbedded. On the east side the roadways rest upon solid rock. The masonry is in a forward state. In consequence of this break in the communication, passengers are obliged to be transferred from one train to the other across the river, which is rapidly performed by omnibuses. These vehicles are compelled to take a sweep of two miles, which distance will be lessened one half by the line. When finished to Milford, the railway will be 163 miles in length.—*Times.*

PATENT LAW CASE.

Case of Alleged Infringement by a Foreign Ship-Captain of an English Patent.

VICE-CHANCELLORS' COURTS,
December 1, 1851.

(*Before Sir* G. TURNER.)

CALDWELL V. ROLFE.

MR. ROLT (with whom was Mr. Amphlett) moved for an injunction to restrain the defendant, who was described in the bill as the master of the Dutch ship *Fyenoord*, from commanding or assisting in the sailing of the ship with the screw propeller constructed according to Lowe's patent, or in imitation thereof. The defendant had put in his answer, and although he ignored the title of the plaintiff to the patent, he had adduced no evidence impeaching that title, and the title itself was proved in this case in the same way as in the numerous other cases in which the Court had granted the injunction. The only point which appeared to be raised by the defendant's answer was, that he is now the captain of the ship; but he did not allege that he was not the captain of the ship when the bill was filed, or that it was intended that he should resume the command when the case, with regard to the injunction, should be disposed of. They submitted that the plaintiff was entitled to the order.

Mr. BACON (with whom was Mr. Miller), for the defendant, argued that the case against the present defendant, a subject of the King of Holland, having no English domicile, and having no connection with this country, except from the accident of his coming hither with a foreign ship, was perfectly novel, and clearly distinguished from all the cases in which the Court had granted an injunction against English subjects. It was not to be denied that no subject of this country could go into a foreign State and construct an article which, if made in England, would be an infringement of an English patent, and bring such article hither; but that case was perfectly different from the case now before the Court. It had never been decided that a foreigner, making in his own country, as he lawfully might, without the infringement of any law, an article which in England was the subject of a patent, might not bring that article to this country, not, indeed, for sale, but for his own especial use. Take the case of a patent axle of a carriage; a foreign workman might even, without any knowledge of the English invention, construct a carriage with a similar axle, and he might sell that carriage to a fellow-subject of his own, who might afterwards visit England and bring his carriage with him, not for sale, but for the purposes of travelling in this country. Was it to be contended that this Court would grant an injunction to restrain such a use of the carriage, which had been made and purchased without contravening any law? Was the Court prepared to restrain the sailing of a foreign vessel, having possibly a valuable cargo of perishable commodities on board, on such a ground as this? The case raised a grave question of international law, which must be tried and determined before the Court would interpose by the exercise of its jurisdiction.

The SOLICITOR-GENERAL, as *amicus curiæ*, observed, that though not in th[i]
he might state to the Court that he a[p]
to oppose a motion for an injunct[ion]
which notice had been given for th[e]
but which he was informed would
moved until the next seal, in whi[ch]
question of the right of the plaintiff[s]
injunction to restrain the sailing of
ships would be raised.

His HONOUR suggested that the r[...]
the present motion, and the furthe[r]
ment, had better stand until the nex[t]
when the question would be fully dis[...]
and both motions disposed of; an ar[...]
ment to which the parties assented,
most convenient.

SPECIFICATIONS OF ENGLISH PATEN[TS EN]ROLLED DURING THE WEEK, E[NDING] DECEMBER 2, 1851.

JOHN FIELDING EMPSON, of Bi[rming]ham. *For improvements in the ma[nufac]ture of buttons.* Patent dated M[...] 1851.

These improvements consist in ap[...] threads or yarns to strengthen and orn[...] those varieties of die and pressur[e] buttons generally known as "sewn thi[...] buttons.

The method of applying the threads [...] may be of silk, cotton, or other s[...] material) is the same as that ord[...] adopted in the manufacture of "wi[re but]tons" and "leek buttons," and the t[...] may be of different colours from the [mate]rial composing the body of the butt[on] they may be disposed so as to p[...] figures or ornamental patterns of di[...] kinds on the faces of the buttons.

Claim.—The application of thre[...] yarns of silk, cotton, or other suitab[le ma]terial to die and pressure-made butto[ns.]

JOHN HARRISON, of Blackburn. [...] *certain improvements in the manufac[ture of] textile fabrics, and in the preparat[ion of] yarns or threads for weaving.* Patent [...] May 27, 1851.

Claims.—1. The application of a [...] in connection with the work beam of [...] for the purpose of regulating the vib[...] of the lever or levers to which the cl[...] clicks of the ratchet wheel are con[...] Also, the application of similarly-const[...] apparatus for letting off the warp.

2. The arrangement of the drivin[g] or clicks, and the teeth of the ratchet [...] obliquely to each other, whether appl[...] the taking-up of the work or letting [...] warp.

3. The application of a surface [...] against a roughened periphery, for th[e pur]pose of taking up the work or letting [...] warp.

dividing one or more of the beams or of warp-sizing or dressing machines, hich the threads or yarns are wound, reral portions, so that a portion or s thereof may be removed at plea-

Iounting the beams or rollers upon e portions of framework, and also plication of suitable apparatus for any desired portion of the yarns or

Iounting the beams or rollers in bear-pable of being raised or lowered, in clear their threads from those of an t beam or roller.

ie application of a brush, which is so cted as not to act continuously on ns or threads.

IIBALD SLATE, of Woodside Iron-Worcester. *For improvements in ngines and steam boilers, and in the and valves for the induction, educ-d working of fluids.* Patent dated , 1851.

Slate's improvements in steam en-nsist in a particular arrangement of inder, slide valve, and passages, the cylinder is kept hot by the its passage to action, and the area lve and ports is nearly equal to that ntire area of the piston; and the are so constructed as to admit the imultaneously at all points in the rence of the cylinder, whereby the e effect of the steam in driving n is obtained, whatever may be the which the engine is working.

e arrangement described by Mr. hree cylinders, concentric to each e employed. The interior cylinder rorking one; the middle cylinder tes the slide valve; and the external encloses the whole, and forms a round, into which the steam passes to its admission to the working The ends of the slide valve have rings in contact with the external king cylinders, and the valve is by rods passing through stuffing-the cylinder cover. The action of is that of the ordinary slide, and e termed a concentric piston valve. king cylinder has at each end the am passages, and the exterior cylin-lso steam ports in connection with n and exhaust pipes.

ame arrangement of parts is also o the working of fluids and gases; valve, however, in the case of fluids, rmed without any lap; and when o the working of gases, the lap of may be varied, and the valve may d into two parts, each actuated by a separate eccentric to suit the degree of pressure under which the engine may be working.

In constructing his steam boilers, Mr. Slate makes a fire-box of double plate, leaving a water space of three or four inches between the plates, which he connects together at the top and sides of the fire-box with tubular stays closed at their outer ends, but open internally to the action of the flame and heated air in the fire-box, so as to expose an extended surface to the heat, and thus increase the generative powers of the boiler. He also applies other tubes, of the full width of the fire-box, but passing through the two thicknesses of plate, and closed at their outer ends with plugs, which can be removed for the purpose of cleaning the tubes. These tubes are set at a slight inclination, and have oblong passages cut through that part of them which is in the water space formed by the double plate at the sides of the fire-box, in order that a free circulation of water may take place through the tubes from one part of the boiler to another —the heated air and flame from the fire-grate playing around the exterior surface of these tubes, which form the water spaces of the boiler.

Claims.—1. The combination of three cylinders constructed and arranged as described, whereby the working cylinder is at all times kept hot by the steam in its passage to action, and the valves or ports are nearly equal in area to the entire area of the working piston, and so arranged as to admit of the entrance and exit of the steam simultaneously at every point in the circumference of the cylinder.

2. The construction of apparatus, as described, for the working of fluids, whereby the ports and valves for the induction and eduction of the fluids may be equal, or nearly equal, in area to the area of the piston.

3. The construction of steam boilers of double plates, intersected and strengthened by tubular stays.

ALFRED VINCENT NEWTON, of Chancery-lane, mechanical draughtsman. *For improvements in the carbonization of coal, and in the utilization of the products disengaged during that operation, in improving the quality of the products intended for illuminating purposes, and in regulating of the same.* (A communication.) Patent dated May 27, 1851.

The arrangements of apparatus specified under this patent are adapted to the manufacture of coke on a large scale, of a suitable quality, to be used in locomotives, &c., at the same time that gas is produced for illuminating purposes, or for employment as fuel in the furnaces of the apparatus, they include also suitable means for regulating the delivery of the gas to consumers.

The distilling or carbonising chamber of the coking apparatus is placed on an inclination above the furnaces, the heat from which is caused to circulate around it by suitable flues, and the coal is supplied in large quantities from above by a tilt wagon. In the same incline, and forming as it were a continuation of the distilling chamber, is constructed a cooling chamber, into which the carbonised mass from the distilling chamber will descend by its own gravity when the two chambers are put in communication by raising the discharge cover of the one and the feeding cover of the other, which is done by means of a crane travelling on rails laid on the incline. The exterior of the cooling chamber is provided with cold-air channels or passages to facilitate the cooling of the carbonised coal introduced into it. The gas evolved during the carbonising process is conducted through a suitable pipe rising from the top of the distilling chamber to a hydraulic main, and in order to facilitate this part of the operation, an exhauster may, if desired, be connected with the chamber. When, however, the gas is to be employed for heating purposes, the pipe leading from the distilling chamber to the hydraulic main is lowered so as to dip into the water in the main, and thus prevent the exit of the gas from the chamber in that direction. The coke produced by this process is stated to be of great density, arising from the large quantities of coal operated on in mass, by which a general consolidation thereof is produced. The heat from the furnace of one apparatus, instead of passing off to waste, may be conducted through suitable flues to aid in the process of carbonisation in another apparatus.

The arrangements by which the supply of gas is regulated are adapted for application to mains or service pipes as may be required. The apparatus consists of a throttle-valve for regulating the area of the delivery-pipe, a float for giving motion to the machinery and actuated by the gas in its flow, an adjustable counter-weight for regulating the minimum pressure under which the gas is to flow, a weight for regulating the passage of the gas, and a metal casing inclosing the whole of the working parts, and forming one piece with the pipe through which the gas passes.

Claims.—1. The arrangement of the distilling chamber or coke oven with or without a stationary cooling chamber, in combination with a furnace or furnaces and *flues or channels* for the distribution and transmission of caloric from one apparatus to another.

2. Constructing upon a curve the chamber in which the operation of distillation or carbonisation is effected, which curved chamber is placed at the degree of inclination most favourable to the feeding of the coal and discharge of the coke.

3. Arranging the cooling chamber in such manner that it may form a continuation of the distilling chamber.

4. The mechanical arrangements by which the coal is fed to the distilling chamber, together with those by which the distilling and cooling chambers are put in communication to allow the charge to pass from one chamber to the other when the coal has been converted to coke.

5. The arrangements which enable the coal to be operated on in such masses as to obtain coke of greater density than hitherto.

6. The railway and moveable crane placed upon the inclined plane for working the moveable covers of the distilling and cooling chambers.

7. The arrangements for collecting the gas evolved during the process of distilling to be applied to illuminating or heating purposes.

8. The peculiar construction and arrangement of regulator for regulating the flow and distribution of the gas to consumers; and particularly the use of an arm and progressive weight in such apparatus for regulating the pressure under which the gas is to flow.

Henry W. Adams, of Boston, U.S. *For an improved means of generating galvanic electricity, of decomposing water or various electrolytes, of collecting hydrogen, of burning it or atmospheric air, separately or in combination.* Patent dated May 29, 1851.

The first part of this invention consists in constructing galvanic batteries so as to obtain from a given weight of metal the greatest possible quantity of electricity, and to give it the highest degree of intensity.

The method adopted by the patentee in constructing his batteries to obtain these results is as follows:—He takes from thirty to fifty copper or other electro-negative metallic wires, of about one-sixtieth of an inch in diameter, lays them side by side, and insulates them in one bundle or package. He then takes a zinc cylinder about 8 inches long, 4 inches in diameter externally, and 3½ inches in diameter internally, about which he winds the electro-negative metal, in double coils, inside and outside, attaching one end of the coil metalically to the zinc, but leaving a space of a quarter of an inch between the zinc and the coils, to be occupied by the water to be decomposed. The me-is arranged constitute a single cell, a great increase of aggregate surface lectro-negative metal over the elec-ltive metal, instead of the metals rranged with equal, or nearly equal in each cell, as is the case in the ell batteries hitherto constructed.

ing these batteries for decomposing r other liquid or electrolyte, the is immersed therein, and the liquid ms a thin sheet occupying the space the coils and the zinc, by which the full decomposing effect of the ty can be obtained. When water g decomposed, the oxygen of the ombines with the zinc, and forms an of zinc, which should be held in by slightly acidulating the water phuric acid; the hydrogen liberated same time may be collected and if so desired, or applied to other s. When the water becomes super-d with oxide it should be drawn off, fresh supply added. The patentee oceeds to describe and claims an ment of apparatus for collecting the supplying fresh acidulated water or quid, and for removing the used m the battery.

her method of decomposing water aining hydrogen and oxide of zinc. in substituting heat for the acid, and for negative metal of the battery. orts are charged with zinc or ores of d the water or steam is passed them while in a heated state; or ort may be charged with carbona-atter, and a second with zinc, and er passed through each of them vely, by which means hydrogen will ated and an oxide of zinc formed in ond retort. The gas produced is ently carbonised by passing it a benzoling vessel, or it is passed a retort containing decomposed , resin, or carbonaceous matter, h the gas is rendered permanent.

er improvement consists in a method asing the heating power of hydro-other gas by passing it, previous to ion, through or in contact with or charred matter in a pulverised

er part of this invention consists in struction of a vessel for impregnat-spheric air or gas with the vapour ile hydrocarbons such as benzole, acetone, pyroxilic spirit and other as burning fluids which are suffi-volatile to carbonise air or gas at temperatures. This vessel consists rfectly air-tight box, divided hori-into compartments communicating each with the one next it, by means of pans supporting perforated shelves, on which are placed layers of sponge, cotton wicking, or other suitable porous material. The top of the box is provided with a funnel for the introduction of the benzoling or other fluid, which is poured in in sufficient quantities to saturate the layers of sponge, and fill each of the pans; the bottom of the vessel is made double, the false bottom being perforated to admit of the superfluous fluid flowing out, and of the gas ascending to the compartments, each of which it traverses in succession, becoming impregnated during its passage, and finally passing off through an aperture at the top of the vessel to a gas or air-holder, or to the burners. By this arrangement, the air is exposed to a continuous surface of the benzoling fluid; and, in consequence, is thoroughly impregnated without the necessity of adopting means for raising the temperature of the fluids; and without the employment of mechanical pressure for forcing the air, as it readily ascends through these chambers of vapour.

Another improvement consists in the adaptation of burners suitable for the combustion of benzolised air. For this purpose the patentee enlarges the aperture of the burner, whether argand, fish-tail, or otherwise, to such an extent as to enable them to deliver a sufficient quantity of benzolised air to produce good light without resorting to extraordinary pressure, and he places over the burners platinum, palladium, or other analogous metallic cones, which by becoming heated to a white heat, when the air is ignited, produce a light of intense brilliancy.

Another part of the invention consists in constructing air-holders and suppliers. In one arrangement the body of the holder is composed of flexible material, with hoops to retain it in form; the air is supplied to the holder by a pair of bellows, and exhausted by compressing the holder by a weight or other suitable means. For producing a regular supply of air to the burners, the patentee employs a common gas-meter, open to the air, which he sets in rotation by a similar arrangement of gearing, and draws in the air through a pipe in connection with the holder, and forces it out to be supplied to the burner. In order to prevent unsteadiness in the light, he also employs a small holder between the burners and the air supplied.

Claims.—1. The construction of a battery with the electro-negative metal in any one cell thereof, arranged in a bundle of two or more thicknesses, presenting a considerably increased aggregate surface opposite the electro-positive metal, instead of nearly equal surfaces of one metal against the other, as batteries have been hitherto constructed; the greatest capacity of the battery being obtained when a given weight of the electro-negative metal presents the greatest aggregate surface within the smallest space opposite the electro-positive metal.

2. The insulation of either of the metals used in a battery, for the purpose mentioned.

3. The construction of a galvanic battery so as to give to the quantity of electricity generated in a single cell the greatest possible intensity by presenting the greatest possible aggregate surface of the electro-negative metal opposite the electro-positive metal, in an insulated coil or coils, which, to give the greatest intensity should be of the smallest possible size that will conduct the quantity of the electric current without its melting the coils.

4. The use in a galvanic battery of the negative metal in the shape of a wire or wires, or sheet, or plates, in single or double coils or packages within or around, or within and around the electro-positive metal, so as to present in the smallest space opposite the electro-positive metal the greatest possible amount of conducting surface.

5. The insulation of the metallic wires, or sheets, or filaments, or packages of metal in a battery, when wound in a coil or otherwise disposed, in order to generate and conduct the current in a circuit of coils, so as to give to the quantity of electricity in a single cell the greatest possible intensity, and also to give a magnetic power to the coils by virtue of the lateral force of the galvanic current flowing on an insulated surface.

6. The making of the electro-negative metal of a battery in the form of a coil, whether the electro-positive metal be of the same or any other form.

7. The direct immersion of the battery in the water to be decomposed, and the dispensing with conducting wires leading from its opposite poles; the water being either hot or cold—thus completing the circuit, and increasing its effect on the liquid by contact with it.

8. The winding or disposing of the coils or metals so as to leave a thin sheet of water merely between the zinc or other oxydizable metal and the electro-negative metal, for the sake of causing the current to exert on the thin sheet of water its entire decomposing energy without being obstructed by too great a body of water.

9. The employment of the galvano-electric current to influence the chemical affinity of the oxygen of the water for the zinc or other oxydizable metal to combine more rapidly with the zinc or other metal, by the simultaneous exercise of both these powers, and thereby liberating the hydrogen with greater rapidity.

10. The use of this battery for producing motive power.

11. The removal of the oxydized water in batteries arranged and operated as aforesaid, to be converted into products of commercial value by any known chemical means.

12. The charging of retorts with zinc, or the ores of zinc, entirely or in combination with charcoal, coke, or other analogous carbonaceous material, when decomposing water and manufacturing oxide of zinc, as described.

13. The employment of charcoal, or other analogous carbonaceous material, to improve the heaty properties of hydrogen or other gas or mixture of gases.

14. The arrangement of a series of basins and perforated shelves, with the funnel and draw-off cocks, in combination with sponge, or other similar porous or absorbent materials, for the purpose of benzoling common air and other non-luminous gases with any of those hydrocarbons which are sufficiently volatile to enable air or non-luminous gases to burn with a clear white light when passed through chambers filled with their vapours at common temperatures, and at a distance from the benzoling reservoir.

15. The conversion of argand, fish-tail, bat's-wing, and union-jet burners, where all the parts are immoveable, into air-burners.

16. The use of any construction of platinum, palladium, or other analogous metallic cones, wicks, or conductors, to be placed in such contact with the flame produced by the burning of common air, imbued with the vapour of any volatile hydrocarbon, as to be heated to a white heat, and thereby to increase the intensity and brilliancy of the light.

17. The construction of air-holders and suppliers in manner set forth.

John Pegg, of Leicester, manufacturer. *For improvements in producing corrugated surfaces on leather.* Patent dated May 29, 1851.

These improvements consist in producing corrugated surfaces on leather by employing, in conjunction with grooved or fluted rollers or crimping surfaces, strips or bands of India-rubber.

In carrying his invention into effect, Mr. Pegg employs bands or strips of India-rubber which has been subjected to the vulcanizing process, or some other process by which it is rendered permanently elastic. These strips or bands he stretches in a frame, and then applies on one side thereof the to be corrugated, and causes the ıbber bands to adhere thereto by cement; and he then applies on the ide of the stretched India-rubber ı piece of leather or woven fabric with cement, and he passes the whole ı rollers to cause the two surfaces of her, or the surface of the leather and the woven fabric, to adhere together, g the India-rubber strands between a stretched state. He then subjects ticated leather to the action of suita- ved or fluted rollers or surfaces, for ng corrugations thereon, using such or surfaces in a heated state to facili- ir action; after which, he soaks the ted leather in warm water, at about ır., by which it is rendered soft and

ı.—The employment of strands or f India-rubber, in combination with by means of grooved or fluted rol- crimping apparatus, so as to corru- ; leather as described.

r Ashworth, of Bristol, Manager Great Western Cotton-works. *For improvements in the method of ing and removing incrustation in oilers and steam generators.* Patent Iay 29, 1851.

ı improvements consist in the em- nt of a certain compound of coal- eed-water, plumbago, or black-lead, tile or other soap, for the purpose of ing the formation on the interior of ive, marine, and other boilers of the ıe incrustation caused by the deposi- lime or other impurities contained ion in fresh-water, or of the saline , such as sulphate of magnesia and ı of sodium, with which sea-water gnated. The following proportions en found by the patentee to answer practice:—33 gallons of coal-tar, ons of linseed-water (prepared by with the aid of steam, 14lbs. of lin- water, and then straining to remove ds and other impurities), 5lbs. of go or black-lead, and 8lbs. of cas- p (for which the common black or ı may be substituted, but not with so lvantage.) These ingredients, when ly mixed together, form a composi- about the consistence of cream; of or a thirty-horse power boiler, one ı introduced twice a week (the steam been previously blown off) through -hole, or any other suitable aperture. t of the composition upon new boilers vent permanent incrustation,—the es contained in the water forming a nd thin coating on the interior sur- face of the boiler, which scales off and falls to the bottom of the boiler, from whence it can be swept out, or otherwise removed; while in the case of old incrusted boilers the deposit on the interior surface thereof soon becomes loosened and detached, and may then be easily removed by sweeping.

Claim.—The use of a mixture or composition of coal-tar, linseed-water, plumbago, or black lead, and castile or other soap, for the purpose of preventing and removing incrustation in steam-boilers, but without restriction to the exact proportions of the ingredients named, or to the precise method of using the same above described.

William Crane Wilkins. *For certain improvements in railway-buffers.* Patent dated May 29, 1851.

These improvements have relation to auxiliary buffers to be employed in conjunction with buffing apparatus of any ordinary nature, for the purpose of receiving and as it were absorbing some of the force resulting from violent collisions or concussions, and preventing as much as possible its transmission from the carriage struck or striking to the others in a train.

These buffers are constructed of double thicknesses of sole-leather, filled with sea-shore sand, or granulated silica, in a perfectly dry state, and are attached to the fore-part of railway-carriages, trucks, or wagons, between or on each side of the ordinary buffers, and are intended to come into action only when the ordinary buffers have been driven back to a greater extent than would occur under any other circumstances than those of violent concussion or collision.

Claim.—The use of sand interposed as a medium for the reception of some of the force resulting from violent concussion, with a view to prevent as much as possible the transmission of such force through the train.

Joseph Reynolds, of Vere-street, card-maker. *For improvements in the manufacture of cards usually denominated "playing cards."* Patent dated May 29, 1851.

These improvements consist in manufacturing playing cards with their edges punched and perforated to any desired ornamental pattern; the pattern must be the same on both sides and ends of the cards, and so punched as to coincide in all points when the cards are made up in packs.

Claim.—The application to playing cards of edges such as described, punched and perforated by any description of tools adapted for that purpose.

WEEKLY LIST OF NEW ENGLISH PATENTS.

William Exall, of Reading, Berks, engineer, for improvements in certain agricultural implements, and in steam engines and boilers for driving the same. December 1; six months.

George Laycock, late of Doncaster, York, but now of Albany, New York, America, dyer, for improvements in unhairing and tanning skins. December 1; six months.

William Grayson, of Henley-on-Thames, Oxford, watch and clock-maker, for an odometer or road-measurer, to be attached to carriages for showing distances over which the wheels pass. December 1; six months.

Thomas Burstall, of Lee-crescent, Edgbaston, Warwick, civil engineer, for certain improved machinery for manufacturing bricks an articles from clay alone, or mixed wit materials. December 1; six months.

John Macintosh, of Berners-street, Mi civil engineer, for improvements in steam in rigging and propelling vessels, and fac their progress through water. December

William Wood, of Oxford-street, Mi carpet manufacturer, for improvements manufacture and ornamenting of carpet and other fabrics. December 4; six mont

James Thompson, and Frederick Al Compton-street, Brunswick-square, bak certain improvements in the means of a ratus for heating ovens. December 5; six

WEEKLY LIST OF DESIGNS FOR ARTICLES OF UTILITY REGISTERED.

Date of Registration.	No. in the Register.	Proprietors' Names.	Addresses.	Subjects of Desi
Nov. 26	3030	Dent, Allcroft, and Co..	Wood-street, Cheapside	European collar faste
27	3031	Samuel Hemming	Bristol	Combined lactometer vessel.
,,	3032	William Hodgson	Bradford	Spool motion.
28	3033	Robert Adams	King William-street	Projectile.
29	3034	Moses Wright	Yorkshire	Shuttle.
Dec. 1	3035	William Marr	Cheapside	Improved girder.
2	3036	G., A., and F. Ferguson	Poplar	Compresser for gas-
,	3037	John Gillam	Woodstock	Seed-cleanser and se
3	3038	T. B. W. Gale	Homerton	Boring tool.
,,	3039	Thomas Paris	Greenwood, Barnet	Brick.
,,	3040	Thomas Paris	Greenwood, Barnet	Brick.
4	3041	John Sanders	Birmingham	Adjusting lock-furn
,,	3042	Wolf and Baker	Basinghall-street	Condensing tobacco-p
,,	3043	Richard Garrett	Saxmundham	Reciprocating knife ing machine.
,,	3044	James Slipper	Leather-lane	Bronchita tube.

WEEKLY LIST OF PROVISIONAL REGISTRATIONS.

Nov. 27	332	George Levalson	Cavendish-square	Valved catheter-plug
28	333	H. W. Atkins	Birmingham	Burglary-preventive furniture.
29	334	B. Clark, M.R.C.S.	Hampstead	Sewer-stopper.
	335	G. P. Cooper	Pall-Mall East	Elliptic waistband.
,,	336	James Leetch...	York-road	Revolving curtain ri

CONTENTS OF THIS NUMBER.

LONDON: Edited, Printed, and Published by Joseph Clinton Robertson, of No. 166, Fle in the City of London— Sold by A. and W. Galignani, Rue Vivienne, Paris; Machin Dublin; W. C. Campbell and Co., Hamburg.

Mechanics' Magazine,

MUSEUM, REGISTER, JOURNAL, AND GAZETTE.

No. 1479.] **SATURDAY, DECEMBER 13, 1851.** **[Price 3*d*., Stamped, 4*d*.**

Edited by J. C. Robertson, 166, Fleet-street.

AMERICAN STEAM AND GAS FIRE-ENGINE.

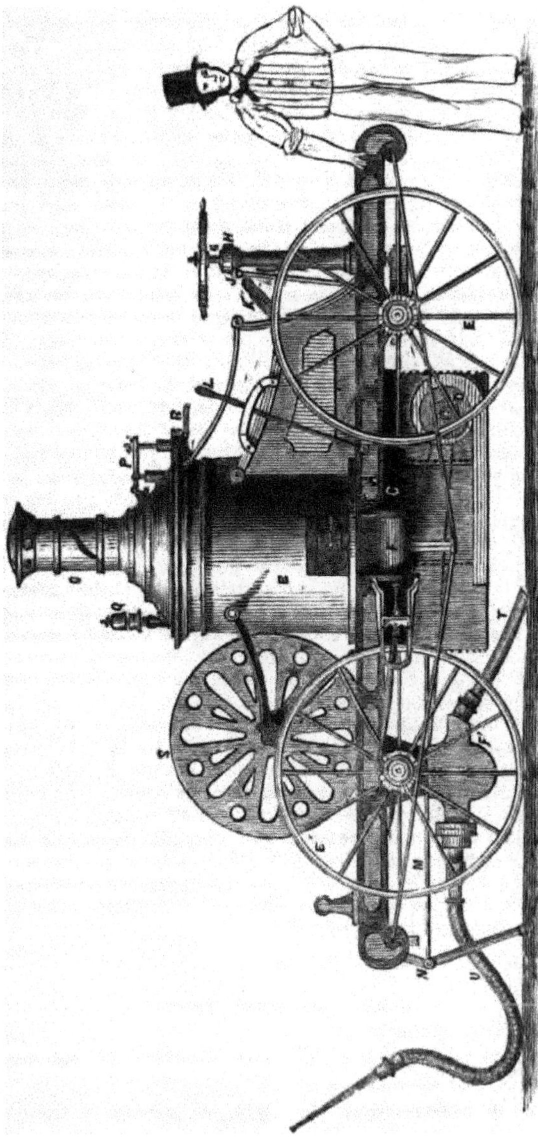

AMERICAN STEAM AND GAS FIRE-ENGINE.

[OUR first page exhibits an engraving of a new fire-engine, which has been brought out in Philadelphia, by a Mr. W. L. Lay, and is highly spoken of by the American press. We can see no difference, however, between it and that which Messrs. Braithwaite and Ericsson built, and exhibited repeatedly at work in London some twenty years ago (see *Mech. Mag.*, vol. xii., p. 488), except this—that the engine is here kept constantly provided with a charge of carbonic acid gas to put it in motion while the steam is getting up, so that not a single moment need be lost in moving it towards any spot where its services may be required. The description, which we subjoin, is that given by the *Scientific American.*]

A is the truck-frame; B is a strong steam tubular boiler; C is the water-tank for supply of boiler, and D is the blower for the fire. EE′ are the wheels; F is the steam cylinder, and F′ is the pump to throw water on the fire; this pump is a rotary one, and occupies but a small space. T is the suction hose, and U is the discharge hose, with the nozzle on the same; G and G′ is the steering gear; it consists of a wheel above, having a vertical shaft with a pinion on its lower end gearing into a segmental rack to guide the wheels, and make them turn easily. H is a circular head with indentations round it to receive the catch-rod I, which is pressed into the indentations by a spring below, to keep the pinion of the steering apparatus secured from moving as required. KL are levers; P is the balance on the valve; R is the lever for operating the valve of the steam whistle Q. A is a telescopic smoke-pipe, which can be elevated or lowered at pleasure; S is the hose carriage. MN exhibit a combination of levers to raise up the back wheels off the ground when the machine is set to working; to do this, the attendant operates the lever L, which draws back the rod M, and, acting upon the joint N, lifts forward the support below, which raises the back wheels F′, and holds up the back end of the engine, thereby allowing the wheels to act the part of fly-wheels to the crank of the piston-rod. The rotary pump has two cog-wheels, like Stewart's engine, and they are driven by cog gearing attached to the inside of the axle of the driving wheels: this gearing is not represented, but to those acquainted with mechanical devices, the mere mention of it is enough. The engine is operated by a lever to open the throttle-valve in the usual way.

When the engine is standing in the engine-house, the boiler always contains a sufficient quantity of water to get up steam, and at the same time is charged with carbonic acid gas by suitable apparatus, until it contains sufficient to work the engine for ten minutes, in which time steam can be raised to take its place when exhausted. The kindling and fuel is laid in the fire-box ready to be ignited in an instant. When an alarm of fire is given, the engineer mounts his seat, and, by opening the throttle valve, the engine will instantly propel itself in the direction of the fire, while at the same time the kindling in the fire-box is ignited, and, the blower being in motion, will raise steam in time to work the engine before the gas is used up. When the engine arrives at the fire, by merely choking the fore wheels, and pulling the lever connected with the standards, the hind wheels will be raised from the ground, and act as fly-wheels when the rotary pump is put in motion by letting on the steam. The pump will force three or four hundred gallons of water one hundred and fifty or two hundred feet high per minute, which will extinguish any ordinary fire in a very few minutes. It is intended to use two 3-horse power engines to do the work. The whole will weigh about one and a half tons.

LOWE'S PROPELLER PATENT.

A Petition has been presented to the Privy Council for the prolongation of the term of this Patent, and appointed to be heard on the 2nd of February next. In consequence of the hot war which has been for some time past waged against ship-builders and engineers for alleged infringements of this Patent, this attempt to obtain a prolongation of the term of the monopoly, has produced an extraordinary degree of excitement; and no fewer than nine different parties have given notice of their a to oppose it. We give, on account great interest attaching to the sub- s Petition entire :—

Mr. Lowe's Petition.

Queen's Most Excellent Majesty in Council.

humble Petition of James Lowe, of No. 4, Charlotte-place, Grange-road, Bermondsey, in the county of Surrey, Mechanic.

reth,—That your Petitioner in the 11, entered into the employment of lward Shorter, who, in the year 1800, d Letters Patent for an invention of chine, or engine, for working and ; the progressive motion of ships and of every description, and of every ehicle in, or on which, men or goods ried on the water, as well as on the on lakes and inland navigation, or without the assistance of sails or

your Petitioner, in the year 1813, prenticed to him, and remained in mployment till the year 1819, and that time had his attention directed subject of the propulsion of vessels.

. your Petitioner entered into an en- nt at sea, in the year 1819, in the Sea whale fishery, and having been uccessful, he returned to England in and subsequently became a partner e said Edward Shorter, and contri- o the expense of the attempts then to render the invention of the said d Shorter available for the propul- f vessels, but all your Petitioner's was sunk and lost in the said ex- nts.

; your Petitioner was then obliged to e said Edward Shorter; but, in the 34, returned into the employment of d Edward Shorter, who was still en- in his attempts at improvements in pulsion of vessels.

; the said invention of the said Ed- Shorter was repeatedly tried, but and was of no practical utility; and mpts were unsuccessful.

: your Petitioner, in the year 1834, d certain improvements in propel- essels; but his pecuniary circum- were such as wholly to prevent his g his said invention into public notice t assistance.

; in the autumn of the same year, your Petitioner having been employed . Webster Flockton, of the Spa-road, ndsey, *aforesaid*, turpentine manu- , *to execute* and adapt certain ry, *for a somewhat similar purpose*, to a model of a boat on the premises of him, the said Mr. Webster Flockton; your Petitioner applied to him to be allowed to make use of the said model for the purpose of making an experiment of his aforesaid invention, which the said Mr. Webster Flockton permitted your Petitioner to do.

That your Petitioner, encouraged by the success of the experiment, was induced to apply to the said Webster Flockton for assistance to bring the same forward, and to some of his own friends to supply the necessary funds for obtaining a Patent for his said invention; but was at that time unable to obtain more than sufficient to pay the expenses of entering a caveat against any Patent being granted for the said invention to any other person than your Petitioner, the necessary funds for which were supplied almost entirely by the said Webster Flockton.

That your Petitioner having been unable to obtain any of his friends to join him in obtaining the said Patent, your Petitioner again applied to the said Webster Flockton, who shortly afterwards succeeded in inducing his brother, Thomas Metcalfe Flockton, then of Potter's-fields, in the county of Surrey, turpentine manufacturer, since deceased, and John Addis, of Shad Thames, in the said county, corn-factor and granary keeper; William Miskin, of Broad-street, Horslydown, in the said county of Surrey, surgeon; and Arthur Davis, of Deptford, in the county of Kent, gentleman, since deceased; to join in and share the expense of making the necessary experiments, and procuring Letters Patent for the said invention.

That, in pursuance of such arrangement, the said last-named parties in the month of February, 1837, purchased a large steam-boat, called the *Wizard*, and went to considerable expense in fitting it up with an engine, and adapting such steam-boat to your Petitioner's machinery; and they also incurred, further, very large expenses in making experiments, and otherwise, for the purpose of forwarding such invention.

That in the month of February, 1838, the said last-named parties were so satisfied of the success of the invention, as to agree to supply the funds for taking out Letters Patent for the same.

That on the 24th day of March, 1838, your Majesty granted unto your Petitioner Letters Patent for his said invention under the title of "Improvements in Propelling Vessels," for the term of fourteen years.

That your Petitioner, in accordance with the proviso in the said Letters Patent in that behalf contained, and within six calendar months next, and immediately after the date thereof—namely, on the 24th day of September, 1838,—by an instrument in writing, under his hand and seal, did particularly describe and ascertain the nature of the said invention, and in what manner the same was to be, and might be, performed, and did cause the same to be enrolled in the High Court of Chancery, at Westminster, in the same day.

That the said instrument in writing so enrolled, described your Petitioner's invention in substance to consist in propelling vessels by means of one or more curved blades set, or affixed on a revolving shaft below the water-line of the vessel, and running in the direction from stem to stern of the vessel, each blade being a section or portion of a curve or screw, which, if continued, would produce a screw.

That the effect of employing blades, being sections or portions of a screw, is that each blade becomes a propelling instrument, which allows the water to pass away in all directions except at that point where the instrument is in full action; hence there is no choking or holding the water towards the centre of motion, which is the case in using one or more than one turn of a screw.

That, although the attention of scientific men on the continent, as long ago as the year 1727, and in this country during more than fifty years had been directed to improvements in the propulsion of vessels—specially, in the first instance, with the object of providing vessels of war with the means of motion by propellers under water, out of the reach of the enemy's shot, and capable of being used in all weathers—and latterly, with the view of dispensing with the cumbrous paddle-boxes and wheels, and other apparatus connected therewith, so as to obtain in such vessels adequate space for a complete broadside fore and aft, and thereby affording, both to the royal and mercantile navies, a means of propulsion safe from injury from without, and free from the impediments to sailing from the swell and surge created by the paddle-wheels, and the effects of the wind and sea upon such paddle-wheels and boxes, so that the power of steam might be conveniently adapted to sailing vessels, and applied as an auxiliary power with great economy and advantage, it was not until several years after the invention of your Petitioner that the success of such means of propulsion was established.

That your Petitioner's invention has been applied to two ships, named the *Adventurer and the Reliance*, which were sent out by *the Lords Commissioners* of the Admiralty *to the arctic regions*, in search of Sir John Franklin's expedition, and the propellers were found to answer admirably under circumstances which would have been fatal to steam-ships with paddle-wheels and boxes.

That numerous attempts had been made, and several patents obtained, prior to the invention of your Petitioner, all of which either wholly failed or were only partially successful, until after the invention of your Petitioner.

That, under these circumstances, the attention of scientific men, and of the engineers of this country, with some few exceptions hereinafter referred to, was, about the time of your Petitioner making his invention and obtaining the said Letters Patent, almost wholly directed to paddle-wheels and the improvements therein, as the means of propulsion, specially with the view of obviating the loss of power from backwater, and of preventing the vibratory action and reducing the surge produced by the paddle-wheel.

That, on or about the 22nd day of March, 1832, Letters Patent were granted to Bennet Woodcroft, for certain "improvements in the construction and adaptation of a revolving spiral paddle for propelling boats and other vessels on water" for fourteen years, the term of which Letters Patent was prolonged by your Majesty for six years on the recommendation of the Judicial Committee of your Majesty's most honourable Privy Council.

That, on or about the 31st day of May, 1836, Letters Patent were granted to Francis Pettit Smith "for an improved propeller for steam and other vessels" for the term of fourteen years, the term of which Letters Patent was also prolonged by your Majesty for five years on the recommendation of the Judicial Committee of your Majesty's most honourable Privy Council.

That, on or about the 31st day of July, 1836, Letters Patent were granted to John Ericsson "for an improved propeller applicable to steam navigation" for the term of fourteen years, the term of which Letters Patent was also prolonged by your Majesty for five years on the recommendation of the Judicial Committee of your Majesty's most honourable Privy Council.

That, on or about the 28th day of November, 1840, Letters Patent were granted to George Blaxland for "an improved method of propelling ships and vessels at sea, and in navigable rivers" for the term of fourteen years.

That, notwithstanding the greatest exertions made by your Petitioner, and the proprietors of the several Patents for inventions having an object similar to the invention of your Petitioner, and from other causes over which your Petitioner and the persons asso-

ith your Petitioner had no controul, ircumstances hereinafter mentioned, invention of your Petitioner made le progress, and, in fact, no remu- could be obtained from it.

in the year 1843, your Petitioner ed that the said Patent so granted to titioner as aforesaid had been in- by Mr. John Penn, of Greenwich, ; and application was made to him ompensation to your Petitioner for ringement, and to take a license for of the said invention; but the said n declined to accede to such appli- nd thereupon an action was brought him by your Petitioner in your Ma- Court of Queen's Bench at West- on the 17th day of November, recover damages for the said in- nt, and to establish the right of titioner to the said Patent, and such as tried before the Lord Chief Jus- man, at the Guildhall of the City of on the 16th day of December, hen a verdict was found for your r on all the issues raised therein.

in the following Hilary Term the Defendant in the said action for and obtained a rule *nisi* for a , and the said rule *nisi* was standing new trial paper for argument until day of May, 1846, when the same ed, and judgment was deferred until day of June, 1846, when the Court e rule absolute for a new trial.

notice of trial in the said action was ds given on the 25th day of Novem- 6, and the said action was in the r trial; but, owing to the state of in the said Court, was not tried 23rd and 24th days of February, hen a verdict was again found for titioner on all the issues.

rule *nisi* for a new trial was again or and obtained on the part of the nt in the said action in the following 'erm, and the argument thereon did place until the 21st day of Novem- 48, when judgment was deferred 22nd day of January, 1849, when was made absolute for a new trial ent of costs; but the said rule has bsequently abandoned, and the ver- l final judgment entered for your r.

on the 1st day of May, 1849, a *ias* was issued out of the Petty Bag- f the Court of Chancery, in the your Majesty, on the prosecution as Henry Maudslay, of the firm of *y and Field*, the eminent engineers, *the said Letters* Patent; and the *facias* did not come on for trial until the 15th day of February, 1850, when it was tried in the Court of Exchequer at Westminster, before Sir Frederick Pollock, Knight, Lord Chief Baron of that Court.

That, at the trial of the said *scire facias*, most of the numerous abortive attempts between the year 1727 and the date of the said Patent of your Petitioner were brought forward in order to impeach the novelty of the invention of your Petitioner and the validity of his said Letters Patent.

That the said Lord Chief Baron having, in the course of the case for the prosecution, intimated his opinion that, in order to impeach the validity of your Petitioner's said Patent, it must be shown that the precise combination and arrangements claimed by your Petitioner had been published to the world or in public use, the counsel for the prosecution consented to a verdict being taken against the Crown on all the issues joined, thereby establishing the validity of the Patent of your Petitioner, each party paying his own costs and agreeing to certain arrangements in reference to the pending and threatened litigation for the infringement of your Petitioner's Patent.

That, as part of such arrangements, it was agreed that the verdict obtained by your Petitioner in the said action of Lowe *versus* Penn should not be disturbed, but that each party should pay his own costs; and final judgment has been accordingly signed for the Plaintiff in that action.

That your Petitioner's said Patent-right, previous to the arrangement just referred to, met with every possible opposition at the instance or on the part of the engineers and the said Mr. John Penn, against whom the said action was brought, and the said Thomas Henry Maudslay, as representing the general body of engineers of the metropolis who had commenced or were about to commence the construction and application of propellers according to the invention of your Petitioner, and of steam engines specially adapted for such propellers.

That your Petitioner, and others associated with him in endeavouring to introduce the said invention into general use, have always been ready and willing to grant licenses for the use of the said invention on payment of a reasonable sum for the same.

That, in consequence of the opposition offered to and the adoption of the said invention without license, several other actions were necessarily commenced for such infringements, and your Petitioner and the others associated with him incurred very considerable expense therein, all which actions, however, have been settled, or are in the course of settlement, in pursuance of the arrangement hereinbefore and hereinafter referred to.

That numerous persons, associated together for the purpose of introducing the improved system of propulsion herein referred to, had become interested in the several Letters Patent granted to Bennet Woodcroft, Francis Pettit Smith, John Ericsson, your Petitioner, and George Blaxland, as aforesaid, and at the time of the arrangement above referred to questions had arisen as to an interference and conflict of right under the said five Letters Patent, and the Patents for Scotland and Ireland for the same inventions; and it was deemed advisable to unite the interests of all parties interested in your Petitioner's said Patent, including your Petitioner, with the interests of the parties interested in the said other Letters Patent, for other reasons, and especially as the Lords Commissioners of the Admiralty, and certain engineers who were opposing your Petitioner's Patent, and the engineers generally, would not treat with the parties interested in your Petitioner's Patent alone, in consequence of the said parties so interested in the above-mentioned other Patents having claimed certain portions of the merit of your Petitioner's said invention, but more especially as the Lords Commissioners of the Admiralty refused to entertain the claims of those parties interested in your Petitioner's said Patent while such claims on the part of other inventors were pending.

That, in consequence thereof, an arrangement was proposed to be entered into, and which was afterwards accordingly entered into between your Petitioner and all the other parties interested in your Petitioner's said Patent, and the other parties interested in the said four other Letters Patent; that the apportionment of the money to be derived from the sale of, or licenses under all or any of the said five Letters Patent, should be referred to William Carpmael, Esq., who should ascertain and determine what each party should be entitled to as a class, and apportion the same accordingly.

That, since the termination of the proceedings in *scire facias* in February of the last year, and the arrangements consequent thereon, several sums of money have been paid into the common fund, agreed to be constituted for apportionment as aforesaid; but the amount of the sum so received hitherto, is insufficient to pay the expenses and other outgoings, which, by the arrangement above referred to, was made a first charge on such fund.

That the amount expended on the first experiments, the Letters Patent, the legal *expenses, and models*, and other matters in respect of your Petitioner's said invention and Patent alone, in the whole exceeds the sum of 10,000*l.*, and the whole amount of the proceeds hitherto received in respect of the same does not exceed 1,000*l.*

That the parties interested in the said four other Patents have expended a very large sum of money upon the said inventions, and the attempts necessary to introduce the same into use, for which they are at present wholly without remuneration.

That your Petitioner respectfully submits that by the conjoint efforts of the said Bennet Woodcroft, Francis Pettit Smith, John Ericsson, George Blaxland, and your Petitioner, and the other persons associated with them, a new system of propulsion has been fully established, whereby they have conferred a national benefit, the value and importance of which, both as a primary and as an auxiliary power in the saving of time, in the certainty with which long voyages are made, and in the consequent saving of capital, it is difficult to estimate.

That your Petitioner and the said other persons who have been instrumental in bringing forward and perfecting the said invention of your Petitioner, are quite unremunerated for their exertions.

That the absence of remuneration, and the great loss which has been hitherto sustained, have resulted from causes over which neither your Petitioner, nor either of the said other parties, had any control.

That the said Letters Patent so granted to your Petitioner, have, during nearly the whole term thereby granted, yielded no return. And your Petitioner submits that, having regard to the expense and labour incurred in introducing the invention, an exclusive right in using and vending the invention of your Petitioner for the further period of seven years in addition to the term in the said Letters Patent mentioned, will not suffice for the reimbursement and adequate remuneration of your Petitioner.

That your Petitioner has caused advertisements to be inserted, in pursuance of the statute in that behalf provided.

Your Petitioner therefore humbly prays that your Majesty will be graciously pleased to refer the matter of his Petition to the Judicial Committee of your Majesty's most Honourable Privy Council, to consider and report to your Majesty thereon; and that your Petitioner may be heard before such Committee by his Counsel, Agents, and witnesses, and that your Majesty will be graciously pleased to grant new Letters Patent for a term of fourteen years, from and after the expiration of the term of fourteen years granted by the said Letters Patent so granted to your Petitioner as aforesaid, or for such term as to your Majesty shall seem cording to the statutes in such case and provided.

And your Petitioner will ever most humbly pray,

James Lowe.

'AIN ADDISON'S RAILWAY SIGNAL.

las been long since universally ad- that the invention of a signal by the passenger travelling in a railway e should be enabled at once to com- ate with the guard or with the engine- was a *desideratum*. Many instances occurred in which the establishment a communication would have pre- the occurrence of much mischief. It frequently happens that the friction iron axletrees of the wheels of a rail- rriage produces ignition of the wood- forming its flooring, and whilst the proceeding it is impossible, save by option of the signal system, that the an be made aware of the occurrence of until the carriage is actually in flames, hich time probably the passengers be burned. "It is not until a bishop ed that any measures will be adopted vent the recurrence of these acci- ' said the late witty Canon of St. ; and shortly afterwards the Bishop of was nearly burned to death from the n of the floor of the carriage in which travelling on a railway, and his life ved only from the fortunate circum- of the train stopping at a station the fire had spred. An accident of a nature placed in jeopardy the lives Paxton and his friends during the of their recent journey to Derby, to the festival which the inhabitants of own had given in honour of their ien, the constructors of the Crystal It is not long since that during a ourney from Birmingham, a gentle- avelling in one of the carriages was d, and nearly murdered, by his fellow ger, who was suddenly seized in a fit ng madness, and who having nearly ed his companion, leaped through dow of the carriage whilst the train oceeding at full speed, and, marvel- relate, escaped unhurt. Only last wo young men leaped from a train it was in motion, one of whom was and the other severely injured; but ellow-passengers had no means of nicating the occurrence to the guard, he arrival of the train at the next , whilst had such communication , the train might have been stopped, and the wounded man conveyed at once to a place of safety, instead of being left, as he was, lying senseless on the road, and in consequent danger of being crushed by the next train traversing the line of rails. Scarcely a week passes that the reports from our Police-courts do not inform us of the punishment by fine,—a very inadequate punishment we admit, of some cowardly mis- creant for assaulting and insulting, females, his fellow-passengers, in a railway carriage, and so common has this offence become, as to amount to a virtual prohibition of the tra- velling of females alone in a railway carriage after evening has set in. All these facts tend to prove how very desirable it is that there should be established a means of instan- taneous communication between the pas- sengers in a railway carriage and the guard attending the train. Several different plans have been proposed to effect this desirable object, but the greater number of them are open to the great, and, in the opinion of almost all persons connected with the prac- tical working of our railway system, insu- perable objection, that whilst they alarm the guard they at the same time alarm the whole of the passengers, and by the confusion and commotion thus produced actually endanger the safety of the whole. This objection ap- plies with greater force to those signals which may be called noisy signals than to any others; and we believe it is to the fact that until very recently it has been deemed impossible to invent a signal which shall alarm the guard without at the same time alarming the passengers, that the objection of the Directors of the great majority of the Railway Companies to permit experiments to be tried on their lines is to be attributed. That objection, however, has been com- pletely obviated by the invention by Cap- tain Addison of a signal which alarms the guard—and the guard alone; which is of me- chanism so simple as to be capable of being always preserved in an efficient state, which can only be made by a person travelling in the carriage, which cannot be made by him without the full cognizance of his fellow passengers, and which indicates clearly and unmistakeably the carriage from which it proceeds. Thanks to the liberality and kind- ness of the Directors of the North Kent line of railway, Captain Addison was afforded an opportunity, on Monday last, of testing the merits of his simple yet admirable in- vention: and nothing could be more suc- cessful than the results of those experiments. The day signal consists of the springing out from the side of the carriage of a large iron flag, somewhat similar to that carried by some of our omnibuses, which is thrown out on an iron arm to such a distance from the carriage as to be distinctly visible to the guard, whose business it is to keep his eye constantly along the whole line of carriages. During the course of the journey, this signal was given at least a dozen times, and was always acknowledged within two seconds of the time of its exhibition. The experiment had been performed several times before any of the passengers by the other carriages were at all aware of what was going on, and had it not been that they heard the conversation on the subject which took place at the different stations where the train stopped, they would have completed their journey without being aware that any experiments whatever had been tried. The experiment with the night signal was equally successful. In the night signal, at the same time that the iron flag flies out, a hammer falls upon a blue light, fitted up with a percussion apparatus, which at once ignites the light, and thus points out the carriage. The success of these experiments was complete, and we congratulate Captain Addison on having invented a safe and simple signal, which meets all the objections that have been made to the signal system, whilst it realises all its advantages.—*Evening Sun.*

SIR SAMUEL BENTHAM'S EXPERIMENTAL VESSELS.

Entertaing the opinion that we are ignorant of some of the most important points of consideration in naval construction,* it will not be supposed that I wish to convey the idea that Sir Samuel Bentham thought the experimental vessels he had constructed were models of perfection; but as the form of those vessels was referred to in his communications to the Admiralty as having been successfully adopted, as the draughts of them, as given by Charnock, are in many respects inaccurate, and as their form still differs materially from that of other vessels, short descriptions of their form, taken from the original draughts of them, are now furnished.

By these draughts, it appears that the *Dart* and the *Arrow* were 128 feet 8 inches long, whilst their breadth in midships, moulded at the deck, was only 33 feet; though the sides of these vessels tumbled out, instead of retreating inwards, thus differing essentially in the proportions of length to breadth from vessels as theretofore constructed, but which afforded examples now very generally followed in vessels of every description. Indeed, the advantage in point of speed gained by great length in proportion to breadth, has been so marked that, at the present day, the proportions exemplified in those sloops are frequently much exceeded.

Another remarkable peculiarity of the experimental vessels was their *sharpness* at and towards the head; this innovation has also been frequently adopted. The sharpness, however, extended no farther than to within about 12 feet of the middle in length of the vessel; from thence it took a form barely curvilinear, and continued nearly straight to the middle, from whence the form slightly receded for another 12 feet aft, at which point it much resembled that of the fore part, though not quite so sharp.

By giving great breadth for so considerable a length in the middle of the vessel, internal space was afforded for stowage, for accommodations for the crew, much deck-room for rigging and manœuvring the vessel, and for working the guns. How far flatness of form for a great length fore and aft of the middle traverse-line may contribute to speed, or may retard it, remains yet to be decided. The question was to have been an early object of the experiments commenced in 1830; besides what is said in Sir Samuel's papers, in Nos. 1476 and 1477, experiments for its determination are indicated by the sketches engraved in No. 1295 of the Magazine. These sketches exhibit different lengths of straight-sided models, intended to be interposed between other models representing different forms of heads and sterns of vessels. His experimental vessels, which were nearly straight for so considerable a part of their length, were swift, and remarkably good sea boats, evincing the greatest superiority in bad and blowing weather: his serpentine or vermicular vessels, constructed in Russia, were remarkable for speed, though the middle links were all of them straight, those at the head and stern being the only sharp ones; it must, however, be noticed that the serpentine vessels were used only in rivers, he not having remained long enough in the south of Russia to try the effect of such a form in agitated seas.

* *Mech. Mag.*, March 30th, 1850.

Dart, the *Arrow*, and the *Netley* ke in their transverse sections, ottoms in midships nearly por- circles; they were originally d with sliding keels, and when re removed a few inches of false re given them; the *Millbrook*, not designed for sliding keels, ch of the same form; the top all of these vessels tumbled s; so did those of the *Red-* nd *Ealing*, but these schooners of being nearly flat bottomed, y sharp towards their keels.

erence so very great in the form wer part of those vessels might n expected to have marked it- heir respective degrees of speed, other properties; yet no diffe- their qualities as fast-sailers sea-boats was observed during the many years they were constantly employed in severe service. Their power of resisting strains from the unintermitted firing of ordnance of great calibre has been already noticed in No. 1325 (p. 635), of the *Mechanics' Magazine;* so have many examples of their other good qualities been specified in the first part of the *United Service Journal* for 1830, p. 337. Many other still existing documents might be added to the extracts given by that journal in proof of the superiority of those vessels; but it seems needless to do so, since those may suffice, and its pages be so easily referred to; in this instance, it may be said, much profit might be derived from them by persons seeking examples on which to ground improvements in naval architecture.

A diagram, showing a midship section

Midship Section of "Redbridge" and "Ealing."

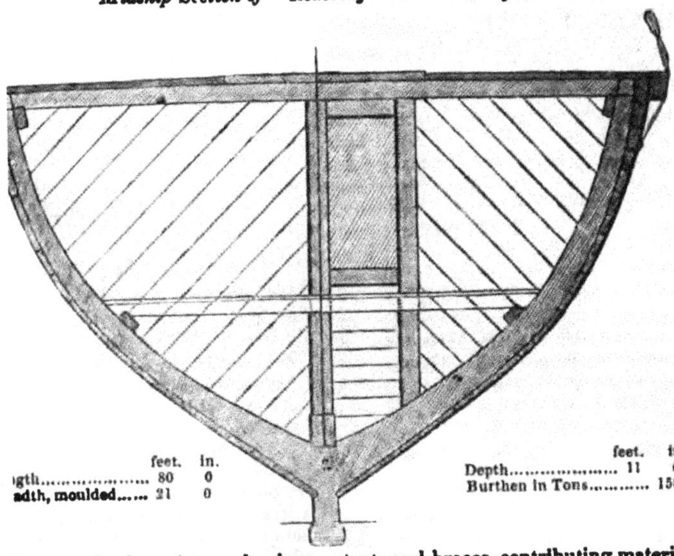

	feet.	in.
ngth	80	0
adth, moulded	21	0
Depth	11	4
Burthen in Tons	158	

Dart and the *Arrow* having been given in the *Mechanics'* s to show their nearly semi- form of hull, a midship of the *Redbridge* and the is now furnished to exhibit pness of the bottom of these s, differing therefore widely sloops in the form of their re- hulls. The present diagram rds a notion of the manner in e fixed bulkheads in all the ves- formed, whereby they become struts and braces, contributing materially to the strength, particularly against racing.

It must be observed, however, that an exhibition of the advantages obtainable by the introduction of the principles of mechanics in naval architecture was the chief object of planning these experimental vessels. The example they afforded has, doubtless, led to the great improvements in point of strength that have since taken place very generally in the construction of navigable vessels;

and as to the economy exemplified in these vessels, it cannot be more forcibly evinced than it was by the offer of a shipbuilder habitually contracting with the Navy Board; he was willing to build a frigate on Sir Samuel's plan for 8*l.* 10*s.* a ton; two-thirds of the then contract price;* and a vessel for carrying water in bulk planned by Sir Samuel, but which was strong enough to admit of being heavily armed, was actually built by contract at Plymouth for 6*l.* 10*s.* per ton.

M. S. B.

ON THE CONSTRUCTION OF LIFE-BOATS.

Sir,—I have read with much gratification the Report upon life-boats, occasioned by the patriotic and humane steps taken by the Duke of Northumberland to promote the means of saving human life in cases of shipwreck, and the encouragement given by his Graces's munificent premium offered for the best model. The Report appears well and ably drawn up in adverting to the many points to be considered in a life-boat, and directing attention to these considerations. I have read, also, a letter in the *Hampshire Advertiser*, of the 18th October, signed "A Voice from the Goodwin," with much interest from the well-meant and many sound remarks contained in it. Some of the observations appear a little at variance with what have been considered requisites,—such as buoyant sides: and the construction I am about to call attention to—namely, the flat floor,—is particularly objected to. These objections I consider to arise from not having studied the qualities of the build proposed, and the means that may be provided to obviate the grand objection to a flat floor—namely, that of leeway.

I would observe, that the bend construction appears to be the chief or only style of build thought of for life-boats. No one will gainsay that such construction is adapted for boats generally for lightness, strength, and safety, and ordinary purposes. But for life-boats we have to consider the form best qualified for that particular service; and I beg leave to submit the advantages of a boat flat-floored similar to a Thames punt.

Such a construction possesses the greatest power of flotation from its extent of surface; and its bearing is the greatest possible, all the breadth surface being brought to rest on the water; and from the same cause, also, there is the least liability to upset.

It may be observed, that there is scarcely an instance of a punt being upset on the river. This is a very different thing, however, from the waves of the sea and the rough weather of a storm; but of these more presently. I will now first point out the advantages of the flat-floor construction for the purposes of a life-boat.

Fig. 3.

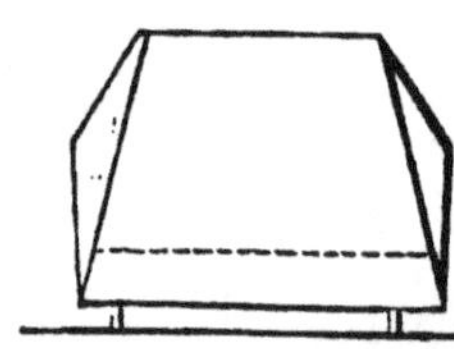

The flat floor offers, besides its stability, the best form for internal flooring of cork,—which is the most desirable material to form the bottom, not being endangered by staving, and acting as floating ballast against air-vessels above, at the sides, ends, and under the thwarts.

The cork floor being flat, will be placed with the greatest advantage both for floating and ballast, from its exposing a large extent of surface, and its weight being placed at the lowest position possible. A cork* floor of 6 inches depth would not immerse with the crew more than or near that depth, and therefore either with scuppers, or delivering tubes, would enable the vessel to free herself easily of water when shipped. The quantity of cork should be regulated by what would float the weight required of the boat and crew.

* "United Service Journal," 1830, p. 336. The whole of this communication of Sir Samuel's, "On the diminution of expenditure without impairing the efficiency of the Naval and Military Establishments" seems well worthy of perusal.

The articles that have appeared in the *Mechanics' Magazine* relative to Sir Samuel Bentham's ideas *as to naval construction* will be found in the following *Numbers:—No.* 1294, 1295, 1305, 1306, 1323, *1325, 1327, 1330,* 1331, 1337, 1363, 1365, 1371, 1390, *1392, 1400, 1401,* 1404, 1405, 1407, 1409, 1419, 1476.

* Weight of cork taken at 12 lbs. to the cubic foot would give near 100 cubic feet, and 72 cubic feet would be sufficient for the support of a crew of eight men.

It is impossible to avoid considerable weight in the construction of a life-boat, and cork appears to be the lightest and safest substance for the purpose. The employment of water as ballast appears to me a great mistake, as it is a considerable weight to move, while cork is very light; and it would always be more easy to move a boat filled with cork than with water. Water tanks, moreover, must weaken the boat.

Oars or paddles alone, and no sails,

Fig. 1. Fig. 2.

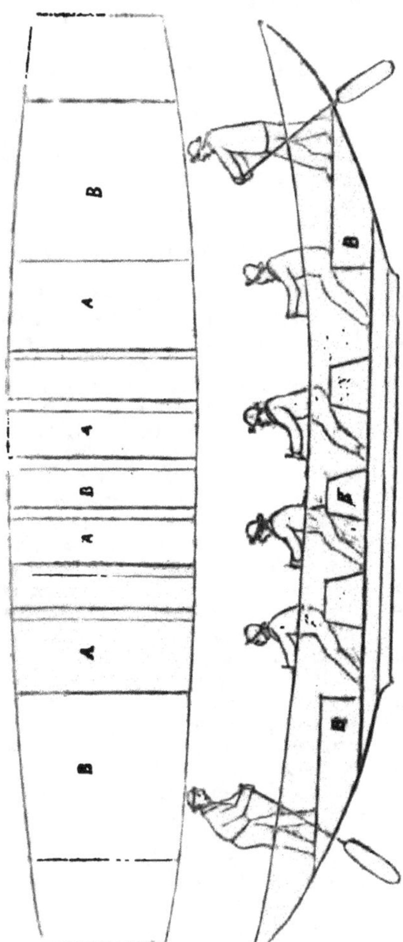

Fig. 4. Fig. 5.

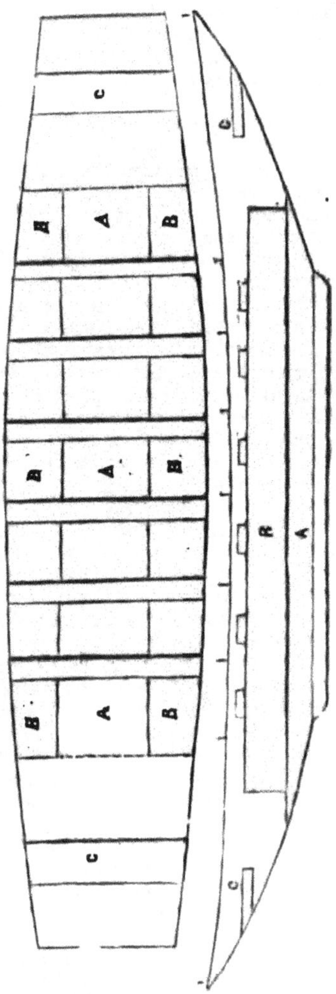

should ever be used on the occasions where a life-boat is required; with the exception, perhaps, of where a long distance may have to be reached,—as in *the case of the Goodwin Sands; and where sails are used, a lug or sprit-sail will be found most suitable.*

A life-boat, with the properties essential for such, is in a great measure unfitted for possessing good sailing qualities; to ensure which, heavy ballast, placed at the lowest position possible for small craft, is an essential.

A life-boat, therefore, must, it is obvious, be mainly, if not entirely, dependent for propulsion on her oars or paddles; and the flat-floor construction will, by drawing the least possible water, give the greatest facility in passage through the water, as offering the least resistance for weight.

The stability of the flat floor is beyond denial, and is demonstrable both practically and theoretically. We have a good practical example of this in the case of the Thames punt, before alluded to, which is rarely, if ever, known to be upset, even in very rough weather. The possibility of such a casualty cannot be denied; but all practical men that I have put the question to, declare to the firmness of the punt, and the improbability of accident from upsetting. We cannot insure certainty in any case, as accidents unforeseen may occur; but our business is not the consideration of providing against impossibilities, but of ascertaining what construction of life-boat is the most firm and safest, and that in which the fewest casualties are known to occur,—and this appears to be the case with the flat-floor construction of the punt; we may instance, also, the flat floor of the sailing barge and sailing lighter of the dockyards. Whoever heard of the capsizing of one of these vessels? And here we may observe that the trial of the flat floor is witnessed in the roughest weather, and that it has to encounter waves of the heaviest character of our coast; which is evidence that the flat floor, while firmest and safest on the smooth water of the river, is equally so in the waves of the sea. We do not say, we repeat, that the upsetting of the flat-floor build is impossible. We are not considering impossibilities, but what form of vessel is known in practice to possess the greatest amount of stability, and to be the frees from the danger of upsetting.

This firmness in a flat floor is attended with other advantages well worthy of consideration. It gives a greater bearing on the side while hauling-in persons immersed in the water, and while stepping into the boat, than any other form.

As regards the comparative advantages and preference of the flat-floor build over all others, it may be observed, that the weight of such a construction for the particular service of a life-boat, is no objection. It may be heavy from the strength required in the timbers and planking of the floor; but even here the weight of the floor is in some measure an advantage in point of acting as ballast; and weight in a life-boat is not of itself an objection, as, for the purposes required, it is unavoidable; and if that weight is so placed that, from its position, it affords least resistance to progressive motion, it becomes the most advantageous in point of utility that can be attained. The flat floor resting on the water, offers scarce any comparative resistance to progression, and may be said to slide over, more than divide the water; and, with efficient propelling force, gives less resistance for weight than any other form.

Boats or vessels built with bends are lighter in construction, and stronger for the quantity of material used; and being, from their lightness, more easily moved, they consequently row better than flat-floored punts. But, again, I may observe, we have not to consider lightness and fast motion from that cause alone; we have stability, bearing, and fastness for weight, to consider; and, for the reasons assigned, I submit that the flat floor in these points exceeds all other constructions for the purposes of a life-boat. And, under these considerations, I do not hesitate to say, that a boat built on the bend construction, with the weight of the additions required to render her safe, with the qualifications as a life-boat, will not row lighter than the flat-floor construction, but, on the contrary, will pull heavier, from the circumstance that she will be immersed deeper, by which the greatest opposition of the water is incurred; while the flat floor, by resting with least depth on the surface of the water, will, weight for weight, pull far more easily than a life-boat built on the bend construction.

There is yet another reason to be assigned in favour of the flat floor as compared with the rounded bearing or sharp-built row-boat, which is, that while the rounder bottom rolls with the motion of the wave till the side bearing begins to act, the flat floor from the first, by its peculiar bearing, begins to act in restoring the equilibrium* of the disturbed

* I would not have this expression taken in the sense of quieting the waves, but that a flat surface keeps the liquid down, and smoother than a

of the wave, and, therefore, from more firmly and smoothly in a ea than a rounded bottom or sharp s not so likely to be turned over. may, nevertheless, by an extra-y effect of a breaker, or by being in an unpropitious moment, be over by a wave, or turned end on the Shields' boat unhappily was. dimensions are the next matter sideration; and I would suggest a of not less than 30 feet by 6 feet; 5½ feet beam; depth from 2½ 3 feet midships, with a sheer of feet, according to length; and to e-quarter of her length at each form a run underneath for head ance into the water. The sides I should have a gentle bow or om end to end, to avoid a straight d better to meet concussion either ve or when falling alongside a Both ends flat—that is, not to a point, but tapering off fine run; and to row single-banked paddles of fir, short in length, led from a pin and gummett, and eered by an oar, and no rudder. derneath part of the flat of the propose having two parallel stout of oak lengthwise, to act as sup-r running her down a beach, and ng ground, and to save the bot-d act in keeping her to the wind loat. The appliance of the oars lles on the lee-side in sailing e another means of keeping the the wind.

ir-boxes, I suggest two modes— have them built-in as thwarts he vessel above the space for the or, with water-tight bulkheads, ing the sheer to form a bulwark ty; which plan will have the ad- that the sea shipped in one com-t will not flow to the others, and ble her the more readily to rise herself of water; or, according her mode, two longitudinal cases the whole length, joined by and ng the thwarts—the cases being in parts by close-fitting inside ces, giving strength and render-rtight divisions of the air-cases. The advantage of this mode would be to give the vessel greater power of floatation on her sides, and render the cork more efficient as ballast. Both kinds of these air-boxes would give an extra power of floatation equal to about 20 cwt. or 30 cwt., or between 1 and 2 tons. These modes of construction are shown in the diagrams annexed. In figs. 1, 2, and 3, A represents the cork floor, and B the air-boxes as thwarts and bulkheads; and in figs. 4 and 5, A is the cork floor, B air-boxes supporting thwarts, and C seat or foot-board for steersman.

The observation that I will here finally make is—risk and danger are unavoidable, if we would render assistance to those shipwrecked in a storm; but humanity dictates to offer every aid in our power to our fellow-creatures in distress; and God forbid that the feeling should ever be subdued in the heart of an Englishman, or of any human being. All we have in this case to consider, is the best and safest means of construction of a vessel that will accomplish the laudable purpose to which our attention has been attracted, and to which so much public attention has been directed by the example of the nobleman at whose instance the Life-boat Committee was appointed.

G. G. V.

London, Dec. 1, 1851.

urface; and so far as any given space is y the flat floor, the effect of it is to keep down and smooth, which, in a degree, its effect even in the small space occu-ost.

ROYAL MAIL STEAM-SHIP "AMAZON."

On Wednesday last, a trial took place down the river of the *Amazon*—a splendid steam-ship built by Messrs. Green, of Blackwall, for the Royal Mail Steam-packet Company, and fitted with engines by Messrs. Seaward and Capel.

The *Amazon* is the largest steam-ship which has been yet built on the Thames, being of no less than 2,500 tons burthen. Her length over all, 310 feet; beam, 42 feet; depth of hold, 34 feet; draught forward, 19 feet—aft, 19 feet 9 inches. The saloon is most capacious, and comfortably and tastefully fitted up; and abaft of this is a ladies' cabin, of a still more sumptuous character. We were much pleased with the arrangements for ventilation; but it seemed to us that there was a deficiency of light in the saloon.

The *Amazon* is propelled by two beam engines of the collective (nominal) power of 800 horses, driving reefing paddle-wheels each of 39 feet 6 inches diameter. The cylinders are 96 inches diameter, and are fitted with Seaward's patent induction and eduction valves; the length of stroke is 9 feet. At the measured mile, with a slight tide against her, the *Amazon* steamed at the rate of full ten miles an hour, the engines making 14 strokes per minute. Mr. Mills, the able and experienced engineer of the company, expressed no doubt that, after the engines settle in their bearings, the ship will steam upwards of twelve miles an hour.

PATENT LAW.—THE CASE OF USER BY A FOREIGNER OF AN ENGLISH PATENT INVENTION IN ENGLISH WATERS.—(SEE ANTE P. 453.)

Vice-Chancellors' Court, Tuesday, December 9.

(Before Sir G. TURNER.)

Caldwell v. *Rolf.*—*Caldwell* v. *Vandlissengen.*—*Caldwell* v. *Verbeek.*

The SOLICITOR-GENERAL and MR. BAGGALLAY proceeded with their arguments against the application for an injunction to restrain the defendants in the second of the above causes from sailing their vessels with the screw-propellers made in conformity with or in imitation of Lowe's patent. They contended that Lowe's patent was invalid—first, because the invention was not novel; secondly, because the specification claimed too much; and thirdly, because it did not sufficiently describe the nature of the invention. It was not novel, for it appeared that Bramah's invention in 1785 was on the same principle. It was assumed that the screw-propeller had no analogy to the vanes of a windmill or of a smoke-jack, which was Bramah's invention; but this assumption was quite unfounded, and not supported by the testimony of any scientific witness. Upon comparison of the models, it was very difficult to say that the principle was no perfectly analogous. Then the specification was too extensive; he claimed a patent for the curved blades of his screw, but he did not limit it to curved blades of a screw-like form; that was only one mode of making the screw. The specification did not claim a patent for sections of a screw; but, according to the claim, the invention might be a perfect unbroken screw. The description went further, but it had been repeatedly held that the description could not be imported into the claim. The claim, however, if not distinct, was too broad. It claimed not only the blade, but the axis of the screw. Now, it was clear that the axis was not new. The specification was also defective in not giving such a description of the invention as would enable a workman to construct it. The invention was described as "the section of a screw, which if continued would make a perfect screw." Was it possible to say that any artificer from this description alone, could, with any certainty, have constructed a machine having the properties of the propeller in question? Next to the objection to the validity of the patent, came the question of user or possession. The trials which had taken place had in no respect established the right of the plaintiffs. In fact, they rather tend to throw doubt upon that right; for in every case where a verdict had been obtained, affirming the novelty of the invention, the Court had set aside the verdict as against evidence, and granted a new trial, and, in every case in which the specification had been under the consideration of the court of law, that court had expressed doubts upon its sufficiency; and it appeared that the plaintiffs, instead of inviting any ultimate decision on these points had in all cases avoided such a decision by settling their claims with the defendants. Many persons had, no doubt, submitted to the payment of a small royalty rather than incur the expense, delay, and annoyance of a suit in Chancery. The engineers, Maudslay and others, had willingly submitted to an arrangement by which all the claims of the several patentees, Woodcroft, Smith, Ericsson, B'axland, as well as that of Lowe, were settled; and, for a certain payment amounting to £1 per horse-power, they were allowed quietly to prosecute the manufacture;* for the same reason they had paid over to the plaintiffs, as representing all those patentees, the guarantee fund which they had received from the Government. It was quite obvious that this might be a very wise and prudent arrangement for the engineers, without in any measure affirming the validity of Lowe's patent, any more than it established any one of the other patents which were vested in the plaintiffs. The next point was the fact that the defendants had been for two or three years in the habit of using this invention, publicly, without any concealment; and, believing themselves to be justified in so doing, they had, as the affidavits showed, expended large sums of money in fitting their ships with these screws. The fourth objection to the motion was one of great public importance,—whether the Court would restrain the departure of a foreign ship, owned and manned by foreigners, because they happened to have fitted to the ship an invention which was lawfully placed there in the country in which it was built, but for which a patent had been obtained in this country? The principle did not apply to ships; if it could be maintained, it must be extended to every article of manufacture. Take, for example, Minter's chairs. Suppose a person to buy chairs according to that form and model made in Paris, and that there was no French patent to prevent them from being made by any person, might not the purchaser bring the chairs which he had so bought to this country, and use them in his own house, or in a coffee-house, if he was a proprietor of such a place, without being subject to any injunction? The case of a carriage made abroad had been suggested in the former argument, and a multitude of similar cases might be supposed. If this right were to be insisted upon as against foreigners, there was no part of a foreign ship of which the owners might not be deprived of the use. One patentee might claim to restrict the use of the sails, another of the cordage, another of the rudder, and another of the anchor; the ship might, as soon as it was unfortunate enough to enter a British port, be dismasted and broken up by all these several claimants. The consequence would be that foreign countries would retaliate. They would grant patents to their own people, which would operate equally in the restraint of our shipping in their ports. The defendants in these cases would be advised to procure patents in Holland for these screws, and the result of the plaintiffs' success would be that they would prevent the use of the screw by the defendants in this country, but they would only exemplify the fable of the dog in the manger, for they might in like manner be restrained in Holland from using it themselves. Where would the interference of the Court end? Suppose a foreign vessel fitted with the screw were stranded on our shores! the accident would make the parties amenable to the law as an infringement of the patent. In the case cited of "the University of Oxford v. Richardson," Lord Eldon had expressed himself to the effect that there might, from the necessity of the case, be a user of a patent article which would not amount to an infringement. The words of the patent did not extend the restric-

* For some explanation of this combination of parties, see Lowe's Petition for prolongation of his Patent, given in another page.

eign subjects; it was addressed "to all
dies politic and corporate, and other Her
subjects;" and the Court would not sanc-
empt to give the jurisdiction so wide an
at least until the right had been formally
by a trial at law.
is was heard for the defendant Verbeek,
nt in the third cause.
r, in reply, contended that the argument
y lawfully acquiring a patented article
ght use it in this country, was founded
lacy,—that although it was true to say
ully acquired abroad, if made or bought
e purpose of being used abroad, it was
y acquired abroad if intended for use in
ry, where the rights of the patentee
ted. The suggestions of inconvenience
d to foreigners were readily answered.
when they came to this country, were
the laws of this country. They were,
e, subjects, amenable to the civil and
risdiction of England. It had been
reigners might have been using a par-
ention for many years, but, on bringing
untry, it might be found to be the sub-
atent, and that it might even happen
iginal inventor would be himself re-
m using his invention here. If the law
ntry gave this privilege to patentees, it
y of the Court to administer that law,
re any hardship in it.
our said, that the foreigner was de-
thing he had before. If he had never
vention in England, he could not coming prevented from using it here, and,
sed it here before, there would be no

r then proceeded, by comparison of the
drawings, to show that Lowe's patent
ecifications on which it was founded,
able to the objections which had been
on by the defendants. The defendants
did not pretend that any Dutch work-
icer had invented the machine; they
how they came by the first model The
erefore, had a right to impute to the
the knowledge of the source of the
nd that it had emanated from Lowe;
owledge they might have had if they
proper to seek it. They did not pre-
e been deceived by any persons, and
believe they had not adopted the
f Lowe. The proceedings in the
his country had been open, public,
o the last, until the parties had
eless to contest any longer the right
tentee. The compromise with the
no doubt, was made upon the con-
hat they should be indemnified against
f all the other patentees; but the right
patentees had nothing to do with the
t of the *scire facias*; the trial in that
upon the validity of Lowe's patent
that patent was thereby as clearly
as the submission of the defendants
it. With regard to the difficulties
been suggested as to the departure of
e would not enter into that question.
s would so far concede the point. The
d be allowed to depart, and not only
ews, but should be allowed to use the
he purpose of leaving the English seas.
for the injunction only to restrain
again coming within the jurisdiction
ented propeller.
-CHANCELLOR.—The defendants have
e the propeller out of the jurisdiction
rt; can I say the ship shall not have it

—*We submit that you* may. It is the
y which the patent can be protected.
ble to follow the ship and ascertain
when the propeller is used, if the parties are allowed to fix it for use. The temptation to use it, in cases of necessity, would be so great that it cannot be supposed it would not be resorted to.

The VICE-CHANCELLOR.—Can I act upon the supposition of such a temptation? Your right would be to an action for damages; and I can only protect the legal right.

Mr. ROLT submitted that the existence of the screw propeller on board the ship would be a case to go to a jury, and in which a jury would give damages, presuming that it had been used.

At the conclusion of the argument,

His HONOUR said the case was one of such general importance and novelty, that he should not determine it without a few days' consideration.

SPECIFICATIONS OF ENGLISH PATENTS ENROLLED DURING THE WEEK, ENDING DECEMBER 8, 1851.

ROBERT WILLIAM SIEVIER, of Upper Holloway, civil engineer. *For improvements in weaving and printing textile fabrics.* Patent dated May 29, 1851.

Mr. Sievier's improvements in weaving have relation to the production of certain fabrics to be employed as substitutes for "Kidderminster," or "yard-wide" carpets, and consist in the employment of an addition warp-thread, (which may be of any thickness, and composed of waste cotton, or other inferior material, and which the patentee calls the "dead warp," from its never appearing on the face of the fabric,) in combination with the binder warp, and with one or two wefts, according as the opposite faces of the fabric are required to be of similar or of dissimilar qualities. When two wefts are employed the dead-warp never moves from its horizontal position, and it then forms a race for the upper shuttle to travel on, the lower shuttle running against the under side of the dead-warp, and the shed being divided into two parts by the dead-warp, and each of the shuttles being thrown simultaneously from one side of the loom to the other. The fabrics thus produced are of considerable substance, but from their being quite plain require to be printed in order to render them applicable to the purposes for which they are intended.

The improvements in printing have relation to certain arrangements for operating on such fabrics as have just been described, or any others requiring to be so printed. Of these arrangements, Mr. Sievier describes three modifications. The first of these consists principally of two flat rings, or circular tables, one above the other, the upper one carrying the printing-blocks, and the lower table the colour-troughs and serving-rollers for applying the colour to the printing surfaces. These rings are caused to rotate so as to bring each block successively over each of the printing-tables in connection with the apparatus on which are laid the fabrics to be printed, and of which there may be any desired number, and while in this position the blocks or printing surfaces are subjected to pressure from a platten over the table actuated by cranks on a revolving shaft, driven by the same machinery as is employed to rotate the tables. The fabrics after being successively printed on by each of the blocks in the series, so as to produce on them a perfect pattern with any number of colours therein, are removed by means of a triangular frame, which has a suitable intermittent motion, and is so constructed as to remove from the table at each partial revolution a length of fabric equal to the repeat of the pattern.

The second arrangement differs from that just described in being of a rather simpler construction; in this arrangement the printing tables, instead of being fixed and acted on by pressing surfaces, are caused themselves to press the fabric against the printing block or engraved plate during its passage in juxtaposition to the fabric.

In the third arrangement, the printing is effected by the employment of engraved rollers, which are supplied with colour by other rollers dipping in'o colour troughs, and the pressure necessary to produce an rmpression is obtained from a third set of rollers, between which and the printing-sollers the fabric is caused to pass on a uitable blanket or carrying band.

Claims.—1. The making or production of carpets and similar fabrics by means of the peculiar mode of weaving described.

2. The adaptation of engraved plates, locks, or printing surfaces, together with colour rollers and troughs, and their appendages, to moveable rings or tables, which are made to rotate round a centre, and bring each of the said engraved plates, blocks, or printing surfaces consecutively over the fabric to be operated on, and whereby the different colours, or shades of colours, may be successively applied thereto until the pattern is completed.

3. The arrangement of apparatus described for printing carpets and other similar fabrics, or any mere modification thereof.

William Bridges Adams, of Adam-street, Adelphi, engineer. *For certain improvements in the construction of roads and ways for the transit of passengers, of materials and of goods; also, in buildings and in bridges, and in locomotive engines and carriages, parts of which improvements are applicable to other like purposes.* Patent dated June 3, 1851.

Claims.—1. The application to the permanent way of railways of stone-block sleepers with bolts passed through them to hold down the rails or chains, or both, and with or without the intervention of timber, or other similar elastic material. Also, the application of longitudinal or transverse sleepers secured to stone blocks by bolts passing between the blocks and secured to timber or other material by bolts passing between the blocks. Also, any similar arrangements substituting masses of brickwork or cement, or concrete, or masses of cast iron or metallic slag, or loaded boxes of timber or iron in place of stone blocks, to secure the rails firmly, and when required elastically, to a sufficient mass of weighty ballast. Also, the use of side clamps to secure the rail ends bolted below the rails, and forming a chair if preferred. Also, certain sectional forms of deep single T iron rails and timber combined for permanent way, and of laterally corrugated or grooved rails. Also, the construction and application of various forms of rails for the permanent way of railways to produce lateral and vertical stiffness, formed of cast or wrought metal, or of both combined, and combined with wood and formed of one piece, or of several pieces combined or bolted, or riveted together, and laid in concrete, or in ballast, or with or without sleepers beneath them, and which are called "Girder rails," such girder rails having a continuous vertical or partly vertical support lower than the horizontal, or partly horizontal bearing surface on the sleepers or ballast. Also, the application to girder rails of cheek plates, and angle cheek plates on one or both sides, or tongue plates, or saddle or channel plates, or girder bearers, or break joint, for the purpose of connecting the girder rails, or parts of them, together, and making them continuous, and combining them in any new form required. Also, a mode or modes of making curved rails by forming the holes of the separate horizontal plates in curved lines. Also, the application of hollow girders, or rails as water pipes, or perforated as drain pipes, and of the various other forms capable of this arrangement, by adding bottom plates. Also, the application of elastic girder rails, and other analogous constructions, to permit elastic yielding of the rail without disturbing the ballast below. Also, the application of longitudinal or other timber bearings beneath the broad horizontal surfaces or plates of metal, to give vertical stiffness. Also, the construction and application of channel rails and of steel, or steeled rails, to form such girder rails as are described. Also, the application of zinc or tin, or other metallic or galvanic covering, or asphalte, or bituminous substance, or mineral, or other paint to these girder rails, or parts of them, to preserve them from on, applicable to other rails also. be construction and application of le railways, or agricultural or other ilways or tramways. Also, the ap-on of a notched bar with staple holes. the application of forked staples, or and holes, or studs and holes, to t portable or other rails and cross gether, with or without notches. Also, astruction and application of elastic 's to diminish noise and vibration in or on arches, particularly adapted for g railways through towns.

'he application to streets or roads, or walks or park walks, or roads, or to station platforms, or to markets or places open, or partly open, to the r, or for like purposes in railway s of perforated surfaces of wrought or cast iron, or other metal, stone, slate, earthenware, or other similar l, supported on legs or ledges, above face of the ground, or of the ordinary ent, so as to permit dust, mud, or or other extraneous matter to pass below, each perforation being either cut holes of any convenient form, nings at the edges, or longitudinal gs, either cast in one piece, or in e pieces, placed side by side, or in convenient forms. Also, the applica-f such surfaces, or of any common laid in contact with sand, gravel or porous soil, so as to give an unyield-et porous surface. Also, the ap-on of metal frames perforated, with s recessed and fitted with wooden , or asphalte, or bitumen, or cement, rcelain, or glass, or tiles, or mosaic or artificial stone or similar matters, l either as ornaments, or for the pur-f preventing slipping, being likewise able to the internal floors of buildings, , shops, and passage ways, or to metal non-perforated. Also, the applica-o perforated metal or other foot sur-or to the openings of cool vaults, or as or door steps of advertisements g open letters, or devices to be lighted ht by gas or other light. Also, the ation for similar purposes of semi-arent porcelain, or of glass. Also, e application of similar advertisements r stamped, or graven on metal plates erforated. Also, the construction of der surface for carriage or horse roads lag stones, or slabs, or sheets of iron n concrete, and constituting an under e, on which the perforated iron grat-re to be laid, or, if preferred, wood or or asphalte or bitumen. Also, the lising or covering with a metallic coat-ing or bitumen, or paint, the wrought or cast iron used in these foot or horseways to prevent oxidation.

3. The application of a chain tension network, and intersecting corresponding girders, either in a square or circular, or other form for flat or slightly-arching roofs or floors, thus distributing the strain in various directions. Also, the application of elastic cords, or piping of gutta percha, caoutchouc, or other elastic material to make watertight joints, instead of putty, and so as to expand or contract with the other materials. Also, this application as a like purpose to stonework or woodwork, generally to prevent the passage of moisture. Also, the application of glass bricks, either plain or coloured, to building purposes, such bricks to be dovetailed, joggled together, or fileted, or tongued and grooved, or similarly connected.

4. The application of double lattices or diagonal framings, to form rigid girders for buildings and bridges in connection with tension chains. Also, the application of shear legs, to erect such or similar girders in their places without a scaffold, by drawing the chains to a horizontal level. Also, the construction and application of channel plates to form girder bridges. Also, the application of cast iron frames with tension bars or chains, to form girders for bridges and other purposes. Also, the construction and application of links for chains for suspension or other purposes, in which that part of the link which forms the eye, is made thicker than the other part of the link, in order to obtain a greater bearing surface without increasing the diameter of the bolts connecting the links.

5. The construction and use of double tank engines, to be used without a separate tender for fuel and water, such engines being provided with tanks and coke-boxes, and being capable of separation when required, and provided with moveable buffers, so that they may serve for two light loads, as two separate tank engines, and the double engines may be constructed with two fire-boxes, and two smoke-boxes, or with a fire-box and smoke-box together. Also the application to locomotive engines of double chimneys formed of double plate, with air between to economise the heat of draught; applicable also to other engines. Also the construction and application to the piston-rods of four-wheel tank engines of slide-bars working outside the cylinders, parallel to the piston-rods, and connected with them to propel all four wheels. Also the combination of a tank-engine carrying its fuel and water for short distances, with a separate

tender for fuel or water to travel long distances when required, the tank of the separate tender being connected either to the tank of the engine or to the boiler direct. Also the use and application of two tank engines, carrying their own water, on four or six wheels, each connected together, and with a moveable foot-plate over the opening between them, so that one man may drive both the tank engines, the fire-boxes being both together; these improvements in connecting and disconnecting engines being also applicable to locomotive engines propelled by other power than steam.

6. The application of coupling together two or more four or six-wheeled carriages or engines by buffer-rods, or dowel-rods, passing from one carriage-frame or engine-frame into the other for the purpose of keeping them at one horizontal level, and giving vertical support, being a variation of the mode of coupling the double carriages. Also the application to railway carriages or wagons of double headstocks and rigid truss frames for the purpose of preventing damage by buffing or traction. Also the application of screw-bolts to connect the parts of iron under frames of railway carriages together, to give facility of separation in case of repairs, or for facility of transport to render them conveniently portable. Also the construction and combination of the patentee's registered coal-hopper wagon to form double wagons. Also the combined construction of railway carriage wheels with the patentee's patent ribbed tires, and with single disc plates and false rib or double disc plates combined with central cast or wrought-iron cheek-plates, all riveted or bolted together without piercing holes in the tread of the tire. Also the application of angle iron riveted to an ordinary rail tire to produce a rib to which to rivet the disc or discs. Also the application of radial flat spokes in such wheels instead of discs. Also the application of tin or other metallic coverings to the separate plates of steel, which form laminated springs for carriages on railways, so that they may be hermetically sealed from rust. Also the application to railway carriages and locomotive engines of improved sledge-breaks, either hinged to the axle-boxes or made to slide on bars inclining to the rails, so that the motion of a screw or other means may cause them to impinge on the rails; and the same sledges will act as safety bearers on the rails in case of the breakage of a wheel or axle. Also the application of water-tank wagons to water railways and prevent them being damaged, and causing dust and wear. Also the application of moveable buffer-heads to the ends of carriages and engines, so that the carriages may be worked either in pairs or separately.

Thomas Parker, of Leeds, broker. *For improvements in machinery for opening, cleaning, and preparing fibrous substances, and for manufacturing felted fabrics.* Patent dated June 3, 1851.

The first of these improvements has relation to a machine for opening, cleaning, and partially dressing tangled, confused, and dirty fibrous substances, such as tow, or tow waste, from hemp, &c., and for opening and cleaning woollen and silk waste.

The machine consists mainly of an endless travelling sheet, on which the material is supplied, to be operated on by the teasing action of rods or pins, suspended above the travelling sheet, or to the smoothing and straightening effect of certain combs or brushes. The endless sheet is composed of bars of wood, iron, or other suitable material, either smooth or having their surfaces provided with sharp or obtuse pointed pins or needles, for the purpose of holding the fibrous materials while under the action of the teasing surfaces; but where a greater amount of retention is required to be exercised on the fibres, the travelling sheet is made to receive clamps (which may be removed at pleasure), which admit of the carrying sheet being filled and emptied continuously, and which serve to retain the longer and straighter fibres when the shorter and waste portions are removed and swept away by the action of the brushes or combs.

The second improvement has relation to a machine for felting wool, hair, or other feltable materials, in the which the same, while undergoing the felting operation, are caused to pass between the flat surfaces of two travelling sheets, composed of belts or straps, and bars of wood, or other suitable material, square or flat, which are laid side by side, close together, so as to cover the whole length of the straps or belts. The sheets are carried, one above the other, by rollers, and when in a horizontal position are parallel to each other. The bars which form the surface of the lower sheet are permanently attached thereto, but those of the upper sheet are connected with suitable means for giving them a rapid reciprocating movement in a direction transversal to the progressive motion of the sheets, and it is by the action of these bars that the felting of the materials under operation is effected. The wool, fur, hair, or other material, having been carded and doffed in the usual manner, is wound on a roller simultaneously with a sheet of linen, or other convenient material. The speed of the take-up roller and sheet is so regulated with reference to that of the doffer, that the felting materials shall be delivered therefrom in a bat, of a thickness suitable for that of the felt to be produced, and the receiving sheet will of course be of a length corresponding to that of the felt to be made. The roller with the sheet and bat wound upon it are then removed to the felting machine, where the carrying sheet and bat, after having been passed through water, or other liquid or vaporous medium, are operated on by the travelling sheet in the manner described, and the felt thus manufactured.

The third branch of the invention has relation to the ordinary paper-making machinery, and consists in certain additions or improvements to adapt the same to the manufacture of a species of felt by the mixture of fibrous materials with glutinous or bituminous substances, or with pulverizable materials, such as saw-dust, founders' sand, &c., for the purpose of covering houses, ship's bottoms, &c., and preserving goods in packing cases. The fibrous materials employed in manufacturing felt for the latter purpose may be of the lowest description, such as burring machine waste, and other woollen refuse, as well as tow-waste, &c.; but when the felt is to be applied to agricultural and building purposes, such as the formation of copings, gutters, and mouldings, a mixture of pulverizable materials is to be employed.

The improvements consist, firstly, in the application of heat, either externally or internally, by means of steam or other heating medium circulating through pipes around or within the machine, or by blowing steam direct into the mixture of materials therein to the rag machine, for the purpose of keeping the glutinous or bituminous materials employed in as liquid a state as possible, in order to facilitate the incorporation and amalgamation therewith of the fibrous substances employed; secondly, in the use of an additional sheet of wire-cloth, placed in a slightly inclined position over the ordinary pulp receiving-sheet for the purpose, in paper-making, of more thoroughly felting the pulp, and, when glutinous and other materials are employed, of spreading and equalizing the thickness of the mass; thirdly, in the application of a roller covered with rings of cloth, or other absorbent spongy material, to the interior of the wire web aforesaid, or to a cylinder of the same material, which may be employed in lieu of the wire web, for the purpose of absorbing the moisture expressed from the pulpy mass by the web or cylinder, which moisture is removed from the sponge covered roller by means of a guttered doctor in a slightly inclined position, against the edge of which the roller revolves; and fourthly, in the use of a roller, covered with cork, for carrying the upper end of the inclined sheet of wire-cloth.

Claims.—1. The use of oscillating pins or teasers in consecutive ranks, while the material is fed and passes off in endless supply; the use of clamps for holding the material and the combination of two clamps, by which a change is effected of the part by which the material is held; the application of vibrating pins or teasers to the feeding sheet of batting machines or cards; and finally, the use of brushes of obtusely-pointed wires or pins.

2. The use of travelling flat surfaces which simultaneously advance and felt the bat; the application of a sheet to support and guide the bat through water or other medium to aid its felting properties, without requiring the bat to be first partially felted; the giving of an oscillating movement to the flat or square felting bars; and the method of arriving at and equalizing the desired thickness of the bat.

3. The use and application of heat to the rag engine, of the additional inclined sheet, of the spongy roller, even if used without the additional sheet or wire cylinder, and of the cork roller.

John Hopkinson, of Oxford-street, pianoforte manufacturer. *For improvements in pianofortes.* (A communication.) Patent dated June 3, 1851.

The improvements claimed under this patent comprehend—

1. Certain modes of communicating motion from the keys to the hammers, applicable to horizontal and upright pianos.

2. The application of sponge as a covering for the hammer heads.

Specification Due but not Enrolled.

Cornelius Alfred Jaquin, of New-street, Bishopsgate, mechanist. *For improvements in the manufacture of nails, pins, tacks, screws, and other similar articles.* Patent dated June 3, 1851.

WEEKLY LIST OF NEW ENGLISH PATENTS.

Joseph Harrison, of 10, Oxford-square, Hyde-park-gardens, engineer, for certain improvements in steam engines. December 8; six months.

Peter Armand Lecomte de Fontainemoreau, of South-street, Finsbury-square, for improvements in the apparatus for kneading and baking bread and other articles of food of a similar nature. (Being a communication.) December 8; six months.

Richard Archibald Brooman, of the firm of J. C. Robertson and Co., of Fleet-street, patent agent, for certain improved modes of applying electrochemical action to manufacturing purposes. (Being a communication.) December 8; six months.

Richard Archibald Brooman, of the firm of J. C. Robertson and Co., of Fleet-street, patent agent, for improvements in the manufacture of sugar, in the preparation of certain substances for such manufacture, and in the machinery or apparatus employed therein. (Being a communication.) December 8; six months.

Isaac Alexander, of 112A, High Holborn, Middlesex, biscuit baker, for a mode of preparing and treating certain kinds of cheese, whereby to render the same applicable to a variety of culinary and other domestic purposes. December 8; six months.

Perry G. Gardiner, of New York, civil engineer and machinist, for improvements in the manufacture of malleable metals into pipes, hollow shafts, railway wheels, or other analogous forms, which are capable of being dressed, turned down, or polished in a lathe. December 8; six months.

Charles Cowper, of Southampton - buildings, Chancery-lane, for improvements in separating coal from foreign matters, and in apparatus for that purpose. (Being a communication.) December 8; six months.

William Pidding, of the Strand, gentleman, for improvements in the treatment, manufacture, and application of materials or substances for building purposes. December 8; six months.

John Lake, of Apsley, Hertford, civil e for improvements in propelling on cana rivers. December 8; six months.

Thomas Restell, of the Strand, Middlesex maker, for improvements in locks or fas December 8; six months.

John Frearson, of Birmingham, for i ments in cutting, shaping, and pressing m other materials. December 10; six mont

James Webster, of Leicester, for impro in dyeing gloves and other articles of December 10; six months.

Etienne Alexander Armand, of Paris, provements in the modes of distilling and organic substances and bituminous matter the treatment of their products, together apparatus used for the said purposes. I 10; six months.

Alfred Vincent Newton, of Chancery-l chanical draughtsman, for improvements textile fabrics. (Being a communicati cember 10; six months.

Thomas Masters, of Regent-street, con for improvements in obtaining and dr aërated and other liquids, and in chargin with gaseous fluids, applicable to vessels ing solid matters, and also as a fastening sils and apparatus, and in holders for cig cember 11; six months.

WEEKLY LIST OF DESIGNS FOR ARTICLES OF UTILITY REGISTERED.

Date of Registration.	No. in the Register.	Proprietors' Names.	Addresses.	Subjects of Desi
Dec. 5	3045	George Sant	Fulham	Milk-tester.
,,	3046	Francis Whisshaw	John-street, Adelphi	Telekouphonon.
,,	3047	Maurice Moses	Tower hill	Janus coat.
6	3048	J. C. Evans	King William-street	Revolving curtain r
,,	3049	E. N. Fourdrinier	Sunderland	Pen-holder.
8	3050	John Hicks	Dorchester	Dorset stove.
,,	3051	W. Flataw and Co.	Mansell-street	Shoe.
9	3052	Henry Stephens	Stamford-street	Parallel ruler.
,,	3053	Cripps and Lindup	Worthing	Duobus coat.
10	3054	—— Lyon	Windmill-street	Sausage-meat cutter.
,,	3055	Charles Clarke	Birmingham	Casement stay and fa
11	3056	D. Thornton and Son	Birmingham	Glass gauge-tube fo engines.

WEEKLY LIST OF PROVISIONAL REGISTRATIONS.

Dec. 10	337	John Sutton	Newington	Bulbous root hol flower-support.
,,	338	Marcus Davis	Strand	Geometer.
11	339	Cyrus and J. Clark	Somerset	Expanding boot.

Errata.—In the List of Designs Registered last week, p. 460, *for* "G. A. & F. Ferguson, *read* "C. A. & T. Ferguson."

CONTENTS OF THIS NUMBER.

LONDON: Edited, Printed, and Published by Joseph Clinton Robertson, of No. 166, Flee in the City of London— Sold by A. and W. Galignani, Rue Vivienne, Paris; Machin a *Dublin*; W. C. Campbell and Co., Hamburg.

Mechanics' Magazine,

MUSEUM, REGISTER, JOURNAL, AND GAZETTE.

No. 1480.] SATURDAY, DECEMBER 20, 1851. [Price 3*d*., Stamped, 4*d*.

Edited by J. C. Robertson, 166, Fleet-street.

BERTHON'S PATENT BOATS, SOUNDING INSTRUMENTS, TIDE GAUGES, AND CURRENT INDICATORS.

Fig. 2.

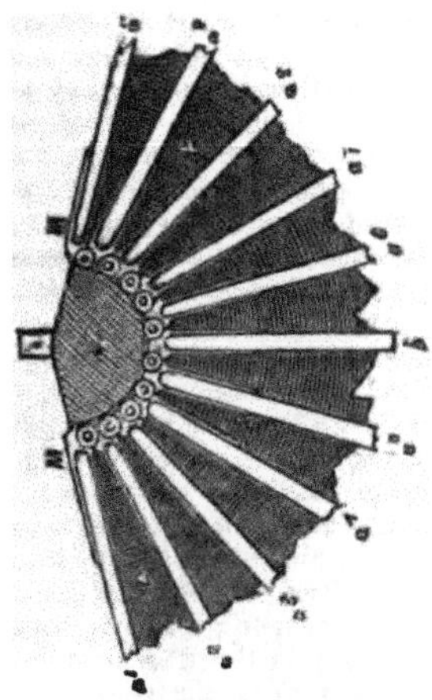

Fig. 1.

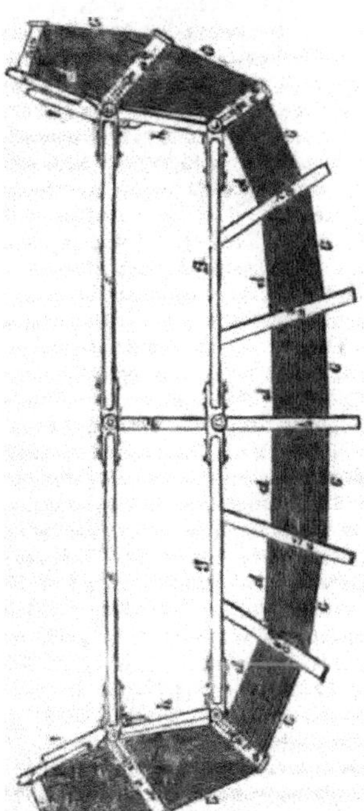

Fig. 4.

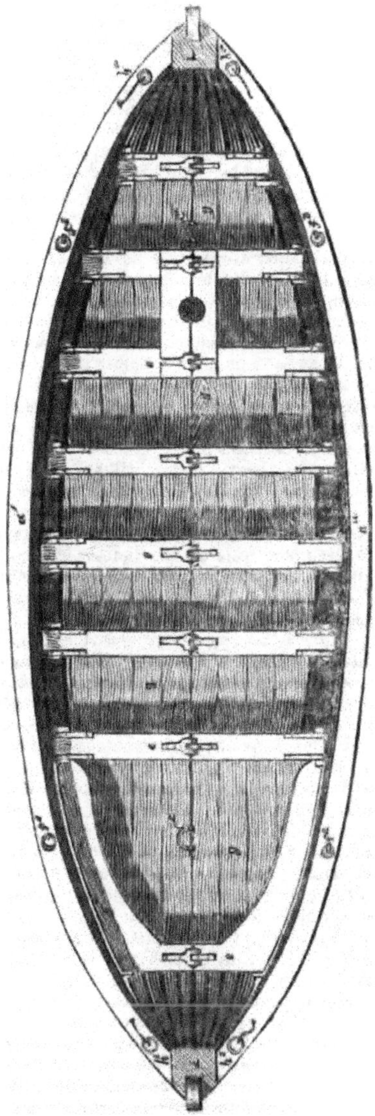

BERTHON'S PATENT BOATS, SOUNDING INSTRUMENTS, TIDE GAUGES, AND CURRENT INDICATORS.

(Patent dated June 12, 1851. Specification enrolled December 12, 1851.)

Specification.

Firstly. My invention has relation to the construction of boats generally, and more especially to the class of boats called life-boats and to boats for the transport of ordnance and other heavy bodies across rivers (sometimes called pontoons or floating bridges), and consists in constructing them in such manner as to combine adequate capacity and strength with the property of being collapsible at pleasure into a small compass, and so easier of transport from one place to another. Figures 1, 2, 3, 4, 5, 6, 7, 7ª, 8, 8ª, represent the details of a life-boat, built on this principle, 30 feet in length and 10 in breadth. Fig. 1 is a midship section; a^1, a^2, a^3. a^4, a^5, a^6, a^7, a^8, a^9, a^{10}, are a series of longitudinal timbers, and *b* is the keel. The timbers are all made flat and broad—say two inches in thickness, and from eight to fifteen inches in breadth; and they are jointed together at the ends (both of which are alike) by chain hinges of the construction separately represented in figs. 2 and 3, and hereafter described; *e, c, e, e, c, c, c, c, c, c,* is an outer sheathing or envelope, composed of any flexible waterproof material, which is attached by nails and cement to the outer edges of the timbers, and *d, d, d, d, d, d, d, d,* an inner lining of similar material, which is similarly attached to the inner edges of the timbers. The sides and bottom of the boat are thus divided into ten distinct longitudinal compartments or cells, which extend lengthwise from stern to stem, and are intended to contain air drawn in through openings at the extreme ends while the boat is in the act of being expanded; *e e* is one of the thwarts, formed in two pieces, jointed together in the middle, one of which pieces is also attached by a joint at its outer end to the timber a^2, and the other is similarly attached to the timber a^9. *f* is a stanchion fixed upon the keel *b* for the thwart *e e* to rest upon; for each thwart there is a stanchion projecting upwards through the flooring; *g g* is the flooring or platform, which is jointed, like the thwarts, along the midship line and along the sides by three or four hinges (as shown in figs. 8 and 8ª) to the timber a^3 on one side, and a^8 on the other. From what has been stated, it will be readily understood that, when the boat is collapsed, the timbers will lie in parallel planes, but assume, as the framework is expanded, positions more or less divergent from the centre line, or line of the keel. When the boat is shut, this flooring rises in the middle, and stands up like the roof of a house; and when open, it drops down, and rests on the timbers a^4, a^5, a^6, a^7, and on the keel *b*. The thwarts also rise and fall like the flooring, and all these together perform the part of extenders when the boat is open. *h h* are moveable stanchions under each thwart, resting upon the flooring, to give additional strength. The gunwale pieces a^1, a^{10} are supported by stanchions I I, and are set up very taut by a lever of the kind shown in fig. 6. Fig. 2 is a portion of one of the ends inside. L is the stem-block, being a strong semi-cylindrical block of wood, which is firmly attached to the stem or end of the keel-piece turned up. Around this stem-block the ends of all the timbers are made to abut when the boat is open, by the directing action of the chain-hinge M M, which then takes a semicircular curve; but when the boat is shut, the chain-hinge assumes a horizontal position, or that of a straight line at right angles to the stem, allowing the timbers to fall into parallel planes, as shown in fig. 3. The chain-hinge M M, figs. 2 and 3, is made like the chain of a watch, but has two straps from each link to grasp the timbers upon which they are riveted; the centre link, which is made fast to the stem, is the strongest, and forms the fixed point for the parts on each side to work upon. Fig. 4 is a gunwale plan or view, showing the internal arrangements; *e, e, e, e,* are the thwarts; *g, g, g, g,* the platform or flooring; c^2 c^2 are two strong metal shackles placed upon two of the hinges of the flooring, so that the pins of the hinges pass through their eyes. To these shackles the falls for hoisting up the boat are hooked. f^2, f^2, f^2, f^2, are other ring bolts on the gunwales, to which the spans are hooked, for expanding and lowering the boat; L L are the stem-blocks before mentioned, and h^2, h^2, h^2, h^2, are openings, with plugs fitted, through which the air enters in opening the boat, and escapes on shutting it. The plugs are put in when the boat is in use, to keep the water out. Fig. 5 shows an end view of the boat as it appears when collapsed and hoisted up under the davits, closely frapped to, and lashed to the ship's bulwarks outside with strong gripes.

Fig. 6 shows, on an enlarged scale, one of the gunwale stanchions with its lever; I^2 is the stanchion; I, the lever, made of tough wood; a^1 the gunwale-piece, a^2 the second timber; *e, a thwart.* The lower end of the stanchion I^2 being placed in the notch of the lever I, and the lower end of that lever on the end of the thwart, then when pressure is made in the direction of the arrow so as to bring the two pieces I^2 and I together, a very great expanding force is obtained, whilst, immediately on passing the centre or straight line, the lever I flies home against the stanchion I^3, where it remains. Figs. 7 and 7^a show the sort of hinge used for the thwarts and flooring along the centre line; the pin *a* is withdrawn, and the two parts *b* and *c* separated; each part has two straps to grasp the wood which is put between them; they are then riveted together.

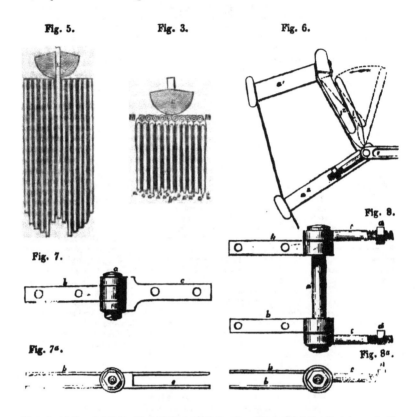

Fig. 5.

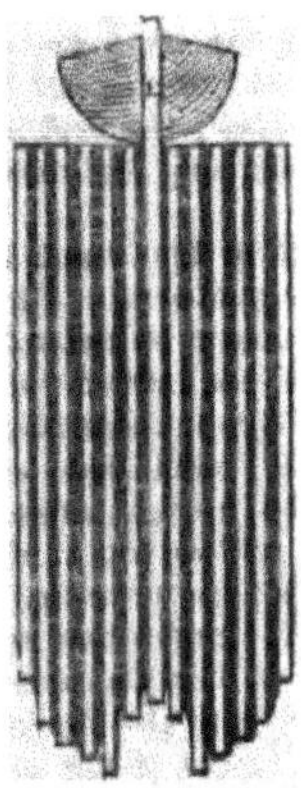

Fig. 3.

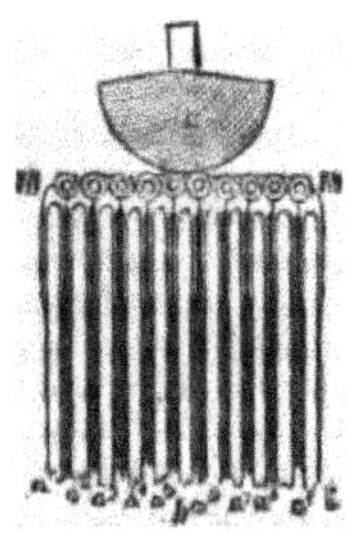

Fig. 6.

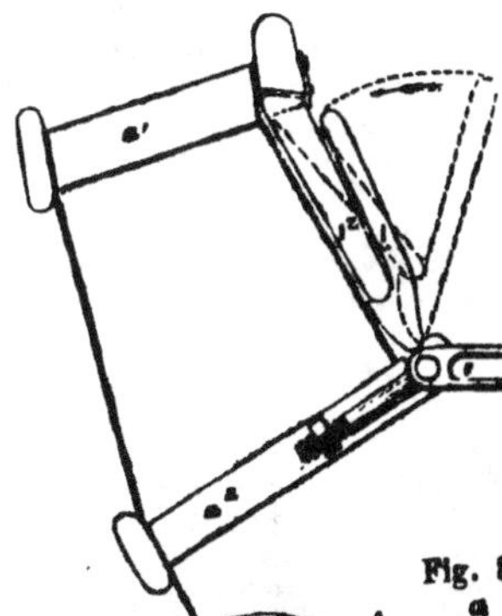

Fig. 8.

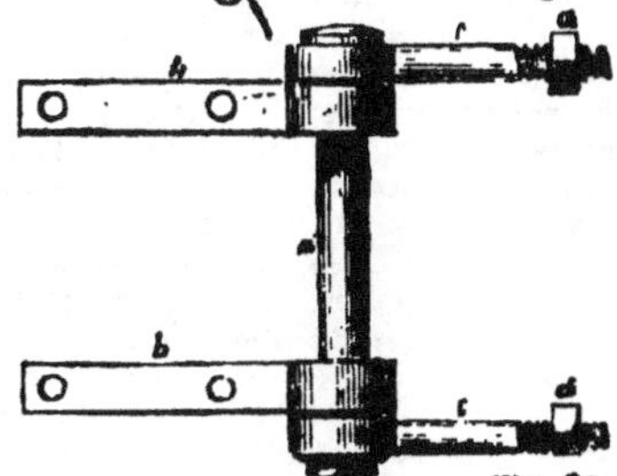

Fig. 7.

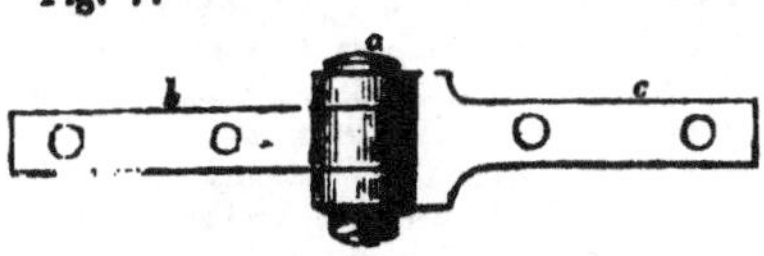

Fig. 7^a.

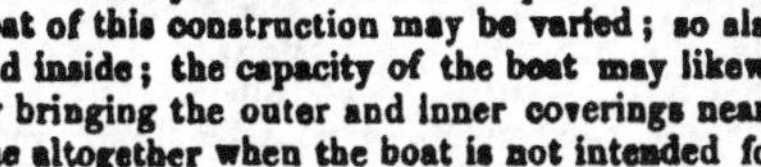

Fig. 8^a.

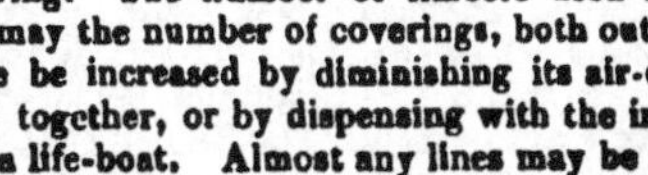

Figs. 8 and 8^a exhibit the kind of hinge used for the outer ends of the thwarts; *a* is the pin; *b b* the straps, which are riveted on to the ends of the thwarts at their edges; *c c* the two eye-bolts, which are screwed into the nuts *d d*, imbedded in the timbers. A similar plan to that just described is adopted for the flooring. The number of timbers used in a boat of this construction may be varied; so also may the number of coverings, both outside and inside; the capacity of the boat may likewise be increased by diminishing its air-cells by bringing the outer and inner coverings nearer together, or by dispensing with the inner one altogether when the boat is not intended for a life-boat. Almost any lines may be followed in respect of the general form. As before stated, any flexible watertight material may be used for the covering; but I give the preference to strong thick felt, coated with India-rubber. Strong bands or webs may be passed round the framework in the direction of the ribs, to give additional strength. The keel should, however, be added after the webs and cloth are put on. The boat is furnished with two or more bilge-keels, or pieces put on outside the covering on the edges of the timbers a^5, a^6, a^4, a^7, &c. (fig. 1.) The edges of the timbers a^1, a^2, a^{10}, a^9, are covered with battens of wood, about three-quarters of an inch thick, and all the edges of the timbers upon which the covering is fastened are defended by strips of copper outside all.

It remains to explain the mode of lowering a boat of this description into the water, and of hoisting it up. And *first, to lower the boat:* Suppose it to be frapped to the ship's bulwarks outside, and firmly lashed thereto with the gripes, which would then bear the whole weight, then the falls or tackles of the davits being rigged in the usual way are attached to spans hooked into the gunwale ring-bolts ready to receive the weight of the boat; the gripes are then cast off, when the boat falls on to the spans, and by its own weight expands. When it thus flies out and open, the thwarts, the flooring, and the stanchions all at once drop into their places, and air is at the same instant drawn into the cells; the boat may then be immediately lowered into the water.

To hoist up the boat, the process is as follows: The boat being brought alongside the vessel, with the masts and oars lashed to the thwarts, the plugs of the air-holes are removed, and the gunwales lowered by unshipping their stanchions, and tricing all the stanchions under the thwarts; the tackles are then hooked to the shackles; the first pull raises the centre of the flooring, on which the whole boat collapses, and may be hoisted up shut.

A modification of the preceding method of construction is represented in figs. 9 and 10, whereby the same may be adapted to the purpose of floating bridges or pontoons. Fig. 9 is a transverse section of a boat constructed according to this modification in its extended state; and fig. 10 an end view of it when collapsed. *a a* is the platform, one half of which is made fast to the gunwale timbers on each side; when the boat is collapsed, these pieces stand apart and parallel to each other, as in fig. 10; when expanded, they come into one plane, as in fig. 9, and their inner ends touch each other, and rest on the upright support *b*, which projects upwards from the midship timber b^1, being secured to the top of this support by a strong staple *d*, which projects from it between them; *e* is a bar or wedge of wood or iron which passes through the staple, and keeps both ends down. *c c* are two supports which abut against the midship timber *b*, and are secured laterally to the platform at d^2 d^2, these supports serve to stiffen the boat and keep it extended; they may also be prolonged upwards in a diagonal direction, so as to abut against similar pieces of the next adjoining boat. At the extreme ends of the platform there are strong hooks and eyes to unite it to those of the adjoining boats. The length of platform carried by each boat would be about 15 feet, the breadth of each boat, when expanded, about 8 feet, and its length 12 feet or 14 feet; the breadth of platform 8 feet. The thickness of the platform being 3 inches in its side rails, and that of each of the other timbers being about 1½ inch; the thickness of the whole when collapsed would be about 16 inches, so that four of them would be carried vertically upon a cart or wagon 5 feet 4 inches wide.

The coverings should be made of the same material as the life-boat, first before described, and may be one, two, or more in number. These boats would be expanded by hand, say by three men at each end of the platform, after which they might be launched. A keel e^2 and bilge-pieces *f f* should be put on to defend the boat when on shore. The platform may also be made detached from the boats, and put on when they are launched. The same arrangement of chain-hinge, stem-block, and air-holes, would be suitable for pontoons, as adopted in the case of the life-boat, first before described.

Secondly My invention, in so far as relates to instruments for sounding, consists in the improved instrument for the purpose represented in fig. 11. *a a a a* is a gutta-percha tube of about one quarter inch external diameter, and about one-sixteenth of an inch internal diameter, and drawn to these dimensions when cold. This tube may be of any required length, and serves both as line and as a communicating pipe between the lead C and indicator B. The lead C is of any form externally, and within it there is a chamber open at bottom, which contains a bag *b* of India rubber attached to, and communicating with, the tube *a, a, a, a*. The indicator B consists of a frame of wood, to which is attached a glass tube *e* and a graduated scale *f*, the tube *a, a, a, a*, communicating with the lower end of the tube *e*, through a small stop-cock *g*.

Now the bag *b* of the lead, and the whole of the tube *a, a, a, a*, are filled with a fluid lighter than sea-water (which may be obtained by diluting sea-water with fresh). Supposing the specific gravity of the water in the instrument to be to that of sea-water as 72 : 73, I place the zero point of the scale of the indicator at the level of the sea, or at the water-line of a vessel. The lead is then hove, and as it descends, the water will rise in the indicator (assuming the stop-cock to be open) one inch for every fathom, because one fathom (or 72 inches) of sea-water outside the bag and tube, will support 73 inches of the lighter water inside. Thus, for every fathom of depth, there is an inch of water in the indicator. But as an indicator more than 4 feet long (the length for 48 fathoms) would be inconvenient, one of more suitable dimensions is shown in fig. 12.

Here the glass-tube $e'e'$ is turned up below, and terminates in a bulb b, like a common rometer, the bulb and bend of the tube contain mercury; the tube a, a, a, a, as before, fastened to the neck of the bulb at c, all air is excluded, and the water rests upon the

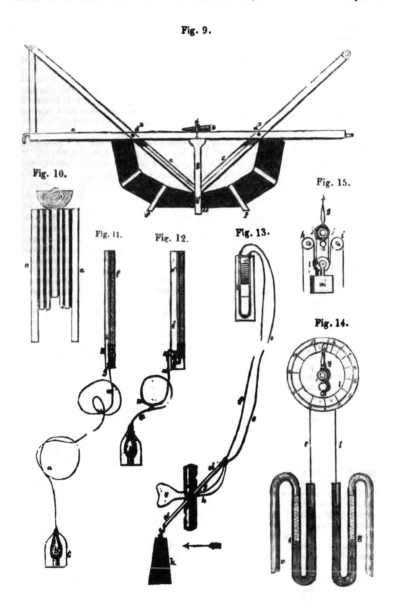

Fig. 9.

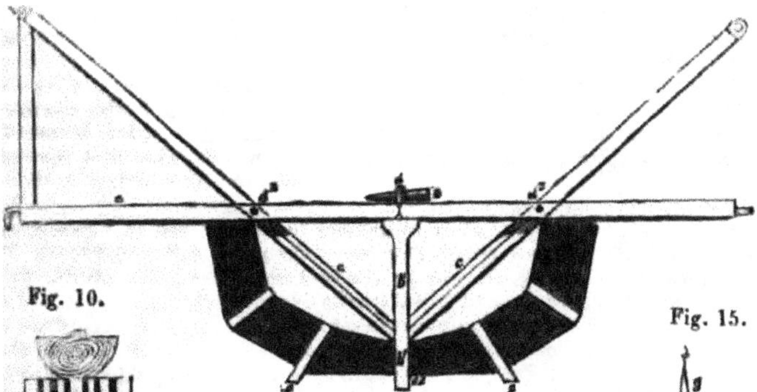

Fig. 10.

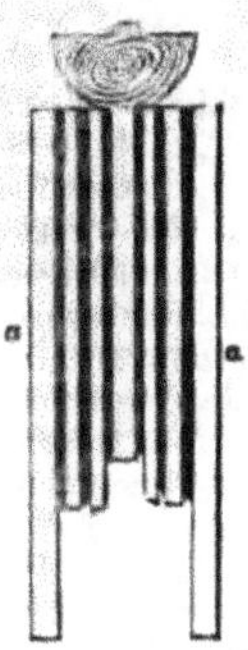

Fig. 15.

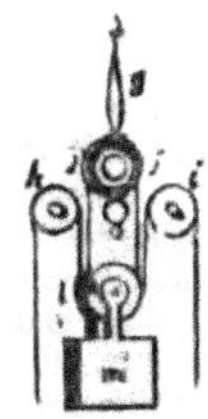

Fig. 11.

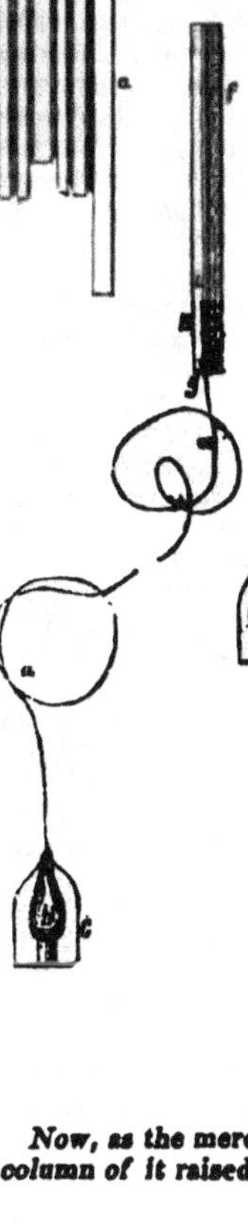

Fig. 12.

Fig. 13.

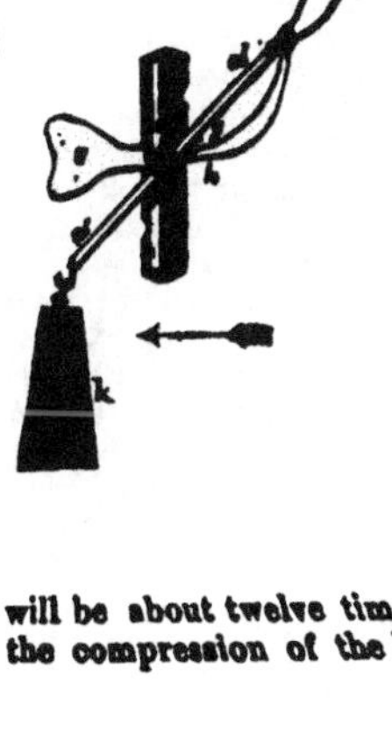

Fig. 14.

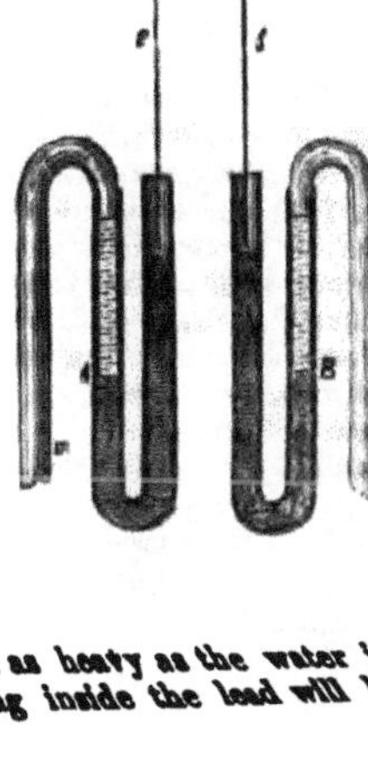

ercury. *Now, as the mercury will be about twelve times as heavy as the water in the be, the column of it raised by the compression of the bag inside the lead will be only* one-twelfth as high as in the former case, and thus one inch of mercury will indicate an immersion of twelve fathoms, or an indicator 4 feet long will measure any depth, not exceeding 576 fathoms; but, if required, another tube can easily be added above that on the indicator by slipping a small piece of vulcanised India-rubber tube over both ends when greater depths are to be sounded.

Another arrangement that will be found extremely convenient for moderate depths is to use the same apparatus with an indicator like fig. 12, with only *air* in the bag *b*, and tube *a, a, a, a*. As the bag descends, the air will be compressed in the tube, and force up the mercury in the indicator. This has the great advantage of being independent of position; that is, the indicator need not be placed at the water-line.

Thirdly.—My invention consists of an improved instrument for indicating the rise and fall of water. It is constructed on the same principle as the instrument last described, the indicator being fixed either on shore or in a moored vessel and used as a tide gauge, the scale being graduated proportionally for salt or fresh water, according as the case may be.

Fourthly.—My invention consists of an instrument for ascertaining and indicating the rate of currents, being an adaptation to that purpose of the perpetual log,* for which former letters patent for England were granted to me of date 19th of December, 1849. Fig. 13 is a side elevation of this instrument. A is a strong metal cylinder about 10 inches long, closed at both ends and divided into two chambers, *b* and *c*, by a diaphragm; *d d* is a bow-piece of a long oval form, which carries the cylinder and has eyes at the sides to receive two small projecting trunnions. The lower chamber *c* is partly filled with lead or other metal, which serves to keep the cylinder always in a vertical position, while a flat plate, or tail-piece *e*, keeps it always in one direction when sunk in a current. The lower chamber of the cylinder has an aperture in the after part, viz., the side from which the tailpiece projects, and the upper chamber has a similar aperture directly opposite to it. Close to the diaphragm there is a small nipple projecting from each chamber, and to these nipples are attached short pieces of flexible hose, *g* and *h*, which at the arch of the bow-piece *d*, are respectively connected with two gutta-percha tubes, *e e* and *f f*, like that used in the sounding instrument before described. These tubes constitute the communication between the cylinder and the indicator, as well as answer the purpose of lowering and raising. I, the indicator, is a glass tube bent into a U-shape, and half filled with mercury, each link being filled up with fresh water. The gutta-percha tubes and the chambers being also filled with water, the whole of the parts are connected together by inserting the ends of the tubes, armed each with a piece of vulcanized rubber, into the open ends of the glass. The chambers, *b* and *c*, have a piece of sponge in each, saturated with water, to prevent the entrance of air when not in use, and keep the tubes always full. The two gutta-percha tubes are "parcelled" together. The scale on the indicator is graduated like the scale of the perpetual log before referred to, and is made to slide, so as to be placed with its zero mark level with the lower surface of the mercury.

To use this instrument the indicator may be fixed or held in the hand while the cylinder is hove into the current; it sinks of its own weight, or a lead *k* may be added if necessary below. The cylinder remains vertical, and the tailpiece keeps the aperture of the chamber *b* in the direction *meeting* the current, and that of *c* in the contrary direction. Thus we have a positive force in chamber *b*, and a negative one in chamber *c*, which acting along the tubes move the mercury and indicate the rate of the current. To ascertain the rate of current at any given depth, it is only necessary to combine this instrument with either of the sounding instruments, as before described.

Fig. 14 represents another form of apparatus for indicating rates of current adapted more particularly for vessels when moored. The same construction of way-tube with its positive and negative water pipes is used in this instrument, as in my perpetual log, but instead of the water pipes *a* and *b* being connected above with air-vessels, they are united respectively with two strong inverted syphons of glass, A and B, half filled with mercury. One limb of each syphon is filled up with water over the mercury, where the water tubes, *a* and *b*, likewise full of water, are connected with them. Upon the other surfaces of the mercury are two glass weights, *c* and *d*, with small lines or strings *e* and *f* attached to them, which pass round the sheaves of the indicator I, which is provided with a circular dial and index-hand *g*. The arrangement of sheaves is shown in fig. 15; *h, i* are sheaves moving on fixed axes; *j j* is the sheave on the pivot of the index-hand, *k* and *l* are two sheaves, side by side, attached to a compensation-weight *m*. The string passes over *h* under *l*, over *j*, under *k*

---

* Described in vol. III., page 501.

er i. Thus, any force which acts equally and similarly on the mercury in both syphons same time, must affect both the strings equally, and the compensation weight m, will fall accordingly without moving the index-hand; but the current acting in contrary upon the two chambers of the way-tube, and therefore upon the two weights, the hand is turned according to the rate of current. The scale is proportioned to the the sheave on the pivot of the index-hand.

having now described the nature of my said invention in the several parts thereof, what manner the same is to be performed, I declare that the improvements which I ire as follow:—

t.—I claim the constructing of boats in the manner represented in figs. 1, 2, 3, 4, , 7a, 8, 8a, 9 and 10, and before described, that is to say, in so far as regards the framework, and the arrangements and combinations whereby the same is made to and collapse' as required, and in expanding or collapsing to fill or empty the air-where such cells are used).

nd.—I claim the improved sounding instrument represented in fig. 11, and before ed, in so far as regards the employment of the difference in specific gravity between -water and that of a lighter fluid contained in a tube communicating therewith, to t all times the depth of such sea-water; as also the modifications of the same repre-in fig. 12, and before described.

d.—I claim the improved instrument for indicating the rise or fall of water, or tide as before described.

th.—I claim the instrument for ascertaining and indicating the rate of currents, so the modification thereof, respectively represented in fig. 13, and figs. 14 and 15, fore described.

TELEGRAPHIC CASE OF MORSE v. BAIN. (SEE ANTE, P. 435.)

quote the following judicious and : observations on the decision in this om the *Scientific American*:—

e action was for an infringement of s patent by the Telegraph line from lphia to Baltimore. This line has ermed the 'Bain Line,' because a al telegraph was employed on it. was also a local arrangement of bat-e invention of Mr. Rodgers, used on e complainants alleged that all the of Morse were infringed by the de-s; viz., electro-magnetic action, a ttery, and Morse's Chemical Patent. have looked over the evidence given; s two huge volumes; and we cannot that, in relation to the practical de-ent and discovery of the principles ed in the Electro-magnet Telegraph essor Morse, our country is more in-to Professor Joseph Henry than any ving man, and he has neither received blic credit nor honour, which are is due, much less any remuneration invaluable discoveries. He was the n in the world who moved machinery Electro-magnet; and he is the inven-he 'Electro-magnet' to do this,—and this Morse's Telegraph would yet blivion.

e decision rendered amounts to this, e made the first 'Recording Tele-' therefore every *recording* telegraph fringement of Morse's patent. We different opinion, and believe that we can prove, by good logic and plain facts, that the said opinion of Judge Kane is incorrect. Let us quote his opinions fairly:—

"'Morse's patent of 1840, in all its changes, asserts his title to two distinct patentable subjects; the first, founded on the discovery of a new art; the second, on the invention of the means of practising it.

"'That he was the first to devise and practise the art of recording language, at telegraphic distances, by the dynamic force of the electro-magnet, or, indeed, by any agency whatever, is, to our minds, plain upon all the evidence.

"'The third patent is for the chemical telegraph. We do not propose to enter on the discussion of this. The subject of it is clearly within the original patent of Mr. Morse, if we have correctly apprehended the legal interpretation and effect of that instrument.'

"The Chemical Telegraph of Bain and the Electro-magnet Telegraph of Morse are totally different inventions, and, in our opinion, the Chemical Telegraph did not, does not, and cannot infringe Morse's patent. The invention of Mr. Morse consists in this, that he transmits messages to a distance using the mechanical action of an electro-magnet to do so, by making marks. It consists in nothing more, and is no less, and is a beautiful invention, and we would not ruffle a single plume which justly belongs to its inventor. The Chemical Telegraph consists in transmitting messages to a distance, not using mechanical action, but chemical action to do so, by making marks. The one telegraph cannot do what the other does at all. Morse's telegraph may be compared to the action of chiselling out letters on a plate; Bain's to etching them out. Morse's telegraph is indebted to the Electro-magnet to make the marks: Bain's uses no magnet at all. Morse's marks are made, not by the direct current of galvanism from a battery, but the secondary current force of a magnet: Bain's marks are made with the current direct, using no secondary current force. The batteries of the two are also different. We cannot conceive how any man, possessing the best scientific skill, can fail to perceive that the two telegraphs are as different in essence, principle, action, construction, operation, and the effects produced, as light and darkness. The great error in the decision, in our opinion, consists in overlooking the fact that the *Recording Telegraph* is not an art in the general sense, but only a branch of it. *Telegraphing* is an art, and signalling and making telegraphs, of which there are many, are but branches; the decision rendered, makes the recording telegraph tantamount to the *whole art*, it therefore over-rides all the testimony adduced, and hence the two huge volumes of evidence might as well have been kept in the drawers of the defendant's counsel, without submitting it at all; in fact, the evidence is shabbily treated, and former dicisions of other courts, totally different, are jauntly passed over. The plain error of the decision, to our view, lies in the first paragraph we have quoted. There can be no such a thing as an art apart from a process, and the very word *recording*—this *adjective*—relates to the process, it qualifies the *act*, and lawyers should always have the organ of comparison large enough to distinguish the difference between *the* act and *an* act. What is an art? Simply a process or manner of doing a thing; recording messages, without any reference to the means of doing so, is a mere abstraction—like an abstract soldier without a gun, blade, bayonet, or any kind of arms whatever. The common and true understanding of the term "*art*" is the manner of doing a thing. Thus we have the *Art* of printing in general; but it, like the different telegraphs, embraces different processes, all of which are distinct in themselves, and entirely different inventions. We have the art of wooden block printing (the oldest), the art of moveable-type printing, copperplate printing, and lithographic printing. These are all totally *separate* and *distinct* arts, but still they are *all embraced in* 'the art of printing generally considered.' Judge Woodbury, in his decision in the Morse and House trial, in Boston, 1850, held an *art* to be just as we have expressed it—*a* process or means of doing a thing, not a mere abstraction, as in the recent decision, raised up into a principle, and which, if once admitted into our Federal Courts, will destroy every principle of equity in them whatever.

"There are two patentable principles in Morse's patent: one is the art, process, or means (we use the word *art* as it is understood in common usage, viz., to be the way of doing a thing) of sending telegraph messages; the other the product of the art, *the* recorded message, which is the same as the word 'manufacture,' in the old laws.

"Our definition of Morse's legal claims is radically different from that expressed in the decision quoted. Judge Kane defines the product or manufacture to be the art; we, the process: hence he makes the manufacture or product cover different processes and other products—whereas a product, in the eye of all law, is specific and inflexible, the least variation from which is a different product (manufacture); and this is what we believe of the recorded messages of the Magnetic Telegraph, and every other. He considers the product or messages produced by the Morse telegraph, to be patentable—so do we; for we believe the word 'manufacture,' in the old patent law, covers this. But neither the action nor the message product or manufacture of the Chemical Telegraph are like those of Morse's telegraph; they are entirely different. There is a greater difference between the two telegraphs, in every point, than there is between the two printing arts or processes of moveable type printing and lithographic printing—both recording arts, but distinct inventions.

"It is the duty of our courts to judge every question upon its real merits; the legal rights of any man, if they are not a day old, are just as sacred as those of one hundred years old; and if our courts do not view questions in this light, then law, with them, is a mere question of privilege, rather than of right and justice.

"There is not the least resemblance, in any respect, between the inventions of Morse and Bain, and surely it cannot be equity to take away from one man that which he has invented, entirely distinct and different, and give it to another who never invented a principle of it; yet this is what the recent decision has done. In respect to the complaints, we could not conscientiously feel easy, in being awarded property that did not belong to us; but, with the author of the 'Bridgewater Ethical Treatise,' we think this is one of the questions which, man and man—the complainant defendant—will yet be settled before r tribunal than that of an earthly

feel deeply for those against whom lsion has been rendered, for we and conscientiously believe, with- disparagement to Professor Morse's n, that the inventor of the Chemical Patent has been deeply wronged, and his property, in every sense of the word, has been awarded to those who have not the least moral right to it. We could not, in conscience, feel easy with such a decision, if we were in the complainant's place. The decision does not affect the Merchants' Bain lines in this State."

NEW MODE OF SECRET CORRESPONDENCE.

Bristol, November 20th, 1851.

ear Sir,—It has long been a desi- ı to obtain what Lord Bacon states requisites for a good cipher for 'riting; namely, simplicity, faci- ts use, and above all perfect im- bility. The transmission of de- ı and messages by means of the : Telegraph, where the writing, :r may be its character, must be . to the inspection of many per- pears at this moment to make the more than ever important. I iven much attention to secret and long since hit on a plan I to be, in its most important part, , and think will fulfil these nents. Unless a letter can be in cipher without much trouble, ıs easily translated by those who ; key, it will seldom be used; ss it will defy the greatest inge- nd the most intense application her, like Bramah and Chubb's will be comparatively valueless. plan I propose, the *Times* news- ill be the best possible vehicle ransmission of *secret despatches*. requires often special messen- :f sent by mail, it is liable to violation, especially in foreign post-offices, and is always in danger of suppression or loss. Twenty newspapers may be sent by as many conveyances, some one of which is likely to reach its destination. That paper may contain the most important instructions to a foreign minister, no other human being having the ability to read them, he alone possessing the key. Every commander of a ship of war may have a key for *general orders*, and another for despatches designed for him alone. The general order will be understood by all, but the private one only by the individual for whom it is intended. A letter may be placed in the hands of a clerk, and the key given him for the first line: he will be as unable to decipher the second, without further instructions, as he would have been the first without the key.

By way of illustration I will give you two or three lines, which, if you will give place to in your Magazine, I think I may venture to predict, even were I to offer the £200 reward, would defy the best *picklock* in the kingdom.

Trial Specimen.

5.16.25.10 2.14.22.8.26.3.7 9.21 10.5.3 15.23.4,6.15.13.8.2 11,1 23.16.10.1.21—
6.2.11 418 2.7 10.23.6.16.18.8 19.25.21.5.3.1.11 23.25.26.5.1721,7.22.23.23.15.
25.22.10.13.4.1.11.21.8.3.9.22.13.9.26.11.9.15.24.24.7.7—

'ollowing is precisely the same, with another key:

6.23.26.10.4.3.16.7.1.23.9.3.21.23.1.17.17.23.15.22.11.8.19.15,24.14.14,8.12.16.
.15.8.13.20.8.24.1. 21.19.6.16.5.21, 21,4,23,8,16,14,12,23,1,8,23,12,8,10,14,8,22,
26 25,23,26.26.11.16.6.1.21.25.25.22.10.14.17.21.12.6.5.20.2.17.1—

re, therefore, two keys to the lock: I could furnish a hundred—nay, any

of the keys to these lines is con- ı this letter; as may always be the any other, or in any newspaper ıication. It is susceptible of infi- iety. Although you may have ce enough in me not to suspect a *imposition of* any kind, some of your readers may entertain doubts. I shall, therefore, after a time, give you the translation to publish, and a key for your own private use. I am, Sir, &c.,

F. B. O.

[We recommend the above to the attention of the curious in the art of deciphering.

though, if what our correspondent says of his system be correct, it is beyond the power of any one, without the help of the proper key, to make anything of it. Having ourselves been favoured with a loan of the key, we can unhesitatingly bear witness to the method being at least exceedingly simple. We insert the communication with the more pleasure that, in as far as regards the writer's mode of conveying the secret message, it is nearly identical with a plan which we had ourselves devised for the purpose some thirty years ago. That was, however, when the rates of postage were high, and the savings to be effected by the use of a public newspaper for *private* correspondence would have been of large amount.—ED. M. M.]

PLAN OF A CHEAPER METHOD OF PROPULSION FOR RAILWAYS—SUITABLE FOR BRANCH LINES.

Among the various causes which have tended so greatly to reduce the value of railway property, perhaps none has had so great an effect as the somewhat undue extension of this means of communication into parts of the country but thinly inhabited. It must, however, be admitted that their introduction into these parts has been a matter of great comfort and convenience to the inhabitants, while to the country in general it is a benefit that all parts of it should be of easy access. Any plan, therefore, that should tend to reduce the working expenses of such lines, may reasonably be expected to engage a share of attention; nor has the subject been neglected, various plans having been proposed, among which may perhaps be mentioned the atmospheric system, which has been tried on one or two lines, but has now, I believe, been finally abandoned. The attempt to work a train by horses acting on a platform in a van has also been made, but has not come into any kind of use. A third plan that has been suggested, and is, I believe, patented, is to work the locomotives by compressed atmospheric air instead of steam—the air to be condensed into large receivers by stationary engines at the principal stations, and the locomotives to take it in, somewhat as they now take in water. This plan would be cheaper than steam; but it is found that air will not work an ordinary steam engine well, and that it requires the condensing property of steam for the rapid action of the piston. This plan seems, however, the most promising; and the object of this paper is to explain a kind of locomotive suitable for this new moving power to act upon.

One of the earliest accounts of motion having been obtained by steam, is that of the well-known instrument called the æolipile, fig. 1:—"The æolipile is formed by a globular metallic vessel, which rests on pivots at B C, and on which it can revolve with perfect facility. Two tubes proceed from this ball at right angles to the pivots, shut at the extremities, but with a small aperture at the side, whence steam may escape. The pivots are the extremities of tubes, connected with a boiler below (D), as marked in the sketch. On the boiler being heated, steam passes by the pivot tubes C B into the cylinder, from which it issues by the little aperture F in the cylinder tube E. As the steam escapes, it rushes out with great force, and as it acts on the side opposite the aperture, it forces it and the cylinder to move round in the contrary direction." Several steam engines of considerable power—20-horse power and upwards—have lately been constructed on this simple principle, and are said to work with great efficiency; and it is this kind of engine I propose to adapt to railway locomotives, by enclosing it in hollow driving wheels, in the following manner:

Fig. 2 represents a transverse section of the locomotive, together with a pair of the driving-wheels. These wheels are each of them formed by two circular discs or plates of iron, a little apart, and connected by a rim of the same metal, so that their appearance is that of a solid wheel, like a grindstone, only that they are in reality hollow The hollow axle passes from the boiler, or rather air-holder, quite through the wheel through both discs. Midway in the axle, between the two discs or sides of the wheel, there is a plate fixed, which I may call a stop-plate, SP. The arms or spokes of the engine are then fitted on to the hollow axle on the air-holder side of the stop-plate, and are also fastened to the side-plate of the wheel at the points a and a. These arms are, as usual, hollow, except at the extremities, which are closed, and each of them has an aperture about a quarter of an inch in diameter on one side, a little from the end, one of which may be seen at *f*, the one on the other spoke being on the other side. There are openings in the hollow axle on the far side of the stop-plate, as represented by dotted lines, to admit of the escape of the air after issuing from the apertures of the arms.

Fig. 1.

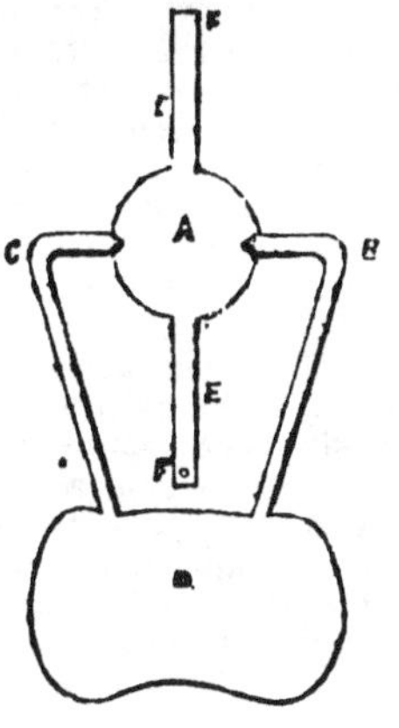

Fig. 2.

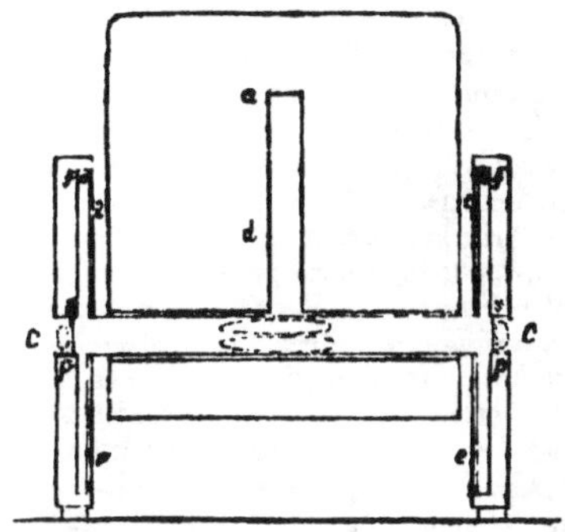

Fig. 3.

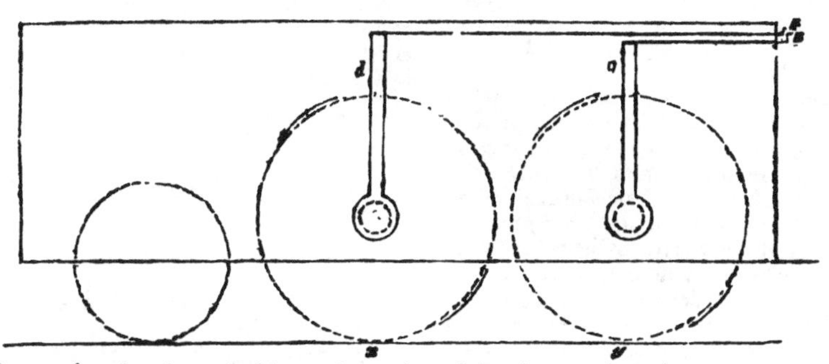

The mode of action of this engine would be as follows. On opening the valve *a* by means of the driving-handle, the air would pass down the vertical pipe *d*, and enter the hollow axle of the wheels. It would then rush along this in both directions, until it met with the obstruction offered at either extremity by the stop-plates. These would cause it to pass into the arms of the engine, and issuing out of the apertures near their extremities, causing thereby their rotation, as also that of the wheel to which they are connected. The air, after filling the hollow interior of the wheel, *would pass through* the openings in the *hollow axle on the far side* of the stop-*plate, and finally* pass out at C, whence it might be conducted to any part thought desirable.

This engine is, in fact, merely the ordinary rotatory engine incased in hollow driving-wheels, and connected with them: and it may be worked fully as well by compressed atmospheric air as by steam.

Fig. 3 represents a longitudinal section; the situation of the driving-wheels being represented by dotted lines. If the handle be turned, it will admit the air into the vertical pipe *d*, and which will consequently act on the wheels *x*, which are constructed to revolve in the direction of their arrows, and consequently to bear the engine forward. If, now, this handle be closed, and the handle B be opened, the air will pass into the vertical pipe *g*, and act on the wheels *y*, which are constructed to rotate in the direction of *their* arrows, and consequently to bear the engine backwards. Thus, there are two driving-handles, and no reversing one, which is a simpler arrangement than that of the ordinary steam locomotive, as the motion can be reversed by two actions instead of three. To reverse the ordinary locomotive, it is necessary; first, to shut off the steam; second, to draw the reversing handle;—and third, to put on the steam again. The second of these actions may be said to be equal to the other two put together; so that the manner of reversing the locomotive which I am proposing, would enable this operation to be performed in half the usual time; a circumstance of no small importance, as it would place the engine more readily under the control of the driver, and consequently tend to the greater safety of the whole train. This mode of checking the engine might also be pretty freely had recourse to, as it would not strain or injure it much more than a break. The two pairs of driving-wheels may be coupled in the ordinary manner without any detriment to the action of either.

Besides the greater safety to which I have alluded, this engine would possess, among others, the following advantages: The prime cost would be much less, probably not more than one-half of an ordinary locomotive (the only parts requiring any particular skill, being where the axle revolves in the sides of the air-holder, which it ought to do, air tight), while it would last at least twice as long: it would also require but one attendant, viz., the driver, the guard of the train having his break-van next the engine, so as to be able to render assistance in case of illness. There would be considerable outlay in the first instance, in erecting large Cornish engines at the principal stations to condense the air, together with large air-holders to contain it, and from which the locomotives would supply themselves much in the same way as they now take in water. But these engines, it is well known, last a considerable time with very little repair, and the air-holders would be still more durable. They would also be worked with coal, and at low pressure, which would be cheaper than a greater number of smaller engines, worked with coke and at high pressure. With the ordinary locomotives there is great loss in having to light so many fires, the heat also has to be maintained during the intervals of work, and when the steam gets too high, it has to be let off, all which is loss and waste. But one of these atmospheric locomotives, when charged, might remain for hours, or even days, without expense, and yet be ready for work at any moment. These are, doubtless, further advantages; but the attempts that have been made to introduce an atmospheric system into railway management sufficiently prove that this mode of propulsion is already appreciated. The atmospheric engine here proposed is different from any yet attempted; and though improvements may, no doubt, be made upon it, I believe its adoption, even as it is, would be a step in the right direction, and might effect a saving of some per cent. in the working of a line of railway.

M. G.

THE MATHEMATICAL INQUIRY.—(SEE ANTE P. 435.)

Sir,—The problem proposed by your correspondent, "A Constant Reader," is certainly not an easy one to be solved in a practical form.

If A = the given area,
d = semi-chord,
x = required radius,

the equation connecting these quantities is of the form

$$A = x^2 \operatorname{Sin.}^{-1}\frac{d}{x} - d\sqrt{x^2 - d^2} \quad (1.)$$

There is no known method of solving this equation directly. But it has occurred to me that the "Table of Segments" usually published with other Tables, might be made use of practically for the purpose required. By those Tables the *areas*, corresponding to *given*

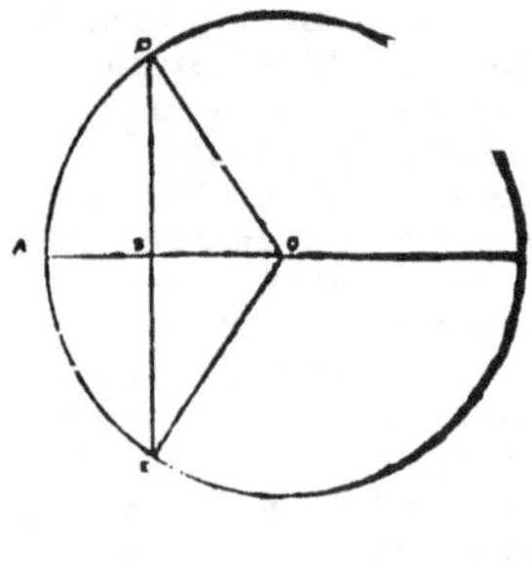

s of the segments may be ascer-. The values registered are the ponding values of

$$\frac{AB}{2\,AC} \text{ and } \frac{\text{area DAEB}}{4\,AC^2}$$

v let m and n be any correspond-lues in these Tables. Then

$$\frac{AB}{2\,x} = m \text{ and } \frac{A}{4\,x^2} = n.$$

and $AB = AC - CB.$

$$= x - \sqrt{x^2 - d^2}$$

$$\text{or } m = \tfrac{1}{2} - \tfrac{1}{2}\sqrt{1 - \frac{4\,n\,d^2}{A}},$$

$$\text{or } \frac{d^2}{A} = \frac{m.(1-m)}{n} \ldots (2).$$

it would not be difficult to form containing the values of

$$\frac{m.(1-m)}{n}$$

rresponding values of m and n. were done, the practical solution be easy, suggested by equation

ide the square of the semi- by the given area. Search the for this number, or the number t to it; and observe the correspond-lue of n. Divide the area by this of n, and the result is the square e diameter required.—Example. the diameter when the area is 8·283 yards, and the chord 12 yards.

s $d = 6$ $A = 8{\cdot}283$;

$$\therefore \frac{d^2}{A} = 4{\cdot}346.$$

along the column of values of

$$\frac{m(1-m)}{n}$$

is number; and the corresponding of

$$n = {\cdot}006391;$$

$$\therefore \text{diameter})^2 = \frac{A}{n} = 1296;$$

$$\therefore \text{diameter} = 36 \text{ yards}.$$

rouble of finding the values of

$$\frac{m.(1-m)}{n},$$

abulating them would be in reality trifling.

I am, Sir, yours, &c.,

J. A. C.

mber 2, 1851.

Another Answer.

Sir,—The following may, perhaps, meet the case proposed by "A Constant Reader" in your last Number. The formula is taken from Elliott's *Mensuration*, page 179, and may also be found in Baker's *Mensuration*, Weale's Series, page 45:

Put c = chord; h = height of segment; A = area. Then

$$\frac{2}{3}h\sqrt{(c^2 + \frac{8}{5}h^2)} = A,$$

very nearly; whence

$$h^4 + \frac{5}{8}c^2h^2 = \frac{45}{32}A^2;$$

a quadratic from which h may readily be determined. When this is done, a perpendicular $= h$ may be erected on the middle point of the chord, and then *three* points are given through which the circle will pass, and its radius is determined by *construction*. But if a *calculation* be preferred, we have

$$\frac{c^2}{4h} + h,$$

for the *diameter*, and

$$\frac{c^2}{8h} + \frac{h}{2}$$

for the *radius* of the required circle.

I am, Sir, yours, &c.,

W.

December 2, 1851.

TRIAL TRIP OF THE PENINSULAR AND ORIENTAL COMPANY'S STEAMER "RIPON."

On Monday last this vessel proceeded down the Solent to Cowes Roads, and thence to the measured mile in Stokes-bay. She performed the first run with the tide in 5 min. 8 sec., equal to 11·726 knots, engines making 19¾ revolutions. The same distance back against the tide occupied 5 min. 30 sec., equal to 10·909 knots, giving an average of 11·317. The *Ripon* is fitted with Berthon's patent log, or indicator, which showed a speed of 11·500, or 11½ knots. The new boilers with which the *Ripon* has been fitted, and which produce an abundant supply of steam, are the new patented boilers of Messrs. Lamb and Summers, of Southampton.* So little vibration of the engines was felt that a full glass of water placed in the binnacle on deck re-

* See *Mech. Mag.*, vol. l., p. 553.

mained there without a drop being spilt. The speed of the *Ripon* has been increased full 2½ knots an hour by the new boilers and feathering paddle-wheels, and her performance has consequently given the greatest satisfaction.

THE WEST INDIA MAIL PACKET COMPANY'S STEAMERS.

(From the Southampton Correspondence of the *Times*.)

The *Amazon* (see *ante* p. 473), under the command of Captain Symonds, arrived here last evening (December 16) from the Thames, and is now in the Southampton tidal dock. She is the finest steam-ship that has ever appeared in these waters, and externally is a great credit to her builders, the Messrs. Green, of Blackwall, as well as to the Royal Mail Steam-packet Company, her full and lofty rig and immense spread of canvas, together with her taut and warlike appearance, giving her at a distance more the look of a formidable war-steamer than that of a packet-ship to be engaged in the peaceful business of conveying mails, merchandise, and passengers. The voyage from the Thames has been much retarded by the dense fogs which have latterly prevailed, and on several occasions the *Amazon* was obliged to anchor in consequence of the thickness of the weather. She started from the anchorage below Gravesend on Saturday morning, but shortly afterwards brought up for the night, having made but little progress. Considerable interest is attached to the performance of the vessel, as she is the first of the new main line of steamers about to be established by the West India Mail Company between Southampton and the Isthmus of Panama. The several trials of speed were tested by the following gentlemen, who were on board, viz.:—Mr. Hughes, the Government Engineer; Mr. George Mills, the Superintending Engineer of the Royal Mail Steam Company; Mr. Brunet, of the firm of Seaward and Co., the makers of the engines; and by Mr. Austen, appointed by the Royal Mail Company to superintend the trials.

We select a few of the particulars of the several results, with a view of giving an idea of the *Amazon's* performances. On Monday morning the *Amazon* ran from the Maplin Light to the Sunk, a distance of 19 miles, in 1 hour and 30 minutes; equal to a speed of 13·6 knots, but from this should be deducted the influence of a 2-knot tide, giving an actual speed through the water of 11·6 knots, by both patent and common logs; the pressure of steam was at this time 12 lbs., the engines *making 13¾ to 14* revolutions, the consumption of coal being 53 cwt. 1 qr. 16 lbs. per hour. Subsequently the *Amazon* ran from the Kentish Knock to the North Sand Head, a distance of 20¾ miles, with no tide, in 1 hour 44 minutes; equal to a speed of 11·88 knots, or nearly 13½ miles an hour. During the night of Monday, the weather being thick, it was deemed advisable to reduce the speed of the ship, and an opportunity of trying the effect of the expansive apparatus was afforded. After running through the Downs the engines were accordingly worked on the third grade of expansion, and with this 11¾ revolutions of the engines were made, with a consumption of 27 cwt. 2 qrs. 25 lbs. of coals per hour, and a speed of 10½ knots; with the steam shut off, the engines making 9½ revolutions, and an expenditure of 25 cwt. per hour, the speed was 8·6 knots. On Tuesday morning the *Amazon* ran through Spithead, and proceeded to the measured mile in Stokes Bay. Six trials were there made, with the following results:

	Min.	Sec.		Knots.
1st run, with the tide...	5	7	= a speed of	11·728
2nd run, against tide ...	5	44	......	10·465
3rd run, with tide	4	45	......	12·632
4th run, against tide ...	5	41	......	10·657
5th run, with tide	4	42	......	12·766
6th run, against tide ...	5	33	......	10·811

The mean result of these six runs is therefore, 11·492 knots, equal to an average of 13·242 statute miles per hour. This performance is, on the whole, considered satisfactory; but a somewhat greater speed is anticipated on the official trial trip, when the stiffness of the machinery will be got rid of, and when the *Amazon* will be put in better trim than she is on the present occasion, her forward spardeck near the bows being now lumbered up with two immense surf boats and some heavy cylinder covers and machinery brought round from the Thames. The quantity of coals on board was between 500 and 600 tons; the regular quantity for sea is, however, 1,000 tons. The *Amazon* steered very easily, and with great precision for so large a ship, and in turning, the segment of the circle described was so small as to surprise the gentlemen on board. The ponderous machinery of the *Amazon* is most creditable to the manufactory of Messrs. Seaward and Capel; its beauty of finish, great strength, and admirable working are much admired by all who have had an opportunity of inspecting it. It is expected that the *Amazon* will be appointed to take out the West India and Pacific mails of the 17th of January.

The *Oronoco*, built by Pitcher, of Northfleet, the engines by Maudslay, Sons, and Field, the second ship of the line, is nearly completed, and will shortly be brought round to Southampton from the Thames.

gdalena and *Parana* will soon after-
ollow, those vessels being in a forward
Neither the builders nor engineers
of these magnificent vessels have
:tered by any conditions on the part
Company that would be likely to in-
vith the speed of the different vessels
aterial degree; the consequence is,
most eminent shipbuilders and engi-
this country have been set against
ier, and it will at some future time
which builder and engineer have
d the fleetest ocean steam-ship.
lerstand that probably the *Amazon*
Oronoco will be sent out on a trial
company, as far as Ushant, and it
e most important for the interests of
building and engineering science if
le four ships could be sent out for a
me as a sort of squadron of evolu-
order that the palm might be
to those to whom the result should
t would be justly due. The Royal
eam-packet Company are just about
act for the construction of an iron
hip of 2,500 or 2,800 tons, into
e engines intended for the unfortu-
merara (recently stranded at Bris-
d which are now nearly completed,
e placed. She will be ready for sea
onths, and the builders will proba-
Messrs. Caird and Co., of Greenock.
ship is also to be built, of immense
take the place of the *Demerara*, but
: be ready for perhaps 18 months.
lmiralty have permitted the Royal
mpany to construct one iron steam-
an experiment, for the West India
rvice; but this concession is only
, in consequence of the Company
ng more vessels than the letter of
ntract requires. It is intended to
e that the two new ships shall be of
ater speed than the four now pre-
for sea.

CATIONS OF ENGLISH PATENTS EN-
LED DURING THE WEEK, ENDING
EMBER 16, 1851.

c HAZLEHURST, of Marton, Dal-
eel refiner. *For certain improve-
s the manufacture of iron.* Patent
une 3, 1851.

invention consists of an improved
of operating upon pig iron in the
g furnace, so as to obtain the metal-
in a spongy state, and admit of its
ground or pulverised previous to
oing the second puddling operation.
n produced by this process is pecu-
adapted for the manufacture of spades, shovels, sickles, scythes, hooks, rivets, boiler-plate, sheet-iron, tin and block plates, wire, and other purposes where iron of a superior quality is required.

In working according to his improvements, the patentee introduces into the ordinary puddling furnace the usual charge of pig iron, or pig and refined iron, or pig iron with a mixture of iron ores and carbonaceous matters,—such as ground coal, coke, charcoal, or sawdust; he then melts the iron, and brings it to as thin and liquid a state as possible for the purpose of clearing the metal, and conducting the process of puddling in the usual manner up to this point. He then lowers the damper until the iron begins to thicken, when it is again boiled and kept very hot, in order to bring it again to a thin and liquid state. The draught of the furnace should then be lowered, until the iron becomes malleable and is ready to "ball." The iron is then drawn out of the furnace in pieces of any required size, but without balling; and the lumps are placed in barrows or other receptacles, where they can be shut in and kept from the external air until cool. When cold, the lumps will be found to be of an open, spongy, and honeycombed texture, and must be ground or pulverised with rollers, or stampers, and the bad iron or crude, or imperfectly worked metal, dirt, and other impurities which would injure the quality of the iron, removed therefrom by picking and sorting. The manufacturer then takes sufficient of this ground iron to produce a bloom or bar of the size intended to be made, which is introduced into the same, or any other suitable furnace having a sand or cinder bottom, and balled at a low heat; after which, the metal is finished under the hammer, or in squeezers, or rolls of the ordinary description; as iron produced by this process requires no cutting or piling.

The iron thus manufactured will be found peculiarly applicable to the purposes above mentioned; but when it is to be polished after being made up, the ground iron should be scoured before being subjected to the second puddling operation. The ground iron may also be used for sinking in the charcoal fires, and for making iron to be converted into steel,—for which purpose charcoal pig iron, or best scrap iron, have hitherto been employed; and by this means, not only is a considerable economy in the cost of the iron effected, but iron of equally good quality produced.

In conclusion; the patentee observes that although he has in his description mentioned certain processes in the manufacture of iron, which are well known, he does not intend to claim the same, nor does he claim the employment of carbonaceous matters, as above mentioned; but what he does claim is—Stopping the puddling process at a certain point, so as to obtain the iron in a spongy or honeycombed state, and then grinding or pounding the same while in that state, so as to admit of the picking or sorting operation taking place, in order that previous to any subsequent working the crude iron, or dirt, or other impurities which would injure the quality of the iron, may be removed therefrom.

JAMES BANISTER, of Birmingham, brass founder. *For improvements in the manufacture of metallic tubes for steam boilers and other uses.* Patent dated June 7, 1851.

These improvements consist in causing the overlapping edges of metallic tubes to be soldered or brazed in a muffle or oven. The usual practice has hitherto been to effect this operation by passing the tube through a fire, and in consequence of the great heat to which some parts of the tube are thus exposed, and the injury to the metal resulting from this cause, this plan is considered objectionable. Mr. Banister's oven or muffle, is so constructed, that the tube to be soldered is subjected successively to different degrees of heat, and the greatest amount of such heat takes effect immediately on the edges to be joined together. The muffle is constructed with two furnaces, one at a lower level than the other; the first of these furnaces heats the end of the muffile at which the tubes are introduced, by radiation, and the flame and heat therefrom are conducted through a flue to the second furnace, which is on a level with and alongside of the muffle, but at its further end, and issue thence, together with the products of combustion from that furnace, into the muffle near the end at which the tubes are withdrawn, returning towards the opposite end of the muffle; but just beyond the point where the flame enters the muffle, and opposite to it, there is a flue leading to the chimney, by which part of the heat escapes, while the rest passes on towards the entrance end, and is conducted away through a flue to the chimney. It will be seen that the muffle is thus of three different degrees of heat, the lowest of which serves to heat the tube; the next degree of heat causes the solder to begin to melt; while the greatest heat effects the union of the portions of the tube by melting the solder, after which the tube is immediately withdrawn. The solder which the patentee prefers to employ consists of a mixture of two solders melting at different heats; the first is composed of 40 parts of spelter to 36 parts of copper, and the second of 40 parts of spelter, and 42 of copper. These are prepared as if for use separately, and equal quantities of them are mixed together, and, with borax, constitute a solder which effects a more perfect junction of the edges of the tube than that ordinarily employed.

Claim.—The mode described of uniting the edges of metallic tubes.

ROBERT ALEXANDER KENNEDY, of Manchester, cotton spinner. *For improvements in machinery applicable to engines for carding cotton and other fibrous substances.* Patent dated June 10, 1851.

These improvements consist,—Firstly, in arranging the rollers for the endless belts or flexibles of carding engines, which carry the sheet or web of wool, cotton or other material, in such manner, that although less longitudinal space shall be occupied by the belt, it shall be of equal length with those of the ordinary carding engines, and this is effected by employing several rollers instead of two only, over and under which the belt is passed; and secondly, in constructing the endless belts of such machines of cotton, or other suitable fabric, woven in one piece without a seam.

Claims.—1. The arrangements and combinations of the rollers for the flexibles or endless belts of carding engines as described, or any modification thereof.

2. The application of woven fabrics for making the said endless belts, whether employed in machinery arranged as described, or in any machine of the ordinary construction.

FELIX CHARLES VICTOR LEON LEVACHER D'URCLE, of Paris, farmer. *For improvements for increasing the produce of autumn wheat.* Patent dated June 12, 1851.

These improvements are based on a supposed discovery of the patentee's, that autumn wheat is, contrary to the generally received opinion, a biennial plant, and it is to develope its natural capabilities, and rescue the plant from the state of degeneracy to which a long course of improper management has reduced it, that is the object proposed to be attained by him. The ground in which the wheat is to be sown must be tilled and well manured, and the sowing is directed to take place between the 20th of April and the 10th of May; it may be a few days earlier or later, but somewhere between those dates is, the patentee says, the proper moment. The field having been divided into squares of about a quarter of an acre each, diagonal rows of holes are dug at a distance of from 15 to 20 inches apart, in each of which are deposited four grains of wheat arranged in a circle, or otherwise at a little distance from each other. This done, the holes are filled in, and when the plants have grown to a height of about 4 inches,

; of the four plants which are sup-
have sprung up from the seed,
pulled up, leaving one plant only,
the strongest and most healthy,
uce of wheat from which, when it
maturity, will be very considerably
over the usual yield. By follow-
his course of treatment the quality
ain will be increased each succeed-
until it arrives at a state of perfec-
which, under the ordinary system
osed to be quite incapable.
—Developing the biennial proper-
utumn wheat by the process de-
by which its produce will be very
reased.

AM HENRY FOX TALBOT, of Lay-
bey, Chippenham. *For improve-
photography.* Patent dated June

st part of this invention consists
ing photographic images on plates
repared by the following means:—
f glass should be selected having a
nd well-polished surface; and in
btain a photographic picture the
proceeds as follows:
takes albumen, or white of egg,
s the most liquid portions thereof
; the rest) with an equal quantity
, and having spread the mixture
and evenly over the surface of the
ows it to dry spontaneously, or
t a fire.
mixes an aqueous solution of
silver with a large proportion of
o that the mixture shall contain
ree grains of the nitrate to each
liquid. (This proportion may be
om one to six grains in the ounce
; but three grains is considered to
st proportion.)
dips the prepared plate for a few
into this mixture, then withdraws
it by a gentle heat, or allows it to
taneously.
dips the plate into distilled water,
e any superfluous nitrate of silver.
applies a second coating of albumen
me way as above directed, and dries
by the application of gentle heat,
the use of too much heat, by which
te of silver might be decomposed.
takes an aqueous solution of prot-
iron, containing 140 grains of the
le to the ounce of water. A small
of free iodine in the solution, by
colour would be rendered slightly
will be found to be of advantage.
measure of the solution he adds one
acid and ten of alcohol, and allows
re to stand for a few days previous

7. He dips the plate into the solution, or allows the liquid to pass over the whole of its surface in a continuous stream. It is then dried, when it should be of a pale yellow colour, very clear, and uniformly transparent; and this completes the preparation of the plate. All the preceding operations may be performed in moderate daylight, but avoiding exposure to too strong a light, or to sunshine.

8. When it is desired to obtain a photographic picture, the operator takes a solution of nitrate of silver containing one hundred grains of nitrate of silver to an ounce of water, and, having mixed two measures of the same with two of acetic acid and one of water, he dips the albuminized plate therein once or twice, for a few seconds each time (performing the operation in a darkened room or by candlelight), for the purpose of rendering it sensitive. If the weather is cold, the plate should be slightly warmed before so dipping it. He then removes it to the camera without loss of time, as the plate ought to be used a few minutes after taking it out of the solution; and when a sufficiently strong photographic image is supposed to be obtained, the plate is transferred from the camera to the dark chamber or operating room.

9. It is then immersed in a solution of sulphate of iron, composed by mixing one measure of a saturated solution thereof in water with two measures of water (but the solution may be stronger or weaker, at the discretion of the operator), by which the previously invisible images will be rapidly rendered perceptible.

10. The plate is then washed, and dipped in a rather strong solution of hyposulphite of soda in water, which, generally, in about a minute renders every part of the image more distinct and visible. The picture is then washed in distilled water, and the surface of the plate may be cleansed from any particles of dust, or other impurities, by rubbing it gently with cotton dipped in water; and if the above-described operations have been properly performed, the surface of the plate will not be at all injured by this cleaning. The picture is then dried, and the operation is finished. For the purpose of better preserving the picture, the plate may be covered with a coating of albumen or fine transparent varnish.

Although throughout the above processes certain proportions of chemical substances have been named, they may be varied very considerably, as is also the case in photographic operations generally.

The images obtained by his improved method, Mr. Talbot calls "Amphitypes," because they appear either positive or negative, according to the circumstances of light under which they are viewed. Thus, if held against a bright light, or against a sheet of white paper, they appear negative, and the reverse when held against a black surface and seen in obliquely reflected light. It is in the power of the operator, by varying the proportions of the chemicals employed, to obtain at pleasure positive images more or less distinct in comparison with the negative images. When it is intended to copy the image upon paper, it is desirable to obtain as strong a negative as possible on the glass plate, which is then copied on the paper, to produce thereon a positive image in the usual manner; but when the operator wishes to have a picture on the glass, he should endeavour to obtain a strong positive image. When this is obtained to his satisfaction, it may be preserved from injury and from contact of the air, by pouring black paint over the pictured side of the plate, and then by turning the glass the picture will be seen correctly, and not reversed as regards the right and left sides. This method of blacking one side of the plate is not however any part of the present invention. Throughout the specification the words negative and positive are made use of in the sense in which they are generally employed by photographers, viz., a positive image is that in which the lights and shades of the object are represented by lights and shades on the photograph, and a negative image is that in which a reverse effect is produced.

The method of operating just described is that which Mr. Talbot recommends when the object is close at hand, and the operator is in the vicinity of a darkened room, to which he can retire for the purpose of rendering his plates sensitive; but under circumstances where the object is at a distance, and when the operator is on a journey or otherwise removed from any house or place where such conveniences exist, the following method of procedure may be adopted:—The operator constructs a glass cell with equal and parallel sides, open at the top and closed at the bottom and sides, and quite water-tight, of a size just sufficient to receive one of the photographic plates, but not much greater, in order that there may be no waste of the chemicals employed. The posterior glass of the cell has one of its sides ground or unpolished, and the cell, when in use, is placed at the hinder part of the camera, so that, when directed towards an object, the unpolished or ground surface may answer the purpose of the sheet of ground glass introduced in cameras to place the objects in their true focus. Allowance must, of course, be made for the unusual position occupied by the ground glass in this case. The top of the cell is provided at one corner with a funnel for the introduction of liquid, and the bottom is furnished with a stop-cock and waste-pipe terminating in a caoutchouc tube, which may be moved by hand from one to the other of two vessels which are provided to receive the used liquors escaping from the camera: the nitrate of silver solution is too expensive to be wasted, but the other ingredients, when once used, may be thrown away. These preparations made, the operator pours into the cell a quantity of liquid sufficient to fill it nearly full when it contains one of the photographic plates, and notes the quantity required. He then provides four bottles of that capacity, one of which he fills with solution of nitrate of silver, prepared as before directed under operation 8; the second bottle is to contain a solution of sulphate of iron, as directed under operation 9; the third bottle is filled with water, and the fourth with a strong solution of hyposulphite of soda. These quantities are sufficient for obtaining a single photographic picture, and when they are used the bottles must be filled again. Having prepared a number of glass plates by means of the processes before described up to No. 7 inclusive, they are to be packed in a box ready for use: the operator, when he desires to obtain a photographic picture of an object, takes one of the plates from the box (which he can do without injury to it, as the plates in this condition are not sensitive to light), and places it in the camera, the focus of which he adjusts to the object. He then closes the front lens or object-glass, lowers a curtain over the camera box, leaving exposed only the funnel at the top (and care should be taken to guard against any light entering through this) and the waste pipe at the bottom of the cell, and pours into the cell through the funnel the contents of the first bottle (nitrate of silver solution), for the purpose of rendering the plate sensitive to light. He may then proceed in two different ways. That is, he may open the front lens, and obtain the image while the plate is immersed in the solution; or, before opening the front lens, he may allow the nitrate of silver solution to escape through the waste pipe, and he will then obtain an image on the plate while the liquid is adhering to its sides. In the latter case, or after allowing the solution to escape, if the former method is adopted, he closes the stop-cock, and successively pours into the cell the contents of the second and third bottles, allowing each to remain in for about half a minute; and, finally, he pours in the hyposulphite of soda solution, after which the plate is removed, and the image being now nd not liable to injury from expo- air, the plate is washed and placed x to be finished and varnished when rs' operations are completed. An- nethod, but one which is less simple, se four bottles of larger size than above described, but containing the quids. These bottles are placed on above the camera, and from each n descends a tube of India-rubber ed with two stop-cocks, which are at such distances apart, that the l of tube between them shall be of a y equal to that of the cell when it s a plate. These tubes dip into a which communicates by a suitable ith the funnel leading to the cell. quids are successively supplied to the l from the bottles, and the method of ng according to this system is the same just described. The images obtained s by these means may be copied on er in the usual manner. In fixing ages on paper, it is recommended, ashing them, to immerse the paper ot solution of iodide of potassium, dipping in the solution of hyposul- of soda, by which means a better n of the image will be obtained.

ler this branch of his invention Mr. claims the mode of preparing the plates, especially the use of a weak n of nitrate of silver immediately e first coating of albumen; also the nt use of protiodide of iron and sul- of iron upon albuminized glass plates; so the simultaneous production upon plates of images, which are both e and negative according to the n which they are viewed. (In the cation of a patent granted to Messrs. e and Talbot, the 19th December, a method is described of producing nages which differs from the present prior formation of the negative which is afterwards converted into a e one.) Also, the apparatus described used along with a camera, enabling erator to work without the necessity kening the apartment in which he or of employing a tent or other ance for working in the shade when photographic pictures at a distance ny house. The form of the appara- ay be considerably varied, but the al point is that the glass plate is in the cell in a partly prepared state, h it is insensible to light, and is not d from the cell until the photogra- cture is finished, with the exception final washing and drying. The paten- s not claim as new the mere use of a all containing nitrate of silver, into which the photographic plate is dropped previous to, or during the formation of the image; but he claims the addition of the stop-cock and waste-pipe, and the general arrangements which render unnecessary the removal of the plate from the cell before the picture is finished. He states, also, that he believes the arrangement of four vessels furnished with tubes and stop-cocks for pouring measured quantities of different fluids into the glass cell to be a new one.

The second part of the invention consists of a method of obtaining, under certain circumstances, the photographic picture of objects which are in rapid motion. An electric battery of the greatest power which can be conveniently obtained is arranged in a darkened room, and, supposing the moving body whose picture is required, is a wheel revolving upon its axis, the camera is placed at a convenient distance from it, and adjusted so as to have the image of the object in its focus. A glass plate is then taken, which has been previously prepared in the way described above, and it is rendered sensitive with nitrate of silver in the way also above described; it is then placed in the camera, and the electric battery is discharged, producing a sudden flash of light, which illuminates the object; the image thus taken on the glass plate is then rendered visible, and the process finished as before directed. If the process is properly conducted, a distinct positive image of the moving body will be seen upon the glass, the rapidity of the motion not affecting the accuracy of the delineation.

What is claimed under this head of the invention is the use of the instantaneous light of an electric battery in such a way as to obtain the photographic image of a body illuminated thereby.

WILLIAM BIRKETT, of Bradford, agent. *For improvements in obtaining soap from wash-waters.* Patent dated June 12, 1851.

In carrying out these improvements soap-suds or wash-waters are subjected to evaporation (which may be done in any convenient vessel), until they are reduced to a pasty or cream-like consistency; a quantity of common salt or other soluble salt is then added, and the addition of such salt is continued during the boiling, until the fatty matters of the soap in the wash-waters are separated from the water and salt to such an extent that (on being allowed to settle for a time) the fatty matters may be skimmed off. They are then removed to another pan, where caustic soda or potash ley is added to them, and hard or soft soap produced according to the nature of the alkali employed. Or, the product of the evaporation is combined with fatty matters or rosin, and converted into soap. The ley or refuse resulting from the above process is allowed to settle, and the clear liquor evaporated, so as again to be used in place of the salt solution, above described, for separating the remainder of the water from fresh quantities of soap-suds or wash-waters.

The patentee makes no specific claim, but states in conclusion, that the novelty of his invention consists in treating suds or wash-waters by evaporation, and separating the remainder of the water and refuse matters by a suitable soluble salt, and then employing alkali to convert such products again into soap.

JOHN CHATTERTON, of Birmingham, agent. *For certain improvements in protecting insulated electro-telegraphic wire, and in the methods and machinery used for the purpose.*

The particulars of this invention will be given in an early Number.

Specification due, but not Enrolled.

JOHN EMANUEL LIGHTFOOT, of Broad Oak, Accrington, calico-printer, and JAMES HIGGINSON, of Cobourg-terrace, Manchester, chemist. *For improvements in treating and preparing certain colouring matters to be used in dyeing and printing.* Patent dated June 12, 1851.

WEEKLY LIST OF NEW ENGLISH PATENTS.

Thomas Twells, of Nottingham, manufacturer, for certain improvements in the manufacture of looped fabrics. December 15; six months.

Frederick William Norton, of Paisley, Renfrew, North Britain, manufacturer, for certain improvements in the manufacture or production of plain and figured fabrics. December 16; six months.

John Gedge, of 4, Wellington-street, Strand, Middlesex, for improvements in the treatment of certain substances for the production of manures. (Being a communication.) December 16; six months.

James Souter and James Worton, of Birmingham, for improvements in the manufacture of papier-mache and in articles made therefrom, and in the manufacture of buttons, studs, and other articles where metal and glass are combined. December 17; six months.

WEEKLY LIST OF DESIGNS FOR ARTICLES OF UTILITY REGISTERED.

Date of Registration.	No. in the Register.	Proprietors' Names.	Addresses.	Subjects of Design.
Dec. 12	3057	Edward John Dent	Strand	Prismatic balance.
13	3058	James Neighbour	High-street, Windsor	Geometrical fimbria, or shirt, with graduating corset.
16	3059	Charles Rowley	Birmingham	Lead and slate pencils, and crayon sharpener.
,,	3060	Williamson & Roberts	Heaton Norris, Lancaster	Apparatus for taking up the cloth in looms.
,,	3061	Edward Kesterton	Long-acre	Frame for carriage-windows.
17	3062	Joseph Welch and John Margetson	Cheapside	Oxonian shirt front.
,,	3063	Samuel Whitfield	Birmingham	Fastening for metallic bedsteads.

CONTENTS OF THIS NUMBER.

LONDON: Edited, Printed, and Published by Joseph Clinton Robertson, of No. 166, Fleet-street, in the City of London— Sold by A. and W. Galignani, Rue Vivienne, Paris; Machin and Co. Dublin; W. C. Campbell and Co., Hamburg.

Mechanics' Magazine,

MUSEUM, REGISTER, JOURNAL, AND GAZETTE.

No. 1481.] SATURDAY, DECEMBER 27, 1851. [Price 3*d.*, Stamped, 4*d.*

Edited by J. C. Robertson, 166, Fleet-street.

CAPTAIN HEPBURN'S PATENT VENTILATED CARRIAGE.

Fig. 1.

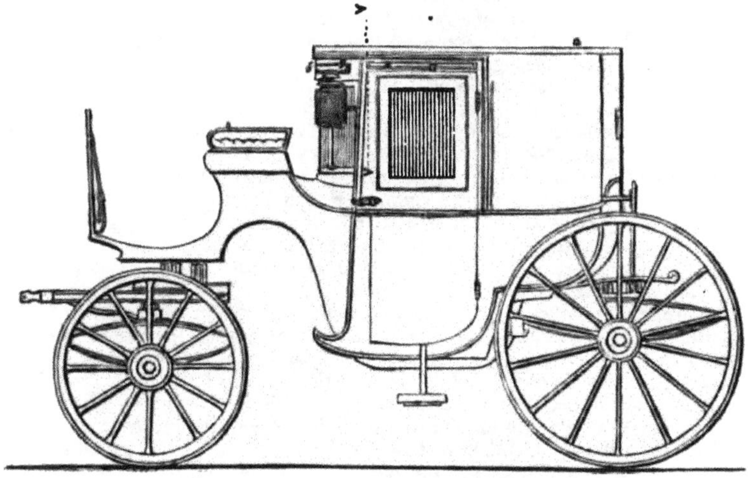

Fig. 4.

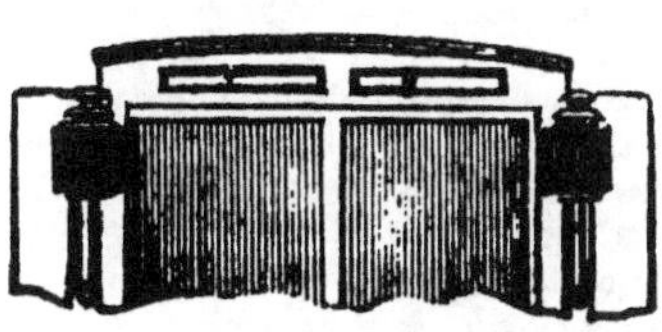

Fig. 6.

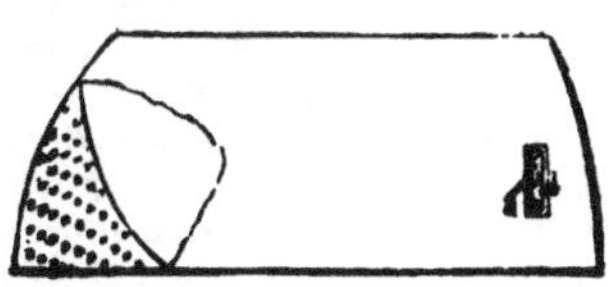

Fig. 5.

Fig. 7.

Fig. 8.

CAPTAIN HEPBURN'S PATENT VENTILATED CARRIAGE.

(Patent dated June 17, 1851. Patentee, Francis John Swaine Hepburn, Captain H. P. Specification Enrolled December 17, 1851.)

Specification.

My invention consists in ventilating carriages and other vehicles and conveyances by constructing them with a second or interior roof or lining, in addition to an ordinary or other exterior roof; the said interior roof or lining being composed of perforated metal, cloth, or some other substance or material permeable to air, or having apertures therein, and the chamber or chambers between the roofs having apertures in its sides, or being otherwise constructed to communicate with the external atmosphere.

And having now set forth the nature of my said invention, I will proceed to describe the manner in which it is to be performed:

Fig. 1 is a side elevation of a carriage commonly known as a "Clarence." Fig. 4 is a front elevation of the upper portion. Fig. 5 is a transverse section of the upper portion taken through the line Y Y in fig. 1. Fig. 6 is an inverted half-plan view of the interior roof or lining. Figs. 7 and 8 are sectional views of one of two apertures (furnished with valves), which form a communication, when desired, between the external atmosphere and the chamber or chambers before mentioned—the former figure showing the valve open, and the latter showing it closed. *a* is an ordinary external roof, supported on the upper part of the hoop-sticks or framing *b*; *c* is the interior roof, supported on the framing *d*, and by the under side of the hoop-sticks. This roof is shown most clearly in the detached view, fig. 6, where it is represented as composed of a piece of perforated zinc covered with cloth, a corner of the latter being turned back to expose the former to view. *e* is the chamber between the two roofs. *f f* are apertures in the sides of the chamber, which communicate with the external atmosphere. *g g* are valves for closing these apertures when desired. These may be opened by pulling the looped cord *h*, with which they are furnished (see figs. 7 and 8) and passing it over the stud *i*. On releasing the loop, the valve will close by the action of the spring *k*. The constructing the interior roof, in the manner set forth and described, effects a division of the current of air into minute streams, which avoids the inconvenience of direct and strong currents; but if in any case it should be desirable to increase the circulation of air to a greater extent, the valves *l l* in the interior roof may be opened for the purpose. *m* is a valve in the lower portion of the front of the carriage, which may be opened when desired. The other parts of the carriage are constructed in the ordinary manner.

I would wish it to be understood that, although I have described perforated zinc covered with cloth as the material I prefer to use for the interior roofing, the same may be made of perforated metal only, or cloth only, or of wire gauze, or of wire gauze and cloth, or of any other suitable substance, material, or combination of substances or materials, which will serve to divide the air into minute streams, and avoid violent or strong currents. The valves may also be shown in any other convenient modes than those represented in the figures. It is also to be understood that although I have only shown my invention as applied to a Clarence, the same may be adapted to a railway carriage or to a passenger canal boat, or to a palanquin, or to an omnibus, or any other description of roofed or inclosed carriage or vehicle.

ON HALBERT'S METHOD OF SOLVING NUMERICAL EQUATIONS.

Sir,—In No. 1476 of this Magazine I gave an extract from Wallace's "Algebra," and requested some possessor of *Halbert's* "Arithmetic" (1789) to furnish you with an example of his method of solving cubic equations. The work appears to be sufficiently scarce, and almost *un-come-at-able*; but my friend, Mr. George Sanderson, of Galashiels, having kindly forwarded me his copy by post, I am in a position not only to gratify myself, but probably some of your readers, by presenting them with Halbert's method entire. It occupies pages 278—286 of his "Arithmetic," and is as follows:

Cubic Equations.

When an equation involves three dimensions of the unknown quantity, viz., the cube, square, and simple power, it is called a *cubic equation*, or though the square be wanting, and it involve only the cube and simple power, it is still called a *cubic*, and in that case is said to want the second term; this happens sometimes, and, when otherwise, it may always be reduced so as to want the second term, and therefore we treat of such in the first instance:

Example 1.

Suppose this equation $a^3+a=10000$ were given, and the value of a were required, we see, that if a the unknown quantity, were deducted from 10000, there would remain just the cube of a, so that, extracting the cube root of the remainder, were it possible to be known, would resolve the equation. Now, this would be effected if, in extracting the cube root, we always subtracted from the resolvent the same figure we put in the quotient; but much easier, if we add 1 to our divisor.—See the work:

$a^3+a=10000$ (21·528874940224...

```
            8 2
1201 }    ------
 123 }     1980
 ----     ------
           1201
             61
          ------
           1262
132400 }  -------
  6375 }   718000
          -------
           662000
            15750
              125
          -------
           677875
13877500 } ---------
   26052 }  40125000
            ---------
            27755000
               25808
            ---------
            27780808
          -------------
1890331200 } 12344192000
  &c.      }    &c.
```

Here the cube root of the first period being 2, *and standing in the place of tens, subtract it from the given quantity*, and at *the same time its cube from the first period*, and you have 1980 for a resolvend; find your divisor as usual* 1,200, to which add 1, and go on to extract the root as far as you choose; but when you have got half the figures you intend, you may find the rest by contracted division, or the greater half may be found so.

As the most laborious part of the work in extracting the cube root when the figures are numerous, is the squaring the quotient for a new divisor, I have found a way to make that part of the work entirely easy, which I reckon a most useful contraction. The squares of any number being known, the square of that number increased by one figure may be had thus:—multiply the increased number by its last figure, and again by the same figure; but the second time omit multiplying the right-hand figure; add these two lines; add also their sum to the former square, shifting two places to the right, and you have the square of the increased number.†

.. ..
.. ..

By finding the divisions this way, we also save the trouble of adding the coefficient of the unknown quantity more than once, *which would be to add to every new divisor found the common way*. Proceed now to another example:

Ex. 2.—$a^3+5a=60$.

If the simple power of the unknown quantity has a coefficient prefixed, your first divisor must be increased by that coefficient; and its product by the first quotient figure must be subtracted from the given quantity, thus:

$a^3+5a=60$ (3·49099....

```
           42
           --
3200)18000
     -----
     12800
      1440
        64
     -----
     14304
     -----
     3696, &c.
```

Carry on the work as far as you please, finding your divisors as taught above:—the cube of 3 here was 27, to which add 3 times the coefficient of a, made 42; which we

* "As usual," that is, "the square of the quotient, multiplied by 300, is your divisor;" as the author remarks in his Rule for the cube root.—T. T. W.

† On this principle the various divisors in the examples are determined, and the author exemplifies the method by giving the detail of the divisors in Ex. 1. These need not be here transcribed.—T. T. W.

subtracted together from the given quantity. Our first divisor was 2700, to which add the same coefficient, made 3200; *for the* 5 *must be added here in the place of hundreds*, as you may understand by considering where its product ought to be subtracted, or, if the given quantity consist of *one* period, this figure must be added *three* places from the right; but if of *two* periods, add it to the *unit* figure.

Example 3.

$a^3+7a=2$ (·28249374....

```
              1·4
             ----
             ·600
             ·008
             ----
 71200 }    ·592000
  1152 }    -------
             569600
               3840
                512
             -------
             573952
             -------
 7235200 }   18048000
    3372 }   --------
             14470400
                 3368
             --------
             14473768
             --------
723857200 }  3574232000
     &c.  }    28954288
               --------
                6788032
                6514714
                -------
                 273318
                 218157
                 ------
                   &c.
```

Our first figure here was a decimal, and its product by 7, subtracted from 2, leaves 6; from this take the cube of ·2, which is ·008, remains 592000; then our divisor is 1200, to which *prefix* 7 the coefficient. This is *really* putting it in the place of hundreds, if you remember your quotient figure was a decimal; then go on with extracting, and when you have gotten three figures, you find other four by contracted division. But if the simple power of the unknown quantity has a *negative* sign, in this case you must *add* where you subtracted before, as it is plain the unknown quantity *wants* the product of the unknown by that figure, to *make up the cube of* the unknown.

Example 4.

$a^3-5a=1$ (2·3300587396..

```
              10
              --
              11
               8
              --
   700 }    3000
   387 }    ----
            2100
             540
              27
            ----
            2677
            ----
108700 }   333000
  4167 }   ------
           326100
             6210
               27
           ------
           332337
           ------
           663000
             &c.
```

Here the product of the coefficient of a, by 2, the first figure in the quote, was 10, to be *added*; and then subtract 8, the cube of 2, annex three ciphers as usual, and find your first divisor 1200; from which take 5, your coefficient in the place of hundreds, as before directed, and go on with the operation, finding the divisor the contracted way; and having got five figures by extracting the root, you may find other five by contracted division.

Example 5.

$a^3-12a=10000$(21·73....

```
             240
             ---
           10240
            8000
           -----
 1188 }     2240
  123 }     ----
            1188
              61
            ----
            1249
            ----
131100)   991000
  &c.        &c.
```

But supposing the equation to contain the *squre* also, which is often the case, then that equation may be reduced to another, which will want the square, or *second* term, by the following rule, which is universally known.*

* The *statement* of the rule is omitted as being sufficiently evident from the subsequent process. —T. T. W.

.. ..
.. ..

Example 6.

$$a^3 - 15a^2 + 63a = 50.$$

To exterminate the second term, make $e + 5 = a$. Then

$$\begin{aligned} a^3 &= e^3 + 15e^2 + 75e + 125 \\ -15a^2 &= -15e^2 - 150e + 250 \\ 63a &= 63e + 315 \end{aligned}$$

$$\therefore\ e^3 - 12e = -15.$$

Now find the value of e, thus:

```
      -15(2·396475......
        24
        --
         9
         8
       ----
 00 }  1000
387 }  ----
       0000
        540
         27
        ---
        567
      -----
38700(433000
     &c.
```

Here the unknown quantity being *negative*, 24 added, makes only 9; from which take 8, the cube of the first quotient figure; our first divisor would be 1200, but 12 our co-efficient, being subtracted in its *proper place*, leaves nothing, &c. Having found the value of e to be 2·396, a being five more, must be 7·397475......, which is gotten with very little trouble, after a little practice. So that I reckon this method a valuable discovery, when compared with the jargon we meet with in other authors, about *Transmutations*, *Limitations*, and *Approximations*, and what never brings us the nearer our purpose."

From an examination of the preceding extract, it will be evident that Mr. Halbert's method is applicable only to equations of the form $x^3 \pm ax \pm b = 0$, and that the *principle* upon which equations of this form are solved is that of *adding to*, or *subtracting from*, $\pm b$ at *each* step, *such a quantity as shall reduce the equation to the form* $x^3 \pm b = 0$. Under this view, the method resolves itself into nothing more than an ingenious extension of the usual rule for the cube root of numbers, and that Mr. Halbert contemplated his rule *as such* is evident from the *formation* of the several terms of his subtrahends, which are obviously the numerical results of b times the *modified* divisors; $3a^2b$; $3ab^2$; and b^3. No indication of a *general* process appears in any of the examples cited, nor would it have been easy, or perhaps possible, to extend the method to equations of the form

$$x^3 \pm ax^2 \pm bx \pm c = 0;$$

and hence Mr. Horner's claims to the *first* discovery of a continuous method of approximating to the roots of equations of all orders are not affected by any thing that appears in Halbert's "Arithmetic." That the *trial* divisors in both processes are *identical* is easily shown, and so far Halbert may be said to have anticipated Horner. His *working form* is nearly the same as that given by Professor Young, for similar equations in pp. 197-8 of his *Cubic and Biquadratic Equations*, but the *principles* upon which these are obtained differ so widely that the one method of solution can scarcely be termed a legitimate anticipation of the other.

I am, Sir, yours, &c.,
T. T. WILKINSON.

Burnley, Lancashire,
December 20, 1851.

NOTES ON THE THEORY OF ALGEBRAIC EQUATIONS. BY JAMES COCKLE, ESQ., M.A., BARRISTER-AT-LAW.

(Concluded from page 460.)

Third and Concluding Series.

IX.—CONCLUSION.

(1.) Among other expedients, which have suggested themselves to me for the solution of the equation of the fifth degree, is that of employing expressions with unreal exponents. Without expressing any opinion as to the prospects of success attending such a course of inquiry, but venturing a caution against rashly entering into these speculations, I may state that, after trying a multiple of an unreal root of unity as an exponent, I have long since abandoned the further prosecution of inquiries in this direction.

Investigations, however, like the following rest upon a surer basis of utility. They had their origin in my perusal of Mr. JERRARD'S *Reflections*, at pp. 545—574 of vol. xxvi of the Third Series of the *Philosophical Magazine*. In the consideration of the METHOD OF SYMME-

TRIC PRODUCTS, we have seen how important may be the discussion of linear functions of the roots of equations; of this we shall have a further instance in the following result, which it will be interesting to compare with those given in the first section of Mr. JERRARD's *Reflections*.

Let a, b, c, d, and e be the roots of the quintic

$$x^5+Ax^4+Bx^3+Cx^2+Dx+E=0,$$

and let p, q, r, and s be undetermined quantities, and make

$$a+c+pb=P,$$
$$d+e+qb=Q,$$
$$a+c+rd=R,$$
$$d+e+sc=S.$$

We hence arrive at

$$ade+cde+pbde=Pde,$$
$$acd+ace+qabc=Qae,$$
$$abe+bce+rbde=Rbe,$$
$$abd+abe+sabc=Sab;$$

whence, by addition, transposition, &c., we obtain

$$\left.\begin{array}{l}\Sigma\,(abc)=Pde+Qac+Rbe+Sab\\ +(1-p-r)\,bde+(1-q-s)\,abc.\end{array}\right\}$$

So, from the previous group of equations, we see that

$$\left.\begin{array}{l}2\Sigma.\,a=P+Q+R+S\\ +(2-p-q)\,b-rd-sc.\end{array}\right\}$$

The following, among other consequences, flows from these results. If we assume that

$$\left.\begin{array}{l}P=0,\ Q=0,\ R=0,\\ \text{and also}\\ \qquad p+q=1,\end{array}\right\}\ldots\ldots(A.)$$

then, *necessarily*,

$$S=0,\ r+s=0,\ \text{and}$$
$$rd+sc=b.$$

The last relation is readily obtained by adding the equations $R=0$ and $S=0$; and we now see that, since

$$A=\Sigma.\,a=0=\Sigma\,(abc)=C,$$

the given quintic, under the conditions (A.), takes the form

$$x^5+Bx^3+Dx+E=0.$$

It would seem that, when

$$pq=rs=-1,$$

we have also

$$5\,D=B^2;$$

in other words, the given quintic takes the solvible form of DE MOIVRE. The converse of this proposition is discussed by Mr. JERRARD, in his *Reflections*.

(2.) Let us apply, under a different notation, the process of Art. 6 of those *Reflections* to the biquadratic

$$y^4-t^4=0,$$

whose roots we will denote by a, b, c, and d respectively. Then, uniting

$$0=\lambda\,(a+b+c+d)$$

with

$$T=\alpha a+\beta b+\gamma c+\delta d,$$

on dividing the result by $\alpha+\lambda$, and designating

$$\frac{\theta+\lambda}{\alpha+\lambda}\ \text{by}\ \theta',$$

we shall find

$$\frac{T}{\alpha+\lambda}=a+\beta' b+\gamma' c+\delta' d.$$

Following the line of argument adopted in the *Reflections*, let us suppose that $\delta'=0$, or $\delta+\lambda=0$, then

$$\frac{T}{\alpha-\delta}=(1+i\beta'+i^2\,\gamma')\,t,$$

i being an unreal fourth root of unity. We may hence infer that

$$\gamma'=-(i^2+i^3\beta').$$

Finally, on making the requisite substitutions, we shall obtain

$$\left.\begin{array}{l}\dfrac{T}{\alpha-\delta}=a+\beta' b\\ \qquad -(i^2+i^3\beta')c;\end{array}\right\}$$

and we see that $T=0$ is equivalent to

$$a-i^2c+(b-i^3c)\,\beta'=0,$$

which gives, independently of β',

$$a-i^2c=0,\ b-i^3c=0.$$

In the above investigation it is assumed that

$$a=t,\ b=i\,t,\ c=i^2t,\ d=i^3t;$$

so that we see at once that the preceding relations hold, as also

$$b-i^2d=0,$$

and a variety of similar ones which it is needless to discuss. But it is very desirable, by making

$$y=x^3+p_1x^2+p_2x+p_3$$

where x is the root of a general biquadratic, to conduct the argument side by side with that of Mr. JERRARD, to trace its failing point—for fail it most undoubtedly will,—and then, guided by the analogy thus afforded, to detect the fallacy (if fallacy there be) in Mr. JERRARD's argument. I may observe that

ting $i^a t$ for t in the above discus- here will arise two new systems of ns, making altogether three pairs, equations, derived in a manner onding to that by which the ten entioned in Art. 4 of the *Reflec-* re obtained. For some further s on quintics, I would refer to my in the *Diary* for 1851 (p. 76) and p. 84), and the places therein ned.

If, at p. 114 of vol. i. of the *matician*, we correct a typogra- error in the article numbered 3, write

$$f^2(3) = h_1^2 + h_2^2.$$

this last expression may be de- ed into factors, both of which are d in the formula

$$\left\{c + (c^2 - 4a)^{\frac{1}{2}}\right\} z_2 + \tfrac{1}{2}(dz_3 + f)$$

$$-4a)^{-\frac{1}{2}}\left\{(2e - cd)z_3 + 2g - cf\right\},$$

ning to z_2 the proper values (*Ib.* Art. 3). The two values of the ormula are given by the double plicitly included in the radical , and they can never both take n

$$z_1 + \beta z_2,$$

e have, simultaneously,

$$dz_3 + f = 0, \text{ and}$$

$$(2e - cd)z_3 + 2g - cf = 0.$$

liminating z_3 from the last two s, we have

$$ef - dg = 0,$$

less the last condition is satis- e at least of the factors is of the

$$z_1 + \beta z_2 + \gamma,$$

is different from zero.

preceding investigation may be adapted to the case in which the nt of z_1 is a quantity (r) dif- rom unity. We have only to , b, ... i each by r, and sim- e result, which belongs to that ent termed by me the METHOD ISHING GROUPES.

omparing this latter discussion t which I gave at pp. 132-3, of *iii. of S. 3 of the Philosophical e*, and adopting Mr. JER- *tation*, we see that

$$r = \tfrac{1}{2}tf\lambda'.\lambda' - s(f\lambda')^2$$

$$rd = (1, 3) = tf\lambda\lambda''' - 2sf\lambda' f\lambda'''$$

$$rg = (2, 4) = tf\lambda''\lambda^4 - 2sf\lambda'' f\lambda^4$$

$$rf = (1, 4) = tf\lambda\lambda^4 - 2sf\lambda' f\lambda^4$$

$$re = (2, 3) = tf\lambda''\lambda''' - 2sf\lambda'' f\lambda'''.$$

It will be noticed that λ^4 here repre- sents λ''''. In connection with the present topic, I beg to refer the reader to my remarks in the Arts. 15, 16, pp. 32, 33, of the Supplement to vol. iii. of the *Mathematician*. But I shall add here that although in general

$$(a, b) = tfab - 2sfafb,$$

yet that, when $a = b$, we have

$$(a^2) = \tfrac{1}{2}tfa^2 - s(fa)^2.$$

(4.) Let $X = 0$ denote the general equation of the nth degree, and let the function

$$(x - p_1)X = \phi(1) = 0$$

be developed as follows,

$$\begin{array}{l} x^{n+1} + p_2 x^{n-1} + p_3 x^{n-2} + \ldots \\ \quad - p_1^2 x^{n-1} - p_1 p_2 x^{n-2} - \ldots \\ \hline x^{n+1} + q_1 x^{n-1} + q_2 x^{n-2} + \ldots \end{array}$$

In the above operation, as well as in those given or indicated below, the terms in x^n are omitted because they necessarily disappear from the result.

If we develope in like manner

$$(x - q_1)\phi(1) = \phi(2),$$

omitting for convenience the various powers of x, we obtain

$$\begin{array}{l} 1 + q_2 + q_3 + q_4 + \ldots \\ \quad - q_1^2 - q_1 q_2 - q_1 q_3 - \ldots \\ \hline 1 + r_1 + r_2 + r_3 + \ldots \end{array}$$

which indicates that

$$x^{n+2} + r_1 x^{n-1} + r_2 x^{n-2} + \ldots = 0.$$

It is needless to give further illustra- tions of the law of coefficients; but the following is the working process,

$$\begin{array}{l} -r_1 - r_2 - r_3 - r_4 - \ldots \\ \quad r_1 \\ \hline -r_1^2 - r_1 r_2 - r_1 r_3 \ldots - r_1 r_n \\ \quad r_2 + r_3 \quad + r_4 \ldots \\ \hline \end{array}$$

Let the coefficient of x^r in $\phi(x)$ be represented by (r, z), then we have

$$(r, z) = (r+1, z-1) - (1, z-1)\ (r, z-1).$$

It may be as well to mention that

$$\phi(z) = \{z - (1, z-1)\}\ \phi(z-1),$$

of which the development is of the form

$$x^{n+z} + \ldots = 0;$$

and also that

$$(n, z) = -(1, z-1)\ (n, z-1).$$

Let u denote the reciprocal of x; divide X by $p_n\, x$; the result (X′) may be written thus,—

$$u + v_1 + v_2\, x + v_3 x^2 + \ldots$$

Form the function

$$(u - v_1)\ X' = \psi(1),$$

and develope it thus

$$u^2 + v_2 + v_3 x + v_4 x^2 + \ldots$$
$$- v_1{}^2 - v_1 v_2\, x - v_1 v_3 x^2 - \ldots$$

The term in u vanishes and is omitted, and the working process is the same as in the last case; and, if we make

$$u^z + \Sigma\{(r+1, z) x^r\} = 0,$$

the equation connecting the coefficients of one step with those in the next will be the same as before. Both are in fact equations in finite differences, and might readily be put under such a form. It will not be forgotten that

$$\{u - (1, z-1)\}\psi(z-1) = \psi(z),$$

which presents a resemblance to a former relation.

(5.) The following investigation will furnish the means of forming at pleasure convenient examples of cubic equations. In my solvible form of cubic

$$x^3 + a\, x^2 + b\, x + \frac{b^2}{3a} = 0,$$

let

$$b = \frac{a^2}{3}(1 - m^3),$$

then x is given by the three values of the expression

$$\frac{b}{\{(1)^{\frac{1}{3}} m - 1\}\, a},$$

one of which is real. Values of a and m may be assumed, and b be thence determined. The advantage of this is that if we form the cubic whose root is $e + x$, e being arbitrary, the application of my Rule (lii., 488) becomes exceedingly simple, in consequence of surds disappearing from the process.

Let the real root of the above solvible cubic be a whole number (N); then we have

$$\frac{b}{\sqrt[3]{a(a^2 - 3b)} - a} = N,$$

and hence we infer

$$(b + aN)^3 = N^3\ (a^3 - 3\, a\, b),$$

or

$$b^3 + 3\, b^2 a\, N + 3\, b a^2\, N^2 + 3ab\, N^3 = 0,$$

which gives on dividing by b

$$b^2 + 3a\, N.\, b + 3\,(a^2 N^2 + a\, N^3) = 0,$$

and b will be rational if

$$12\,(a^2 N^2 + a N^3) - 9\, a^2\, N^2 = 0,$$

or

$$a = -4\, N,$$

in which case we have

$$b + \frac{3\, a\, N}{2} = 0, \text{ or } b = 6\, N^2.$$

James Cockle.

2, Pump-court, Temple,
December 13th, 1851.

(End of the Series.)

CLARK'S RAILWAY MACHINERY.*

We have had many works on Railway Engineering, but not one which embraces the whole field so completely as that of which some specimen numbers are now before us. Some have treated of the modes of setting out or laying down railways only; others, of the locomotive machinery employed on them only; but the plan of Mr. Clark's work includes "the principles and construction of (both) rolling and fixed plant in all its departments." Well designed, the work gives every promise of being also well executed. Mr. Clark is a careful collector of facts, and shows generally great judgment and discernment in bringing out their salient points. He gives proofs, too, of being thoroughly conversant,

---

* "Railway Machinery: A Treatise on the Mechanical Engineering of Railways, embracing the Principles and Construction of Rolling and Fixed Plant in all Departments. Illustrated by a Series of Plates on a large scale, and by numerous Engravings on Wood. By Daniel Kinnear Clark, Engineer. Parts I. and II. 1851. Blackie and Son."

both theoretically and practically, with all the branches of his subject; and if, occasionally, we miss in his descriptions that clearness which is the usual result of fulness of knowledge, we are inclined to ascribe this wholly to inexperience in literary composition—a fault which may be expected very surely to disappear with the progress of the undertaking. For one thing Mr. Clark is especially to be commended, and that is, his entire freedom from those personal and party leanings by which most preceding works on railways have been disfigured. In his judgments he is bold and free—sometimes, to say the truth, rather flippant, but always (to all seeming) honest and conscientious.

Chapter I. gives the general history of the locomotive from 1784 till 1831; and chapter II. brings it forward from 1831 to 1849. The following is the author's own summary of the events comprehended within these periods:

"The leading facts are, that on the first introduction of passenger railways, speeds of about 12 miles per hour only were anticipated; the rails thus employed weighed only 35 lbs. per yard, and the engines from 5 to 7 tons. When speeds of 20 and 24 miles were attempted, it was found necessary to have 50 lbs. rails, and engines of 10 and 12 tons. The engines were thus divisible into classes—four-wheeled engines and six-wheeled engines, patronized respectively by Bury and Stephenson and the leading makers. As speed and power—convertible terms—increased, more accommodation was wanted; outside cylinder engines were constructed weighing 15 and 16 tons; inside cylinders continued in vogue, as, by simplification of the machinery, accommodation was provided for them. In other quarters, the gauge of the railway having been increased 50 per cent. upon the previously-existing gauge, the constraint incidental to the inside cylinder was very much relaxed, and still more powerful engines were made, weighing, from first to last, 15 to 35 tons charged. On the ordinary gauge, too, the boilers of the engines were lengthened, and the fire-box increased, and the long boiler engine was the result, weighing from 20 to 22 tons. Finally, as some of the later engines on the common gauge had, owing to peculiarities of construction, acquired a character of unsteadiness at high speeds, and of increased cost of maintenance, Crampton's engine was introduced, in which the wheel-box was extended and more solidly arranged, the driving-wheel being placed behind the fire-box, and the boiler lowered considerably. Meanwhile the rails were increased progressively to 65, 75, and 85 lbs. per yard."

We do not, however, think this at all a good summary, and must beg our readers not to take it as a fair sample of the quality of the work—which it is not. Some of the most striking facts in the progress of the railway locomotive it passes over altogether unnoticed. Take, for instance, the introduction of the chimney blast and multitubular boiler. And, for proof, take what Mr Clark himself elsewhere says of that event:

Introduction of Multitubular Boilers and the Blast Pipe.

"The introduction into France of the imperfect locomotive of the time (1825-30)—imperfect in so far as high speeds were concerned—led to the solution of the question of light and powerful engines. 'The first locomotives, few in number, that were sent to France,' says M. Lobet, 'were made by George Stephenson, and arrived there in 1829, for the Lyons and St. Etienne Railway, of which M. Seguin was the engineer. On trial, their mean velocity did not exceed four miles per hour. To increase the efficacy of her engines, M. Seguin felt the necessity of increasing their evaporative power, and resolved to apply a scheme of his own to the engines he was about to construct (on the model of Stephenson's)—a scheme which he had cherished since 1827, and had patented in February, 1828, which consisted in multiplying the heating surface, by subdividing the current of hot air into streamlets which flowed through a series of tubes immersed in the water of the boiler. The method of the tubes increased amazingly the heating surface, and without the evaporative power, and it is precisely to this evaporation we are indebted for speeds which were before thought impossible. But another difficulty presented itself, the height of the chimney, necessarily limited, was incompetent to maintain the draft, the resistance of which was so much increased by the increase of surface in the new boiler. M. Seguin, therefore, added a circular fan for promoting the draft, and it was partially successful. M. Pelletan, however, completed the solution of the problem by suggesting the steam-jet in the chimney; and, as usual, England appropriated the invention of the two French engineers.'*

* "Lobet Du Chemin de Fer de France." 1845.

"The suggestion and application of the subdivided tube surface is by common consent ascribed to M. Seguin. The steam-jet in the chimney, though no doubt invented independently by M. Pelletan, had been previously applied by Stephenson and Hackworth. It was, however, at the same time but partially employed in this country, as we may infer from the absence of the jet in the sample engines sent to France. The method of the multitubular flue and the steam-jet are parts of one system; they co exist as naturally as the condenser and air-pump of Watt's engine. The locomotive was not ripe for the application of the blast pipe; the large and vacuous cavity of the flue tube, while it presented a very restricted area of heating surface, permitted great power of concentration, and the greater length and surface of Hackworth's doubled flue enabled him on this account probably to employ the blast with greater success than had been done by his predecessors. Again; the tubes of M. Seguin, while they increase the heating surface, increase friction surface of the flue-way simultaneously; and here it was that the aid ot mechanical expedients became more than ever necessary to uphold the requisite rate of combustion. The method of the steam-blast, therefore, of spontaneous invention at home, was in France the child of necessity. The tubes formed the link between the fire-box and the jet, and thus the problem of producing a light and powerful locomotive was solved."

The important influence which these improvements exercised on the progress of the railway system is here correctly enough appreciated; but we must altogether object to the readiness with which Mr. Clark surrenders the claim of his own country to the invention of the jet. The French writer quoted (Lobet) asserts that, "*as usual*, England appropriated the invention of the two French engineers (Seguin and Pelletan);" M. Lobet would have spoken more truly had he said that, "as usual, France, as soon as any important discovery is made in art or science, sets up a claim to the authorship, though, in nine times out of ten, she is but a copyist, or, at best, second-hand discoverer." Mr. Clark makes a distinction between a thing being of "spontaneous invention" and "the child of necessity," which we must confess is of a fineness beyond our comprehension. However begotten, *the fact is absolutely undeniable, that the steam blast, at least, was known and in use in England more than twenty years before a single railway locomotive had been seen on the French soil.* To substantiate this fact, we must quote from Mr. Clark himself:

"Experiments for stimulating the draught of the furnaces were not wanting; great height of chimney being of course inadmissible. Trevethick employed hollow bellows for that purpose in his first engine; in his second, made in 1804, *he turned the waste steam into the chimney*."

Contemporary evidence—the best of all—of Trevethick's right to the invention of the blast, is to be found in *Nicholson's Journal* for September, 1805, where distinct mention of it is made in a communication from Mr. Davies Giddy (afterwards Gilbert, and President of the Royal Society). Mr. G. gives there an "account of certain circumstances observed by himself and others, during the working an engine on Mr. Trevethick's construction at Merthyr Tydvil;" one of which circumstances was "whether, and in what degree, the draft was effected by the *admission of steam into the flue*." "Every one agreed in declaring that the fire brightened each time the steam obtained admission into the chimney, as the engine made no smoke." The good effect of the steam jet in promoting combustion seems never to have been once doubted either by Trevethick or others, but it did not come at the time into general use; because, as Mr. Clark justly observes, "until 1825, nothing but slow speeds were ever contemplated by those engaged in the manufacture of engines, and the unassisted evaporative power of the boiler of the old engines was in general competent to the production of at least as much steam as would do the work of seven or eight horses. This was of itself considered an achievement, and it seemed to have contented mechanical men interested with railways for a quarter of a century."

Not quite "contented," for it is related that, in 1827, Mr. Timothy Hackworth, manager of the Stockton and Darlington Railway, applied the blast-pipe in the chimney to an engine, the Royal George, constructed for that line, and re-arranged by him; which also was prior to the adoption of the plan by the French engineers.

It is supremely idle after this to talk of the claims of Seguin and Pelletan; or rather of Pelletan alone, for what Seguin proposed was confessedly not the blast, but a fan which, too, happened to be a borrowed child; for the same thing was proposed by the Chapmans as early as 1812. Mr. Clark is pleased to say that the steam-jet was "no doubt invented *independently* by Pelletan." But how do we know that? Where is there the least proof of it? History knows no second inventors; there can be but one, and he the first. The actual circumstances of the case render it fully as probable that Pelletan was aware of the English invention, and profited by it, as that he had recourse to his own wits on the occasion. The two engines first sent from England to France were accompanied by English workmen, and nothing is more likely than that these workmen, or some of them, knew of their countryman Trevethick's device for accelerating the draught; and knowing, would make mention of it. Nicholson's work, moreover, was a journal in universal circulation both at home and abroad, and impartial history will not admit of such well informed men as Pelletan and Seguin being ignorant of its contents.

The claim of the French to the multitubular boiler stands on much better grounds than their claim to the steam-jet, and must, we think, be admitted. Seguin had his patent in 1828, and the first boiler with multitubular flues seen in England, was that of the Rocket, built by Hackworth, which was one of the competing engines at the famous competition on the opening of the Liverpool and Manchester Railway in 1829—a *year after*. "The tubes," says Mr. Clark, "were adopted at the suggestion of Mr. Booth, of the Liverpool and Manchester Railway, to whom the merit of the invention in this country is commonly ascribed." Mr. Booth may possibly have heard nothing of M. Seguin's patent, but we are not at liberty to indulge in any such supposition. It is no more open to an English than a French man of science to plead ignorance of a world-published fact. As long as M. Seguin's patent cannot be displaced from the chronological position which it holds, which we believe to be the case, M. Seguin must be frankly recognised as the inventor of the multitubular system.

Of Timothy Hackworth, referred to in he preceding extracts, and of his services to the railway system, Mr. Clark speaks in terms of high and well-merited praise, which we quote the rather, that Mr. Hackworth has hitherto had but scant justice done to him by writers on railway affairs:

Timothy Hackworth.

"From the foregoing notices it appears that no single individual in this country had up till the year 1830 done so much for the improvement of the locomotive, and for its establishment as a permanent railway moter as Mr. Timothy Hackworth." "He first employed six coupled wheel locomotives; he first applied the waste steam to heat the feed-water; he first employed the eccentrics to work the feed-pumps,—which in many cases is a matter of convenience; he substituted spring balances for weights to the safety-valves; he schemed and first applied the steam chamber in the boiler, a valuable auxiliary for obtaining dry steam; he first placed the cylinders beneath the boiler, and employed the double inside crank axle, coupled directly to the pistons; and he also was the first to employ in railway locomotives a separate crank-shaft hung in bearings fixed to the frame."

The memorable controversy respecting the four-wheel and six-wheel engines is touched on, as we have seen, rather superficially, by Mr. Clark in his historical summary; but there are other passages in which their comparative merits are brought out with sufficient distinctness and circumstantiality:

Six-wheeled Engines, Plain Driving-wheels, and Heavy Rails.

"The increased weight and speed of the locomotive supplied to the Liverpool and Manchester Railway, gave rise to a new and unexpected difficulty. The Rocket class of engine, in working order, did not weigh above five tons. This weight distributed amongst four wheels, did not afford above 30 cwt. to each driving-wheel on a railway that was laid originally with 35-pound rails.

The additional weight of the engines of the Planet class,* 9 tons charged, and loading the rails under the driving-wheels with 5½ tons, or 2¾ tons to each rail, speedily disorganised the permanent way; the rails were for the most part broken or bent under the extra load moving at thirty miles per hour. The cause of the destruction was of a compound character. On the one hand, there was the unequal distribution of the weight, and the mal-arrangement of the wheels, made worse by the increased load laid upon them; on the other hand, the radical defects in the mode of laying the rails had also much to do with it.

"It was impracticable more equally to divide the load over the four wheels; with inside cylinders, the crank axle is necessarily placed in front of the fire-box; a large portion of the engine therefore overlays that axle. The fore wheels, it is true, might have been placed so close to the engines as to divide the weight equally with them; to secure base on the rails, however, they were necessarily placed forward, and were, of course, more remote from the centre of gravity of the suspended mass. Neither was the preponderating load on the driving-wheels required for the purposes of adhesion, as their power of bite was more than ample. In Pambour's experiments on the velocity and load of locomotives on the Liverpool and Manchester Railway, there was not one in which the motion was stopped or slackened for want of adhesion, though there were loads amongst them equivalent to 300 tons on a level.. Again; the wheels fore and hind did not exceed 5 feet apart centres, the axles being placed between the fire-box and the smoke-box; and thus, with overhanging loads and a total length of engine of 15 feet, depressions and the inequalities of the rails gave rise to irregular motions, which were magnified seriously by the length of the machine.

"It was proposed to replace the light rails with 66 lb. rails. Immediate measures were, however, imperative; and it was finally determined to add a pair of wheels to the engine, *behind the fire-box*, constituting it a six-wheel engine, and extending the base to 9 feet. The object of the additional pair was not to relieve the driving-wheels of any fixed part of their load, but to check the pitching of the engine by receiving a portion of the weight at the time of plunging. They were lightly loaded when the engine was empty, merely to bind the springs; and when the engine was charged, the springs had a burden of 5 cwt. each.

"The design was good; the hind wheels remedied the unsteadiness, both vertically and laterally, beyond all expectation; the speeds were maintained without involving any further destruction of the permanent way; and the engineer proceeded diligently to replace the light rails by heavier ones. Since that time the rails have been increased successively to 30, 60, 70, and 75 lbs. per yard lineal.

"Besides deriving from the adoption increased stability on the rails, the six-wheel engine kept firmly together for a longer time; the tubes leaked less; the frame was less disturbed at the bolts and stays; running off the rail became less frequent on the fracture of a crank-axle; this axle itself lasted longer, as it was saved from much of the straining to which, as hind axle, it was previously exposed.

"To preserve the crank-axle of the new engine from the lateral pressure on the wheel flanges on the passage of curves, the driving-wheels were deprived of their flanges, and formed straight on the rims, and were thus left free to suit themselves to the rails. The plane driving-wheel formed the subject of a patent in 1833."

Mr. Bury's Four-wheel Engines.

"A new and original class of engines had been advocated by Mr. Edward Bury, of Liverpool, since 1830, the date of the delivery of his first locomotive on the Liverpool and Manchester Railway. He was appointed locomotive superintendent of the London and Birmingham Railway, opened in 1837, and had opportunities of extensively introducing on that line his peculiar class of locomotives. Though the arguments then advanced by Mr. Bury were deemed irrefragable by a large section of the mechanical men of the time, and gave rise to animated discussion, they have for the most part sunk into unimportance as the arts of construction have advanced. The peculiar features of Bury's engines were the circular fire-box; the arrangement of the tubes in circular arcs; the exclusive inside frame, composed of forged iron bars; and the exclusive number of wheels—four—on which the engine was placed. Mr. Bury's ideas were all of a mathematical complexion. The fancied superiority of the circular fire-box, and the arrangement of the tubes in circular lines for the promotion of equable combustion, and the free circulation of the water in the boiler, is now estimated at

* The Planet was the ninth engine built by Mr Stephenson for the Liverpool and Manchester Railway, and embraced some conspicuous improvements; it was the combination, in fact, of what had previously been known — the multitubular boiler, the blast pipe, the inside horizontal cylinders which were placed inside the smoke-box, and the double crank-axle.—*Mr. Clark*, *ante*

nothing compared with the convenience of the square fire-box as a piece of manufacture, and its value as embodying greater area of grate and heating surface in proportion to its economic bulk. The superiority of the circular arrangement of the tubes has also proved a mere figment; and where tube surface is an object—and, above all, sectional area of tube—the arrangement of tubes in straight lines becomes imperative. Bury's inside frame, composed of bar iron, imparted to his engine a peculiar, wiry air of lightness. Mr. Bury justly conceived that, next to a good boiler, 'the most important point in the construction of a locomotive, was to connect all the parts firmly together by a strong and well-arranged framing, so that they shall retain their relative positions when the engine is in motion.' In this respect, he considered the inside framing much superior to the outside framing, as the connection of the cylinder and crank-axle was more direct; and the boiler was considered to be relieved from much of the strain of the engine incidental to outside frames. Much of this is true, and has been followed up in the works of modern locomotive builders."

On the whole, Mr. Clark has left us very favourably impressed with his qualifications for the important task he has undertaken: and if he but go on as he has begun, we think he may confidently reckon on the general approbation and support of the profession. The work, we must not forget to add, is got up in a style which does great credit to the press of the publishers; the type and paper are excellent; and the pictorial embellishments as splendid as profuse.

GILLAM'S REGISTERED SEED-CLEANSER AND SEPARATOR.

(John Gillam, of Woodstock, Proprietor.)

Fig. 1. Fig. 3. Fig. 2.

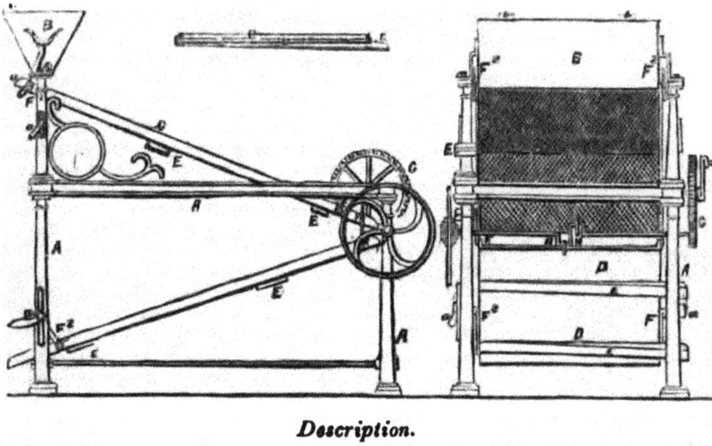

Description.

Fig. 1 is a side elevation; fig. 2 an end elevation of this machine; fig. 3 is a cross section of one of the sieves and the shaking frame in which it is mounted. AA is the frame-wock; B, the hopper; CC, two sieves, each of which is composed of wire gauze, but the wire gauze at the upper end or half is finer in the meshes than that at the lower end, the width of the meshes increasing in regular gradation from the upper end of the top sieve to the lower end of the under one. Below each division of the sieve there is fixed a web of cloth DD, upon which the seeds as they fall through are conducted to the spouts EE, from the mouths of which the different varieties are collected. FF are links, by which the sieve-frames are suspended. F^2 F^2, other links, which may be lengthened or shortened at pleasure, by means of the screws *a a*, so as to give the sieves a greater or less inclination according to the kind of seed operated upon. GG is the gearing by which the frames are put in motion by means of the cranks HH. The hopper is furnished with a shaking apparatus and slide, to regulate the feed.

THE LAW OF PATENTS.—THE CASE OF FOREIGN USER OF AN ENGLISH INVENTION IN ENGLISH WATERS.—(SEE ANTE, PP. 453, 474.)

VICE-CHANCELLOR'S COURT,

Saturday, Dec. 20.

(*Before Sir* G. TURNER.)

CALDWELL *v.* VANVLISSENGEN.—CALDWELL *v.* ROLFE.—CALDWELL *v.* VERBECK.

The VICE-CHANCELLOR gave judgment in these causes. The bills were filed to restrain the use of the screw propeller (constructed according to Lowe's Patent) by the defendants (who are the owners and masters of ships belonging to subjects of the King of Holland) in England, Wales, and the town of Berwick upon-Tweed, and the seas, rivers, and havens of the same. The injunction was resisted on several grounds: first, that the injunction ought not to be granted, for want of novelty in the invention;-- secondly, that there had not been such an enjoyment of the patent as would justify the Court in sustaining it by the injunction; and thirdly, that, supposing the patent to be valid, and the enjoyment to be established, still there appeared upon the affidavits such circumstances as would preclude the interference of the Court. The principles upon which the Court interfered to protect a patent were very simple. It was the duty of the Court to protect the legal right pending litigation; but the party seeking its interposition was bound to show some title, and in the case of a new patent which had not been established in a court of law this title was wanting; but if there had been proceedings at law in which the right of the patentee had not been successfully resisted, and might have been resisted if it had been without foundation, there was a sufficient case for the interference of the Court for the protection of the patentee. The circumstances brought before the Court, as a defence to the application, were stated in the affidavit of one of the defendants in the first cause. The affidavit stated that the ship referred to in that cause, the *Burgomaster Heinkopfer*, was the property of a company in Holland, called the "Amsterdam Steam Screw Schooner Company;" that the company was composed of numerous partners, all of whom were subjects of the King of Holland, and none of whom were English subjects; that the company was entitled by the law of Holland to trade with steam ships, built and fitted up with the propelling power which was the subject of the application; that the screw propellers in their ships were manufactured and fitted by the defendants at Amsterdam; that the defendants were, and always had been, unacquainted with the invention of James Lowe, and that the deponent believed that all the said ships were built and fitted in ignorance of the existence of any such patent; that no patent had been granted to secure the alleged invention in Holland, and that, according to the laws of Holland, it was open to any English subject to apply for and obtain a patent in the kingdom of Holland; that before the vessel in question had been built and fitted with the screw propeller, several other vessels had been built and fitted in the same manner, and had traded between Amsterdam and London, and made many voyages; that the defendants had not, until September last, heard of any objection to their so trading on the ground of the alleged infringement of the patent; that various other vessels had been built and fitted in Holland with propellers on the same principle, and with the same propelling power; and that it would be a great loss to the company, and to both England and Holland, if the trade, which was profitable to both countries, should be restrained by the Court. This affidavit set forth, in clear and distinct language, the grounds on which the case of the defendants was founded. He was of opinion that he could not withhold the injunction on the ground stated. Upon the general principle, foreigners were subject to the laws of the country in which they happened to be. If there were any cases in which they were subject to their own laws in another country, it was not by force of those laws, but of the laws of the country in which they were adopting their laws into its own. This was the doctrine laid down by Mr. Justice Story in his *Conflict of Laws*. The principle in this country did not depend upon the general law. It was the subject of special provision by statute. The statute 32nd Henry VIII. chap. 16, sec. 9, provided "that every alien and stranger born out of the king's obeisance, not being denizen, which now or hereafter come in or to this realm, or elsewhere within the king's dominions, shall, after the 1st day of September next coming, be bounden by and unto the laws and statutes of this realm, and to all and singular the contents of the same." Natural justice, in fact, required that the defendants, when in this country, should be subject to its laws. The question then was, what were the rights of patentees? The Crown had in this kingdom always exercised the right of interfering with the trade of the country, and had at a former period exercised that power very prejudicially. The abuse of this power had estrained by the statute of James. In se of the monopolies reported by Sir 'd Coke, it was held that the Crown ower to grant an exclusive right of g for a reasonable period, and this nited by the statute for the term of en years. The statute did not, how- reate, but control the power of the . to grant patents; and the patentees l their rights not from the statute, m the grant of the Crown. What, ere the words of the patent?—"The thereby gave the patentee, his exe- , administrators, and assigns, special , full power, sole privilege, and au- , that he the said patentee, his exe- , administrators, and assigns, and of them, by himself and themselves, is and their deputy or deputies, ser- r agents, or such others as he the tentee, his executors, administrators, gns, should at any time agree with, others, from time to time, and at all thereafter during the term of years expressed, should and lawfully might use, exercise and vend his said inven- thin that part of the United King- f Great Britain and Ireland called d, the dominion of Wales, and town wick-upon-Tweed, in such manner as the said patentee, his executors, ad- ators, and assigns, or any of them, in his or their discretions seem meet." 'oreigners, as well as British subjects, able to actions for injuries to the civil of British subjects; and there was no why they should not be equally liable n for the infringement of the right ranted. If that were so, there was no reason why the jurisdiction of urt should not be appealed to against The right would in former times een enforced, in aid of the King's by proceedings in the Star-chamber, course of the argument he had in- whether, if a locomotive engine on a , the subject of a patent in England, which no patent had been obtained land, were made in the latter coun- ould be allowed to run into England any objection on the ground of the ment of the English patent; or, if ention had been the subject of a in England, but not in Ireland, the ould be permitted to trade between and Liverpool without any such n? The answer given to this was, prior use of a patent in Scotland e fatal to a patent obtained in Eng- ut that such would not be the case if or use were in a foreign country. as not, however, an answer to the tion. *In one case* the result would depend on the previous knowledge of the invention—in the other case, on the effect of the patent. The remarks of Lord Eldon, in the case of the Bibles—"Richardson *v.* the University of Oxford," had been referred to on the cases of necessity which arise for allowing a user of the subject of a patent, and it was said that this was such a user as the Court would not restrain. There might, no doubt, be such cases of necessity, and perhaps the case suggested of a foreign ship stranded on the English coast might be such a case. It must be remembered that foreigners were at liberty to apply for and obtain patents in this country with the same privileges as British subjects. If foreign inventors did not take this step, they, to that extent at least, withheld the use of their invention from the subjects of this country; and, if they were restrained from using their own inventions in this country, such inventions being the subjects of patents granted to other persons, they had nothing taken from them by that restraint, for, if the patent were valid, the right of using their inventions in this country was one which they had never enjoyed. It had been argued that any interposition of this Court might be met by similar restraints on our ships abroad; but this question resolved itself into one of national policy. It was a proper subject for the consideration of the Legislature; but it was the duty of this Court to administer the law, and not to make it. He was of opinion that the facts stated did not afford a sufficient ground for refusing the injunction. In the second case, the master of the ship having stated that he was not now the master, and there being no suggestion of any intention to reappoint him, the injunction would not be granted; and in the other cases it must be qualified by restraining the defendants, unless and until they should have obtained the license of the plaintiffs so to do, from using or exercising, or permitting to be used or exercised in the limits of that part of the United Kingdom of Great Britain and Ireland called England, the dominion of Wales, or Berwick-on-Tweed, the invention of James Lowe in the bills in the causes mentioned, or any mode or process for the propulsion of steam vessels merely colourably differing therefrom, and restraining the defendants, unless and until so licensed as aforesaid, from propelling the vessels in the bill mentioned, by sea, or permitting the same to be propelled, within the limits aforesaid, by means of a propeller screw attached thereto, in such seas or ports, or permitting the same to be propelled within the limits aforesaid with or by means of any screw propeller or propellers constituted or applied according to the forms and modes respectively described and mentioned in the said James Lowe's patent, or merely colourably differing therefrom. The plaintiffs must give an undertaking to bring an action. It would be for the benefit of all parties that the question should be tried at law, and the points which had been raised in the argument determined.

INFRINGEMENT OF A PROVISIONAL REGISTRATION.

Marlborough-street.

Mr. Richard Grimes, lamp-maker, No. 83, New Bond-street, was summoned before Mr. Bingham, for having sold a certain improved lamp that had been provisionally registered under the Protection of the Inventions Act of 1851, without the consent of Messrs. Clark and Roskell, lamp-manufacturers, Strand, the registered proprietors of the said lamp,

Mr. Clarkson appeared for the complainants, Mr. Hurlstone for the defendant.

The case was to this effect. The defendant was formerly in Messrs. Clark and Co.'s employment, and recently had commenced business on his own account in New Bond-street. Messrs. Clark and Co., had registered an invention for the purpose of displaying it at the Great Exhibition, which invention consisted of a certain application of machinery to a lamp, whereby the lamps were fitted with cottons without the necessity, as at present, of using the fingers. The invention had been applied and sold with Messrs. Clark's Carcel lamps only, but was capable, under certain circumstances, of being applied to any other kind of lamps. This invention, it was contended, had been pirated by Mr. Grimes, who had effected a registration on his own account of what it was contended was the invention of the complainants.

Mr. Hurlstone said the complainants had divested themselves of the protection of the Registration of Designs Act, because they had sold lamps with the invention attached to them, before the twelvemonth allowed under the Act in question had expired.

Mr. Clarkson denied this. The invention had been sold with Carcel lamps; but this was legal under the special patent. The invention, as registered for other lamps, had not been applied and sold with any other kind of lamp except the Carcel lamp.

Mr. Hurlstone said the reason of this was plain. Messrs. Clark and Co. had not been able to overcome the difficulties of applying the invention to other than Carcel lamps, therefore it was practically useless to them. The defendant had, by certain improvements, overcome all difficulties, thereby making the invention his own, and was therefore entitled to register it.

After a good deal of discussion between the learned counsel, the question left for the Court to decide was, whether the invention of Mr. Grimes was not, in fact, the invention of Messrs Clark and Co., and whether Messrs. Clark and Co. had not divested themselves of legal protection by selling the invention with Carcel lamps?

Mr. Bingham gave his decision on the 15th inst. The Magistrate said—The 9th sect. of the 14th Vic. c. 8, applies to inventions all the provisions of the Designs Act of 1850. The Designs Act, 1850, 13th and 14th Vic. c. 104, s. 4, provides, "that if, during *the continuance of a provisional registration, the proprietor of any design provisionally registered, shall sell any article or thing to which any such design has been applied, such provisional registration shall be deemed null and void.*" Mr. Clark has registered an improvement for lamps—that is, an improved cotton-holder. This cotton-holder, during the continuance of a provisional registration he has sold, applied to a Carcel lamp, of which he is the patentee—he has not sold it applied to a ring-headed lamp. Mr. Grimes has sold the invention during the same period attached to a ring-headed lamp; and it is contended, on behalf of Mr. Clark, that the 14th sec., 13th and 14th Vic. c. 104, has not deprived him of the privilege attached to the registration of his improved cotton-holder, at least as far as regards the application of it to a ring-headed lamp. But the provision seems to me to be so general and unqualified as to leave no room for any such distinction. The inventor loses the privilege if he sells any article to which the registered invention has been applied. Here the complainant has sold such an article, namely, his Carcel lamp, and therefore I feel bound to dismiss the summons.

SPECIFICATIONS OF ENGLISH PATENTS ENROLLED DURING THE WEEK, ENDING DECEMBER 23, 1851.

Frederick Crace Calvert, of Manchester, chemist. *For a new application of certain fluids for manufacturing extracts applicable to the processes of dyeing, printing and tanning, and in the apparatus connected therewith.* Patent dated June 12, 1851.

The first part of this invention consists

employment in the manufacture of to be used in dyeing, printing and of certain fluids not hitherto used, t is essential should possess the two g properties; first, they should be ansformable into vapour and recon- with facility; and, second, they e enabled by their chemical com- to dissolve with facility the colour- tanning matters, or principles from stances in which they are contained, ild also be similarly capable of dis- resins and the resin of gum resins. luids suitable for the purpose are ich are rich in carbon and hydrogen, composed entirely of carbon and n, such as the impure hydrate of f methyle, or as it is commonly wood-naphtha, pyroligneous spirit, or the hydrate of oxide of mesithyle, ne, camphine, the volatile oils ob- by the destructive distillation of the volatile oil of shale coal, and tuminous shale, or schist, also the carburetted hydrogens obtained by llation of coal tar, usually known or mineral naphtha; also those ob- y the distillation of vegetable tars, bog or peat, or from tar and oil from bog or peat, and lastly, alco- ethers, such as the hydrate of oxide , the acetate of oxide of ethyle, or rieties of ethers.

econd part of the invention has re- the apparatus employed for manu- ; extracts from substances contain- uring or tanning principles, or mat- the use of the volatile fluids above ted, or any of them. The appara- ists of two oblong boxes for holding stances to be operated on. These e lined with copper and divided into ments by partitions of about half th; they are also constructed with d false bottoms, beneath which nged steam pipes, and are pro- th doors to admit of the materials troduced and removed. Each of s is in communication with a close pan or evaporator, and also with a (a vessel of the same length as the d placed underneath them), from tter a pipe leads to the boiling pan. e boiling pan rises a vertical pipe in a steam jacket, for the pur- preventing the condensation in of the vapours rising from the which pipe leads to a condenser, bove the operating-boxes and com- ng with a reservoir also above the r the supply of liquid to them e working of the apparatus is ced. A separate condenser is also provided, which communicates by suitable pipes with cones placed in the top of the operating-boxes which act as still heads. The method of operating is as follows:—The materials from which the colouring or tanning matters, or principles, are to be extracted, are supplied to one of the boxes (which to save time are used alternately), either in a dry or moist state, (being moistened, when this method of proceeding is adopted, with water or with solutions of the fixed alkalies, such as potash, soda, or ammonia, or with solutions of alkaline carbonates or bicarbonates, phosphates, borates and sulphates of soda and potash, stannates of soda and potash lime, carbonate of lime, baryta, strontia, and their carbonates, magnesia and its carbonates, or oxide of lead. The alkalies and substances presenting an alkaline reaction may also be added in a dry state. It will be sometimes found necessary to treat substances from which extracts are to be made previous to putting them into the boxes by washing or macerating with one of the above alkalies, or alkaline materials dissolved in water), they are then spread evenly over the surface of the box and the volatile fluid is admitted from the reservoir, or it may be introduced in the state of vapour direct from the boiling pan, and condensed by contact with the materials under operation. After maceration for a few hours, the fluid is let off into the receiver, and thence to the boiler, carrying with it a certain portion of the colouring or tanning matter; the fluid is then volatilised, leaving behind it the portion of tanning or colouring matter which it has extracted, and rising through the vertical pipe is condensed in the condenser and returned to the reservoir for use over again, and the operation is repeated until the whole of the tanning or colouring principle is supposed to be extracted. Should any portion of the volatile fluid remain in the exhausted materials, it is volatilized by steam caused to circulate through the steam pipes in the boxes, and passing away through the conical still-head, is condensed in the detached condenser before mentioned, from which it is removed and supplied to the reservoir to be used over again. Or the operation may be completed by introducing water into the boxes after repeated use of the volatile fluid, for the purpose of combining with any of the said fluid, and any remaining portion of colouring or tanning matter which the substances under operation may contain. The liquid resulting from this operation is conducted to the boiler, and mixed with the extract contained therein. The extract when drained off from the boiler may be diluted with water, or employed at once in the state in which it is so withdrawn.

When sumach is the material operated on, wood naphtha is the fluid which the patentee prefers to employ for extracting its tanning principles; the quantity of naphtha used will be determined by experience and the nature of the extract required, so also when other materials than sumach are operated on, the nature of the material, and the use to which the extract is to be applied, will readily enable any practical chemist to determine what volatile fluid, and what quantity thereof, will be under the circumstances most suitable.

Another method of operating on such substances as above mentioned, consists in employing, instead of the apparatus described and a volatile fluid, an alkali or substance presenting an alkaline reaction. When this method is adopted, the substances containing the tanning principle or matter are treated by maceration or percolation, either with or without the aid of heat. The operation may be conducted in open vessels, but the patentee prefers the use of closed, or partially exhausted vessels, as it is well known that tannin and gallic, or pyro-gallic acids undergo rapid oxidation or destruction by their absorption of oxygen from the atmosphere, when under the influence of alkalies. It will also be found advantageous to add to these extracts a sufficient quantity of some organic or mineral acid to neutralize the quantity of alkali, alkaline earth or carbonate employed, and to give a slight acid reaction to the solution in which such alkaline substances have been used.

In conclusion, Mr. Calvert observes, that the form of apparatus employed is not material, provided it complies with the conditions that the volatile agents used to extract the colouring or tanning matters or principles are made to pass through the substances containing the same, and in passing to extract a portion of the said colouring or tanning matters, from which they are liberated by the application of heat, the agent employed being volatilized and rising through a tube or aperture to be condensed, and then again passing through the substances to be operated on, and the process being repeated as often as necessary.

Claims.—1. The application of the before named volatile fluids for the manufacture of extracts from substances containing colouring and tanning matters or principles, which extracts are applicable in the processes of dyeing, printing, and tanning.

2. The apparatus described for the manufacture of extracts applicable to dyeing, printing, and tanning, by which apparatus the volatile fluids enumerated, or any others possessing the same or similar qualities, are volatilized (leaving the colouring or tanning-matter or principle in the evaporating vessel), and then condensed and employed in the manner described.

3. The use of the before-named alkalies, or substances having an alkaline reaction when mixed with substances containing colouring or tanning matters or principles, previous to the subjecting them to the action of volatile fluids.

4. The manufacture of extracts from substances containing tanning matters or principles by the agency of alkalies, or substances presenting an alkaline reaction.

James Hinks, of Birmingham, manufacturer. *For certain improvements in the construction of metallic reels for winding cotton, silk, and other threads, and in machinery for making the same.* Patent dated June 14, 1851.

Claims.—1. The construction of metallic reels, by forming projections on the ends of the hollow metallic cylinder constituting the body of the reel, and passing the said projections through holes in the discs constituting the end of the reel, and fixing the said discs or ends by turning back the said projections.

2. The construction of metallic reels by forming a series of slits in those edges of the plate of metal which when the said plate is rolled into a cylinder constitute the ends of the said cylinder, and afterwards expressing the said ends into a conical form, and completing the reel by attaching caps to the expressed ends; also, supporting the interior of the said or other reels by a coil of sheet metal placed therein.

3. The construction of metallic reels by passing the ends of the hollow metal cylinder constituting the axis of the reel through holes in the centre of the discs constituting the ends of the said reel, and attaching the said discs to the cylinder by turning back the ends of the cylinder, and soldering the turned-back ends of the cylinder to the external faces of the discs.

4. The construction of metallic reels by constructing the axis of the reel of two hollow cylinders placed one upon the other, the outer one being shorter than the inner one, and affixing the discs constituting the ends of the reel by turning back the ends of the inner cylinder (on which the discs have been placed), and thus holding the discs between the ends of the large outer hollow cylinder and the turned-back ends of the smaller internal hollow cylinder.

5. A machine for capping reels by attaching the caps to the reels by compressing the same between a fixed and moveable die.

John Machin, of Stockport, manufacturer. *For certain improvements in boots and shoes.* Patent dated June 17, 1851.

. Machin proposes (possibly not being
that he is already anticipated) to ma-
ure boots and shoes with revolving
composed of leather or gutta percha,
t when one part of the heel is worn
it may be turned round, and a fresh
presented for wear; and so also that
el may be substituted for another at
easure of the wearer. He also de-
a method of attaching spurs to
with revolving heels, by means of
on the spur, which is clamped
n the fixed and moveable parts of
el.
ms.—1. The manufacture of boots
es, with circular heels, in such man-
it they may be attached or detached
sure.
'he so arranging or adapting the said
ole heels that they may be capable of
readily adjusted so as to present a
point of bearing when they become

FREY ERMEN, of Manchester, cot-
nner. *For certain improvements in
thod of and apparatus for finishing
or threads.* Patent dated June 17,

e improvements have relation spe-
o threads employed for sewing pur-
but are also applicable to other
tions of thread; and the object pro-
by Mr. Ermen is, to impart to such
a lustre, smoothness, and compact-
fibre which is not obtainable by any
ordinary methods of finishing. [The
:us employed for this purpose con-
a stand or framework, in which are
d two rollers, at some little distance
parallel to each other, and capable
istment vertically. The rollers are
together so as to revolve in the same
n. In front of the rollers is placed
-brush, which is caused to rotate by
able means; and a fan is also attached
wer part of the machine, for directing
of hot or cold air against the threads
the finishing operation. The hanks
d, either in a bleached, dyed, or un-
d state, having been starched or sized,
superfluous size removed by squeez-
stretched evenly over the two rollers,
re then set in motion, carrying the
vith them, and exposing every part
urface of the thread to the action of
h, which is caused to rotate in con-
rewith.
s.—1. The production of a lustre
othness, together with perfect adhe-
fibre, by the application of friction,
or threads in the hank, while in a
tension and motion—such friction
roduced by a revolving brush or

2. The arrangement of apparatus described for that purpose.

THOMAS CROOK, of Preston, manufacturer, and JAMES MASON, of Preston, warper. *For certain improvements in looms for weaving.* Patent dated June 17, 1851.

Claims.—1. A method of weaving checked, striped, or figured fabrics, by throwing any suitable number of picks into a shed by the use of a check-rod.

2. The application of tappet or cam-chains, links, or belts, for varying the number of picks in a shed.

3. The application of tappet or cam-chains, or cam-wheels, for arresting the progress of the warp at determined intervals.

4. mode of weaving checked or striped fabrics, wherein the traverse of the warp is stopped at stated intervals, to allow of a firm beat-up.

5. A method of actuating the catch-cords.

6. The use of disengaging cams actuated by cam or tappet-chains.

7. A method of producing pattern or figured surfaces on woven fabrics by a combination of cams or tappets with cam-chains.

8. A method of locking the loose reed at the moment of beating-up, and of stopping the motion of the loom at certain intervals; and the employment for that purpose of pin or tappet-wheels.

9. A method of effecting a change of shuttles without stopping the motion of the loom; and the use of pin or tappet-wheels for that purpose.

10. The application of varnish or other coating to certain parts only of healds (the knot and upper shank being left unvarnished); and the use of healds, of which portions of the working surfaces are varnished, and portions left unvarnished.

Specification Due, but not Enrolled.

RICHARD FLETCHER, of Blackdowns-farm, Ebrington, Gloucester, farmer. *For an improvement in obtaining motive power.* Patent dated June 21, 1851.

WEEKLY LIST OF NEW ENGLISH PATENTS.

William Hirst, of Manchester, manufacturer, for certain improvements in machinery, or apparatus for manufacturing woollen cloth, and cloth made from wool and other materials. December 19. Six months.

Moses Poole, gentleman, for improvements in apparatus for excluding dust and other matters from railway carriages and for ventilating them. (Being a communication.) December 19. Six months.

Henry Clayton, of Atlas Works, Upper Park-place, Dorset-square, for improvements in the manufacture of tubes, pipes, tiles, and other articles, made from plastic materials. December 19. Six months.

Samuel Wilkes, of Wolverhampton, brass founder, for improvements in the manufacture of kettles, saucepans, and other cooking vessels. December 19. Six months.

Joseph Burch, of Craig Works, Macclesfield, for improvements in printing and manufacturing cut pile and other fabrics and yarns. December 19. Six months.

Christopher Rands, of Shad Thames, miller, for improvements in grinding wheat and other grain. December 19. Six months.

James Frederick Lackerstein, of Kensington-square, civil engineer, for improvements in machinery for cutting or splitting wood and other substances, and in the manufacture of boxes. December 19. Six months.

Frederick Bousfield, of Devonshire-place, Islington, gentleman, for a new manufacture of manure. December 19. Six months.

Charles Howland, of New York, engineer, for improvements in apparatus for ascertaining and indicating the supply of water in steam boilers. December 19. Six months.

William Elliott, of Birmingham, manufacturer, for improvements in the manufacture of covered buttons. December 19. Six months.

Rodolphe Helbronner, of Regent-street, for improvements in apparatus used when obtaining instantaneous light. December 19. Six months.

John Thornton and James Thornton, both of Melbourne, Derby, mechanics, for improvements in the manufacture of meshed and looped fabrics and other weavings, and in raising pile and looped fabrics and other weavings. December 19. Six months.

William Emery Milligan, mechanical engineer, of the city of New York, for certain improvements in the construction of boilers for generating steam. December 19. Six months.

Charles La nport, of Workington, Cumberland, ship-builder, for improvements in reefing sails. December 19. Six months.

Richard Archibald Brooman, of the firm of J. C Robertson and Co., of Fleet-street, patent agents, for improvements in sounding instruments. (Being a communication.) December 19. Six months.

John Davie Morries Stirling, of Black-grange, North Britain, Esq., for certain alloys and combinations of metals. December 22 Six months.

Sydney Smith, of Nottingham, for improvements in indicating the height of water in steam boilers. December 22. Six months.

Augustus Applegarth, of Dartford, Kent, for improvements in machinery used for printing. December 24. Six months.

Antonio De Sola, of Madrid, Spain, for certain improvements in the treatment of copper minerals. (Being a communication.) December 24 Six months.

Christopher Nickels, of York-road, Lambeth, and Thomas Ball and John Woodhouse Bagley, of Nottingham, for improvements in the manufacture of knitted, looped, and other elastic fabrics. December 24.

Alfred Vincent Newton, of Chancery-lane, Middlesex, mechanical draughtsman, for improvements in separating substances of different specific gravities. December 24. Six months.

Joseph Stenson, of Northampton, engineer and iron manufacturer, for improvements in the manufacture of iron, and in the steam apparatus used therein, part or parts of which are also applicable to evaporative and motive purposes generally. December 27. Six months.

WEEKLY LIST OF DESIGNS FOR ARTICLES OF UTILITY REGISTERED.

Date of Registration	No. in the Register	Proprietors' Names	Addresses.	Subjects of Design.
Dec. 18	3064	James Haywood	Derby	Stench-trap
20	3065	Charles Lenny	Croydon	Wicker-bodied carriage.
23	3066	J. J. Lane	Bethnal-green	Lozenge-cutting machine.

WEEKLY LIST OF PROVISIONAL REGISTRATIONS.

Dec. 16	340	Cyrus & J Clark	Street, Somerset	Elastic gusset for boots.
17	341	George Pate Cooper	Suffolk-street, Pall-mall	Elliptic gusset.
19	342	Felix P. Rovère	New-inn, Strand	Safety-catch for window-sashes.
23	343	Frederick Bristley	Fitzroy square	Fastening.
,,	344	Moritz Pillischer	Oxford street	Elliptical compasses.
24	345	Francis Higginson, Lieut., R.N.	Pentonville	Apparatus for laying down sub-marine electric telegraph cables, and electric wires, along railways.

CONTENTS OF THIS NUMBER.

END OF VOLUME FIFTY-FIVE.

LONDON: Edited, Printed, and Published by Joseph Clinton Robertson, of No. 166, Fleet-street, in the City of London—Sold by A. and W. Galignani, Rue Vivienne, Paris; Machin and Co., Dublin; W. C. Campbell and Co., Hamburg.

INDEX

TO THE FIFTY-FIFTH VOLUME.

CPSIA information can be obtained
at www.ICGtesting.com
Printed in the USA
BVHW04*2331031018
529052BV00028B/807/P

9 780484 602174